Biology and Ecology of Pharmaceutical Marine Sponges

Biology and Ecology of Pharmaceutical Marine Sponges

Ramasamy Santhanam
Santhanam Ramesh
Anbu Jeba Sunilson

CRC Press
Taylor & Francis Group
Boca Raton London New York

CRC Press is an imprint of the
Taylor & Francis Group, an informa business

CRC Press
Taylor & Francis Group
6000 Broken Sound Parkway NW, Suite 300
Boca Raton, FL 33487-2742

Printed on acid-free paper

International Standard Book Number-13: 978-0-8153-5456-7 (Hardback)

Library of Congress Cataloging-in-Publication Data

Names: Santhanam, Ramasamy, 1946- author.
Title: Biology and ecology of pharmaceutical marine sponges / Ramasamy Santhanam, Santhanam Ramesh, and Anbu Jeba Sunilson.
Description: Boca Raton : Taylor & Francis, 2018. | Includes bibliographical references and index.
Identifiers: LCCN 2018006232 | ISBN 9780815354567 (hardback : alk. paper)
Subjects: LCSH: Sponges. | Sponges--Ecology. | Sponges--Therapeutic use.
Classification: LCC QL371 .S26 2018 | DDC 593.4--dc23
LC record available at https://lccn.loc.gov/2018006232

Visit the Taylor & Francis Web site at
http://www.taylorandfrancis.com

and the CRC Press Web site at
http://www.crcpress.com

Contents

Preface

Sponges (Phylum: Porifera) are the most primitive multicelled animals that have existed for 700–800 million years. Of the approximately 15,000 sponge species, most occur in marine environments and only about 1% of the species inhabit freshwater. These sponges are in great commercial demand as some of them are good sources of bath sponges for the cosmetic industry. Pharmaceutical interest in marine sponges was aroused in the 1950s by the discovery of nucleosides, namely, spongothymidine and spongouridine in *Cryptotethia crypta.* These nucleosides were the basis for the synthesis of Ara-C, the first marine-derived anticancer agent, and the antiviral drug Ara-A. Ara-C is currently used in the routine treatment of patients with leukemia and lymphoma. Marine sponges are the potential sources of many unique metabolites such as nucleosides, sterols, terpenes, alkaloids, fatty acids, peroxides, cyclic peptides, and amino acid derivatives with anticancer, anti-inflammatory, antibiotic or antifouling, antiviral, antimalarial, and immune or neurosuppressive properties.

Manoalide, from the marine sponge *Luffariella variabilis,* has been found to possess antibiotic analgesic and anti-inflammatory properties. The triterpenoid hydroquinones from *Haticlona* sp., latrunculin A from *Latrunculia magnifica* and swinholide A from *Theonella swinhoei* exhibit antitumour properties. Three polyoxygenated sterols from a *Dysidea* sp., simplexides from the Caribbean sponge *Plakortis simplex* and ontignasterol from *Petrosia contignata* show immunosuppressive properties. Although the compound Hamigeran B from *Hamigera tarangaensis* showed 100% in vitro inhibition against both the herpes and polio viruses, weinbersterols A and B from *Petrosia weinbergi* exhibited in vitro activity against feline leukemia virus. Kalihinol A from *Acanthella* sp. and a number of terpenoid isocyanates, isothiocyanates, and isonitriles from *Cymbastela hooperi* display selective in vitro antimalarial activity against *Plasmodium falciparum.*

These marine sponges are responsible for more than 5300 different pharmaceutical products, and some of them have the potential to provide future drugs against important diseases, such as cancer, a range of viral diseases, malaria, and inflammations. Although many bioactives have been discovered in marine sponges, only a few of these compounds have been commercialized. Despite the wide range of functional roles performed by marine sponges, they are still poorly represented in many research, monitoring, and conservation programs.

Though a few books are presently available on marine natural products, a comprehensive book on the biology and ecology of pharmaceutical marine sponges has not been published so far. This publication, the first of its kind contributed by the scientists of both fisheries science and pharmacy disciplines, will fill up this long felt need. The taxonomy, common name, global distribution, habitat, diagnostic features and pharmaceutical compounds (along with their chemical structure and activities) of 450 species of marine sponges are given with suitable illustrations in this book. It is hoped that the present publication will be of great use as a standard reference for researchers, teachers, and students of various disciplines, such as fisheries science, marine biology, biochemistry, biotechnology, and pharmacy; part of the library collections at colleges, universities, and institutions; and as a valuable guide for pharmaceutical companies involved in the development of new drugs from marine sponges.

The authors sincerely thank Dr. K. Venkataramanujam, former dean, Fisheries College and Research Institute, TANUVAS, Thoothukudi, India, for his valuable suggestions. Special thanks must be expressed to all international friends/agencies who were kind enough to provide images of certain species of marine sponges for the book. The secretarial and photographic assistance provided by Mrs. Albin Panimalar Ramesh is gratefully acknowledged.

Suggestions from the readers are welcomed.

Preface

Authors

Ramasamy Santhanam, PhD, is the former dean of Fisheries College and Research Institute, Tamil Nadu Veterinary and Animal Sciences University, India. His fields of specialization are fisheries environments and marine biology. He is presently serving as a fisheries expert for various government and nongovernment organizations in India. Dr. Santhanam has published 20 books on various aspects of fisheries science and 70 research papers. He was a member of the American Fisheries Society, United States; the World Aquaculture Society, United States; the Global Fisheries Ecosystem Management Network (GFEMN), United States; and the International Union for Conservation of Nature's Commission on Ecosystem Management, Switzerland.

Santhanam Ramesh, PhD, professor, Department of Pharmaceutics, Ratnam Institute of Pharmacy, Nellore, AP, India, obtained his PhD degree from Jawaharlal Nehru Technological University, Hyderabad, India in 2011. His fields of specialization are biopharmaceutical products, pharmaceutical nanotechnology, and novel drug delivery systems. Dr. Ramesh has 13 years of teaching/overseas experience. He has authored four books (*Marine Pharmaceutical Compounds*, *Freshwater Phytopharmaceutical Compounds*, *Biology and Ecology of Pharmaceutical Marine Plants*, and *A TextBook of Novel Drug Delivery Systems*) and ten research papers published in nationally and internationally reputed journals. He is presently serving as an associate member of the Academy Pharmacy Group, Royal Pharmaceutical Society, Great Britain, and as a Fellow of Institution of Chemists, India.

Anbu Jeba Sunilson, PhD, associate professor, Department of Siddha Medicine, Tamil University, Thanjavur, India, obtained his PhD degree from the Tamil Nadu Dr. M.G.R. Medical University, Chennai, Tamil Nadu, India in 2006. His fields of specialization are pharmacognostical and phytochemical aspects in herbal drugs. Dr. Sunilson has 20 years of teaching/research experience in KPJ Healthcare University College, Malaysia; Masterskill University College of Health Sciences, Malaysia; and La Trobe University, Australia. He has 75 research papers to his credit. He has served as an associate editor/editor for the *Research Journal of Medicinal Plants*, Academic Journals, USA; *International Journal of Pharmacology*, Science Alert, USA; *Journal of Pharmacology and Toxicology*, Science Alert, USA; *International Journal of Biological Chemistry*, Science Alert, USA; and *Research Journal of Phytochemistry*, Science Alert, USA. Presently he is a member of the Malaysian Pharmaceutical Society.

Authors

1 Introduction

Marine sponges (Phylum Porifera, "pore-bearing"), the simplest of the multicellular invertebrate animals, possess neither nervous system nor internal organs. To date, about 11,000 species of sponges (including 150 freshwater species belonging to the family Spongillidae, class Demospongiae) have been described. Based on the composition of the skeletal elements, these sedentary (sessile) animals are divided into four classes: Calcarea (calcareous sponges), Hexactinellida (glass sponges), Sclerospongiae (coralline sponges), and Demospongiae (horny sponges, siliceous sponges or common sponges). Although the classes Calcarea, Hexactinellida, and Sclerospongiae make up only about 10% to 20% of the known species of sponges, the class Demospongiae comprises the remaining 80% to 90% of the so far described sponge species. Further, the Demospongiae assume greater significance owing to their diversity and pharmaceutical importance. Only the species belonging to the class Demospongia are considered in this book, since this group of Porifera includes almost all the records on natural products—chemical structures and their pharmacological properties—found in the database search constrained to the records between 2010 and 2012.

ECOLOGY

Marine sponges occur at all latitudes and are particularly abundant in Antarctic waters. Marine sponges are found at all water depths, from the tidal zone to the deepest regions (abyss). Members of the Calcarea and Demospongiae are found mainly on the rocky bottoms of the continental shelf, and members of the Hexactinellida are characteristic of the deepest muddy bottoms of oceans and seas. A restricted number of species are adapted to brackish waters. These sponges survive in a variety of conditions including areas where there is scarce light as well as in cold or warm water. Symbiotic relationships between algae and sponges usually occur, especially in strongly illuminated zones; the algae may act as a protective device because they deposit pigments in the superficial cell layers of the sponge.

PHYSICAL CHARACTERISTICS

Size range: Most sponges are only a few centimeters in size, but some urn-shaped or shapeless ones are less than a centimeter. Others, shaped like vases, tubes, or branches, may be 1 to 2 meters tall, and broad-rounded masses may be 1 to 2 meters in diameter. The size of sponges within a species may vary with age, environmental conditions, and food supply.

Morphology: Unlike higher metazoans, the sponges lack true tissues and organs. Further, the poriferans have no body symmetry. However, they have specialized cells that perform specific functions. The shapes of their bodies are adapted for maximal efficiency of water flow into the central cavity. Many sponges have internal skeletons of spongin and/or spicules of calcium carbonate or silica. Primarily, their body consists of a thin sheet of cells over a frame (skeleton).

Diversity of structure: Marine sponges vary greatly in their external appearance. Some are bushy or treelike and have fingerlike projections. Others, particularly in the class Demospongiae, are shapeless, or amorphous, masses that form thin encrustations on objects or are cushion shaped. A few species of Demospongiae have well-defined spherical shapes and others may be cup or fan shaped. Some species of the calcareous sponges may look like tubular sacs, and some members of the Hexactinellida are erect or cylindrical.

Color: Whereas deep-water sponges are brownish, shallow-water sponges are frequently brightly colored ranging from red, yellow, and orange to violet and occasionally black. Most calcareous sponges are white. Some sponges (e.g., the Spongillidae) are often greenish because green algae live in a symbiotic relationship within them; others are violet or pinkish, because they harbor symbiotic blue-green algae. Another variable character in sponges is consistency, which may range from the soft and viscous state of some encrusting species to the hard stonelike quality as in the genus *Petrosia*.

Skeleton in sponges:

Calcareous sponges (Class, Calcarea): Sponges with spicules made of calcium carbonate ($CaCO_3$) in the form of calcite, often three-rayed.

Glass sponges (Class, Hexactinellida): Sponges with silicon dioxide (SiO_2) spicules occurring as long fine hairs, free or woven into a fused scaffold; free spicule are often six-rayed.

Coralline sponges (Class, Sclerospongiae): Sponges with skeleton made of calcium carbonate, either aragonite or calcite.

Horny sponges (or common sponges) (Class, Demospongia): Sponges with silicon dioxide (SiO_2) spicules, and/or sand, and/or fibrillar collagen, and/or fibrous (spongin) collagen.

BIOLOGY

Food and feeding: Sponges create unidirectional water flow throughout their body to facilitate food and oxygen intake and send out the waste. This water enters the sponge through ostia (minute pores) and leaves through a larger opening (osculum). The necessary flow of water is created by the movement of hair cells, that is, choanocytes, which are specific to sponges. They are filter feeders and they feed on microparticles, for example, bacteria, unicellular algae, and organic debris, not exceeding 3 μM. A baseball-sized sponge can filter 50 gallons an hour, and it requires at least one direct daily feeding.

Life cycle: Most sponges reproduce sexually, although asexual reproduction may also occur. Sponges are generally hermaphroditic (that is, male and female germ cells in one animal). However, some sponge species are sequential hermaphrodites (that is, male and female germ cells that develop at different times in the same animal).

Asexual reproduction: Sponges undergo various types of asexual reproduction as described next.

Regeneration: All sponges possess a remarkable ability to regenerate their lost parts. A piece cut from the body of a sponge is capable of growing into a complete sponge.

Budding and branching: In budding, numerous archaeocytes gather near the surface resulting in a small outgrowth on the pinacoderm. The bud thus formed grows outward to produce a small individual, which either remains attached with the parent individual or gets detached and attached to a nearby rock to grow into an independent colony.

Fission and fragmentation: In some sponges multiplication takes place by developing a line of fission and throwing off parts of the body that later can develop into a new sponge.

Reduction bodies: Many marine sponges disintegrate in adverse environmental conditions particularly in winter, leaving small rounded balls called reduction bodies. When favorable conditions return, these reduction bodies grow into complete new sponges.

Gemmules: Marine sponges such as *Ficulina*, *Suberites*, and *Tethya* possess the remarkable ability to produce specialized bodies called gemmules, which survive during unfavorable conditions and germinate to produce new sponges.

Sexual reproduction: Sexual reproduction involves formation of sperms and ova. The sex cells arise either from archaeocytes or choanocytes. Although most sponges are hermaphrodite, cross-fertilization often occurs as the eggs and sperms are produced at different times. Oocytes are produced inside the body and they remain inside mesogloea waiting for fertilization. In sperm formation, archaeocyte and trophocyte cells are involved and in demospongiae choanocytes form sperm and leave the body of the sponge through osculum in large numbers. Sperm from water enter the body of another sponge through canal system and reach the flagellate chambers, where choanocytes trap them. Acting as nurse cells, choanocytes transport the sperm body without tail to the mature ova that wait in the mesogloea. The sperm nucleus then fuses with the nucleus of ovum, ensuring internal fertilization. The merging larva swims for a period of time that may vary from a few hours to a few days before it descends to find a surface suitable for attachment. After attachment, the larva metamorphoses into a young sponge.

ASSOCIATIONS

Associations with other organisms: Various plants and animals may live on the surface of the sponge or inside its canals and cavities. In some cases the associations are specific, for example, the coral *Parazoanthus axinellae* grows on the sponge *Axinella*. The organisms that live in the cavities of sponges include crustaceans, nematode and polychaete worms, ophiuroid echinoderms (brittle stars), and bivalve molluscs. Some inhabit a sponge for occasional shelter or nourishment and others establish more intimate associations as parasites or predators. Some organisms that live on (called epibionts) and in (called endobionts) sponges act as parasites. Cyclopoid copepods are the most important parasites of marine sponges

Symbiotic associations: The most important symbiotic associations of sponges occur with single-celled and multicellular algae. The sponge protects the algae from enemies, from unfavorable environmental conditions, and from their own metabolic waste products. Further, the sponge uses the algae as a source of oxygen as a mechanism for eliminating its metabolic wastes, and as a food source (consuming both algal waste products and dying algae). In some boring clionid sponges (*Cliona viridis*) of the class Demospongiae, some single-celled brown algae are constantly present. Some marine sponges may also harbor multicellular blue-green algae (e.g., *Oscillatoria*), red (Rhodophyceae), and green (Chlorphyceae) algae. Red and green algae sometimes provide skeletal support for certain sponges.

Association with microorganisms: The marine sponges gained great interest due to their association with a wide variety of microorganisms. These microorganisms are known to be a rich source of secondary metabolites that exhibit a broad range of bioactivities such as inhibition of enzyme activities and cell division, and antiviral, antimicrobial, anti-inflammatory, antitumor, cytotoxic, and cardiovascular properties.

Predators of marine sponges: Marine sponges are often eaten by other organisms, for example, mollusks (gastropods such as snails and nudibranch slugs, prosobranchs such as *Patella* and *Littorina*, and chitons), some crustaceans, and some fishes (especially on coral reefs).

Diseases of sponges: Sponges may be attacked by diseases of epidemic character, the agents of which are not well known. The commercial sponges of the West Indies once were nearly completely destroyed by a fungus-like microorganism.

AQUACULTURE

Fishing: Commercially valuable sponges, which may be found from tidal level to a depth of about 70 m, are usually harvested by hooking or harpooning in shallow waters, by skin diving, or by deepwater fishing. Although the most valuable sponges are found in the eastern Mediterranean area, they are also harvested off the west coast of Florida and the Florida Keys, in the West Indies, off Mexico and Belize, and, to a limited extent, off the Philippines.

Marine sponge culture: Considering the future need of sponge biomass for new drug development, the sponge culture may become a viable alternative and possibly the emerging new field of marine biotechnology. The method most commonly used for sponge culture is the suspension method, which is based on threading sponge fragments on ropes.

Conservation of marine sponges: Sponges, which are potential underwater treasure chests of future medical discoveries, are important for maintaining ecosystem function and integrity of marine benthic communities worldwide. Attention is also focused upon species that have been of recent commercial interest as sources of metabolites with biomedical potential, such as *Dysidea avara*, which is considered endangered due to a pressing demand for the industrial extraction of chemicals it produces (e.g., Avarol). Sponge culture in situ is suggested as a useful approach that may allow eco-compatible management and conservation of this natural resource. Of the described sponge species, twenty have been reported to be on the threatened species lists, and all such species belong to the northeastern Atlantic and Mediterranean. However, there has been no detailed assessment of their current global conservation status. Parameters such as their status, patterns of temporal variation in sponge populations and assemblages, the number of sponges on threatened species lists, and the impact of environmental pressures need to be studied urgently to safeguard their species diversity and biomass.

IMPORTANCE OF SPONGES

Historical importance of sponges: The soft elastic skeletal frameworks of certain species of the common sponges (e.g., *Spongia officinalis, Hippospongia communis, Spongia zimocca, S. graminea*) have been familiar household items since ancient times. In ancient Greece and Rome, sponges were used to apply paint, as mops, and by soldiers as substitutes for drinking vessels. During the Middle Ages, burned sponge had therapeutic value especially in the treatment of various diseases. Natural sponges now are used mostly in arts and crafts such as pottery and jewelry making, painting and decorating, and in surgical medicine.

Pharmaceutical values of marine sponges: The relationship between marine sponges and medicines goes back to Alexandrian physicians and was described by Roman historian Plinius. Physicians used sponges that were saturated with iodine to stimulate coagulation of the blood. Sponges were soaked with pure wine and put on the left part of the chest in case of heartaches and soaked in urine to treat bites of poisonous animals. They were also used against all kinds of wounds, bone fractures, dropsy, stomachaches, and infectious diseases. At present "stodal," a syrup containing roasted *Spongia officinalis* is used for homeopathic treatment of dry and asthmatic cough in the Western world. The unusual nucleosides spongouridine and spongothymidine, isolated from the marine sponge *Tethya crypta* in the 1950s, served as lead structures for the development of the commercially viable antiviral drug Ara-A and the anticancer drug, Ara-C. Ara-C is also currently used in the routine treatment of patients with leukemia and lymphoma. One of its fluorinated derivatives has also been approved for use in patients with pancreatic, breast, bladder, and lung cancer. Marine sponges have the potential to provide future drugs against important diseases, such as cancer, a range of viral diseases, malaria, and inflammation. The following quote from the Bible describes the importance of marine sponges:

> He will wipe away every tear from their eyes. There will be no more death or mourning or crying or pain for the old order of things has passed away.
>
> **—Book of Revelations 21:4**

Bioactive Compounds/Secondary Metabolites in Marine Sponges

Diversity of sponge natural products: Among the marine organisms, the marine sponges are considered to be the gold mine with respect to the diversity of their bioactive compounds/secondary metabolites discovered during the past 50 years. The following table testifies to this.

Number of New Compounds Isolated from Different Marine Groups during 2001–2010

Marine Source/Phylum	Compounds
Marine microorganisms (including phytoplankton)	1750
Green algae	100
Brown algae	250
Red algae	450
Sponges	2400
Cnidarians	2000
Bryozoans	40
Molluscs	300
Tunicates	350
Echinoderms	450

Source: Mehbub, M.F. et al. 2014, *Mar. Drugs*, 12: 4539–4577.

Present status of sponge natural products as pharmaceuticals: In the past decades, marine life-forms have been the origin of a remarkable number of novel secondary metabolites. The number of isolated compounds from marine sources has increased steadily, from an annual number of approximately 20 in 1984 to an annual number of more than 1000 in 2010. The total number of approved drugs from the marine environment is steadily increasing, from four in 2010 to seven in 2014. The first U.S. Food and Drug Administration (FDA)–approved marine-derived drug reached the marked in 1969. It was the anticancer pyrimidine analog cytarabine (Cytosar-U®), isolated from the Caribbean sponge *Cryptotheca crypta*. Since then, six more marine natural products have been approved as drugs (one of which is the anticancer sponge derived macrolide eribulin mesylate (Halaven®). Of the 18,000 marine natural products described, over 30% are from sponges, and of the antitumor natural product patent registrations in recent years, over 75% are from sponges. During 2001–2010 Japan had the highest number of marine sponge–based bioactive compounds (332) followed by Indonesia (235), Korea (211), Australia (187), and China (146). The Bahamas, Mexico, Palau, Papua New Guinea, Philippines, and Vanuatu are the other countries that were sources of more than fifty new sponge-based compounds (Mehbub et al., 2014).

Composition and activities of sponge secondary metabolites: Most bioactive compounds from sponges can be classified as anti-inflammatory, antitumor, immunosuppressive or neurosuppressive, antiviral, antimalarial, and antibiotic. The chemical diversity of sponge products is remarkable. Marine sponges contain nucleosides, bioactive terpenes, sterols, cyclic peptides, alkaloids, fatty acids, peroxides, and amino acid derivatives (which are frequently halogenated). The number of new compounds isolated from marine sponges with different bioactivities during 2001–2010 is given next.

Bioactivity	Compounds
Anticancer/Cytotoxicity	828
Antibacterial	145
Antifungal	114
Anti-HIV	41
Antituberculosis	31
Antimalarial	27
Anti-inflammatory	9
Anti-Alzheimer's	9
Antiviral	4
Miscellaneous	407
Total	1615

Source: Mehbub, M.F. et al. 2014, *Mar. Drugs*, 12: 4539–4577.

Sponge secondary metabolites in clinical trials: The clinical status of certain secondary metabolites (as FDA-approved agents) of marine sponges are given next.

Compound	Species	Disease/Molecular Target	Clinical Status
Cytarabine (Ara-C)	*Cryptotethya crypta*	Cancer/DNA polymerase	FDA approved
Vidarabine (Ara-A)	*Cryptotethya crypta*	Antiviral viral/DNA polymerase I	FDA approved
Eribulin Mesylate (E7389)	*Lissodendoryx* sp.	Cancer/microtubules	FDA approved
Hemiasterlin derivative (E7974)	*Cymbastella* sp.	Cancer/microtubules	Phase I
Discodermolide	*Discodermia dissoluta*	Cancer/microtubules	Phase I
Bengamide derivative (LAF389)	*Jaspis* sp.	Cancer	Phase I
Spongistatin 1	*Hyrtios erecta*	Cancer/microtubules	Preclinical
Manoalide	*Luffariella variabilis*	Inflammation/ inhibition of phospholipase A2	
Salicylhalimide A	*Haliclona* sp.	Cancer/microtubules	Preclinical
Laulimalide	*Cacospongia mycofijiensis*	Cancer/microtubules	Preclinical
Peloruside A	*Mycale hentscheli*	Cancer/microtubules	Preclinical
Halichondrin B	*Halichondria okadai*	Cancer	Phase II/III
Hemiasterlin (E 7974)	*Hemiasterella minor*	Cancer	Phase I
IPL576*	*Petrosia contignata*	Inflammation/asthma	Phase II

Sources: Agrawal, S. et al, 2016, *Front. Pharmacol.*, 7: 333; Newman, D.J., and G.M. Cragg, 2016, *J. Nat. Prod.*, 79, 629–661; Mayer, A.M. et al., 2010, *Trends Pharmacol. Sci.*, 31: 255–265; Chapter I: Bioactive metabolites of marine organisms, n.d., http://shodhganga.inflibnet.ac.in/bitstream/10603/28761/3/03_chapter%201.pdf.

* Synthetic analog of the steroid contignasterol.

Constraints: Despite the considerable amount of new bioactive metabolites that originated from marine sponges, very few compounds have reached advanced stages of clinical trials. One of the reasons for this status is that many of the sponge bioactive metabolites are very toxic, thus leading to a low therapeutic index. A second major obstacle disturbing the pharmaceutical use of sponge natural products is the supply problem. Approval of any new compound for human therapeutic use requires a steady supply of material. Sponge tissues usually bear only trace amounts of the secondary metabolites of interest. Therefore, obtaining sufficient amounts of the bioactive metabolites from natural sponge populations requires collection of adequate amounts of sponge biomass. It is very clear that for most sponge species, natural populations are unable to sustain such heavy exploitation. Different methods have therefore been proposed to overcome the supply problem. The most direct, simple method is by chemical synthesis of the target compound. Most natural products are not amenable to such synthesis due to the high complexity of their chemical structure. Even for those compounds that can be synthesized in laboratory settings, the scaling up of their production is usually not feasible. Sponge tissue culture (primmorphs) may solve this constraint to a greater extent. In this regard, collaborative, intensive research by marine biologists, pharmacists, and organic chemists is urgently required for utilizing the plethora of marine sponge species and their derivatives for the benefit of human kind. Conceptually, it is clear that although marine sponges offer a huge potential in the naturally based pharmacopoeia of this century, an unfavorable balance between discovery and the very small number of candidates incorporated for clinical evaluation exists. So it appears that a better and more pragmatic approach is urgently needed in order to translate innovative discoveries into active clinical therapeutics.

2 Marine Sponges: Biology and Pharmaceutical Aspects

FAMILY: AGELASIDAE

Agelas clathrodes (Schmidt, 1870) (= *Agelas flabellrformis*)

Class: Demospongiae Subclass: Heteroscleromorpha
Order: Agelasida Family: Agelasidae

Common name: Orange elephant ear sponge

Distribution: Western Atlantic and Western Central Pacific; Florida, Bahamas, and Caribbean

Ecology: It is a reef-associated inhabiting rocky bottoms at depths of 2–15 m. It prefers areas with some water movement

Biology: The orange elephant ear sponge rises from a narrow base. Its form is massive and compressed. Larger species have a broad fan shape (broadly flabellate). Sometimes they have several coalescent fronds. The plane of the fronds is usually twisted. The surface is covered with a variable degree of fleshy skin. The color of the body is brilliant red-orange, somewhat paler internally. Like other demosponges, it sponge is a filter feeder. Water circulates through the sponge, entering through fine pores and leaving through the larger holes. Oxygen and food particles (mostly bacteria) are removed from the water as it passes through the sponge, and waste products are carried away. It is a hermaphrodite, but it also reproduces asexually. When pieces of sponge become detached, they may later reattach themselves to the substrate. It attains a maximum length of 10 cm.

Compound(s) and Activities

Cytotoxic activity: The hydroethanolic extracts of this species have shown cytotoxicity activity against tumor cell lines at a concentration of 20 mg/mL. The values of mean inhibitory concentration (IC50) are shown next (Ferreira et al., 2007).

IC50 (μg/mL) Values of Tumor Cell Lines

HL-60	HCT-8	MDA-MB435	SF-295
48.5	85.5	58.0	62.4

AChE inhibition activity: The undiluted methanol extracts from this species were found to be strongly hemolytic and they showed substantial anti-acetylcholinesterase (AChE) activity (Sepčić et al., 2010). The enzyme acetylcholinesterase has been reported to cause several diseases such as abdominal pain, lack of appetite, yellowed skin, dizziness, slow heartbeat, sudden or substantial weight loss, and weakness in man.

Antibacterial activity: Several compounds, namely, (–)-agelasidine A, zooanemonin, 1-carboxymethylnicotinic acid, hymenidin, mukanadins A and C, monobromodispacamide, agelasidines A, 2-amide-4-bromopyrrole, and O-methyltryptophan have been isolated from this species. Among them, agelasidine A has shown both antibacterial and antifungal activities (Medeirosa et al., 2006).

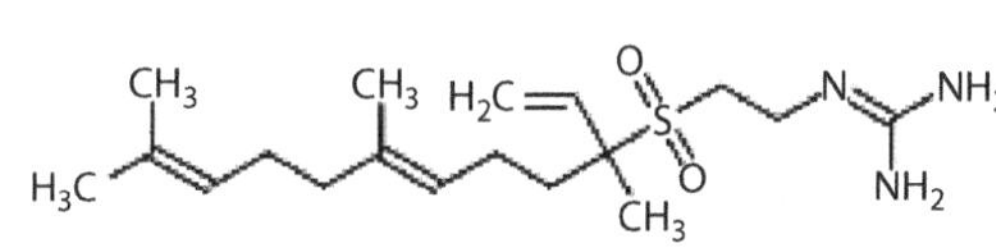
Agelasidine A

The MIC (minimum inhibitory concentration) values recorded for agelasidine A are given next.

Species	MIC (μg/mL)
B. subtilis	50
E. coli	100
M. smegmatis	100
S. aureus	50
S. faecium	100
S. mutans	100

Source: Medeirosa, M.A. et al., 2006, *Z. Naturforsch C.*, 61: 472–476.

Immunosuppressive activity: The compound 4a-methyl-5a-cholest-8-en-3~-ol and 4,5-dibromo-2- pyrrolic acid of this species have shown immunosuppressive activity (Perdicaris et al., 2013).

Agelas linnaei (De Voogd, Parra-Velandia, and van Soest, 2008)

Order: Agelasida Family: Agelasidae

Common name: Not designated

Distribution: Indo-Pacific, off Jakarta, West Java, Indonesia

Ecology: It inhabits rocky and sandy areas of coral reef; overgrowing other reef invertebrates.

Biology: This species is roundly lobate to thickly flabellate. It is consistency very soft and spongy. Color of the animal is bright orange at the surface to cream-orange internally. Surface is with dense conules (1–3 mm in height) supported by tips of ascending fibers covered by a bright easily distinguishable membrane. Small apertures (<2 mm) are found scattered on the surface and bigger pores (2–3 mm) are connected to internal axial canals.

Compound(s) and Activities

Cytotoxicity: The compounds agelanesin A, agelanesin B, agelanesin C, and agelanesin D isolated from this species showed prominent activity for cytotoxicity against the murine L1578Y mouse lymphoma cell line with IC50 values of 9.6, 9.3, 16.8, and 13.1 μM, respectively (Putra and Jaswir, 2014).

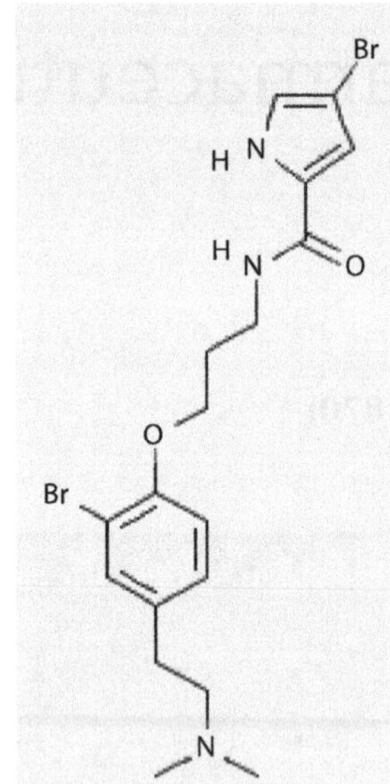

Agelanesin A

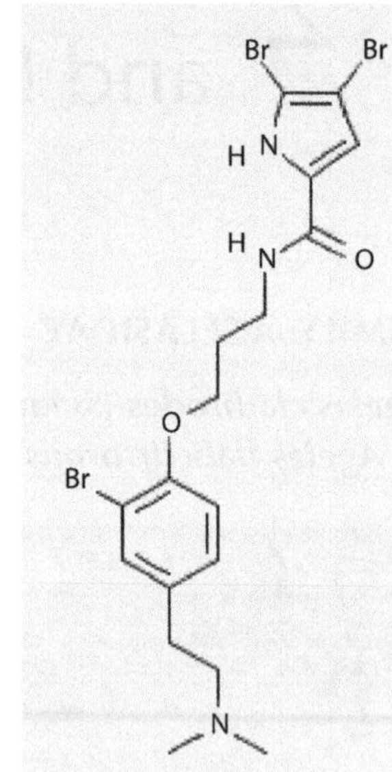

Agelanesin C

Agelas mauritiana (Carter, 1883)

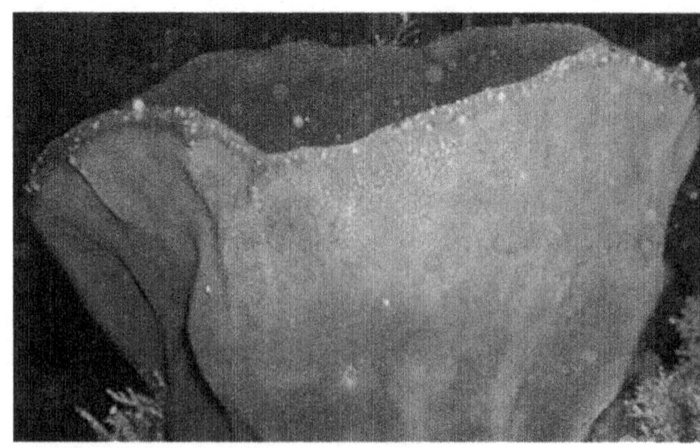

Order: Agelasida Family: Agelasidae

Common name: Pocked cup sponge

Distribution: Indo-West Pacific, Port Elizabeth to Mozambique

Ecology: It is sessile inhabiting subtidal reefs. Sessile at depths of 10 to 50 m.

Biology: This species, which is about 30 cm tall forms large, often irregular grayish to purple cups. External surface is with indentations and small projections. The wall is thin in relation to the height of the cup. Surface is tough and not compressible. Color of the body is greenish brown or grayish blue with an olive tint. This species can build large colonies on flattish reefs. It is a filter feeder.

Compound(s) and Activities

Antimicrobial and antileishmanial activities: Four new alkaloids, namely, (–)-8′-oxo-agelasine D (2), ageloxime B (3),

(+)-2-oxo-agelasidine C (4), and 4-bromo-N-(butoxymethyl)-1H-pyrrole-2-carboxamide (5), and a known compound (−)-ageloxime D (1) have been isolated from this species. Among these compounds, compounds 1 and 3 showed antifungal activity against *Cryptococcus neoformans* and antileishmanial activity against *Leishmania donovani in vitro*. Compound 3 also exhibited antibacterial activity against *Staphylococcus aureus* and methicillin-resistant *S. aureus in vitro* (Perdicaris et al., 2013; Yang et al., 2012).

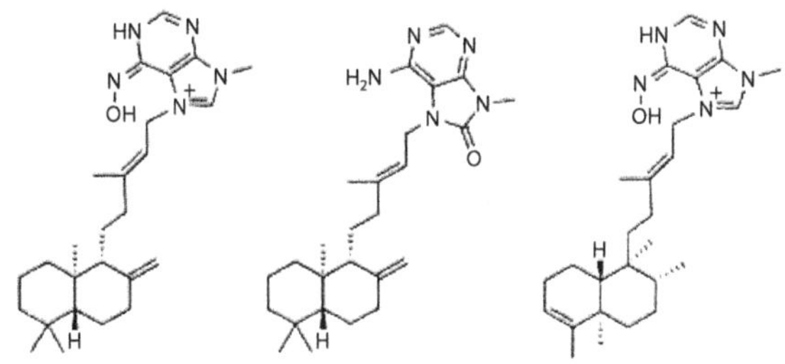

1 2 3

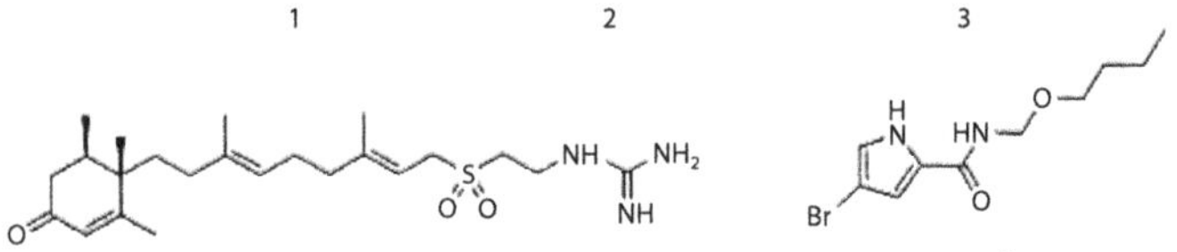

4 5

(1) (−)-ageloxime D; (2) (−)-8′-oxo-agelasine D; (3) ageloxime B; (4) (+)-2-oxo-agelasidine C; (5) 4-bromo-N-(butoxymethyl)-1H-pyrrole-2-carboxamide

Antitumor activity: The compound agelasphin (KRN7000), a galactosylceramide isolated from this species, has shown antitumor activity as an NKT cell activator (Anjum et al., 2016; Sipkema et al., 2005).

Cytotoxicity: Two bicyclic diterpenoides, agelasimine-A, and agelasimine-B; a new bromine-containing alkaloid, 5-debromomidpacamide; along with midpacamide and methyl N-methyl-4,5-dibromopyrrole-2-carboxylate have been isolated from this species. Of these compounds agelasimine-A and -B exhibited a wide range of biological activities such as cytotoxicity, inhibition of adenosine transfer into rabbit erythrocytes, Ca2+-channel antagonistic action, and α1 adrenergic blockade (Fathi-Afshar and Allen, 1988).

Cytotoxicity and antibacterial activity: Five new diterpene alkaloids, namely, (+)-80 -oxo-agelasine B (1), (+)-agelasine B (2), (+)-80 -oxo-agelasine C (3), agelasine V (4), and (+)-80-oxo-agelasine E (5), along with two known compounds, ()-80 -oxo-agelasine D (6), and agelasine D (7) have been isolated from the ethanol extract of this species. Of these compounds, compounds 2 and 7 not only exhibited moderate cytotoxicity toward the cancer cell lines PC9, A549, HepG2, MCF-7, and U937 with IC50 values of 4.49–14.41 μM, but also showed potent antibacterial activities against a panel of methicillin-resistant *Staphylococcus aureus* (MRSA), clinical isolates with MIC90 values of 1–8 mg mL^{-1} (Hong et al., 2017a).

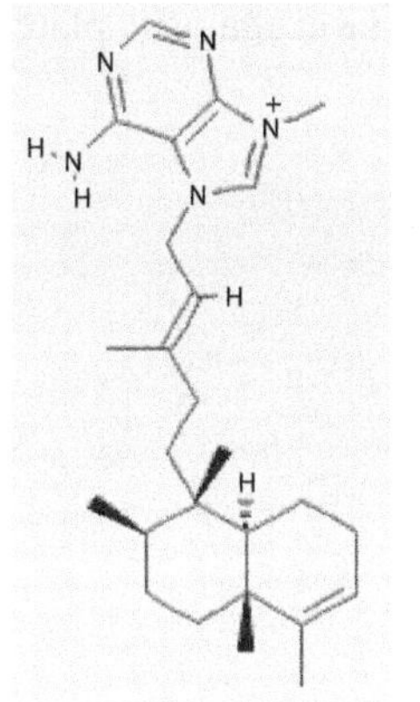

Agelasine B

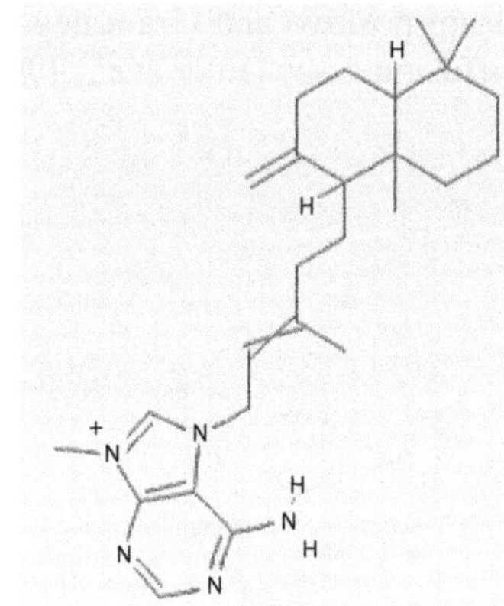

Agelasine D

Antifungal activity: The compounds ageloxime B and (−)-ageloxime D of this species have shown antifungal activity against *C. neoformans* (Cheung et al., 2014).

Antidiabetic and antitumor activities: The alpha-galactosylceramide (alpha-GalCer), natural killer isolated from extracts of this species, showed antidiabetic activity. Further, this compound exhibits profound antitumor activities (Kitamura et al., 1999; Lauritano and Ianora, 2016).

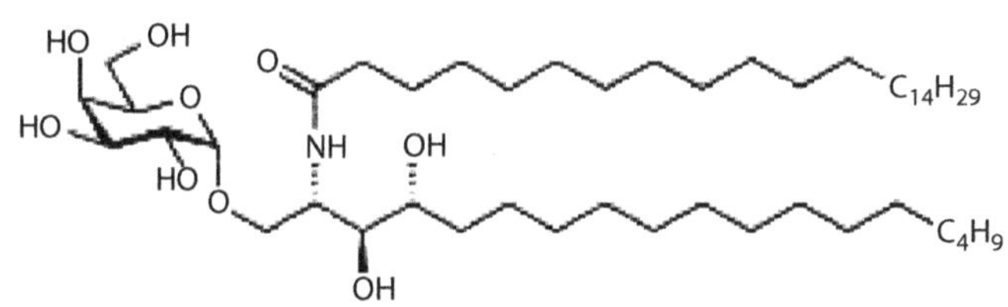

Alpha-galactosylceramide

Agelas nakamurai (Hoshino, 1985)

Order: Agelasida Family: Agelasidae

Common name: Okinawan marine sponge, South China Sea sponge

Distribution: Indo-Pacific; Zamami Island, Ryukyus, Japan; Japanese Exclusive Economic Zone

Ecology: It inhabits shallow water of 20 m depth.

Biology: It is a massive, rounded, or thickly encrusting species. Color of the body is orange rufous to brick red at the surface in dry specimens. Surface is smooth, uneven, with irregularly meandering surface grooves. Consistency is firmly spongy and is resilient when wet; and is very hard, hardly compressible when dry. Ectosomal skeleton is armed by tangential acanthostyles, which are irregularly distributed.

Compound(s) and Activities

Antibacterial activity: Two new dimeric bromopyrrole alkaloids, nakamuric acid (1) and its corresponding methyl ester (2), have been isolated from this species along with the known metabolites sceptrin (3), debromosceptrin (4), and ageliferin (5). All these compounds inhibited the growth of several

Gram-positive and Gram-negative bacteria in the agar plate diffusion assay (Eder et al., 1999).

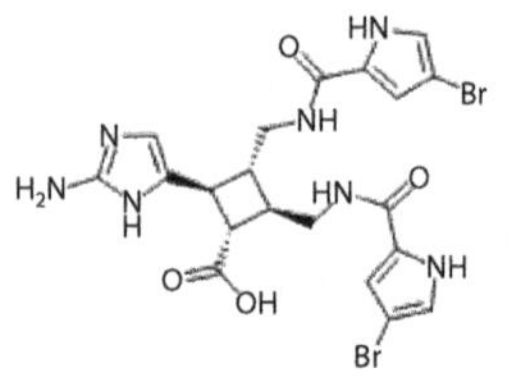

Nakamuric acid

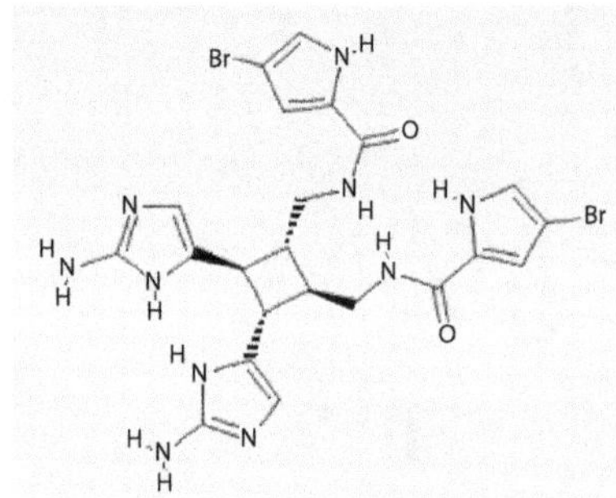

Sceptrin

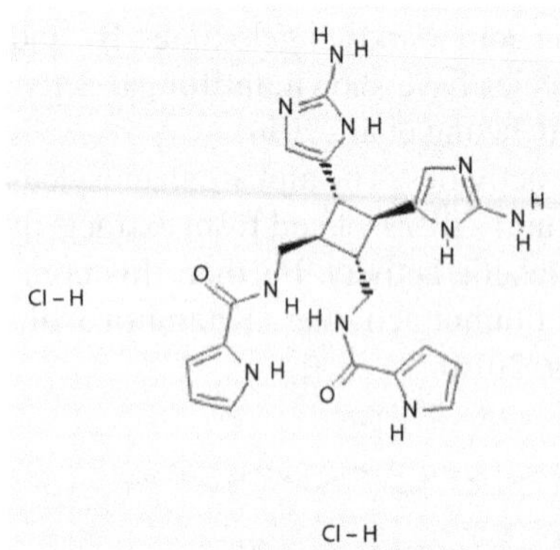

Debromosceptrin

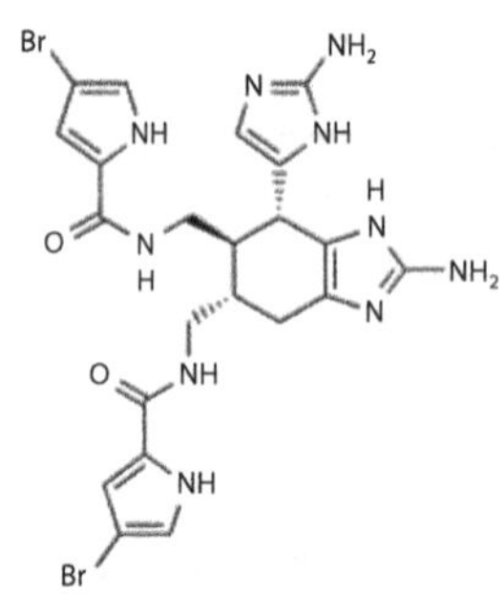

Ageliferin

Antimycobacterial and protein tyrosine phosphatase inhibiting activities: Three new N-methyladenine-containing diterpenes, 2-oxoagelasines A (1) and F (2) and 10-hydro-9-hydroxyagelasine F (3); and eight known agelasine derivatives, 2-oxoagelasine B (4), agelasines A (5), B (6), D (7), E (8), F (9), and G (10), and ageline B (11) have been isolated from this species. Of these compounds, compounds 3 and 5–11 inhibited the growth of *Mycobacterium smegmatis* with inhibition zones of 10, 14, 15, 18, 14, 20, 12, and 12 mm at 20 μg/disc, respectively. All these compounds were, however, inactive (IC50 >10 μM) against Huh-7 (hepatoma) and EJ-1 (bladder carcinoma) human cancer cell lines. Compounds 1, 2, and 4 exhibited markedly reduced biological activity against *M. smegmatis*. On the other hand, compound 10 inhibited protein tyrosine phosphatase 1B (PTP1B) activity with an IC50 value of 15 μM/mL (Abdjul et al., 2015).

Anti-TB activity: The three methoxy analogs of the compound, agelasine E of this species showed high anti-TB activity against *Mycobacterium tuberculosis* H37Rv with MIC values of 3.1, 1.6, and 3.1 μg/mL, respectively (Rogoza et al., 2011).

Cytotoxicity and antimicrobial activities: Two pairs of new non-brominated racemic pyrrole derivatives, (±)-naka murine D (1) and (±)-nakamurine E (2); two new diterpene alkaloids, isoagelasine C (16) and isoagelasidine B (21); together with thirteen known pyrrole derivatives ((±)-3–15), five known diterpene alkaloids (17–20, 22) have been isolated from this species. An unidentified diterpene alkaloid of this species showed cytotoxicity against HL-60, K562, and HCT-116 cancer cell lines with IC50 values of 12.4, 16.0, and 19.8 μm, respectively. Six diterpene alkaloid compounds showed potent antifungal activity against *Candida albicans* (MIC values ranging from 0.6 to 4.7 μg/mL) and four such compounds exhibited moderate antibacterial activity against *Proteusbacillus vulgaris* (MIC values ranging from 9.4 to 18.8 μg/mL) (Chu et al., 2017).

Putra and Jaswir (2014) reported that the diterpene alkaloids of this species, namely, (–)-agelasine D and (–)-ageloxime D, exhibited cytotoxicity against L5178Y mouse lymphoma cells (IC50 4.03 and 12.5 μM, respectively) (Putra and Jaswir, 2014).

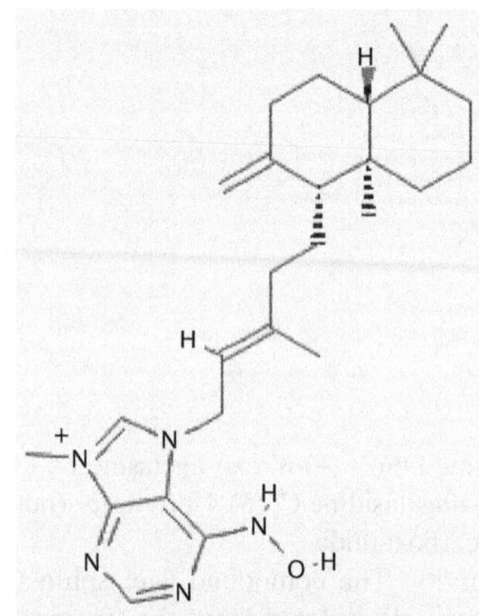

Ageloxime D

MMP-2 inhibiting activity and antiangiogenic effect: The compound ageladine A of this species showed inhibitory activity against the enzyme matrix metalloproteinase-2 (MMP-2) with an IC50 value of 2.0 μg/mL. MMP-2 plays a key role in endometrial menstrual breakdown, regulation of vascularization, and the inflammatory response. Ageladine A has also been reported to inhibit several other MMP isoforms including MMP-1, 8, 9, 12, and 13 with IC50 values of 1.2, 0.4, 0.8, 0.3, and 0.5 μg/mL, respectively. The ageladine A, which is also a matrix metalloproteinase inhibitor, has also shown antiangiogenic effect by inhibiting vascular formation at 10 μg/mL (Fujita et al., 2003a; Sullivan et al., 2009).

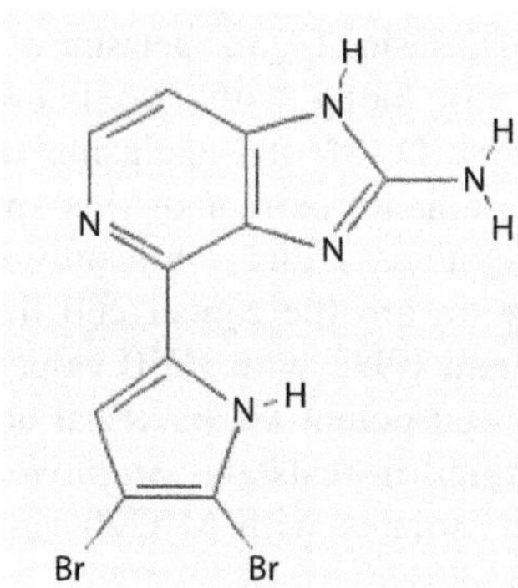

Ageladine A

Agelas oroides (Schmidt, 1864)

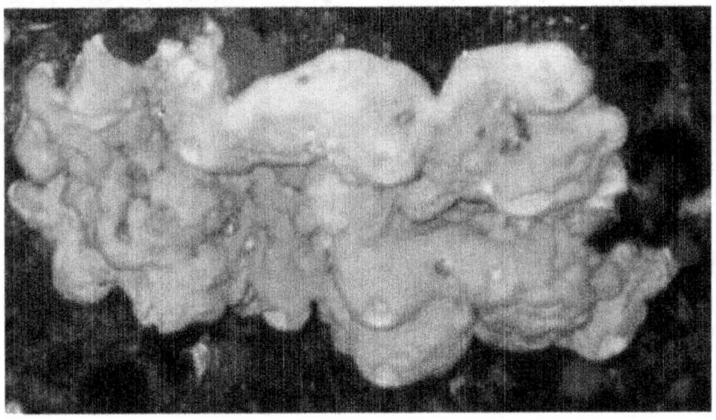

Order: Agelasida Family: Agelasidae

Common name: Orange sponge, Mediterranean sponge, Turkish marine sponge

Distribution: Mediterranean, Atlantic, English Channel, and North Sea

Ecology: This sponge lives fixed on the hard substrates of the circalittoral stage. It is often found in cave entrances up to 50 m deep.

Biology: This sponge is massive, with irregular lobes that can reach 20 cm in height. Its texture is original, elastic but very strong and tear-proof because its spongine skeleton is very developed. Its surface appearance, which is blistered, has tiny protuberances. Its color varies from yellow to orange. Morphologically it has very varied forms, which are also very irregular. Like most sponges, it is an active filtering animal that can inhale large amounts of water through myriad pores. This sponge is common in places rich in its food organisms, namely, plankton and bacteria. Reproduction may be either asexual, by budding, or sexual, by production of male and female gametes by the same individual. However, fertilization occurs only with two different individuals. Reproduction is oviparous. The emerging larvae are ciliated, and they swim for some time before settling. This sponge can contain a large amount of symbiotic bacteria.

Compound(s) and Activities

Antioxidant activity: The extracts of this species showed 48% DPPH radical scavenging activity (inhibition) at concentration of 2000 μg mL^{-1} (Orhan et al., 2012).

IL-2 inhibitor: The pyrrole-imidazole alkaloid compound, taurodispacamide A of this species has been reported to inhibit IL-2, which is related to blood-related diseases (Sipkema et al., 2005).

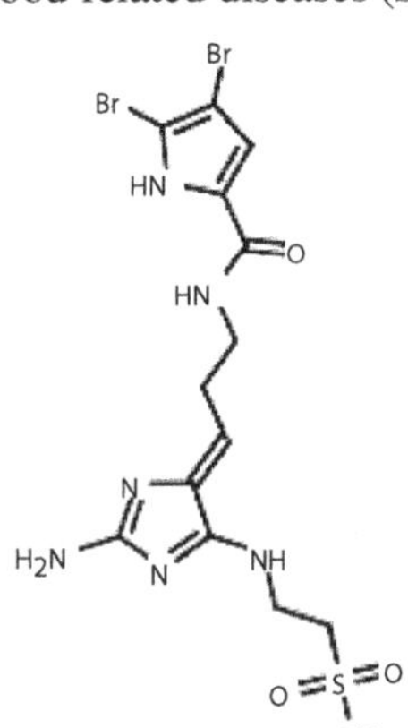

Taurodispacamide A

Antimalarial activity: The (E)-Oroidin alkaloids of this species have shown antimalarial activity against *P. falciparum* (D6 clone) with an IC50 value of 0.30 μg/mL (Anjum et al., 2016).

Antimicrobial activity: The organic extracts of this species showed antibacterial activity against Gram-positive bacteria and antifungal activity, and the values of MICs (μg mL^{-1}) are shown next (Orhan et al., 2012).

MIC (μg mL^{-1})				
S. aureus	*E. faecalis*	*B. subtilis*	*C. albicans*	*C. parapsilosis*
128	128	32	16	16

Source: Orhan et al., 2012, *Rec. Nat. Prod.*, 6: 356–367.

AChE inhibitory activity: The organic extracts of this species showed acetylcholinesterase (AChE) inhibitory activity with a percentage inhibition value of 49.8 at a concentration of 200 μg mL^{-1} (Orhan et al., 2012).

Cytotoxicity: The methanolic extract of this species showed cytotoxicity of human neuroblastoma cell lines, LAN5 and SK-N-BE(2)-C, in a time- and dose-dependent manner. The cell death caused by this species was calpain- and caspase-dependent and of oxidative nature (Ferretti et al., 2007).

The compounds agelanesin A, agelanesin B, agelanesin C, and gelanesin D isolated from this species showed prominent activity for cytotoxicity against the murine L1578Y mouse lymphoma cell line with IC50 values of 9.6, 9.3, 16.8, and 13.1 μM, respectively (Putra and Jaswir, 2014).

Antimalarial, antimycobacterial, and antibacterial activities: The aqueous MeOH extracts of this species yielded six metabolites, namely, 24-ethyl-cholest-5alpha-7-en-3-beta-ol (1), 4,5-dibromopyrrole-2-carboxylic acid methyl ester (2), 4,5-dibromopyrrole-2-carboxylic acid (3), and (E)-oroidin (4), 3-amino-1-(2-aminoimidazoyl)-prop-1-ene (5), taurine (6), and minor compounds, namely, FAMA, consisting of a 1:2 mixture of (5Z,9Z)-5,9-tricosadienoic (7) and (5Z,9Z)-5,9-tetracosadienoic (8) acids, and FAMB composed of 8, (5Z,9Z)-5,9-pentacosadienoic (9) and (5Z,9Z)-5,9-hexacosadienoic (10) acids. Among these compounds, (E)-Oroidin (4) showed antimalarial activity and was found to be a more potent *Plasmodium falciparum* FabI inhibitor with an IC50 value of 0.30 μg/mL. The FAMA and FAMD (mainly consisting of methyl-branched fatty acids) inhibited *Mycobacterium tuberculosis* FabI with IC50 values of 9.4 and 8.2 μg/mL, respectively, and *Escherichia coli* FabI with IC50 values of 0.5 and 0.07 μg/mL, respectively. The majority of the above compounds also exhibited *in vitro* trypanocidal and leishmanicidal activities (Tasdemir et al., 2007).

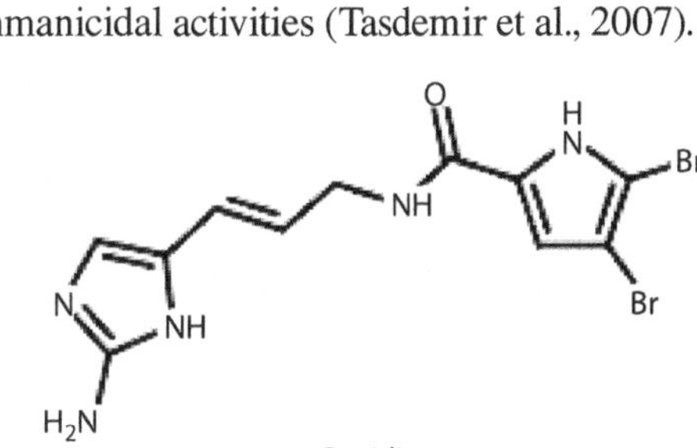

Oroidin

Agelas sp.

Compound(s) and Activities

Cytotoxic activity: The hydroethanolic extracts of this species showed cytotoxicity against tumor cell lines at a concentration of 20 mg/mL (Ferreira et al., 2007). The values of mean inhibitory concentration (IC50) recorded for the different cell lines are given next.

IC50 (µg/mL) Values of Tumor Cell Lines

HL-60	HCT-8	MDA-MB435	SF-295
62.5	63.3	52.9	59.4

Source: Ferreira, E.G. et al., 2007, *Porifera Res.: Biodiversity, Innovation Sustainability*, 313–318.

Neurosuppressive and muscle relaxant activity: The pyrrole-guanidine alkaloid compound, keramadine isolated from this unidentified species showed neurosuppressive and muscle relaxant activity by serving as a serotonergic receptor antagonist (Sipkema et al., 2005).

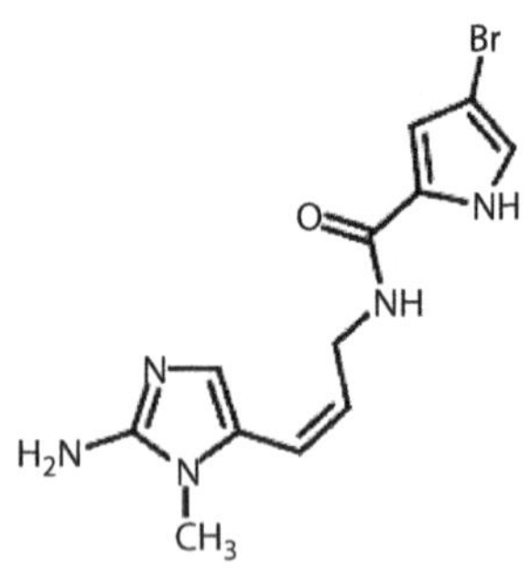

Keramadine

Antimicrobial and cytotoxic activities: The compounds agelasines and agelasimines of this unidentified species have shown antimicrobial and cytotoxic activities. It was particularly active against the fungal pathogen, *Candida krusei* with an MIC value of 15.6 µg/mL (Anjum et al., 2016).

Antibacterial activity: The alkaloids, agelasines O–R and T, showed antibacterial activity against *S. aureus* and *B. subtilis* (MIC 8.0–32.0 µg/mL). Further, the bromopyrroles, agelamadins A and B, of this species displayed antibacterial activity against B. subtilis (MIC 16 µg/mL each) and Micrococcus luteus (MIC 4.0 and 8.0 µg/mL, respectively). On the other hand, its 2-bromokeramadine showed weak activity against E. coli (Gribble, 2015).

Antifungal activity: The compounds of this species, namely agelasines Q and R, showed varying degrees of antifungal activity against the fungi *C. albicans, Aspergillus niger, Trichophyton mentagrophytes* and *Cryptococcus neoformans*, with the greatest activity toward the latter fungus (IC50 8.0 µg/mL each). Its bromopyrrole compound, mukanadin G, was active against *C. albicans* and *Cryptococcus neoformans* (IC50 16 and 8.0 µg/mL, respectively). Further, agelamadins A, B, C and E of this sponge displayed antifungal activity against *Cryptococcus neoformans* with IC50 values of 8, 4, 32, and 32 µg/mL, respectively. Other compounds isolated from it with antifungal activity against *Candida albicans, Trichophyton mentagrophytes, Cryptococcus neoformans*, and *Aspergillus niger* include nagelamides U–Z (Gribble, 2015).

FAMILY: AXINELLIDAE

Auletta elongata (Dendy, 1905) (= *Acanthella elongata*)

Order: Axinellida Family: Axinellidae
Common name: Not designated
Distribution: Gulf of Mannar, India
Ecology: It is a coastal, reef-associated species.
Biology: The body of the sponge is stalked with tubular branches, which grow in a bushy pattern or in one plane. The size of branches may vary from 4 to 9mm dia. Color of the sponge varies from pink to orange when alive. The texture is firm but compressible with good resiliency. The oscules are terminal on branches, circular, and 3–5 mm in diameter.

Compound(s) and Activities

Mosquito larvicidal activity: At a concentration of 10%, the methanol extract of this species showed high larvicidal activity (95%) with low LC value (0.066 mg/mL) (Sonia and Lipton, 2012).

Antibacterial activity: The crude methanol extracts of this species showed antibacterial activity against *Aeromonas hydrophila, Pseudomonas aeruginosa, Vibrio alginolyticus, V. anguillarum, V. fischeri, V. fluvialis*, and *V. vulnificus* with inhibition zones of 10–25 mm (dia.) (Sonia et al., 2008).

Auletta sp.

Order: Axinellida Family: Axinellidae

Compound(s) and Activities

Antitumor activity: An unusual tripeptide, hemiasterlin isolated from this species has shown antitumor activity through the stabilization of microtubules (Sipkema et al., 2005).

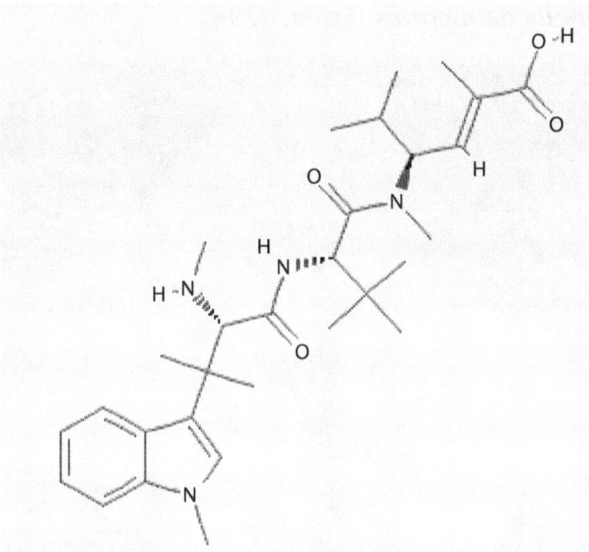

Hemiasterlin

Anticancer activity: The compound milnamide C isolated from this sponge showed significant anticancer activity against MDA-MB-435 breast cancer cells with IC50 values of 3.2×10^{-1} μg/mL (Agrawal et al., 2016).

Axinella cannabina (Esper, 1794)

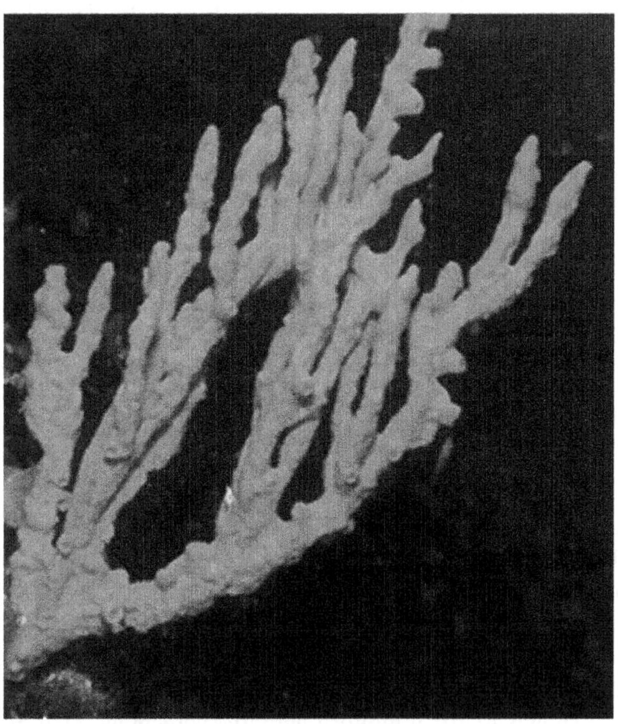

Order: Axinellida Family: Axinellidae
Common name: Tree sponge, barrel sponge
Distribution: North Atlantic Ocean, Mediterranean
Ecology: Sublittoral rocky substrates

Biology: This species is often with one or, at most, two trunks that branch out early from which few branches randomly come out. In general, both along the main axis, there are many side segments bearing oscula. It may show numerous lamellar structures that characterize this species. On the seabed these sponges are often isolated or otherwise well-spaced thinned to form small groups. The main branches of some specimens may exceptionally reach a meter of length. More often this species may have ramification or erect axe, with height ranging from 5 cm to 30 cm.

Compound(s) and Activities

Antimalarial activity: The compound (+)-axisonitrile-3 isolated from this species has shown strong antimalarial activity against *Plasmodium falciparum* (chloroquine-sensitive strain D6: IC50 = 142 ng/mL; chloroquine-resistant strain W2: IC50 = 16.5 ng/mL). On the other hand, the compound, (+)-axisothiocyanate-3 of this species was found to be 500-fold less active than the axisonitrile (strain D6: IC50 = 12340 ng/mL, strain W2: IC50 = 3110 ng/mL) (Emsermann et al., 2016).

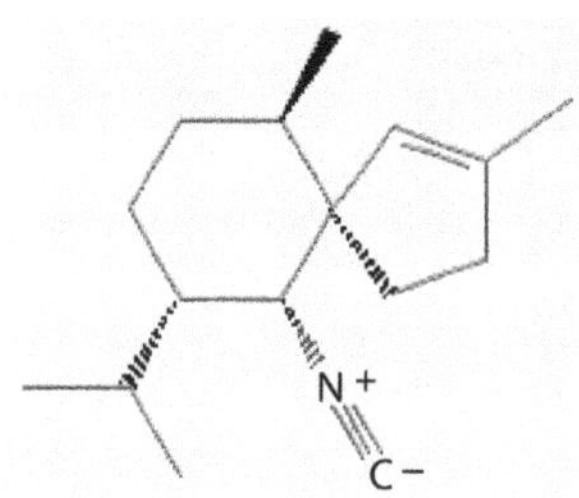

Avxisonitrile-3

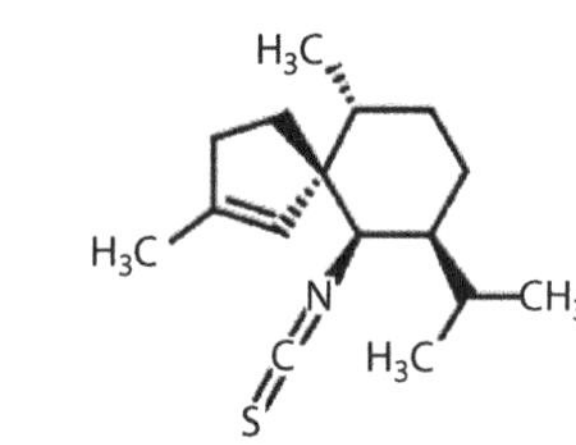

Axisothiocyanate-3

Antimicrobial activity: The organic extracts of this species have shown antibacterial activity against Gram-positive bacteria and antifungal activity, and the values of MICs (μg mL^{-1}) recorded for the different species of microorganisms are given next (Orhan et al., 2012).

MIC (μg mL^{-1})				
S. aureus	*E. faecalis*	*B. subtilis*	*C. albicans*	*C. parapsilosis*
128	128	64	16	16

Source: Orhan, I.E. et al., 2012, *Rec. Nat. Prod.*, 6: 356–367.

Antioxidant activity: The organic extracts of this species have shown DPPH radical scavenging activity (inhibition, 17%) at a concentration of 2000 μg mL^{-1} (Orhan et al., 2012).
AChE inhibitory activity: The organic extracts of this species have shown AChE inhibitory activity (percentage of inhibition, 39) at a concentration of 200 μg mL^{-1} (Orhan et al., 2012).
Others: This species has been reported to yield 5%–13% (dry weight of sponge) of collagen, which finds a plethora of applications, primarily in the sectors of cosmetics, pharmaceuticals, and medical care products (Tziveleka et al., 2017).

Axinella corrugata (George and Wilson, 1919)

Order: Axinellida Family: Axinellidae

Common name: Not known

Distribution: Indian and Pacific Oceans, Caribbean and Western Atlantic

Ecology: It inhabits coral reefs in shallow benthic zone at a depth range of 30–40 m.

Biology: It is a branching, bushy, or tubelike species. Surface is very corrugated. Consistency is tough and rubbery. Color of the body is orange to orange-yellow, yellow, or red-orange.

Compound(s) and Activities

Antiprotozoal activity: The associated fungal strain, *Hypocrea lixii*, of this species yielded an unidentified compound that displayed antiprotozoal activity against *Trichomonas vaginalis* fresh isolate, ATCC 30236, and metronidazole-resistant LACM2 with an IC50 value of 250 μg/mL (Indraningrat et al., 2016).

Others: The aqueous extracts of this species showed the presence of lectin. The biotinylated lectin has the capacity to stain the tumor cells such as that of breast (T-47D, MCF7), colon (HT-29), lung (H460), ovary (OVCAR-3), and bladder (T24). Further the lectin is used as potential metastatic markers and are crucial for developing successful tools for therapy against cancer (Dresch et al., 2013).

Axinella damicornis (Esper, 1794)

Order: Axinellida Family: Axinellidae

Common name: Yellow sponge

Distribution: British Isles, Mediterranean

Ecology: It grows on vertical cliff face at 15 m in sheltered location, and on rock reefs at 30 m.

Biology: It is branching-erect with branches compressed and "webbed" together. It is variable in appearance and is growing up to 10 cm in height, with short stalk. The fusion of the branches tends to result in irregular growth forms in taller specimens. In lower growing, squatter specimens, the lamellae become fused and convoluted. The terminal branches are about 3 mm in diameter. The surface has a characteristic "mealy" appearance. The surface is velvety, with projecting spicules of uneven length. Wide canals run up the sponge to arrive radially at "vents" on the distal edges of the lamellae. Color of the animal is bright to deep yellow, verging sometimes toward deep orange at the margins.

Compound(s) and Activities

Antioxidant activity: The organic extracts of this species have shown DPPH radical scavenging activity (inhibition, 38%) at a concentration of 2000 μg mL^{-1} (Orhan et al., 2012).

Aktas et al. (2013) reported that the organic extracts of this species have shown DPPH radical scavenging activity (inhibition, 71%; IC50, 519.5 μg/mL) at a concentration of 800 μg mL^{-1}.

AChE inhibitory activity: The organic extracts of this species have shown AChE inhibitory activity (percentage of inhibition, 52) at a concentration of 200 μg mL^{-1} (Orhan et al., 2012).

Antimicrobial activity: The organic extracts of this species have shown antibacterial activity against Gram-positive bacteria and antifungal activity and the values of MICs (μg mL^{-1}) recorded for the different species of microorganisms are given next (Orhan et al., 2012).

MIC (μg mL⁻¹)				
S. aureus	*E. faecalis*	*B. subtilis*	*C. albicans*	*C. parapsilosis*
128	128	32	16	16

Source: Orhan, I.E. et al., 2012, *Rec. Nat. Prod.*, 6: 356–367.

Cytotoxicity and neuroprotective effect: The fungal strain *Aspergillus niger* isolated from this sponge yielded a compound, 3,3′-bicoumarin (bicoumanigrin A), which showed moderate cytotoxicity against human cancer cell lines *in vitro*. Another compound of this fungal strain, aspernigrin B, displayed a strong neuroprotective effect by significantly reducing the increase of intracellular calcium concentration in rat cortical neurons stimulated with glutamic acid or quisqualic acid (Thomas et al., 2010).

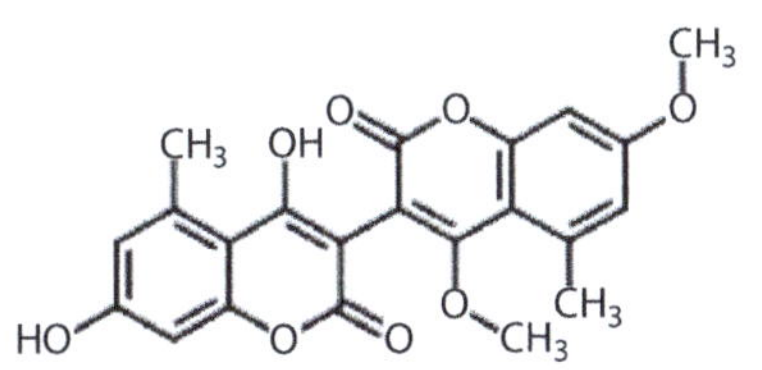
Bicoumanigrin A

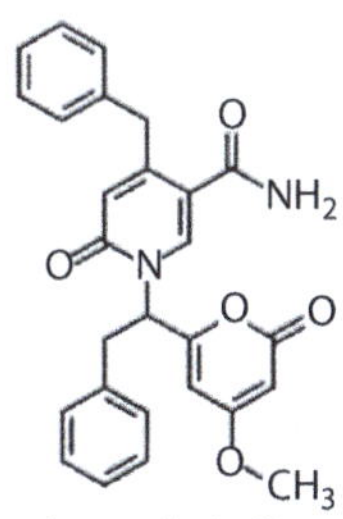
Aspernigrin B

Neuroprotective properties: A known pyrrole alkaloid agelongine (1) and a new natural product daminin (2), the bromine-free analog of 1, have been isolated from this species. Compound 2 showed significant neuroprotective properties (Aiello et al., 2005).

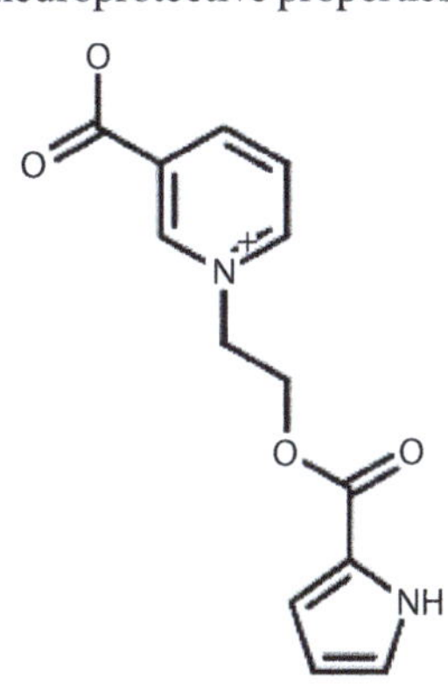
Daminin

Axinella donnani (Bowerbank, 1873)

Order: Axinellida Family: Axinellidae

Common name: Not known

Distribution: Western and Northern Madagascar, South India and Sri Lanka, Seychelles, Gulf of Guinea West, Indian Ocean, Kenya, Mozambique Channel

Ecology: It is a coastal, reef-associated species.

Biology: It is a dark, thick, cup-shaped sponge with undulated margin. It has a dark brownish color. Specimens may also appear orange. Body is distinctly pedunculate. The spicules are fairly stout, gradually sharp-pointed, and are with curved styli, which may be small and slender and scarce.

Compound(s) and Activities

Mosquito larvicidal activity: At methanol extract concentration of 10%, this species showed larvicidal activity (60%) with low LC value (1.7 mg/mL) (Sonia and Lipton, 2012).

Axinella polypoides (Schmidt, 1862)

Order: Axinellida Family: Axinellidae

Common name: Antler's sponge

Distribution: Western Mediterranean Sea, and on the Atlantic coast of Europe and Africa from the Bay of Biscay to Cape Verde

Ecology: It lives on rocky bottoms at depths of 20 to 100 m.

Biology: It has a cylindrical main shaft, which is usually erect. Sponges can form branches or fans, which are oval at the end. Its branches are rounded and surface is smooth with a yellow or orange color. The body length usually varies between 20 cm and 25 cm. This species can be flexible at the top part but hard at the lower section.

Compound(s) and Activities

Antiprotozoal activity: The associated action bacterial strain *Streptomyces axinellae* Pol001T of this species yielded the compounds Tetromycin 1–4, which displayed antiprotozoal activity against *T. brucei brucei* with IC50 values of 26, 40, 23, and 32 μg/mL respectively. Similarly the compound Tetromycin B produced by the above strain and valinomycin of its *Streptomyces* sp. 34 strain displayed similar activity with IC50 values of 17 and <0.1 μg/mL, respectively (Indraningrat et al., 2016).

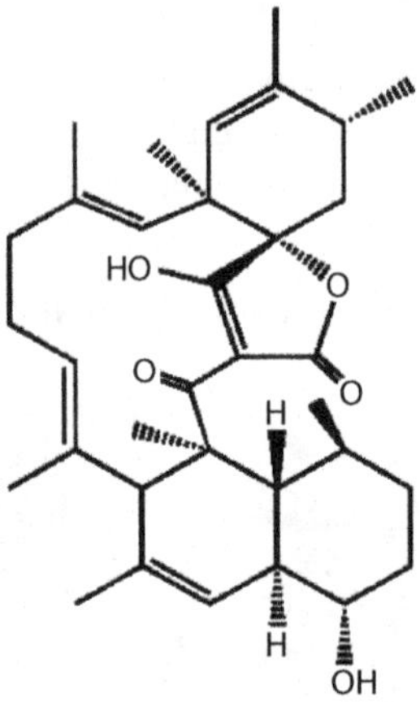

Tetromycin B

Axinella verrucosa (Esper, 1794)

Order: Axinellida Family: Axinellidae

Common name: Mediterranean sponge

Distribution: Mediterranean

Ecology: It grows on cliffs and on pebble bottom at a depth range of 10 to 75 m.

Biology: It is an arborescent species quite variable in shape Occasionally it is club-shaped, with cylindrical branches. The surface is grooved and hispid. It grows to a size of 10 cm high and the branches are up to 1 cm in diameter. Consistency is stiff. The styles of the spicules are often modified to tylostyles. Color of the animal is ocher yellow.

Compound(s) and Activities

Anti-inflammatory activity: The brominated alkaloid, hymenialdisine isolated from this species has shown anti-inflammatory activity by inhibiting IL-1 induced prostaglandin E2 production with an IC50 value of 0.6 μM (Keyzers, 2005).

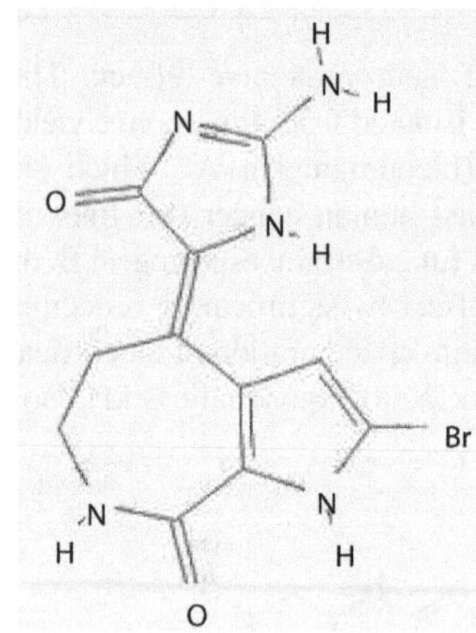

Hymenialdisine

Antiproliferative and antifungal activities: The ethyl acetate extract of *Penicillium* sp., derived from this sponge species yielded the compound communesin B and its new congeners communesins C and D, and the known compounds oxaline, griseofulvin, and dechlorogriseofulvin. The communesins exhibited moderate antiproliferative activity on different leukemia cell lines (Jadulco et al., 2004; Thomas et al., 2010).

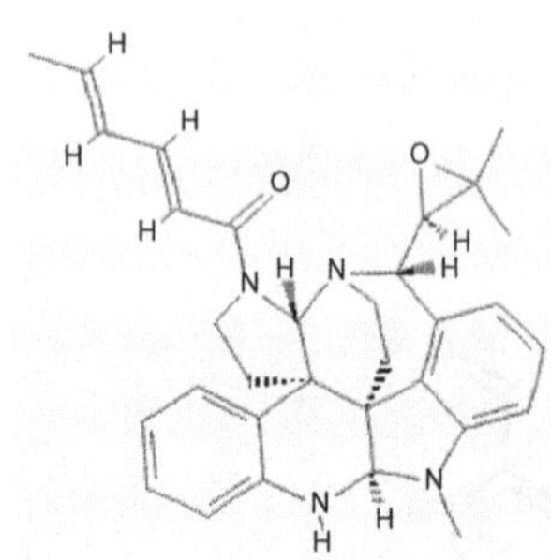

Communesin B

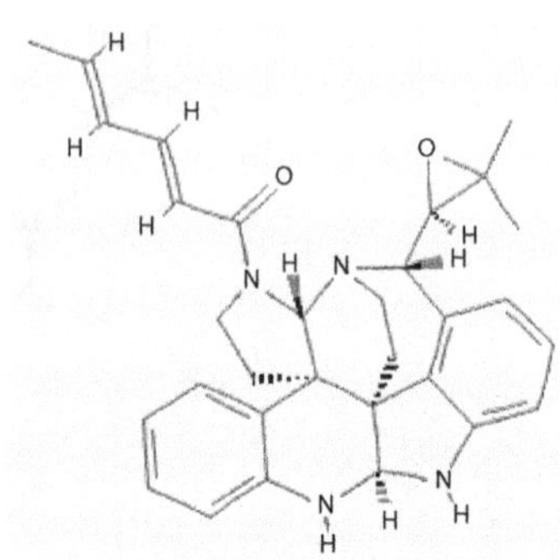

Communesin C

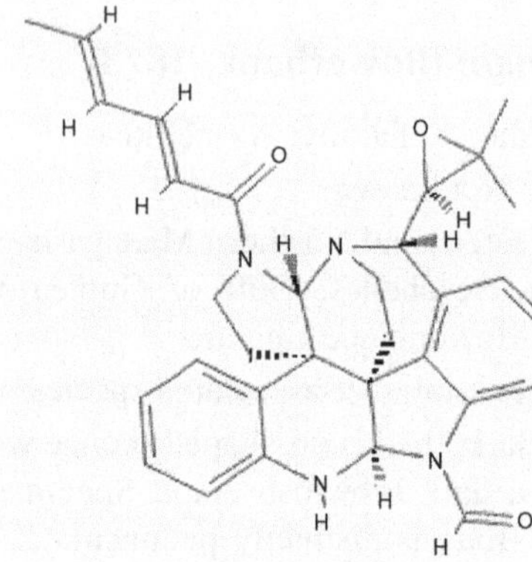

Communesin D

Neuroprotective activity: The bioactive compounds isolated from this species have shown neuroprotective activity against the agonists serotonin and glutamate *in vitro* (Aiello et al., 2006).

Axinella weltnerii (Lendenfeld, 1897)

Order: Axinellida Family: Axinellidae
Common name: Crumpled sponge
Distribution: Zanzibar
Ecology: Not reported
Biology: Not studied

Compound(s) and Activities

Cytotoxicity: A special group of triterpenoids, namely, sodwanones A–I and sodwanones K–W have been isolated from this species. Sodwanones G, H, and I have been found to have cytotoxic activity against cell cultures of P-388 murine leukemia, A-549 human lung carcinoma, HT-29 human colon carcinoma, and MEL-28 human melanoma. Sodwanones G, H, and I showed high specificity toward human lung carcinoma cell line A-549. The cytotoxic triterpenes sodwanones K, L, and M were found to be cytotoxic to P-388 murine leukemia cells. Sodwanone V inhibited both hypoxia-induced and iron chelator (1,10-phenanthroline)-induced HIF-1 activation in T47D breast tumor cells (IC50, 15 μM), and sodwanone V was the only sodwanone that inhibited HIF-1 activation in PC-3 prostate tumor cells (IC50, 15 μM). Sodwanone A and sodwanone T inhibited hypoxia-induced HIF-1 activation in T47D cells (IC50 values, 20–25 μM), and sodwanone V showed cytotoxicity to MDA-MB-231 breast tumor cells (IC50, 23 μM) (Li et al., 2013).

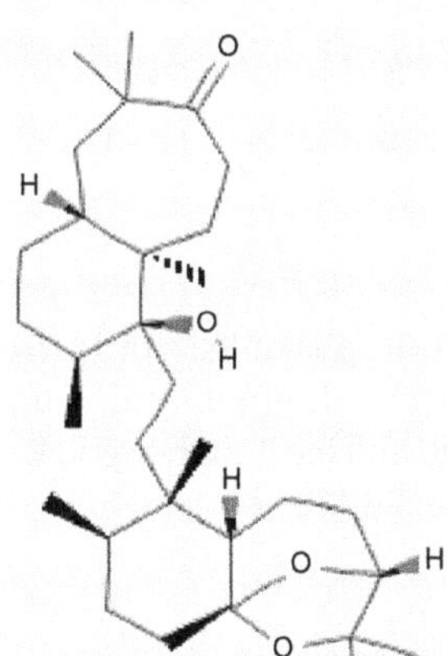

Sodwanone M

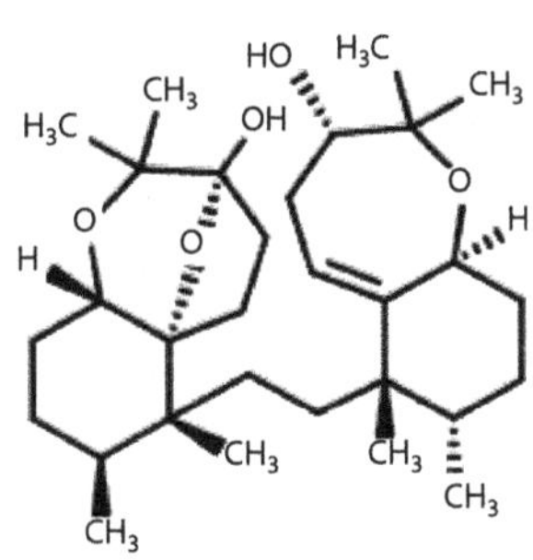

Sodwanone V

Axinella sp.

Order: Axinellida Family: Axinellidae

Compound(s) and Activities

Cytotoxicity: Sodwanone-derived compounds, 3-epi-sodwanone K (1), 3-epi-sodwanone K 3-acetate (2), and 10,11-dihydrosodwanone B (3), have been isolated from this unidentified *Axinella* sp. While compounds 1 and 3 inhibited hypoxia-induced HIF-1 activation in T47D cells (IC50 values, 20–25 μM), compound 2 was cytotoxic to T47D cells (IC50, 22 μM) (Li et al., 2013).

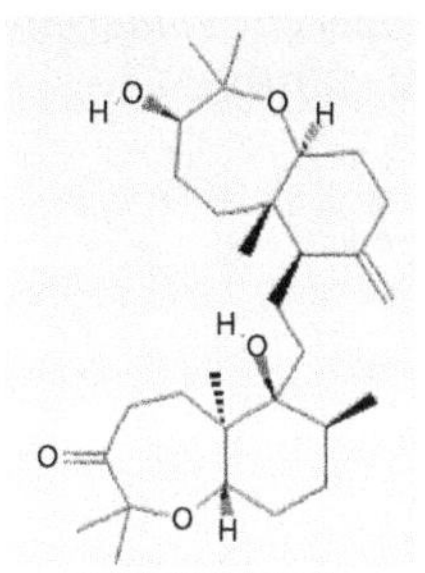

3-epi-sodwanone K

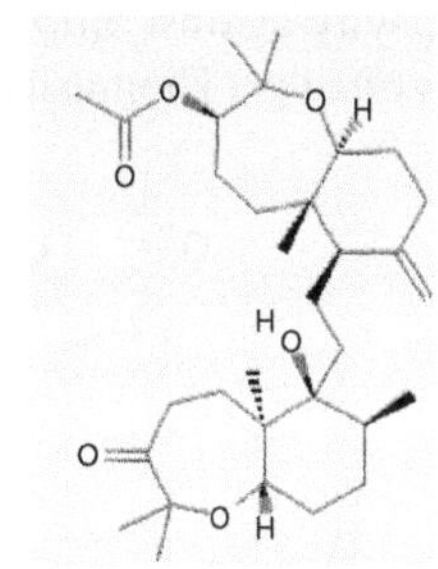

3-epi-sodwanone K 3-acetate

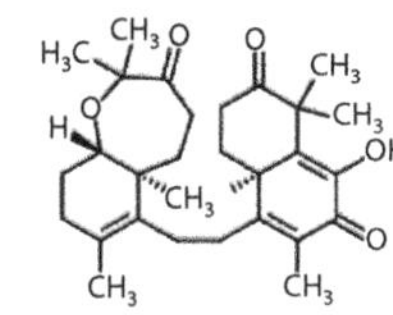

10,11-dihydrosodwanone B

The crude extract from a small-scale culture of the fungus *Acremonium* sp. 021172C derived from this sponge displayed potent cytotoxicity against leukemia and solid tumor murine and human cancer cell lines (Thomas et al., 2010).

Antibacterial activity: The imidazole alkaloids, namely, axinellamines B–D, isolated from this unidentified species have shown antibacterial activity against *M. luteus* with an MIC value of 16.7 μg/mL (Anjum et al., 2016).

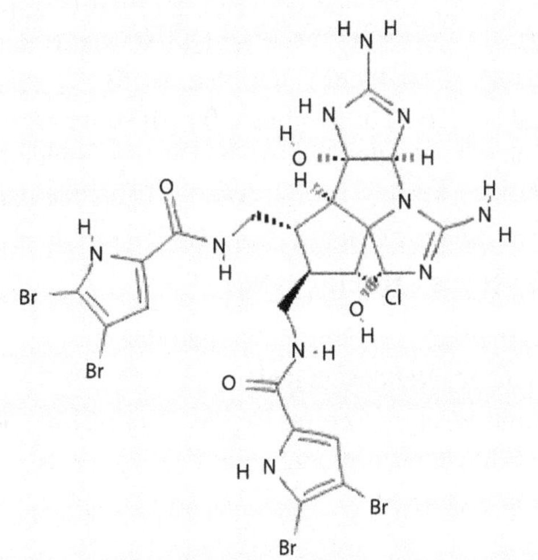

Axinellamine B

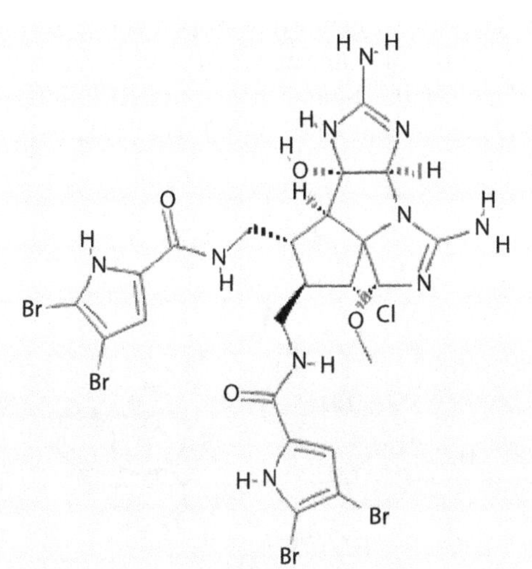

Axinellamine D

The unique steroids isocyclocitrinol A and 22-acetylisocyclocitrinol A isolated from the extract of a saltwater culture of this sponge-derived fungus *Penicillium citrinum* exhibited weak antibacterial activity against *Staphylococcus epidermidis* and *Enterococcus durans* (Thomas et al., 2010).

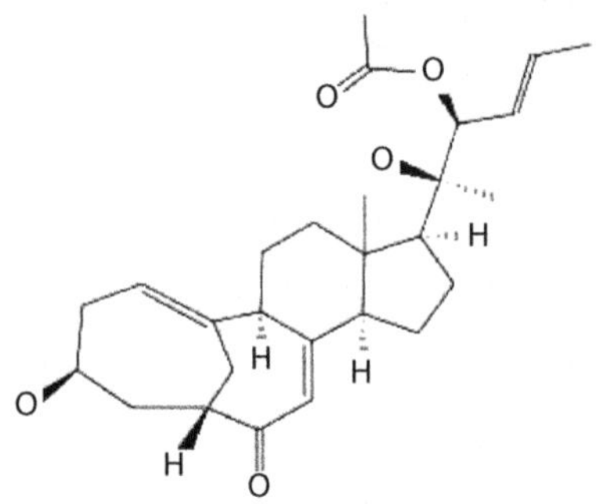

22-acetylisocyclocitrinol A

Antibacterial activity: The chlorine-containing metabolite, 3-O-methyl massadine chloride, of this species exhibited antibacterial activity against the Gram-positive bacteria *Staphylococcus aureus* (ATCC 9144 and 25923; IC50 3.7 and 4.2 μM, respectively) and *B. subtilis* (ATCC 6051 and 6633; IC50 2.6 and 2.2 μM, respectively), and the Gram-negative bacteria *E. coli* (ATCC 11775; IC50 4.4 μM) and *P. aeruginosa* (ATCC 10145; IC50 4.9 μM) (Gribble, 2015).
Antifungal activity: The symbiotic fungal strain *Myrothecium* sp. JS9 isolated from this unidentified species yielded antifungal metabolites, namely, roridin A and D (Thomas et al., 2010).

Phakellia fusca (Thiele, 1898)

Order: Axinellida Family: Axinellidae
Common name: Not designated
Distribution: Western Central Pacific, South China Sea
Ecology: It inhabits coral reef areas.
Biology: Not reported

Compound(s) and Activities

Cytotoxic activity: A new cyclic heptapeptide phakellistatin 13 isolated from this species was found to be significantly cytotoxic against the human hepatoma BEL-7404 cell line with an ED50 values of <10^{-2} μg/mL (Li et al., 2003).

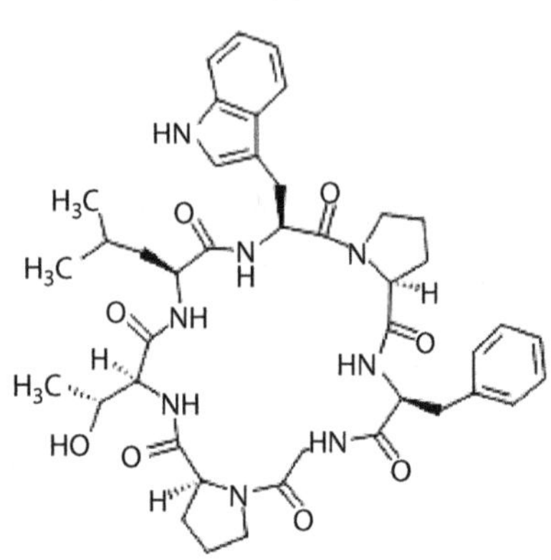

Phakellistatin 13

Antibacterial and antioxidant activities: Two new furan derivatives, hypofurans A and B (1 and 2), and three new cyclopentenone derivatives, hypocrenones A–C (3–5), along with seven known compounds (6–12), were isolated from a marine fungus *Hypocrea koningii* PF04 associated with this sponge species. Among them, compounds 1, 10, and 12 all showed modest antibacterial activity against *Staphylococcus aureus* ATCC25923 (MIC, 32 μg/mL). Further, compounds 1, 10, and 11 exhibited moderate DPPH radical scavenging capacity with IC50 values of 27.4, 16.8, and 61.7 μg/mL, respectively (Ding et al., 2015).

Phakellia stelliderma (Lévi and Lévi, 1989)

Order: Axinellida Family: Axinellidae
Common name: Not designated
Distribution: Western Central Pacific, Philippines
Ecology: It is a sessile species.
Biology: Not available

Compound(s) and Activities

Cytotoxic activity: Two new carotenoid neoplasm inhibitors, 19-hexanoyl-oxymytiloxanthin and 19-butanoyloxymytiloxanthin isolated from this species showed mild cytotoxic activity against P388 mouse leukemia cells (Kilimnik et al., 2016).
The carotenoids, 19-hexanoyloxymytiloxanthin and 19-butanoyloxymytiloxanthin isolated from this species were the neoplasm inhibitors and showed mild cytotoxic activity against P388 mouse leukemia cells (Dembitsky et al., 2006).

Phakellia sp.

Order: Axinellida Family: Axinellidae

Compound(s) and Activities

Cytotoxicity: A new cycloheptapeptide, phakellistatin 14, isolated from this sponge showed cancer cell growth inhibitory activity (ED50, 5 μg/mL) against the murine lymphocytic leukemia P388 cell line (Pettit and Tan, 2005).

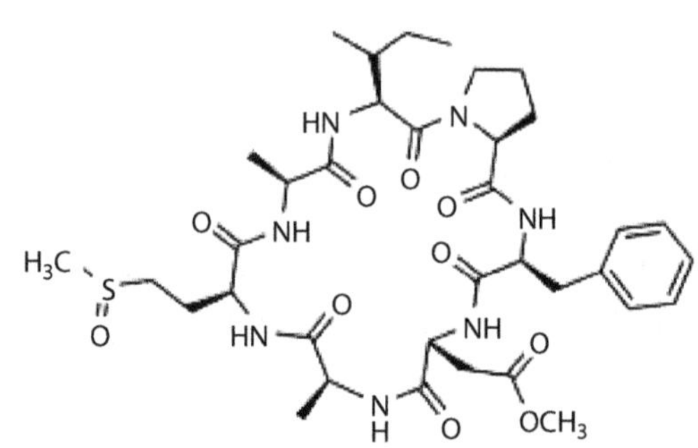

Phakellistatin 14

Anticancer/cytotoxic activity: The cyclic peptide compounds phakellistatins 2 and 4 isolated from this species showed cancer cell growth inhibitory properties (Pettit et al., 1995).

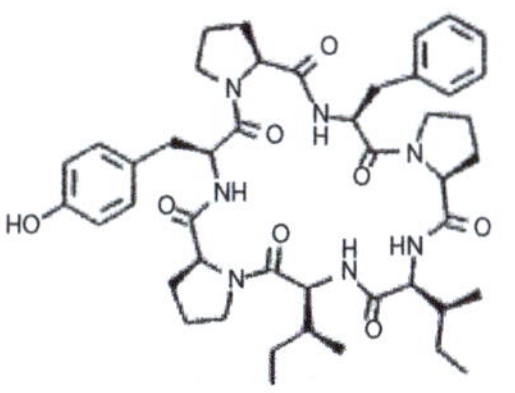

Phakellistatin 2

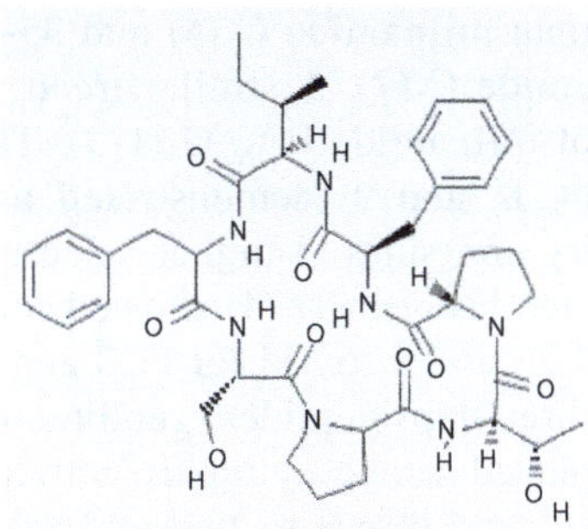

Phakellistatin 4

The cycloheptapeptide compound, phakellistatin 1 isolated from this species showed cell growth inhibitory property against P-388 murine leukemia with an ED50 value of 7.5 μg/mL (Pettit et al., 1993d).

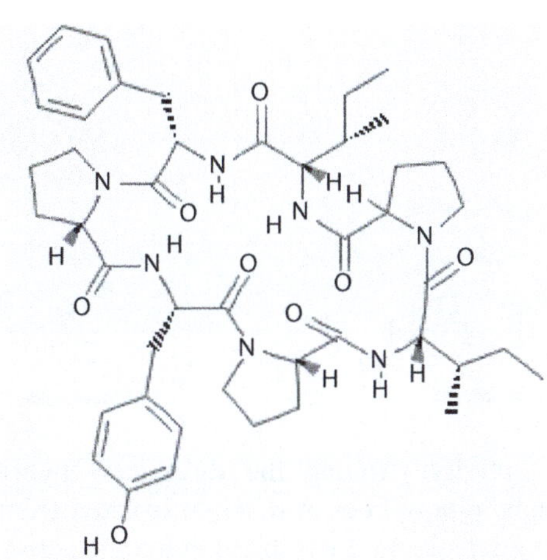

Phakellistatin 1

Galzitskaya and Caflisch (1999) reported on the isolation of three cyclic decapeptides—phakellistatins 7, 8, and 9—with cancer cell growth-inhibitory potentials.

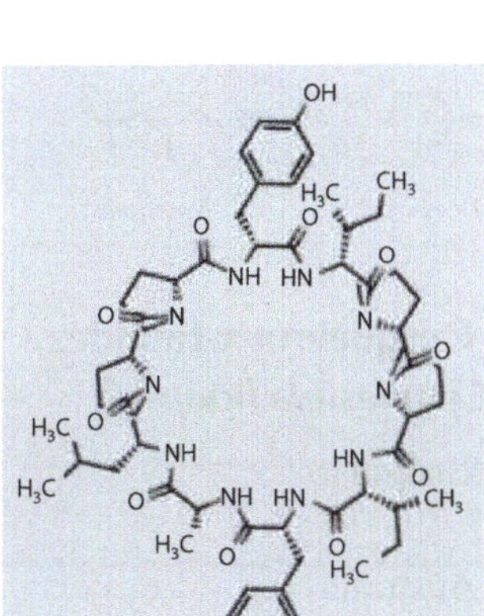

Phakellistatin 7

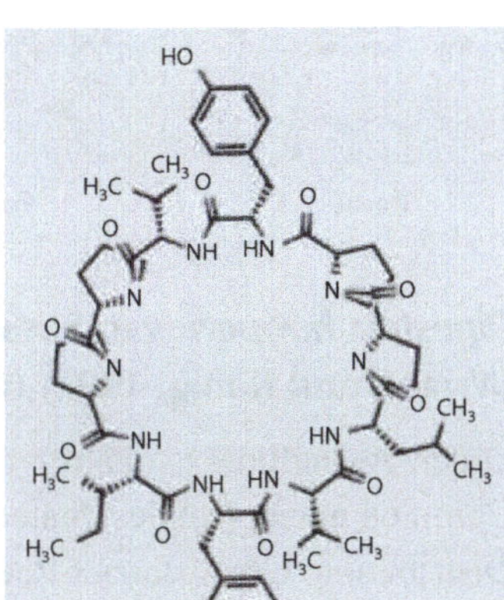

Phakellistatin 9

The compound phakellistatin 6 of this species was found to significantly inhibit growth of certain human cancer cell lines (GI50, 0.1 to 0.01 μg/mL) (Pettit et al., 1994b).

Phakettia trachys (Laubenfels, 1954) (= *Ptilocaulis trachys*)

Order: Axinellida Family: Axinellidae
Common name: Not designated
Distribution: Northwestern Gulf of Mexico
Ecology: It is a shallow water species inhabiting at a depths of 25–60 m.
Biology: Not documented

Compound(s) and Activities

Antifungal activity: The cyclic depsipeptide majusculamide C isolated from this species exhibited antifungal activity (Thomas et al., 2010).

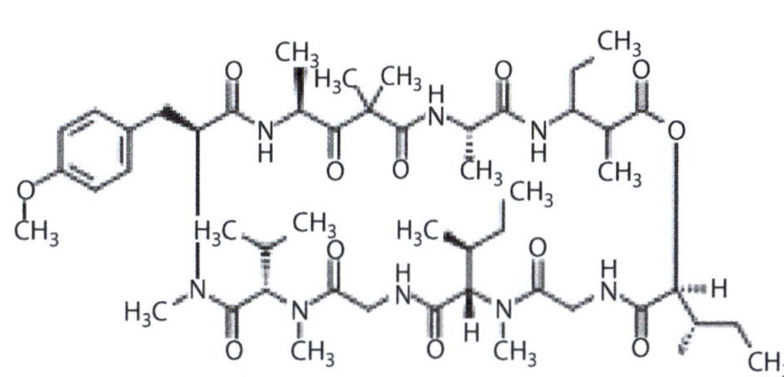

Majusculamide C

Pipestela candelabra (Alvarez, Hooper, and van Soest, 2008)

Order: Axinellida Family: Axinellidae

Common name: Bob Marley sponge, Australian marine sponge

Distribution: Western Central Pacific, Papua New Guinea, Solomon Islands, Vanuatu, Coral Sea Territories and Great Barrier Reef, Australia

Ecology: This sessile species is generally found growing on slopes, walls, overhangs, or caves of coral reefs in clear water (10–30 m visibility), between 5–50 m deep.

Biology: This sponge is erect and 15–40 cm tall on solid stalk of variable length and diameter. Growth form is arborescent, with cylindrical branches, lobes, hollow digits or tubes of 0.5–3 cm in diameter. These tubes are with walls of 3–5 mm thick and are with or without bulbous swellings along longitudinal length; sometimes they are fused together into a lamellar mass. Color of the animal is orange, yellow, or tan in life, and yellow beige in spirit. Oscula (3–5 mm dia.) are apical, large, conspicuous, and are on ends of tubes and on sides or tips of branches; sometimes they are covered with thin membrane. Texture is flexible, moderately soft, velvety, rubbery, tough, and may be sticky. Surface ornamentation: It is optically smoot and, fleshy. It is slightly translucent and membranous with subdermal drainage channels.

Compound(s) and Activities

Antiproliferative activity: Two consecutive prefractionated fractions this sponge extract were found to to be selectively active on the human prostate cancer cells (PC3) compared to the human neonatal foreskin fibroblast noncancer cells (NFF). Twelve metabolites—milnamides E–G (1–3), hemiasterlin D (4), milnamide A (5), milnamide C (6), milnamide D (7), hemiasterlin (or milnamide B) (8), hemiasterlin A (9), geodiamolide D (10), geodiamolide E (11), and geodiamolide F (12)—isolated from this species inhibited selectively the growth of prostate cancer cells (PC3 and NFF) with IC50 values in the picomolar to sub-micromolar range. Among the compounds, milnamide A (5) exhibited the strongest activity (IC50 = 11.0 nM) and was three-fold more potent than milnamide C (6) and 35-fold more potent than milnamide C (7). A similar trend was observed in the activity of milnamide E to G (1–3). The compounds hemiasterlins (4, 8, and 9) demonstrated much more potent cytotoxicity and showed higher selectivity toward noncancer cells. Hemiasterlin D (4) showed inhibition with IC50 values of 2.20 and 8.16 nM for PC3 and NFF cells, respectively. The three depsipeptides, geodiamolides D–F (10–12), also displayed strong cytotoxic activity against the PC3 cells and showed three- to five-fold selectivity with the NFF cells. Geodiamolide D (10) showed the strongest activity with an IC50 value of 33.1 nM, which was four-fold more potent than geodiamolide E (11) and five-fold more active than geodiamolide F (12) (Tran et al., 2014).

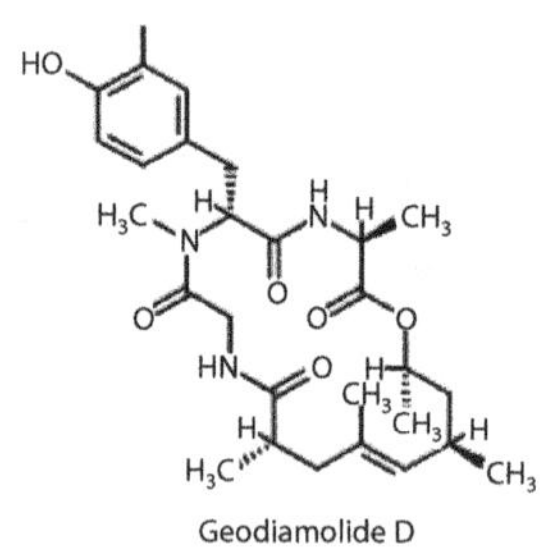

Geodiamolide D

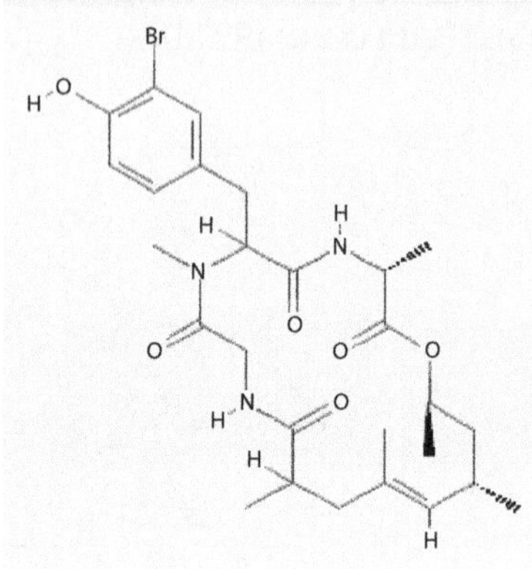

Geodiamolide E

Cytotoxic activity: Among the three new hybrid macrolides, namely, pipestelides A–C (1–3) isolated from this species, compounds 2 and 3 exhibited cytotoxic activities in the micromolar range (Sorres et al., 2012).

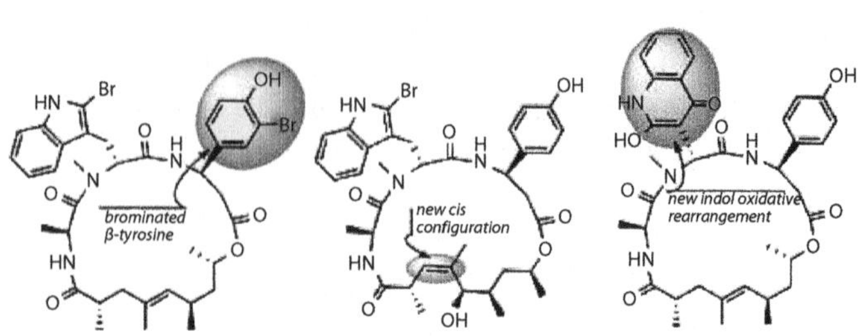

Pipestelide A Pipestelide B Pipestelide C

Pipestela hooperi (van Soest, Desqueyroux-Faúndez, Wright, and König, 1996) (= *Cymbastela hooperi*)

Order: Axinellida Family: Axinellidae

Common name: Not designated

Distribution: Great Barrier Reef, Australia

Ecology: It occurs in coral reef habitats and dead coral rocks at depths of 6–9 m.

Biology: This sponge irregularly spreads over the substrate and is stalkless. It may take partly an encrusting shape, but it is a modified cup-shaped, thickly lamellate sponge, with uneven, undulating incurved margins. The basal flattened part is large, about 30 cm diameter. Thickness of the lamella is about 10–30 mm. Both surfaces are similar. They are rough, slightly microhispid. They bear characteristic rounded elevations and are covered overall by a translucent collagenous membrane. Numerous and evenly distributed pores are visible on both surfaces. Oscules are small (2 mm dia.) and are irregularly distributed. They are usually placed on the top of the rounded elevations or on surface ridges present on both surfaces of the sponge. Consistency of the sponge is hard, stiff, and inflexible. Color of the sponge is yellowish to brown in life and beige in alcohol. Skeletal structure is of exclusively thin oxeas (encased in lightly developed spongin fibers).

Compound(s) and Activities

Antiplasmodial activity: The alkaloid, manzamine A isolated from this species has shown activity against Chloroquine-resistant *P. falciparum* with an IC50 value of 4.5 ng/mL. Similarly, the tetracyclic diterpene diisocyanoadociane of this species showed antiplasmodial activity against *P. falciparum* (D6 clone) with an IC50 value of 0.005 μg/mL (Anjum et al., 2016).

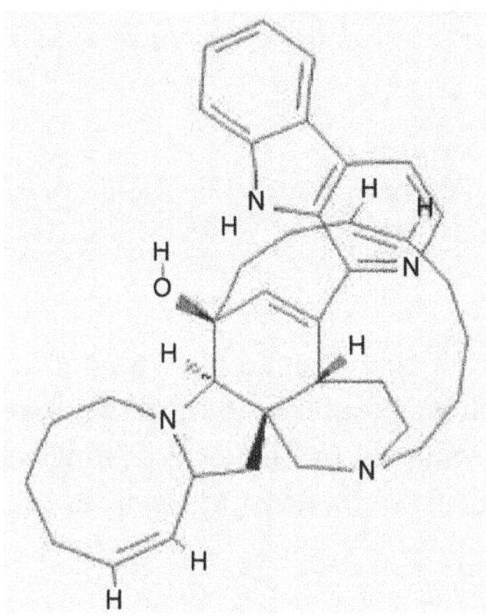

Manzamine A

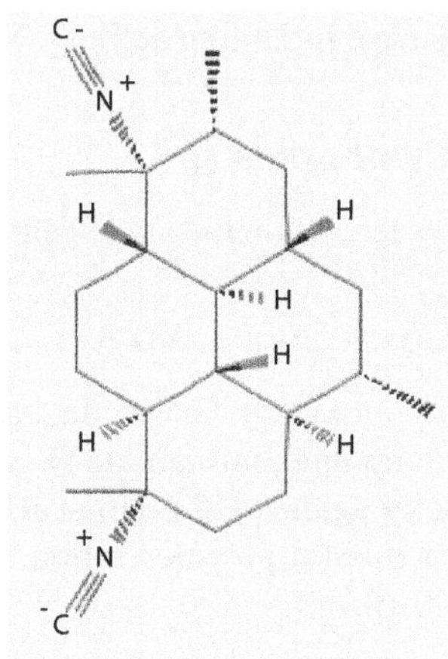

Diisocyanoadociane

Antibacterial and antitubercular activities: A total of nine diterpene isonitriles (1–9), and a sesquiterpene axisonitrile-3 (10), isolated from this sponge showed antibacterial (Gram positive and negative), antifungal, and antitubercular activities (Perdicaris et al., 2013; Wright et al., 2011).

Antiplasmodial activity: A total of 15 diterpenes (1–15) that contain isocyanate, isothiocyanate, and isonitrile functionalities have been reported from this species. All isolates with the exception of compound 1 demonstrated significant and selective *in vitro* antimalarial activity (Konig and Wright, 1996).

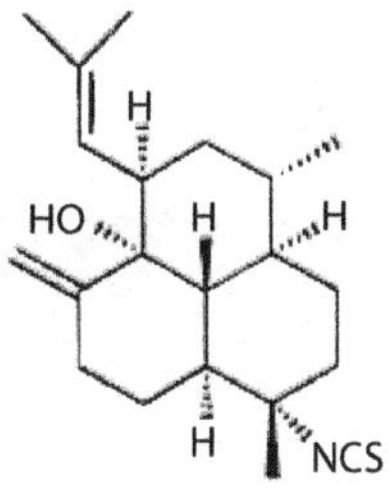

Diterpene 14

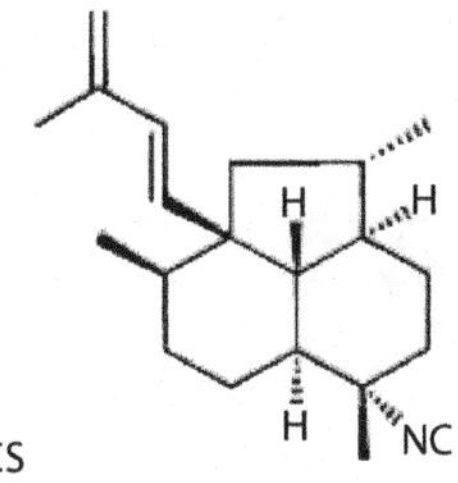

Diterpene 15

Among the five new diterpene formamides, namely, (1S, 3S, 4R, 7S, 8S, 11S, 12S, 13S, 15R, 20R)-7-formamido-20-isocyano-isocycloamphilectane (1), (1S, 3S, 4R, 7S, 8S, 11S, 12S, 13S, 15R, 20R)-7, 20-diformamido-isocycloamphilectane (2), (1S*, 3S*, 4R*, 7S*, 8S*, 12S*, 13S*)-7-formamido-cycloamphilect-11(20)-ene (3), (1R*, 3S*, 4R*, 7S*, 8S*, 12S*, 13S*)-7-formamido-amphilecta-11(20), 14-diene (4), (1S*, 3S*, 4R*, 7S*, 8S*, 12S*, 13S*)-7-formamido-amphilecta-11(20), 15-diene (5), isolated from this species, compound 1 containing both formamide and isonitrile was found to have moderate antiplasmodial activity (IC50, 0.5 μg/mL) (Wright and Lang-Unnasch, 2009).

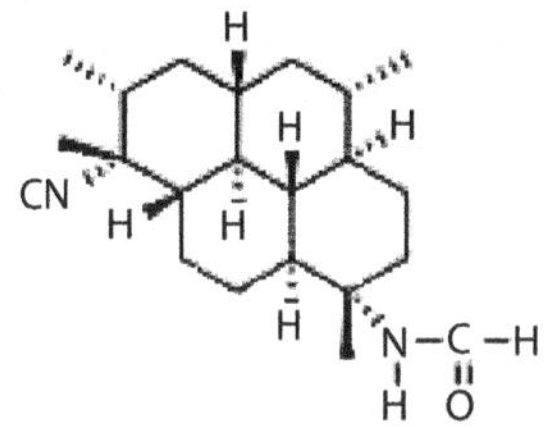

Formamide 1

Pipestela terpenensis (Fromont, 1993) (= *Amphimedon terpenensis*)

Order: Axinellida Family: Axinellidae

Common name: Not designated

Distribution: Australia, Coral Sea, Torres Strait, Northern Great Barrier Reef

Ecology: It is more commonly found in lagoons and more sheltered waters where it has been found to grow to a reasonably large size.

Biology: This sponge has tall erect lamellate and/or coalescing branches with the appearance of flattened organ pipes. The color is red brown alive with cyanobacterial symbionts. Oscula (2 mm in dia.) are located on the edges, raised ridges, and nodes of the sponge. Skeleton is reticulated with paucispicular ascending tracts, encased in lightly developed spongin fibers, and connected by short tracts and ending in narrow fans at surface. Spicules are exclusively oxeas, which are curved with fusiform or round tips and most with axial channel.

Compound(s) and Activities

Antimalarial activity: The diterpene isonitrile derivatives (including di-isocyano-adociane) of this species demonstrated significant *in vitro* antimalarial activity (Soest et al., 1996).

Cymbastela cantharella (Lévi, 1983) (= *Pseudaxinyssa canthranella*)

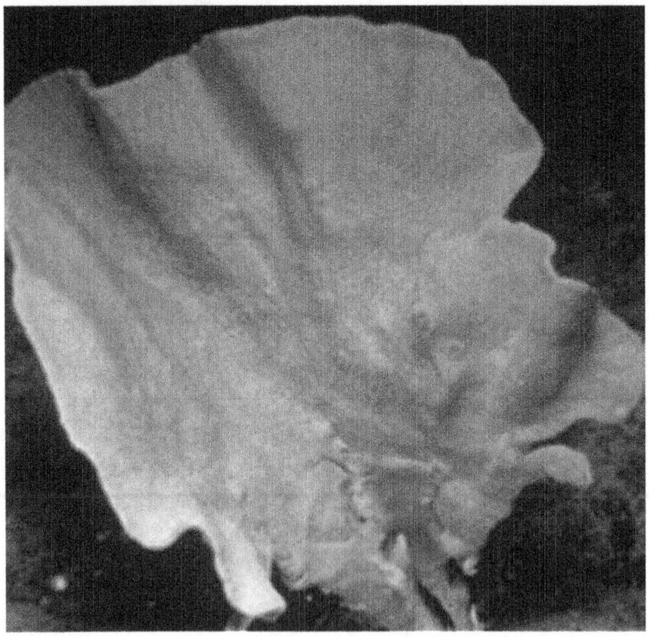

Order: Axinellida Family: Axinellidae

Common name: New Caledonian marine sponge

Distribution: Only from New Caledonia.

Ecology: It occurs at depths ranging from 10 to 40 m.

Biology: It is a short, erect, cup-shaped or semivasiform lamellate form (80–150 mm high and up to 170 mm maximum dia.) with a short stalk (30–40 mm long and 10–17 mm dia.) and is usually with convoluted margins. Occasionally this sponge has secondary cups or lamellae growing inside the primary cup, often with buttresses and secondary projections. The surface of the lamellae is predominantly smooth, although distinct differences between interior (inhalant) and exterior faces of cups exist. The exterior of the lamellae is microscopically smooth, optically uneven, and the interior of the cup has numerous small oscules (0.5–2 mm dia.). Oscules are approximately 2 mm apart. Color of the sponge is pale orange in life and beige in ethanol. Texture is firm, flexible, and slightly compressible. The exterior of the lamellae has a sandpaper-like texture and the interior is more compressible. Ectosome is membraneous, without specialized skeleton but with a heavy layer of slightly darker collagen. Oxeas of the spicules are relatively short, slender, slightly curved at the center, occasionally asymmetrical, tapering, fusiform, and sharply pointed. These projecting oxeas occur individually or in paucispicular plumose bundles, and are regularly dispersed across the surface, arising from ascending radial tracts in peripheral skeleton.

Compound(s) and Activities

Cytotoxic activity: The compound girolline isolated from this species showed cytotoxic activity and inhibited exponentially growing leukemic cells. Further, this compound had significant activity in several models of grafted murine tumors (Sullivan et al., 2009).

Antiproliferative activity: The compound girolline (or girodazole) isolated from crude ethanol extracts of this species has demonstrated potent antiproliferative activity *in vitro* on several cell lines such as P388 murine leukemia and P388/dox, KB naso-pharyngeal, and T24 bladder carcinoma human cell lines. The half maximal inhibitory concentration (IC50) values were found to be 0.06 μM, 0.08 μM, 0.21 μM, and 0.19 μM, respectively. Moreover, girodazole also has antitumor activity *in vivo* on the P388/dox cell line (Motuhi et al., 2016).

Antiplasmodial activity: The compound girolline has also shown promising results for the treatment of malaria, as this compound has displayed potent antiplasmodial activities against four *Plasmodium falciparum* strains with IC50 values ranging from 77 to 215 nM (Motuhi et al., 2016).

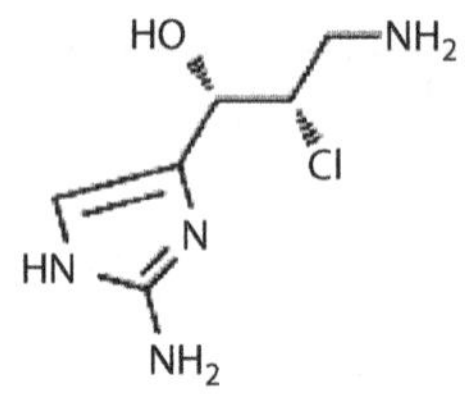

Girolline

Enzyme inhibitory activity: The dihydrohymenialdisine derivatives isolated from this species have shown improved kinase inhibiting activity (Sauleau et al., 2011).

Cymbastela sp.

Order: Axinellida Family: Axinellidae

Compound(s) and Activities

Cytotoxicity: The crude extract of this sponge gave a peptide compound, milnamide D, which was found to exhibit cytotoxicity against HCT-116 at IC50 value of 66 nM and inhibition of tubulin polymerization at IC50 of 16 μM (Agrawal et al., 2016).

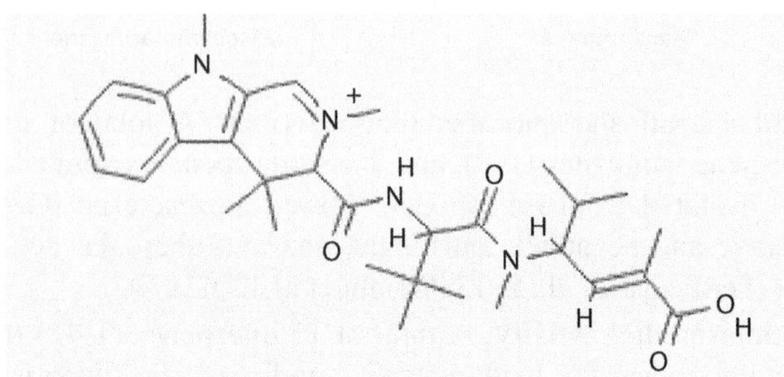

Milnamide D

The extracts of this sponge have yielded cytoloxic peptides, namely, geodiamolide G, hemiasterlin A, hemiasterlin B, criamide A, and criamide B, which showed cytotoxicity (Coleman et al., 1995).

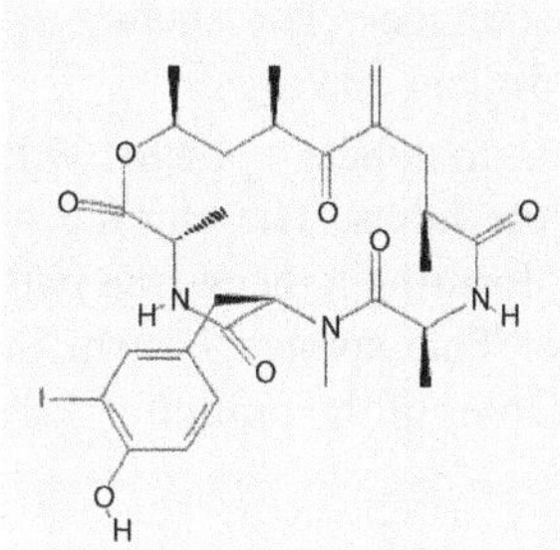

Geodiamolide G

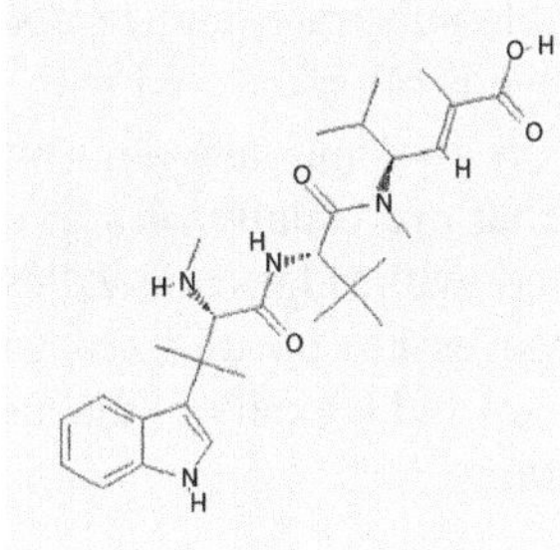

Hemiasterlin A

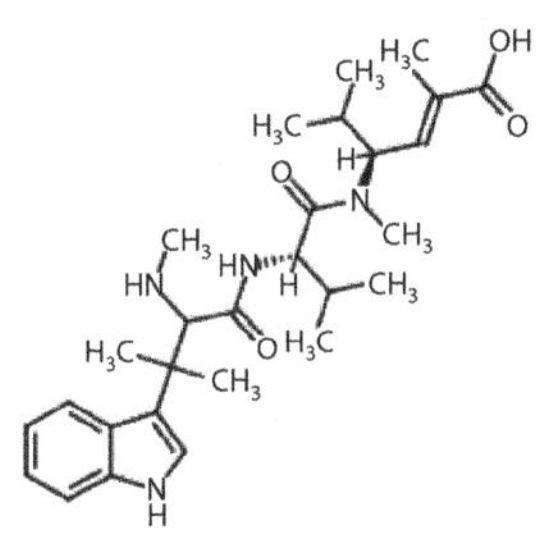

Hemiasterlin B

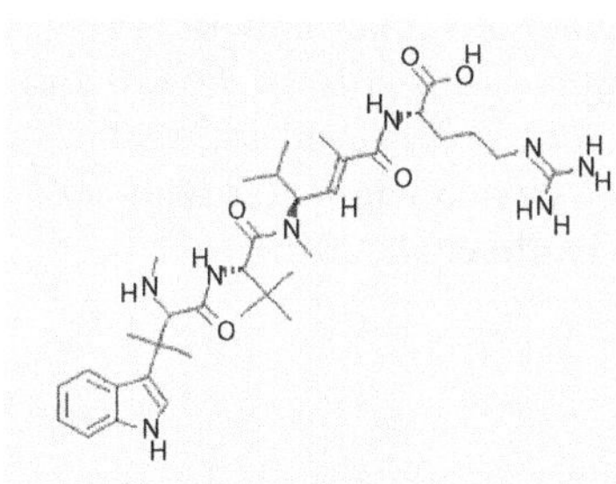

Criamide A

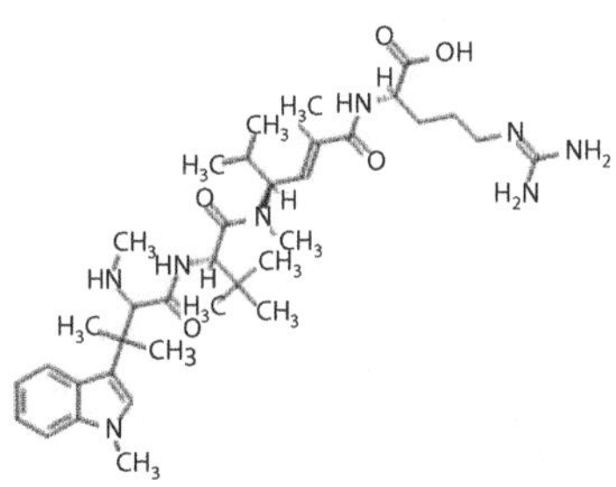

Criamide B

FAMILY: HETEROXYIDAE

Didiscus oxeatus (Hechtel, 1983)

Order: Axinellida Family: Heteroxyidae

Common name: Not known

Distribution: Eastern Brazil

Ecology: Not reported

Biology: This sponge is lobate with a size of 2.5 cm (height) and 4 × 2.5 cm. It is soft and easily crumbled. Color of the body is light tan. Cortex is papery with about 0.5 mm thickness. There are two oval oscules of 1–3 mm size placed apically. Surface of the sponge is even, rough to touch, and microgranular.

Compound(s) and Activities

Antiyeast, cytotoxic, and enzyme inhibitory activities, and antidermatophytic effect: The compounds (+)-curcuphenols and (+)-curcudiols have been isolated from this species. Whereras (+)-curcuphenol inhibited the growth of the fungi *Absidia ramosa, Aspergillus niger, Botrytis cinerea, Cladosporium cucumerinum, Fusarium oxysporum, Penicillium expansum, Rhizopus oryzae, Trichoderma harzianum, T. koningii*, and *T. mentagrophytes*, (+)-curcudiol only partially inhibited the growth of *A. ramosa* at a concentration of 200 μg/disc. Further, (+)-curcuphenols also displayed cytotoxicity against P-388 murine leukemia, A-549 lung, HCT-8 colon, and MDA-MB mammary cancer cell lines; and strongly inhibited the activity of gastric H, K-ATPase. Furthermore, (+)-curcuphenols also displayed an antidermatophytic effect against the human pathogenic fungus *Trichophyton mentagrophytes*. The dermatophytoses are the most common form of fungal infections found in most countries, affecting skin, hair, and nails (Gaspar et al., 2004).

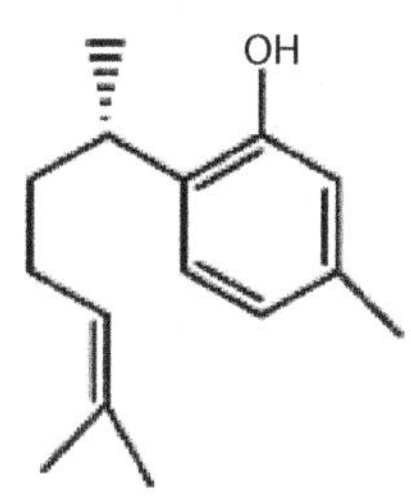

Curcuphenol

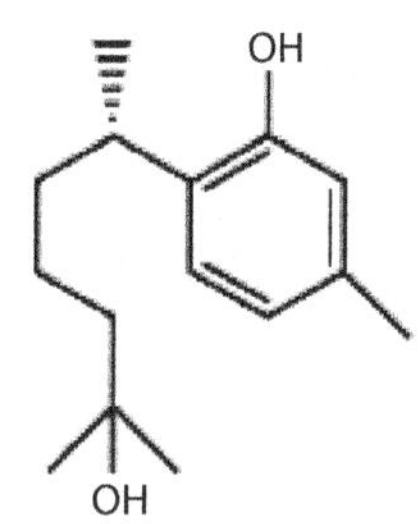

Curcudiol

Myrmekioderma granulatum (Esper, 1794)

Order: Axinellida Family: Heteroxyidae

Common name: Convoluted orange sponge

Distribution: Indo-West Pacific; Bay of Bengal, India

Ecology: It is found in sublittoral areas to a depth of 20 m. It inhabits fringing coral platforms, coral pools in sand and silt, beach rock, and dead coral substrates.

Biology: It is a massive-amorphous to lobate sponge with oscules strewn in spongin-enforced plumose tracts, which run radially and oblique to the surface. Ectosome consists of a palisade of small spicules with vertically packed arrangement. This sponge has been reported to compete for space with epizootic didemnid ascidian and other sessile marine invertebrates and plants.

Compound(s) and Activities

Antimicrobial activity: Fatty acid, volatile compounds, and sterols present in the lipophilic extract of this species were found to be active against several pathogenic microorganisms (Fernández et al., 2013; Mishra et al., 2009).

Myrmekioderma gyroderma (Alcolado, 1984) (= *Epipolasis reiswigi*)

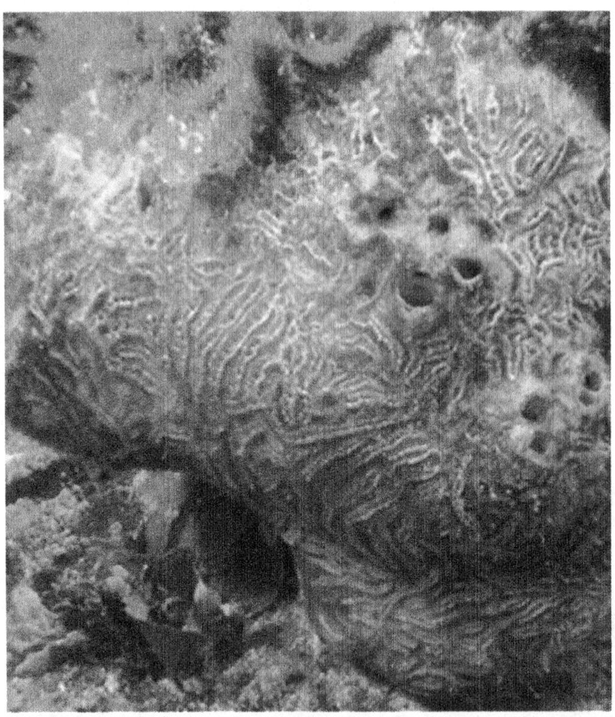

Order: Axinellida Family: Heteroxyidae

Common name: Convoluted orange sponge, Caribbean marine sponge

Distribution: Widespread in the Caribbean Sea; Jamaica

Ecology: It is common in deep areas of coral reefs (20–80 m) in protected locations with little water movement, in sand-rubble substrate, and in coral crevices. The surface of this sponge is often covered with sediment and algae.

Biology: It is a massive-amorphous to lobate sponge with oscules that are usually on top of the lobes. The surface is convoluted, with ridges and valleys, forming a maze-like pattern. Grooves are up to 1 cm wide. This sponge is firm in consistency and it exudates mucus. Color of the animal is yellow to orange.

Compound(s) and Activities

Antitubercular activity: The 27 diterpenes called cyanthiwigins of this species showed anti-TB activity against *M. tuberculosis*. However, the best result was the moderate activity of cyanthiwigin (C) with 50% of inhibition at 6.25 μg/mL (Jyothi et al., 2016).

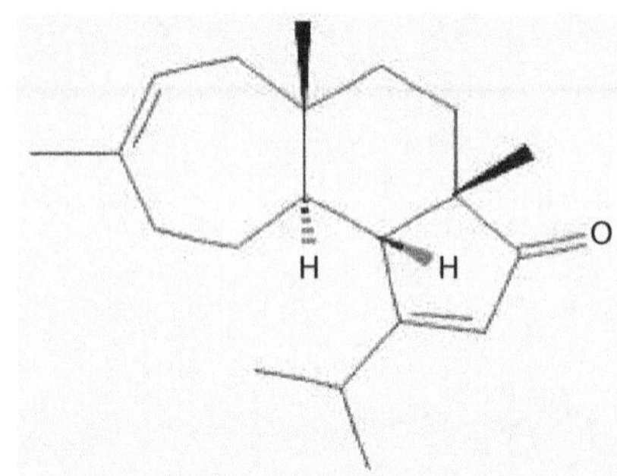

Cyanthiwigin

Antitumor activity: The organic extract fraction of this species showed antitumor activity by inhibiting MDA-MB-231 and HT29 cell line growth (IC50, <1 μg/mL) at 5 and 25 μg/mL concentration, respectively (Fernández et al., 2013).

Cytotoxic activity: The compounds cyanthiwigin A–D isolated from this species showed cytotoxic activity (Green et al., 1992).

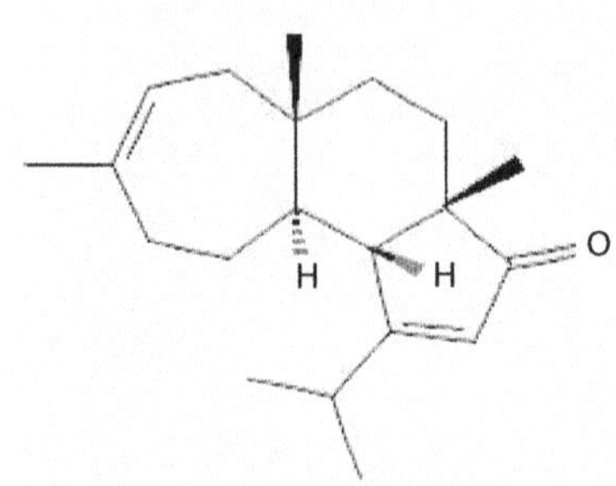

Cyanthiwigin A

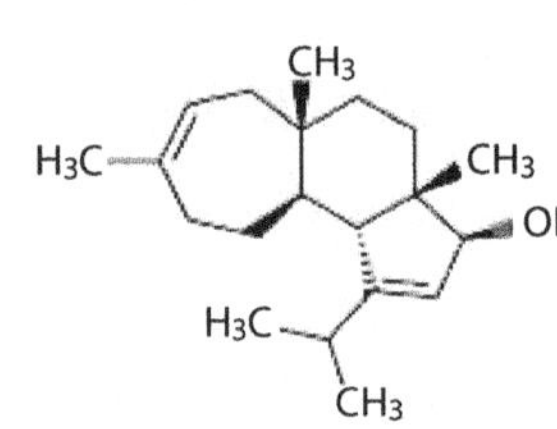

Cyanthiwigin C

Antiproliferative effect: The fatty acids of the extract fraction of this species totally inhibited breast cancer (MDA-MB-231) and colon carcinoma (HT29) cell line growth at 5 and 25 μg/mL concentrations, respectively (IC50, <1 μg/mL) (Fernández et al., 2013).

Myrmekioderma rea (Laubenfels, 1934) (= *Myrmekioderma styx*)

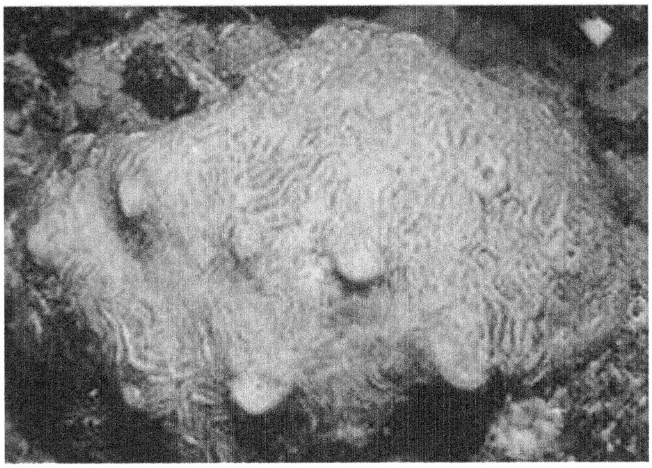

Order: Axinellida Family: Heteroxyidae

Common name: Not designated

Distribution: Pacific and Atlantic Oceans, Gulf of Mexico, Jamaica

Ecology: It occurs in shallow water habitats (0–30 m) including barrier reefs.

Biology: It is an amorphous, lumpy, and orange sponge with a height of 7 cm and diameter of 14 cm. Surface is whitish due to numerous spicules and foreign debris. Consistency is mediocre. Surface is extremely irregular with hundreds of hemispherical lumps or tubercles of 3 mm diameter. Several saucer-shaped depressions (12 mm dia.) are also seen.

Compound(s) and Activities

Antiviral and anti-TB activities: Diterpenes, cyanthiwigins A–AA, isolated from this sponge showed activity against hepatitis B virus (HBV); *Mycobacterium tuberculosis* (Mtb) strain H37Rv at 43 mg/mL EC50, and 50% inhibition at 6.3 mg/mL; the human immunodeficiency virus (HIV-1) with 42 mM EC50; and against human primary tumor cells at 3–18 mM IC50 (Fernández et al., 2013).

Myrmekioderma sp.

Order: Axinellida Family: Heteroxyidae

Compound(s) and Activities

Antitumor, antimalarial and enzyme inhibitory activities: The mono-O-alquyl-glycosyl glycerols of this species caused reverting in tumor cell morphology. Its sesquiterpenes (bisabolene type) displayed inhibiting activity of the gastric ATPase. Further, its (+)-curcuphenol showed inhibiting activity against cancer cell lines and parasites, such as *P. falciparum* (Fernández et al., 2013).

FAMILY: RASPAILIIDAE

Ectyoplasia ferox (Duchassaing and Michelotti, 1864)

Order: Axinellida Family: Raspailiidae

Common name: Brown encrusting octopus sponge, Caribbean marine sponge

Distribution: Western Central Atlantic; Bahamas, Brazil, Florida, Caribbean, Trindade and Martin Vaz Islands

Ecology: It occurs on coral reefs, hard bottoms, lagoons, and mangroves at a depth range 2 to 25 m. It hosts alpheid shrimp in its canals.

Biology: This sponge grows in two forms: (1) thickly encrusting, with oscules on top of hemispherical or low conical projections, up to 5 cm thick, with a spread of 30 cm or more; and (2) massive, erect (10 cm), long often compressed, thickly flabellate, with oscules on lateral, horizontally projecting elevations, and on terminal erect chimneys. The oscules in both types are 2 to 7 mm wide. The surface has a felt-like texture and the consistency is spongy, compressible, resilient, and easily torn. The color of this sponge is brownish orange externally and yellow orange internally.

Compound(s) and Activities

Cytotoxic/antitumor activity: The hydroethanolic extracts of this species showed cytotoxic activity (IC50, >100 μg/mL) against tumor cell lines HL-60, HCT-8, MDA-MB435, and SF-295 at a concentration of 20 mg/mL (Ferreira et al., 2007). The compounds, feroxosides A and B, isolated from the polar extract of this species were found to be moderately cytotoxic (IC50, 19 μg/mL) against murine monocyte-macrophage cell line (Campagnuolo et al., 2001).

Two novel norlanostane triglycosides, ectyoplasides A and B, isolated from the methanolic extracts of this species showed moderate cytotoxic activity when tested *in vitro* toward various cell lines (Cafieri et al., 1999).

Three new triterpene glycosides, named ulososide F (1), and urabosides A (2) and B (3), together with the previously known ulososide A (4) have been isolated from this species. No significant cytotoxicity against two cell lines (Jurkat and CHO cells) was detected below 50 μM for any of the above compounds. Only compound 3 exhibited a low cytotoxicity, with an IC50 value of 100 μM. Other compounds were considered inactive, with IC50 >100 μM, and with cytotoxicities below 30% up to 100 μM. Studies on the bioactivities of triterpene glycosides reported moderate cytotoxicity against J774 (murine monocyte-macrophage), WEHI164 (murine fibrosarcoma), and P388 (murine leukemia) cell lines with IC50 values ranging from 8.5 to 19 μg/mL (Colorado et al., 2013). The fungus *Phoma* sp. associated with this sponge species yielded cytotoxic compounds epoxyphomalin A and B. Among these, compound A showed superior activity against various human tumor cell lines (Thomas et al., 2010).

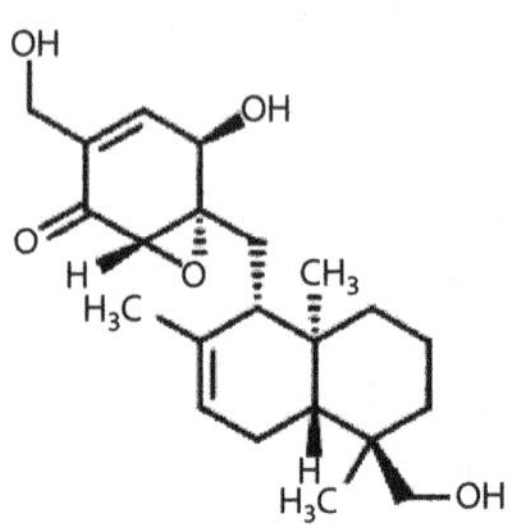
Epoxyphomalin A

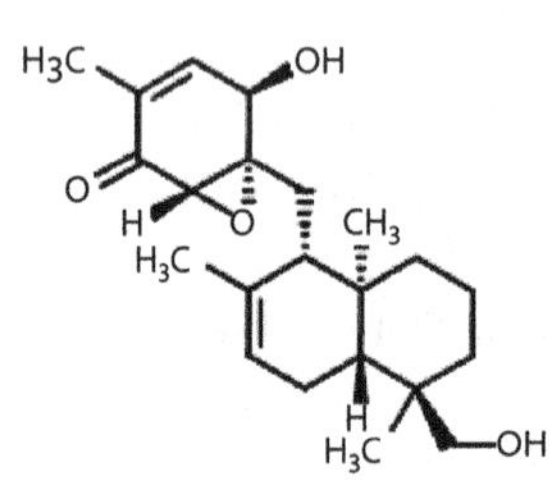
Epoxyphomalin B

Anticancer activity: The fungal strain *Spicellum roseum* 193H15 derived from this sponge was found to produce trichothecenes such as trichodermol and 8-deoxytrichothecin. These compounds inhibited the activity of LacCer synthase (which has a greater role in oncogene expression and cell proliferation) in neuroblastoma cells (Thomas et al., 2010).

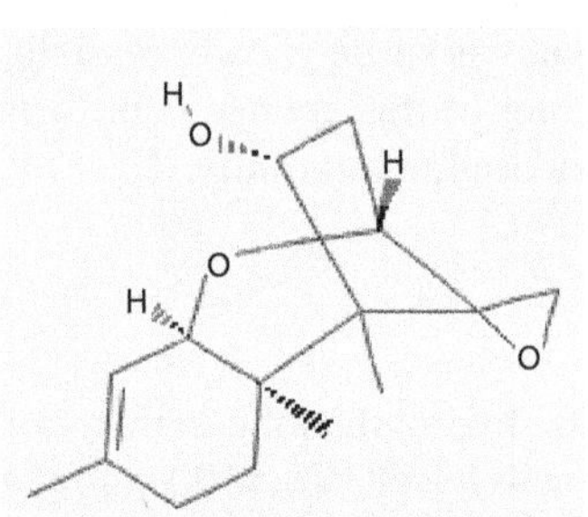
Trichodermol

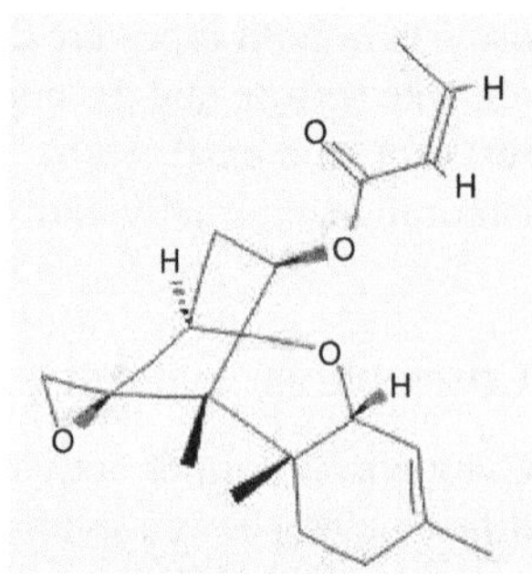
8-deoxytrichothecin

Antimicrobial activity: The associated fungal strain *Coniothyrium* sp. 193477 of this sponge species yielded antimicrobial compounds such as (3S)-(3′,5′-dihydroxyphenyl) butan-2-one, 2-(1′(E)-propenyl)-octa-4(E),6(Z)-diene-1,2-diol, (3R)-6-methoxymellein, (3R)-6-methoxy-7-chloromellein, and cryptosporiopsinol. Among these, cryptosporiopsinol demonstrated significant antimicrobial activity (Thomas et al., 2010).

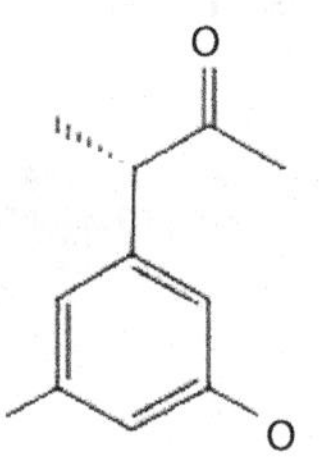
(3S)-(3′,5′-dihydroxyphenyl)butan-2-one

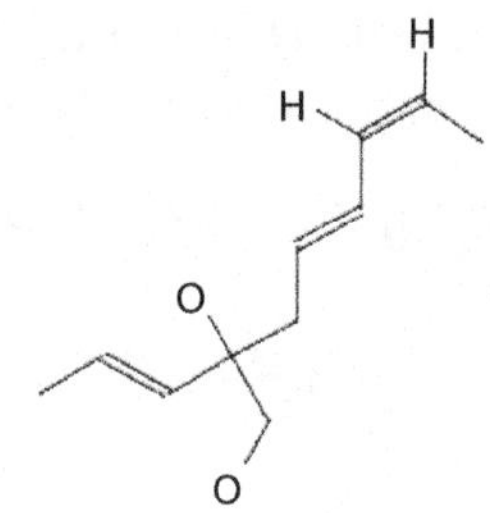
2-(1′(E)-propenyl)-octa-4(E),6(Z)-diene-1,2-diol

Lithoplocamia lithistoides (Dendy, 1922)

Order: Axinellida Family: Raspailiidae

Common name: Not designated

Distribution: Panama, Mauritius, Seychelles

Ecology: It occurs at depths of 80 to 200 m.

Biology: It is a massive, encrusting sponge with uneven, nodular surface that is minutely rough. Vents are minute and are inconspicuous. Texture is very hard, compact, and is incompressible. Color in spirit is pale grayish yellow. The body has a size of 16 mm height, 3.5 mm breadth, and 10 mm thickness. The main skeleton is dense, subisodictyol reticulation of acanthostrongyla, in which the spicules are stout, short, and slightly curved with strong, sharp spines. The spicules make up the entire skeleton, but there may be ill-defined bands of smooth styli can be seen running toward the surface.

Compound(s) and Activities

Antitumor activity: The tubulin-binders, namely, polyketides PM050489 (1) and PM060184 (2), of this species showed excellent growth inhibition against human tumor cells, including HT-29 (colon), A-549 (lung), and MDA-MB-231 (breast), with GI50 values of 0.5, 0.4, and 0.5 (1) and 0.4, 0.6, and 0.7 (2) nM, respectively (Gribble, 2015).

Prosuberites laughlini (Díaz, Alvarez, and van Soest, 1987) (= *Eurypon laughlini*)

Order: Axinellida Family: Raspailiidae

Common name: Caribbean marine sponge

Distribution: Southern Caribbean Sea, West Indian region, Bahamas

Ecology: It occurs as a thin film over dead coral; elevation from the substrate is up to 4 mm.

Biology: This encrusting (0.3–4.0 cm thick) sponge is dull orange to yellow externally and lighter internally. Consistency is soft. The surface is visually smooth in thin specimens and rugose on thicker ones. Microscopically, the surface is hispid, due to the outward projection of single spicules (mainly tylostyles). Some specimens however, show coarse elevations made up by spicule tracts. The oscules (1–6 mm dia.) have a collar up to 5 mm in height and are not visible as they are generally contracted. The ectosome is thin organic skin, which is pierced by large spicules.

Compound(s) and Activities

Anticancer activity: The cyclic heptapeptide rolloamide A (which exists in two conformers, major and minor) exhibited growth suppressive activity against a panel of histologically diverse cancer cell lines (Williams et al., 2009).

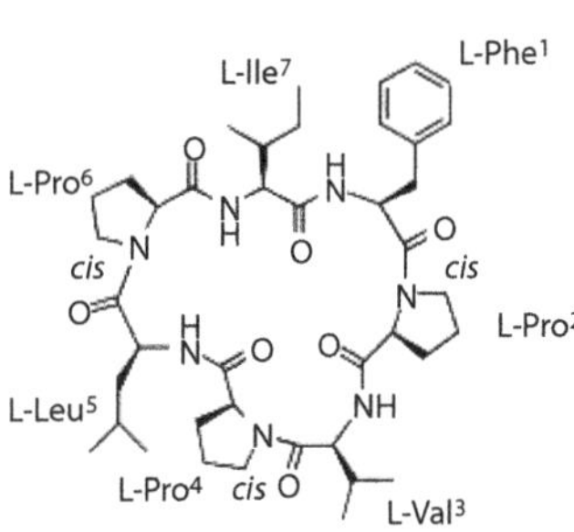

Rolloamide A (major conformer)

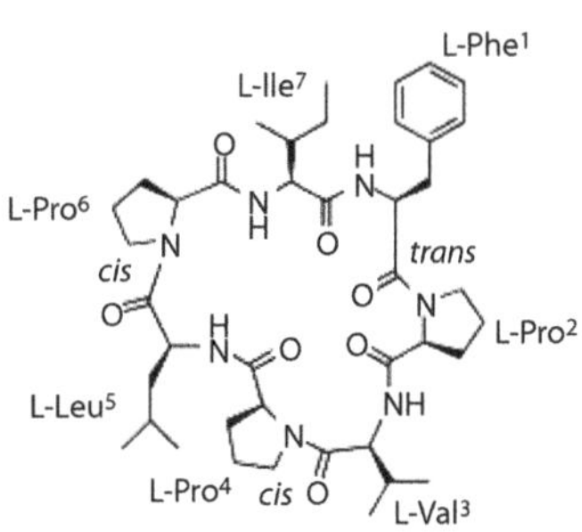

Rolloamide A (minor conformer)

Cytotoxicity: The cyclic heptapeptides euryjanicin A–D (1–4) and the cyclic octapeptide dominicin (5) isolated from this species have shown weak cytotoxicity against non-small cell lung cancer (NCI-H522), melanoma (LOX IMVI), ovarian cancer (IGROV1), and renal cancer (UO-31) (Vera et al., 2009).

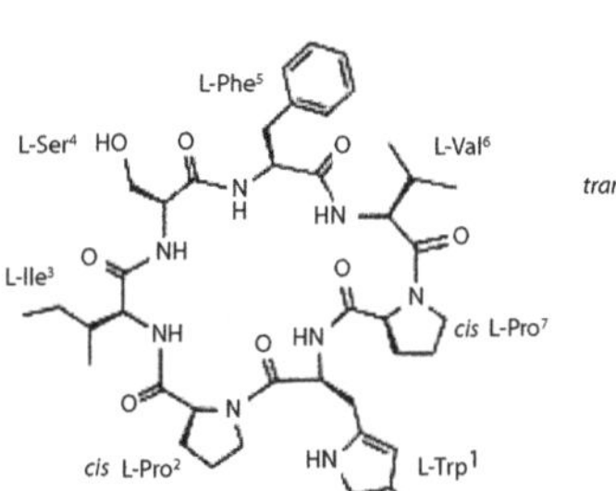

1

Euryjanicin A

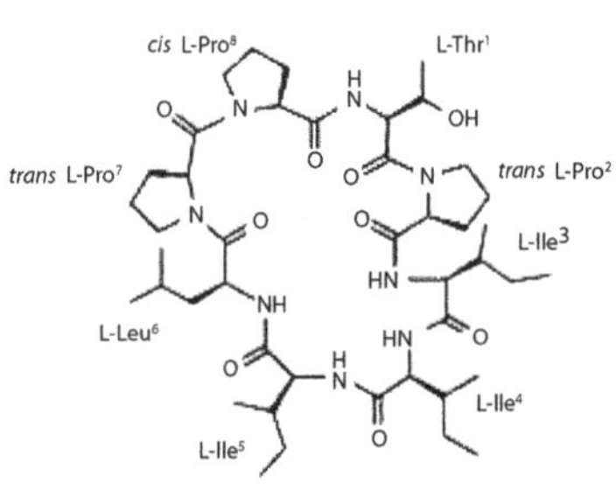

2

Euryjanicin B

The percent growth values recorded on the different human cancer cell lines treated with the above compounds are given below.

Percent Growth Values Recorded on Human Cancer Cell Lines Treated with Euryjanicins A–D (1–4) and Dominicin (5)

	Compounds				
Cell line	1	2	3	4	5
NCI-H522	72	84	95	95	100
LOX IMVI	89	89	88	88	100
IGROV1	83	91	100	100	100
UO-31	74	76	83	83	33

Source: Vera, B. et al., 2009, *J. Nat. Prod.*, 72: 1555–1562.

Raspaciona aculeata (Johnston, 1842)

Order: Axinellida Family: Raspailiidae

Common name: Red sponge

Distribution: East coast of England, Galicia; Mediterranean

Ecology: It occurs in the sublittoral zones on stones and shells at a depth of 15 m.

Biology: It is an encrusting to massive sponge with a thickness from about 2 mm to 2 cm, and its lateral size is up to 6 × 2 cm. It is cushion shaped and has a strongly indented, almost honeycombed surface. Sponges with fused conical masses are also seen. Surface color is red or bright rose and is very hispid. Spicules are tylostyles or styles that are smooth, slightly curved and elongated but with a distinct head. Some spicules are oxeotes. Skeleton is of long megascleres that form a dense palisade at the substrate in thin crusts, but form strongly plumose bundles.

Compound(s) and Activities

Cytotoxicity: Raspacionin triterpinoids, namely, raspacionin, raspacionins A, raspacionins B, 21-deacetylraspacionin,

10-acetoxy-21-deacetyl-28-hydroraspacionin, 10-acetoxy-21-deacetyl-4-oxo-28-hydroraspacionin, 10-acetoxy-15,21-dide acetyl-4-oxo-28-hydroraspacionin, 10-acetoxy-15-deacetyl-4-oxo-28-hydroraspacionin, 10-acetoxy-4-acetyl-15-deacetyl-28-hydro raspacionin, 10-acetoxy-15-deacetyl-4-21-dioxo-28-hydroraspacionin, 10-hydroxy-4,21-dioxo-28-hydroraspacionin, 21-oxo raspacionin, 15-deacetyl-21-dioxoraspacionin, 4,21-dioxo-ras pacionin, 10-acetoxy-4,21-dioxo-28-hydroraspacionin, 10-acetoxy-4-acetyl-21-oxo-28-hydroraspacionin, 10-acetoxy-4-acetyl-28-hydroraspacionin, 10-acetoxy-28-hydroraspacionin, and 10-acetoxy-21-deacetyl-4-acetyl-28-hydroraspacionin, have been isolated from this species. All these compounds showed cytotoxicity against the MCF-7 tumor cell line with IC50 values between 4 and 8 μM (Li et al., 2013).

Antitumor activity: The renieramycin M of this species has shown antitumor activity by inducing apoptosis in lung cancer cells (Perdicaris et al., 2013).

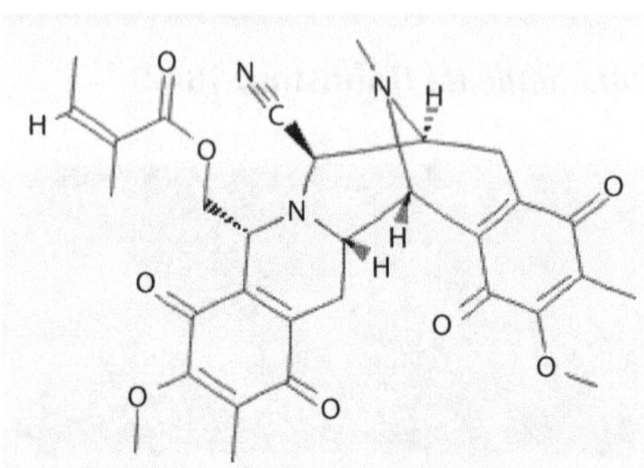

Renieramycin M

FAMILY: STELLIGERIDAE

Paratimea sp. (= *Halicortex* sp)

Order: Axinellida Family: Stelligeridae

Compound(s) and Activities

Antiviral activity: The indole alkaloid dragmacidin F isolated from this sponge showed antiviral activity against HIV-1 (Anjum et al., 2016).

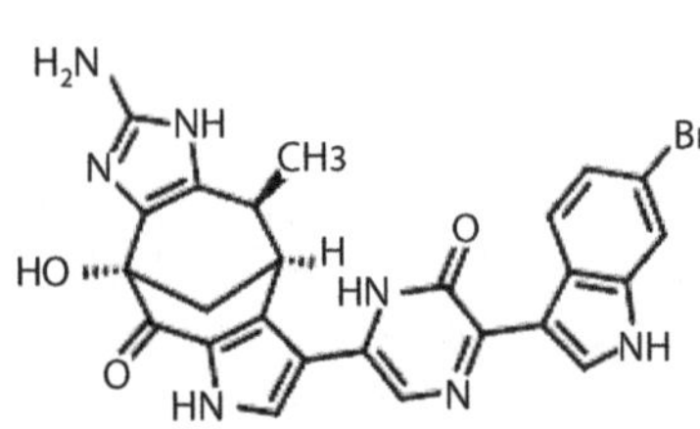

Dragmacidin F

FAMILY: BIEMNIDAE

Biemna ehrenbergi (Keller, 1889)

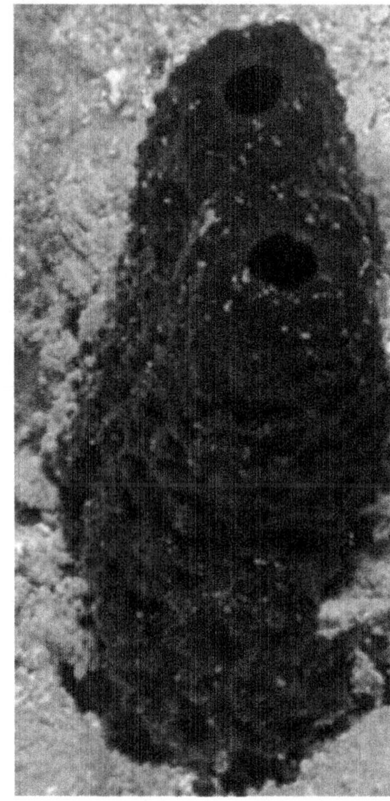

Order: Biemnida Family: Biemnidae

Common name: Red Sea sponge

Distribution: Caribbean Sea, Colombia, Eritrea, Southern Red Sea, Western Arabian Sea

Ecology: It is found exclusively in soft-bottom lagoons, usually in the shallowest part. The sponge is attached to the buried beach-rock, which reduces the risk of dislodgment during storms.

Biology: In this species, most of the sponge body is covered by sediment, but the chimney-like siphons protrude from the sediment surface. The transcellular ostia of this sponge are mostly located on the buried surface of the sponge. The oscula, however, are located on top of the siphons; their elevated position and their ability to close combine to prevent the filtering system outflow from clogging. The sponge maintains a large population of bacteria. When the latter disintegrate, they may be consumed by the sponge.

Compound(s) and Activities

Antimicrobial and anticancer activities: The organic extract of this species yielded the bioactive compounds, namely, (22E)-ergosta-22-ene-8,14-epoxy-3,7-dione (1),

(E)-tetracos-8-en-5-ynoic acid (2), (22E)-ergosta-5,8,22-trien-7-one-3β-ol (3), 32,35-anhydrobacteriohopanetetrol (4), (24R)-ergosta-6,22-diene-5,8-epidioxy-3-ol (5), melithasterol B (6), thymidine (7), and 2′-deoxyuridine (8). Among these compounds, compound 1 displayed an inhibition zone of 20 mm at 100 μg/disc against *Escherichia coli* in the disc diffusion assay. Similarly, compounds 2 and 4 displayed inhibition zones of 20 and 18 mm, respectively, against *Candida albicans* at the same concentration. Compounds 1–3 exhibited weak cytotoxic activity against human colon adenocarcinoma (HCT-116) cancer cell line (Youssef et al., 2015a).

Biemma fortis (Topsent, 1897)

Order: Biemnida Family: Biemnidae

Common name: Indonesian marine sponge

Distribution: Papua New Guinea; Straight of Malacca, Dead Sea; Indonesia; Palau and Ponapé; Philippines

Ecology: It is found at 9–10 m in two habitats (coral reef and sandy substrate with coral patches). It thrives well in areas with high siltation.

Biology: It is a massive sponge (15 cm long and 9 cm in dia.) with chimney-like projections. Its base is buried in the substrate and sometimes only the tubular projections are visible at the surface. The sponge is rugged and hispidulous, and ostia are not visible, while the oscula (3–8 mm) are terminally located at each projection. Texture is woody and corklike. In live specimen, the portion buried to the ground is yellowish green to yellowish brown, while the top of the projection is dark green to gray. This color variation is largely due to accumulated debris.

Compound(s) and Activities

Neuroprotective activity: The pyridoacridine alkaloid named labuanine A isolated from this species induced multipolar neuritogenesis in more than 50% of cells at 0.03–3 μM concentrations (Putra and Jaswir, 2014).

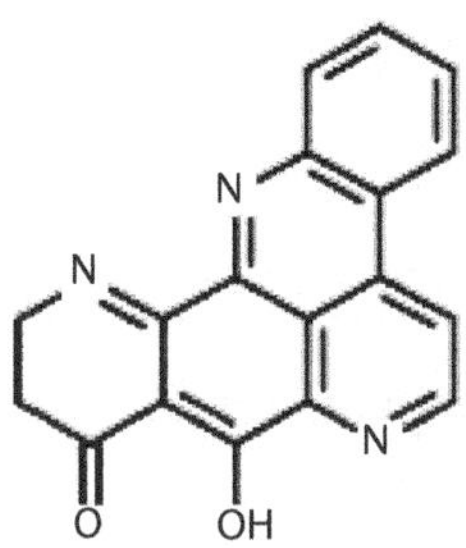

Labuanine A

Biemna tubulata (Dendy, 1905)

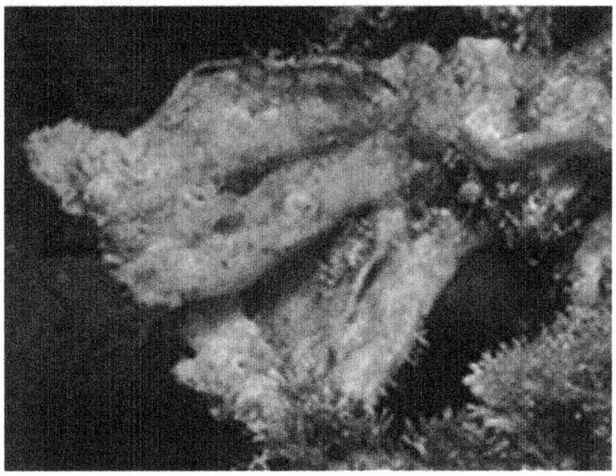

Order: Biemnida Family: Biemnidae

Common name: Not designated

Distribution: Florida to Curacao; Sri Lanka

Ecology: It grows on mangrove roots and in the reef environment.

Biology: It is encrusting to massive, amorphous or with some tubular processes. The surface of the sponge is corrugated conulose and its consistency is soft and delicate. Color of the animal is yellow.

Compound(s) and Activities

Antibacterial activity: The butanol extracts of this species showed maximum antimicrobial activity against *E. coli* with MIC of 0.091 mg/mL (Govinden-Soulange et al., 2014).

Neofibularia nolitangere (Duchassaing and Michelotti, 1864)

Order: Biemnida Family: Biemnidae

Common name: Touch-me-not sponge, Jamaican sponge

Distribution: Western Atlantic; Florida, Bahamas, Caribbean

Ecology: It is occurring in the reef environment, but often in areas with some disturbance; occasionally it is also a member of the fouling community.

Biology: It is a cake-shaped or massively encrusting sponge with an apical depression, which may lead to a cup shape. Oscules are conspicuous and are apical or inside the apical depression. Surface is smooth or shaggy, areolate. Size is up to about 30 cm high and 30 cm wide. Consistency is corky to compressible, fragile and crumbly. Color of the sponge is brownish to dark red.

Compound(s) and Activities

PP1 inhibitory activity: The organic extracts of this species caused protein phosphatase 1 (PP1) inhibition. The concentrations necessary for total enzyme inhibition ranged from 54 to 790 μg/mL. Malfunctions of PP1 inhibitors have been linked to a variety of diseases, including cardiovascular disease and cancer (Sepčić et al., 2010).

Antileishmanial and antimalarial activities: The fractions (Y4 and Y5) of the crude methylene chloride extract of this species demonstrated notable antimalarial properties giving percentage inhibitions of 87% and 78%, respectively, for fractions Y4 and Y5 at 20 μg/mL. Fractions Y4 and Z4 showed remarkable antileishmanial activity inhibiting the growth of the pathogen by 93.31% and 91.77%, respectively, at 20 μg/mL (Thompson and Gallimore, 2013).

FAMILY: DESMANTHIDAE

Petromica (Chaladesma) ciocalyptoides (van Soest and Zea, 1986) (= *Monanthus ciocalyptoides*)

Order: Bubarida Family: Desmanthidae

Common name: Not designated

Distribution: Southwestern Caribbean, eastern Brazil

Ecology: It grows in sand bottom beyond coral formations and in rubble flats in wave-exposed bays at depths of 10–35 m; and also in shallow reefs.

Biology: Its basal mass (1 cm thickness) is found buried in sand with prominent tapering fistules, which are thick-walled 10 cm long. All apices are closed. Fistules may be anastomosing into flattened groups of two or three. Consistency is rather elastic, fairly soft and crumbly inside. Surface is slightly rough. Color of the sponge is pale yellow orange above. It is white and semitransparent in spirit. It is a confused, halichondrial skeleton forms a distinct skin.

Compound(s) and Activities

Antibacterial/antiviral activity: The compound halistanol sulfate A isolated from this species displayed the most potent antimicrobial effect against *Streptococcus mutans*. This compound also inhibited the initial oral bacteria colonizers, such as *S. sanguinis*, but at much higher concentrations. The results obtained indicate that halistanol sulfate A may be considered a potential scaffold for drug development in *S. mutans* antibiofilm therapy, the main etiologic agent of human dental caries (Lima et al., 2014). Guimarães et al. (2013) reported that this compound displayed antiviral activity.

Cytotoxic activity: The hydroethanolic extracts of this species showed cytotoxic activity against tumor cell lines, namely, HL-60, HCT-8, MDA-MB435, and SF-295 (IC50 values >100 μg/mL) at a concentration of 20 mg/mL (Ferreira et al., 2007).

Petromica (Chaladesma) citrina (Muricy, Hajdu, Minervino, Madeira, and Peixinho, 2001)

Order: Bubarida Family: Desmanthidae

Common name: Yellow sponge

Distribution: Pacific Ocean, Tropical West Atlantic, Southeastern Brazil

Ecology: It grows in seamounts and knolls.

Biology: It has confused reticulation of long oxeote spicules with ascending spicule tracts.

Compound(s) and Activities

Antibacterial activity: The aqueous extract was fractionated and the compound, halistanol trisulfate (241,25-dimethylcholestane-2b,3a,6a-triol trisodium sulfate) isolated from this species displayed antibacterial activity against *S. aureus*

(MIC, 512 mg/L), *Staphylococcus epidermidis, Enterococcus faecalis, Mycobacterium fortuitum*, and *Neisseria gonorrhoeae*. Further, this compound showed cytotoxicity against cancer cell line L929 (fibrosarcoma) (Marinho et al., 2012; Perdicaris et al., 2013).

Antibacterial activity: The aqueous extracts of *Petromica citrina* showed a large action spectrum over resistant-bacteria such as *Staphylococcus aureus*, coagulase-negative staphylococci, and *Enterococcus faecalis*. This fraction demonstrated a bactericidal effect on *S. aureus* cells at the MIC of 16 μg/mL (Marinho et al., 2010).

Antiviral activity: The organic extracts of strains B555, B584, and B616 of the associated *Bacillus* sp. of this species showed activity against Bovine viral diarrhea virus (BVDV) with IC50 values of 27.4, 10.2, and 47 μg/mL, respectively (Indraningrat et al., 2016).

Aqueous extracts of *Petromica citrina* showed a large action spectrum over resistant-bacteria such as *Staphylococcus aureus*, coagulase-negative staphylococci, and *Enterococcus faecalis*. Aqueous extract of *P. citrina* was fractioned and the aqueous fraction showed the greatest inhibitory activity on *Staphylococcus* strains. In addition, this fraction demonstrated a bactericidal effect on exponentially growing *S. aureus* cells at the MIC (16 μg/mL) (Marinho et al., 2010).

Cytotoxicity and antiherpetic activity: The compounds halistanol sulfate (1) and halistanol sulfate C (2) of this species have shown cytotoxicity and antiherpetic activity against HSV-1 (KOS strain) (Guimarães et al., 2013).

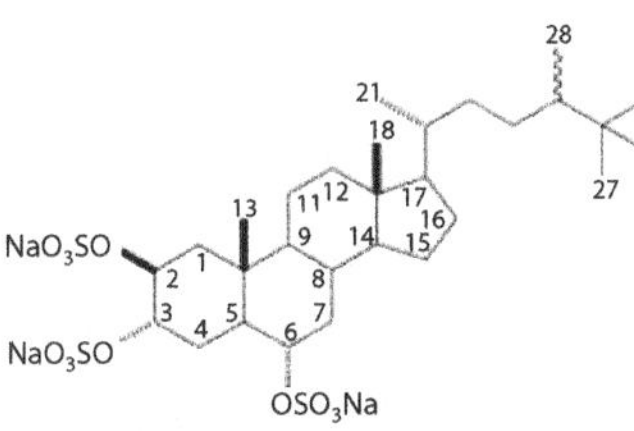

Halistanol sulfate

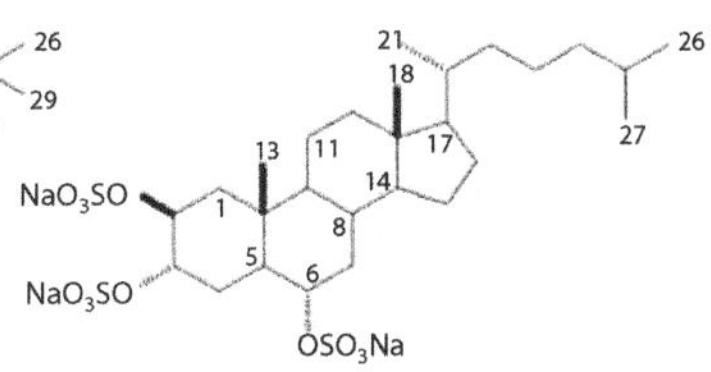

Halistanol sulfate C

The CC50 (50% cytotoxicity concentration) and IC50 (50% viral inhibitory concentration) values are given next.

	CC50 μg/mL	IC50 μg/mL
Compound 1	13.8	5.6
Compound 2	11.9	6.1

Source: Guimarães, T.R. et al., 2013, *Mar. Drugs*, 11: 4176–4192.
Note: CC50, 50% cytotoxicity concentration; IC50, 50% viral inhibitory concentration.

FAMILY: DICTYONELLIDAE

Acanthella acuta (Schmidt, 1862)

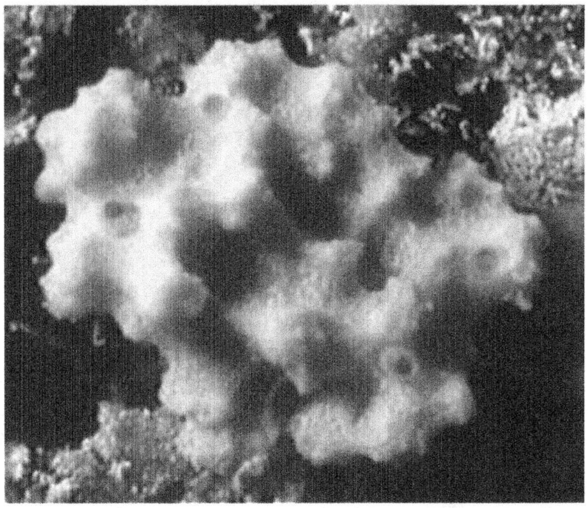

Order: Bubarida Family: Dictyonellidae

Common name: Orange spiny sponge, orange lumpy sponge

Distribution: Mediterranean, Atlantic, English Channel, North Sea

Ecology: This species lives on semidark rocky bottoms (cliffs, caves) and coralligenous of the infralittoral floor, about 10 m up to about 70 m depth.

Biology: This massive sponge, attached to the substrate by a short peduncle, has many small, flattened outgrowths arranged in fans more or less intertwined with each other. The spiky or hispid surface has many conic and pointed papillae giving it the appearance of a cactus. With cartilaginous consistency, its color is orange or more or less dark, and its size can reach about 10 cm in diameter. The nudibranch mollusk *Phyllidia flava* is often associated with this sponge and is the main source of food.

Compound(s) and Activities

Antitumor and cytotoxic activity: The secondary metabolites isolated from the associated bacterial strain (*Bacillus pumilus*) of this species, namely, diglucosyl-glycerolipid and diglucosyl-glycerol, strongly inhibited the growth of the tumor cell lines HM02 and Hep G2 (Thomas et al., 2010).

The dichloromethane/methanol (1:1) extract of this species showed cytotoxicity against two human malignant cell lines, namely, human cervix carcinoma (HeLa) and human colon carcinoma (LS174), and also a normal fetal lung fibroblast cell line (MRC5). The IC50 values in the MTT assay in LS174 and HeLa cells were found to range from 9.9 to 29.5 μg/mL (Stanojković et al., 2015).

Antimicrobial activity: Among the isonitrile sesquiterpenes isolated from this species, the compound acanthellin-1 has shown antimicrobial activity (Minale et al., 1974).

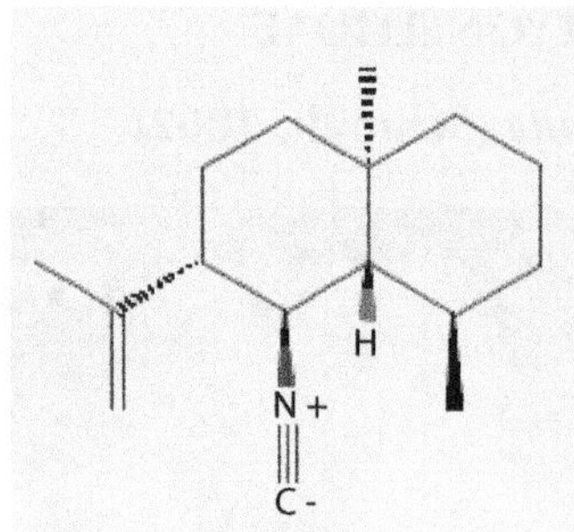

Acanthellin-1

Acanthella cavernosa (Dendy, 1922)

Order: Bubarida Family: Dictyonellidae

Common name: Yellow or red lattice sponge, cactus sponge

Distribution: Indo-West Pacific Ocean

Ecology: This sessile sponge inhabits shallow reefs (up to 20 m) and is usually found encrusting various substrates near or in caves where water currents are fairly swift. Depending upon its area of origin, various color varieties and shapes are available. It must have strong water movement and requires a shaded area. It will not tolerate algae growth in its surface.

Biology: The shape of the species is variable with three basic morphotypes: massive to semispherical; bushy with flat-spiky branches arising from multiple bases or from short peduncle, with crowded and complex branching, or in honeycombed reticulation; and lobate-tubular with thick lobes joined at base. The oscules are irregularly distributed, and are up to 5 mm in diameter. Ostia, which are minute, are aggregated in surface grooves. The surface of the species is clathrate, rubbery, cartilaginous or membranous skin with blunt or spiky short conules. Color of the animal is generally orange or beige. It is a delicate filter-feeder, feeding on plankton, bacteria, and suspended detritus and will generally stay static. It is a hermaphrodite.

Compound(s) and Activities

Antifungal and cytotoxic activities: The compounds isokalihinol B and kalihinene of this species have shown antifungal activities against *Mortierella romannicus* and *Penicillium chrysogenum*. Further, these compounds exhibited cytotoxicity against P388 murine leukemia cells with IC50 values of 0.8 μg/mL and 1.2 μg/mL, respectively (Emsermann et al., 2016).

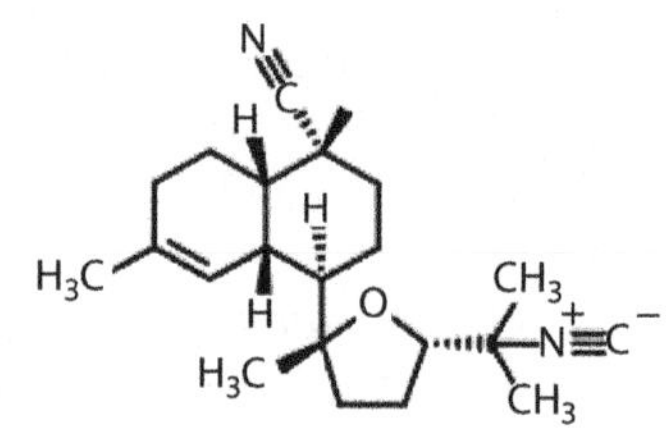

Kalihinene

Cytotoxicity and antifungal activities: The following compounds of this species showed cytotoxicity against human cell lines (IC50 μM) (Gribble, 2015; Xu et al., 2012).

Human Cell Lines

Compound	HCT-116	A549	HeLa	QGY-7701	MDA-MB-231
Kalihipyran A	>50	13.1	11.2	13.5	>50
15-formamido-kalihinene	>50	17.5	4.7	16.4	>50
10-formamido-kalihinene	>50	7.0	13.3	14.5	6.8
Kalihinene X	12.3	8.6	10.6	13.0	15.2

Source: Gribble, G.W., 2015, *Mar. Drugs*, 13: 4044–4136.

Further, the compound 15-formamido-kalihinene showed weak antifungal activity against *Trichophyton rubrum* and *Microsporum gypseum* with MIC values of 8 and 32 μg/mL, respectively. The compound 10-formamido-kalihinene also displayed weak antifungal activity against the fungi *Candida albicans*, *Cryptococcus neoformans*, *T. rubrum*, and *M. gypseum* with MIC values of 8, 8, 4, and 8 μg/mL, respectively (Xu et al., 2012).

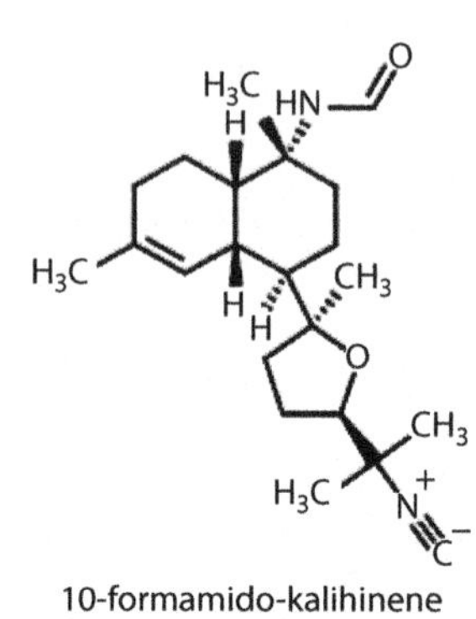

10-formamido-kalihinene

Tumor suppressive activity: The compounds mirabilin G and netamine M isolated from this species inhibited cellular degradation of PDCD4 with EC50 values of 1.8 μg/mL and 2.8 μg/mL, respectively. These compounds have been reported to stabilize PDCD4 under tumor promoting conditions (Grkovic et al., 2014).

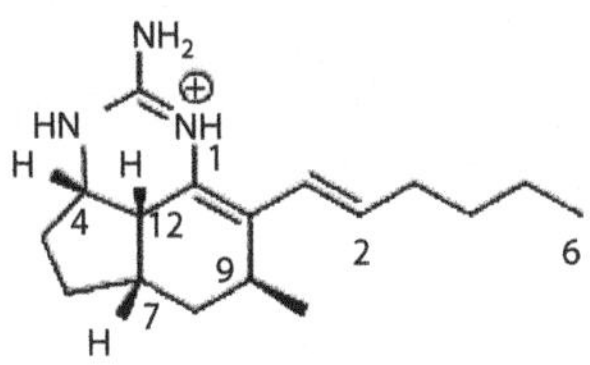

Mirabilin G

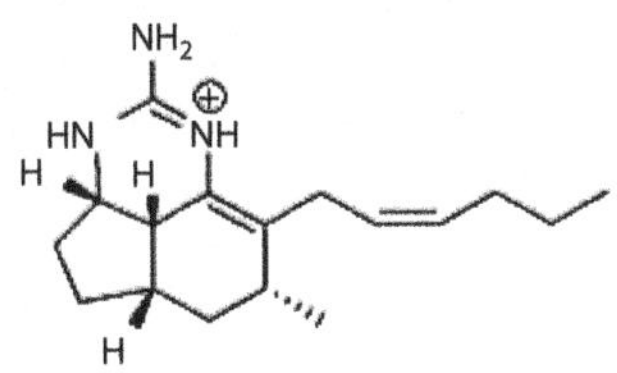

Netamine M

Antimalarial activity: The compound 4-isothiocyanatoeudesm-11-ene of this species has shown moderate antimalarial activity against *Plasmodium falciparum* (chloroquine-sensitive strain D6: IC50 = 4000 ng/mL; chloroquine-resistant strain W2: IC50 = 550 ng/mL) (Emsermann et al., 2016).

Anthelmintic/antiparasitic activity: Among the diterpenes obtained from *A. cavernosa* included known kalihinols X [1] and Y [2] and new kalihinols J [3] and I [4], isolated from this species. Compounds 2, 3, and 4 showed anthelmintic/antiparasitic activity (Alvi et al., 1991).

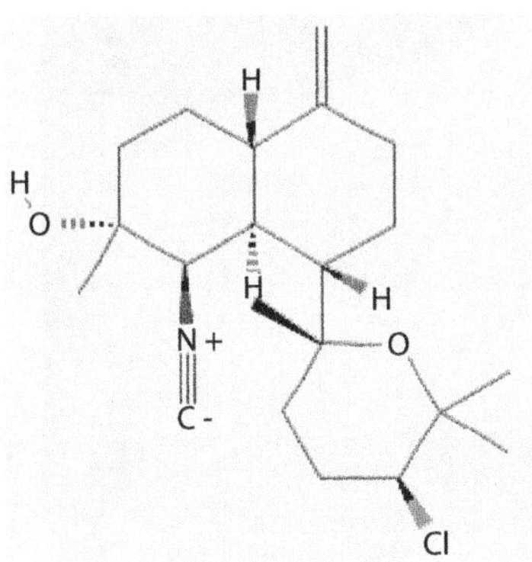

Kalihinol Y

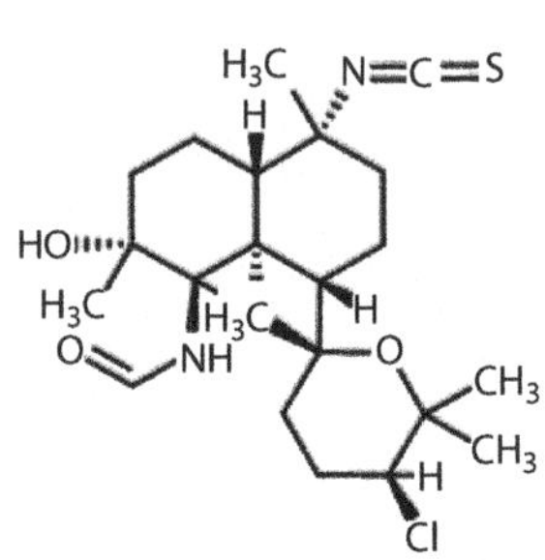

Kalihinol J

Acanthella klethra (Pulitzer-Finali, 1982)

Order: Bubarida Family: Dictyonellidae

Common name: Not known

Distribution: Western Indian Ocean: Kenya and Madagascar

Ecology: This sessile species inhabits depths between 10 and 20 m.

Biology: It is hermaphroditic. The zygote develops into parenchymella larva (free-swimming) before settling down on a substrate where it grows into a young sponge.

Compound(s) and Activities

Antimalarial and anti-TB activities: The sesquiterpenoid compound axisonitrile 3 of this species showed varying degrees of cytotoxic activity with cultured *Plasmodium falciparum*. This compound may therefore be a promising antimalarial agent (Angerhofer et al., 1992; Sipkema et al., 2005). This compound has also shown antitubercular activity with an MIC value of 1.6–2.0 μg/mL (Jyothi et al., 2016; Zhang and Demain, 2005).

Antimalarial activity: The compound 4-isothiocyanatoeudesm-11-ene of this species has shown moderate antimalarial activity against *Plasmodium falciparum* (chloroquine-sensitive strain D6: IC50 = 4000 ng/mL; chloroquine-resistant strain W2: IC50 = 550 ng/mL) (Emsermann et al., 2016).

Cytotoxic activity: The diterpene isonitriles kalihinene and isokalihinol B of this species have shown cytotoxicity (Fusetani et al., 1990).

Acanthella pulcherrima (Ridley and Dendy, 1886)

Order: Bubarida Family: Dictyonellidae

Common name: Not designated

Distribution: New Caledonia, Sahul Shelf (Australia); French Polynesia, Tuamotu

Ecology: It inhabits outer reef slope.

Biology: It is a small, club-shaped sponge (7.5–12 cm long, 6–7.5 cm breadth), with short cylindrical stalk (1.2–2.2 cm long, 0.7–1.0 cm wide), enlarged basal holdfast, and several very thin, leaflike, flattened branches (10–30 mm maximum width, 4–8 mm thick), with even margins. Color of the animal is pale orange brown or brown alive.

Compound(s) and Activities

Antimalarial activity: The sesquiterpenoid compound 11-isothiocyano-7β-H-eudesm-5-ene isolated from this species showed moderate antimalarial activity against *Plasmodium falciparum* (chloroquine-sensitive strain D6: IC50 = 2240 ng/mL; chloroquine-resistant strain W2: IC50 = 610 ng/mL) (Emsermann et al., 2016).

Acanthella sp.

Order: Bubarida Family: Dictyonellidae

Compound(s) and Activities

Antimalarial activity: The isonitril compound (containing kalihinane diterpenoid) kalihinol A of this unidentified species has shown antimalrial activity against *Plasmodium falciparum* with an IC50 value of 0.0005 µg/mL (Anjum et al., 2016; Sipkema et al., 2005).

Antioxidant activity: The extract of this species showed moderate DPPH antioxidant activity with IC50 values of <100 µg/mL (Abdillah et al., 2013).

Dictyonella incisa (Schmidt, 1880)

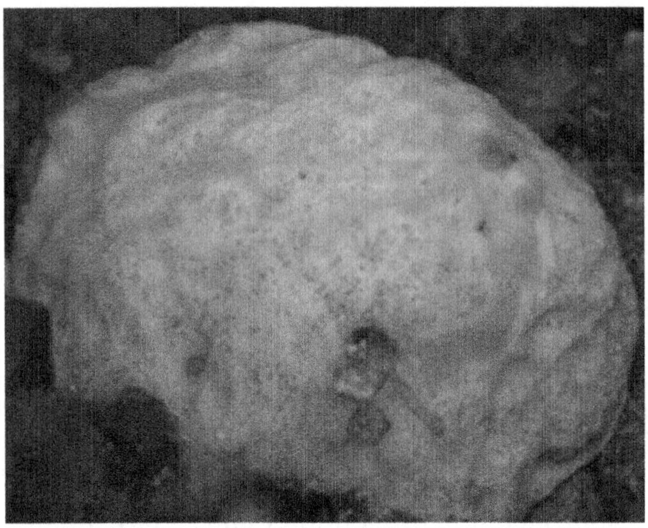

Order: Bubarida Family: Dictyonellidae
Common name: Not designated
Distribution: Mediterranean Sea
Ecology: It dwells in shallow seas of less than 50 m depth.
Biology: Not available

Compound(s) and Activities

Cytotoxic activity: An unusual degraded sterol, incisterol, has been isolated from this species (Pietra, 2002). This compound has been reported to serve as agonists of PXR (pregnane X receptor). The PXR has implications in several important physiological and pathological conditions such as bile acid–induced hepatotoxicity.

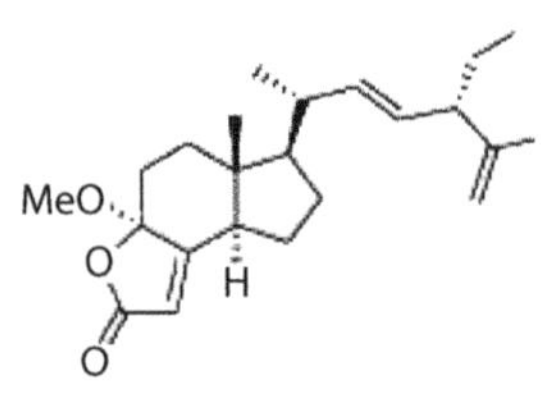

Incisterol

The methanol extract of this species has been reported to possess cytotoxic activity against HeLa (human cervix epithelial cancer cell) with an IC50 value of 159.0 µg/mL (Kutluay et al., 2016).

Dictyonella sp.

Order: Bubarida Family: Dictyonellidae

Compound(s) and Activities

Cytotoxic activity: The hydroethanolic extracts of this species showed cytotoxicity of tumor cell lines at a concentration of 20 mg/mL. The IC50 values recorded for the different cell lines are given next.

IC50 (µg/mL) Values of Tumor Cell Lines

HL-60	HCT-8	MDA-MB435	SF-295
39.0	23.8	54.6	33.1

Source: Ferreira, E.G. et al., 2007, *Porifera Res.: Biodiversity, Innovation Sustainability*, 313–318.

FAMILY: CLIONAIDAE

Cliona celata (Grant, 1826)

Order: Clionaida Family: Clionaidae
Common name: Sulfur boring sponge
Distribution: North Sea, Western Europe, South West Britain
Ecology: The boring form is very common in oyster and mussel beds such as *Ostrea edulis*, *Crassostrea gigas*, *Crepidula fornicate*, and *Arctica islandica* where it causes damage to shellfish farming. It is visible as small yellow or orange nipples on shells or limestone. It uses its acid to drill tunnels in the calcium material.
Biology: The massive form of this species has raised, rounded ridges up to 40 cm across. Large oscules with raised rims are seen along the tops of the ridges. This sponge often forms a

thick platelike structure standing on its edge with large specimens growing up to 1 m across and 50 cm high. The surface is evenly covered by tuberculate inhalant papillae. It is often bright to deep yellow, becoming darker out of water. Sometimes red coloration due to algae may be observed around oscular (exhalent) openings. The whole sponge shows a noticeable decrease in size when removed from the water. The consistency of this sponge is firm and inflexible with a tough outer layer.

Reproduction: This boring sponge can reproduce asexually and sexually. It can simply release sperm into the water and this fertilizes the female's eggs. This sponge may also attach its larvae onto mollusks like clams and mussels. This usually results in the death of the host. It then will begin to grow and colonize.

Compound(s) and Activities

Mosquito larvicidal activity: The methanol extract of this species showed the highest larvicidal activity at 500 ppm against the fourth instar larvae of the mosquito *Anopheles stephensi*. The LC50 and LC90 values were found to be 80.6 and 220.8 ppm, respectively. High ovicidal activity of 91.2% was recorded at 500 ppm concentration of the methanol extract. The hexane extract, on the other hand, was found to be the most effective protectant against the adult female mosquitoes of this species. The mean protection time recorded in hexane extract was up to 245 min at 5 mg/cm^2 dosage against the adults (Reegan et al., 2015).

Antibacterial and cytotoxic activities: The compounds clionamide, celenamides, and storniamides A–D isolated from this species exhibited antibiotic activity against several Gram-positive bacteria. Further, these compounds possessed cytotoxicity against leukemia cell lines (Sugumaran and Robinson, 2010).

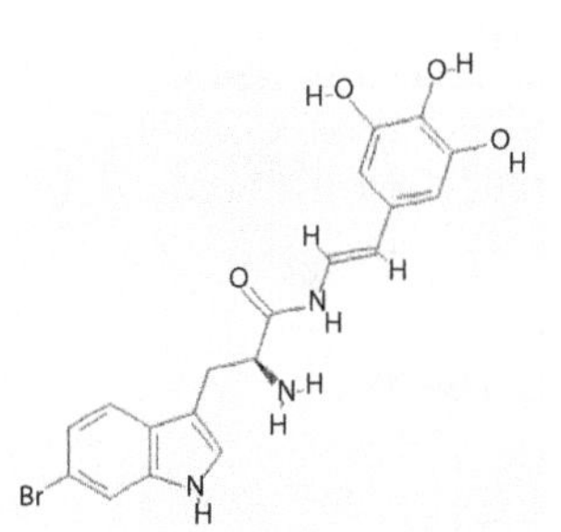

Clionamide

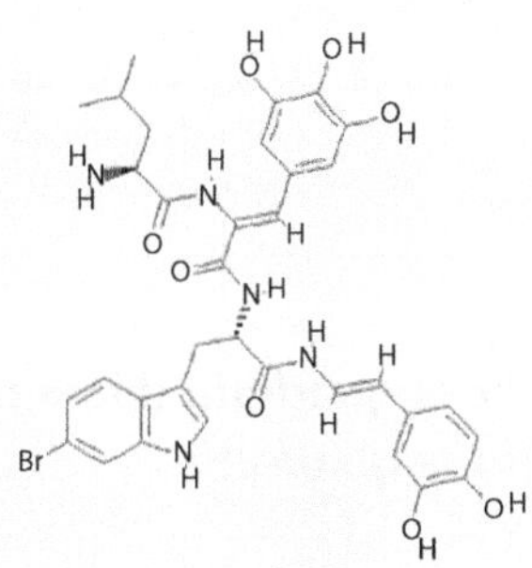

Celenamide

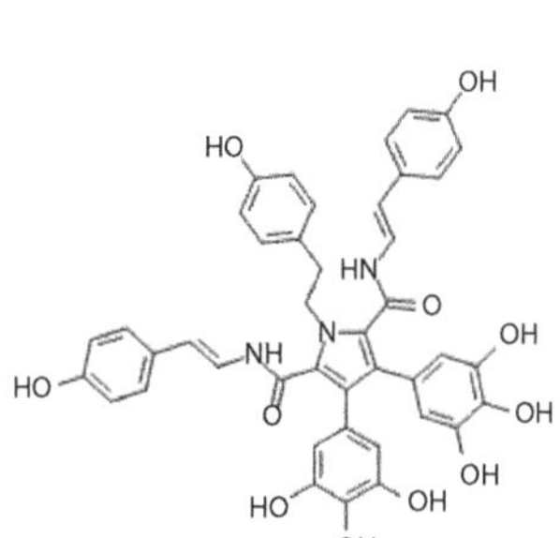

Storniamide A

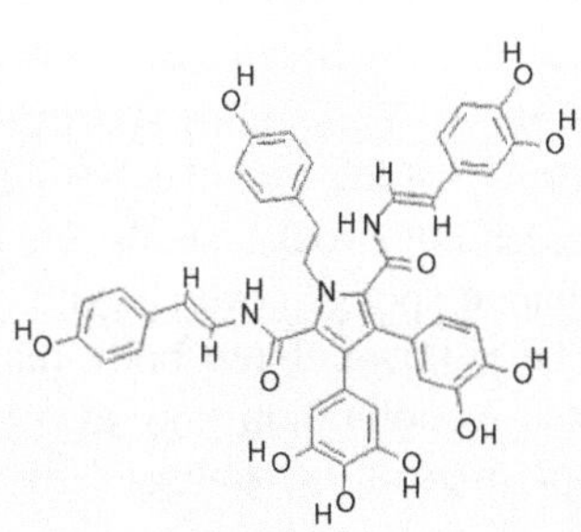

Storniamide B

Cliona varians (Duchassaing and Michelotti, 1864)

Order: Clionaida Family: Clionaidae

Common name: Coral-boring sponge, brown variable sponge

Distribution: Western Central Atlantic and Indo-West Pacific

Ecology: It is common on coral reefs, reef flats, seagrass beds and lagoons, and burrows in dead corals. It is also found in mangrove ponds and brackish waters.

Biology: Thick encrustations of this sponge may cover several square meters of substratum. Oscules are relatively large, paler, and slightly elevated. This sponge may be irregularly massive, hemispheric, globular, or partly divided into irregular lobes. Encrusting patches may cover 1 to 2 m in a round or semicircular pattern. Its smooth and nonporous surface may often partly be covered by sediment. Color of the animal is yellow, tan, green, or brown. Thin encrustations may be dark green. Internally it is pale yellowish, with a dark brown hemispheric form with an orange interior. Spongin fibers are absent. Presence of small anthosigma spicules is a unique and diagnostic characteristic feature of this species.

Compound(s) and Activities

Antibacterial and antileishmanial activities: The lectin of this species has displayed a cytotoxic effect on Gram-positive bacteria, such as *Bacillus subtilis* and *Staphylococcus aureus*.

Further this compound showed antileishmanial activity by agglutinating the *Leishmania chagasi* promastigotes. These findings are indicative of the physiological defense roles of the lectin of this species and its possible use in the antibiosis of bacteria and protozoa pathogenic (Moura et al., 2006).

Anjum et al. (2016) reported that the lectin isolated from this species showed antibacterial activity against *S. aureus* with an MIV value of 16 μg/mL.

Cytotoxic, antileukemic effect: The lectin extracted from this species inhibited the growth of human leukemia cells, with IC50 values of 70 and 100 μg/mL for K562 and JURKAT cells, respectively. The lysosomal protease cathepsin B has played a key role in mediating the cell death (Queiroz et al., 2009).

Cliona viridis (Schmidt, 1862) (= *Cliona nigricans*)

Order: Clionaida Family: Clionaidae

Common name: Green boring sponge, green clione

Distribution: Mediterranean, Atlantic, English Channel and North Sea, Indo-Pacific

Ecology: It is present in the calcareous substrates, and it can also be found in the rhizomes of posidonia, on sedimentary bottoms, and occasionally on molluscan shells. The deeper the depth (about 350 m), the more it is found on calcareous algae.

Biology: The surface of this sponge is optically smooth, and papillae may be up to 8 mm in diameter and up to 15 mm high. The consistency is tough, almost hard. The different color tinges (whitish [rare], green yellow, brownish, or even blackish) of the animals of this species are due to variations in the amount of zooxanthellae (a photosynthetic symbiont), stage of development, and the depth at which the specimens are found. The visible parts are only the oscules (exhalant openings) and the warts (which are in the form of flat sieves, circular, or ovoid) with the ostioles (inhalant pores). In some cases, this sponge can cover the substrate and get the form of a brownish-green colony.

Food and feeding: Like other sponges, it draws water in through its ostioles and filters out food particles such as bacteria and other microorganisms. During this process oxygen diffuses into the tissues and waste products diffuse out. The water is then expelled through the osculi.

Reproduction: It is oviparous, that is, the eggs are fertilized internally as the sperm are drawn into the interior of the sponge with the incoming water flow. The eggs are then ejected into the water column via the outgoing water stream. The larvae are planktonic and have limited dispersal ability. When they settle on the seabed, they develop into juvenile sponges in about two weeks.

Compound(s) and Activities

Antioxidant, antibacterial and antifungal activities: The organic extracts of this species showed the abundance of sterols and saponins, and moderate levels of tannins, free quinines, flavonoids, and alkaloids, which showed DPPH antioxidant and antimicrobial activities (Bary et al., 2016).

Anticancer activity: The extract of this species showed anticancer activity against A549, HeLa, and HT-29 cancer cell lines with the following GI50 values.

GI50 (μg) Values of Cancer Cell Lines

A549	HeLa	HT-29
12.4	7.6	16.3

Source: Sponge crude extraction and primary screening, n.d., http://shodhganga.inflibnet.ac.in/bitstream/10603/64131/10/10_chapter%205.pdf.

Cliothosa quadrata (Hancock, 1849) (= *Cliona quadrata*)

Order: Clionaida Family: Clionaidae

Common name: Not designated

Distribution: South India and Sri Lanka

Ecology: Sediments are its primary habitat.

Biology: This yellow boring sponge occurs in two distinct forms: yellow papillae, which are sticking out of limestone (calcareous rocks, shells, etc.); and as a large massive, wall-shaped sponge covered with characteristic flattened papillae. The yellow-colored body may become darker out of water. Red discoloration may also be seen surrounding the ocular openings of this sponge.

Compound(s) and Activities

Anticancer activity: The extract of this species showed anticancer activity against A549, HeLa, and HT-29 cancer cell lines with the following GI50 values.

GI50 (µg) Values of Cancer Cell Lines

A549	HeLa	HT-29
22.4	17.9	24.4

Source: Sponge crude extraction and primary screening, n.d., http://shodhganga.inflibnet.ac.in/bitstream/10603/64131/10/10_chapter%205.pdf.

Spheciospongia inconstans (Dendy, 1887) (= *Spirastrella inconstans*)

Order: Clionaida Family: Clionaidae
Common name: Cork sponge
Distribution: Indo-West Pacific, Eastern India
Ecology: It is a sessile sponge inhabiting depths up to 100 m.
Biology: This large sponge is a hemispherical and cake-like mass (30 cm dia.), which is attached to calcareous debris with its very broad base. Upper surface is uneven and is furrowed by numerous, closely placed, very deep meandering grooves or elongated pits of 2–3 mm broad. Small pits are also seen between the grooves. Surface is very minutely hispid. Color of the sponge is light brownish orange. Texture is hard, woody, and incompressible.

Compound(s) and Activities

Antibacterial activity: The ethanolic extract of this species exerted antibacterial activity against cerebrospinal fluid (CSF) disease (neurological disease) causing pathogenic bacteria (Sujath et al., 2014). The values of the inhibition zone diameter for different species of bacteria are given next.

Inhibition Zone Diameter (in mm)

E. coli	5
S. aureus	12
H. influenzae	No inhibition
N. meningitides	10
K. pneumoniae	No inhibition
Streptococcus agalactiae	12
E. faecalis	9

Source: Sujatha, S. et al., 2014, *Pharma Innovation*, 3: 81–86.

Kumar et al. (2012) reported that the chloroform extract of this species displayed antimicrobial activity against Methicillin-resistant *Staphylococcus aureus* with a maximum of 30 mm diameter of zone inhibition
Antimicrobial activity: The crude extract of this species showed antimicrobial activity against *S. aureus, B. subtilis, E. coli, V. cholerae*, and *C. albicans*, and the values of zone of inhibition recorded were 8.1, 8.3, 9.5, 8.9, and 7.8 mm, respectively (Putra and Hadi, 2017).
Anticancer activity: The extract of this species showed anticancer activity against A549, HeLa, and HT-29 cancer cell lines with >100 GI50 values ("Sponge crude extraction," n.d.).
Mosquito larvicidal activity: The extract of this species displayed larvicidal activity against the fourth instar larvae of *Aedes aegypti* (yellow fever mosquito vector) with an LC50 value of 112.0 ppm (Rao et al., 2008).

Spheciospongia purpurea (Lamarck, 1815) (= *Spirastrella purpurea*)

Order: Clionaida Family: Clionaidae
Common name: Not designated
Distribution: Atlantic Ocean, Mediterranean, South West Australia, Japan, South Africa, China
Ecology: It inhabits reef areas, at depths of 10 to 320 m.
Biology: This sponge grows over masses of ascidians, polyzoans, and fragments of shells. Its greatest thickness is 12 mm. Surface is even and is finely hispid. Skeleton consists of a dense reticulation of closely packed tylostyles. At surface, the tylostyles, which are shorter, are arranged in vertical brushes and their points project slightly beyond the dermis. Tylostyles have slightly curved shafts and rounded heads.

Compound(s) and Activities

Antifungal activity: Four known (1–4) and two new (5 and 6) lysophospholipids have been isolated from this sponge. The isolated compounds displayed various moderate *in vitro* antifungal activities against four fungi (*Cryptococcus neoformans, Candida glabrata, Trichophyton rubrum*, and *Aspergillus fumigatus*) (Lin et al., 2015).

Spheciospongia vagabunda (Ridley, 1884)

Order: Clionaida Family: Clionaidae

Common name: Chocolate sponge, vagabond boring sponge, Red Sea sponge

Distribution: Indonesia; Indo-West Pacific, Fiji Islands; Palau; Papua New Guinea; Cebu to Mindoro, La Union, Ilocos Norte, Ilocos Sur, and San Francisco, Philippines

Ecology: It inhabits shores on muddy bottoms of lagoons near the midwater mark; found at 6 m depth in areas with coral rubbles and sandy substrata. There may be a broad zone of these sponges near coral reefs.

Biology: This smooth, velvety sponge is irregular in shape (10 cm long, 6 cm wide, and 1.5–3 cm thick). There are three forms based on the shape: an irregular mass with full of little holes, an irregular mass with many tall stout cones with holes at the tips, and an irregular mass with globular lumps attached on a stalk. These globular lumps may break off to form new sponges. All three forms have similar internal structures (skeleton form and spicules). Surface is hispid and has steep-sided conical projections (8–12 mm high and 5–9 mm wide). The grooves are filled with calcitic materials. Ostia are not visible, and the oscula located at the apex of the conules could not be easily seen out of water. Texture is hard, corky, and is not readily compressible. Its color in life and in alcohol is brown. The top of conules is darker than the other parts of the sponge due to heavy concentration of pigments. Megascleres are tylostyles of two size categories, with terminal or subterminal heads, which are slightly curved at the anterior half and sharply pointed. Microscleres are finely spined spirasters. Spicules are tightly packed and confused, crisscrossing each other with some protruding to the surface.

Compound(s) and Activities

Antiprotozoal activity: The associated actinobacterial strain *Actinokineospora* sp. EG49 of this sponge species showed antiprotozoal activity against *L. major* and *T. brucei* with 48% growth inhibition (Indraningrat et al., 2016).

Antimicrobial activity: The associated actinobacterium *Micrococcus* sp. EG45 of this species yielded the compound microluside A, which showed antibacterial activity against *S. aureus* NCTC 8325 and *E. faecalis* JH212 with MIC values of 12.4 μg/mL and 9.6 μg/mL, respectively. Similarly *Nocardiopsis* sp. RV163 yielded 1,6-dihydroxyphenazine (1,6-phenazinediol) which was active against *Bacillus* sp. P25 and *Actinokinespora* sp. EG49 with DOI (diameter of inhibition) values of 11 mm and 15 mm, respectively. Further, another actinobacterium *Actinokineospora* sp. EG49 yielded an unidentified compound that showed antifungal activity against *Candida albicans* with DOI value of 12 mm (Indraningrat et al., 2016).

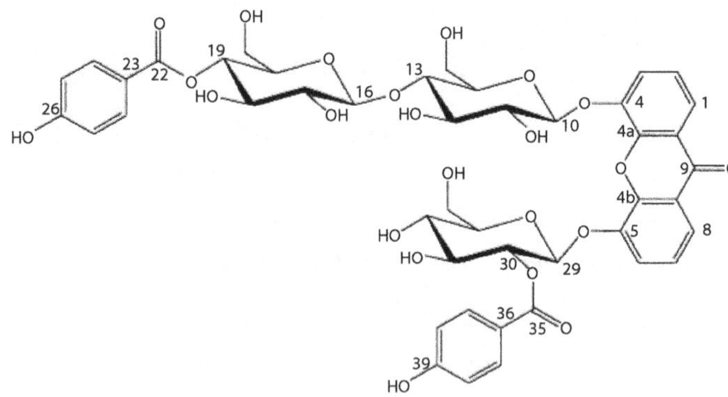

Microluside A

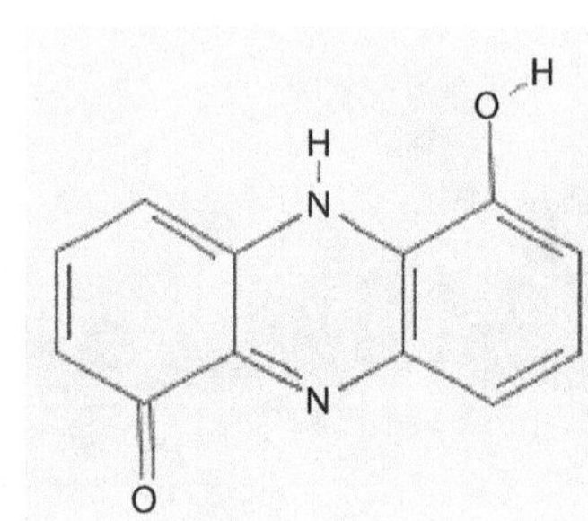

1,6-Dihydroxyphenazine

Antibacterial activity: The unidentified fungus associated with this sponge species yielded a polyketide—14,15-secocurvularin—which was mildly antibiotic against *Bacillus subtilis* (Thomas et al., 2010).

Cytotoxicity: The extract of this species yielded three new ceramides: N-[(2S,3S,4R)-1,3,4-trihydroxytetradecan-2-yl] tridecanamide (1), (R)-2′-hydroxy-N-[(2S,3S,4R)-1,3,4-trihy droxypentacosan-2-yl] octadecanamide (2), and (R,Z)-2′-hydroxy-N-[(2S,3S,4R)-1,3,4-trihydroxytricosan-2-yl) nonadec-10-enamide (3). Of these, compounds 2 and 3 displayed high potential cytotoxicity against HepG2 (IC50 24.7 and 21.3 μM, respectively) and MCF-7 (IC50 26.8 and 29.8 μM, respectively) (Ibrahim et al., 2015; Nadeem et al., 2015).

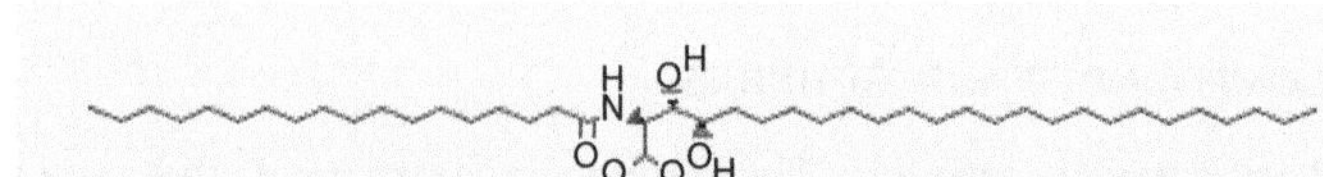

(R)-2′-hydroxy-N-[(2S,3S,4R)-1,3,4-trihydroxypentacosan-2-yl] octadecanamide

FAMILY: SPIRASTRELLIDAE

Spirastrella abata (Tanita, 1961)

Order: Clionaida Family: Spirastrellidae

Common name: Not designated

Distribution: Japan, Korea
Ecology: Not reported
Biology: Not available

Compound(s) and Activities

Cytotoxic, antimicrobial, and enzyme inhibitoryactivities: Three sphingosine 4-sulfates (1–3) and a lysophosphatidylglycerol (4) have been isolated from this species. These compounds exhibited significant cytotoxicity against K562 cell line with an LC50 values of 4–8 μM, weak antimicrobial activity (1), and weak-to-moderate inhibitory activity against isocitrate lyase and Na(+)/K(+)-ATPase (Jang et al., 2012). Wang et al. (2002) reported that the compound sphingosine 4-sulfate 5 of this species showed cytotoxic effects on several human tumor cell lines.

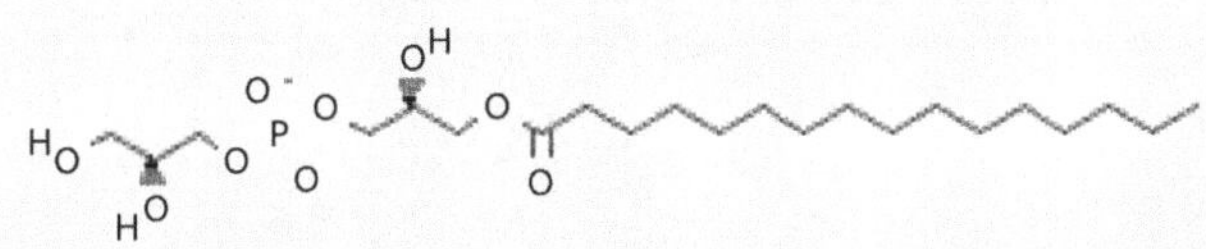

Lysophosphatidylglycerol

Hypocholesterolemic activity: The lyso-PAF analogs and lysophosphatidylcholines of this species possessed hypocholesterolemic activity by inhibiting cholesterol biosynthesis (Perdicaris et al., 2013).

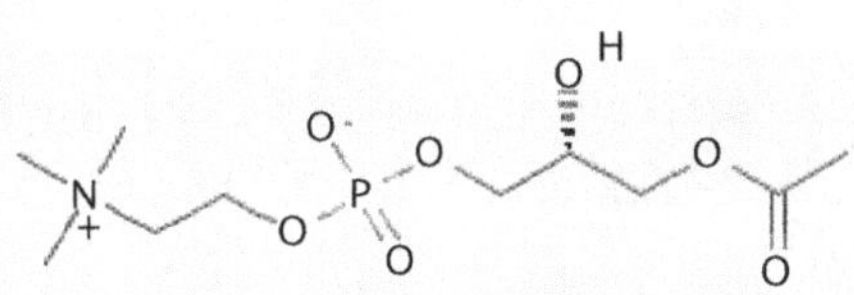

Lysophosphatidylcholine

Platelet activating: Alam et al. (2013) reported on the isolation of four new lyso-PAF (Platelet activating factor) derivatives from this species.

Spirastrella coccinea (Duchassaing and Michelotti, 1864)

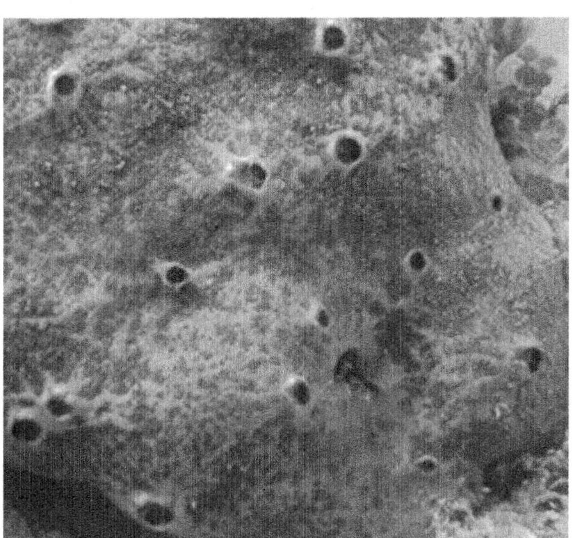

Order: Clionaida Family: Spirastrellidae
Common name: Caribbean marine sponge
Distribution: Tropical Atlantic America, North Carolina, Eastern Caribbean, Japan, Bahamas, United States
Ecology: It inhabits rocks; on coral reefs and hard bottoms.
Biology: It is an encrusting sponge of about 1 mm thick. Color of the animal is dull orange or brown in life and is a pale grayish-brown or pale brown in alcohol. Consistency is tough and slightly compressible. Surface is smooth. In the living sponge, riverlike systems of subdermal canals lead to small oscules. There is no detachable skin in ectosome. Bands of microscleres enclose dermal pores. The endosome has numerous tylostyles that are strewn in confusion. A few spicule tracts may be present. The tracts are often obliquely oriented and they end in plumose tufts at the surface. Spirasters, mostly of large size, are abundant in the flesh. Spicules are tylostyles, which are pinlike, with long, straight, gradually pointed shafts and distinct heads.

Compound(s) and Activities

Antimitotic activity: Williams et al. (2003) isolated a novel macrolide, spirastrelloide A methyl ester, from this sponge that showed potent antimitotic activity in a cell-based assay.

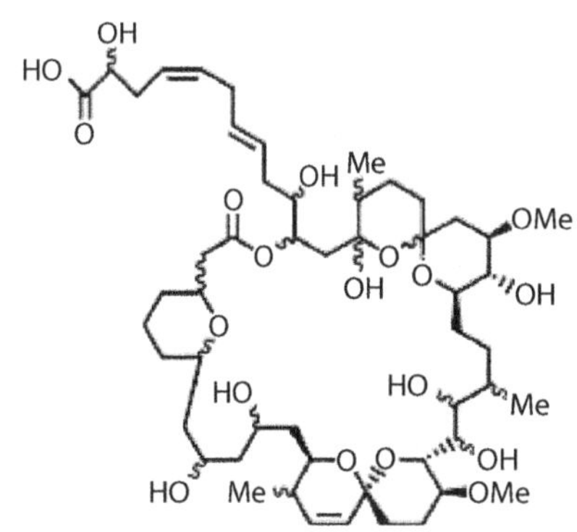

Spirastrelloide A

Spirastrella cunctatrix (Schmidt, 1868)

Order: Clionaida Family: Spirastrellidae

Common name: Orange ray sponge, orange encrusting sponge

Distribution: Western Mediterranean

Ecology: It inhabits rocky substratum.

Biology: Live sponges of this species have dull pinkish-red crusts and are 5–10 cm or more in diameter and 1–5 mm thick, depending on the state of contraction. They are soft and show veinlike, branching surface canals converging toward elevated oscula. The ectosomal membranes that cover the canals have a pattern of radiating grooves. Preserved sponges are leathery tough, with rough surface texture, and about 1 mm thick. Bundles of a few tylostyles extend from the base to the surface of the crust. The smaller spirasters are condensed near the surface. Tylostyles have rounded to oval, pronounced heads from which they taper gradually to a sharp point.

Compound(s) and Activities

Antiprotozoal activity: The associated actinobacterial strains *Modestobacter* sp. SBT362 and *Rhodococcus* sp. SBT367 of this sponge species showed antiprotozoal activity against *Trypanosma brucei brucei* with IC 50 values of 19.34 µg/mL and 19.97 µg/mL, respectively (Indraningrat et al., 2016).

Spirastrella hartmani (Boury Esnault, Klautau, Bézac, Wulff, and Solé-Cava, 1999)

Order: Clionaida Family: Spirastrellidae

Common name: Not designated

Distribution: Western Central Atlantic, Caribbean, Japan

Ecology: It is common between and under coral rubble, in shallow reefs, and in seagrass environments. This sessile sponge is also seen in brackish waters; depth range 1 to 10 m.

Biology: It is a thick encrusting sponge with a maximum thickness of about 4 cm. It is dull reddish to brown externally, lighter color internally. Surface is smooth and leathery to the touch. It is compressible in consistency.

Compound(s) and Activities

Cytotoxicity: Martin et al. (2005) isolated two new halogenated derivatives, chlorine and bromine, of helinane from the 2-propanol extract of this sponge. Among them, the chlorine-substituted helinane showed in vitro cytotoxicity against the human tumor cell lines A549, HT29, and MDA-MB231.

Spirastrella mollis (Verrill, 1907)

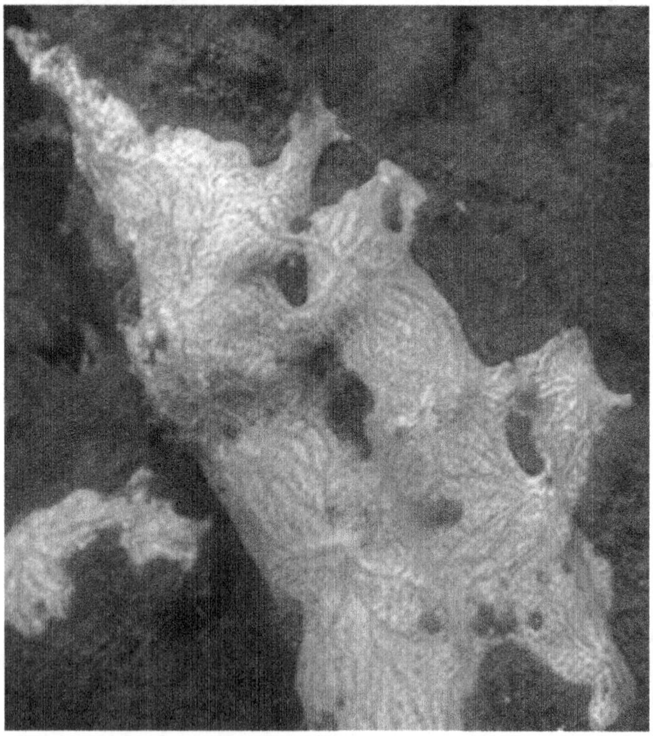

Order: Clionaida Family: Spirastrellidae

Common name: Not designated

Distribution: Bahamas, Bermuda

Ecology: It is seen on dead, cavernous corals.

Biology: It is an encrusting sponge forming soft films with 0.5–3 mm thickness. Surface is smooth. Oscules and ostia are not visible. Interior is without canals. Skeletal spicules are long, slender tylostyles that are scattered and in groups. These tylostyles have well-rounded or ovate/elongate heads. Microscleres are large spined spirasters that have 3 to 4 whorls of sharp conical spinules. Some of these spirasters are curved.

Compound(s) and Activities

Antitumor activity: The mollenyne compound chlorodibromohydrin of this species was cytotoxic toward HCT-116 (human colon cancer cells) with an IC50 value of 1.3 µg/mL (Gribble, 2015).

The brominated lipid mollenyne A isolated from this sponge showed *in vitro* cytotoxic activity against the human colon tumor HCT-116 with IC50 values of ≤0.1 μg/mL (Morinaka and Molinski, 2011).

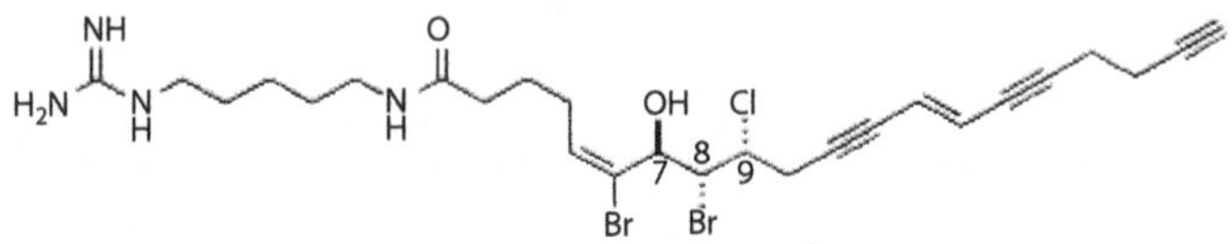

Mollenyne A

Spirastrella pachyspira (Lévi, 1958)

Order: Clionaida Family: Spirastrellidae

Common name: Not known

Distribution: Red Sea and Indian Ocean, Japan

Ecology: It inhabits coral rocks; intertidal zones.

Biology: It is a small encrusting sponge with a size of only 4 × 2 × 1 mm. Surface is uniform. It is pale white in spirit. Consistency is fleshy. Oscula and pores are not traceable. There is no differentiation between the ectosome and endosome. Endosome is quite dense due to interlocking of the modified spirasters. Tylostyles are set in a radial pattern, and their tips are projecting beyond the surface giving a characteristic roughness to the surface. Spicules are tylostyles that are straight or slightly curved. Their head is well developed, globular, or rounded. The spirasters are of four types: branching, robust, slender, or small with curved axes.

Compound(s) and Activities

Antioxidant activity: The ethyl acetate extract of this species showed a promising nitric oxide radical scavenging activity (80%) followed by the hexane extract at a concentration of 1000 μg/mL. Further, at the highest concentration (1000 μg/mL), the values of the DPPH radical scavenging activity of the hexane, chloroform, ethyl acetate, and methanolic extract were found to be 85.7%, 73.2%, 80.4%, and 77.2%, respectively (Morais and Chitra, 2016).

Antiangiogenic activity: The methanolic extract of this species showed antiangiogenic activity at a concentration of 62.5 μg/mL (Kumar and Gopalkrishnan, 2014).

Spirastrella sp.

Order: Clionaida Family: Spirastrellidae

Compound(s) and Activities

Antiprolific and antitumoral activities: The compounds, spongistatins, isolated from this species were found to be cytotoxic and inhibited the cancer cell proliferation with antitumoral activity (Sepčić et al., 2010).

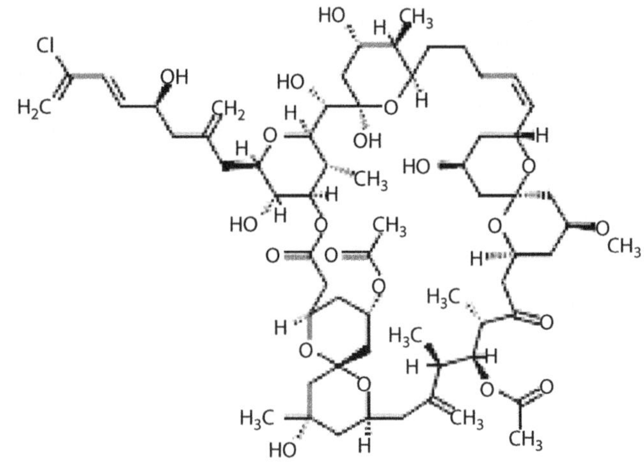

Spongistatin

FAMILY: CALLYSPONGIIDAE

Arenosclera brasiliensis (Muricy and Ribeiro, 1999)

Order: Haplosclerida Family: Callyspongiidae

Common name: Not known

Distribution: Brazil

Ecology: Not reported

Biology: It is a massive, lobate sponge, with up to 50 lobes or tubes 5–20 mm in diameter by 22–30 mm high, fused laterally. The whole sponge reaches 12 cm in height and 30 cm in width. Each lobe bears a circular oscule at the top, 1–10 mm in diameter, surrounded by a translucent oscular rim. Surface is rough to the touch, filled with abundant debris and sand particles. Color is whitish, drab, or cream, both alive and in spirit. Consistency is soft, compressible, with abundant mucus.

Compound(s) and Activities

Antibacterial activity: The alkyl piperidine alkaloid compounds—arenosclerins A, B, and C—isolated from this species have shown antibacterial activity (Sipkema et al., 2005).

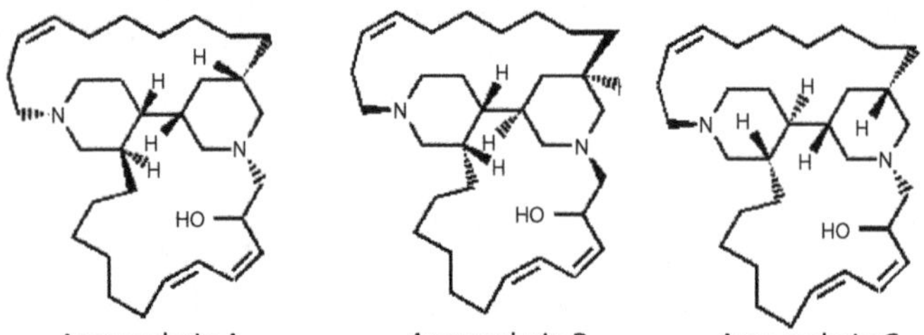

Arenosclerin A Arenosclerin B Arenosclerin C

Anjum et al. (2016) reported that the alkyl piperidine alkaloid haliclonacyclamine E isolated from this species showed antibacterial activity against *S. aureus* with an MIC value of 8 μg/mL.

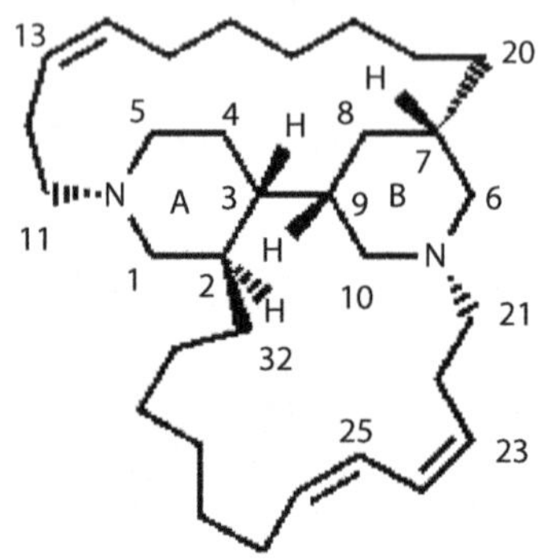

Haliclonacyclamine E

Antibacterial and antitubercular activities: The alkyl piperidine alkaloids—arenosclerins A, C—of this species have also shown activity against *S. aureus* and *M. tuberculosis* with MIC values of 16 μg/mL and 30 μg/mL, respectively (Anjum et al., 2016).

Cytotoxic and antimicrobial activities: The alkyl piperidine alkaloids—arenosclerins A–C and haliclonacyclamine E—isolated from this species displayed antibacterial activity against *Staphylococcus aureus, Escherichia coli,* and 12 antibiotic-resistant bacteria (isolated from hospital environment). Haliclonacyclamine E and arenosclerins A and C were active against a larger number of bacteria strains than arenosclerin B. However, arenosclerins B and C presented more potent antibacterial activity. These compounds also showed cytotoxic activity against human HL-60 (leukemia), L929 (fibrosarcoma), B16 (melanoma), and U138 (colon) cancer cell lines at concentrations between 1.5 and 7.0 μg/mL (Torres et al., 2002).

Antimutagenic activity: The crude extract of this species has shown a significant antigenotoxic effect, which suggests that this species has the potential to protect DNA from the action of 4NQO, 2-aminofluorene, sodium azide, and mitomycin C (Stankevicins et al., 2008).

Callyspongia (Cladochalina) aerizusa (Desqueyroux-Faúndez, 1984)

Order: Haplosclerida Family: Callyspongiidae

Common name: Tube sponge, Indonesian sponge

Distribution: Great Barrier Reef, Australia; Indonesia; New Caledonia; Tanzania; Papua New Guinea; Palau; Philippines

Ecology: It is found attached to coral reefs at a depth of 10 m.

Biology: It is a tubular and erect form (178 mm long, 17.5 mm wide, and walls 2.5 mm thick), with clusters that are attached to the substrate by a common base. Internal surface of tubes is smooth with plenty of small pores, while the external surface is laden with tapering and distally directed spinelike projections. Ostia are not visible but the oscula (5 mm in dia.) are terminally located at each tube. Texture is soft, spongy, compressible, and easy to tear. The color of the animal is blue-green to green in life.

Compound(s) and Activities

Antibacterial activity: The metabolite acetyl Sumiki's acid isolated from *Cladosporium herbarum,* the fungal isolate of this sponge species has shown antibacterial activity against *Bacillus subtilis* and *Staphylococcus aureus* at a concentration of 5 μg/disc (Thomas et al., 2010).

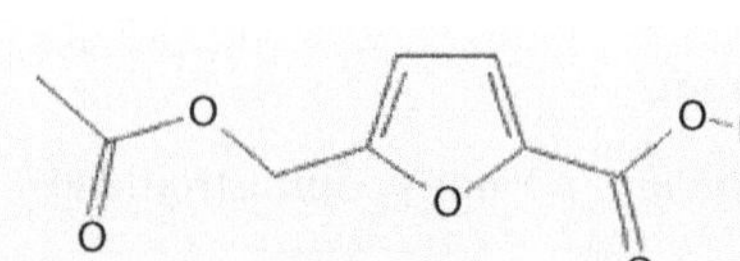

Sumiki's acid

Antifungal activity: The compound callyaerin A of this species showed strong inhibitory properties towards *Candida albicans* (Ibrahim et al., 2010).

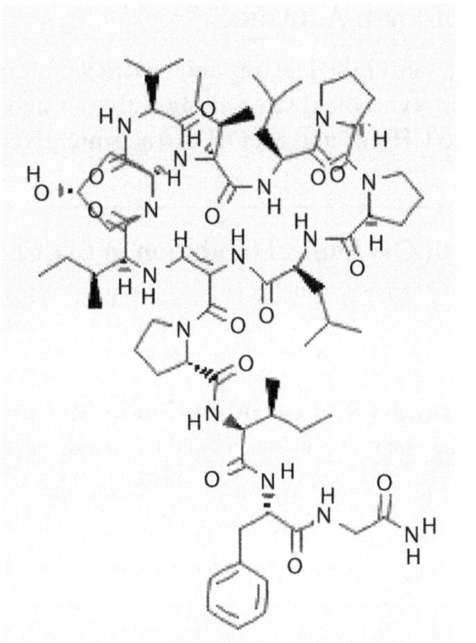

Callyaerin A

Anti-TB activity: Five new cyclic peptides, callyaerins I–M (1–5), along with the known callyaerins A–G (6–12) have been isolated from this species. Among these compounds, callyaerins A and B showed potent anti-TB activity against *Mycobacterium tuberculosis* with MIC90 values of 2 and 5 μM, respectively. Callyaerin C (8) was found to be less active, with an MIC90 value of 40 μM (Daletos et al., 2015).

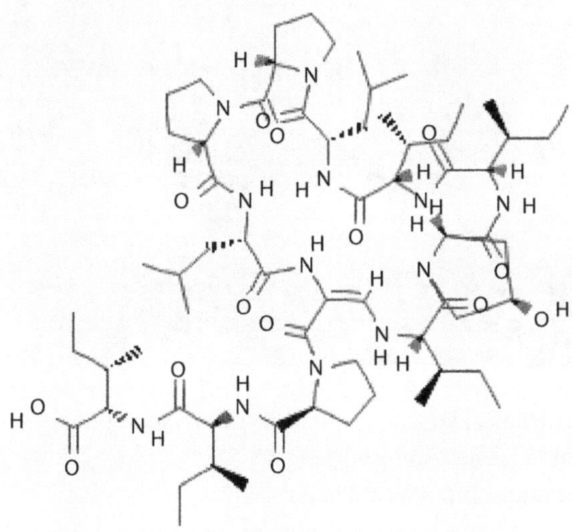

Callyaerin B

Cytotoxic/anticancer effect: The methanol extract of this species showed an anticancer effect on cell lines TE-8 (esophagus), Hep-G2 (liver), and Mia PaCa2 (pancreas) and in noncancerous cells (HET-1A). The IC50 values recorded for 24 h in these cell lines were 3.1 ng/mL, 10.6 ng/mL, 10.7 ng/mL, and not detected, respectively; and for 48 h, 1.06 ng/mL, 17.65 ng/mL, 0.51 ng/mL, and not detected, respectively (Hadisaputri, 2017).

Cytotoxicity: The compound callyaerin G of this species was found to be cytotoxic toward the mouse lymphoma cell line (L5178Y) and HeLa cells with ED50 values of 0.53 and 5.4 ug/mL, respectively, (Ibrahim et al., 2008).

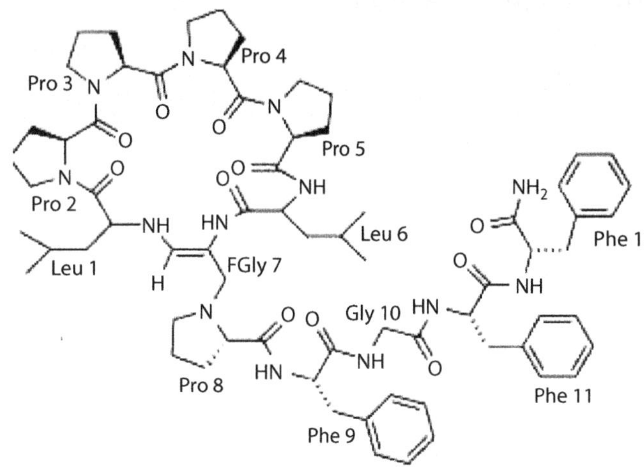

Callyaerin G

The compounds callyaerins E and H exhibited strong cytotoxic activity against the L5178Y cell line with ED50 values of 0.39 and 0.48 μM, respectively (Agrawal et al., 2016; Ibrahim et al., 2010).

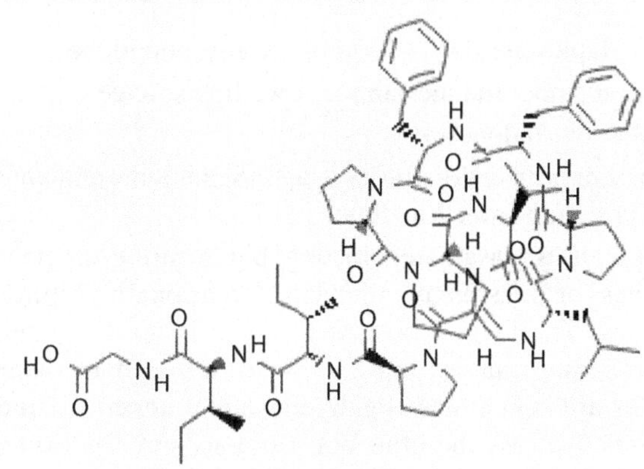

Callyaerin E

***Callyspongia (Cavochalina) bilamellata* (Lamarck, 1814)**

Order: Haplosclerida Family: Callyspongiidae

Common name: Not designated

Distribution: Australia

Ecology: It is attached to solid substrates in benthic habitats.

Biology: It is a pedicellate, bilamellate sponge doubled up like a bivalve shell with a long, hard stem, ending in a root-like expansion. It is 28 cm in height and 15 cm in length. Consistency is leathery. Color of the animal is pale pinkish brown. Outer surface is irregular, corrugated, and wrinkled, and reticulated with numerous microconules. Internal surface is even, smooth, microhispid, and concentrically lined. There are small oscula that are regularly scattered over the internal surface.

Compound(s) and Activities

Nematocidal activity: The depsipeptides phoriospongin A and B isolated from this species displayed significant nematocidal activity against *Haemonchus contortus* with LD99 values of 100 and 194 μg/mL, respectively (Agrawal et al., 2016).

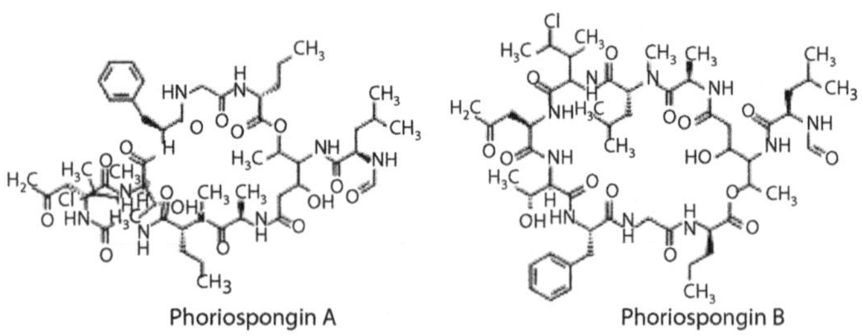

Callyspongia (Euplacella) biru (de Voogd, 2004)

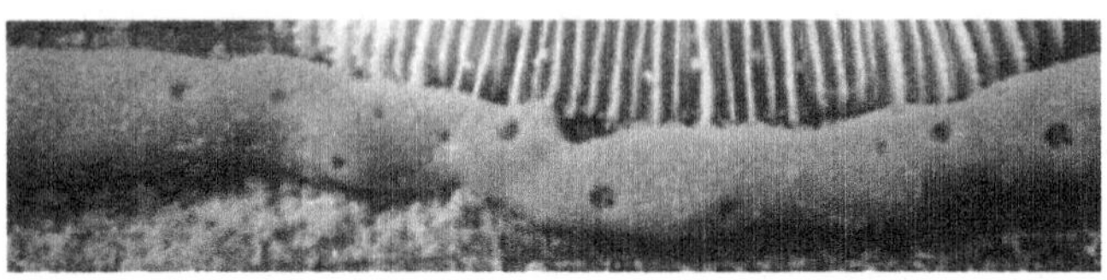

Order: Haplosclerida Family: Callyspongiidae
Common name: Indonesian reef-dwelling sponge
Distribution: Indonesia
Ecology: It is an encrusting sponge in seamounts and knolls at depths ranging from 2 to 10 m.
Biology: It is basally encrusting but forming single erect branches or masses of rounded, occasionally bifurcating branches. The longest observed branch is up to 40 cm and the branching masses may be 10 to 150 mm in diameter. Oscules are sometimes slightly elevated, numerous, and are distributed across the branches. Copious slime is exuded in this species when pressed or cut. Color of the body is bright blue. Surface of the sponge is optically smooth. Consistency is spongy. The high growth potential and high survival rate of this sponge species suggest that it is a promising candidate for mariculture development. It is a toxic sponge too.

Compound(s) and Activities

Antibacterial activity: The amphitoxin of this species possesses antibacterial activity (de Voogd et al., 2005).

Callyspongia clavata (Keller, 1889)

Order: Haplosclerida Family: Callyspongiidae
Common name: Not designated
Distribution: Eritrea, Southern Red Sea, Persian Gulf
Ecology: It grows at depths of 5 to 20 m.
Biology: Not studied

Compound(s) and Activities

Antioxidant activity: The organic extracts of this species showed hydroxyl radical scavenging activities and the percent values in the CH_2Cl_2 and MeOH extracts are given next.

Percent of OH Radical Inhibition in CH_2Cl_2 Extract

Concentrations in milligrams of dried sponge extract

0.5	0.25	0.125	0.062
22.4	46.3	46.6	47.6

Percent of OH Radical Inhibition in MeOH Extract

Concentrations in milligrams of dried sponge extract

0.5	0.25	0.125	0.062
66.2	68.4	70.4	73.2

Source: Seradj, H. et al., 2012, *Iranian J. Pharm. Sci.*, 8: 249–255.

Callyspongia (Cladochalina) diffusa (Ridley, 1884)

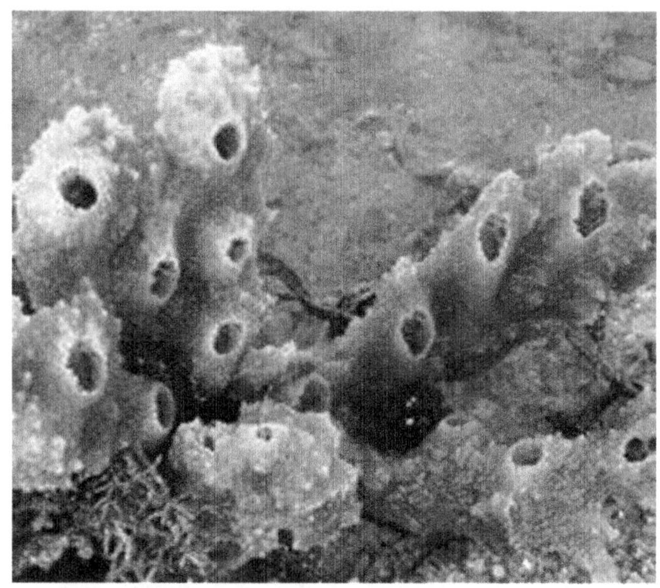

Order: Haplosclerida Family: Callyspongiidae
Common name: Not reported
Distribution: Indo-West Pacific
Ecology: This sessile species is growing on coral rubble at a depth up to 100 m. This group of sponges is sometimes seen on some southern shores, growing on coral rubble.
Biology: These are small (10–15 cm) sponges with rough thick walls and colored a dull champagne pink or purplish pink (lilac). Texture is of a fine network. Forms are variable. Some have thick cylindrical branches, long or short. These branches may fuse to form upright, fanlike, or other irregular shapes. There may be a large circular hole (0.5 cm) at the tip of a branch, but the hole is not very deep. There may also be holes of regular size and spaced along the length of the branch.

Compound(s) and Activities

Immunosuppressive activity: The associated bacterial strains (including *Virgibacillus* sp.) of this species were immunosuppressive and showed antagonistic activity against clinical bacterial pathogens (Perdicaris et al., 2013).

Antimicrobial and anti-TB activities: The organic extracts of this species showed significant antibacterial activity against *Kleibesella pnumoniae, Escherichia coli*, and *Enterococcus feacalis* with the lowest MIC values (0.2–0.4 µg/mL). Similarly, these extracts also showed significant antifungal activity against *Candida albicans* (MIC, 0.2–0.8 µg/mL) and *Aspergillus niger* (MIC, 0.2–0.4 µg/mL). A significant antifungal activity was observed against *Candida albicans* (0.2–0.8 µg/mL) and *Aspergillus niger* (0.2–0.4 µg/mL). Further, the chloroform fraction and ethyl acetate fraction of these extracts have shown significant antitubercular activity with an MIC value of 6.3 µg/mL (Warad et al., 2017).

Antibacterial activity: The aqueous extract of this species showed antibacterial activity against *Vibrio anguillarum* (zone of inhibition, 12 mm), *V. harveyi* (11 mm), *S. aureus* (16 mm), *Bacillus subtilis* (14 mm), *E. coli* (10 mm), *Klebsiella pneumoniae* (10 mm), *P. vulgaris* (10 mm), and *Salmonella enterica typhi* (14 mm) (Mol et al., 2016).

Antibacterial activity: The associated bacterial communities derived from this species displayed antibacterial activity against the pathogenic bacteria, namely, *S. aureus, S. typhi, S. paratyphi, K. oxytoca, E. coli, P. mirabilis, P. vulgaris, K. pneumoniae, V. cholerae*, and *A. tumefaciens*. The values of zone of clearance for different pathogens against each associated bacterial species are given next.

Pathogens Tested (Zone of Clearance in mm)

Associated Bacteria	*S. aureus*	*S. typhi*	*S. paratyphi*	*K. oxytoca*	*E. coli*	*P. mirabilis*	*P. vulgaris*	*K. pneumonia*	*V. cholera*	*A. tumefaciens*
A. faecalis	–	4	1	5	–	3	3	5	1	1
A. hydrophila	8	5	3	–	–	–	5	3	6	2
B. licheniformis	5	5	7	1	4	6	6	3	5	–
B. subtilis	2	5	5	3	4	7	6	7	9	8
P. aeruginosa	12	11	11	7	8	4	4	–	2	12
P. fluroscens	10	11	17	9	12	5	16	11	8	14
P. putida	10	6	6	8	9	–	5	6	6	8

Source: Vasanthabharathi, V., and S. Jayalakshmi, 2012, *Afr. J. Biotechnol.*, 11: 7500–7511.

Antifungal activity: The crude extracts of this species showed antifungal activity against *Aspergillus niger, Pencillium notatum*, and *Candida albicans*, and the values of zone of inhibition at different concentrations of the extract are given next.

Zone of Inhibition (diameter in mm)

Aspergillus niger		*Pencillium notatum*		*Candida albicans*	
50 µl	100 µl	50 µl	100 µl	50 µl	100 µl
1.2	1.2	1.3	1.5	0.8	1.8

Source: Dhinakaran, D.I. et al., 2012, *J. Appl. Pharm. Sci.*, 2: 52–55.

Mosquito larvicidal activity: At a methanol extract concentration of 10%, this species showed larvicidal activity (50%) with low LC value (1.4 mg/mL) (Sonia and Lipton, 2012).

Rao et al. (2008) reported that the extract of this species displayed larvicidal activity against the fourth instar larvae of *Aedes aegypti* (yellow fever mosquito vector) with an LC50 value of 142.2 ppm.

Callyspongia (Cladochalina) fibrosa (Ridley and Dendy, 1886)

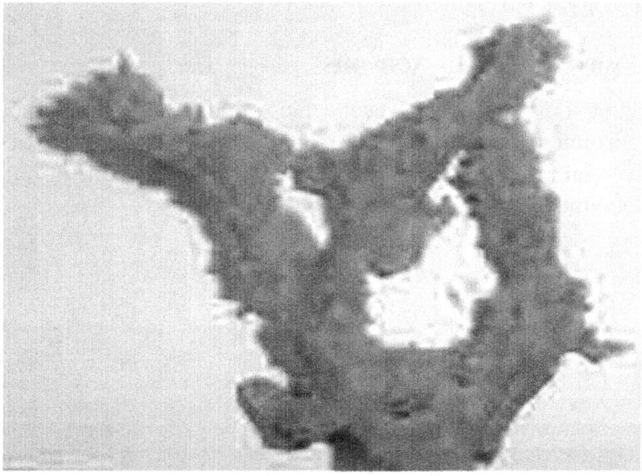

Order: Haplosclerida Family: Callyspongiidae
Common name: Not known
Distribution: Southern Vietnam
Ecology: It is a shallow water sponge.

Biology: This sponge is composed of finger-shaped or flattened branches. Surface may or may not have strong conules. Conules are prominent at growing tips. Oscules are irregularly distributed and may be terminal or marginal, rounded or elliptical, shallow, and compound with diameters up to 4 mm. Color of the sponge is pale yellow.

Compound(s) and Activities

Antimalarial activity: Among the four new polyhydroxylated sterols isolated from this species, the major steroid 4a showed antimalarial activity against *Plasmodium falciparum* on the chloroquine-resistant stain better than on the chloroquine-sensitive strain (Rao et al., 2010).
Antibacterial activity: The ethanolic extract of this species exerted antibacterial activity against cerebrospinal fluid (CSF) disease (neurological disease) causing pathogenic bacteria (Sujatha et al., 2014). The values of inhibition zone diameter (in mm) for different species of bacteria are given next.

Inhibition Zone Diameter (in mm) of Bacteria in *Callyspongia fibrosa* (Ms-4)

E. coli	*S. aureus*	*H. influenzae*	*N. meningitides*	*K. pneumonia*	*Streptococcus agalactiae*	*E. faecalis*
No inhibition	8	9	13	No inhibition	10	11

Source: Sujatha, S., 2014, *Pharma Innovation*, 3: 81–86.

Callyspongia fistularis (Topsent, 1892)

Order: Haplosclerida Family: Callyspongiidae
Common name: Red Sea sponge
Distribution: Red Sea
Ecology: It occurs in shallow coastal waters.
Biology: Not reported

Compound(s) and Activities

Cytotoxicity: The polyacetylenic amide callyspongamide A isolated from this species showed a moderate cytotoxicity against HeLa cells with an IC50 value of 4.1 μg/mL (Dembitsky et al., 2006).

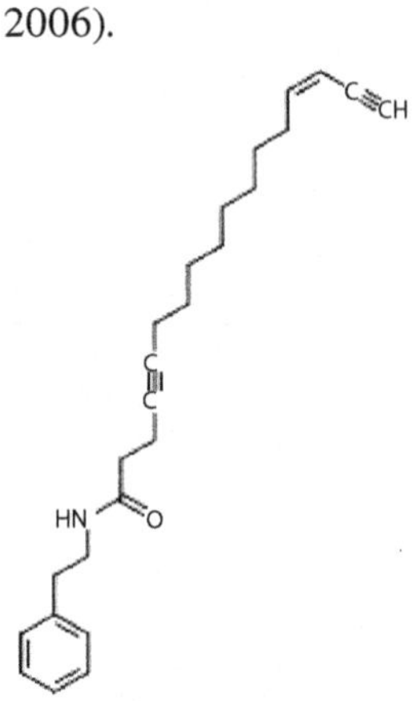

Callyspongamide A

Callyspongia implexa (Topsent, 1892) (= *Ceraochalina implexa*)

Order: Haplosclerida Family: Callyspongiidae
Common name: Red Sea sponge, tube sponge
Distribution: Djibouti, Egypt, Eritrea, Saudi Arabia, Sudan, Yemen
Ecology: It is a coral reef–associated species.
Biology: Not available

Compound(s) and Activities

Antichlamydial (antibacterial) activity: Among the compounds, 26,27-bisnorcholest-5,16-dien-23-yn-3β,7α-diol, gelliusterol E (1) and C27-polyacetylene, callimplexen A (2), and β-sitosterol (3) isolated from this species, gelliusterol E (1) acted against *Chlamydia trachomatis*, an obligate intracellular Gram-negative bacterium and inhibited the formation and growth of chlamydial inclusions in a dose-dependent manner with an IC50 value of 2.3 μM. This bacterium has been reported as the leading cause of ocular and genital infections worldwide (Abdelmohsen et al., 2015; Nadeem et al., 2015).
Anti-TB activity: The methonolic extract of this species found potent bioactivity against TB (Cheng, 2012).

Callyspongia (Callyspongia) reticutis (Dendy, 1905)

Order: Haplosclerida Family: Callyspongiidae
Common name: Not known
Distribution: Seychelles, South India, and Sri Lanka
Ecology: It grows in shallow coastal waters.
Biology: Not studied

Compound(s) and Activities

Antifungal activity: The crude extracts of this species showed antifungal activity against *Aspergillus niger, Pencillium notatum*, and *Candida albicans*, and the values of zone of inhibition at different concentrations of the extract are given below.

Zone of Inhibition (Diameter in mm)

Aspergillus niger		*Pencillium notatum*		*Candida albicans*	
50 μl	100 μl	50 μl	100 μl	50 μl	100 μl
0.7	1.0	0.5	0.9	1.0	1.5

Source: Dhinakaran, D.I. et al., 2012, *J. Appl. Pharm. Sci.*, 2: 52–55.

Antibacterial activity: The ethyl acetate extracts of the associated bacterial isolate GUVFCM derived from this species were found to possess antibacterial activity against human bacterial pathogens *S. typhi, S. marcescens, S. vitreus*, and *P. vulgaris* with the values of zone of inhibition (dia. mm) of 2, 4, 3, and 6, respectively (Velho-Pereira and Furtado, 2012).

Callyspongia (Callyspongia) siphonella (Lévi, 1965) (= *Siphonochalina siphonella*)

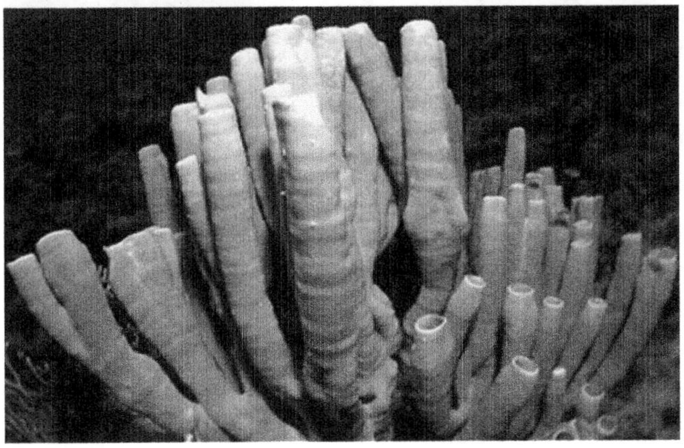

Order: Haplosclerida Family: Callyspongiidae

Common name: Colonial tube sponge, Red Sea sponge

Distribution: This species is endemic to the Red Sea; widely distributed throughout the Gulfs of Aqaba and Suez.

Ecology: It dwells from shallow water but mostly deeper than 5 m.

Biology: This sponge forms clusters of its pale lavender long tubes, which reach a size up to 60 cm.

Compound(s) and Activities

Cytotoxicity: This species is a rich source of sipholane triterpenoids including sipholenols (A, C–L); sipholenones (A, E); and siphonellinols (C, D, E). Among these compounds, sipholenol A was found to have increased the sensitivity of resistant KB-C2 cells. Sipholenol A, sipholenol I, sipholenol L, sipholenone A, sipholenone E, siphonellinol C, and siphonellinol D showed potent reversal of multidrug resistance in cancer cells that overexpressed P-glycoprotein (Li et al., 2013).

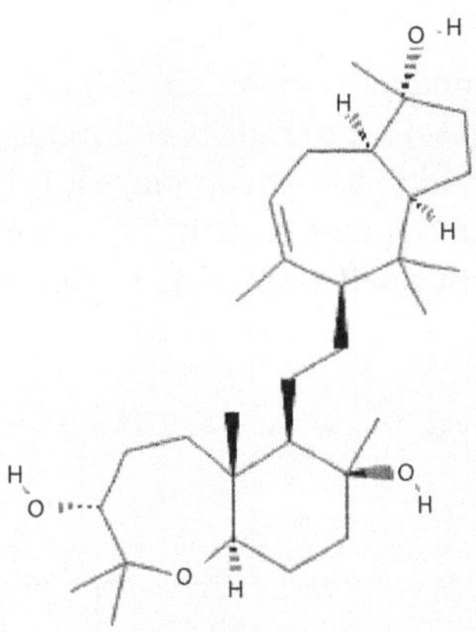

Sipholenol A

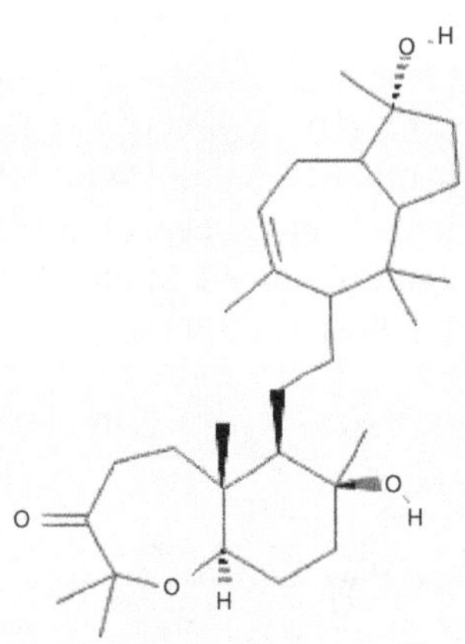

Sipholenone A

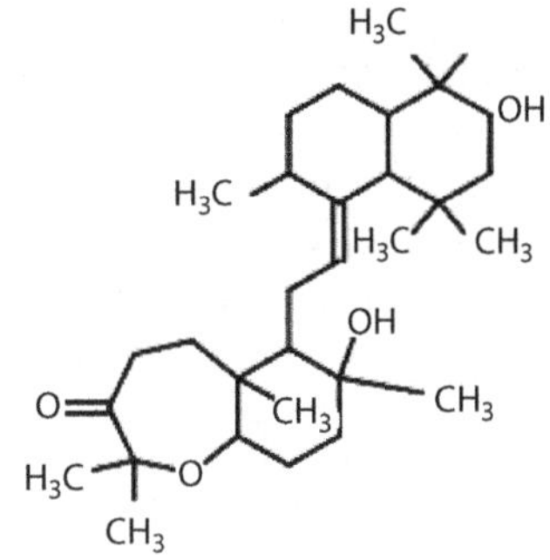

Sipholenone E

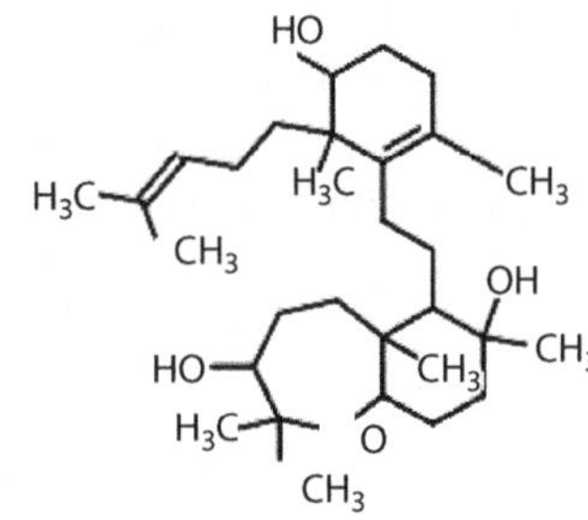

Sipholenone D

Six metabolites—callyspongenol D (1), callyspongendiol (2), ergosta-5, 24 (28) dien-3-ol (3), sipholenone A (4), sipholenol A (5), and sipholenol L (6)—have been isolated from this species. Among these, compounds 4–6 exhibited cytotoxicity on various cell lines with IC50 values (μM) of 95.8, 14.8, and 19.8, respectively. Compounds 1–3 showed IC50 values more than 100 μM. Siphenone A (4), sipholenol A (5), and sipholenol L (6) showed moderate activity against HCT-116 cells with IC50 values of 14.8 and 19.8 μM, respectively, compared to 6.0 μM for doxorubicin B (Sobahi et al., 2017).

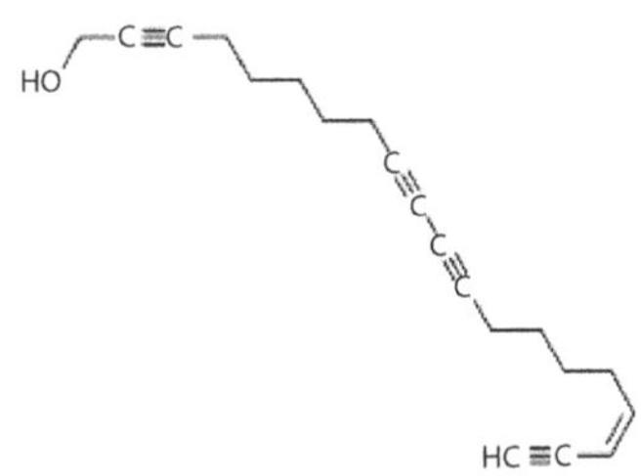

Callyspongenol D

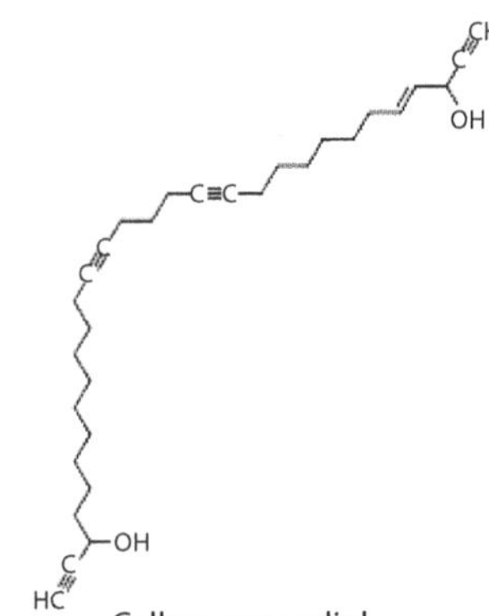

Callyspongendiol

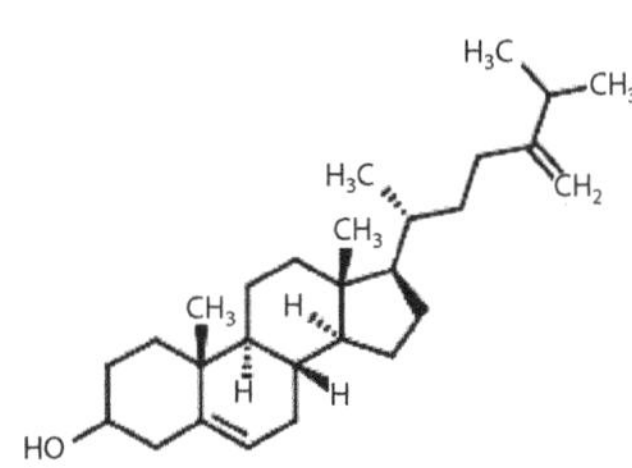

Ergosta-5, 24 (28) dien-3-ol (3)

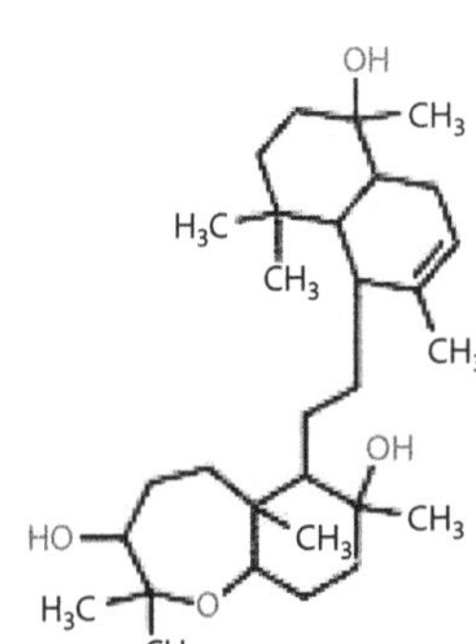

Sipholenol L

Antiproliferative activity: A new compound, namely, neviotine-C (1), along with four known metabolites (2–5) have been isolated from this species. All compounds, except 5 showed antiproliferative activity against MCF-7, PC-3, and A549 cell lines with IC50 in range 7.9 to 87 μM, while compound 3 showed potent activity against PC-3 and A549 with IC50 = 7.9 and 8.9 μM, respectively (Angawi et al., 2014). Similarly its compound sipholenone A showed activity against MCF-7 and HepG-2 cell lines with IC50 values of 3.0 and 2.8 μM, respectively (Nadeem et al., 2015).

Anti-inflammatory activity: The chloroform fraction of methanol extracts of this species yielded a compound, callysterol (ergosta-5,11-dien-3beta-ol), which showed anti-inflammatory activity (Youssef et al., 2010).

Antioxidant activity: Seradj et al. (2012) reported on the hydroxyl radical scavenging activities of this sponge extracts.

Percent of OH Radical Inhibition in CH_2Cl_2 Extracts

Concentrations in milligrams of dried sponge extract

0.5	0.25	0.125	0.062
71.4	72.3	69.3	72.1

Percent of OH Radical Inhibition in MeOH Extracts

Concentrations in milligrams of dried sponge extract

0.5	0.25	0.125	0.062
48.4	46.6	40.3	35.3

Source: Seradj, H. et al., 2012, *Iranian J. Pharm. Sci.*, 8: 249–255.

Callyspongia (Cladochalina) subarmigera (Ridley, 1884)

Order: Haplosclerida Family: Callyspongiidae

Common name: Not known

Distribution: Palau, Palawan/North Borneo, Western and Northern Madagascar

Ecology: Not reported

Biology: Body of this species is tubular and flabellate with branches that are 5 to 10 mm in diameter. Color of the animal is yellowish brown. This sponge has compressible consistency with good resiliency. Oscules of 2.5–5.0 mm diameter are arranged along the upper surface.

Compound(s) and Activities

Mosquito larvicidal activity: At methanol extract concentration of 10% this species showed high larvicidal activity (90%) with low LC value (0.4 mg/mL) (Sonia and Lipton, 2012).

Callyspongia truncata (Lindgren, 1897)

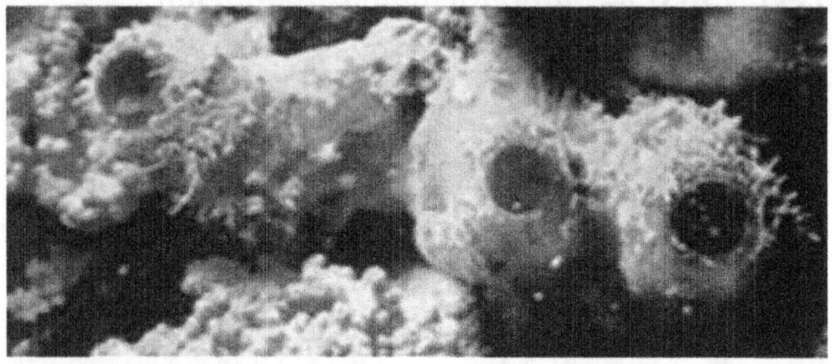

Order: Haplosclerida Family: Callyspongiidae

Common name: Not designated

Distribution: Vietnam, Japan

Ecology: It grows in islands.

Biology: Not reported

Compound(s) and Activities

Cytotoxicity: A polyketide compound, callystatin A, isolated from this species exhibited potent cytotoxicity against KB cells at IC50 0.01 ng/mL (Kobayashi et al., 1997; Sipkema et al., 2005).

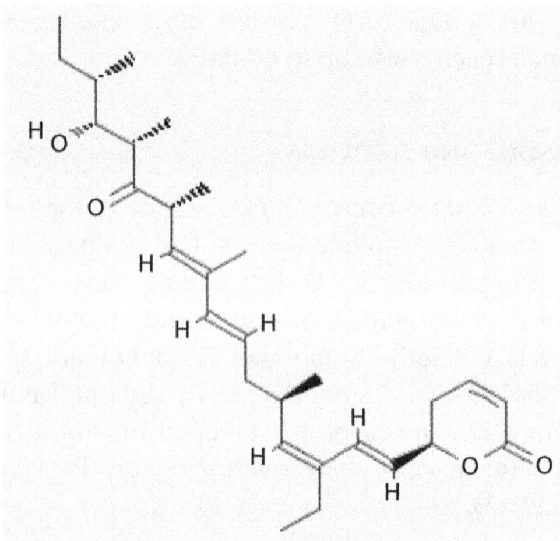

Callystatin A

Anticancer activity: The compound callysponginol sulfate A isolated from this species serves as a membrane type 1 matrix metalloproteinase (MT1-MMP) inhibitor, which plays a role in reducing cancer cell motility and tumor growth. This compound inhibited MT1-MMP with an IC(50) of 15.0 μg/mL (Fujita et al., 2003b).

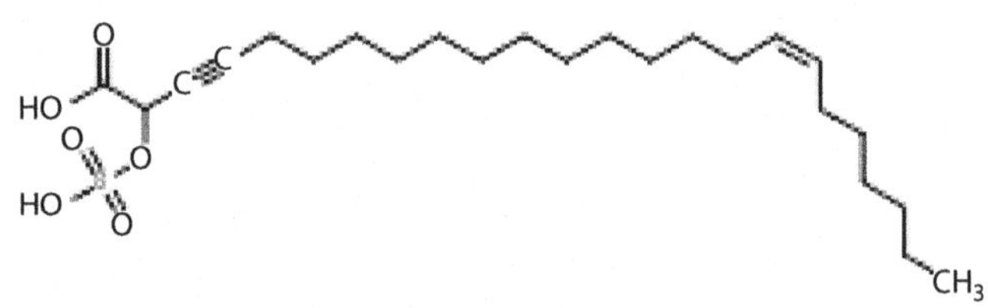

Callysponginol sulfate A

Antiviral activity: The polyacetylene compound callyspongynic acid of this species has potential antiviral activity. Further, it also inhibits glucosidase, an enzyme responsible for blood-related diseases (Sipkema et al., 2005).

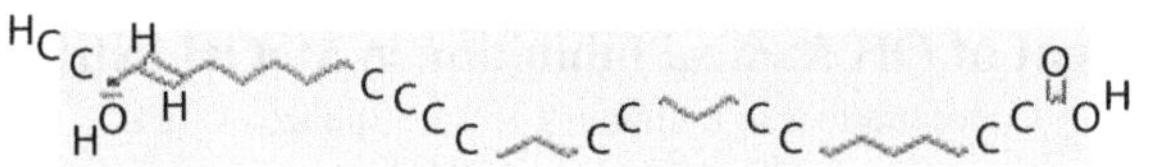

Callyspongynic acid

Antidiabetic activity: The polyacetylenic acid, callyspongynic acid, of this species serve as a α-glucosidase inhibitor, which interfere with the hydrolysis of glycogen, keeping the glucose concentration in the blood at a lower level. This compound may be of great use in the treatment of patients with type 1 and 2 diabetes (Nakao et al., 2002; Senthilkumar and Kim, 2013).

Callyspongia vaginalis (Lamarck, 1814)

Order: Haplosclerida Family: Callyspongiidae

Common name: Branching vase sponge, tube sponge, South China Sea sponge

Distribution: It is a native species in Trinidad and Tobago, and the Caribbean, and is distributed in South Florida, the Bahamas, and the Caribbean.

Ecology: It inhabits shallow and midrange reefs, walls, and rocky areas. This species is often infested with sponge zoanthids (*Parazoanthus parasiticus*) and sponge brittle stars (*Ophiothrix suensonii*).

Biology: The shape of this species is a cluster of long erect tubes, tapering only slightly. The tubes have a wide, thin-walled vent, and their outer walls are strongly conulose. The tubes show a considerable tendency to anastomose and this may lead to hollow, fan-shaped specimens. Tubes are up to 25 cm high and 5 cm in diameter. Fan-shaped forms are up to 18 cm high, 21 cm in the widest expansion, and up to 3 cm in thickness. Vents are up to 2.5 cm in diameter. Conules are about 3 mm high and 5 mm apart. Surface of outer walls is smooth between the conules. The consistency is very spongy. Color of the animal is very variable from lavender to brownish gray to greenish gray and occasionally light tan.

Food and feeding: These sponges acquire their food heterotrophically by suspension-feeding, and feed on plankton and other microorganisms that are small enough to filter out of water currents that are moved through its anatomy.

Reproduction: This species would normally release its larvae in the daytime between 9 a.m. and 5 p.m. at varying intervals and numbers but can release an average of 50 larvae per day per individual. When the larvae are developed enough for release they may leave by squeezing through ostioles and secondary oscules (narrow openings) in the parent's body.

Compound(s) and Activities

Antimicrobial/anticancer activity: The associated fungal strain of this sponge species, namely, *Ulocladium botrylis* 193A4, yielded a new tyrosine kinase inhibitor that is used in cancer therapy. The antimicrobial compound ulocladol together with the antifungal agent 1-hydroxy-6-methyl-8-(hydroxyl- methyl) xanthone have also been isolated from the said fungal strain (Thomas et al., 2010).

Carotenoids: The bacterial symbionts of this species have been reported to produce carotenoid pigments such as diadinoxanthin, fucoxanthin, neoxanthin, dinoxanthin, and diadinochrome. The therapeutic values of these pigments for degenerative diseases of the retina are well known (Abfa et al., 2017).

Callyspongia sp.

Order: Haplosclerida Family: Callyspongiidae

Antiviral activity: The associated *Epicoccum* sp. JJY40 of this unidentified sponge species has yielded pyronepolyene compounds, namely, C-glucoside iso-D8646-2-6 and 8646-2-6, which showed antiviral activity against H1N1 with IC50 values of 56.1 and 62.1 μg/mL, respectively (Indraningrat et al., 2016).

Cytotoxicity: A new 1,4-diazepine, callysponine (1) together with four known proline-based diketopiperazines—cyclo-(S-Pro-R-Leu) (2), cyclo-(S-Pro-R-Val) (3), cyclo-(S-Pro-R-Ala) (4), and cyclo-(SPro-R-Tyr) (5)—have been isolated from this unidentified species. Among these compounds, compound 1 showed a marginal activity against a small panel of three human tumor cell lines: HepG2 (hepatoma carcinoma cell), A549 (lung carcinoma cell), and HeLa (cervical cancer cell). The inhibition ratio of these cell lines were lower than 10% at concentration of 100 μg/mL with the IC50 values of 13.7, 2.1, and 3.8 μg/mL, respectively (Huang et al., 2010).

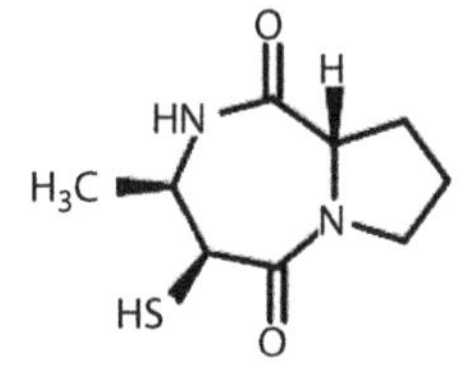

Callysponine

The compounds, polyacetylenic alcohols, callyspongenols A–C, and dehydroisophonochalynol isolated from this species showed moderate cytotoxicity against P388 and HeLa cells (Dembitsky et al., 2006).

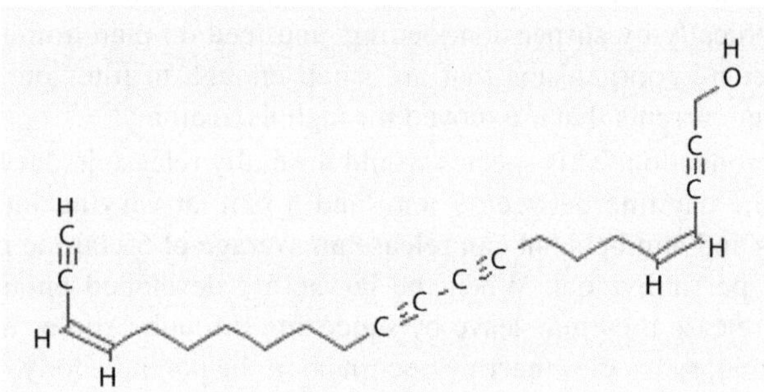

Callyspongenol B

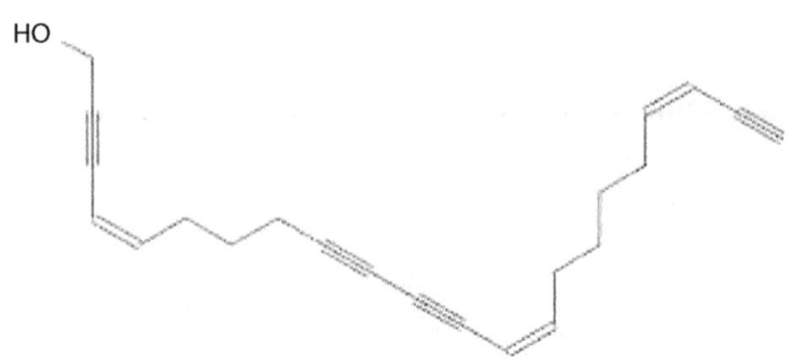

Callyspongenol C

Antitumor activity: The compound callyspongiolide of this species exhibited potent cytotoxicity against L5178Y mouse lymphoma cells, human Jurkat J16 T and Ramos B lymphocytes with IC50 values of 320, 70, and 60 nM, respectively (Gribble, 2015).

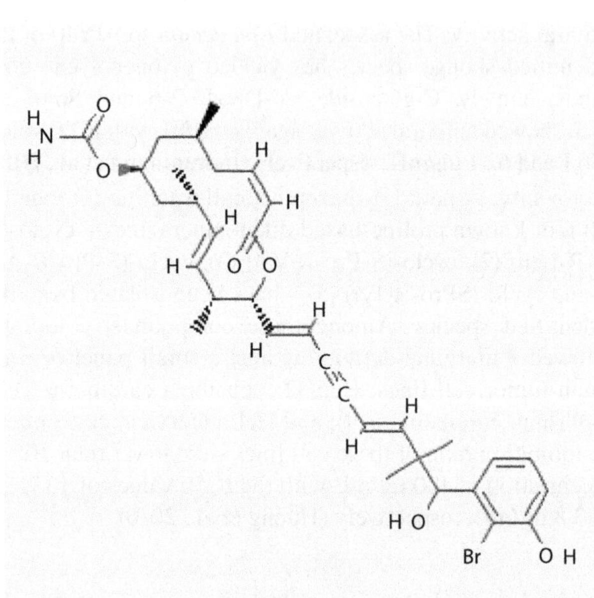

Callyspongiolide

Antioxidant activity: Seradj et al. (2012) reported on the hydroxyl radical scavenging activities of this sponge's extracts.

Percent of OH Radical Inhibition in CH_2Cl_2 Extracts

Concentrations in milligrams of dried sponge extract

0.5	0.25	0.125	0.062
9.1	9.8	10.8	−2.2

Percent of OH Radical Inhibition in MeOH Extracts

Concentrations in milligrams of dried sponge extract

0.5	0.25	0.125	0.062
33.5	22.0	17.0	−2.7

Source: Seradj, H. et al., 2012, *Iranian J. Pharm. Sci.*, 8: 249–255.

Enzyme inhibiting activity: The compound bromotyrosine of this species showed weak ability to increase ApoE from human astrocytoma cells (CCF-STTG1) at a concentration of 40 μM (Gribble, 2015).

Siphonochalina sp.

Order: Haplosclerida Family: Callyspongiidae

Compound(s) and Activities

Cytotoxicity: The compound dehydroisophonochalynol isolated from this species showed cytotoxicity against HeLa and P388 cells (Dembitsky et al., 2006).

FAMILY: CHALINIDAE

Adocia sp.

Order: Haplosclerida Family: Chalinidae

Compound(s) and Activities

Antibacterial activity: The compound o-aminophenol isolated from the associated bacterial isolate of this species has shown antibacterial activity by inhibiting the growth of *Staphylococcus aureus* and *Bacillus subtilis* (Oclarit et al., 1994).

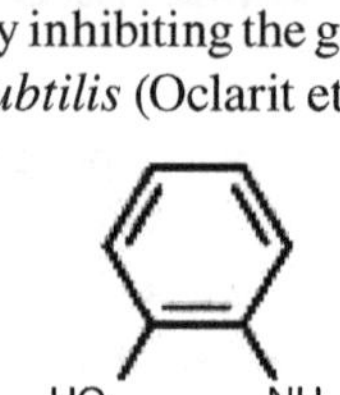

o-Aminophenol

Antitumor activity: The triterpenoid hydroquinone compounds, namely, adociasulfates, isolated form this unidentified species have shown antitumor activity by inhibiting A1,3-fucosyltransferase and kinesin motor protein (Anjum et al., 2016).

The compounds adociacetylenes A–D (polyacetylenes-adociacetylenes A, C, and D) of this species exhibited inhibitory activity in the *in vitro* endothelial cell-neutrophil leukocyte adhesion assay. All these compounds were highly cytotoxic to P388, A-549, HT-29, and MEL-28 melanoma cells (Dembitsky et al., 2006).

Haliclona (Reniera) cribricutis (Dendy, 1922)

Order: Haplosclerida Family: Chalinidae

Common name: Indian bright blue sponge

Distribution: Indian Ocean

Ecology: Not reported

Biology: Sponge of this species is irregular, tuberous and is attached by many points to the substratum. Surface is slightly conulose. The ridges connecting the adjacent conules and the projecting lateral walls of the subdermal cavities together give a characteristic appearance to the surface. Dermal membrane is transparent. Diameter of the body is 1.2 cm. Color of the animal is pale gray when dry. Consistency is slightly compressible with poor resiliency. Oscules are numerous and are irregularly distributed on the general surface. Oscular rim slightly elevated. Pores are minute. Oxeas of the spicules are slightly curved, sharply and gradually pointed.

Compound(s) and Activities

Mosquito larvicidal activity: The extract of this species displayed larvicidal activity against the fourth instar larvae of *Aedes aegypti* (yellow fever mosquito vector) with an LC50 value of 31.5 ppm (Rao et al., 2008).

Cytotoxicity: The compounds ceramides isolated from this species showed cytotoxicity by serving as important intracellular second messengers that play a role in the regulation of cell growth, differentiation, and programmed cell death (Tilvi et al., 2005).

Haliclona (Gellius) cymaeformis (Esper, 1794) (= *Sigmadocia symbiotica*)

Order: Haplosclerida Family: Chalinidae

Common name: Not designated

Distribution: Australia, India, Singapore, Madagascar, Sri Lanka

Ecology: It occurs in subtidal, shallow coastal waters and semidark coral reefs off island where the water is calm and turbid.

Biology: The morphology of this sponge varies according to water-movement intensity. In sheltered positions, the sponge specimens have a bushy form, with flattened, elk horn–shaped branches that offer wider surfaces to solar radiation. In relatively more exposed areas, sponge specimens become cushion shaped to lobate, devoid of branches. The oscules (1.5 to 3 mm in dia.) are round or elongated, flush, or slightly raised, and are always located in the upper part of the sponge. The color of the sponge in life is dark green. The spicules may be arranged in parallel to the algal bundles without forming the usual reticulation. Oxeas of the spicules are fusiform or slightly angulate, with both extremities pointed. The alga Ceratodictyon–Haliclona symbiotic association is very common and is quite remarkable. At first glance, this sponge is not immediately identifiable because ~70% of the biomass of each clump consists of the said algal tissue.

Compound(s) and Activities

Anti-inflammatory activity: The cyclic heptapeptide compounds, cis,cis- (1) and trans,trans-ceratospongamide (2) and the minor trans,trans- conformer (3), have been isolated from this sponge. Among them, compounds 1 and 3 inhibited expression of secretory PLA2 with ED50 of 32 nM (Keyzers, 2005).

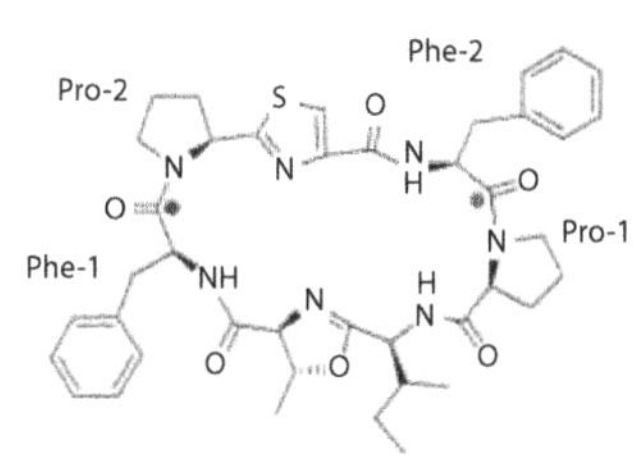

Cis, cis-ceratospongamide

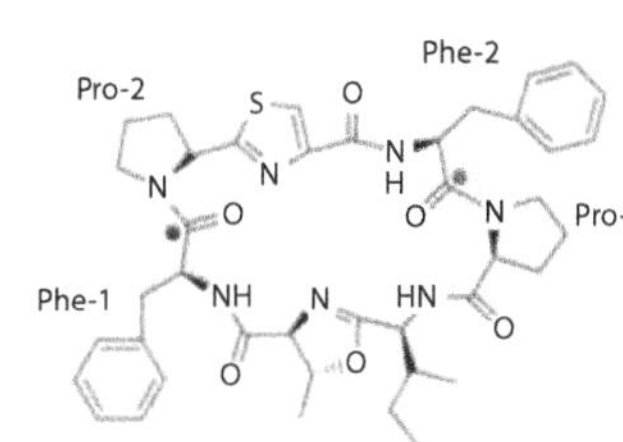

Trans, trans-ceratospongamide

Haliclona (Reniera) fascigera (Hentschel, 1912)

Order: Haplosclerida Family: Chalinidae

Common name: Large tube-sponge

Distribution: Indo-West Pacific Ocean

Ecology: Inhabits hard substrates on reef flats.

Biology: It is a purple-colored sponge. Its surface is quite smooth, fragile, and is easily torn.

Compound(s) and Activities

Antimicrobial activity: The ethyl acetate extracts of the fungi isolates derived from this sponge showed antimicrobial activity against *Staphylococcus aureus*, *Escherichia coli*, and *Candida albicans*. The values of zone of inhibition (mm dia.) for eight isolates against *Staphylococcus aureus* were HF2 isolate (14.5 mm), HF8 isolate (12.5 mm), HF9 isolate (14 mm), HF13 isolate (16.5 mm), HF14 isolate (11 mm), HF19 isolate (14.5 mm), HF21 isolate (14.5 mm), and HF22 isolate (13 mm). One isolate was active to *C. albicans*: HF16 isolate (12.5 mm) (Handayani et al., 2015).

Antibacterial activity: The ethyl acetate extracts of symbiotic bacteria from this sponge showed antibacterial activity against the multidrug resistant organisms such as Vancomycin-resistant *Enterococcus* sp. (VRE), extended-spectrum beta (Î²)-lactamase Gram-negative organisms (ESBL), Carbapenem-resistant Enterobacteriaceae (CRE), and multi-resistant *Acinetobacter baumannii* (MRAB). Nine extracts showed significant activity against VRE and ESBL. The strongest inhibition activity was found in extract of *Micrococcus sp.* with MIC values of 0.625% against VRE (Handayani et al., 2016a).

Haliclona (Gellius) fibulata (Schmidt, 1862) (= *Sigmadocia fibulata*)

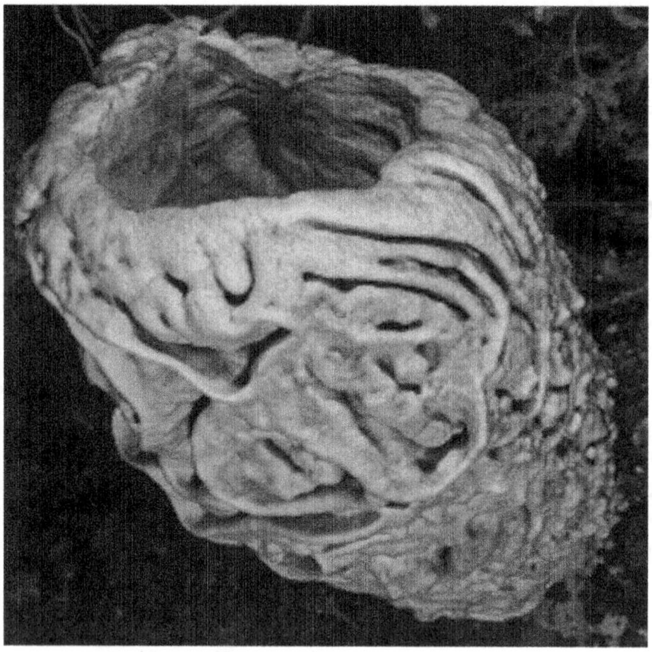

Order: Haplosclerida Family: Chalinidae

Common name: Not designated

Distribution: British Isles, Atlantic coasts of France, Spain, Mediterranean, Greenland

Ecology: It occurs normally in the infralittoral, to about 60 m on stones, conglomerates of shells, calcareous algae, and sponges. Also reported from a depth of about 500 m.

Biology: It is a cushion-shaped sponge that may become a laterally spreading mass, occasionally with fistular proliferations. It may reach a size of 10 cm or more. Oscules (1–2 mm) are few and they flush with the surface. Surface is smooth and is strongly reticulated. Consistency is rather firm, but compressible. Oxeas of the spicules are straight or slightly curved, with a rather long and sharp point. The ectosomal skeleton is irregular. Color of the sponge is old rose or pale yellow when alive and in alcohol it is cream, yellowish white.

Compound(s) and Activities

Antibacterial activity: The ethanolic extract of this species exerted antibacterial activity against Cerebrospinal fluid (CSF) disease (neurological disease) causing pathogenic bacteria (Sujatha et al., 2014). The values of inhibition zone diameter for different species of bacteria are given next.

Inhibition Zone Diameter (in mm) of Bacteria in *Callyspongia fibrosa* (Ms-4)

E. coli	*S. aureus*	*H. influenzae*	*N. meningitides*	*K. pneumonia*	*Streptococcus agalactiae*	*E. faecalis*
No inhibition	12	9	No inhibition	11	6	12

Source: Sujatha, S., 2014, *Pharma Innovation*, 3: 81–86.

Mosquito larvicidal activity: The extract of this species displayed larvicidl activity against the fourth instar larvae of *Aedes aegypti* (yellow fever mosquito vector) with an LC50 value of 97.3 ppm (Rao et al., 2008).

Haliclona (Flagellia) flagellifera (Ridley and Dendy, 1886) (= *Haliclona (Gellius) flagellifera*)

Order: Haplosclerida Family: Chalinidae

Common name: Not designated

Distribution: Circumglobal southern ocean localities

Ecology: It occurs in a depth range limited largely to continental and upper bathyal waters.

Biology: It is a massive encrusting sponge with a size of 2.9 cm in lateral expansion and 1.6 cm in thickness. Color of the animal is pale grayish (in alcohol). A single oscule is present. Surface is shaggy. Consistency is soft and brittle. Oxeas of the spicules are slightly curved, fat, cigar shaped, and gradually pointed.

Compound(s) and Activities

Antibacterial activity: The extracts of this species showed antibacterial activity against the laboratory, commensal, and clinically relevant bacterial strains, namely, *Bacillus subtilis* EXB-V6 and *Macrococcus* 1F with the MIC values of 65 and 650 μg/mL, respectively (Turk et al., 2013).

The extracts of this species showed antibacterial activity against the environmental bacterial strains *Exignuobacterium* sp., *Pseudoalteromonas* sp., and *Janthinobacterium svalbardensis* with the MIC values 70 μg/mL, respectively (Turk et al., 2013).

Haliclona (Halichoclona) fulva (Topsent, 1893) (= *Reniera fulva*)

Order: Haplosclerida Family: Chalinidae

Common name: Mediterranean sponge

Distribution: North Atlantic Ocean, Western Mediterranean, Aegean Sea, Azores, Canaries, Madeira, Black Sea, Gulf of Guinea West

Ecology: It is found on molluscan (mussel) shells and other hard substrata in shallow water.

Biology: This coralligenous species has massive and thickly encrusting sponges. The body of the sponge has an orange color when alive and greenish in alcohol; fragile and hard. Oscules are few, small (2–4 mm), slightly elevated, and circular. Ectosomal and choanosomal skeleton consists of primary multispicular and secondary unispicular tracts. Spicules include only slender and fusiform oxeas. This species is normally fed by the dotted sea slug *Peltodoris atromaculata* (Doridoidea: Nudibranchia).

Compound(s) and Activities

Antibacterial activity: Nine linear polyoxygenated acetylenes, fulvynes A–I, isolated from the butanolic extract of this species were found to be active against a chloramphenicol-resistant strain of *Bacillus subtilis* (Nuzzoa et al., 2012).

Cytotoxicity: A new long-chain acetylene compound fulvinol isolated from this species exhibited cytotoxicity against four tumor cell lines (ED50 = 1 μg/mL) (Ortega et al., 1996).

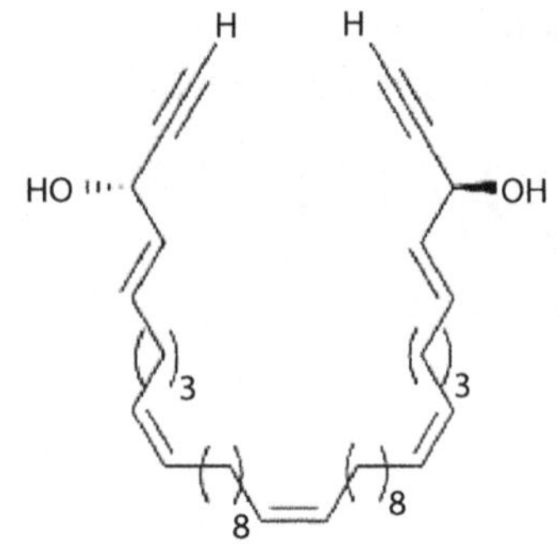

Fulvinol

The compound, fulvinol of this species was found to be cytotoxic to P-388 mouse lymphoma, A-549 human lung carcinoma, HT-29 human colon carcinoma, and MEL-28 human melanoma with ED50 1 μg/mL each (Dembitsky et al., 2006).

Haliclona (Gellius) lacazei (Topsent, 1893) (= *Raphisia lacazei*)

Order: Haplosclerida Family: Chalinidae

Common name: Mediterranean marine sponge

Distribution: Eastern Atlantic, Western Mediterranean

Ecology: It inhabits coral and sandy bottoms.

Biology: This sponge has a broad-based tube (2.5 cm high, 1 cm dia.) with a thick-walled oscule of 0.5 cm diameter. Surface is smooth and fine-grained. Consistency is rather soft and fragile. Color of the animal is light yellow in alcohol. Ectosome is represented by some tangential spicules. Choanosome has an irregular unispicular reticulum of megascleres. Microscleres are extremely abundant. Spicules are megascleres that are robust and slightly curved.

Compound(s) and Activities

Cytotoxicity: The bisindole alkaloids topsentin B1 and B2 isolated from this species inhibited human broncopulmonary cancer cells NSCLC-N6 (IC50 12.0 and 6.3 μg/mL, respectively) (Gul et al., 2005).

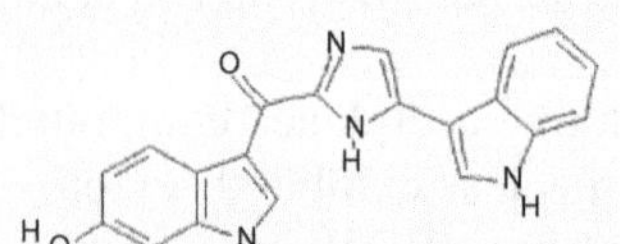

Topsentin B1

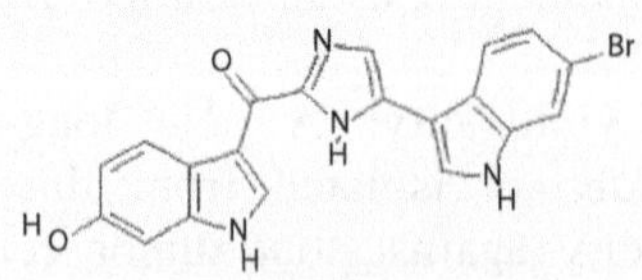

Topsentin B2

Haliclona (Soestella) mucosa (Griessinger, 1971)

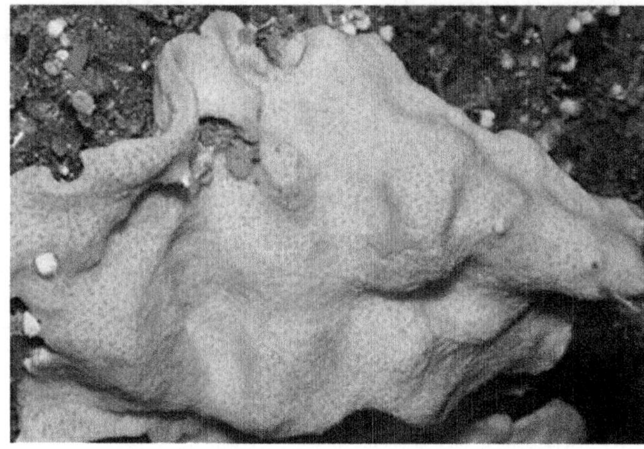

Order: Haplosclerida Family: Chalinidae

Common name: Mediterranean sponge

Distribution: Western Mediterranean

Ecology: It is a coralligenous (coralligenous framework results from a multistratified accretion due to the accumulation of crustose coralline algae and animal builders) species almost covering the live rock.

Biology: The growth form of this massive, encrusting species is with a mixture of yellow and pale pink color.

Compound(s) and Activities

Anticancer activity: The compound panicein A hydroquinone isolated from this species inhibited the doxorubicin efflux activity of patched, and enhanced the cytotoxicity of chemotherapeutic agent on melanoma cells *in vitro*. Panicein A hydroquinone strongly increased cell mortality induced by dxr on MEWO cells derived from a melanoma metastatic site (lymph node tissue; 5 to 8 times) and on A375 cells derived from a human malignant melanoma (2 to 3 times). The IC50 values of the said compound were about 5 μM and 20 μM on MEWO and A375 cells, respectively, in the presence of dx, while in the absence of dxr, this molecule was weakly cytotoxic with IC50 values of more than 30 μM. The use of such inhibitors of patched drug efflux activity in combination with classical chemotherapy could represent a novel approach to reduce tumor drug resistance, recurrence, and metastasis (Fiorini et al., 2015).

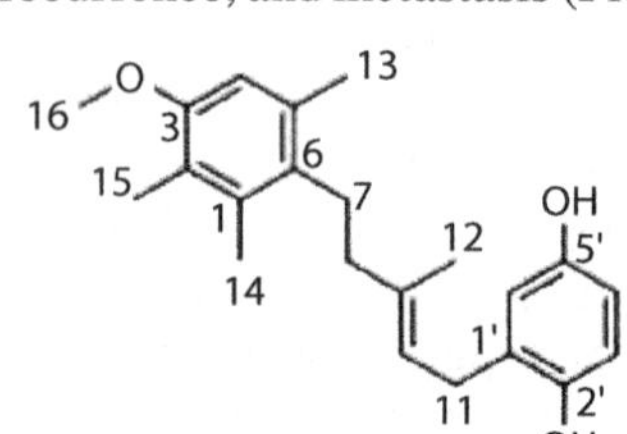

Panicein A hydroquinone

Haliclona nigra (Burton, 1929)

Order: Haplosclerida Family: Chalinidae
Common name: Not designated
Distribution: Western Central Pacific, Indonesia
Ecology: It inhabits protected environments such as undersides of rocks and crevices, both in intertidal and subtidal habitats.
Biology: Not available

Compound(s) and Activities

Cytotoxicity: The hexapeptides, namely, haligramides A and B isolated from this species showed cytotoxicity against Cancer/A-549 (lung), HCT-15 (colon), SF-539 (CNS), and SNB-19 (CNS) with IC50 values of 5.2–15.6 μg/mL and 3.9–8.8 μg/mL, respectively (Rashid et al., 2000).

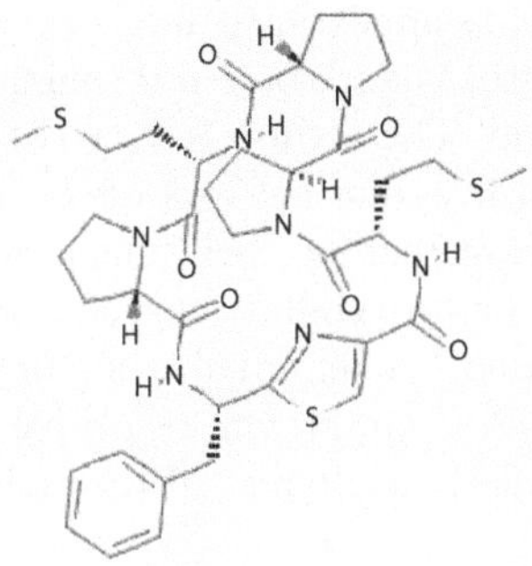

Haligramide A

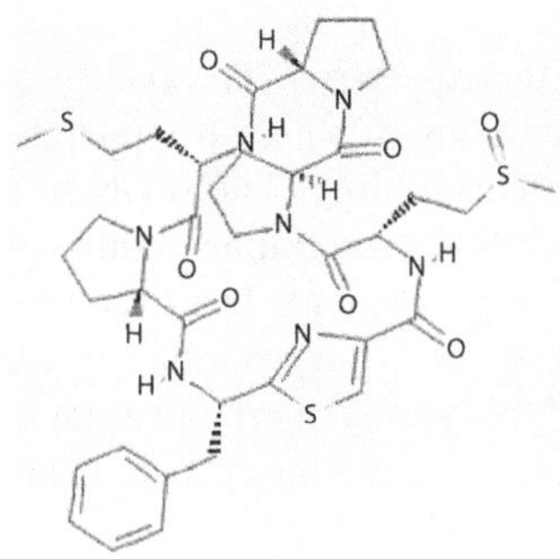

Haligramides B

Haliclona (Reniera) osiris (de Laubenfels, 1954) (= *Prianos osiros*)

Order: Haplosclerida Family: Chalinidae
Common name: Not designated
Distribution: Micronesia, Australia
Ecology: Not available
Biology: Not available

Compound(s) and Activities

Antitumor activity: The acetylenic carotenoid compound, (3R,3′R,5S)-3,3′,5,19′-tetrahydroxy-7′,8′-didehydro-γ, ε-carotene-8-one isolated from this species was found to be cytotoxic toward cultured human colon tumor cell HCT 116 (IC50, 4.38 μg/mL) (Rogers and Molinski, 2005).

Cytotoxicity: The polyacetylene compounds osirisynes A–F exhibited moderate cytotoxicity against a human leukemia cell line (KS62), with LC50 values of 25, 48, 52, 25, 20, and 22 μM for 106–111, respectively. In addition, osirisyne C, E, and F inhibited Na+/K+-ATPase and reverse transcriptase at concentrations of 1 μg/mL (Dembitsky et al., 2006).

The compound, acetylenic carotenoid, (3R,3′R,5S)- 3,3′,5, 19′-tetrahydroxy-7′,8′-didehydro-γ,ε-carotene-8-one isolated from this species was cytotoxic toward cultured human colon tumor cell HCT 116 (IC50 4.38 μg/mL) (Dembitsky et al., 2006).

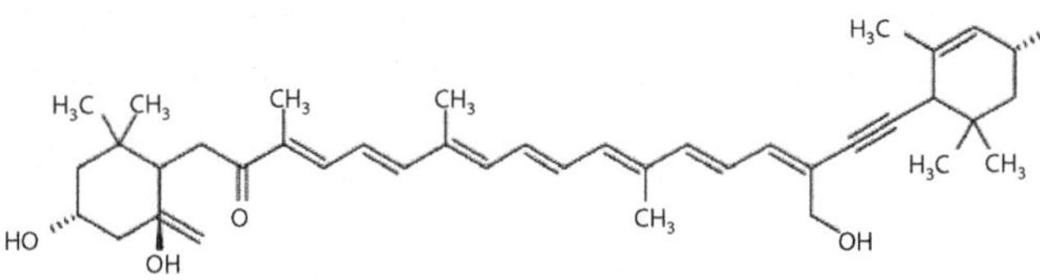

(3R,3′R,5S)- 3,3′,5,19′-tetrahydroxy-7′,8′-didehydro-γ,ε-carotene-8-one

Haliclona (Haliclona) oculata (Linnaeus, 1759)

Order: Haplosclerida Family: Chalinidae
Common name: Encrusting turret sponge, mermaid's glove, finger sponge, eyed finger sponge
Distribution: Atlantic coast of Canada and North America, Portugal, France, the Netherlands, British Isles, Shetland, Faroe, Denmark, Sweden, Norway, Spitsbergen, Bear Island, White Sea, Kara Sea; in the entire arctic and subarctic area
Ecology: It occurs among a rich faunal turf of hydroids and bryozoans in the infralittoral, to about 100 m, and on rocky and sandy bottoms, exposed slopes of silty cobble and pebble subject to strong to moderate tidal streams. It can tolerate low salinity and turbid water with suspended silt.
Biology: This sponge consists of solid branches growing from a short stalk that is attached to the substratum with a pedicel. There is a considerable variation in the number and the degree of fusing of the branches. The branches may be isolated along their entire length or fuse to become flabelliform. The oscules are small (1–3 mm) and are arranged on the narrower sides of the compressed branches. They may be slightly elevated. In the flabelliform sponges, the oscules are situated on the margins of the blade. This sponge may reach a considerable size, and it may be between ca. 10 cm and 30 cm

in height. Surface is very smooth, velvety, and slightly hispid. Consistency is rather soft but elastic. The stalk is very firm and incompressible.

The spicules are short oxeas with short and sharp points. Ectosomal skeleton is mostly absent, but occasionally some dermal spicules may be seen. The color of the sponge is light brown or dark yellow, often with a greenish tinge, sometimes reddish.

Reproduction: The larvae (parenchymella) in this species are produced in summer and fall (July to November), and are elongate, white, and have a bare posterior pole. Gemmules, which are of about 0.5 to 1 mm in diameter, are dark-orange colored and are normally present in dense clusters at the base of the stalk. These gemmules are probably not functional. But new sponges may appear from these gemmules.

Compound(s) and Activities

Antileishmanial activity: The n-chloroform fraction of this species showed the highest *in vitro* (>75%) and *in vivo* (84.3%) antileishmanial activity with IC50 values of 54 μg/mL and 61 μg/mL against promastigotes and intracellular amastigotes, respectively (Lakshmi et al., 2015).

Antifungal property: The organic extracts of this specis showed antifungal activity, and the MIC values recorded for the different fungal species in different extracts are given next.

carcinoma following further *in vivo* experiments and clinical trials (Seydi et al., 2015).

Antioxidant and anti-AChE activities: The organic extracts of this species showed weak antioxidant and anti-anticholinesterase activities with an IC50 value of 729 μg/mL (by DPPH method) and <1000 μg/mL respectively (Bianco et al., 2015).

Haliclona pigmentifera (Dendy, 1905)

Order: Haplosclerida Family: Chalinidae

Common name: Not designated

Distribution: Western Central Pacific: Indonesia; India: Gulf of Mannar

Ecology: It inhabits low intertidal zones and sublittoral zones.

Biology: The external form of this species is more or less flabellate, with rounded margin. The vents are about 2 mm in diameter and appear to have been scattered singly along the margin. The surface has a porous appearance and is very minutely hispid. The color (in spirit) is dark brown throughout, sometimes with a purplish tint. The texture is very soft and crumbling. The skeleton is an irregular network of short spicules, sometimes connected together at their ends by spongin cement. The oxea of spicules are slightly curved and gradually sharp pointed at each end. A remarkable feature of this sponge is the presence of granular, brown- or purple-colored pigment cells that are thickly scattered all through the sponge.

In vitro Evaluation of Antifungal Activity (MIC, μg/mL) in Different Extracts

Fungi	MeOH	50% MeOH-$CHCl_3$	Hexane	Chloroform	n-Butanol
Candida albicans	125	500	125	125	500
Cryptococcus neoformans	250	250	500	250	500
Sporothrix schenckii	125	125	31	31	500
Trichophyton mentagrophytes	31	125	63	63	250
Aspergillus fumigatus	500	–	–	250	–
Candida parapsilosis	63	63	31	31	125

Source: Lakshmi, V., and P.K. Shukla, 2013, *Nat. Prod.*, 9: 339–344.

Anticancer activity: The methanolic extracts (200, 400, and 800 μg/mL) of this species increased reactive oxygen species (ROS) formation, mitochondrial membrane potential (MMP), mitochondrial swelling, and cytochrome c release in the mitochondria obtained from cancerous hepatocytes. These extracts also induced caspase-3 activation, which is known as a final mediator of apoptosis, in the hepatocytes obtained only from cancerous rat livers. This species may therefore be a promising therapeutic candidate for the treatment of hepatocellular

Association: The amphipod species *Leucothoe tridens* was found associated with this sponge species.

Compound(s) and Activities

Mosquito larvicidal activity: The extract of this species displayed larvicidal activity against the fourth instar larvae of *Aedes aegypti* (yellow fever mosquito vector) with an LC50 value of 87.9 ppm (Rao et al., 2008).

Haliclona (Haliclona) simulans (Johnston, 1842)

Order: Haplosclerida Family: Chalinidae

Common name: Not designated

Distribution: British Isles, Channel Islands, France, Portugal, Mediterranean, Madeira, and the Canary Islands

Ecology: The encrusting form of this sponge species is common under boulders; it overhangs on the low shore and shallow sublittoral zones. Larger branching colonies occur on bedrock and boulders in areas sheltered from wave action but exposed to moderate to strong tidal streams.

Biology: This species of sponge is very polymorphic. It may grow as a thin sheet or cushion, or as a branching-repent or branching-erect colony, up to about 50 cm across. Its oscules are found scattered along the branches on small mounds. It is usually beige brown in color with patches of orange pigment. The boundaries between the color zones are often sharp. Surface is smooth and even; in certain places, it is slightly hispid. It has a very hard and brittle consistency. The spicules are of short, fat, cigar-shaped oxeas, with rather short and sharp points. The ectosome is a strong, very regular, unispicular and isotropic, tangential reticulation of oxeas.

Compound(s) and Activities

Antimicrobial activities: A total of 52 bacterial isolates belonging to the genera *Pseudoalteromonas*, *Pseudomonas*, *Halomonas*, *Psychrobacter*, *Marinobacter*, *Sulfitobacter*, *Pseudovibrio*, *Salegentibacter*, *Bacillus*, *Cytophaga*, *Rhodococcus*, and *Streptomyces* have been isolated from this species. These isolates showed antimicrobial activities (Thomas et al., 2010).

Antitrypanosomal, cytotoxic, and antimycobacterial activities: Three steroids—24-vinyl-cholest-9-ene-3β,24-diol, and 20-methyl-pregn-6-en-3β-ol,5a,8a-epidioxy, and 24-methylenecholesterol—isolated from this species possessed antitrypanosomal and antimycobacterial activity. All the steroids were also found to possess low cytotoxicity against the Hs27 cell line (Viegelmann et al., 2014).

Haliclona (Rhizoniera) sarai (Pulitzer-Finali, 1969) (= Reniera sarai)

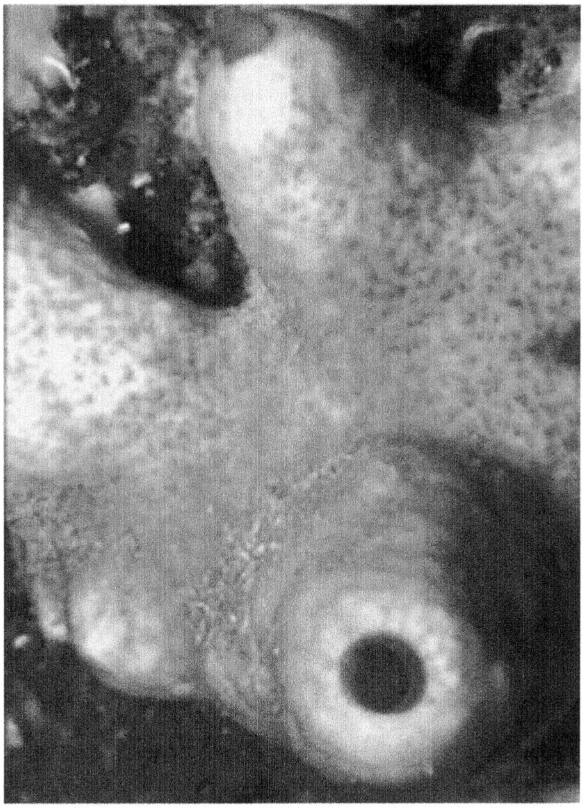

Order: Haplosclerida Family: Chalinidae

Common name: Mediterranean sponge

Distribution: Indonesia; Tremiti Islands, Adriatic Sea

Ecology: It is a shallow water species inhabiting a depth range of 2 to 30 m.

Biology: It is a massive sponge with circular oscules (up to 10 mm dia.) at the summit of the low conical processes. Texture is compact, fragile, and crumbling. Color of the animal is variable from pale cream-yellow to light brown to orange yellow to orange red.

Compound(s) and Activities

AChE inhibitory activity: The alkylpyridinium polymers isolated from this species strongly inhibited AChE and acted as anticancer compounds affecting the cholinergic system expressed in a variety of cancer cell lines (Sepčić et al., 2010).

Hemolytic and cytotoxic activities: The 3-alkylpyridinium polymers isolated from this species were found to be potent anticholinesterase agents. In addition, they showed hemolytic and cytotoxic activities (Sepčić et al., 1997).

Neuroprotective activity: Among the diamide alkaloid sarains (sarains 1, 2, and A) isolated from this species, sarain A was found to be the most promising compound, improving mitochondrial function and decreasing reactive oxygen species levels in human SH-SY5Y neuroblastoma cells at 0.01, 0.1, and 1 μM. Considering the protective effects, sarain A may be a candidate compound for further studies in neurodegenerative diseases (Alvariño et al., 2017).

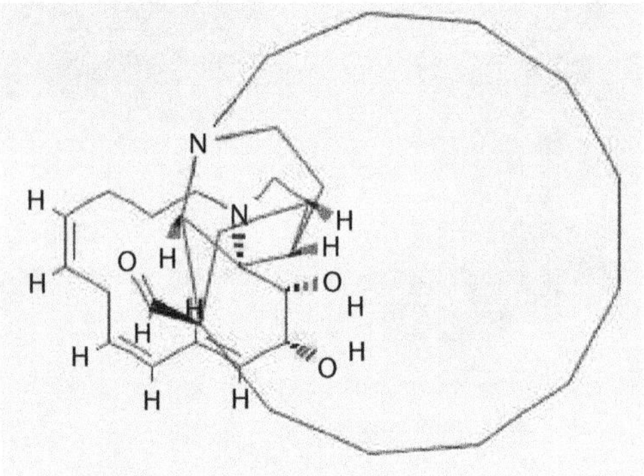

Sarain A

Haliclona (Reniera) tubifera (George and Wilson, 1919)

Order: Haplosclerida Family: Chalinidae

Common name: Not designated

Distribution: Western Central Atlantic, Caribbean.

Ecology: This sessile species is found on reef flats and on rocks in shallow water of depth range 1 to 10 m. It is also very common on lagoon bottoms and on mangrove root. It tolerates brackish water conditions.

Biology: It is a cushion-like sponge with less than 0.5 cm thickness and has oscular chimneys. Size of the animal is 0.5 to 5 cm long. Thin long proliferations (0.1–0.2 cm wide and several centimeters long) are branching off the main body. Color of the sponge is highly variable from pink, to violet or purple externally and internally. Oscules are 0.05–0.1 cm wide. Surface seems to be smooth but it is faintly hispid and distinctly porous. Texture is soft, compressible, and limp.

Compound(s) and Activities

Cytotoxicity and anticoagulant and antioxidant activities: The compounds C18 sphingoid bases—(2R,3R,6R,7Z)-2-aminooctadec-7-ene-1,3, 6-triol (halisphingosine A) and (2R,3R,6R)-2-aminooctadec-1,3,6-triol (halisphingosine B)—isolated from this species were found to be cytotoxic in glioma (U87) and neuroblastoma (SH-SY5Y) human cell lines. These compound have also been reported to possess anticoagulant and antioxidant activities (Biegelmeyer et al., 2015).

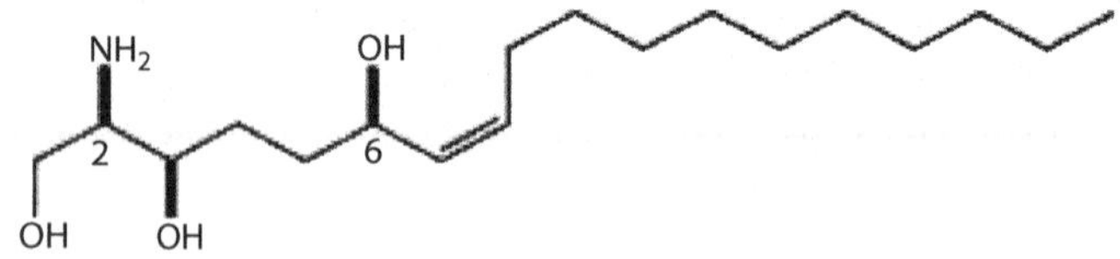

Halisphingosine A

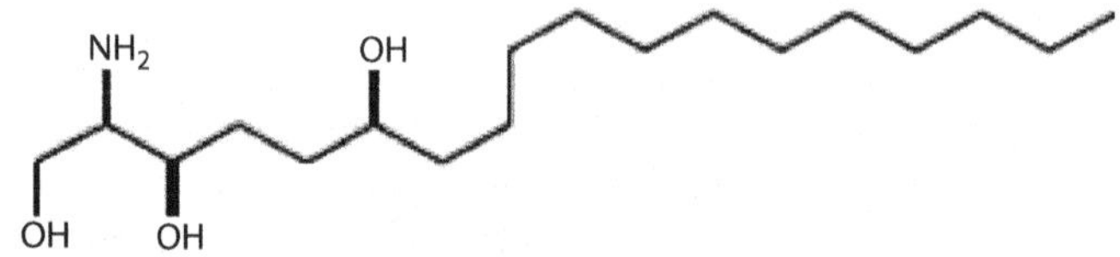

Halisphingosine B

Haliclona tulearensis (Vacelet, Vasseur, and Lévi, 1976)

Order: Haplosclerida Family: Chalinidae

Common name: Orange encrusted turret sponge, Madagascan sponge

Distribution: Western and Northern Madagascar

Ecology: It grows on rocky subtidal reefs.

Biology: This bright yellow sponge has conical turrets and a pocked surface.

Compound(s) and Activities

Cytotoxic activity: The bisquinolinylpyrrole compound, halitulin isolated from this species exhibited cytotoxicity with an IC50 value of 25 μg/mL against P-388 murine leukemia cell line, 12 μg/mL against A-549 human lung carcinoma cells, 12 μg/mL against HT-29 human colon carcinoma cells, and 25 μg/mL against MEL-28 human melanoma cell line (Kashman et al., 1999; Sugumaran and Robinson, 2010).

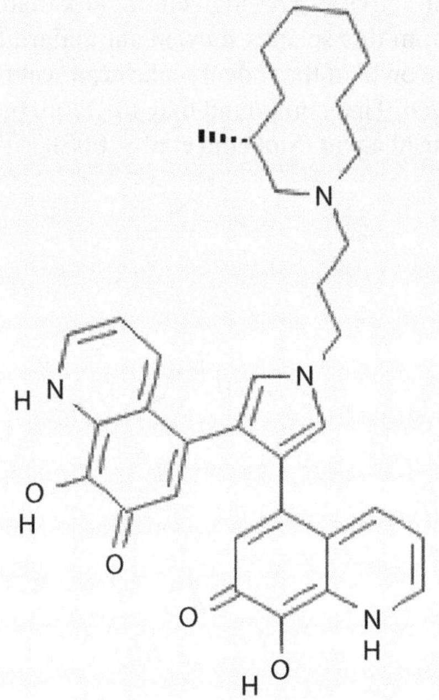

Halitulin

Haliclona (Soestella) valliculata (Griessinger, 1971)

Order: Haplosclerida Family: Chalinidae

Common name: Not designated

Distribution: Western Mediterranean

Ecology: It flourishes in protected environments as undersides of rocks and crevices, both in intertidal and subtidal habitats.

Biology: Not available

Compound(s) and Activities

Antiproliferative activity: The fungal strain *Emericella variecolor* associated with this species produced a novel anthraquinone, evariquinone, which revealed a strong antiproliferative activity against KB (ATCC CCL17, human cervix carcinoma) and NCI-H460 (NCI 503473, non-small cell lung cancer) cells (Thomas et al., 2010).

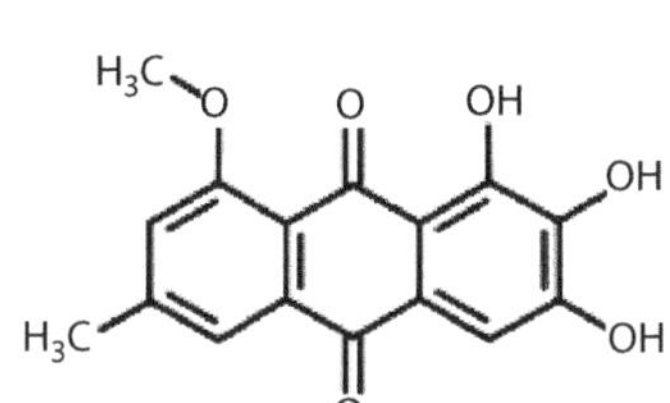

Evariquinone

Haliclona (Rhizoniera) viscosa (Topsent, 1888)

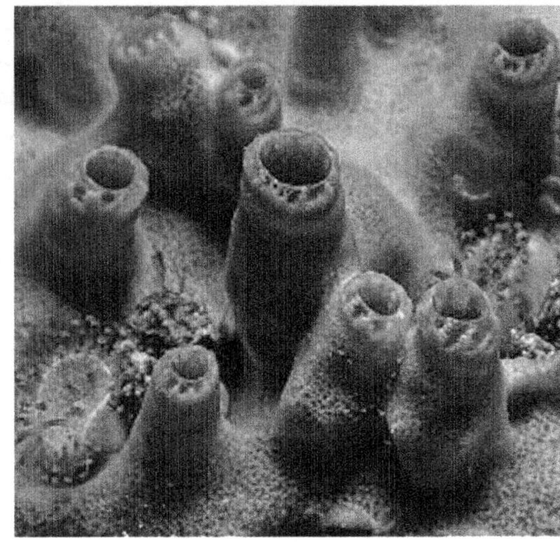

Order: Haplosclerida Family: Chalinidae

Common name: Not designated

Distribution: Shetland Islands, British Isles, Belgium, France, Portugal, Mediterranean

Ecology: It occurs in the infralittoral zone to a depth of 50 m. Resides on vertical or horizontal sides of rocks in places with strong currents and low turbidity.

Biology: This sponge forms purple, pink, or brownish spreading cushions with rows of large oscular chimneys. The surface is distinctly punctate and it may be rugose in larger specimens. The consistency is firm. The species is slimy and when broken it produces a lot of slime. Color of the sponge is grayish purple, commonly verging to yellow toward the base. The color may turn to blackish brown when the sponge is exposed to the air.

Compound(s) and Activities

Antifungal activity: The compound haliscosamine ((9Z)-2-amino-docos-9-ene-1,3,13,14-tetraol), which is a sphingosine derivative isolated from this species, showed antifungal activity against the species of *Candida* and *Cryptococcus* (Cheung et al., 2014).

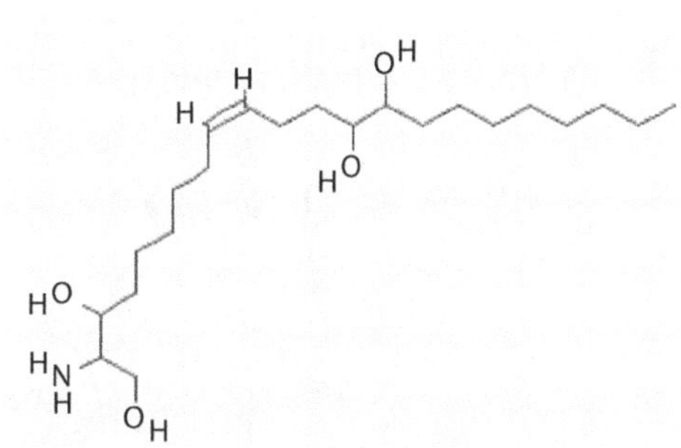

Haliscosamine

Cytotoxic and antibacterial activities: The cyclic bis-1,3-dialkylpyridiniums and cyclostellettamines isolated from this species had cytotoxic and antibacterial activities against A549 cell line and Gram-positive strains, respectively (Perdicaris et al., 2013).

Antitumor activity: The salicylate macrolide salicylihalamide A isolated from this sponge possessed antitumor activity by serving as a v-ATPase inhibitor (Sipkema et al., 2005).

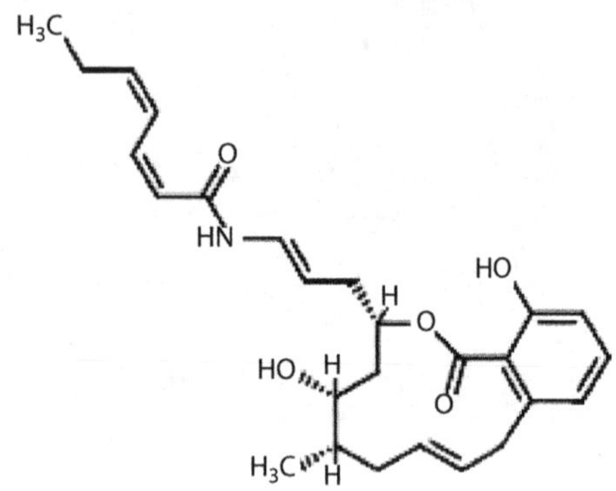

Salicylihalamide A

Antibacterial activity: The unidentified fungus associated with this sponge species produced hirsutane sesquiterpenes such as hirsutanols A–C and ent-gloeosteretriol. Among these compounds, hirsutanol A and ent-gloeosteretriol exhibited mild antibiotic activity against *Bacillus subtilis* (Thomas et al., 2010).

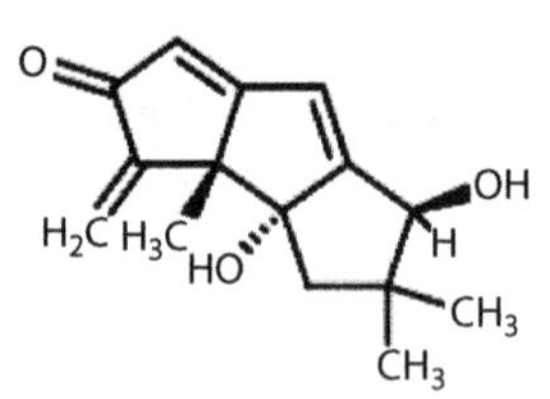

Hirsutanol A

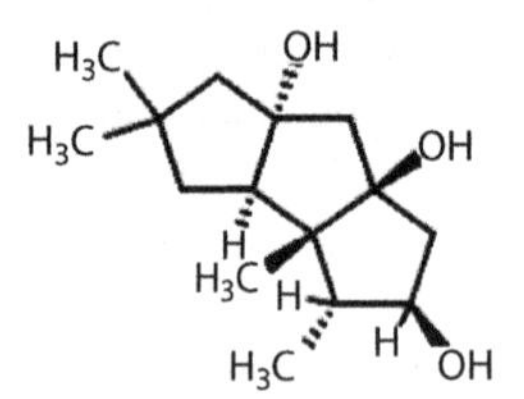

Ent-gloeosteretriol

Antifungal activity: The isolated metabolite of this species, haliscosamine, showed a significant antifungal activity against *Candida neoformans, C. albicans*, and *C. traopicalis* and the values of growth inhibition diameter (mm) for these species were 28, 25, and 25, respectively (El-Amraoui et al., 2013).

Anti-inflammatory activity: Among the peptides halipeptins A and B isolated from this unidentified sponge, the former inhibited mouse paw edema in a dose-dependent manner with 60% inhibition at 300 μg/kg (Keyzers, 2005).

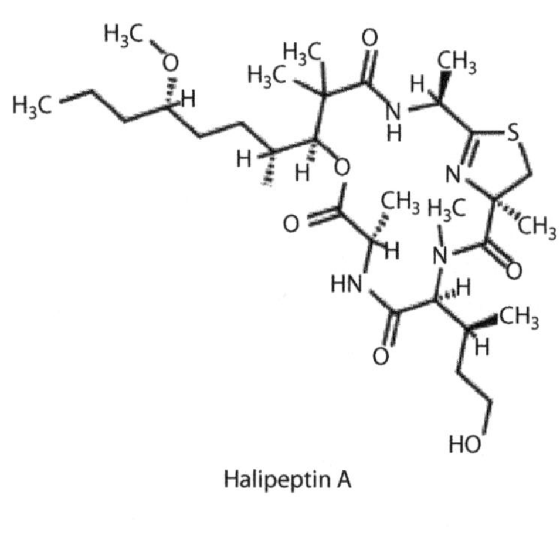

Halipeptin A

Antimalarial activity: The manzamine alkaloid manzamine A isolated from this species showed antimalarial activity and inhibits the growth of the rodent malaria parasite *Plasmodium berghei in vivo*. This compound may therefore be a promising new antimlarial agent (Sipkema et al., 2005).

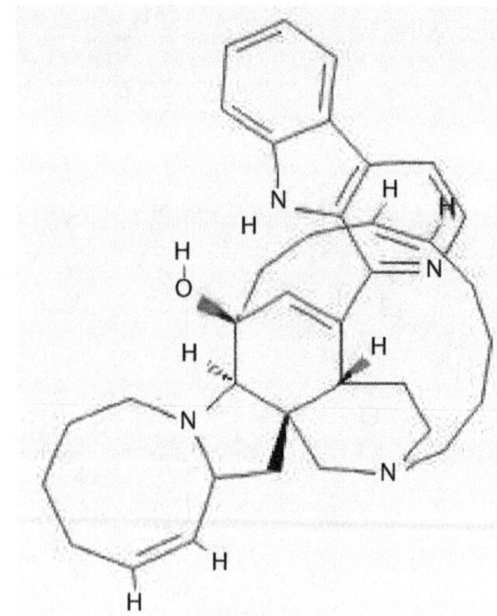

Manzamine A

The unidentified bacterial isolate THB14 associated with this sponge showed antiplasmodial activity against *P. faciparum* with an IC50 value of 12 μg/mL (Indraningrat et al., 2016).

Immunomodulatory effects: Two new merohexaprenoids halicloic acids A (1) and B (2) have been isolated from this sponge. These two compounds are indoleamine 2,3-dioxygenase (which plays a key role in the immunomodulatory effects) inhibitors (Williams et al., 2012).

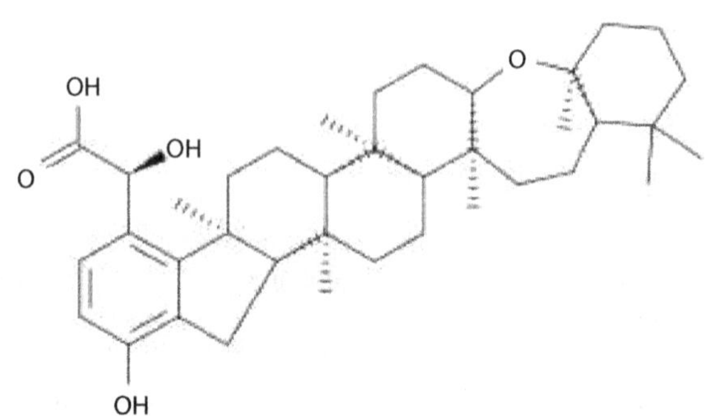

Halicloic acid A

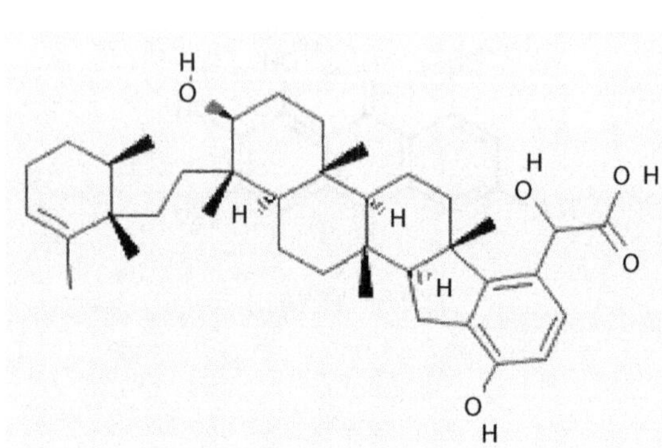

Halicloic acid B

Haliclona (Gellius) sp.

Order: Haplosclerida Family: Chalinidae

Compound(s) and Activities

Cytotoxicity: The indole alkaloid compound, gelliusines A and B of this sponge showed cytotoxicity and inhibited KB, P-388, HT-29, and NSCLCN-6 cell lines with IC50 values of between 10 and 20 μg/mL (Gul et al., 2005).

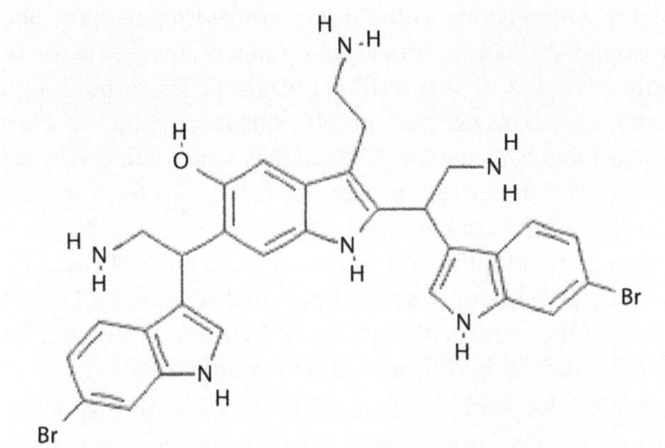

Gelliusine A

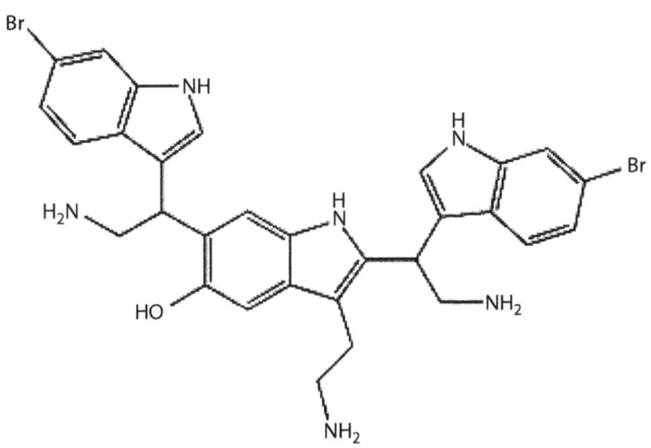

Gelliusine B

The compounds gelliusterols A, B, and C have been isolated from this species. Of them, gelliusterols A and B exhibited moderate cytotoxic activity, with IC50 values greater than 1 μg/mL. An activity of 0.5 μg/mL was observed with gelliusterol C against HT-29, while the other cell lines gave IC50 values above 1 μg/mL (Dembitsky et al., 2006).

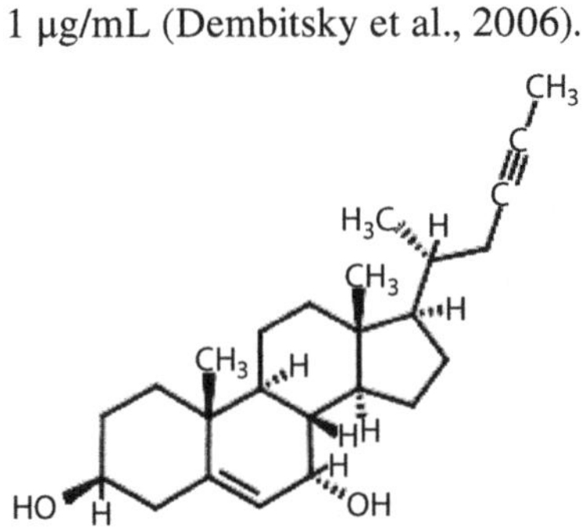

Gelliusterol A

The compounds halicynones A and B; triangulyne A and triangulyne E; and pellynols A, B, C, D, and I isolated from this species showed cytotoxicity (Dembitsky et al., 2006).

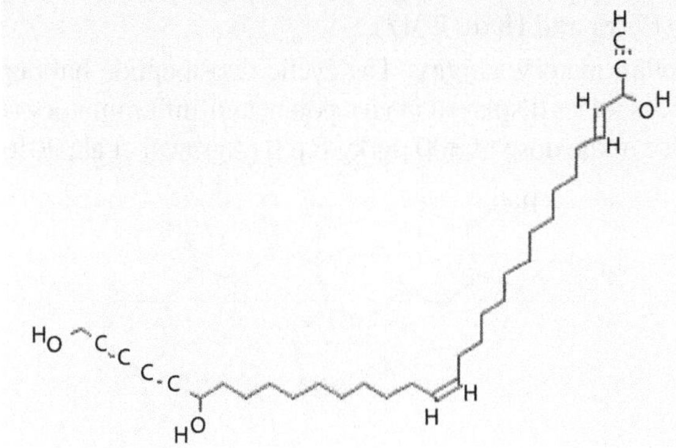

Pellynol A

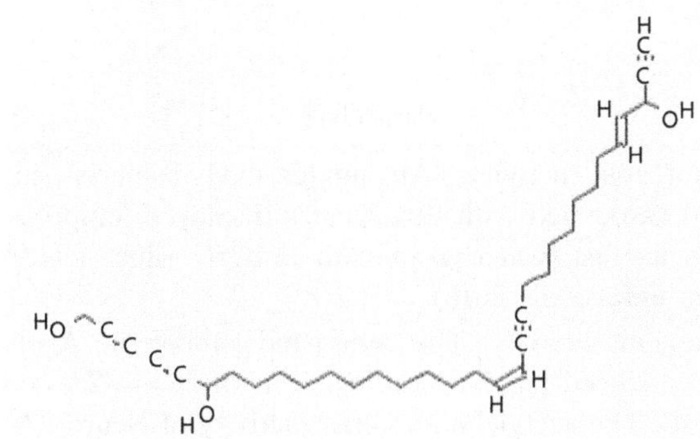

Pellynol C

Cultured HCT-116 cells were found to be very sensitive to compoun, pellynol A (IC50 0.026 μg/mL), pellynol B (0.12 μg/mL), pellynol C (0.127 μg/mL), pellynol D (0.103 μg/mL), and pellynol I (78 μg/mL). Pellynols A–D, I, and F, having a terminal 1-yn-3-ol, showed strong cytotoxicity against several melanoma and ovarian cancer cell lines (IC50 0.08–2.0 μM/mL). Pellynic acid inhibited inosine monophosphate dehydrogenase with an IC50 of 1.03 pM/mL (Dembitsky et al., 2006).

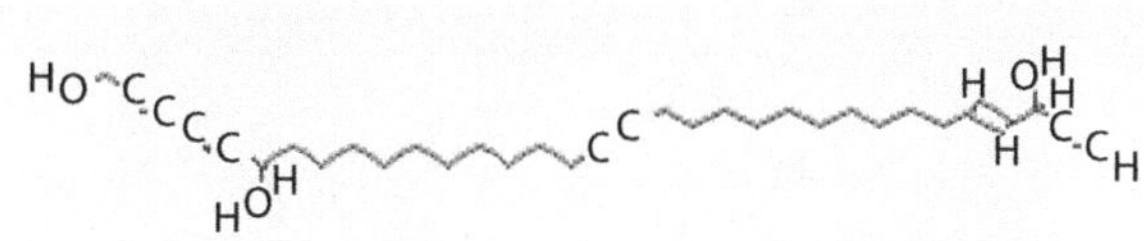

Pellynol F

The methanol extract of this sponge on a species gave a linear peptide, kendarimide A, which reversed P-glycoprotein-mediated multidrug resistance in mammalian cells (Agrawal et al., 2016).

Antitumor activity: The triterpenoid hydroquinones, adociasulfates, isolated from this sponge showed antitumor activity by serving as A1, 3-fucosyltransferase, and kinesin motor protein inhibitors (Anjum et al., 2016).

Two brominated acetylenes of this species were found to be cytotoxive toward MCF-7 human breast cancer cells with IC50 values of 32.5 and 50.8 μM, respectively (Gribble, 2015).

The compounds halicynones A and B isolated from this unidentified sponge showed antitumor activity against the human colon tumor cell line HCT-116 (Mokhlesi et al., 2017).

Antimicrobial activity: The crude extract of this species showed antimicrobial activity against *S. aureus*, *B. subtilis*, *E. coli*, *V. cholera*, *and C. albicans*, and the values of zone of inhibition recorded were 8.7, 9.5, 8.3, 6.6, and 9.2 mm, respectively (Putra and Hadi, 2017).

Anti-inflammatory activity: The cyclic depsipeptide halipeptins A of this species displayed *in vivo* potent anti-inflammatory activity (mice at the dose of 300 μg/kg [i.p.]) (Agrawal et al., 2016).

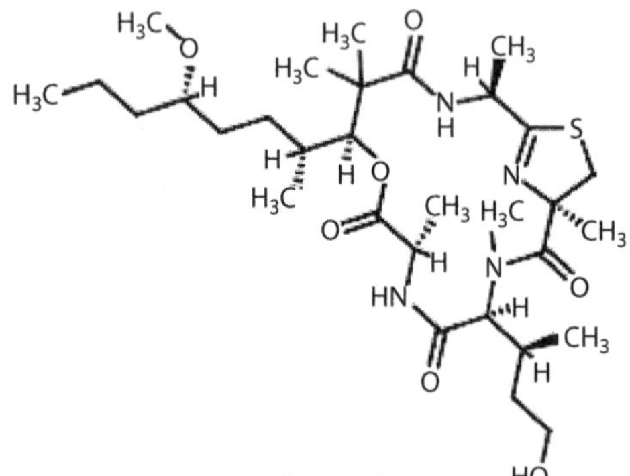

Halipeptin A

Antiprotozoal activity: An unidentified bacterial isolate THB14 associated with this species displayed antiprotozoal activity against *P. faciparum* with an IC50 value of 11.98 μg (Indraningrat et al., 2016).

Neuritogenic activity: The compound lembehyne A of this species induced bipolar neuritogenesis of Neuro-2A cells at 1 μg/mL. The acetylcholinesterase activity of Neuro-2A was also increased by treatment with this compound. Further lembehynes B and C also exhibited neuritogenic activity against a neuroblastoma cell line Neuro-2A (Dembitsky et al., 2006).

FAMILY: NIPHATIDAE

Amphimedon compressa (Duchassaing and Michelotti, 1864) (= *Haliclone compressa*)

Order: Haplosclerida Family: Niphatidae

Common name: Erect rope sponge, red tree sponge, red tubular sponge, red sponge, scattered pore sponge

Distribution: Western Central (tropical) Atlantic, southern Florida, Caribbean Sea, and Bahamas

Ecology: This sponge grows above the substrate, on the top of coral reefs, occasionally on the reef wall between 5 and 20 m. It is found more rarely in the mangrove and in brackish waters. It prefers waters that are not agitated (but clear).

Biology: The body of the animal has an optically smooth surface, which is due to the rarely protruding primary fibers. Ectosomal skeleton is with regular tangential reticulation with rounded meshes. Color in life is bright green throughout the "tissue." This sponge has a soft consistency, and is elastic. Growth form has cushion-like patches which are thick, about 104 cm. The finger-like processes, which are from the cushions are up to 20 cm long, and up to 3 cm diameter. Surface of the sponge is smooth, with ridges on some parts of the surface of certain specimens. Oscules are found scattered all over the surface of the sponge, about 1.2 to 2.5 cm apart, with diameter 2.8 to 4.3 mm. It is commensal with amphipods.

Food, predator, and associated species: Like most sponges, it feeds by filtering microscopic particles in the water. It often has the sponge brittle star (*Ophiothrix suensoni*) living on its surface. These sponges are often eaten by sea stars.

Reproduction: The reproduction in this species takes place all year. But in the warm season, the greatest activity is observed. The fertilization of the gametes takes place inside the sponge. The larvae are incubated to a size of about 0.5 mm, then released into the water and lead a short pelagic, planktonic life before settling. This species also has the ability to multiply asexually, by "cuttings" of fragments torn off during a hurricane.

Compound(s) and Activities

Cytotoxic activity: The hydroethanolic extracts of this species showed cytotoxicity against tumor cell lines at a concentration of 20 mg/mL. The values of MIC (IC50) for the different cell lines are given next.

IC50 (μg/mL) Values of Tumor Cell Lines

HL-60	HCT-8	MDA-MB435	SF-295
1.6	1.6	1.7	3.0

Source: Ferreira, E.G. et al., 2007, *Porifera Res.: Biodiversity, Innovation Sustainability*, 313–318.

Antibacterial activity: A new compound, 8,8′-dienecyclostellettamine, isolated from this species showed vigorous antibacterial activities (Xu et al., 2007).

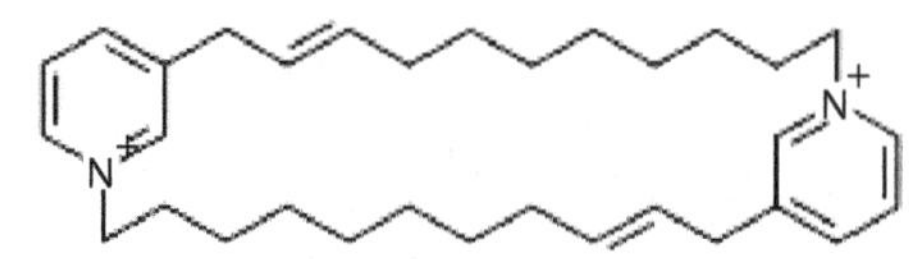

8,8′-dienecyclostellettamine

Antitrypanosomal activity: The hexane and methanol extracts of this species showed antitrypanosomal activity and the IC50 values recorded for these extracts at 48 hr were 52.1 and 6.0 g/mL, respectively (Tabares, 2011).

Amphimedon navalis (Pulitzer-Finali, 1993)

Order: Haplosclerida Family: Niphatidae
Common name: Not designated
Distribution: East Africa
Ecology: It is a coral reef–associated species.
Biology: Not available

Compound(s) and Activities

Neuroprotective activity/cholinesterases (ChEs) activity: The ethyl acetate extracts of this species rich in terpenoid compounds showed strong AChE inhibition (96%) at a concnentration of 0.1 mg/mL (Grosso et al., 2014).

Amphimedon paraviridis (Fromont, 1993)

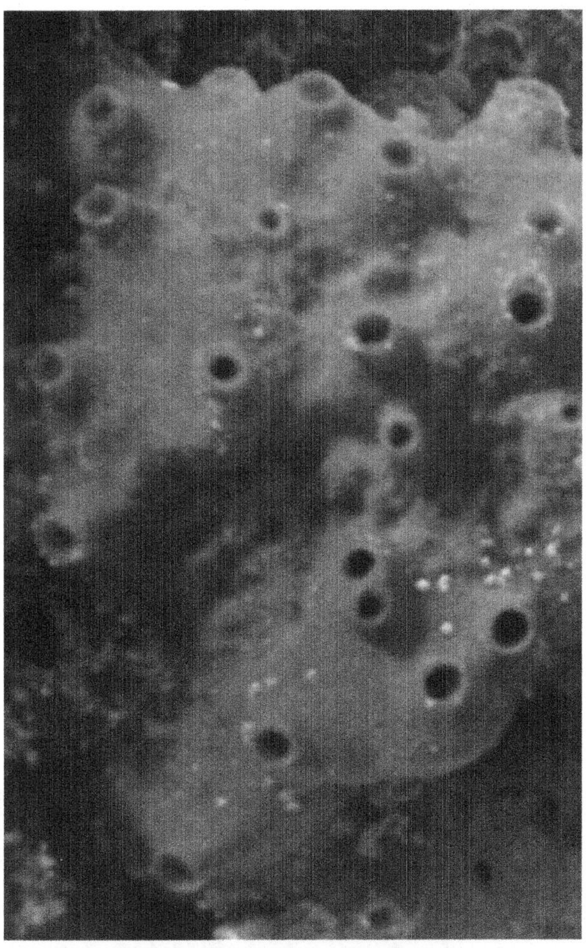

Order: Haplosclerida Family: Niphatidae
Common name: Not designated
Distribution: Australia
Ecology: Not reported
Biology: Owing to the high growth potential and high survival rate of this species, it is considered a promising candidate for mariculture development (de Voogd, 2007). No other features are available.

Compound(s) and Activities

Antiangiogenic activity: The antiangiogenic purine analog, 1,3-dimethylisoguaninium, isolated from the ethanol (EtOH) extract of this species exhibited specific inhibition of the basic fibroblast growth factor (bFGF)-induced proliferation of bovine aorta endothelial cells (BAECs). Moreover, this compound reduced the tube formation of BAECs in a time-dependent manner (Jeong et al., 2003).

Amphimedon sp.

Order: Haplosclerida Family: Niphatidae

Compound(s) and Activities

Cytotoxicity: The indole alkaloid compounds, namely, manzamine B, E, and F, isolated from this unidentified species have shown cytotoxicity against P-388 murine leukemia cells with IC50 values of 6, 5, and 5 μg/mL, respectively (Gul et al., 2005).

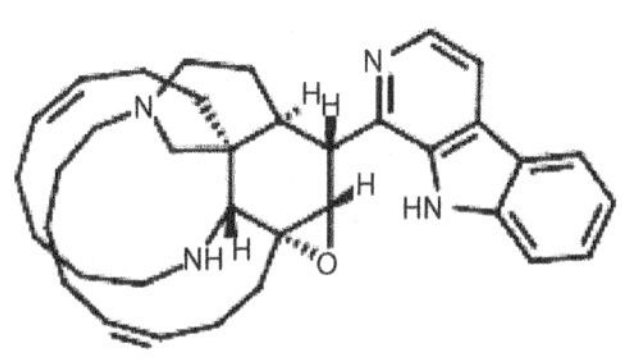

Manzamine B

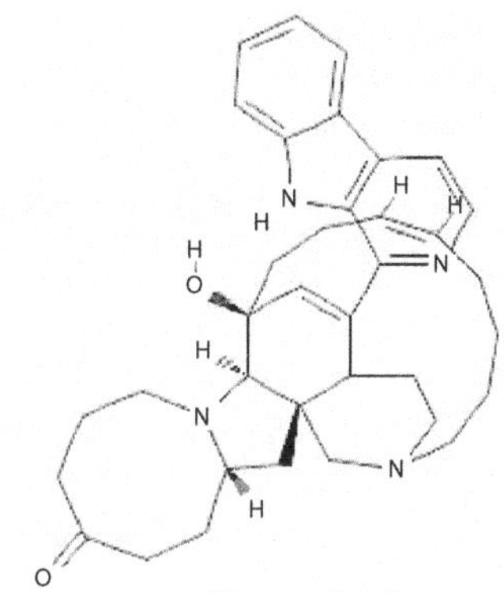

Manzamine E

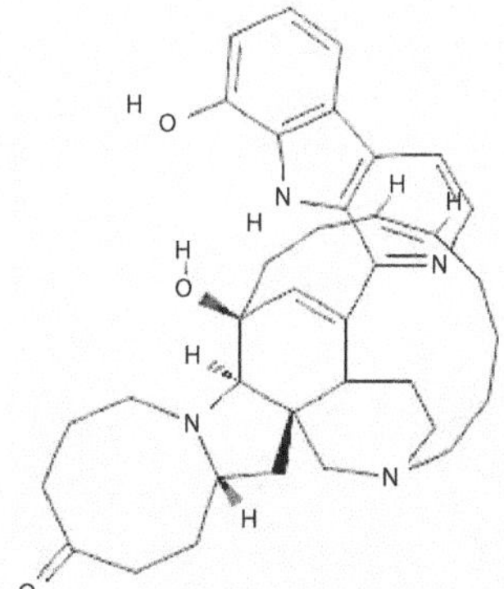

Manzamine F

Hemigellius bidens (Topsent, 1901)

Order: Haplosclerida Family: Niphatidae

Common name: Antarctic sponge

Distribution: Antarctica, Bellinghausen Sea, Graham Land, Victoria Land, Weddell Sea

Ecology: It is a deepwater species at depths of 30 to 550 m.

Biology: It is an erect sponge with dimensions of 4.2 × 1.2 cm and 0.3 cm thickness. It has a light rugose surface, which is hispid to the touch. Preserved material is friable in consistency. Color of the animal is light brown with clearer regions. Choanosome is with a reticulation composed by paucispicular tracts. Microscleres are irregular or randomly occur between the tracts. The oxeas of the megascleres are smooth, straight, or slightly curved.

Compound(s) and Activities

Antibacterial activity: The extracts of this species showed antibacterial activity against the laboratory, commensal, and clinically relevant bacterial strains with the MIC values.

Bacterial Strain	MIC (µg/mL)
Staphylococcus aureus (MRSA) S-943 A	400
S. pseudintermedius (MRSP) S-053 A	400
S. pseudintermedius (MRSP) S-043 A	400
Listeria monocytogenes	500
Staphylococcus epidermidis EXB-V55	80
Staphylococcus aureus 10F	400
Bacillus subtilis EXB-V68	80
Macrococcus 1F	6
Micrococcus 2F	60
Escherichia coli HB101	400

Source: Turk, T. et al., 2013, *Mar. Drugs*, 11: 1126–1139.

The extracts of this species showed antibacterial activity against the environmental bacterial strains *Exignuobacterium sp.*, *Pseudoalteromonas sp.*, *Alteromonas sp.*, *Vibrio ruber*, and *Janthinobacterium svalbardensis* with the MIC values of 120, 110, 200, 120, and 100 µg/mL, respectively (Turk et al., 2013).

Cribrochalina dura (Wilson, 1902)

Order: Haplosclerida Family: Niphatidae

Common name: Not designated

Distribution: Southeast Pacific and Western Central Atlantic, Easter Island

Ecology: It occurs in the offshore waters at depths of 300 to 400 m.

Biology: This sponge is ramose with crooked, knotty, and compressed branches, which are sprawling or arching from one holdfast to another. The branches have thickness of 0.5 to 2 cm. The size of the sponge is 15 × 30 cm in diameter and 15 cm in height. The presence of oscules and dense ectosome are also the chacterisitic features of this species. Color of the animal is purplish brown to olive.

Compound(s) and Activities

Cytotoxicity: The compound duryne isolated from this species inhibited the growth of both mouse and human tumor cell lines *in vitro*. Further, its compound, vasculyne exhibited modest differential cytotoxicity toward the melanoma and colon tumor cell-line subpanels with average G150 and LC50 values of 0.2 and 6.7 µg/mL, respectively (Dembitsky et al., 2006).

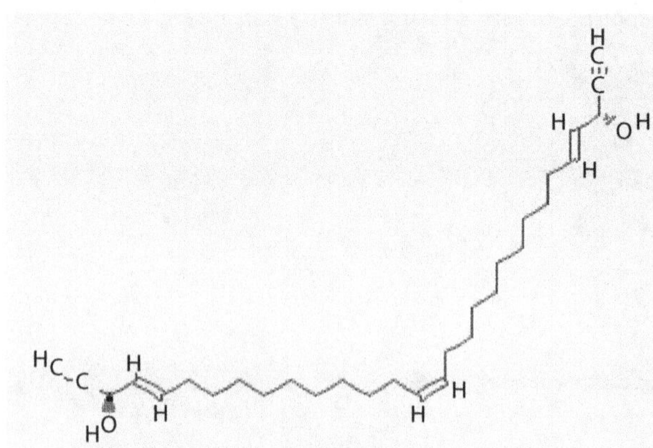

Duryne

Cribrochalina vasculum (Lamarck, 1814)

Order: Haplosclerida Family: Niphatidae

Common name: Brown bowl sponge

Distribution: Eastern Caribbean, Floridia, Greater Antilles, Southwestern Caribbean, Western Caribbean, Western Atlantic Ocean, Bahamas

Ecology: It inhabits coral reef slopes.

Biology: It is a ramose sponge with crooked, knotty, compressed branches that are sprawling or arching from one holdfast to another. It measures 15 by 30 cm in diameter and 15 cm in maximum height wth an apical hole. The spicules are arranged radially. Color of the sponge is purplish brown to olive.

Compound(s) and Activities

Antitumor activity: Two acetylene alcohols (compound 1 and compound 2) have been reported to inhibit phosphorylation of IGF-1Rβ as well as reduced its target signaling molecules IRS-1 and PDK1. These compounds may therefore serve as novel therapeutic agents targeting IGF-1R signaling for antitumor treatment (Zovko et al., 2016).

Antitumor/cytotoxic activity: A total of eight acetylenic alcohols, namely, 3-hydroxyeicos-(4)-en-1-yne, 3-hydroxydocosa-(4E, 15E)-dien-1-yne, 3-hydroxy-16- methyleicos-(4E)-en-1-yne, 3-hydroxy-19-methyleicos-(4E)-en-1-yne,3-hydroxy-21-methyldocosa-(4E, 15E)-dien-1-yne, 3-hydroxy-14-methyldocosa-(4E)-en-1-yne(knowncompounds),3-hydroxy-heneeicos-(4E)-en-1-yne, and 5-hydroxy-16-methyleicos-(3Z)-en-1-yne (new metabolites, minor components), have been isolated from this species. All these compounds showed selective *in vitro* antitumor activity against the H-522 non-small cell lung line and the IGROV-1 ovarian line (Hallock et al., 1995).

Antiproliferative activity: Two acetylenic compounds isolated from the organic extracts of this species showed strong tumor-specific toxicity in non-small cell lung carcinoma (NSCLC) cells and small-cell lung carcinoma cells, and less prominent toxicity in ovarian carcinoma. These acetylenic compounds were found to cause a time-dependent increase in activation of apoptotic signaling involving cleavage of caspase-9, caspase-3, and PARP, as well as apoptotic cell morphology in NSCLC cells. These acetylenic compounds may therefore serve as novel therapeutic agents for NSCLC and other tumor malignancies (Zovko et al., 2014).

Cytotoxicity: The acetylenic alcohols of this species showed selective *in vitro* antitumor activity against the H-522 non-small cell lung line and IGROV-1 ovarian line. Further, these compounds displayed cytotoxicity against P388 leukemia cells, human lung cancer cells (A549), and colon (HT-29) tumors, with IC50 values ranging from 0.86 to 90 μg/mL (Dembitsky et al., 2006).

A new C43 acetylenic alcohol, vasculyne, of this species yielded average G150, TGI, and LC50 values of 0.2, 0.7, and 6.7 μg/mL, respectively, and exhibited modest differential cytotoxicity toward the melanoma and colon tumor cell-line subpanels when tested against the NCI's 60-cell antitumor screening panel (Dembitsky et al., 2006).

Cribrochalina sp.

Order: Haplosclerida Family: Niphatidae

Compound(s) and Activities

Antibacterial activity: The alkaloid compounds of this unidentified species, namely, cribrostatin 3 and cribrostatin 6, have shown antibacterial activity against *N. gonorrheae* (antibiotic resistant strain) and *S. pneumonia* (anitibiotic resistant strain), respectively. Cribrostatin 6 showed an MIC value of ≤2 μg/mL (Anjum et al., 2016).

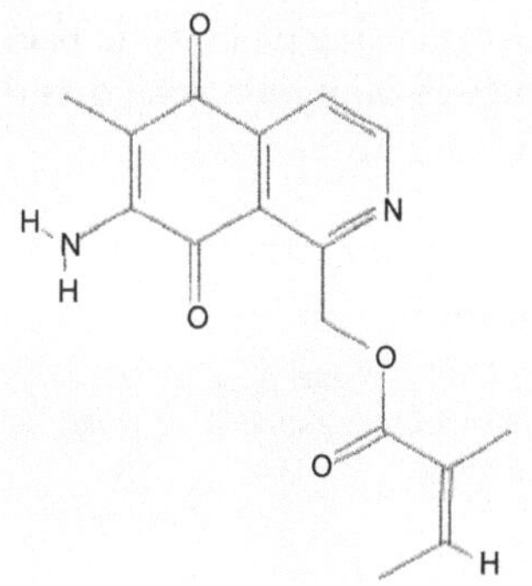

Cribrostatin 3

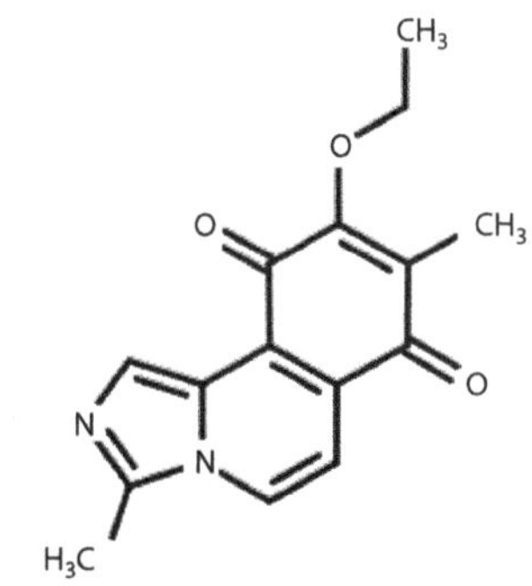

Cribrostatin 6

Gelliodes carnosa (Dendy, 1889) (= *Sigmadocia carnosa*)

Order: Haplosclerida Family: Niphatidae

Common name: Not designted

Distribution: South China Sea.

Ecology: It inhabits island coastal waters.

Biology: This sponge is erect, sessile, and more or less lamellar proliferous. Sometimes it rises into short tubular processes. Its size is 18 cm high and 20 cm broad. Oscula (2–3 mm dia.) are numerous and are presnt on the margins of the sponge and on isolated papillae. Surface is smooth but uneven with a rather fleshy appearnce. Texture is tough and resilient. Color of the sponge is gray in life and brownish gray when dry.

Compound(s) and Activities

Antibacterial activity: The ethanolic extract of this species exerted antibacterial activity against cerebrospinal fluid (CSF) disease (neurological disease) causing pathogenic bacteria (Sujath et al., 2014). The values of inhibition zone diameter for different species of bacteria are given next.

Inhibition Zone Diameter (in mm)

E. coli	*S. aureus*	*H. influenzae*	*N. meningitides*	*K. pneumoniae*	*Streptococcus agalactiae*	*E. faecalis*
9	12	12	20	No inhibition	14	No inhibition

Source: Sujatha, S. et al., 2014, *Pharma Innovation*, 3: 81–86.

Antibacterial activity: The associated bacterial communities derived from this species displayed antibacterial activity against the pathogenic bacteria *S. aureus, S. typhi, S. paratyphi, K. oxytoca E. coli, P. mirabilis, L. vulgaris, K. pneumoniae, V. cholerae, and A. tumefaciens.* The values of zone of clearance for different pathogens against each associated bacterial species are given next.

Pathogens Tested (Zone of Clearance in mm)

Associated Bacteria	*S. aureus*	*S. typhi*	*S. paratyphi*	*K. oxytoca*	*E. coli*	*P. mirabilis*	*L. vulgaris*	*K. pneumoniae*	*V. cholerae*	*A. tumefaciens*
B. cereus	10	2	–	–	4	4	5	3	2	1
B. macerans	4	13	11	5	–	11	3	–	3	–
B. megaterium	8	5	3	–	–	–	5	3	6	–
B. brivis	5	3	3	2	3	1	1	–	–	–
L. casei	6	3	9	5	4	9	1	1	3	1
S. marcescens	1	2	7	6	4	–	–	–	9	7
P. aeruginosa	11	6	2	–	4	10	3	11	2	8
P. putida	7	2	3	6	6	–	3	10	7	–

Source: Vasanthabharathi, V., and S. Jayalakshmi, 2012, *Afr. J. Biotechnol.*, 11: 7500–7511.

Bactericidal and fungicidal activities: The different extracts of this sponge species inhibited the growth of bacteria (70%) more frequently compared to fungal strains (26%). Gram-negative bacteria were more sensitive (72%) compared to Gram-positive bacteria (65%). Considerable antimicrobial activity was exhibited against *Bacillus subtilis* (MIC 203 μg/mL), *Klebsiella pneumonia* (MIC 203 μg/mL), *Escherichia coli* (MIC 407 μg/mL), *Saprolegnia parasitica* (MIC 500 μg/mL) and *Fusarium solani* (MIC 500 μg/mL) (Khakshoor and Pazooki, 2014).

Antidermatophytic activity: The methanol extract of this species showed antidermatophytic activity against the pathogenic fungi *Microsporum gypseum*, *Epidermophyton floccosum*, and *Trichophyton rubrum*. The values of MIC were found to be 125, 250, 250, and 250 μg/mL, respectively. The dermatophyte infections in humans are among the most common forms of skin diseases worldwide. This superficial fungal infection damages the keratinized tissues such as the stratum corneum, nail, and hair (Dhayanithi et al., 2012).

Gelliodes petrosioides (Dendy, 1905) (= *Sigmadocia petrosioides*)

Order: Haplosclerida Family: Niphatidae

Common name: Not designated

Distribution: South India and Sri Lanka

Ecology: It occurs at a depth ranging from 10 to 25 m.

Biology: This sponge is massive, depressed, and cushion shaped. It is flattened below, where it has apparently been attached by a broad base. Further, it is evenly rounded and is strongly convex above. Surface is coarsely granular and is not hispid. Vents are not visible. Pores are numerous and are scattered in the thin dermal membrane, which roofs over the numerous small, rounded subdermal cavities. Color of the sponge (in spirit) is pale yellowish gray. Texture is compact, hard. and almost stony. The greatest size of specimen, which is irregularly rounded in outline, is about 2 cm in diameter. The main skeleton is a very dense but quite irregular reticulation of very coarse, stout, densely spicular fiber, with a great many loose megascleres scattered between. The spicules are short, stout, fusiform, and with slightly curved oxea.

Compound(s) and Activities

Anticancer activity: The extract of this species showed anticancer activity against A549, HeLa, and HT-29 cancer cell lines with >100 μg GI50 values (Sponge crude extraction, http://shodhganga.inflibnet.ac.in/bitstream/10603/64131/10/10_chapter%205.pdf).

Gelliodes pumila (Lendenfeld, 1887) (= *Sigmadocia pumila*)

Order: Haplosclerida Family: Niphatidae
Order: Haplosclerida Family: Adociidae
Common name: Not known
Distribution: Indian Ocean, Australia
Ecology: It occurs at depth ranging from 10 to 25 m.
Biology: Body of this sponge is finger shaped, with an oscule which has a diameter of 6 mm at its terminal. Color is pale gray. Consistency is tough and is slightly compressible. The dermal skeleton is polygonal. The oxeas of the spicules are slightly curved.

Compound(s) and Activities

Anti-inflammatory/analgesic/antipyretic activity: The methanol extract of this species showed anti-inflammatory activity at the concentrations of 100 mg/kg and 200 mg/kg. The experimented rats showed significant decrease in the paw thickness at the fifth hour of extract administration. This extract also exerted more analgesic activity. As for the antipyretic activity during the second and third hour, the rats showed significant reduction in temperature with the treatment of the extract at 100 mg/kg and 200 mg/kg dosages. This sponge species may therefore be used to produce novel drugs (Dhinakaran and Lipton, 2016).

Mosquito larvicidal activity: The sponge extracts of this species showed larvicidal activity against the third instar stage of *Aedes egypti*. The extracts produced 100% mortality during 24 h at 5.5% concentration, during 48 h at 5.5% concentration, and at 60 h with 4.5% concentration. The LC50 value recorded for 60 h was 3.3 mg/mL (Collection, extraction of compounds, http://shodhganga.inflibnet.ac.in/bitstream/10603/27919/8/08_chapter%201.pdf).

Antimicrobial activity: The methanol extracts of this species exhibited high inhibitory activity toward human pathogens of Gram-negative organisms such as *E. coli* (inhibition zone diameter, 16 mm), *Klebsiella pneumoniae* (14 mm), *Pseudomonas putida* (15 mm), and *Serratia liquefaciens* (16 mm). They were also active against the Gram-positive *Staphylococcus aureus* (2–12 mm). The methanol extracts of this species also showed moderate antifungal activity against *Candida albicans* and *Penicillium chrysogenum*. In *Candida albicans*, the diameter of the inhibition zone was 6 to 10 mm, and for *Penicillium chrysogenum* it was 3 to 6 mm (Dhinakaran and Lipton, 2012). The methanol extracts of this species showed antibacterial activity against Gram-positive bacterial strains *Bacillus thuringiensis* MTCC 4714, *Enterococcus faecalis* MTCC 439, *Listeria monocytogenes* MTCC 1143, *Staphylococcus aureus* MTCC 737, and *Proteus vulgaris* MTCC 426; and Gram-negative strains *E. coli* MTCC 443, *Klebsiella pneumoniae* MTCC 109, *Pseudomonas putida* MTCC 1688, and *Serratia liquefaciens* MTCC 3039. The extracts also showed antifungal activity against *Trichoderma viride* MTCC 2047, *Fuzarium* sp. MTCC 284, *Aspergillus niger* MTCC 1344, *Candida albicans* MTCC 183, *Penicillium chrysogenum* MTCC 5108, and *Aspergillus flavis* MTCC 873 (Collection, extraction of compounds, http://shodhganga.inflibnet.ac.in/bitstream/10603/27919/8/08_chapter%201.pdf).

Anticancer activity: The organic extracts of this species showed anticancer activity against human epidermoid cancer cells (Hep2) and human breast adenocarcinoma cell line (MCF7) with an IC50 value of 1.25 mg/mL (Bhimba et al., 2013). The anticancer activity of the extract on these cancer lines at different concentrations are shown next.

Anticancer Activity against MCF7 and Hep2 Cell Lines

	Cell Viability (%)	
Extract Concentration (mg/mL)	MCF7	Hep 2
5.0	27.3	31.4
2.5	41.8	37.3
1.25	52.7	47.1
0.63	61.8	56.9
0.16	74.5	68.6
0.08	80.0	94.1
0.04	94.5	98.0

Source: Bhimba, B. V., 2013, *J. Pharm. Res.*, 6: 401–403.

Antiviral and antitumor activities: The compounds, benzamidines and isobenzofuran isolated from this species exhibited antitumor activity. Further, its cinnamaldehyde exhibited antiviral and antitumor activities.

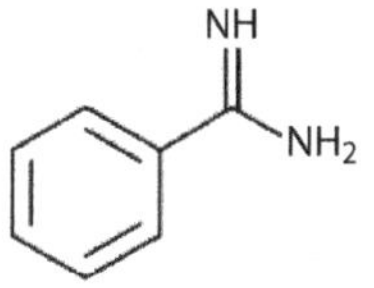

Benzamidine

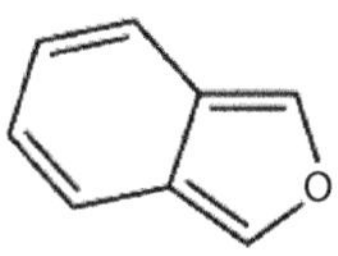

Isobenzofuran

The percentage inhibition values of the extracts associated with the *in vitro* antiviral activity (against HSV-1) are given next.

Extract Concentration (μg)	% Plaque Inhibition
70	83
60	75
50	68

Source: Antiviral activity, http://shodhganga.inflibnet.ac.in/bitstream/10603/27919/12/12_chapter%205.pdf.

Pachychalina alcaloidifera (Pinheiro, Berlinck, and Hajdu, 2005)

Order: Haplosclerida Family: Niphatidae
Common name: Not designated
Distribution: Southeastern Brazil
Ecology: The species is known from 5 to 25 m depth.

Biology: It is an encrusting or massive, cushion-like sponge with or without lobate. It may also be digitiform or more frequently volcaniform projections. Specimens can cover nearly 1 m^2 of substrate, but are frequently less than 1 cm^2 in area. It is friable in consistency. Oscula are apical. Colors, when alive, vary among white, beige, and gray, which may be a bit yellowish. The ectosomal skeleton appears as an irregular paratangential reticulation, mostly unispicular. The choanosomal skeleton is (plumo)reticulated with primary ascending tracts. Spongin is not visible. Spicules are of a single category, namely, megascleres of the oxea type, and they are less stout and less curved.

Asociation: Common associated organisms of this species include polychaetes, hermit crabs, amphipods, isopods, hydroids, anemones, ophiuroids, tunicates, and bivalves.

Compound(s) and Activities

Cytotoxic activity: The bis-piperidine alkaloids, madangamine F, haliclonacyclamine F, and arenosclerins D and E isolated from this species displayed cytotoxic activity against different cancer cell lines (de Oliveira et al., 2007).

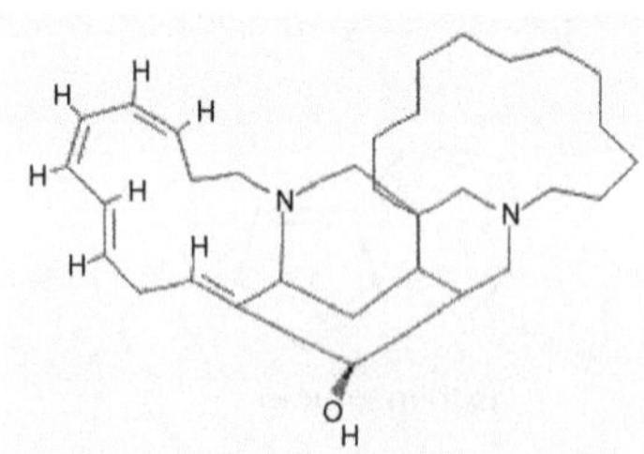

Madangamine F

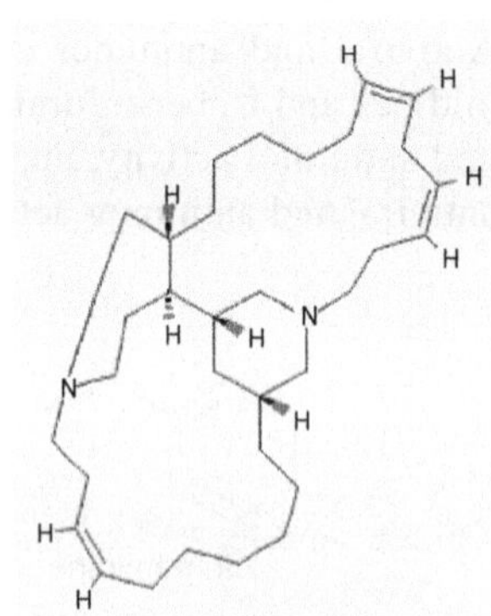

Haliclonacyclamine F

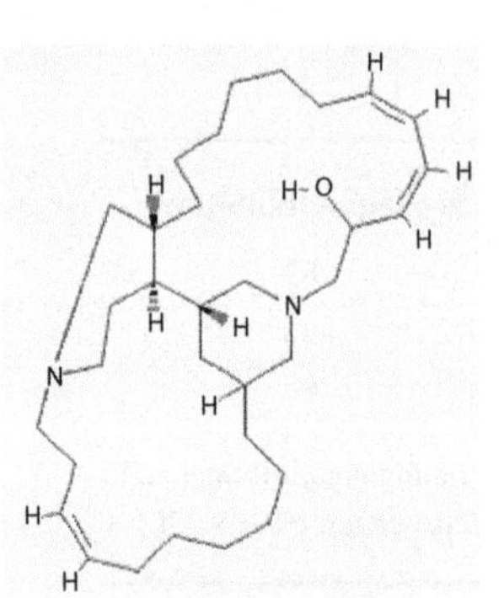

Arenosclerin D

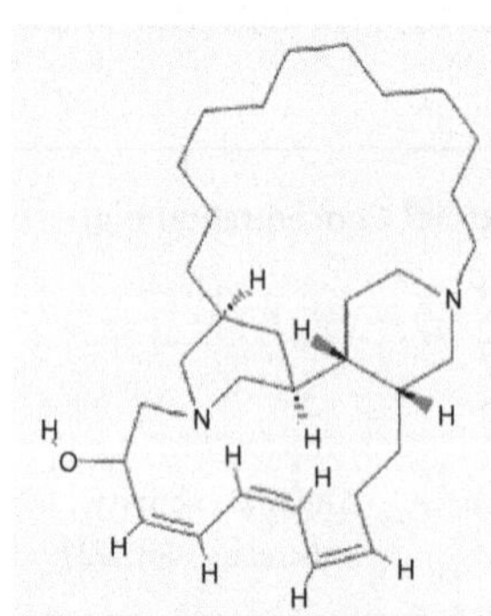

Arenosclerin E

The compound ingenamine G, an alkaloid isolated from this species, displayed a moderate cytotoxic activity against human proliferating lymphocytes with an IC50 value of 15 µg/mL, and antitubecular activity against *M. tuberculosis* (H37Rv) at 8 µg/mL (Cavalcanti et al., 2008).

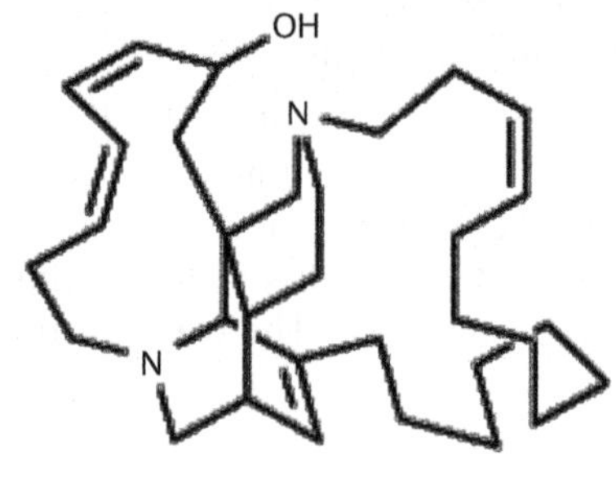

Ingenamine G

Pachypellina sp.

Order: Haplosclerida Family: Niphatidae

Compound(s) and Activities

Antitubercular and antimalarial activity: The compounds (+)-8-hydroxymanzamine A isolated from this unidentified sponge showed antitubercular activity against *M. tuberculosis* (CH37RV) with an inhibition of 98%–99% and MIC values of less than 12.5 µg/mL. Further, these compounds had antimalarial activity against *P. berghei* with a single dose of 100 µM in *P. berghei*-infected mice in 9 to 12 days (Asif, 2017; Fattorusso and Taglialatela, 2008).

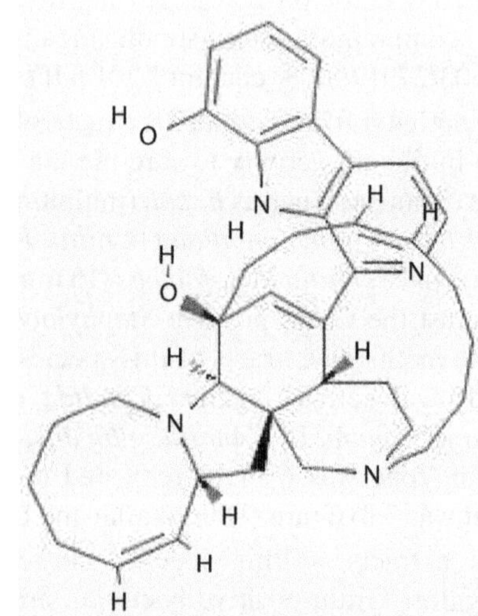

8-hydroxymanzamine A

Niphates digitalis (Lamarck, 1814) (= *Dasychalina cyathina*)

Order: Haplosclerida Family: Niphatidae

Common name: Vase sponge

Distribution: Gulf of Mexico, Florida

Ecology: It dwells in coral reefs.

Biology: These sponges are tapering cups, up to 19 cm high and up to 14 cm in diameter. The apical vent, with sharp rim, is mostly laterally flattened. Outer surface is fairly smooth to extremely conulose or spinous. Oscules are present on the inside of the cup. Consistency is firm and spongy. Color of the sponge is light bluish to grayish green, and occasionally purplish gray. Ectosome is of paratangential reticulation of fibers often ill developed. In choanosome, the skeleton is irregularly radial plumose. In this sponge, the fiber cells are among thin sponging A fibers, which closely resemble vertebrate reticular fibers occurring in loose connective tissue.

Association: Brittle stars are often seen in this sponge

Compound(s) and Activities

Cardioactive activity: The marine nucleosides adenosine and 2′-deoxyadenosine of this species serve as cardioactive agents. These nucleosides have also been reported to display antiviral, anticancer, vasodilator, muscle relaxant, and hypertensive activities (Weinheimer et al., 1978).

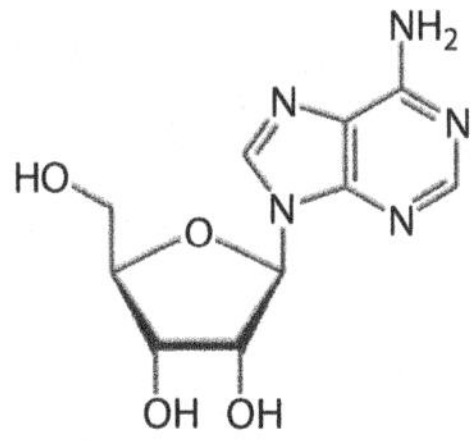

Adenosine

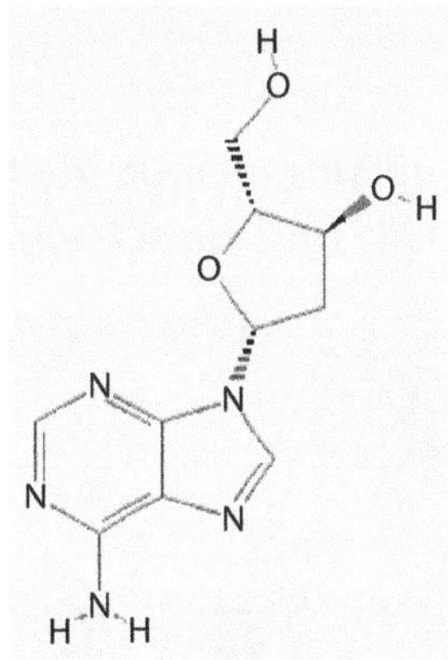

2'-deoxyadenosine

Niphates furcata (Keller, 1889)

Order: Haplosclerida Family: Niphatidae

Common name: Not reported

Distribution: Eritrea, Red Sea

Ecology: It occurs in benthic habitats including coral reefs and seagrass.

Biology: Not available

Compound(s) and Activities

Antibacterial activity: The methanol extracts of this species have shown antibacterial activity against Gram-positive and Gram-negative bacteria. The values of MIC recorded were 5 mg/mL for *Bacillus subtilis spizizenii*, 10 mg/mL for *Staphylococcus aureus aureus* and *Bacillus cereus*, 30 mg/mL for *Escherichia coli*, and 20 mg/mL for *Serratia marcescens* (Nazemi et al., 2017a).

Antioxidant activity: Seradj et al. (2012) reported on the hydroxyl radical scavenging activities of this sponge's extracts.

Percent of OH Radical Inhibition in CH_2Cl_2 Extracts

Concentrations in milligrams of dried sponge extract

0.5	0.25	0.125	0.062
82.9	80.9	79.7	74.6

Percent of OH Radical Inhibition in MeOH Extracts

Concentrations in milligrams of dried sponge extract			
0.5	0.25	0.125	0.062
71.6	75.4	75.1	80.5

Source: Seradj, H. et al., 2012, *Iranian J. Pharm. Sci.*, 8: 249–255.

Niphates olemda (de Laubenfels, 1954) (= *Cribrochalina olemda*)

Order: Haplosclerida Family: Niphatidae

Common name: Tube sponge

Distribution: Palau; Western Pacific Ocean, Malaysia, Philippines

Ecology: It inhabits current swept reefs and very shallow or intertidal, sheltered habitats of 5 m depth.

Biology: It is a fan-shaped sponge and is thicker at the point of attachment than at the edges (3 cm long, 5 cm wide, and 0.5 mm at its thickest portion). Surface is rugged with protruding fiber tracts forming subdermal cavities, while the underside is smoother. Ostia (pores) are not visible, but the oscula (0.5 cm in dia.) are distributed 8 to 15 mm apart. Texture is soft, compressible, and difficult to tear. Color of the sponge is brown with bluish tint when out of water, while in spirit it is light brown. Megascleres are oxeas and microscleres are absent. The skeleton is fibro-reticulate with cored spongin fibers forming irregular meshes; ascending fibers terminate at the protrusions at the surface of the sponge.

Compound(s) and Activities

Cytotoxicity: The compound kapakahine E isolated from this species was found to display moderate cytotoxicity against P388 murine leukemia cells at IC50 of 5.0 μg/mL (Agrawal et al., 2016).

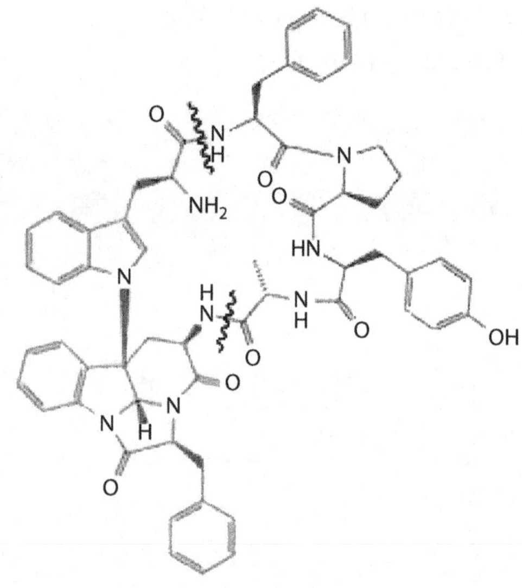

Kapakahine E

Antibacterial activity: The anthraquinone compounds such as lunatin and cytoskyrin A isolated from the associated fungus *Curvularia lunata* derived of this sponge species have shown antibacterial activities against *Staphylococcus aureus*, *Escherichia coli*, and *Bacillus subtilis* (Thomas et al., 2010).

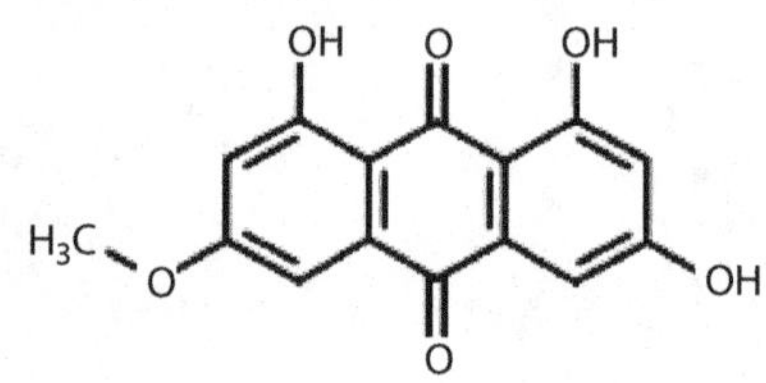

Lunatin

Niphates sp.

Order: Haplosclerida Family: Niphatidae

Compound(s) and Activities

Antiviral activity: The compounds chartarutine B, G, and H produced by the associated fungus *Stachybotrys chartarum* of this seponge species have shown antiviral activity against HIV-1 with IC50 values of 1.8, 2.1, and 2.1. μg/mL, respectively (Indraningrat et al., 2016).

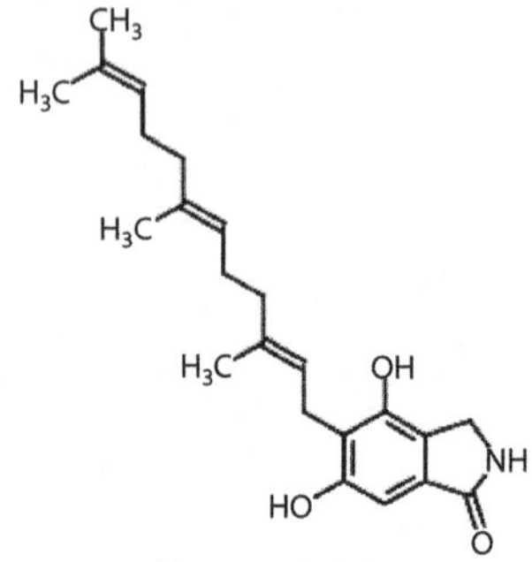

Chartarutine B

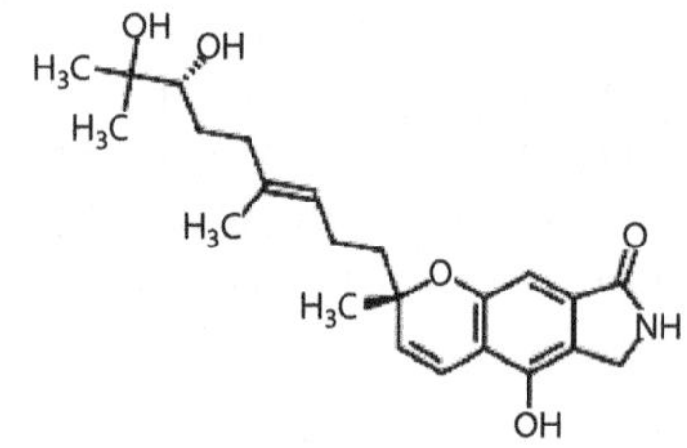

Chartarutine G

Cytotoxicity: The compounds nepheliosyne A and B isolated from this sponge have shown cytotoxicity against K562 chronic myelogenous leukemia, U266 myeloma, SKM1 myelodysplastic syndrome, and Kasumi acute myeloid leukemia human cell lines with IC50 values ranging from 150 to 200 μM (Motuhi et al., 2016).

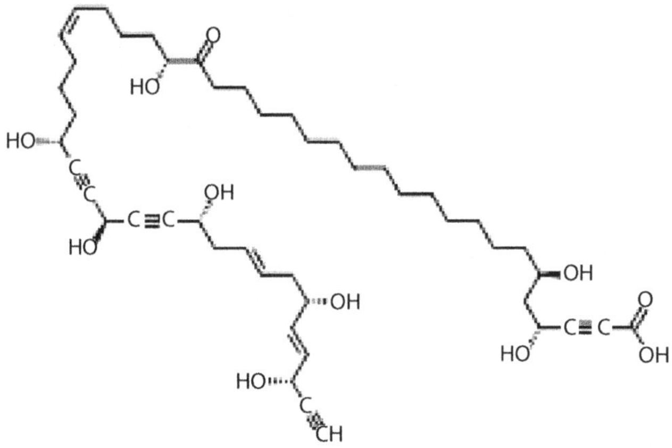

Nepheliosyne A

FAMILY: PETROSIIDAE

Acanthostrongylophora ingens (Thiele, 1899)

Order: Haplosclerida Family: Petrosiidae
Common name: Indonesian marine sponge
Distribution: Indo-Pacific
Ecology: Coastal waters
Biology: Not studied

Compound(s) and Activities

Antiprotozoal activity: The compound manzamine A isolated from the associated bacterial strain *Micromonospora* sp. M42 of this speices has shown antiprotozoal activity against *Plasmodium falciparum* and *P. berghei* at a concentration of 0.0045 and 0.008 μg/mL, respectively (Indraningrat et al., 2016).

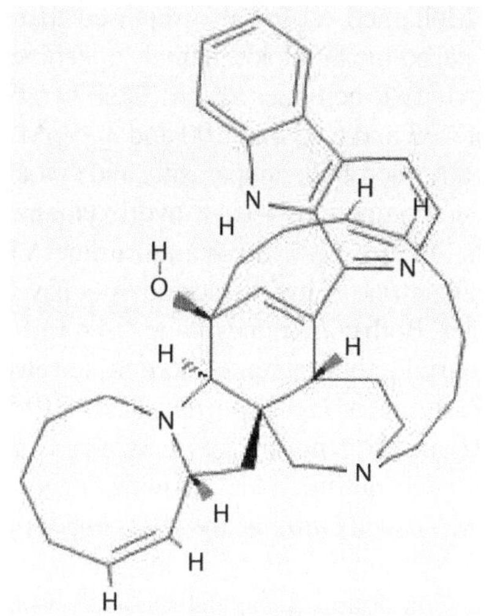

Manzamine A

Cytotoxicity: The compound annomontine (1) and two alkaloids, ingenine A (2) and ingenine B (3), have been isolated from this species. Among them the compounds 1 and 3 showed pronounced cytotoxicity against the murine lymphoma L5178Y cancer cell line with ED50 values of 7.8 and 9.1 μg/mL, respectively. Compound 2, however, showed weak activity (Ibrahim et al., 2015).

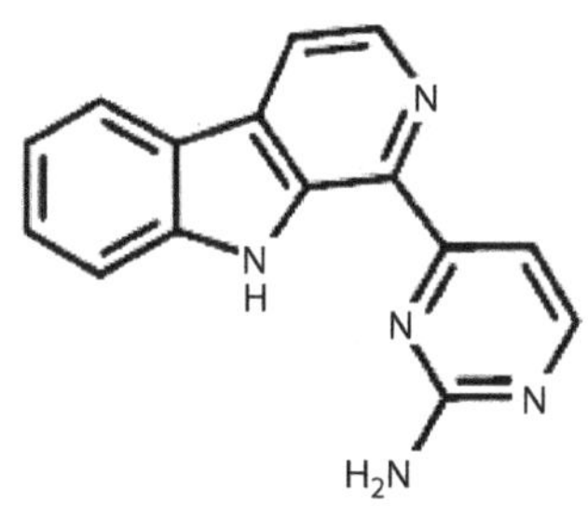

Annomontine

A total of seven metabolites (1–7) including a new manzamine derivative, ircinal E (1), have been isolated from the methanol extracts of this species. Of these, compounds 1–6 exhibited strong to moderate cytotoxicity against the murine lymphoma L5178Y cell line with IC50 values ranging from 2.8 to 21.7 μM (AlTarabeen et al., 2015).

Ibrahim and Mohamed (2017) reported on the isolation of a new β-carboline alkaloid ingenine E (4), along with three known metabolites—annomontine, acanthomine A, and 1,2,3,4-tetrahydronorharman-1-one—from this species. Among them, ingenine E exhibited cytotoxic activity against MCF7, HCT116, and A549 cancer cell lines with IC50 values of 3.5, 0.67, and 2.15 μg/mL, respectively.

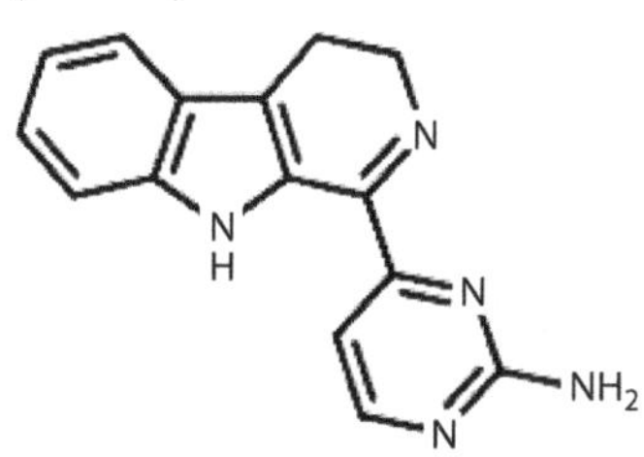

Acanthomine A

Ibrahim and Mohamed (2016) also reported that the two new pyrimidine-β-carboline alkaloids, namely, ingenines C (1) and D (2) exhibited cytotoxic activities toward MCF7 and HCT116 with IC50 values of 4.33 and 6.05 and 2.90 and 3.35 μM, respectively.
Antimalarial, antimicrobial, antiparasitc, and cytotoxic activities: A total of four compounds—(+)-8-hydroxymanzamine A (1), (+)-manzamine A (2), (+)-8-hydroxymanzamine A hydrochloride (3), and (+)-manzamine A hydrochloride (4)—have been isolated from this species. Both compounds 1 and 3 showed equally potent *in vitro* antimalarial activity against chloroquine-sensitive (D6) and -resistant (W2) strains of *P. falciparum* (IC50 = 19.5 and 22.0 ng/mL versus 27.0 and 36.5 ng/mL, respectively). Compounds 1, 3, and 4 showed good antimicrobial activities against methicillin-resistant *Staphylococcus aureus* and *Mycobacterium intracellulare*, and antileishmanial activity against *Leishmania donovani* promastigotes. Compound 2 showed relatively weaker cytotoxic activity toward cancer (HepG2: human hepatocellular carcinoma or hepatoma), and noncancer cells (VERO: monkey kidney fibroblast; LLC-PK11: pig kidney epithelial) (Samoylenko et al., 2009). The values recorded on the different activities of these compounds are given next.

IC50 Values of Antiparasitic Activity of Compounds 1 to 4

Compound	*P. falciparum* (ng/mL)	*L. donovani* (mg/mL)
1	19.5	2.2
2	20.8	11.2
3	27.0	2.5
4	6.1	1.6

Source: Samoylenko, V. et al., 2009, *Nat. Prod. Commun.*, 4: 185–192.

IC50 (μg/mL) Values of Antimicrobial Activity of Compounds 1 to 4

Compound	*C. albicans*	*C. neoformans*	*M. intracellulare*
1	2.7	2.7	0.1
2	*	7.3	0.4
3	3.0	2.7	0.2
4	5.2	1.0	0.1

Source: Samoylenko, V. et al., 2009, *Nat. Prod. Commun.*, 4: 185–192.
* Inactive up to 20 μg/mL.

Antimicrobial Activity (MIC/MFC/MBC [μg/mL]) of Compounds 1 to 4

Compound	*C. albicans*	*C. neoformans*	*M. intracellulare*
1	5 (10)	10 (10)	0.2 (20)
2	*	*	0.6 (ND)
3	5 (10)	5 (10)	0.2 (10)
4	20 (ND)	5 (10)	0.2 (10)

Source: Samoylenko, V. 2009, *Nat. Prod. Commun.*, 4: 185–192.
Notes: ND: No data. Values in barackets, MFC or MBC.
* Inactive up to 20 μg/mL.

IC50 (μg/mL) Values of Cytotoxic Activity of Compounds 1 to 4

Compound	SK-MEL[a]	KB[b]	BT-549[c]	HepG2[d]	LLC-PK11[e]
1	0.5	0.8	0.8	0.5	1.3
2	1.0	1.0	1.1	4.4	2.2
3	0.8	0.7	1.1	1.5	1.9
4	0.6	0.4	0.9	1.6	0.7

Source: Samoylenko, V., 2009, *Nat. Prod. Commun.*, 4: 185–192.
[a] SK-MEL: Human malignant melanoma.
[b] KB: Human epidermoid carcinoma.
[c] BT-549: Human breast ductal carcinoma.
[d] HepG2: Human hepatocellular carcinoma (hepatoma).
[e] LLC-PK11: Pig kidney epithelial cells.

Thomas et al. (2010) reported that the compound manzamine A of this speices possesses both antitumor activity (against mouse leukemia cells) and antimalarial properties (against the rodent malarial parasite *Plasmodium berghei*).
Antiprotozoal and antiplasmdial activities: Indraningrat et al. (2016) reported that the organic extracts of this species showed antiprotozoal activity against *Micromonospora* sp. and antiplasmodial activity against *P. falciparum* with an IC50 value of 0.0045 μg/mL.
Antibacterial activity: The ethyl acetate extract (containing phenolic, terpenoid, and steroid compounds) of the symbiotic fungi of this species showed antibacterial activity. There were eight isolated symbiotic fungi strains from the sponge. The LC50 values of the different fungal extracts were found to be 44.6 ppm (IB121 isolate), 11.5 ppm (IB131 isolate), 662.2 ppm (IB142 isolate), 221.8 ppm (IB161 isolate), 172.7 ppm (IB102 isolate), 58.6 ppm (IB141 isolate), 3.8 ppm (IB101 isolate), and 0.5 ppm (IB151 isolate). The values of diameter inhibition for the tested species, namely, *Bacillus subtilis, Staphylococcus epidemidis, Salmonella typosa* and *Escherichia coli* were 14.4 mm, 11.5 mm, 10.25 mm, and 14.75 mm (at a concentration of 5%), respectively (Handayani and Aminah, 2017).
Neuritogenic activity: Two manzamine-class alkaloids, manzamine A and 8-hydroxymanzamine, isolated from this species showed neuritogenic activity (which relates to the preventive and therapeutic effects of neurodegenerative diseases) against Neuro 2a cells (Zhang et al., 2008).

Acanthostrongylophora sp.

Order: Haplosclerida Family: Petrosiidae

Compound(s) and Activities

Antiviral and anti-TB activities: The alkaloids manadomanzamines A and B isolated from this unidentified species exhibited strong anti-TB activity against *Mycobacterium tuberculosis* with MIC values of 1.9 and 1.5 μg/mL, respectively. These two alkaloids also exhibited significant activity against human immunodeficiency virus (HIV-1) with EC50 values of 7.0 and 16.5 μg/mL, respectively, and moderate activity against several AIDS opportunist infections (Jyothi et al., 2016).

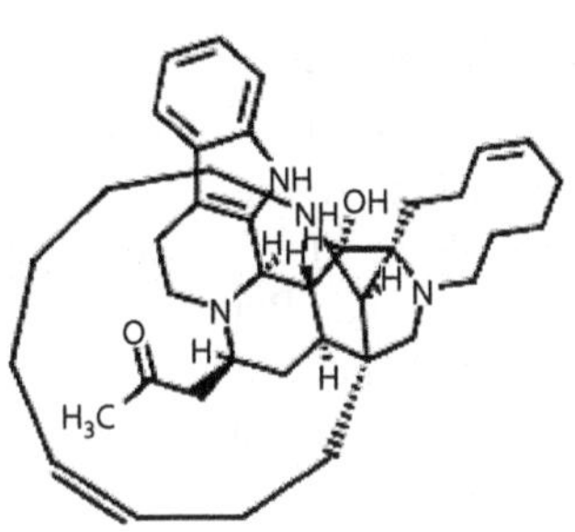

Manadomanzamine A

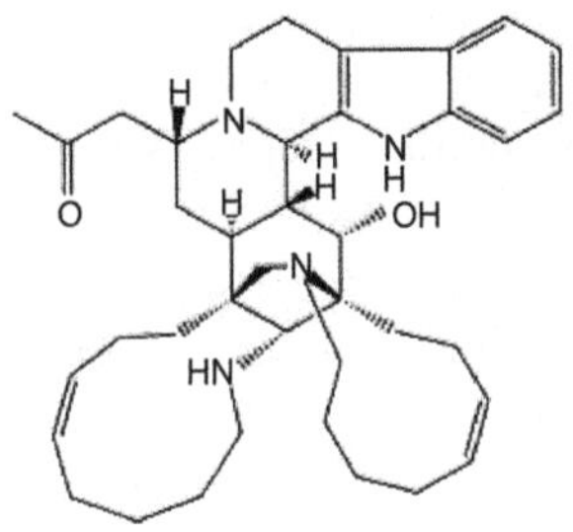

Manadomanzamine B

Anjum et al. (2016) reported that the compound 6-hydroxymanzamine E of this unidentified species showed anti-TB activity against *M. tuberculosis* with an MIC value of 0.9 μg/mL.

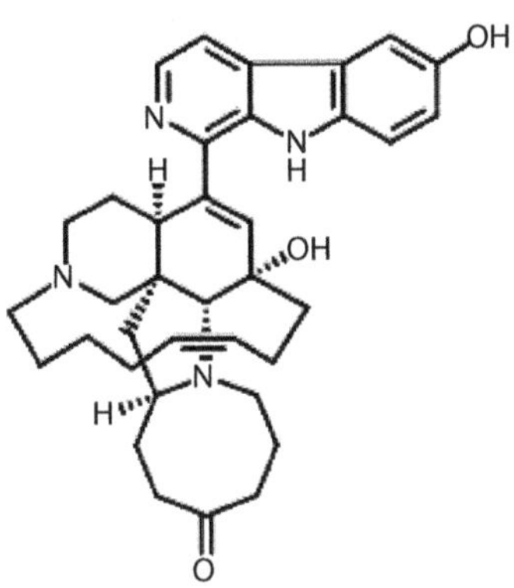

6-hydroxymanzamine E

Cytotoxicity: Of the three alkaloid compounds of this unidentified species—one pyrimidine-β-carboline alkaloid named ingenine A (2) and one pyrimidine-γ-carboline alkaloid named ingenine B (3), along with annomontine (1)—the compounds 1 and 3 have shown pronounced cytotoxicity against the murine lymphoma L5178Y cancer cell line with ED50 7.8 and 9.1 μg/mL, respectively, while compound 2 showed weak activity (Ibrahim et al., 2015).

Neuroprotective activity: The compound manzamine Y isolated from this unidentified speices showed significant inhibitory activity of GSK3, an enzyme implicated in Alzheimer's disease pathology (Rao et al., 2006).

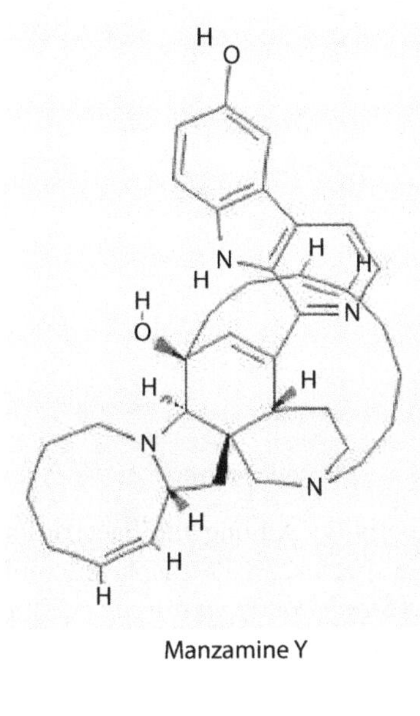

Manzamine Y

Anti-TB and anti-HIV-1 activities: The compounds manadomanzamines A and B of this species exhibited significant activities against *Mycobacterium tuberculosis* (Mtb) and human immunodeficiency virus (HIV-1) and moderate activity against several AIDS opportunistic infections (OI) (Putra and Jaswir 2014).

Petrosia (Strongylophora) corticata (Wilson, 1925)

Order: Haplosclerida Family: Petrosiidae

Common name: Okinawan marine sponge

Distribution: Philippines, Eastern Indonesia, Papua New Guinea

Ecology: It grows in shallow reef.

Biology: This sponge has undulating smooth branches of 3 to 10 cm diameter and is up to 30 cm long. There are numerouss small typical sieve-like oscules (4–8 mm wide) that are scattered across the surface. The ectosome forms a distinct, firm rind and is slightly transparent and easily detachable. The inside is pulpy. Color of the specimen is ocher-greenish tinge and in spirit it is dull brown. The skeleton of the ectosome is a paucispicular reticulation of the larger spicules forming irregular triangular meshes, with perpendicular tufts of the intermediate spicules. The dominating spicules are strongyles, smooth, slightly curved with evenly rounded ends.

Compound(s) and Activities

Antifungal activity: The 2-methoxy acetylenic acids corticatic acids A–E isolated from this sponge showed antifungal activity (Mokhlesi et al., 2017).

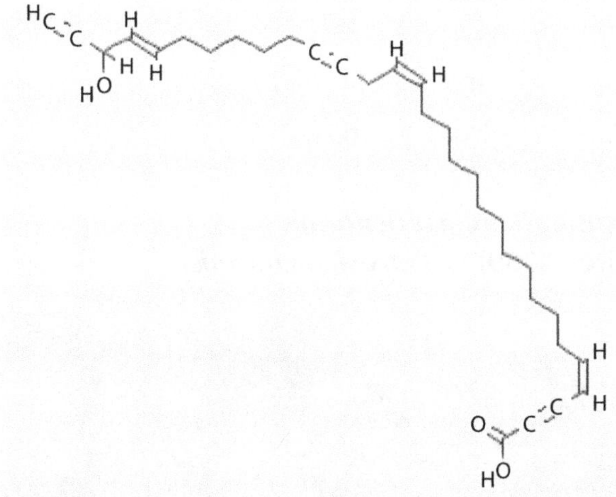

Corticatic acid A

Cytotoxicity: The strongylophorins isolated from this species showed cytotoxicity against HeLa cells (Hoshino et al., 2003).

Anti-invasion activity: A new meroditerpenoid, strongylophorine-26, isolated from this sponge showed anti-invasion activity related to the cancer cells (Warabi et al., 2004).

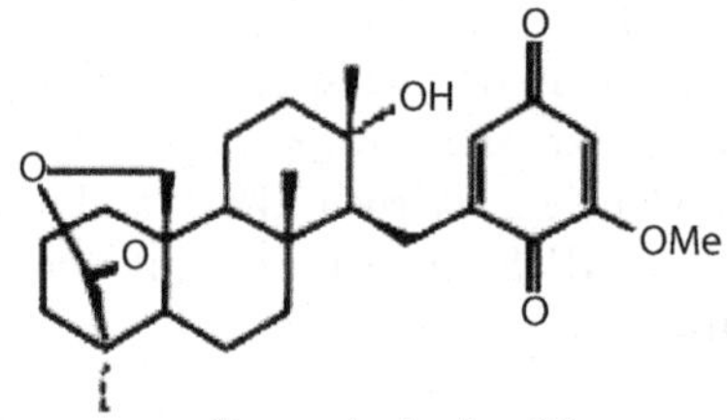

Strongylophorine-26

Petrosia (Strongylophora) durissima (= *Strongylophora durissima*)

Order: Haplosclerida Family: Petrosiidae

Common name: Taiwanese marine sponge

Distribution: South India, Sri Lanka

Ecology: Not reported

Biology: Not available

Compound(s) and Activities

Cytotoxicity: The acetylenic derivative, durissimol B, and duryne of this species showed potent cytotoxicity against human gastric tumor (NUGC) cells. Further, duryne inhibited the growth of both mouse and human tumor cell lines *in vitro* (Dembitsky et al., 2006).

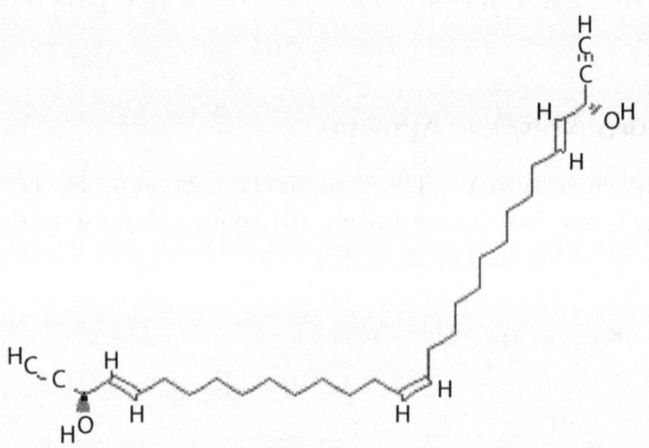

Duryne

Petrosia (Petrosia) ficiformis (Poiret, 1789) (= *Petrosia ficiformis*)

Order: Haplosclerida Family: Petrosiidae

Common name: Stony sponge, Mediterranean sponge

Distribution: Adriatic Sea, Aegean Sea, Azores, Canaries, Madeira, Cape Verde, Ionian Sea, Levantine Sea, Mediterranean Sea, North Atlantic, Tunisian Plateau/Gulf of Sidra, West Africa, and Western Mediterranean

Ecology: It is seen in various rocks, caves, and cracks; generally beyond 5 m.

Biology: This sponge has very variable morphology: crusts massive, reticules, reniform, or irregular. It is several centimeters thick. Surface is more or less smooth, with many oscules that are irregularly spread over the surface. Consistency lasts. Coloring varies from pure white to brown to nuance violets. Interior of sponge is also of clear color. This sponge may also be purple brown in color due to symbiosis with photosynthetic cyanobacteria.

Association and predator: *Petrosia ficiformis* is the main and preferred food of the nudibranch *Peltodoris atromaculata,* which is generally found over the sponge. These nudibranchs accumulate the chemical compounds of the sponge on their digestive tract and use them as one of their defensive strategies.

Compound(s) and Activities

Cytotoxicity: The methanolic extract of this species showed a marked cytotoxic effect on human neuroblastoma cells at any concentration or time of exposure (Ferretti et al., 2007).

The methanolic extract of this species showed a marked cytotoxic effect against two human neuroblastoma cell lines, namely, LAN5 and SK-N-BE(2)-C, at any concentration or time of exposure. A marked cell death of these NB cells was found also after 30 min treatment with this extract (Ferretti et al., 2007).

The compound, petrosynol, a polyacetylene isolated from this species showed antimicrobial, cytotoxic, and antitumoral activities (*Wikipedia*).

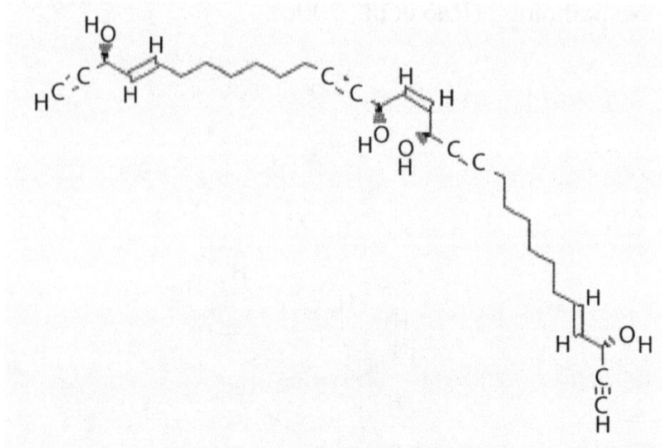

Petrosynol

Antimicrobial activity: Among the bacterial strains isolated from this sponge species, 58.3% displayed antibacterial activity active against *Staphylococcus aureus*, 6.7% against *Bacillus subtilis*, 11.7% against *Enterococcus faecalis*, 11.7% against

Escherichia coli, 38.3% against *Pseudoalteromonas atlantica*, and 33.3% against *Pseudomonas elongata*. The percent values of inhibition observed for *Staphylococcus aureus*, *Bacillus subtilis*, *Enterococcus faecalis*, *Escherichia coli*, *Pseudoalteromonas atlantica*, and *Pseudomonas elongata* were 58.3, 6.7, 11.7, 11.7, 38.3, 33.3, respectively. The high percentage of bioactive isolates derived from this sponge suggests that this sponge-associated microorganisms are promising sources for drug discovery (Chelossi et al., 2007).

Antifungal activity: The associated fungus *Penicillium brevicompactum* of this sponge yielded cyclopentadepsipeptides, petrosifungins A and B, along with the known fungal metabolites brevianamide A, mycophenolic acid (a well-known immunosuppressive agent), and asperphenamate, which showed significant antifungal activity (Thomas et al., 2010).

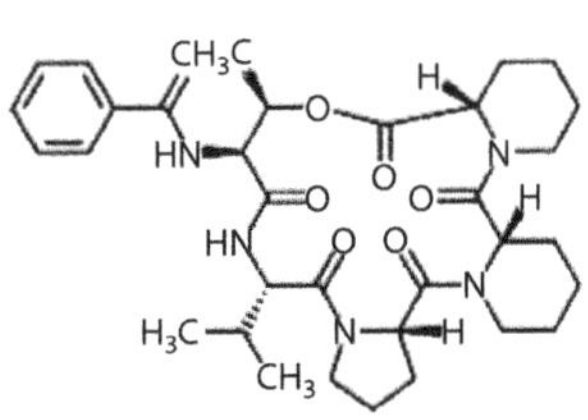

Petrosifungin A

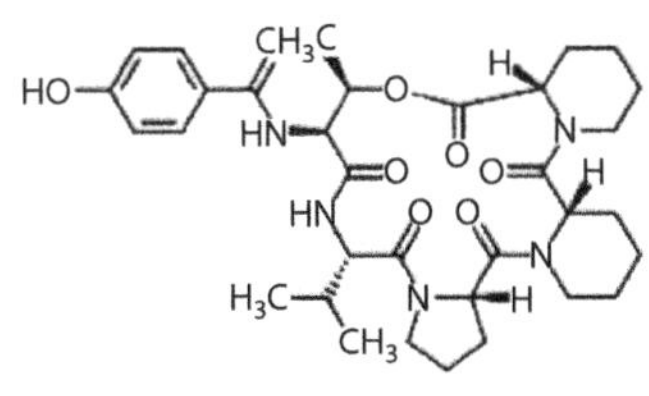

Petrosifungin B

Chelossi et al. (2004) reported that the bacteria *Rhodococcus* sp. and *Pseudomonas* sp. isolated from the epibiotic microbial community of this species showed antimicrobial activity.

Antiprotozoal activity: The associated actinobacterial strains *Streptomyces* sp. SBT348 and SBT344 isolated from this species displayed antiprotozoal activity against *T. brucei brucei* with IC50 values of 16.5 µg/mL and <10 µg/mL, respectivley (Indraningrat et al., 2016).

Petrosia (Petrosia) nigricans (Lindgren, 1897)

Order: Haplosclerida Family: Petrosiidae

Common name: Not designated

Distribution: Indian Ocean, Kenya

Ecology: It grows on seamounts and knolls.

Biology: The shape of this sponge is lamellate, flabelliform, cup shaped, or broadly tubular. The size may be considerable, up to 150 cm in height, 200 cm in diameter, and with a lateral thickness of up to 10 cm. The outside surface of the sponge is undulate or irregularly sharply ridged, almost honeycombed, while the inner side is smooth. Oscules, other than an eventual vent in more tubular specimens are not very obvious, but are 0.5 to 1 cm in diameter and are slightly elevated. Copious slime is exuded when the sponge is cut or put in spirit. Color of the animal is chocolate brown to dark, almost black brown.

Compound(s) and Activities

Cytotoxicity: One new biscembranoid, petronigrione (1), and five known compounds, methyl tortuoate B (2), lobophytone U (3), lobophytone H (4), (24S)-ergostane 3beta, 5alpha, 6beta, 25 tetraol-25-monoacetate (5), and (24S)-ergostane-1beta, 3beta, 5alpha, 6beta, 25-pentaol-25-monoacetate (6), have been isolated from the methanol extract of this species. Among them, compounds 1 and 2 exhibited moderate cytotoxic activities against human cancer cell lines HepG2, KB, LU-1, and MCF-7 on the four human cancer cell lines with IC50 values ranging from 20.7 to 28.9 µg/mL (Nhiem et al., 2013).

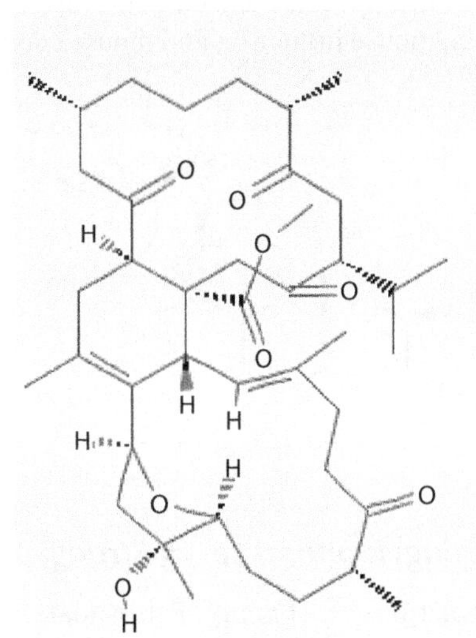

Methyl tortuoate B

Antimicrobial activity: The ethyl acetate extracts of the endophytic fungi of this species showed antimicrobial activity against human pathogenic bacterium *Staphylococcus aureus* with 14 to 24 mm values of zone of inhibition (Handayani et al., 2016b).

Petrosia (Petrosia) weinbergi (van Soest, 1980)

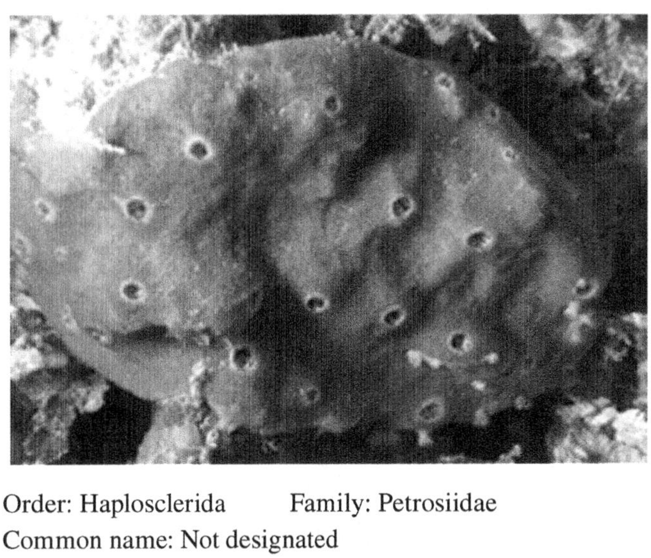

Order: Haplosclerida Family: Petrosiidae

Common name: Not designated

Distribution: Puerto Rico and Curacao; Greater Antilles, Guianan, Southwestern Caribbean, Suriname

Ecology: These sponges are seen in deeper reef slopes as stony masses encrusting corals and hard substrates

Biology: This sponge is scattered with oscules 2 to 5 mm in diameter. The surface is smooth and the consistency is hard and brittle. The color of this sponge is whitish to dirty green. These sponges are often infested with Parazoanthus (Coelenterata: Cnidaria).

Compound(s) and Activities

Antiviral activity: The sulfated sterols weinbersterols A and B isolated from this species showed antiviral activity against feline leukemia, mouse influenza, and mouse corona (Sipkema et al., 2005).

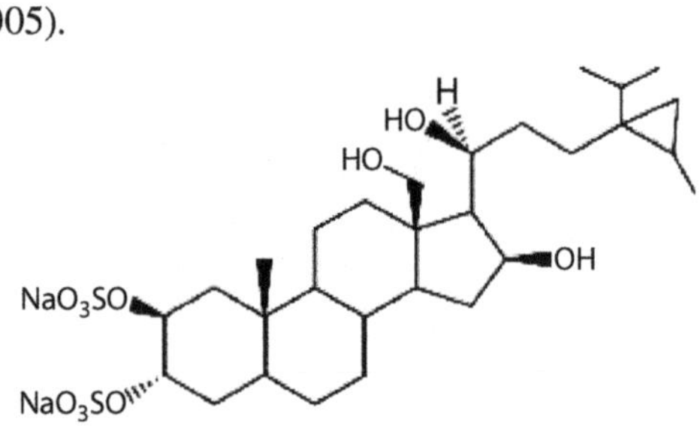

Weinbersterol A

Petrosia (Strongylophora) sp. (= *Strongylophora* sp.)

Order: Haplosclerida Family: Petrosiidae

Compound(s) and Activities

Antiviral activity: Two acetylenic compounds, petrosynol and petrosolic acid, from this sponge inhibited reverse transcriptase of human immunodeficiency virus (Dembitsky et al., 2006).

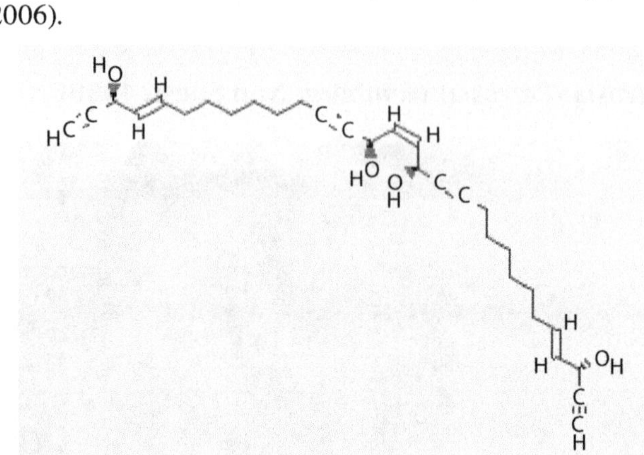

Petrosynol

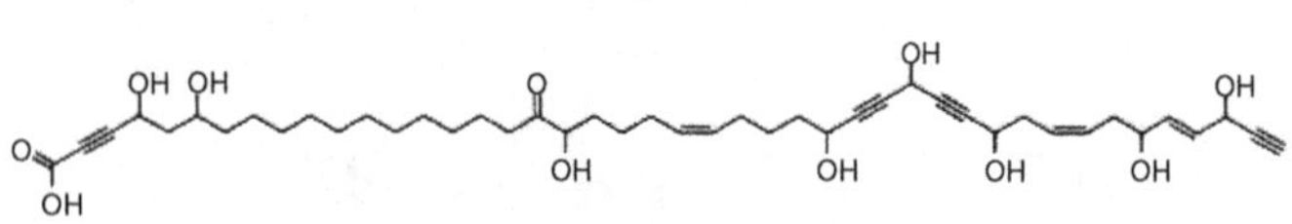

Petrosolic acid

Cytotoxic activity: Polyacetylene compounds dideoxypetrosynols A–D of this species showed cytotoxic activities against human tumor cell lines (A549, SK-OV-3, SK-MEL- 2, XF498, and HCT15). Further, compound A inhibited DNA replication, and a mechanism of its action on cultured human SK-MEL-2 skin melanoma cells has also been suggested (Dembitsky et al., 2006).

The acetylenic alcohols, strongylodiols A–D, isolated from this species showed cytotoxic activity toward human T lymphocyte leukemia (MOLT-4) cells (Dembitsky et al., 2006).

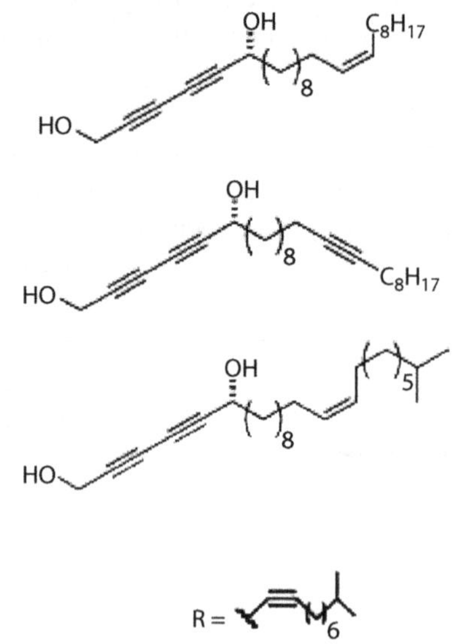

Strongylodiols A-D (top to bottom)

The polyacetylenic alcohols named petrocortynes of this sponge showed different cytotoxic activities human tumor cells (Dembitsky et al., 2006).

Of the acetylenic enol ethers of glycerols (1–6) isolated from this species, compounds 1–3 exhibited weak cytotoxicity against a human leukemia cell line (K-562) with LC50 values of 9.2, 57, and 29 μg/mL, respectively, while compounds 4–6 were found to be less active (LC50, >100 μg/mL) (Dembitsky et al., 2006).

The polyketides such as decumbenones A and B and versiol isolated from its associated fungus *Aspergillus versicolor* showed cytotoxic activity. The compound decumbenone A is also a good inhibitor of melanin (Thomas et al., 2010).

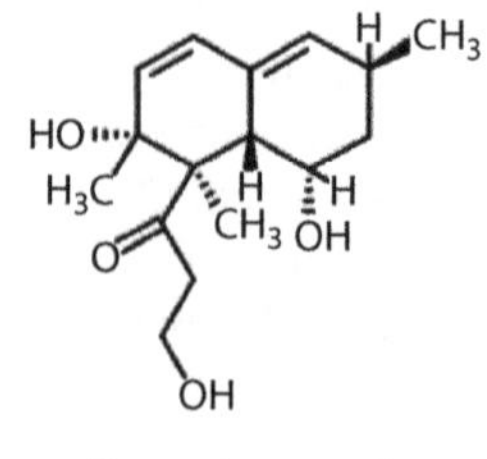

Decumbenone A

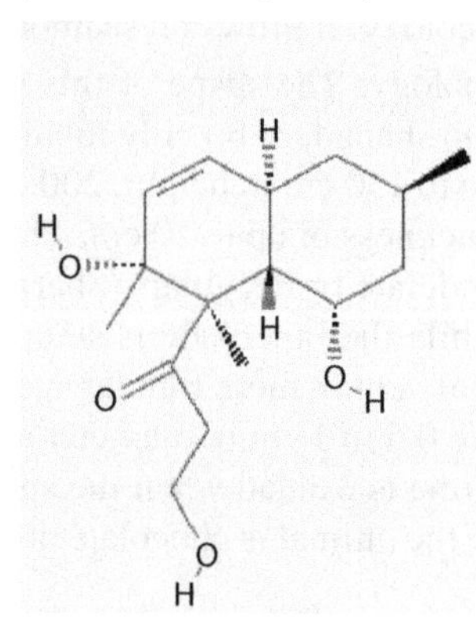

Decumbenone B

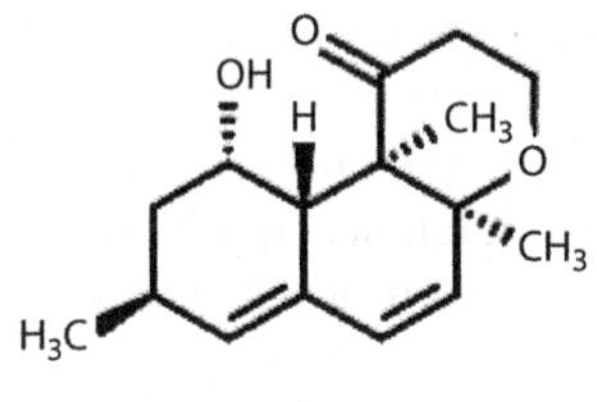

Versiol

The MeOH extract of this unidentified sponge yielded 12 compounds—7α-hydroxyaragusterol I (1), 14-hydroxyaragusterol I (2), 26,27-cyclo-24,27-dimethylcholest-5-ene-3β,7α,12β, 20β-tetraol (3), aragusterol I (4), B (5), and xestokerol B (6), 7-oxopetrosterol (7), 7α,12β,20β-trihydroxycholesta-22E-en-3-one (8), 26,27-cyclo-24,27-dimethylcholestan-3,3-dimethoxy-12β,21,20α,22α-tetraol (9), xestokerol A (10), aragusterol A (11), and aragusteroketals A (12)—which showed cytotoxicity with the following cell lines. The IC50 values recorded for the different cell lines are given next.

Cytotoxicity Data of Pure Compounds from the Marine Sponge *Petrosia* sp.

	Cell Lines (IC50, μM)						
Compound	**MOLT-3**	**HepG-2**	**A549**	**HuCCA-1**	**HeLa**	**MDA-MB-231**	**MRC-5**
1	17.9	12.7	20.0	21.3	ND	ND	40.1
2	32.4	44.8	38.6	37.9	ND	ND	60.6
3	ND	44.6	ND	47.5	ND	ND	ND
4	20.1	54.9	34.7	37.1	ND	ND	63.4
5	16.3	26.5	23.6	25.1	11.2	27.0	ND
6	ND	18.2	32.6	103.4	ND	ND	90.1
7	ND	49.8	52.2	ND	ND	ND	44.8
8	36.6	56.5	54.3	66.1	ND	ND	76.9
9	14.9	12.5	17.9	20.8	ND	ND	37.7
10	24.5	41.4	23.8	34.4	ND	ND	65.9
11	12.8	7.1	37.9	37.6	6.1	18.0	ND
12	14.6	11.3	29.1	34.6	ND	ND	49.6

Source: Pailee, P. et al., 2017, *Mar. Drugs*, 15: 54.
Note: ND, No data.

Nadeem et al. (2015) reported that the following compounds of this unidentified species possessed activity against MCF-7 and HepG2 cancer cell lines. The IC50 values are given next.

Anticancer Agent	**Cell Line**	**IC50 (μM)**
3-,7-,9-trihydroxycholest-5-en	MCF-7	153
3,7-dimethyl-2-(methylamino)-3H-purin-6(7H)-one	MCF-7	130
(7H)-one • (N-((3S,E)-1,3-dihydroxytetracos-4-en-2-yl)stearamide	MCF-7	23
3-,7-,9-trihydroxycholest-5-en	HepG2	230
3,7-dimethyl-2-(methylamino)-3H-purin-6(7H)-one	HepG2	145
N-((3S,E)-1,3-dihydroxytetracos-4-en-2-yl)stearamide	HepG2	20

(*Continued*)

Source: Nadeem, F. et al., 2015, *J. Mol. Biomark. Diagn.*, 7:267.

Antiplasmodial activity: The pentacyclic ingamine alkaloids 22(S)-hydroxyingamine A and dihydroingenamine D, together with the known compound ingamine of this sponge species showed antiplasmodial activity against chloroquine-sensitive (D6) and -resistant (W2) strains of *Plasmodium falciparum* (Perdicaris et al., 2013).

Neuroprotective activity/cholinesterases (ChEs) activity: The ethyl acetate extracts of this unidentified species containing the compound 2-bromoamphimedine showed AChE inhibition with an IC50 value of 300 μM. Similarly another species containg the compound petrosamin showed similar activity with an IC50 value of 91 μM (Grosso et al., 2014).

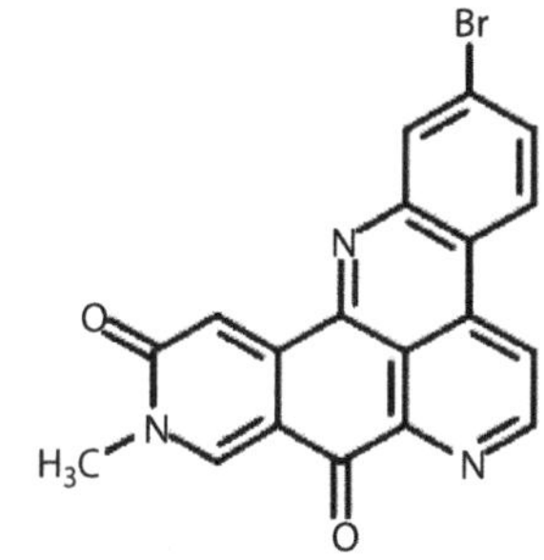

2-Bromoamphimedine

Negombata magnifica (Keller, 1889) (= *Latrunculia magnifica*)

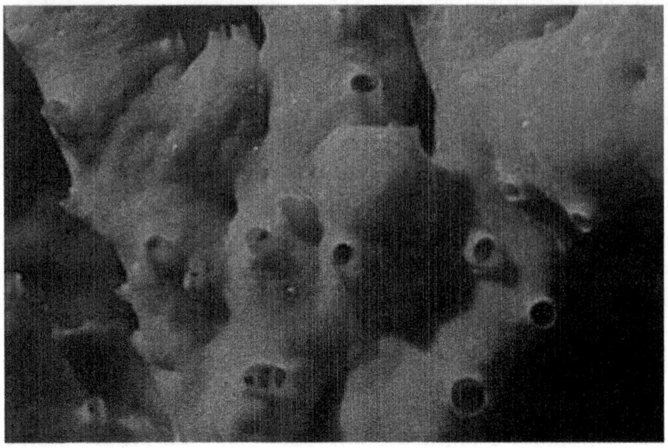

Order: Haplosclerida Family: Petrosiidae

Common name: Toxic finger-sponge, Red Sea sponge

Distribution: Red Sea, Indian Ocean, Sudan, northern Gulf of Aqaba

Ecology: It prefers to grow between corals and rocks, or under them. It lives on shallow coral reefs.

Biology: Its reddish-brown, narrow, crooked branches can grow up to 70 centimeters. The spicules of this species are megascleres styles in addition to strongyles oxeas. This species may be cultured in sea ranching systems for the production of its cytotoxic compound latrunculin B.

Compound(s) and Activities

Cytotoxicity: The two bioactive metabolites, latrunculin A and latrunculin B, of this species showed high cytotoxicity that exert various effects on cell cycle, such as inhibition of the microfilament-mediated process (Hadas et al., 2005).

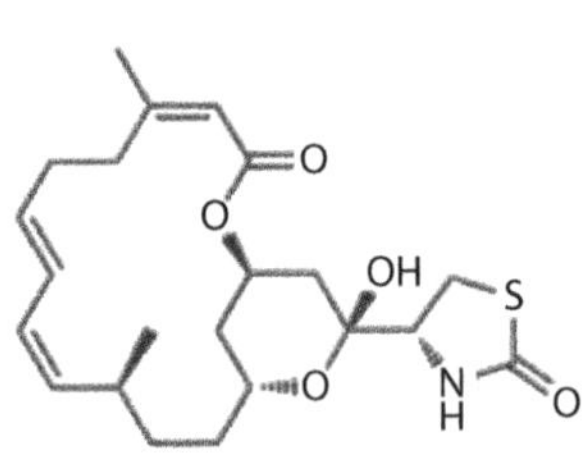

Latrunculin A

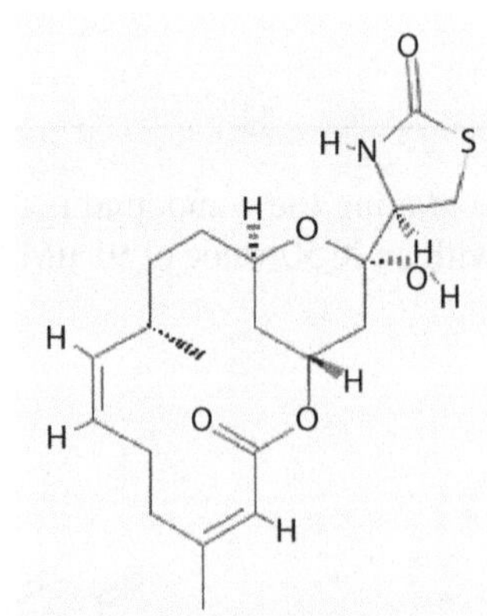

Latrunculin B

Antitumor activity: The thiazole macrolide latrunculin A isolated from this species showed antitumor activity by serving as a stabilizer of monomeric G-actin and polymerization inhibitor (Sipkema et al., 2005).

El-Damhougy et al. (2017) reported that the ethanolic extract of this species showed antitumor activity against the human colon cancer cells with an IC50 value of 1.1 µg/mL. The corresponding value for breast cancer cells (Mcf-7) was 0.4 µg/mL (El-Damhougy et al., 2017).

Antiproliferation activity: The organic extracts of this species showed antiproliferation activity against the human colon adenocarcinoma cell line (Caco-2) with an IC50 value of 5 µl/mL (Rady, 2014).

Sepčić et al. (2010) reported that the compounds latrunculins isolated from this species were found to be cytotoxic and inhibited the cancer cell proliferation with antitumoral activity.

Antiviral activity: The crude extract of this species showed cytotoxic effect on Vero cell with an MIC value of 0.3 µg/mL (El-Damhougy et al., 2017).

Cholinesterase inhibitor: The compound latrunculin A also serves as a cholinesterase inhibitor in the prevention of Parkinson's disease and dementia, which inhibit arousal, attention, learning, memory, and motivation (Nèeman et al., 1975).

BChE inhibitory activity: The hexane extract of this species possessed butyrylcholinesterase (BChE) inhibitory activity. People with no butyrylcholinesterase activity are healthy, fertile, and live to old age (Turk et al., 2013).

Neopetrosia carbonaria (Lamarck, 1814) (= *Xestospongia carbonaria*)

Order: Haplosclerida Family: Petrosiidae

Common name: Encrusting sponge

Distribution: Pacific Ocean and Western Central Atlantic, Caribbean wide

Ecology: This sessile sponge is very common between coral rubble, in shallow reefs, seagrass beds, and mangroves at depths of 1 to 10 m. It can tolerate brackish water conditions.

Biology: It is massive, lobate to repent branches with fistules that easily break off. It is black externally and internally. Smooth surface is scattered with membrane-bearing oscules (3–5 mm in dia.) that are located on top of volcano-shaped elevations. This sponge is brittle to pulpy in consistency.

Compound(s) and Activities

Antimicrobial activity: The CH2Cl2/MeOH extract of this species is a potent antimicrobial extract against the bacterial strains *Oceanobacillus iheyensis*, *Kocuria* sp., *Vibrio harveyi*, and *Bacillus megaterium* with more than 12 mm inhibition zone at 300 μg extract concentration (Qaralleh, 2016).

Aqueous and organic extracts of this species exhibited stronger antibacterial and antifungal activities. The highest activity was obtained for the aqueous extract against the Gram-positive bacteria *B. cereus* (inhibition zone 25 mm and MIC 0.07 mg/mL) and *S. aureus* (17.5 mm and 0.12 mg/mL) and against *C. albicans* (21 mm and 0.32 mg/mL) (Qaralleh, 2016).

The methanol extract of of this species indicated significant antifungal activity *in vitro* against *Candida albicans* (MIC = 7.8 μg/mL), *Cryptococcus neoformans* (MIC = 31.2 μg/mL), *Sporothrix schenckii* (MIC = 31.2 μg/mL), *Trichophyton mentagrophytes* (MIC = 31.2 μg/mL), *Aspergillus fumigatus* (MIC = 31.2 μg/mL), and *Candida parapsilosis* (MIC = 7.8 μg/mL) (Qaralleh, 2016).

The compound araguspongin C of this species showed promising antifungal activity against *Cryptococcus neoformans, Sporothrix schenckii, Trichophyton mentagrophytes,* and *Aspergillus fumigatus* with identical MIC of 50 μg/mL (Qaralleh, 2016).

Antibacterial and antitubercular activities: Halenaquinone, a pentacyclic polyketide isolated from the benzene extract of this species, was reported with antibacterial activity against *Staphylococcus aureus* and *Bacillus subtilis.* The compounds (+)-araguspongine C and neopetrosiamine A were reported with antituberculosis activity against a pathogenic strain of *Mycobacterium tuberculosis* (H37Rv) with MIC values of 3.9 μM and 7.5 μg/mL, respectively (Qaralleh, 2016).

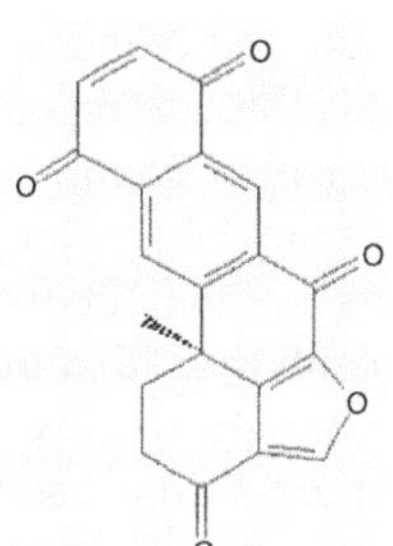
Halenaquinone

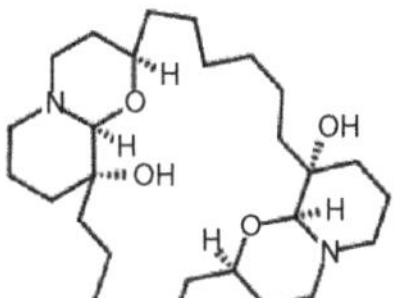
Araguspongine C

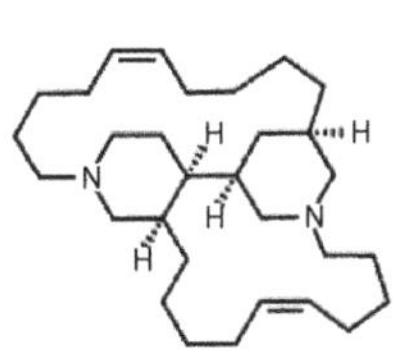
Neopetrosiamine A

Cytotoxic activity: The araguspongines M, B, and D and three 5α, 8α-epidioxy sterols of this species have shown cytotoxicity against the human leukemia cell line HL-60 with IC50 values of 5.5, 5.5, 5.9, 22.4, 9.5, and 9.6 μM, respectively (Qaralleh, 2016).

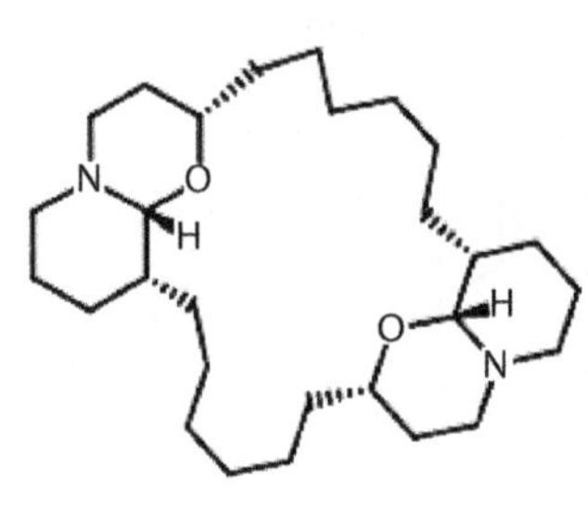
Araguspongine B

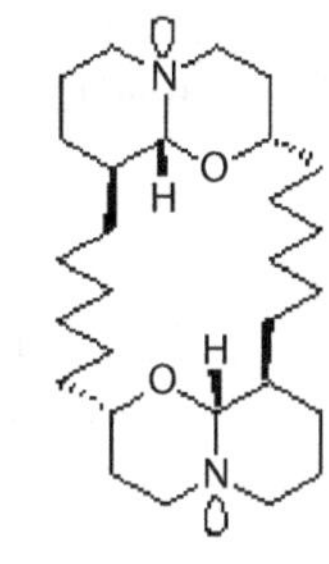
Araguspongine D

The compound demethylxestopongin B isolated from this sponge showed a potent cytotoxic activity against the human tumor cell line A-549 with inhibition ratio of 94.3% at 10 μM (Qaralleh, 2016).

Cytotoxic activity: The pyridoacridine alkaloid amphimedine isolated from this species exhibited potent cytotoxic activity at 30 μM (Qaralleh, 2016).

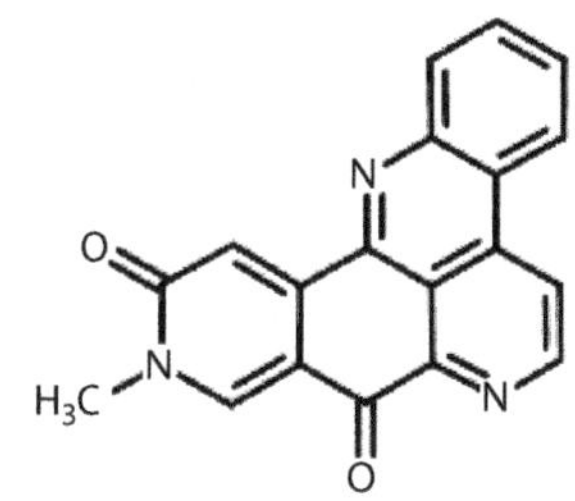
Amphimedine

Antitumor activity: The pyridoacridine alkaloid neoamphimedine of this species showed antitumor activity through the inhibition of the enzyme topoisomerase II (Sipkema et al., 2005).

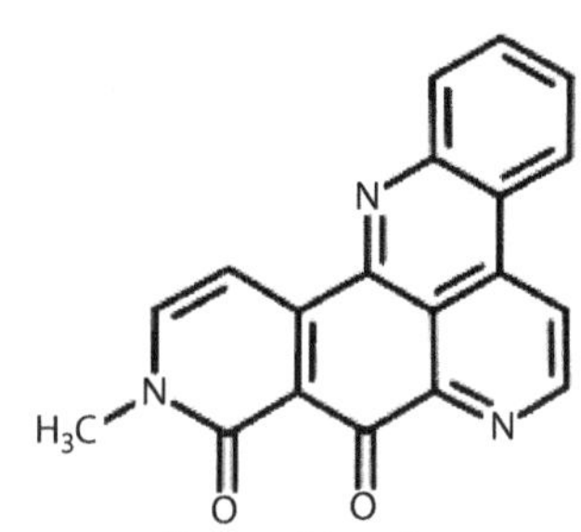
Neoamphimedine

Enzymes inhibiting activity: The compounds halenaquinone and 14-methoxyhalenaquinone have been reported as potent protein tyrosine kinase (PTK) inhibitors (Qaralleh, 2016).

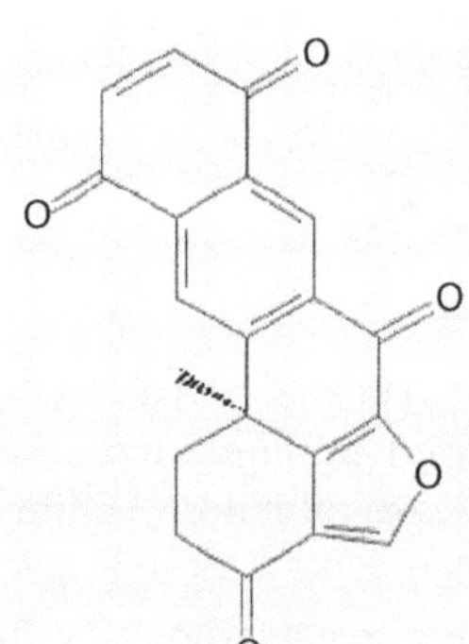
Halenaquinone

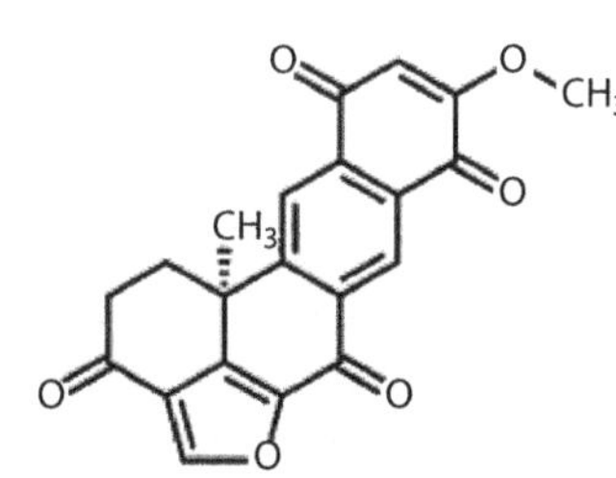
14-methoxyhalenaquinone

Cardioactivity: The compound halenaquinol of this species possesses cardioactivity (Qaralleh, 2016).

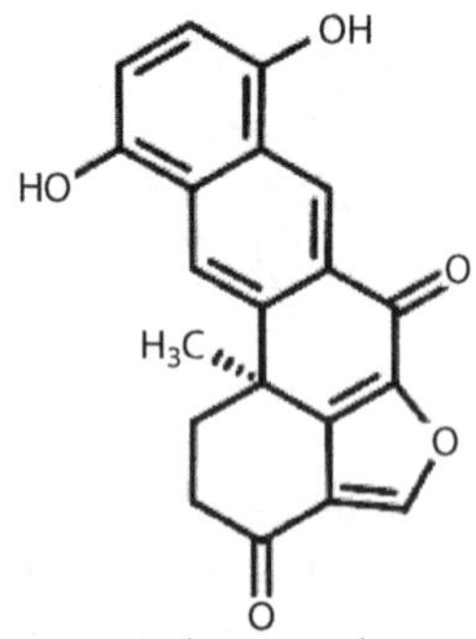

Halenaquinol

Inhibitor of osteoclastogenic differentiation: Its compound halenaquinone was found to be an inhibitor of osteoclastogenic differentiation of murine (Qaralleh, 2016).

Neopetrosia contignata (Thiele, 1899) (= *Petrosia contignata*)

Order: Haplosclerida Family: Petrosiidae

Common name: Not known

Distribution: Indo Polynesian province; Indonesia, Sulawesi Sea/Makassar Strait

Ecology: It occurs in seamounts and knolls.

Biology: Not reported

Compound(s) and Activities

Anti-inflammatory activity: The steroid contignasterol isolated from this species showed anti-inflammatory activity by inhibiting the anti-immunoglobulin E (Keyzers, 2005).

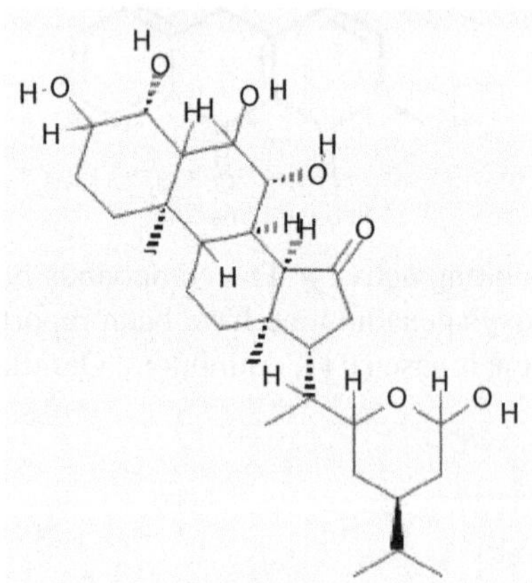

Contignasterol

Immunosuppressive activity: The oxygenated sterol contignasterol of this species also possessed immunosuppressive activity by serving as a histamine-release inhibitor (Anjum et al., 2016; Sipkema et al., 2005).

Cytotoxic activity: The extracts of this species showed cytotoxicity against tumor cell lines HT-29, T47D, and Casky with IC50 values of 78.9, 35.6, and 36.2 μg/mL, respectively (Qaralleh, 2016).

Antioxidant activity: The hydroethanolic extract of this species exhibited moderate DPPH antioxidant activity with IC50 values less than 100 μg/mL (Abdillah et al., 2013; Qaralleh, 2016).

Neopetrosia exigua (Kirkpatrick, 1900) (= *Xestospongia exigua*)

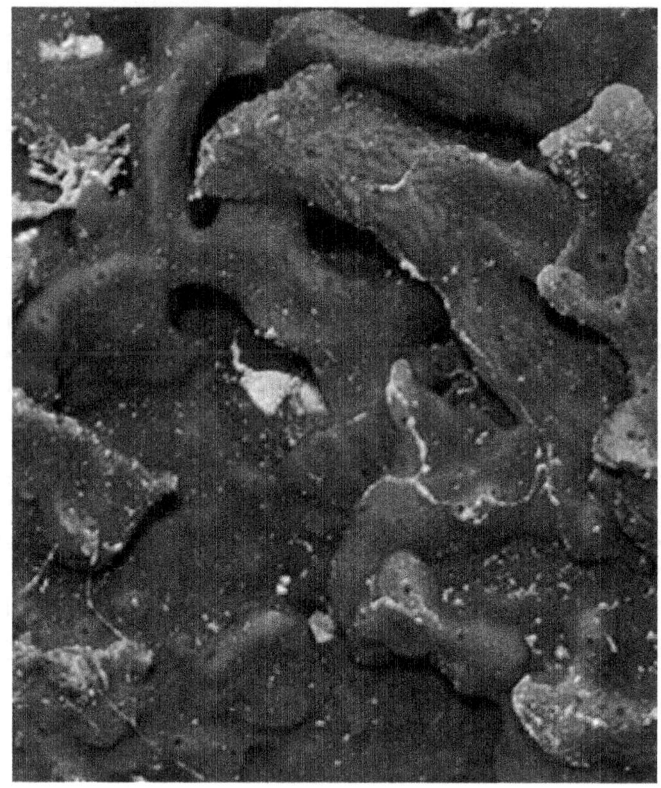

Order: Haplosclerida Family: Petrosiidae

Common name: Indonesian sponge, Thai sponge, Hainan sponge

Distribution: Indo-Pacific; Micronesia

Ecology: This sessile sponge occurs in shallow lagoons and fringing reefs; depth range 0 to 100 m.

Biology: It is an erect sponge with numerous slender branches, which are sometimes anastomose. Specimens that are repent with extremely irregular stout branches are also common. Surface of the sponge is characterized by a reticulate pattern formed by the ascending endosomal columns that fan out distally and support the aspiculous dermal membrane. Texture of the sponge is soft. The skeleton is basically an isodictyal reticulation with unispicular to multispicular ascending tracts developed to varying degrees.

Compound(s) and Activities

Antimicrobial activity: The aqueous and organic extracts from this species exhibited stronger antibacterial and antifungal activities. The highest activity was obtained for the aqueous extract against the Gram-positive bacteria *B. cereus* (inhibition zone, 25 mm and MIC, 0.07 mg/mL) and *S. aureus*

(17.5 mm and 0.12 mg/mL) and the fungus *C. albicans* (21 mm and 0.32 mg/mL) (Majali et al., 2015; Qaralleh et al., 2010). Its methanol extract indicated significant antifungal activity *in vitro* against *Candida albicans* (MIC = 7.8 μg/mL), *Cryptococcus neoformans* (MIC = 31.2 μg/mL), *Sporothrix schenckii* (MIC = 31.2 μg/mL), *Trichophyton mentagrophytes* (MIC = 31.2 μg/mL), *Aspergillus fumigatus* (MIC = 31.2 μg/mL), and *Candida parapsilosis* (MIC = 7.8 μg/mL) (Lakshmi et al., 2010). The extract provided one active compound, araguspongin C, that showed promising activity against the fungi *Cryptococcus neoformans, Sporothrix schenckii, Trichophyton mentagrophytes,* and *Aspergillus fumigatus* with an MIC value of 50 μg/mL (Qaralleh, 2016). The extract also showed moderate antibacterial activity against the fouling bacterial strains *Bacillus cereus, B. pumilus, B. megaterium, Pseudoalteromonas haloplanktis, Pseudomonas chlororaphis, P. putida,* and *P. aeruginosa* (Qaralleh, 2016).

The fungal isolate *Penicillium* cf. *montanens* derived from this sponge yielded three novel decalactone metabolites: xestodecalactones A, B, and C. Among these, xestodecalactone B showed antifungal activity against *Candida albicans.* Further, the compound aspergillitine isolated from its fungus, namely, *Aspergillus versicolor,* showed moderate antibacterial activity against *Bacillus subtilis* (Thomas et al., 2010).

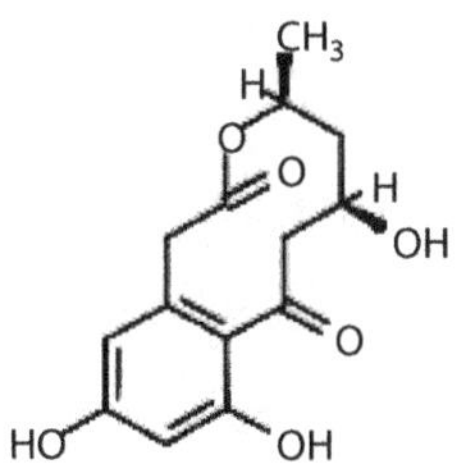

Xestodecalactone B

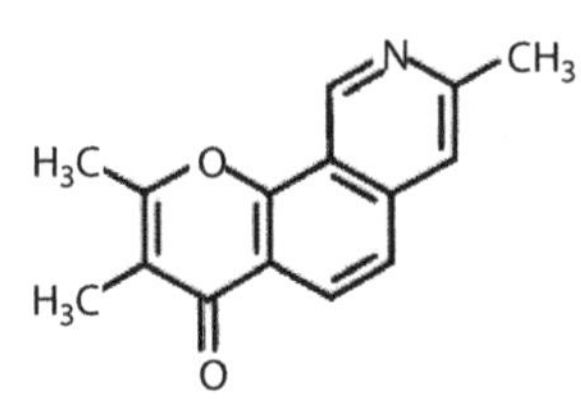

Aspergillitine

Antibacterial activity: The extract of this sponge species showed antibacterial activity against the following values of zone of inhibition: *E. coli*, 10.75 mm, and *P. aeruginosa* MDR, 8.0 mm (Cita et al., 2017).

Antituberculosis and antibacterial activities: The pentacyclic polyketide, halenaquinone, isolated from the benzene extract of this species was reported with antibacterial activity against *Staphylococcus aureus* and *Bacillus subtilis* (Roll et al., 1983). The compound (+)-araguspongine C was reported with antituberculosis activity with an MIC value of 3.9 μM (Orabi et al., 2002). The *in vitro* antituberculosis activity (MIC value, 7.5 μg/mL) against a pathogenic strain of *Mycobacterium tuberculosis* (H37Rv) was found to be due to its other compound, neopetrosiamine A (Weiet al., 2010). A pentacyclic hydroquinone exiguaquinol of this species inhibited *Helicobacter pylori* glutamate racemase (MurI) with an IC50 of 4.4 μM (Qaralleh, 2016).

Anti-TB/anticancer activity: The bis-1-oxaquinolizidine alkaloids, (+)-araguspongine C of this species possessed anti-TB activity with an MIC value of 3.9μg/mL. Further, these compounds induced autophagic death in breast cancer cells and inhibited the growth of *Mycobacterium tuberculosis strain H37Rv* (MIC 1.9 μg/mL) (Asif, 2017; Jyothi et al., 2016).

Antimalarial activity: The ethanol soluble extracts of this species with doses of 400 and 200 mg/kg showed suppression of growth activity against *Plasmodium berghei* by 80.7% and 60.6%, respectively. Further, the compound neopetrosamine A (26) of this species has been reported with antiplasmodial activity against *Plasmodium falciparum* with an IC50 value of 2.3 μM. Furthermore, the compound araguspongine C of this species also exhibited *in vitro* antimalarial activity against *Plasmodium falciparum* with IC50 values ranged from 280 to 670 ng/mL (Qaralleh, 2016).

Antiprotozoal activity: The compound renieramycin A of this species exhibited antiprotozoal activity against *Leishmania amazonensis* with IC50 = 0.2 μg/mL and cytotoxicity with IC50 = 2.2 μg/mL (Qaralleh, 2016).

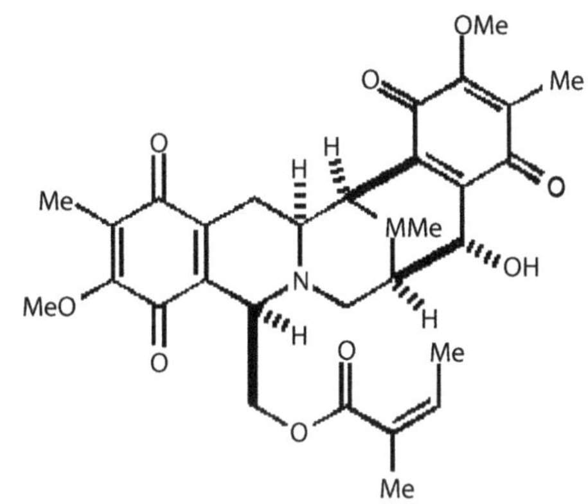

Renieramycin A

The ethanol soluble extracts of this species with doses of 400 and 200 mg/kg showed suppression of growth activity against *Plasmodium berghei* by 80.7% and 60.6%, respectively (Qaralleh, 2016).

Anticancer activity: Jha and Zi-rong (2004) reported that its compounds, motuporins D–F, inhibited the invasion of breast carcinoma cells into new tissues.

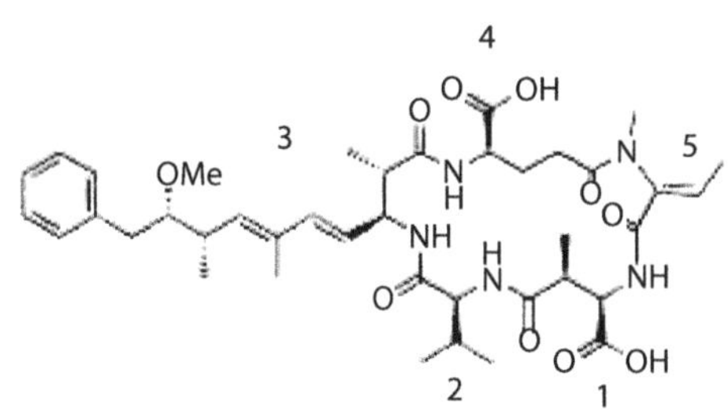

Motuporin

Qaralleh (2016) also reported that the compound halenaquinone of this species was found to exhibit anticancer activity through apoptosis.

Cytotoxic activity: The compounds araguspongine M and araguspongines B and D, and three 5α,8α-epidioxy sterols showed cytotoxicity against the human leukemia cell line HL-60 with IC50 values of 5.5, 5.5, 5.9, 22.4, 9.5, and 9.6 μM, respectively (Qaralleh, 2016).

The compound demethylxestopongin B isolated from this sponge was a potent cytotoxic compound against the human tumor cell line A-549 with inhibition ratio of 94.3% at an extract concentration of 10 μM (Qaralleh, 2016).

Anti-invasion activity: The compounds motuporamines A, B, and C and the mixture of G, H, and I of this species exhibited anti-invasion activity. Roskelley and his colleagues (2001) showed that the compound motuporamine C interfered with the migration of human breast carcinoma, prostate carcinoma, and glioma cells in culture and inhibited angiogenesis (Qaralleh, 2016).

Enzymes inhibiting activity: The compound exiguamine A of this species has been found to be one of the most potent inhibitors of indoleamine-2,3-dioxygenase (IDO) *in vitro*. IDO inhibition can delay tumor growth (Qaralleh, 2016).

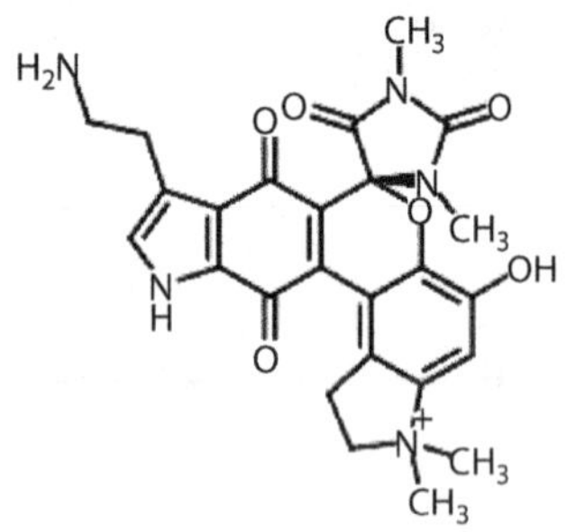

Exiguamine A

The compound araguspongines A and C of this species showed an ability to inhibit rat brain nitric oxide synthase activity *in vitro* with estimated IC50 values of 31.5 and 46.5 mM, respectively. The compounds along with xestospongin B were also reported as potent inhibitors for inositol 1, 4, 5- triphosphate receptor mediated Ca2+ release and endoplasmic reticulum-calcium pump (Qaralleh, 2016).

Antioxidant activity: The hydroethanolic extract of this species exhibited moderate DPPH antioxidant activity with IC50 values of less than 100 µg/mL (Qaralleh, 2016).

Antiamebic activity: The n-hexane and n-butanol soluble fractions of the methanol-chloroform extract of this species showed antiamebic activity with 80% inhibition of trophozoites of *Entameba histolytica* at 900 mg/kg (body weight) dose. On the other hand, the methanol and methanol-chloroform (1:1) extracts of the freshly collected sponge showed MIC values of 125 µg/mL and 250 µg/mL, respectively (Lakshmi et al., 2009).

Vasodilating activity: The compounds xestospongin A–D were found to be active as vasodilator compounds, which induce relaxation of blood vessel *in vivo* (Qaralleh, 2016).

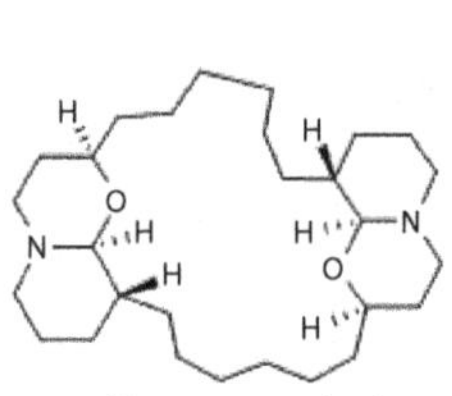

Xestospongin A

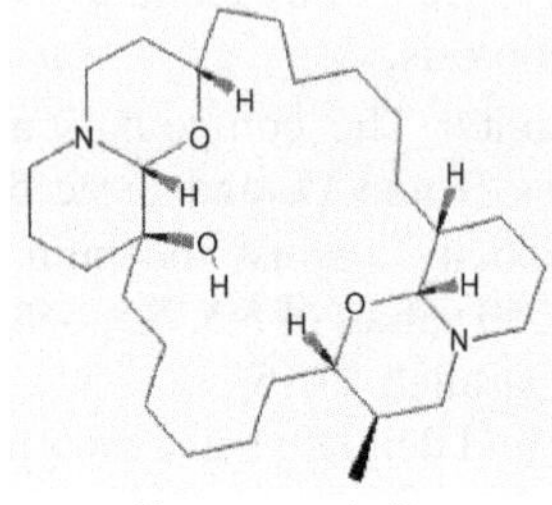

Xestospongin B

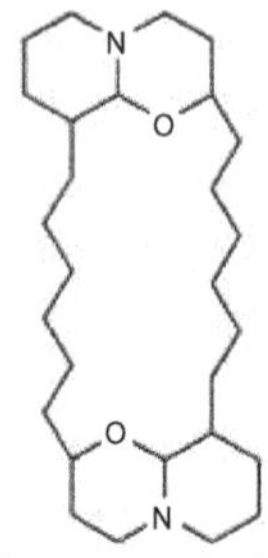

Xestospongin C

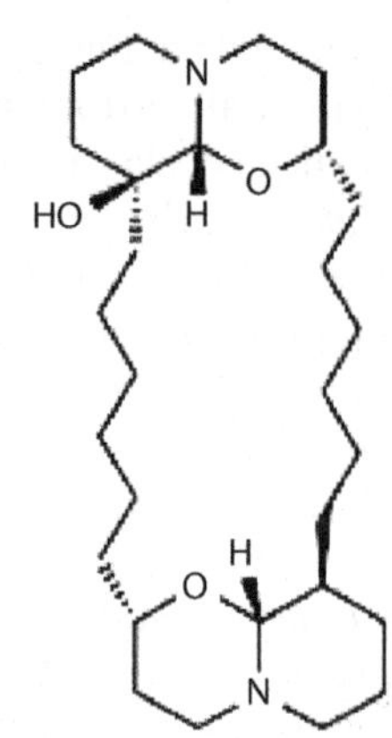

Xestospongin D

Neopetrosia proxima (Duchassaing and Michelotti, 1864)

Order: Haplosclerida Family: Petrosiidae

Common name: None

Distribution: Tropical and subtropical waters of the western Atlantic Ocean

Ecology: This sponge is common on sand and hard bottoms of reef; and on coral rubble, seagrass beds.

Biology: Thes sponges are externally brown, purplish brown, or maroon in coloration, while internally it is light tan. They form thick spreading masses on the substrate (usually coral rubble). The masses are highly variable in shape. They are lobated, with regular to irregularly shaped lobes typically 3 to 5 cm tall. Each lobe has one opening (oscule) of about 2 to 5 mm in diameter and is either located flush on the surface or elevated in small chimneys. The surface is generally smooth or somewhat wrinkled (rugose). The consistency is hard but not brittle. The spicules are composed solely of strongyles, cylindrical in shape with rounded ends.

Association: This sponge serves as a host to symbiotic cyanobacteria, as well as colonies of the eusocial snapping shrimp of the genus *Synalpheus*.

Compound(s) and Activities

Anticancer/antiprolifeartion activity: The compounds neopetrosiquinones A (1) and B (2) isolated from this species showed antiproliferation activity against DLD-1 human colorectal adenocarcinoma cell line with IC50 values of 3.7 and 9.8 μM, respectively, and the PANC-1 human pancreatic carcinoma cell line with IC50 values of 6.1 and 13.8 μM, respectively. Neopetrosiquinone A (1) also inhibited the *in vitro* proliferation of the AsPC-1 human pancreatic carcinoma cell line with an IC50 value of 6.1 μM (Qaralleh, 2016).

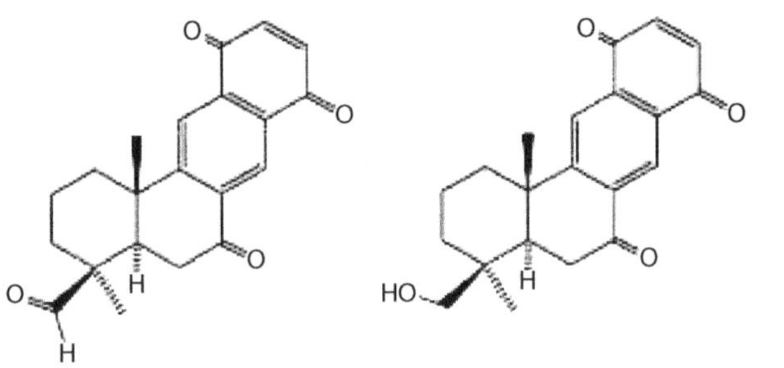

Neopetrosiquinones A and B

The compound neopetrosiamine A of this species exhibited strong inhibitory activity against MALME-3M melanoma cancer, CCRFCEM leukemia, and MCF7 breast cancer with IC values of 1.5, 2.0, and 3.5 μM, respectively (Qaralleh, 2016).

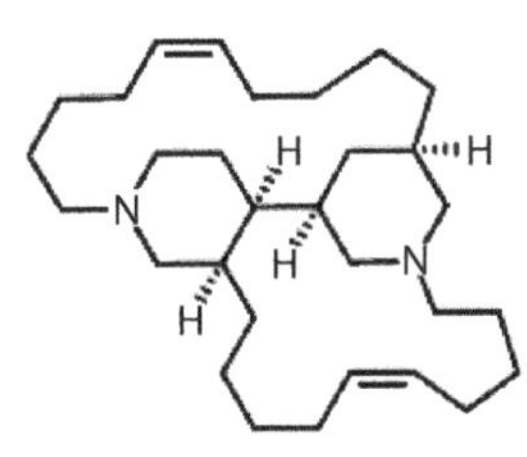

Neopetrosiamine A

Anti-inflammatory activity: The methanolic extract of this species exhibited *in vitro* and *in vivo* anti-inflammatory activities. At a concentration of 100 mg/Kg, the total extract significantly inhibited the paw edema of rats (60%). Further, the deichloromethane and methanol fractions of this species reduced the proinflammatory myeloperoxidase (MPO) enzyme activity (Qaralleh, 2016).

Antimicrobial activity: The organic extract obtained from this species showed *in vitro* antibacterial activity against the Gram-positive *Staphylococcus aureus* and *Streptococcus faecalis*, and antifungal activity against *Candida albicans* (Qaralleh, 2016).

Neopetrosia rosariensis (Zea and Rützler, 1983)

Order: Haplosclerida Family: Petrosiidae

Common name: Not designated

Distribution: Central Atlantic, Panama, Canary Islands, and Colombia; Eastern Caribbean, Greater Antilles, Southwestern Caribbean

Ecology: It is common in patch reef, rock substrate and littoral rock, sandy areas with small corals, and sometimes in the seagrass beds of the genus *Thalassia*. The depth range is 2 to 30 m.

Biology: This sponge appears as cylindrical tubes of 4 to 78 cm high and 2 to 13 cm wide. They emerge from basal mass. Coloration of the sponge is dark reddish brown externally and tan internally. Texture is hard and dense. Surface is smooth to the touch, with rugose and uneven parts. Sometimes it is covered by the cnidarians (coelenterate) zooanthids. Oscules are present on the inner wall of tubes and are of 3.5 cm (max.) wide.

Compound(s) and Activities

Anti-inflammatory activity: The total extracts of this species significantly inhibited the paw edema of rats (72%) at a concentration of 100 mg/Kg. Further, the dichloromethane and methanol fractions of this species reduced myeloperoxidase activity (MPO). On the other hand, the dichloromethane fraction significantly inhibited nitric oxide (NO) (66%), prostaglandin E2 (PGE2) (30.5%), and tumor necrosis factor alpha (TNF-alpha) production (72%) (Franco et al., 2012; Qaralleh, 2016).

The compound clionasterol isolated from this species exhibited potent anticomplementary inhibitor with an IC50 value of 4.1 μM (Qaralleh, 2016).

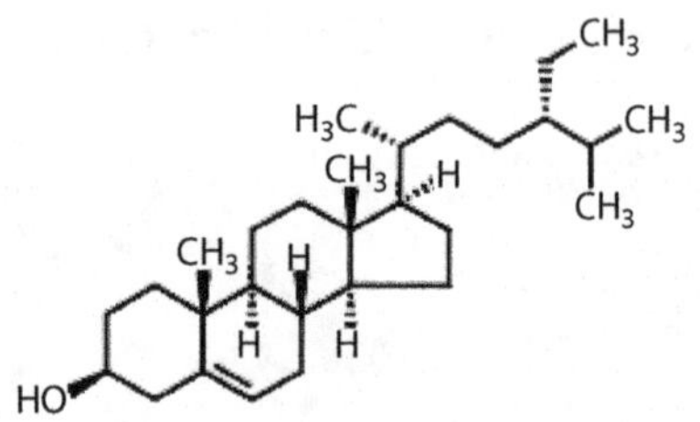

Clionasterol

Neopetrosia sapra (de Laubenfels, 1954)

Order: Haplosclerida Family: Petrosiidae

Common name: Okinawan sea sponge

Distribution: Micronesia

Ecology: It occurs in deepwater habitats.

Biology: It is a massive sponge with a maximum size of 30 cm long, 9 cm diameter, and 6 cm height. Color of the sponge in life is blackish in exterior and olive drab in interior. Consistency is very soft. Surface is slimy and finely cornulose or rugose. Oscula are of 2 mm diameter. There is no ectosome but black pigments are seen. Skeleton of this species consists of oxeas.

Compound(s) and Activities

Cardiotonic activity: The compound xestosaprol C isolated from this species has been reported with cardiotonic activity (Qaralleh, 2016).

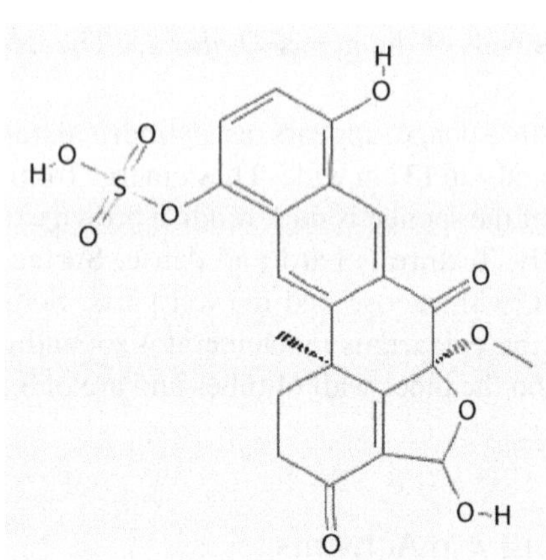

Xestosaprol C

Kobayashi (1991) and Winder (2009) reported on the isolation of a novel cardiotonic agent, xestoquinone, from this species. This compound produced dose-dependent cardiotonic effects on guinea pig left and right atria.

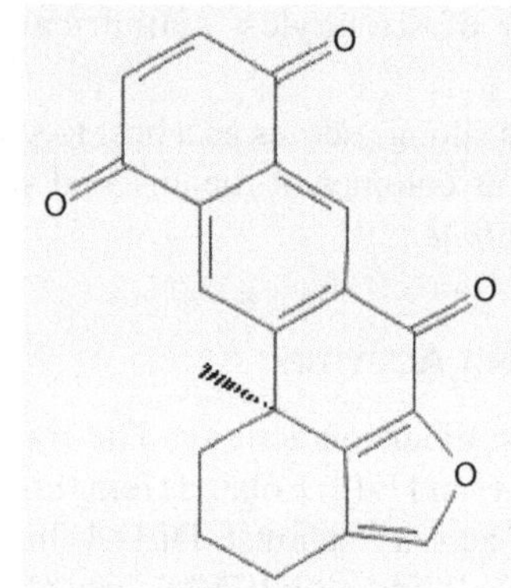

Xestoquinone

Neopetrosia seriata (Hentschel, 1912) (= *Petrosia seriata*)

Order: Haplosclerida Family: Petrosiidae

Common name: Not known

Distribution: Indo-Australian waters; Arafura Sea, Southern Vietnam, Western and Northern Madagascar

Ecology: This sessile sponge is attached to substrata by many points in the subtidal habitats.

Biology: This species has a clathrous appearance with a size of 9 × 3 cm. Color of the sponge is dark brown. Consistency is tough and incompressible. Surface is uniform, but microscopically hispid. Oscules are numerous and are arranged serially in all bits, which have ridges on surface. In certain specimens, oscules (2–4 mm dia.) are found scattered irregularly on hillock-like projections. Pores are minute and they have an average diameter of 0.5 mm. Dermal skeleton is ill-defined at places. Spicules are usually placed horizontally, but they may also be arranged vertically giving a microscopic hispidity to surface. Oxeas of the spicules are uniformly curved and gradually pointed.

Compound(s) and Activities

Anti-HIV activity: Two bis-quinolizidine alkaloids, namely, petrosin and petrosin-A isolated from this species were reported as anti-HIV inhibitors with IC50 values of 41.3 and 52.9 μm, respectively (Goud et al., 2003).

Enzymes inhibiting activity: The compound, halenaquinol isolated from this species was reported as inhibitor for rat brain cortex Na+/K+-ATPase with an I50 value of 1.3×10^{-6} M or 325 nmol per mg of protein (Qaralleh, 2016).

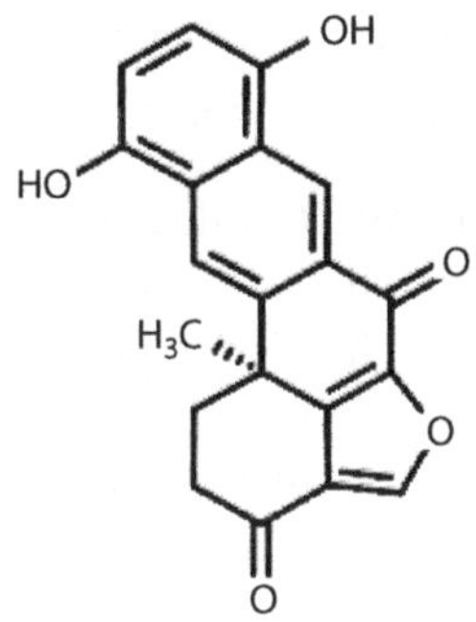

Halenaquinol

- Cardiotonic activity: The compound halenaquinol of this species was reported with cardiotonic activity (Qaralleh, 2016).

Neopetrosia similis (Ridley and Dendy, 1886) (= *Petrosia similis*)

Order: Haplosclerida Family: Petrosiidae

Common name: Not designated

Distribution: Agulhas Bank, Eastern Philippines, Philippines exclusive economic zone, South African exclusive economic zone, South India, Sri Lanka

Ecology: It dwells on reef slopes, from shallow waters down to 40 m.

Biology: Sponge of this species is variable in its external form, that is, repent and branching. Branches are anastomosing, subangular or cylindrical, with large circular oscula (6 mm in dia.) arranged in series on the upper surface. When the branches are erect and lobate, the large circular oscula are present as pits at the summits of the lobes. Pores and round openings are few in the dermal membrane. Size of the specimen measures 9 cm in length by about 1 cm in diameter, and bears nine large circular oscula arranged in a single row along its upper surface. Color of the sponge is yellowish gray in spirit. Texture of the lobate specimen is firm, incompressible, and fibrous. Repent specimens are easily compressible and fibrous but brittle. Surface is smooth but uneven, often with a reticulate appearance.

Compound(s) and Activities

Anti-HIV activity: The bisquinolizidine alkaloids petrosin and petrosin A isolated from this species showed anti-HIV inhibition with IC(50) values of 41.3 and 52.9 μM, respectively. Further, its compounds 1,2-dihydroisoquinoline and isoquinolinequinone have also been reported to possess HIV-1 integrase inhibiting activity and anti-HIV activity, respectively (Goud et al., 2003; Senthilkumar and Kim, 2013).

Neopetrosia vanilla (de Laubenfels, 1930) (= *Xestospongia vanilla*)

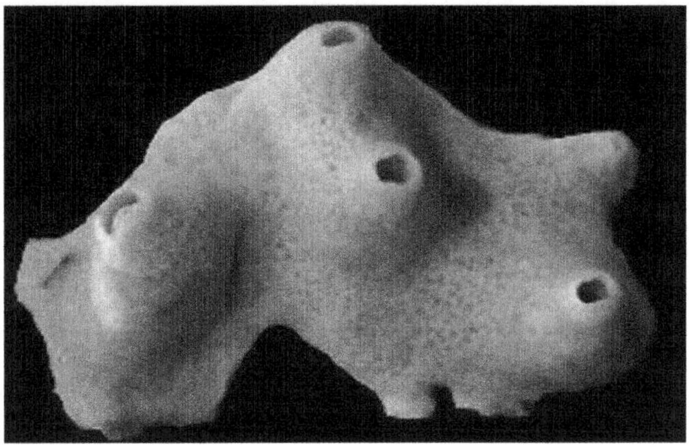

Order: Haplosclerida Family: Petrosiidae

Common name: Northeastern Pacific marine sponge

Distribution: Northeastern Pacific, California

Ecology: It inhabits low to very low intertidal areas of the coral reefs.

Biology: It forms a thin, smooth encrustation that looks like a cake icing. It has a thickness of 5 to 15 mm. Oscules are up to 2 mm across. Color of the sponge is buff, tan, or dull orange brown.

Compound(s) and Activities

Antifungal activity: Among the triterpenoid glycosides xestovanin A and secoxestovanin A isolated from this species, xestovanin A showed antifungal activity against *Pythium ultimum* (Qaralleh, 2016).

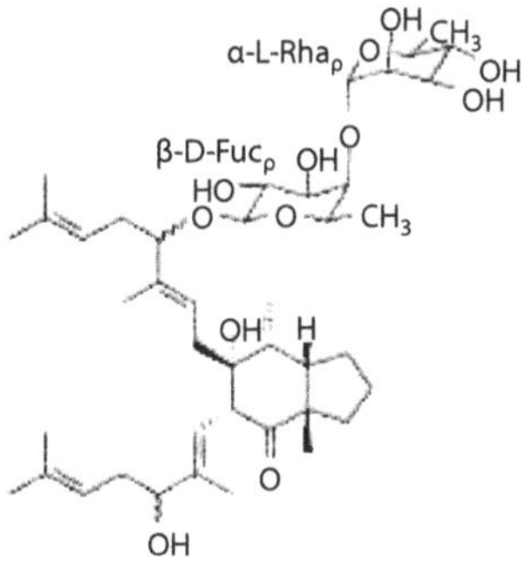

Xestovanin A

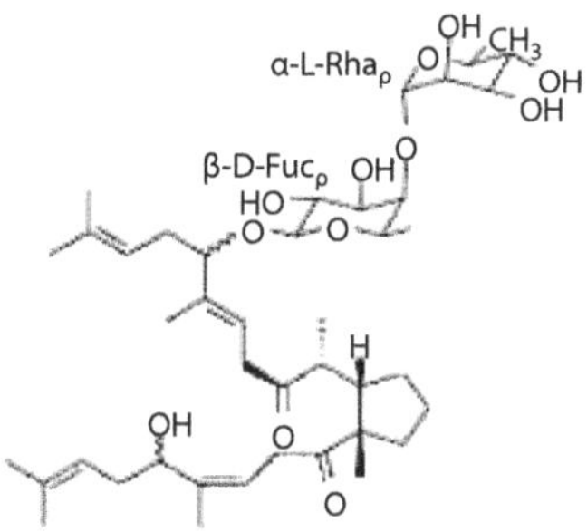

Secoxestovanin A

Neopetrosia zumi (Ristau, 1978) (= *Toxadocia zumi*)

Order: Haplosclerida Family: Petrosiidae

Common name: Not designated

Distribution: California

Ecology: It is seen in semicryptic environments and in exposed areas; depth range 10–20 m.

Biology: It is an erectly, dichotomously branching or lobate (palmate or fan shaped) sponge. The latter is the size of 17 (width) × 5 (thickness) × 18 cm (height). The branching type may have a height of 6.5 cm and the diameter of the branch is 2 cm. This sponge is firm and rigid, and coral-like due to the ectosomal skeleton. Surface is hispid. The spicule bundles are erect on a well-developed tangential layer of oxea. Oscula (10–15 mm dia.) are spaced 5–10 mm apart over the entire surface. Spicules consist of megascleres of oxea.

Compound(s) and Activities

Antibacterial activity: The sterols from this species inhibited the growth of two species of bacteria namely *Staphylococcus aureus* and *Bacillus subtilis* (Datta et al., 2015).

Others: The sterol sulfates of this species possess antimicrobial, cytotoxic, ichthyotoxic, and larvicidal activities (Atta-ur-Rahman, 2017).

Neopetrosia sp.

Order: Haplosclerida Family: Petrosiidae

Compound(s) and Activities

Antimetastatic activity: The compounds neopetrosiamdes A and B isolated from this unidentified sponge have been reported as potential antimetastatic agents that inhibit tumor cell invasion by both amoeboid and mesenchymal migration pathways (Qaralleh, 2016).

Cytotoxicity: The compound renieramycin A of this species exhibited cytotoxicity with IC50 = 2.2 μg/mL. Further its compound renieramycin J was reported with cytotoxic activity against 3Y1, HeLa, and P388 cells with IC50 of 5.3, 12.3, and 0.53 nM, respectively (Qaralleh, 2016).

Anticancer activity: The compound halenaquinone was found to exhibit anticancer activity through apoptosis. This activity was due to the inhibition of phosphatidylinositol 3-kinase activity (Qaralleh, 2016).

Two diastereomeric tricyclic peptides neopetrosiamdes A and B of this species inhibited amoeboid invasion of human tumor cells at 6 μg/mL (Agrawal et al., 2016).

Antiproliferation activity: The compounds neopetrosiquinones A and B of this sponge were found active against the DLD-1 human colorectal adenocarcinoma cell line with IC50 values of 3.7 and 9.8 μM, respectively, and the PANC-1 human pancreatic carcinoma cell line with IC50 values of 6.1 and 13.8 μM, respectively. Neopetrosiquinone A also inhibited the *in vitro* proliferation of the AsPC-1 human pancreatic carcinoma cell line with an IC50 value of 6.1 μM. Neopetrosiamine A of this species exhibited strong inhibitory activity against MALME-3M melanoma cancer, CCRFCEM leukemia, and MCF7 breast cancer with IC values of 1.5, 2.0, and 3.5 μM, respectively (Qaralleh, 2016).

Anti-invasion activity: The compounds motuporamines A, B, and C and the mixture of G, H, and I exhibited anti-invasion activity. The compound motuporamine C also interfered with the migration of human breast carcinoma, prostate carcinoma, and glioma cells in culture and inhibited angiogenesis. Neopetrosiamdes A and B of this sponge were reported as potential antimetastatic agents that inhibit tumor cell invasion (Qaralleh, 2016).

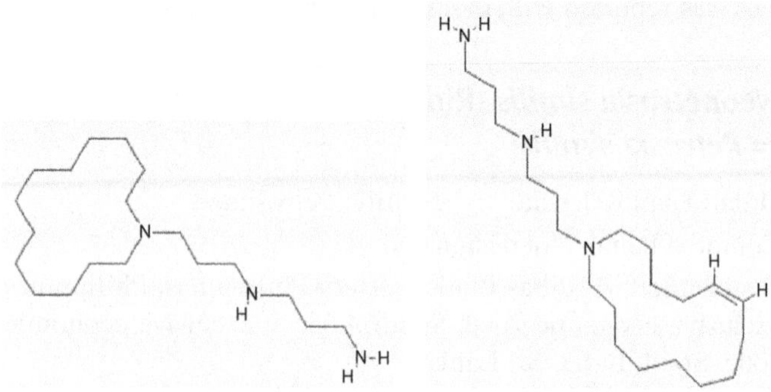

Motuporamine B Motuporamine C

Enzyme inhibiting activity: The compounds araguspongines A and C of this sponge showed an ability to inhibit rat brain nitric oxide synthase activity *in vitro* with an estimated IC50 of 32 and 47 mM, respectively. These compounds along with xestospongin B were reported as potent inhibitors for inositol 1,4,5-triphosphate receptor mediated Ca2+ release and endoplasmic reticulum-calcium pump (Qaralleh, 2016).

The compound exiguamine A of this sponge has been found to be the most potent inhibitor of indoleamine-2,3-dioxygenase (IDO) *in vitro*. IDO inhibition can delay tumor growth (Qaralleh, 2016).

The compounds halenaquinone and 14-methoxyhalenaquinone were reported as potent protein tyrosine kinase (PTK) inhibitors with IC50 values (Qaralleh, 2016).

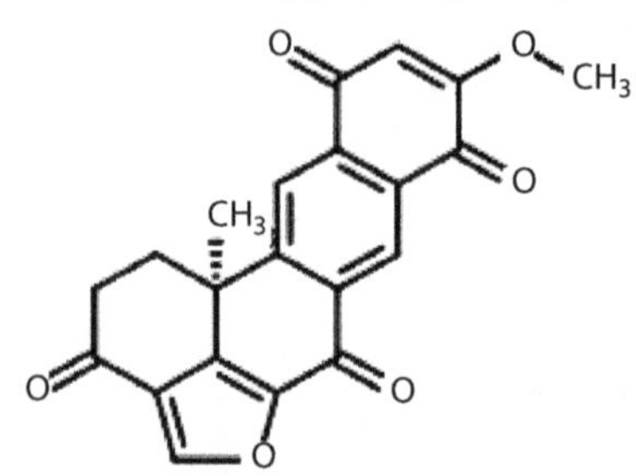

14-methoxyhalenaquinone

Antiprotozoal and cytotoxic activities: The compound neopetrosamine A was reported with antiplasmodial activity against *Plasmodium falciparum* with an IC50 value of 2.3 μM. The compound araguspongine C exhibited *in vitro* antimalarial activity against *Plasmodium falciparum* with IC50 ranged from 280 to 670 ng/mL. The compound renieramycin

A exhibited antiprotozoal activity against *Leishmania amazonensis* with IC50 = 0.2 μg/mL and cytotoxicity with IC50 = 2.2 μg/mL (Qaralleh, 2016).

Vasodilating activity: The compounds xestospongin A, B, C, and D of this sponge were found to be active as vasodilator compounds since they induced relaxation of blood vessels *in vivo* (Qaralleh, 2016).

Xestospongia bergquistia (Fromont, 1991)

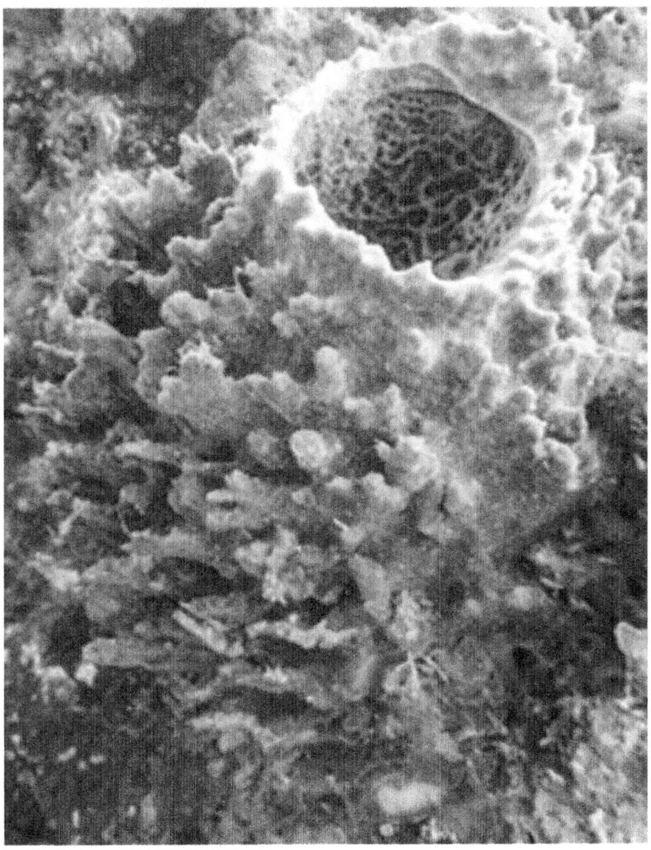

Order: Haplosclerida Family: Petrosiidae

Common name: Not designated

Distribution: Western Central Pacific, New Caledonia.

Ecology: It is found on dead corals and rock substrata in areas of high sediment load and in caves. The depth range is 0 to 100 m.

Biology: It is an erect, cup-shaped sponge with vertical flukes or ridges on its outer surface. These ridges may be flatter and form rounded knobs. Size of the specimens may range from 5 cm to 1 m with the apical cup forming a central hollow up to one-third of the overall height of the sponge. Color of the sponge may vary from maroon to red brown in life and fawn in spirit. Texture is firm and compressible. Exterior surface is smooth. Surface skeleton is an isotropic isodatyal reticulation of spicule tracts forming small meshes. Main spcules are extremely variable in shape and size. They are usually gently curved. Some are centrally bent while others are undulating.

Compound(s) and Activities

Anti-inflammatory activity: The 14b sterols, xestobergsterols A (51) and B (52), of this species showed anti-inflammatory activity (IC50 values, 0.05 μM and 0.10 μM, respectively) by significantly inhibiting histamine release from rat peritoneal mast cells induced by anti-IgE (Keyzers, 2005; Sipkema et al., 2005).

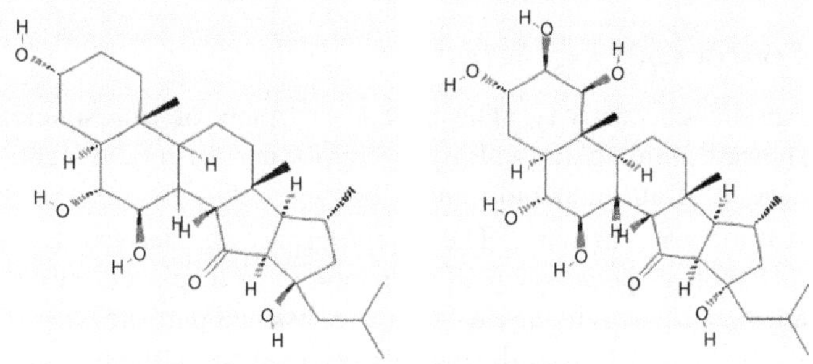

Xestobergsterols A and B

Xestospongia muta (Schmidt, 1870)

Order: Haplosclerida Family: Petrosiidae

Common name: Giant barrel sponge

Distribution: Common in Bahamas and Caribbean, occasional in Florida

Ecology: It inhabits midrange to deep coral reefs and muddy sand bottoms in deeper water.

Biology: It is a cup- or barrel-shaped sponge with a rough, often jagged, stone-hard exterior. Giant specimens may reach a diameter of up to 2 m. Smaller specimens of deeper localities may be cone shaped, that is, with the base broader than the top. Walls on the outside are irregularly ridged and pitted. In between the elevations, the surface is smooth. These sponges are mostly without oscules, but there may be a few small openings (2–3 mm in dia.) that appear to be oscules. Inner wall is rough and irregularly conulose. The central cavity extends to about halfway the height of the cup. The consistency is brittle, incompressible, and easily crumbles or breaks. Colors of the sponges may be shades of gray, brown, red brown, or rose purple. These sponges have been and reported to live for 100 years and they grow only about 1.5 cm a year.

Compound(s) and Activities

Antidiabetic activity: The aqueous extracts of this species showed antidiabetic activity by inhibiting dipeptidyl peptidase IV (Lauritano and Lanora, 2016).

Antimalarial activity: The *Streptomyces* sp. derived from this sponge species yielded a dipeptide that displayed a weak antimalaria activity against the drug-resistant parasite with an IC50 value of 10 µg/mL (Rakotondraibe et al., 2014).

Cytotoxicity: The compound title acid (1) and its methyl ester 33B (2) of this species showed cytotoxicity. The EDs of compound 1 for 50% inhibition *in vivo* and leukemic L1210 cell lines were 24 and 34 µg/mL, respectively, and the corresponding doses of compound 2 were 29 and 34 µg/mL (Dembitsky et al., 2006).

Xestospongia testudinaria (Lamarck, 1815) (= *Petrosia testudinaria*)

Order: Haplosclerida Family: Petrosiidae

Common name: Giant barrel sponge, Red Sea sponge, South China Sea marine sponge

Distribution: Indo-West Pacific; Western and Central Indian Ocean; Philippines, Australia, Indonesia, Malaysia and New Caledonia

Ecology: These sponges attach to rubble in areas with sandy substrate and patches of corals at a depth of 17 m.

Biology: Large encrusting sponges (46 mm long, 22 mm wide, and 15 mm thick) of this species are generally vase- or barrel-shaped, as they have a deep cavity in the center. The inside of the cavity has an uneven and rough texture. The outside may be smooth, bumpy, or with regular fingers, ridges, or "wings." Younger, smaller sponges may be tall, hollow tubes. Often several "vases" of various sizes and shapes are found together, as they are emerging from a common base. Specimens growing on the intertidal areas are about 10 to 20 cm in diameter and about 10 to 20 cm tall. But deeper water sponges can grow to more than 1 m tall. It is maroon to pinkish and the "opening" of the barrel may be paler to white. Surface is micropunctipore, hispid, and conulose. Megascleres are oxeas that are slightly bent with blunt to pointed ends. Microscleres are absent. Ectosomal and choanosomal skeleton are isotropic reticulation of spicules forming thick tracts with meshes. Many spicules are found scattered in confusion obscuring the skeletal network. Spongin is absent.

Compound(s) and Activities

Antiviral activity: The compound, stachybotrin D isolated from the associated fungal strain of *Stachybotrys chartarum* MXH-X73 of this species showed antiviral activity and the EC50 values recorded are given next.

Target	EC50 Value (µg/mL)
HIV-1	3.7
NNRTI* resistant HIV-1 strain 1RT-K103N	3.1
NNRTI resistant HIV-1RT-L100I, K103N	10.5
NNRTI resistant HIV-1RT-K103N, V108I	5.9
NNRTI resistant HIV-1RT-K103N, G190A	6.3
NNRTI resistant HIV-1RT-K103N, P225H	2.7

Source: Indraningrat, A.A.G. et al., 2016, *Mar. Drugs*,14: 87.
* Non-nucleoside reverse transcriptase inhibitor.

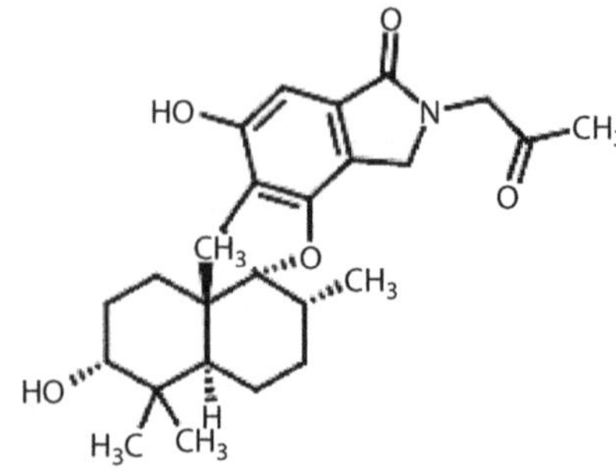

Stachybotrin D

Anticancer activity: The bioactive peptide consisting of 13 amino acids (KENPVLSLVNGMF) identified from this species was found to be selectively toxic to human cervical cancer cells (HeLa) (*Wikipedia*).

Cytotoxicity: The organic extracts of this species yielded the following compounds: xestosterol palmitate (1), xestosterol (2), xestosterol ester of 18′-bromooctadeca-7′E,9′E-diene-7′,15′-diynoic acid (3), (5E,11E,15E,19E)-20-bromoeicosa-5,11,15,19-tetraene-9,17-

diynoic acid (4), 18,18-dibromo-(9E)-octadeca-9,17-diene-5,7-diynoic acid (5), 18-bromooctadeca-(9E,17E)-diene-7,15-diynoic acid (6), 18-bromooctadeca-(9E,13E,17E)-triene-7,15-diynoic acid (7), 16-bromo-(7E,11E,15E)hexadeca-7,11,15-triene-5,13-diynoic acid (8), 2-methylmaleimide-5-oxime (9), maleimide-5-oxime (10), tetillapyrone (11), and nortetillapyrone (12). Among these compounds, the alcoholic and n-hexane extracts containing compounds 5, 6, and 8 showed remarkable cytotoxic activity against human cervical cancer (HeLa), human hepatocellular carcinoma (HepG-2), and human medulloblastoma (Daoy) cancer cell lines. Interestingly, the dibrominated C18-acetylenic fatty acid (5) exhibited the most potent growth inhibitory activity against these cancer cell lines followed by the compounds 6 and 8. Compound 6 was more potent than 8, since the IC50 of 7 was 30.38 and 23.1 μg/mL on HeLa and Daoy cells, respectively, compared to the compound 8, which exhibited IC50 values of 44.41 and 24.57 μg/mL on HeLa and Daoy cells, respectively (El-Gamal et al., 2016).

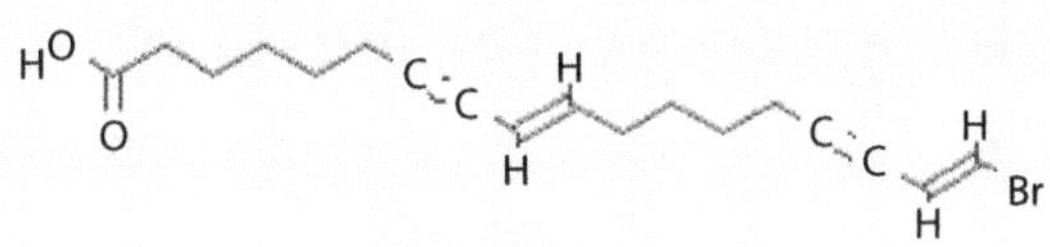

18-bromooctadeca-(9E,17E)-diene-7,15-diynoic acid

Anti-inflammatory, antioxidant, and immunomodulator effects: The methanolic extract of this species exhibited antioxidant activity against DPPH free radicals. Further, this extract (100 mg/kg) showed anti-inflammatory activity by significantly decreasing the percent increase in paw weight measured at 1, 2, 3, and 4 h after carrageenan injection. The extract decreased paw malondialdehyde (MDA) and nitric oxide (NO) and increased the reduced glutathione (GSH), glutathione peroxidase (GPx), and catalase (CAT) activity. It also decreased the inflammatory cytokines, tumor necrosis factor-α (TNF-α), interleukin-1β (IL-1β), and IL-6 (El-Shitany et al., 2015).

Antidiabetic activity: A new steroidal ketone isolated from this species exhibited significant inhibitory activity against protein tyrosine phosphatase 1B (PTP1B), a key target for the treatment of type II diabetes and obesity, with an IC50 value of 4.27 ± 0.55 μM (He et al., 2016).

Antibacterial activity: The extract of this sponge species showed antibacterial activity against with the following values of zone of inhibition.

Zone Inhibition (mm)

S. aureus	*E. coli*	*K. pneumonia*	*S. typhi*	*P. aeruginosa* MDR	*S. aureus* MRSA
20.1	9.5	15.3	15.3	15.3	17.5

Source: Cita, Y.P. et al. 2017, *J. Marine Sci. Res. Dev.*, 7:237.

Enzyme-inhibiting activity: The compound mutafuran of this species showed significant acetylcholinesterase activity (IC50 = 0.64 μM) (Gribble, 2015).

Mosquito larvicidal activity: The extract of this species displayed larvicidal activity against the fourth instar larvae of *Aedes aegypti* (yellow fever mosquito vector) with an LC50 value of 73.8 ppm (Rao et al., 2008).

Xestospongia vansoesti (Bakus and Nishiyama, 2000)

Order: Haplosclerida Family: Petrosiidae

Common name: Not designated

Distribution: Indonesia, Eastern Philippines

Ecology: It inhabits hard substrata of shallow reef habitats.

Biology: It is a thickly encrusting digitate and crusty sponge with dark green to black color. Oscula are many, 1–2 cm apart, and are up to 3 mm diameter. Pores are abundant. The intact surface is smooth. It is rough to touch, hard, and incompressible. Skeleton consists of a meshwork of multipicular tracts of oxea. Spongin is sparse.

Compound(s) and Activities

Antitumor activity: Among the compounds xestolactone A and xestosaprol O isolated from the methanol extract of this species, the latter was found to be 20 times more potent as an inhibitor of indoleamine 2,3-dioxygenase (IDO) IC50 = 4 μM) than the former (IC50 = 81 μM), making it a lead toward inhibitors of tumor immune escape (Blunt et al., 2015).

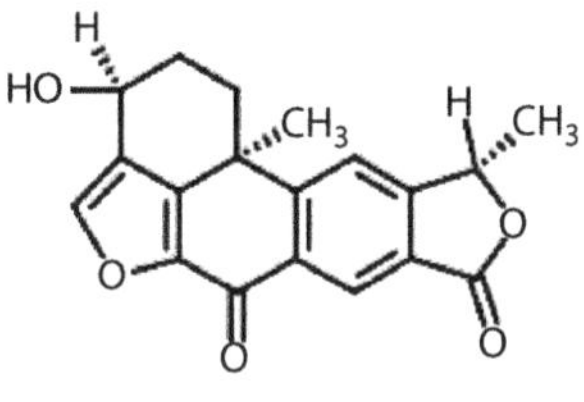

Xestolactone A

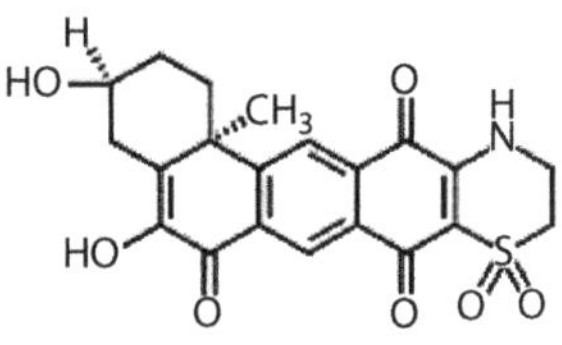

Xestosaprol O

Proteasome inhibitory activity: Salsolinol (1), a tetrahydroisoquinoline alkaloid, was isolated from this species, along with three salsolinol derivatives, namely, norsalsolinol (2), cis-4-hydroxysalsolinol (3), and trans-4-hydroxysalsolinol (4). Among them, the compounds 1 and 2 inhibited the chymotrypsin-like activity of the proteasome with IC50 values of 50 and 32 μg/mL, respectively. Proteasome inhibitors are drugs that block the action of proteasomes, which break down proteins. Thee inhibitors are being studied in the treatment of cancer, and three of them have been approved for use in treating multiple myeloma (Nagasawa et al., 2011).

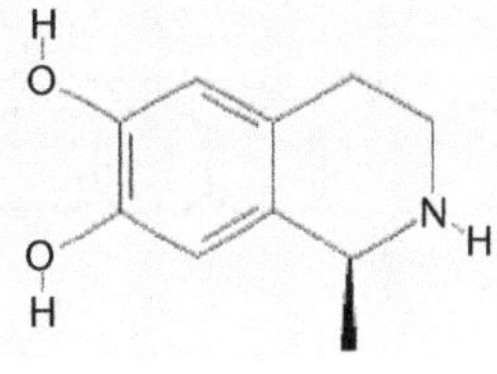

Salsolinol

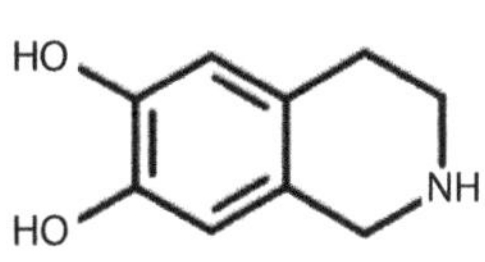

Norsalsolinol

Xestospongia sp.

Order: Haplosclerida Family: Petrosiidae

Compound(s) and Activities

Antibacterial activity: The crude extracts of a bacterial strain *Micrococcus luteus* R-1588-10 derived from this sponge showed potent antibacterial activities (Thomas et al., 2010).

Neurosuppressives and muscle relaxant activity: The macrocyclic bis-oxaquinolizidine and xestospongin C isolated from this species showed neurosuppressives and muscle relaxant activity by acting as an IP3-inhibitor (Sipkema et al., 2005).

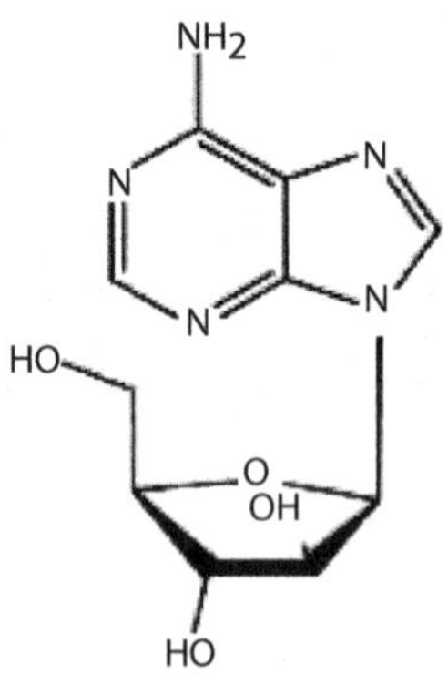
Bis-oxaquinolizidine

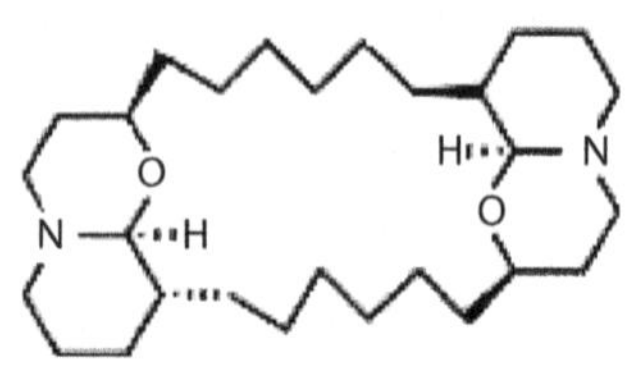
Xestospongin C

Antiviral activity: The sulfamated steroids haplosamates A and B isolated from this species displayed antiviral (anti-HIV-1) activity (Anjum et al., 2016; Sipkema et al., 2005).

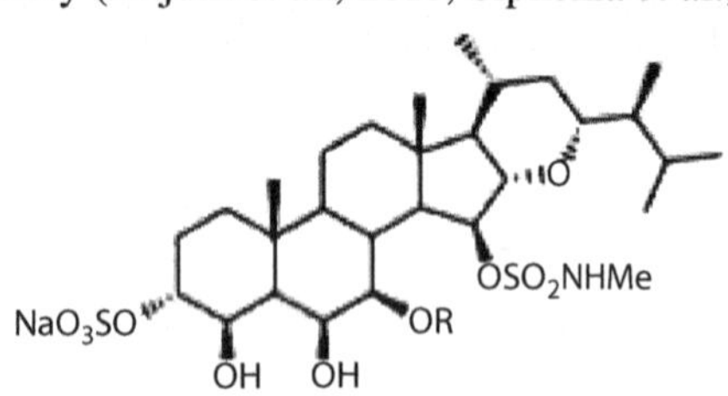
Haplosamates A,B

Anticancer activity: The compounds adociaquinones and xestoquinones of this sponge showed anticancer activity (Concepcion et al., 2014).

Protein kinase inhibitory activity: Seven new adociaquinone derivatives—xestoadociaquinones A and B, 14-carboxy-xestoquinol sulfate, and xestoadociaminals A–D—isolated from this sponge showed protein kinase inhibitory activity (He et al., 2015).

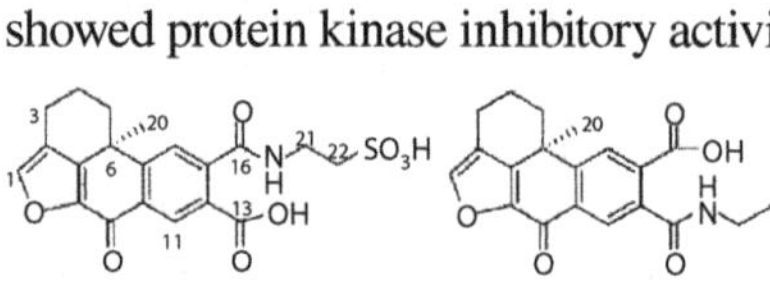
Xestoadociaquinones A,B

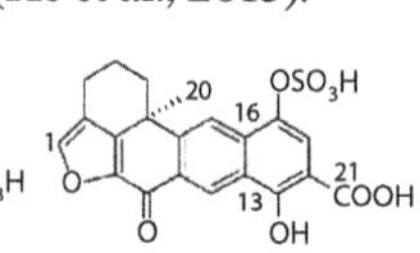
14-carboxy-xestoquinol sulfate

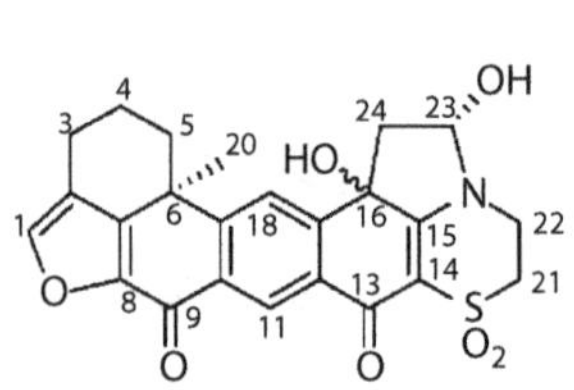
Xestoadociaminals A(16β-OH); B (16α-OH)

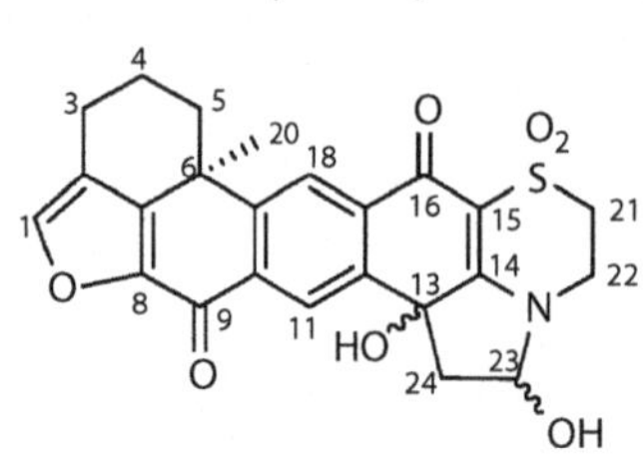
Xestoadociaminals C,D

Cytotoxic activity: The organic dried extract of this species exhibited cytotoxicity against lung fibroblast (V-79) and human colon epithelial (CaCo-2) cancer cell lines at a concentration of 100 μg/mL. The percentage inhibition values recorded for these cell lines were found to be 89 and 95, respectively (Turk et al., 2013)

Cardiotonic activity: The compound xestosaprol C of this sponge was reported with cardiotonic activity (Qaralleh, 2016).

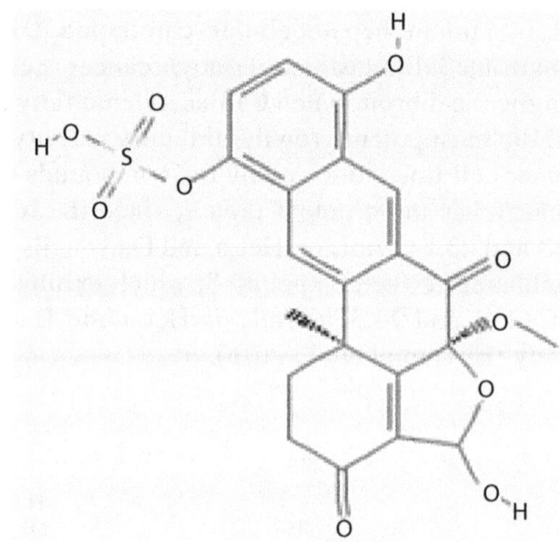
Xestosaprol C

Osteoclastogenic inhibitor: The compound, halenaquinone of this sponge was found to be an inhibitor of osteoclastogenic (osteoclast is a cell that nibbles at and breaks down bone and is responsible for bone resorption) differentiation of murine RAW264 cells (Qaralleh, 2016).

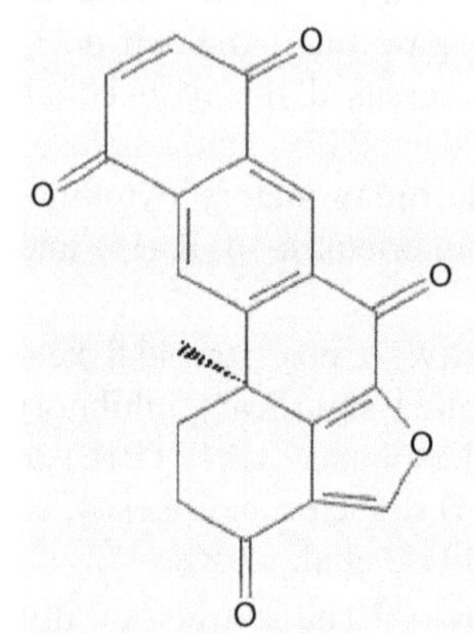
Halenaquinone

FAMILY: PHLOEODICTYIDAE

Oceanapia phillipensis (Dendy, 1895)

Order: Haplosclerida Family: Phloeodictyidae

Common name: Not known

Distribution: South Australia

Ecology: Not reported

Biology: Not available

Compound(s) and Activities

Antifungal activity: The bisaminohydroxylipid glycoside oceanapiside isolated from this species has shown antifungal activity against *Candida glabrata* with an MIC value of 10 µg/mL (Nicholas et al., 1999; Sipkema et al., 2005).

Oceanapia sagittaria (Sollas, 1902)

Order: Haplosclerida Family: Phloeodictyidae

Common name: Pink puffball sponge, maiden's fan

Distribution: Western Central Pacific

Ecology: This sessile sponge is seen on the shores among coral rubble in a depth range 0 to 73 m. It is just the tip of an enormous buried sponge.

Biology: It resembles a small delicate prickly ball of about 1 to 2 cm in diameter. It has a short conical stem (called the capitate structure). The ball has fine sheets between the prickles. It is seen singly, but sometimes a few may be seen growing on coral rubble. This puffy fluff is just a tiny part of the sponge. The bulk of the sponge is actually buried. The main sponge is bulbous and spherical. Color is usually a deep maroon to bright pink.

Compound(s) and Activities

Antitumor activity: A total of seven compounds—24 α-methylcholestanol (1), p-hydroxybenzaldehyde (2), p-hydroxy benzoic acid (3), phenylacetic acid (4), 3-formylindole (5), kuanoniamine A (6), and kuanoniamine C (7)—have been isolated from the ethyl acetate extract of this species. Among them, kuanoniamine A was found to be a potent growth inhibitor of all the human tumor and a nontumor cell lines, whereas kuanoniamine C was less potent but showed high selectivity toward the estrogen dependent (ER+) breast cancer cell line (Kijjoa et al., 2007).

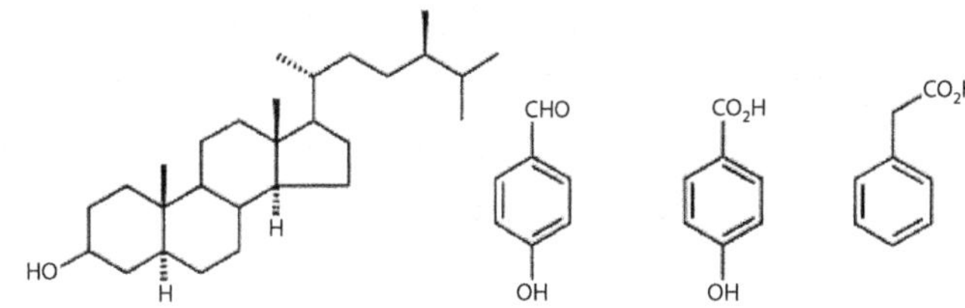

1 2 3 4

1: 24 α-methylcholestanol; 2: p-hydroxybenzaldehyde; 3: p-hydroxybenzoic acid
4: phenylacetic acid

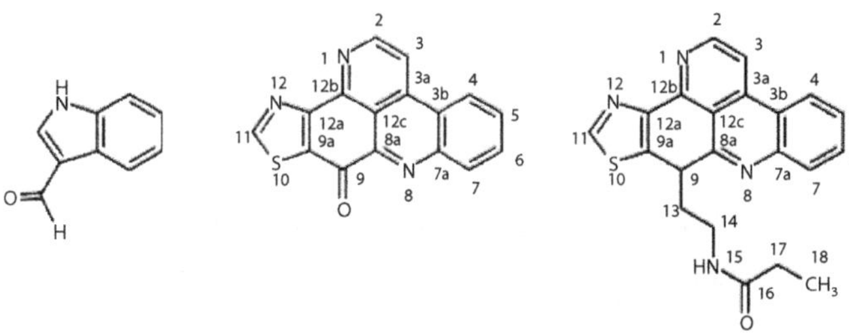

From left: 3-formylindole; kuanoniamine A; kuanoniamine C.

The GI50 values recorded for the different cell lines are given next.

GI50* (µM) Values

Compound	MCF-7	MDA-MB-231	NCI-H460	SF-268	UACC-62	MRC-5
Kuanoniamine A	0.1	0.7	0.9	4.7	1.8	0.6
Kuanoniamine C	0.8	10.2	21.5	33.2	15.8	Not detected

Source: Kijjoa, A., 2007, *Mar. Drugs*, 5: 6–22.

* Fifty percent of cell growth inhibition (GI50) after a continuous exposure for 48 h.

Oceanapia sp.

Order: Haplosclerida Family: Phloeodictyidae

Compound(s) and Activities

Antibacterial activity: The acetylenic acid isolated from this unidentified species showed antibacterial activity (Braga, 2014).

Cytotoxic effect: The pyridoacridine alkaloid sagitol C isolated from this species showed cytotoxic effect against L5178Y, PC12, and HeLa cell lines. It gave 93%, 88%, and 76% growth suppression against the tested cell lines at a concentration of 24.6 µM, and 81%, 74% and 37% at a concentration of 12.3 µM with ED50(s) of 0.7, 0.9, and 2.3 µM, respectively (Ibrahim et al., 2013; Putra and Jaswir, 2014).

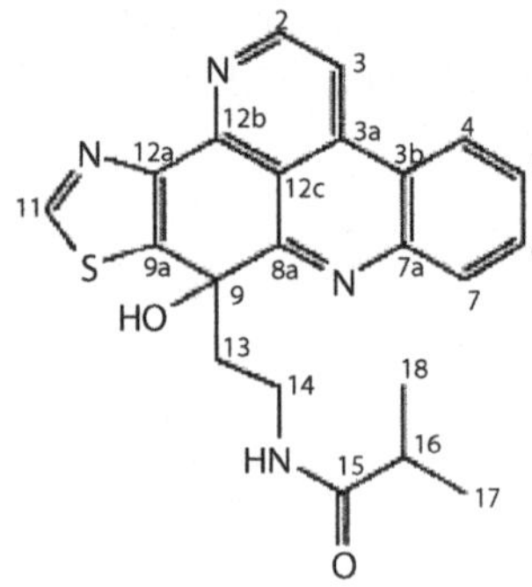

Sagitol C

The brominated fatty acids isolated from this species exhibited mild cytotoxicity toward KB cells (Dembitsky et al., 2006).

Siphonodictyon coralliphagum (Rützler, 1971) (= *Aka coralliphaga*)

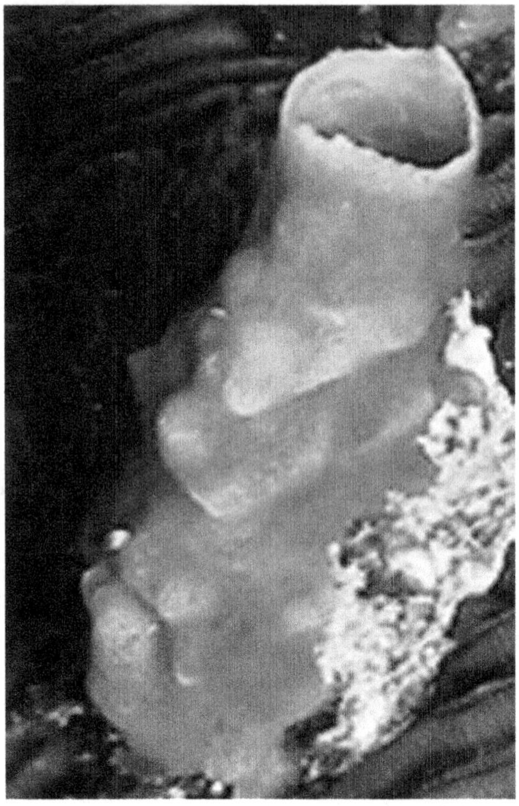

Order: Haplosclerida Family: Phloeodictyidae

Common name: Caribbean sponge, variable boring sponge

Distribution: Western Central Atlantic, Caribbean, Greater Antilles

Ecology: It is a reef-associated species. It is restricted to the lower buttress zone, fore reef, and fore reef slope of fringing barrier-type reefs at depths between 1.5 and 70 m. It is also found in mangroves and brackish waters at a depth range 3 to 5 m.

Biology: This sponge has several tuberculate chimneys protruding from the coral substrate. Circular oscula (2–10 mm dia.), which are unelastically compressible. are found elevated on top of chimneys or tubes. If no osculum, irregular papillae of different sizes are seen. Various epilithic structures are of 5–30 mm diameter and 6–40 mm high. Choanosome is endolithic. Color of the sponge is deep yellow and turns pinkish yellow if exposed to air. Consistency is friable but chimneys are elastic. Surface is finely porous and microhispid. Ostia are found distributed over the entire surface. Skeleton is a more or less a delicate regular network of polygonal meshes. Spicules consists of oxea that are slightly bent.

Compound(s) and Activities

Antidiabetic activity: The bacterial communities of Firmicutes, Actinobacteria, Proteobacteria, and Bacteroidetes associated with this sponge species possess antidiabetic activity by inhibiting the enzyme β-glucosidase (Barde et al., 2015).

Anticancer and antibacterial activities: Four new sulfated sesquiterpene hydroquinones, namely, siphonodictyals E1–E4 (1–4) (derivatives of siphonodictyal E), and cyclosiphonodictyol A (5) were isolated from this species. Among these compounds, siphonodictyal E4 (4) exhibited mild antiproliferation activity against L929 mouse fibroblast, KB-31 epidermoid carcinoma, and MCF-7 breast cancer cell lines with IC50 values of 27, 108, and 54 μM, respectively. Compound E5 acted against MCF-7 with an IC50 value of 175 μM. For antibacterial activity, the compound E3 showed activity against *Staphylococcus aures* (MRSA) and *Micrococcus luteus* with IC50 values of 37 and 74 μM, respectively. The compound 5 acted on these bacterial species with IC50 values of 117 and 58 μM, respectively (Göthel and Köck, 2014).

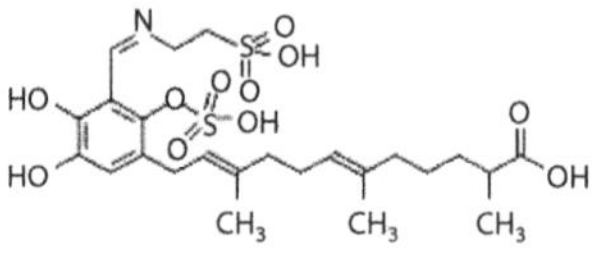

Siphonodictyal E1

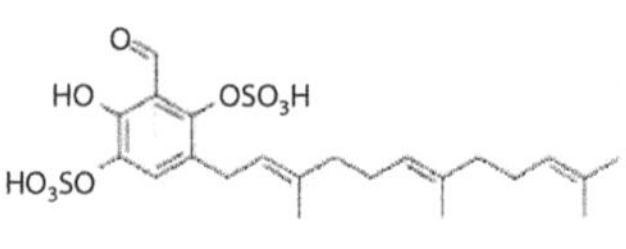

Siphonodictyal E3

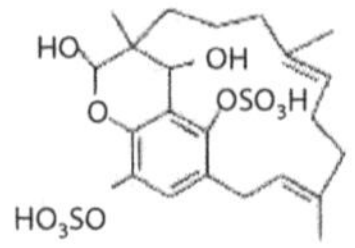

Siphonodictyal E4

FAMILY: HAMACANTHIDAE

Hamacantha sp.

Order: Merliida Family: Hamacanthidae

Compound(s) and Activities

Antifungal activity: The indole alkaloid hamacanthine B isolated from this unidentified species showed antifungal activity against *C. albicans* and *C. neoformans* (Gul et al., 2005).

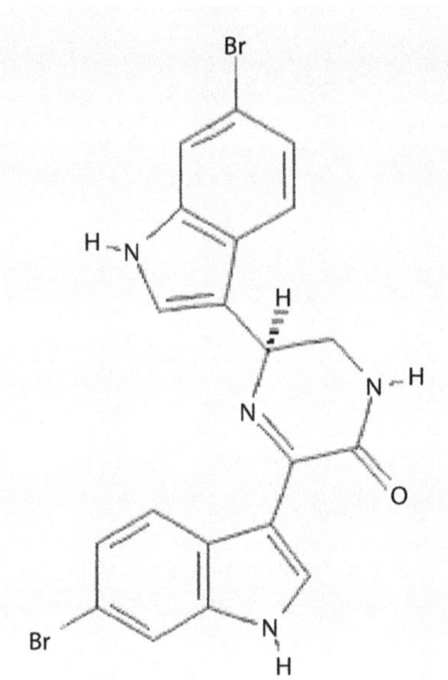

Hamacanthine B

FAMILY: ACARNIDAE

Zyzzya fuliginosa (Carter, 1879) (= *Zyzzya marsailis*)

Order: Poecilosclerida Family: Acarnidae

Common name: Pohnpeian sponge, Philippine sponge

Distribution: Indo-PacificNew Zealand, Eastern Papua New Guinea

Ecology: It inhabits submerged offshore reef and steep sand slope.

Biology: It is a dark brown subspherical mass, with at least one-third of its base lying in a shallow burrow in soft sediment. The upper surface tapers into a single fistule with a terminal sieve plate. The ectosomal skeleton is a tangential layer of tylotes. The choanosomal skeleton contains multispiculartracts of tylotes and an irregular subisodictyal reticulation of single acanthostrongyles and occasional tylotes.

Compound(s) and Activities

Antioxidant activity: The compound makaluvamine F isolated from this species showed antioxidant activity and was potent with a reduction of 54.5% at a concnentration of 1 μM (Alons et al., 2016).

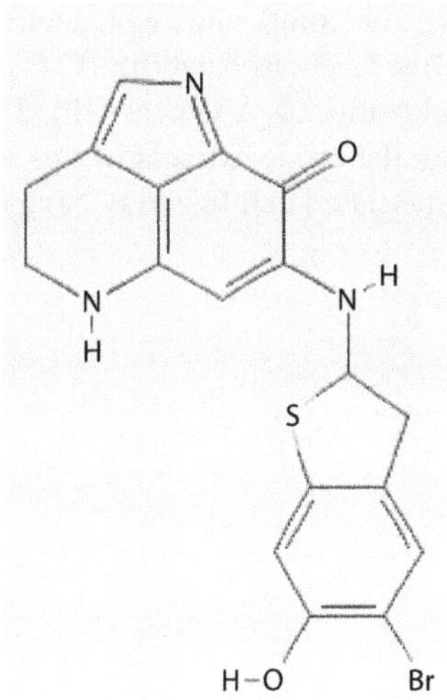

Makaluvamine F

Among the compounds zyzzyanones A–D (1–4) and makaluvamines C (8), E (5), G (6), H (9), and L (7) isolated from this sponge, zyzzyanones A and B displayed moderate antioxidant activities in the ABTS assay and in the autoxidation of linoleic acid (61%–66% inhibition at concentrations of 0.1 mM) (Utkina, 2013).

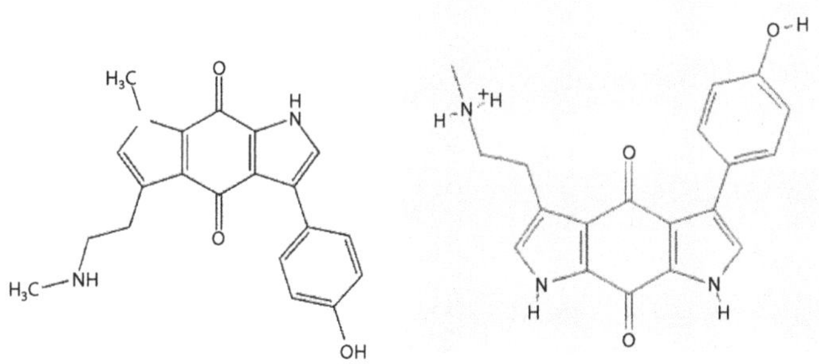

Zyzzyanones A,B

Cytotoxic and antioxidant activities: A new pyrroloiminoquinone alkaloid, makulavamine P, isolated from this species displayed cytoxicity in the micromolar range on KB cells and antioxidant activity (Casapullo et al., 2001).

A dipyrroloquinone, zyzzyanone A, isolated from this sponge exhibited moderate cytotoxic activity against mouse Ehrlich carcinoma cells (Utkina et. al., 2005).

Anticancer activity: The compound makaluvamine A of this sponge showed antitumor activity against colon cancer (HCT-116) cells (Senthilkumar and Kim, 2013).

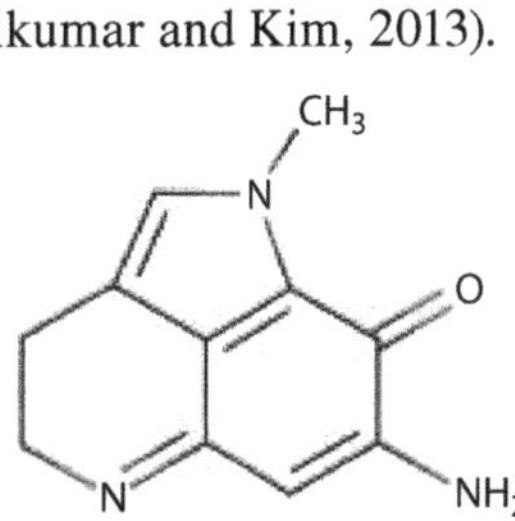

Makaluvamine A

Antiviral activity: The compounds isobatzelline E (1), batzelline D (2), batzelline C (3), isobatzelline C (4), makaluvamine D (5), and makaluvamines A (9), H (10), J (7), K (8), and P (6) isolated from the crude extracts of this sponge inhibited HIV-1 envelope-mediated cell fusion (Chang et al., 2002).

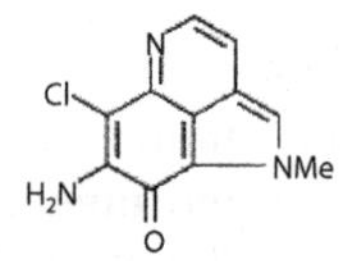
Isobatzelline E

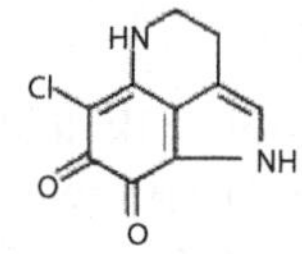
Batzelline D

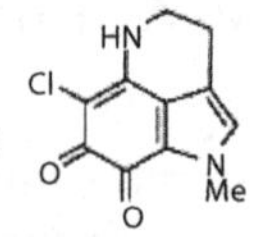
Batzelline C

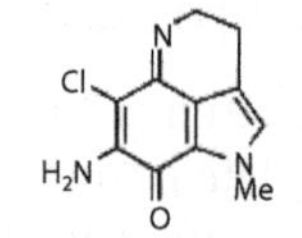
Isobatzelline C

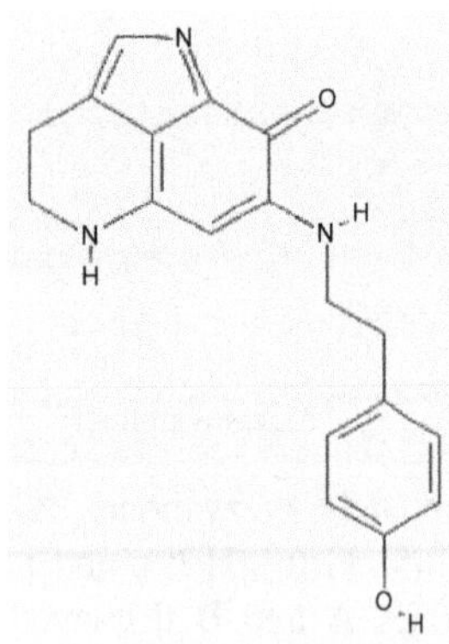
Makaluvamine D

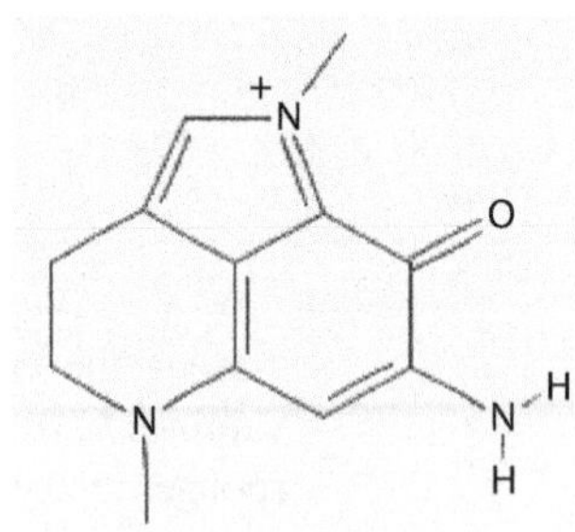
Makaluvamine H

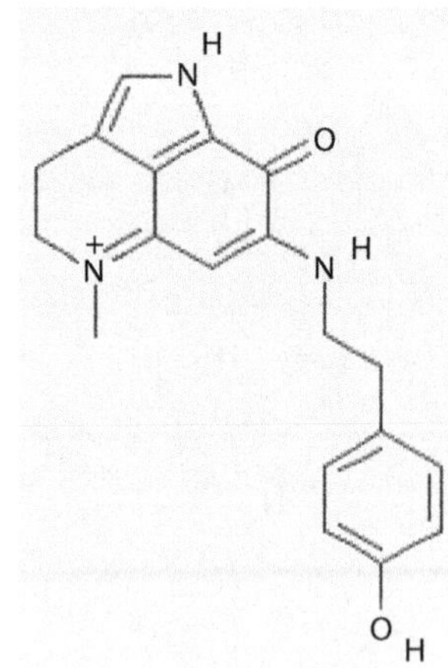
Makaluvamine J

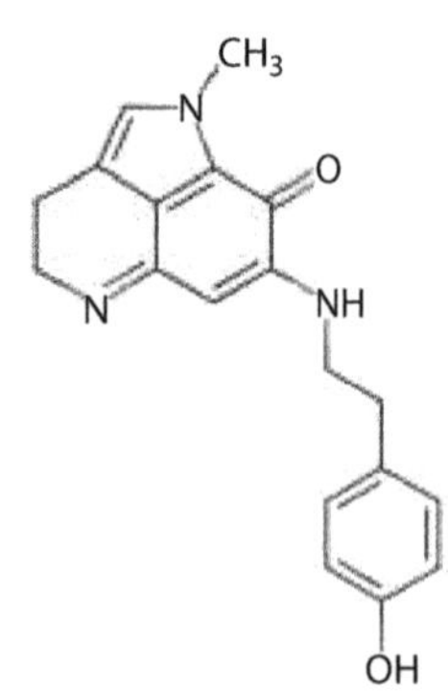
Makaluvamine K

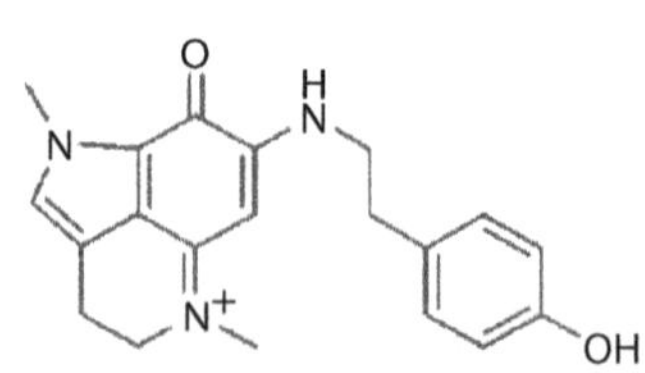
Makaluvamine P

Enzyme inhibitory activity: A new pyrroloiminoquinone compound, makaluvamine N, demonstrated an ability to inhibit the catalytic activity of topoisomerase II (Venables et al., 1997).

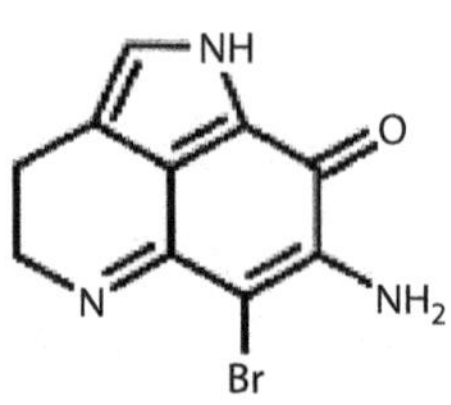
Makaluvamine N

Zyzzya sp.

Order: Poecilosclerida Family: Acarnidae

Compound(s) and Activities

Antimalarial activity: A new bispyrroloiminoquinone alkaloid, tsitsikammamine C (1), isolated from this species displayed potent *in vitro* antimalarial activity with IC50 values of 13 and 18 nM against chloroquine-sensitive (3D7) and chloroquine-resistant (Dd2) *Plasmodium falciparum*, respectively. Among the previously reported compounds of this species, namely, makaluvamines J (2), G (3), L (4), K (5), and damirones A (6) and B (7), compounds 2–4 displayed potent growth inhibitory activity (IC50 <100 nM) against both *P. falciparum* and only moderate cytotoxicity against HEK293 cells (IC50 = 1–4 μM). Makaluvamine G (3) suppressed the parasite growth in *P. berghei*–infected mice following subcutaneous administration at 8 mg kg^{-1} day^{-1} (Davis et al., 2012; Perdicaris et al., 2013).

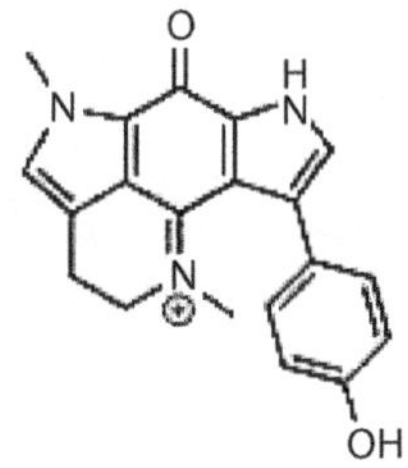

Tsitsikammamine C

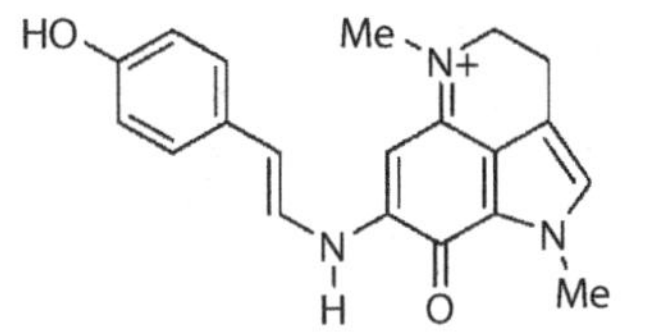

Makaluvamine G

Cytotoxic activity: The polyketides brocaenols A–C isolated from the associated fungal species *Penicillium brocae* showed cytotoxicity (Thomas et al., 2010).

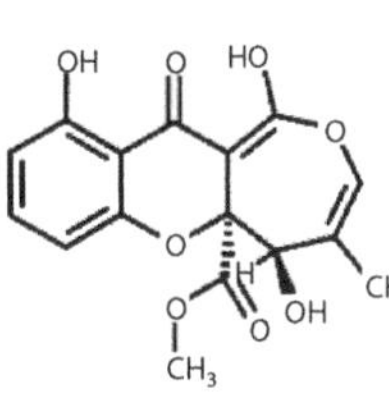

Brocaenol A

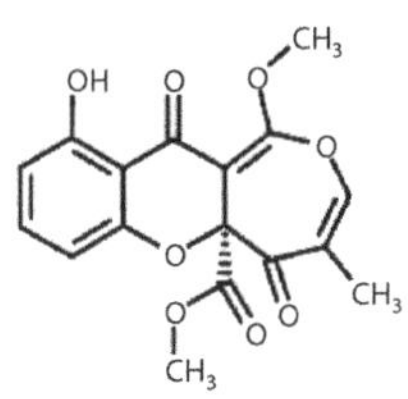

Brocaenol B

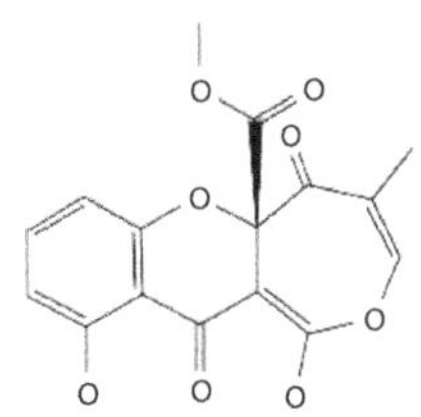

Brocaenol C

Anticancer activity: The pyrroloquinoline compound makaluvamine A of this species showed activity against colon cancer (HCT-116 cells) (Senthilkumar and Kim, 2013).

FAMILY: CRAMBEIDAE

Monanchora arbuscula (Duchassaing and Michelotti, 1864) (= *Monanchora unguifera*)

Order: Poecilosclerida Family: Crambeidae

Common name: Red encrusting sponge

Distribution: Western Atlantic, Greater Antilles, Eastern Caribbean, Florida

Ecology: It is a shallow reef species at a depth of 2 to 60 m. It may encrust dead corals, mollusk shells and gorgonian axes. It is also on hard bottoms and mangrove ponds, and tolerates brackish water conditions.

Biology: This species (up to 12 cm high and wide) is massive or ramose. It displays a variety of different habits. It forms thin encrustations with several rootlike canals from the slightly raised oscules that are of 0.3 to 2 cm in diameter. The species can also form erect, lobate branching bushes with highly penicillate surface projections. It is up to 12 cm high and wide. The projections are 5 to 10 mm long, irregular in outline and spiny. It has soft tissue consistency and tough skeleton. It is brilliant red externally and internally.

Compound(s) and Activities

Cytotoxic activity: The hydroethanolic extracts of this species showed cytotoxicity against tumor cell lines at a concentration of 20 mg/mL. The IC50 values recorded for the different cell lines are shown next.

IC50 (µg/mL) Values of Tumor Cell Lines

HCT-8	MDA-MB435	SF-295
2.0	1.6	1.6

Source: Ferreira, E.G. et al., 2007. *Porifera Res.: Biodiversity, Innovation Sustainability*, 313–318.

Antiparasitic activity: The metabolites batzelladines D, F, L, and nor-L, as well as of synthetic monalidine A, of this species have shown antiparasitic activity against *Trypanosoma cruzi* and *Leishmania infantum* (Santos et al., 2015).

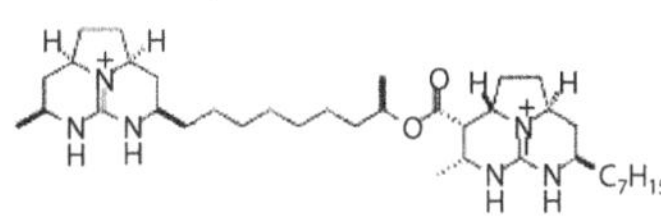

Batzelladine F

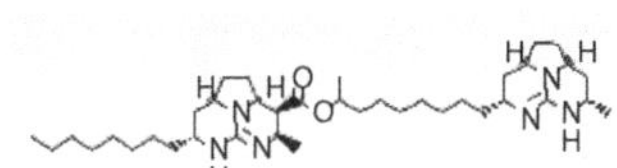

Nor-Batzelladine L

Monanchora pulchra (Lambe, 1895)

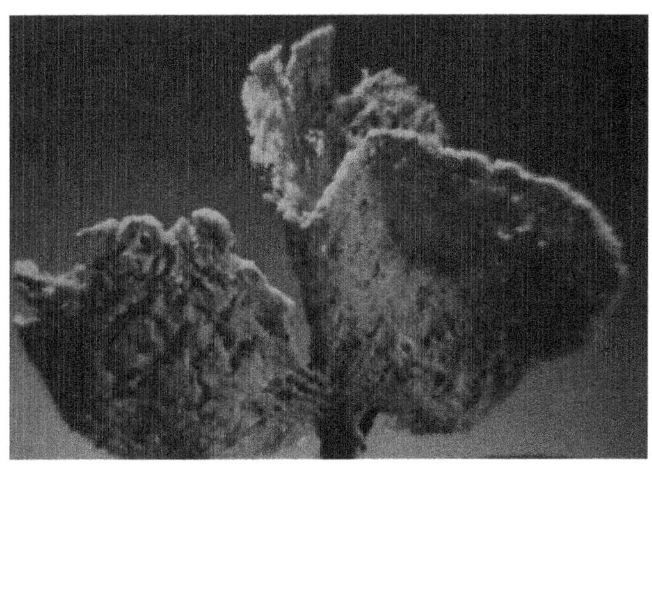

Order: Poecilosclerida Family: Crambeidae
Common name: Far-Eastern marine sponge
Distribution: North Atlantic Ocean, North Pacific Ocean, Canada
Ecology: It is known from seamounts and knolls.
Biology: Not available

Compound(s) and Activities

Antitumor activity: The polycyclic guanidine alkaloid monanchocidin of this species showed antitumor activity by inducing cell death in human monocytic leukemia (THP-1) and human cervical cancer (HeLa) (Perdicaris et al., 2013).

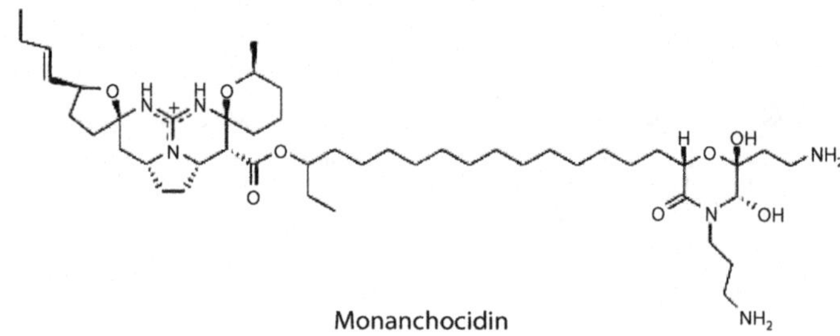

Monanchocidin

Proapoptotic and cytoxic activities: The guanidine alkaloids monanchocidin A (1), monanchocidin B (2), monanchomycalin C (3), ptilomycalin A (4), monanchomycalin B (5), normonanchocidin D (6), urupocidin A (7), and pulchranin A (8) isolated from this species induced cell cycle arrest (apart from 8) and programmed death of cancer cells (Dyshlovoy et al., 2016).

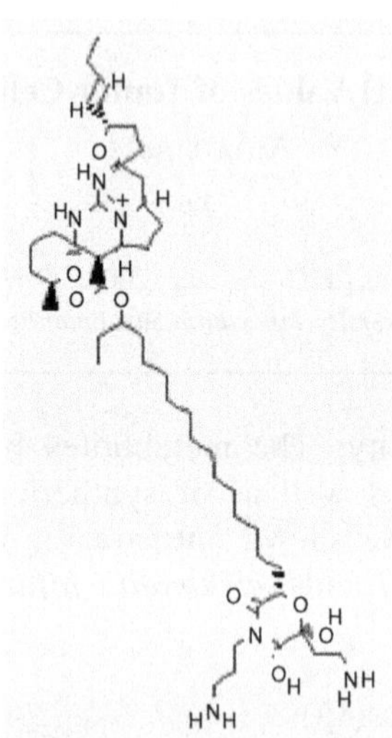

Monanchocidin A

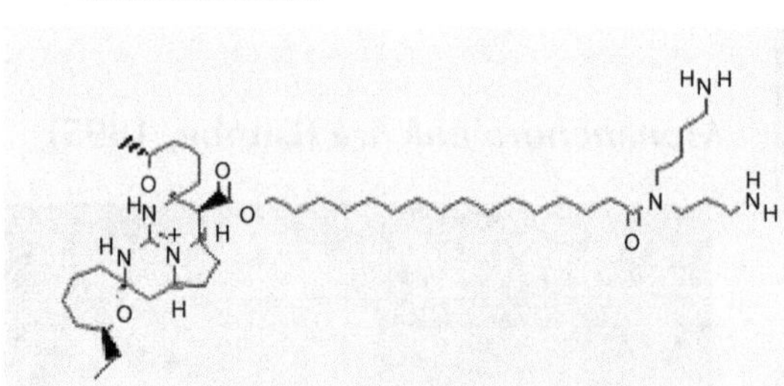

Ptilomycalin A

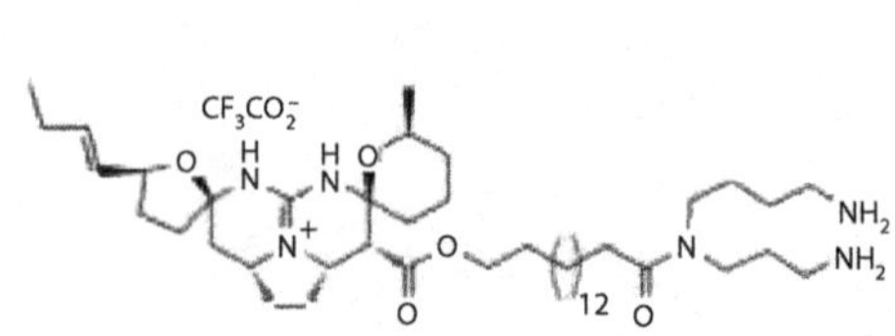

Monanchomycalin B

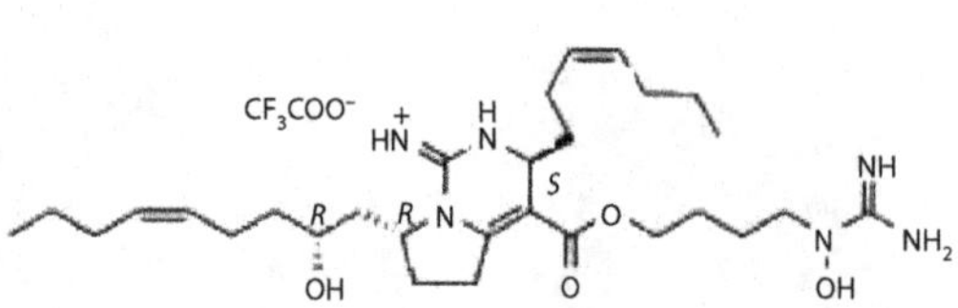

Urupocidin A

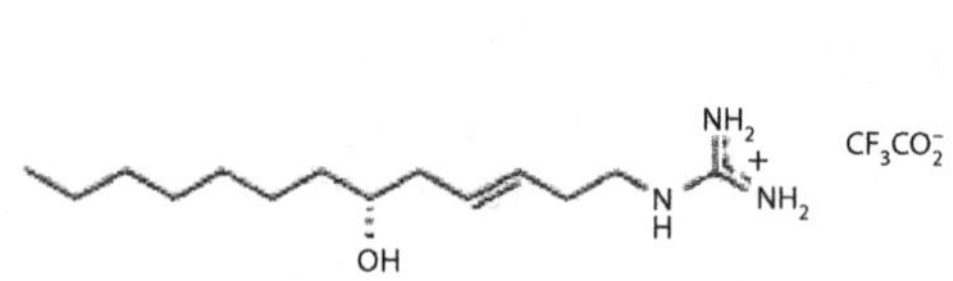

Pulchranin A

Cytotoxic activity: The polycyclic guanidine alkaloids monanchocidins A–E isolated from this species showed potent cytotoxic activities against HL-60 human leukemia cells with IC50 values of 540, 200, 110, 830, and 650 nM, respectively (Makarieva et al., 2011).
The pentacyclic guanidine alkaloids monanchomycalins A and B of this species exhibited potent cytotoxic activities against HL-60 human leukemia cells with IC50 values of 120 and 140 nM, respectively (Makarieva et al., 2012).

Monanchora sp.

Order: Poecilosclerida Family: Crambeidae

Compound(s) and Activities

Anti-HIV activity: The alkaloid crambescidin 826 isolated from this species inhibited HIV-1 with an IC50 value of 1–3 μM (Senthilkumar and Kim, 2013).

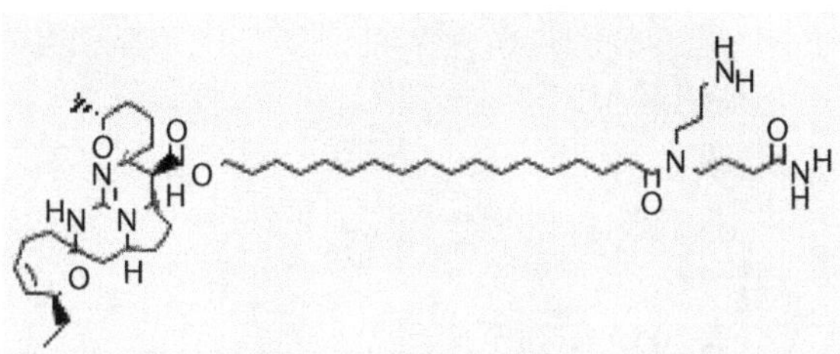

Crambescidin 826

Perdicaris et al. (2013) reported that the compound norbatzelladine L exhibited antiviral activity against herpes simplex virus type (HSV-1).

Crambe crambe (Schmidt, 1862)

Order: Poecilosclerida Family: Crambeidae

Common name: Oyster sponge, orange-red encrusting sponge

Distribution: Kenya, Madagascar, Mozambique Channel, North Atlantic Ocean

Ecology: It commonly occurs in well-lit waters on hard substrate at depths of 5 to 30 m.

Biology: It is an encrusting sponge. The colonies of this species form thin, orange to orange-red plates, rarely lobed, with a very rough surface perforated by raised oscula found along the exhaling channels. These colonies can cover a surface area of about 1 m^2. It feeds by filtering bacteria, microorganisms, and single-celled algae. This species is hermaphrodite. Larvae are planktonic. These sponges often cover the shell of live shellfish (*Arca noae*, *Spondylus*, and various sedentary bivalves). The polychaete worm *Eupolymnia nebulosa* sometimes hides itself on these sponges. This species is also suitable mariculture. The high growth rates of this species observed in continuous flow-through systems (CFTHS) with or without marine microalga (*Phaeodactylum tricornutum*) suggest that this technique is a promising method for culturing this species in closed systems.

Compound(s) and Activities

Antitumor activity: The crude methanol extracts of this species showed highest activity toward pancreatic cell lines MIA-PaCa2 and BxPc-3. It induced efficiently apoptosis, and its treatment overcomes intrinsic and acquired resistance toward chemotherapy in CSC-enriched highly resistant pancreatic tumor cell lines. Further, this extract efficiently and specifically inhibits proliferation and induces cell cycle arrest. This species may therefore serve as a promising source for isolation of new therapeutics for elimination of pancreatic tumors enriched in cancer stem cells (Ottinger et al., n.d.).

The compounds cambescidins 1–4 (pentacyclic guanidine derivatives) of this species showed antitumor activity by serving as a Ca2+ channel blocker (Anjum et al., 2016; Sipkema, et al., 2005).

Antiviral and cytotioxic activity: The organic extracts of this species yielded the pentacyclic guanidines crambescidins 816, 830, 844, and 860, which showed both antiviral and cytoxic activities. Compounds 816, 844, and 860 completely inhibited HSV-1 with diffuse cytotoxicity at 1.3 μg/well and were 98% effective against L1210 murine leukemia cells at a concentration of 0.1 μg/mL (Jares-Erijman et al., 1991).

Cytotoxic activity: Among the compounds isolated from this species, crambescidin 816 was proven to be more cytotoxic against neuronal cell lines. A 24 h of exposure of the cortical neurons to 1 μM crambescidin 816 extract led to an increase in the cytotoxic effect and the almost complete cell death (86.8%). On the other hand, the compound crambescin C1 lowered cellular viability by 22.6% at 1 μM. Further, crambescin C1 and crambescidin 816 caused reductions in the number of cortical living cell at 1 μM (Bondu et al., 2012).

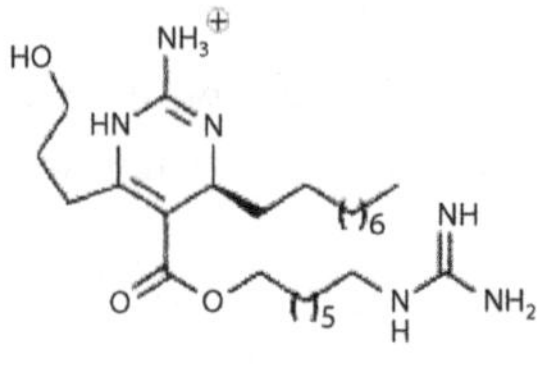

Crambescin C1

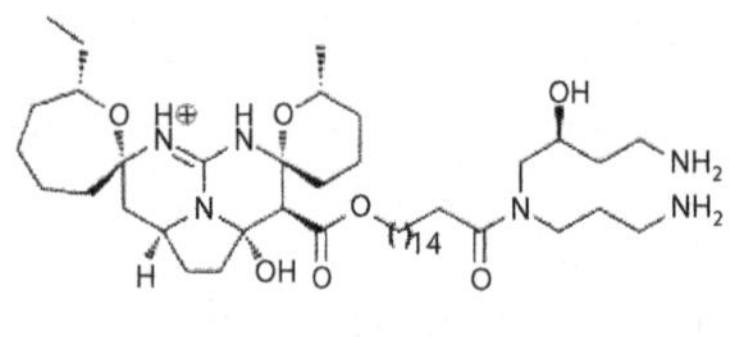

Crambescidin 816

FAMILY: CRELLIDAE

Crella (Grayella) cyathophora (Carter, 1869) (= *Grayella cyathophora*)

Order: Poecilosclerida Family: Crellidae

Common me: Not designated

Distribution: Red Sea, Gulf of Suez

Ecology: This sessile animal spreads on rocks and hard surfaces.

Biology: This sponge has a cuplike body and is massive and spreading. Its surface is undulating, smooth, and is interrupted by the presence of numerous, subcircular, oval or conical, cuplike projections or pores. The spicules are of two kinds: long and short. The long spicules are smooth, straight, slightly fusiform, and cylindrical. The short ones are thickly spinous, slightly curved, fusiform, and sharp pointed. Size of the sponge is 4 cm long and 1 cm thick.

Compound(s) and Activities

Antitumor activity: The ethanolic extract of this species showed antitumor activity against the human colon cancer cells (caco-2) with an IC50 value of 2.1 μg/mL. The corresponding value for breast cancer cells (Mcf-7) was 1.7 μg/mL (El-Damhougy et al., 2017).

Antibacterial activity: The crude extract of this species showed significant antibacterial activity against *P. aeruginosa* (El-Damhougy et al., 2017).

Antioxidant activity: The crude extracts of this species showed DPPH antioxidant activity (3%–53%) in extract concentrations of 25–800 μg/mL. The IC50 value of antioxidant activity with ascorbic acid method was 748 μg/mL (El-Damhougy et al., 2017).

Anti-inflammatory activity: The ethanolic crude extracts of this species showed anti-inflammatory activity with a percentage value of 89.9 (El-Damhougy et al., 2017).

Crella (Yvesia) spinulata (Hentschel, 1911) (= *Crella spinulata*)

Order: Poecilosclerida Family: Crellidae

Common name: Pink sponge

Distribution: Shark Bay

Ecology: Not reported

Biology: This species has a porous body and a hard surface like a stone. In addition, it can absorb oxygen from water through the diffusion process.

Compound(s) and Activities

Antiproliferation activity: The organic extracts of this species showed antiproliferation activity against the human colon adenocarcinoma cell line (Caco-2) with an IC50 value of 8 μl/mL (Rady, 2014).

Cathepsin inhibitory activity: Two dimeric steroid derivatives, shishicrellastatins A and B, isolated from this species displayed inhibitory activity against cathepsin B with an IC50 value of 8 μg/mL each. This cathpsin has been reported to enhance cancer activity (Murayama et al., 2011).

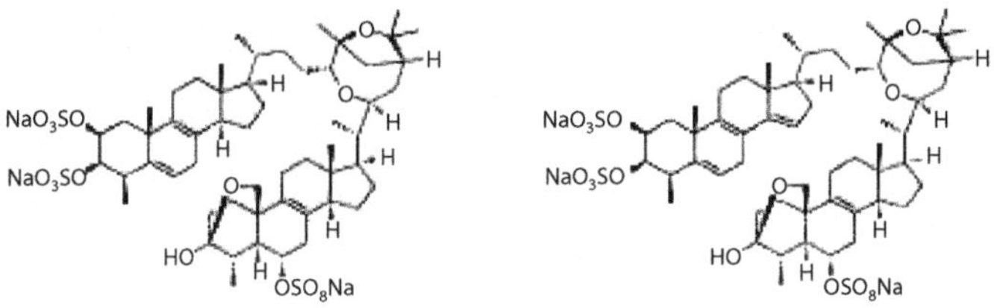

Shishicrellastatin A,B

Crella sp.

Order: Poecilosclerida Family: Crellidae

Compound(s) and Activities

Antitumor activity: Seven new cytotoxic dimeric steroids, namely, crellastatins B–H (2–8), have been isolated from this sponge. Like crellastatin A, crellastatins B–H exhibited *in vitro* antitumor activity against human bronchopulmonary non-small cell lung carcinoma cell lines (NSCLC) with IC50 values in the range of 2 to 10 μg/mL (Zampella et al., 1999).

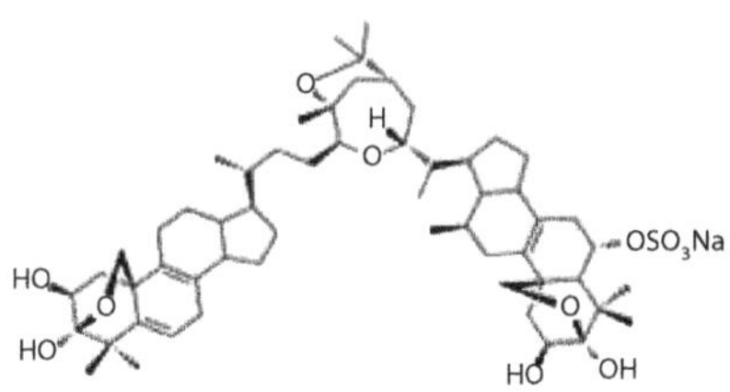

Crellastatin A

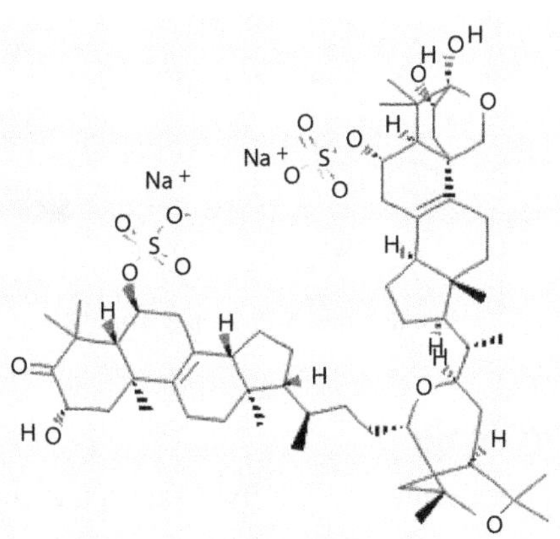

Crellastatin H

FAMILY: CHONDROPSIDAE

Batzella sp.

Order: Poecilosclerida Family: Chondropsidae

Compound(s) and Activities

Antibacterial activity: Three new C-alkylated iminosugars, batzellasides A (1), B (2), and C (3), along with the known halitoxin (4) polymer have been isolated from this species. Among these compounds, compounds 1–3 exhibited antibacterial activity by inhibiting the growth of *Staphylococcus epidermidis* with MIC values of ≤6.3 μg/mL (Segraves and Crews, 2005).

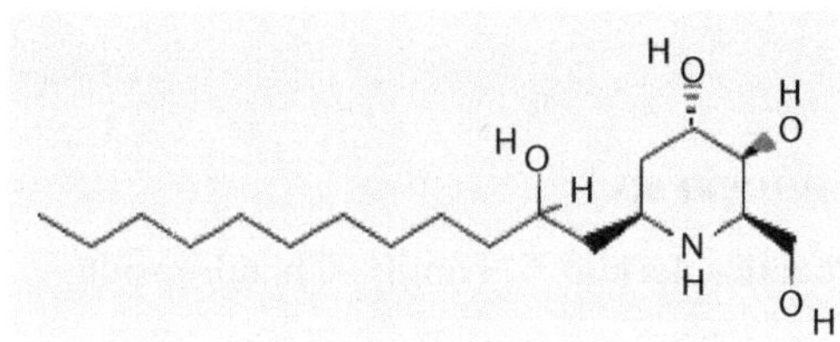

Batzellaside A

Antiviral activity: The guanidine alkaloids batzelladines A and B of this species are the potential inhibitors of HIV, as they inhibited the binding of HIV gp-120 to CD4 (Patil et al., 1995).

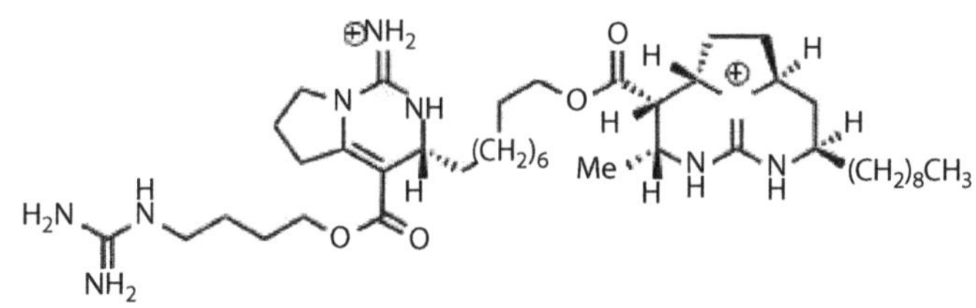

Batzelladine A

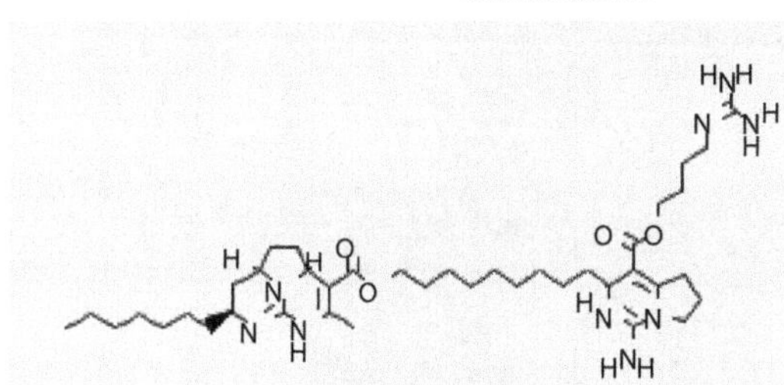

Batzelladine B

Chondropsis sp.

Order: Chondrosiida Family: Chondropsidae

Compound(s) and Activities

Antitumor activity: The compounds chondropsin A and B isolated from this species showed antitumor activity by inhibiting v-ATPase (Sipkema et al., 2005).

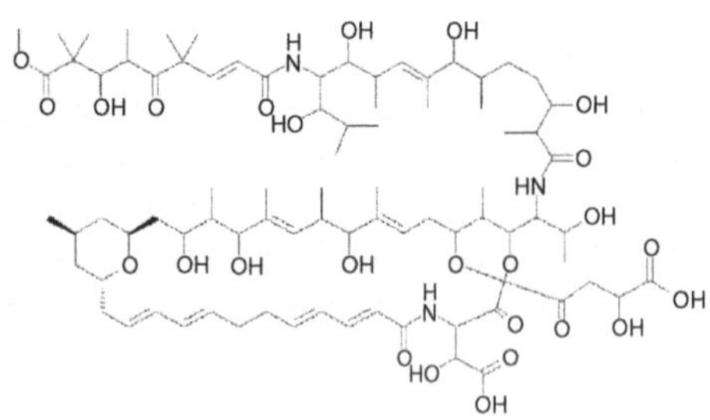

Chandropsin A

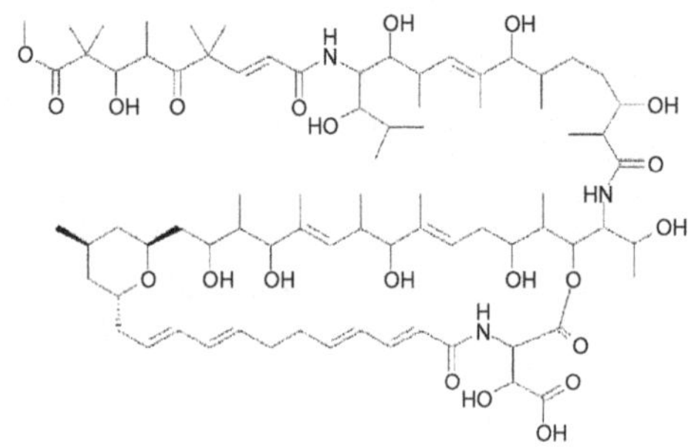

Chandropsin B

Phoriospongia sp.

Order: Poecilosclerida Family: Chondropsidae

Compound(s) and Activities

Nematocidal activity: The depsipeptides phoriospongin A and B isolated form this species displayed significant nematocidal activity against *Haemonchus contortus* with LD99 values of 100 and 194 μg/mL, respectively (Agrawal et al., 2016).

FAMILY: COELOSPHAERIDAE

Lissodendoryx (Acanthodoryx) fibrosa (Lévi, 1961)

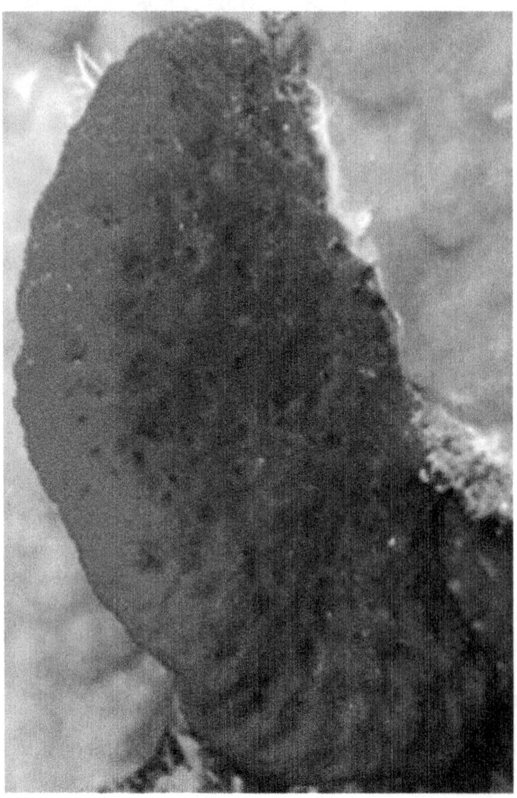

Order: Poecilosclerida Family: Coelosphaeridae

Common name: Not designated

Distribution: Philippines

Ecology: Not reported

Biology: Not documented

Compound(s) and Activities

Cytotoxic/anticancer activity: The dimeric sterols named manadosterols A and B isolated from this species have shown anticancer cytotoxic activity by serving as inhibitors of Ubc (Ubiquitin-conjugating enzyme)13-Uev1A interaction with IC50 values of 0.09 and 0.13 μM, respectively. These compounds have been reported to be approximately 1000 times more potent than the earlier compound with similar activity, namely, leucettamol A, isolated from *Leucetta microrhaphis* (Miosco et al., 2017).

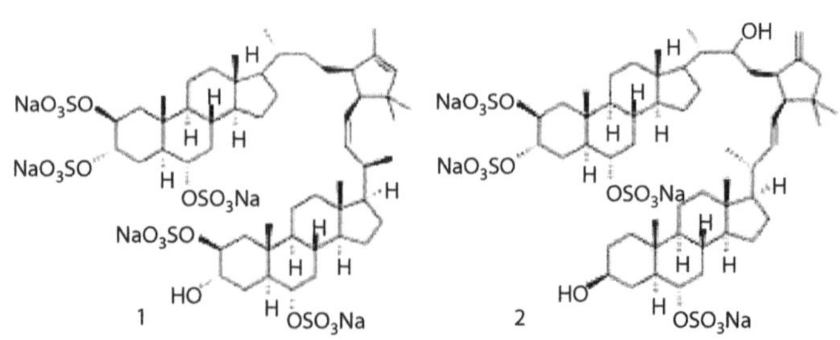

1, Manadosterol A; 2, Manadosterol B

Histodermella sp.

Order: Poecilosclerida Family: Coelosphaeridae

Compound(s) and Activities

Cytotoxicity: The indole alkaloid, makaluvamine G isolated from this sponge was found to be cytotoxic against P-388, A-549, HT-29, MCF-7 (human breast cancer), and KB with IC50 values of 0.5, 0.5, 0.5, 0.5, and 0.4 μg/mL, respectively (Gul et al., 2005).

FAMILY: DESMACIDIDAE

Desmapsamma anchorata (Carter, 1882)

Order: Poecilosclerida Family: Desmacididae

Common name: Lumpy overgrowing sponge, pink rope sponge

Distribution: South Florida and throughout the Caribbean Sea

Ecology: It is common in mangrove, seagrass, and shallow reef habitats to at least 18 m.

Biology: It is a highly variable sponge from thick crusts to massive lobate, or sparsely to profusely branching. The branches are of 1 to 4 cm in diameter. Surface is smooth, with pale barely visible dermal canals toward the oscules. Consistency is soft, compressible and elastic, easy to tear. The interior is crumbly. The oscules are elevated, 0.5 to 2 cm across, with a white conical membrane. Color of the sponge is pink or pale reddish through lavender gray to almost white externally, and always darker red internally.

Compound(s) and Activities

Antioxidant activity: Both the 70% and 100% fractions of this species showed low antiradical efficiency, the values of which ranged between 21.27% and 48.47% (Tovar-Barrios et al., 2016).

Antimalarial and anticoagulating activities: The heparin-like sulfated polysaccharides of this species demonstrated antiplasmodial capacity against *Plasmodium falciparum* with an IC50 ~66 μg/mL. Further, these compounds showed low anticoagulating activity, which could play a dual role as drugs and as potentiators of immune responses (Marques et al., 2016).

FAMILY: HYMEDESMIIDAE

Kirkpatrickia varialosa (Kirkpatrick, 1907)

Order: Poecilosclerida Family: Hymedesmiidae

Common name: Antarctic sponge

Distribution: From Sub-Antarctica and South Georgia to the Antarctic Peninsula and Continent

Ecology: It is found in patchy assemblages at depths of 20 to 600 m.

Biology: This sponge is in the form of a mass of thick, flabellate or digitate fronds arising from a common base. It has a maximum height of 30 cm. There are circular sphinctate oscules that are of 1 cm diameter and are situated at the summits or along the upper edges of the branches. Surface is covered with circular pore areas of about 4 mm diameter each. It is vivid red in color. Consistency is soft and fleshy. Spicules are megascleres, and microscleres are absent.

Predator: It is preyed on by seastars, particularly *Perknaster fuscus* when juvenile, and *Acodontaster conspicuus*.

Compound(s) and Activities

Neuroprotective activity: The extracts of this species have shown neuroprotective activity by regulating the activities of cholinesterase enzyme (it breaks down acetylcholine, the accumulation of which may impair the nervous system and muscles) (Berne et al., 2016).

Antitumor and antiviral activities: The derivatives from this species have been found to have antitumor and antiviral properties, and are being trialed as potential anticancer drugs (British Antarctic Society, n.d.).

Antiviral activity: The pyridopyrrolopyrimidine alkaloid variolin B isolated from this species has shown antitumor and antiviral activities (Perry et al., 1994; Sipkema et al., 2005).

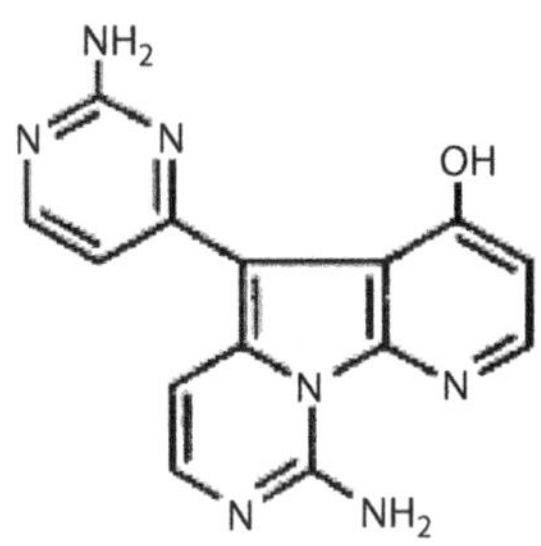

Variolin B

Cytotoxic and antifungal activities: Among the pyridopyrrolopyrimidine alkaloids variolin A and N(3′)-methyl tetrahydrovariolin B isolated from this species, N(3′)-Methyl tetrahydrovariolin B was found to be moderately cytotoxic and showed antifungal activity, while variolin A was weakly cytotoxic (Trimurtulu et al., 1994).

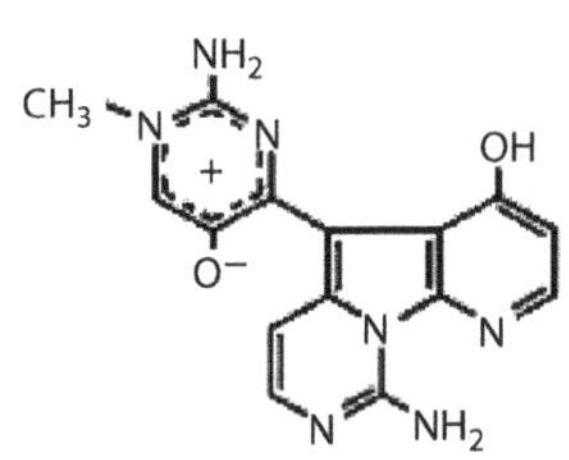

Variolin A

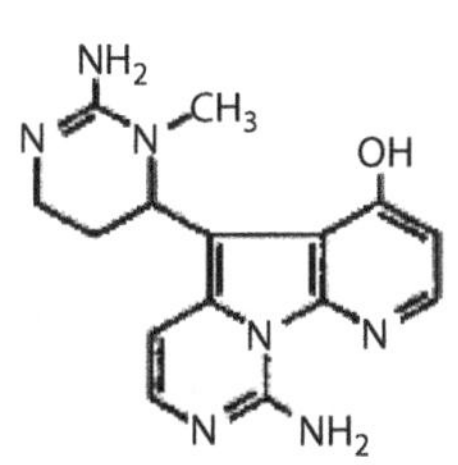

N(3′)-Methyl tetrahydrovariolin B

Phorbas tenacior (Topsent, 1925) (= *Anchinoe tenacior*)

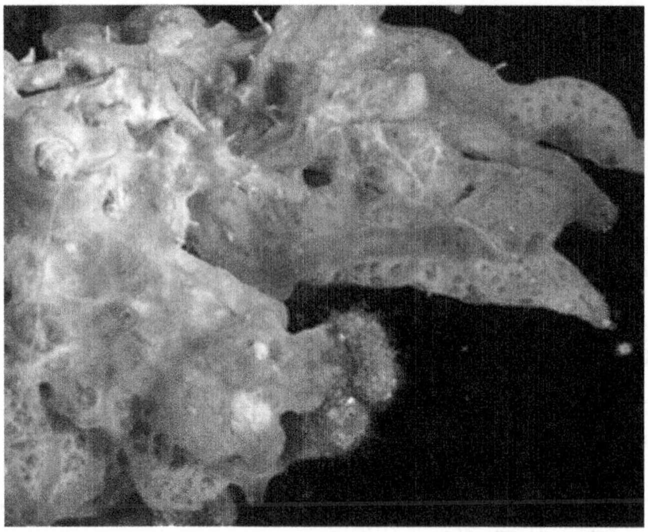

Order: Poecilosclerida Family: Hymedesmiidae

Common name: Not designated

Distribution: Northern Atlantic Ocean, Mediterranean

Ecology: Inhabits shaded reef ledges and crevices in areas where current is strong.

Biology: It is continuously encrusting a large extension of a vertical wall of the cave (more than two square meters). The strongyles of the spicules are constant in shape.

Compound(s) and Activities

Antimicrobial activity: The associated *Citricoccus alkalitolerans* strain of this sponge species showed a moderate antimicrobial activity against many environmental stressors including *V. parahaemolyticus* (Dupont et al., 2013).

Antiprotozoal activity: The associated actinobacterial strain *Micromonospora* sp. SBT687 isolated from this species showed antiprotozoal activity against *T. brucei brucei* with an IC50 value of 14.9 μg/mL (Indraningrat et al., 2016).

Antiplasmodial activity: The associated bacterial species of this sponge showed antiplasmodial activity with 45% inhibition of *Plasmodium falciparum* at 100 μg ml^{-1} (Dupont et al., 2013).

Antioxidant and antibacterial activities: The crude extracts of the associated species of *Vibrio* and *Shewanella* of this sponge species showed antioxidative and antibacterial properties (Dupont et al., 2013).

Phorbas sp.

Order: Poecilosclerida Family: Hymedesmiidae

Compound(s) and Activities

Antitumor activity: The sesterterpenoid phobaketal N of this species showed potent cytotoxicity against the human pancreas cell line (Panc-1) and the human renal cell lines (A498 and ACHN) with IC50 values of 11.4, 18.7, and 24.4 μM, respectively. Further, of the two nonbrominated phorbaketals (L and M) isolated from this sponge, only phorbaketal L showed cytotoxicity (A498, 17.3 μM) (Gribble, 2015).

Hamigera tarangaensis (Bergquist and Fromont, 1988)

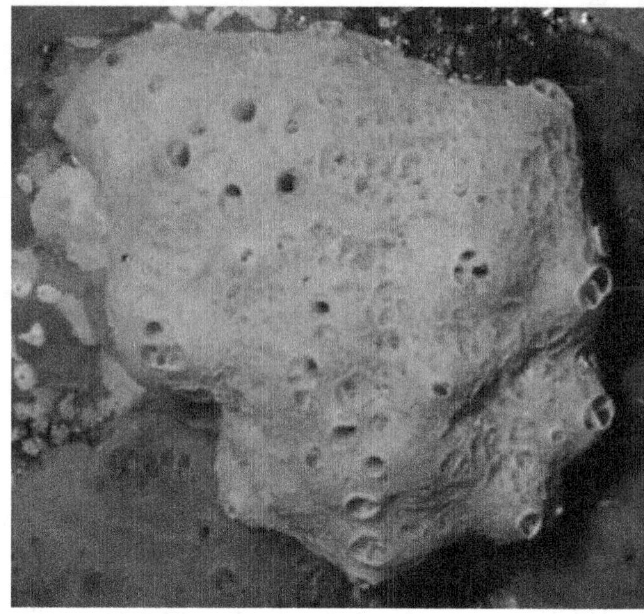

Order: Poecilosclerida Family: Hymedesmiidae

Common name: Not designated

Distribution: New Zealand

Ecology: Not reported

Biology: It is an encrusting sponge with numerous pronounced, round, areolate pore areas raised about 0.5 mm above the surface. Thickness of the sponge is 7 mm. Texture is firm but elastic and is readily compressed. Ectoderm consists of diactional megascleres in a palisade around pore areas. Color of the sponge is orange red to yellow.

Compound(s) and Activities

Antiviral activity: The phenolic macrolide hamigeran B isolated from this species showed antiviral activity against herpes and polioviruses (Anjum et al., 2016; Sipkema et al., 2005).

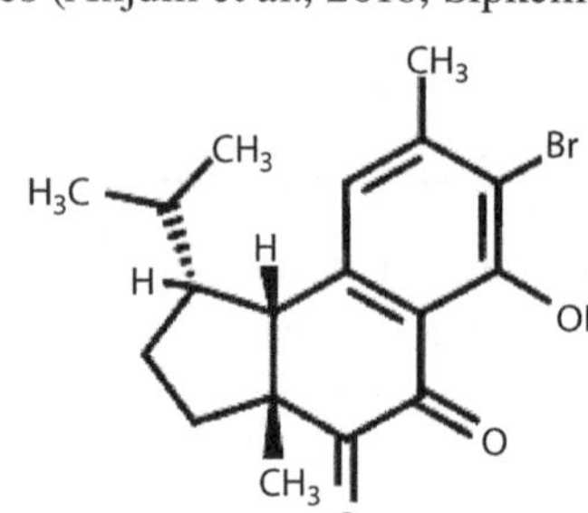

Hamigeran B

Antifungal activity: The terpenoid compound hamigeran G isolated from this species had antifungal activity against *S. cerevisiae* (Cheung et al., 2014; Gribble, 2015).

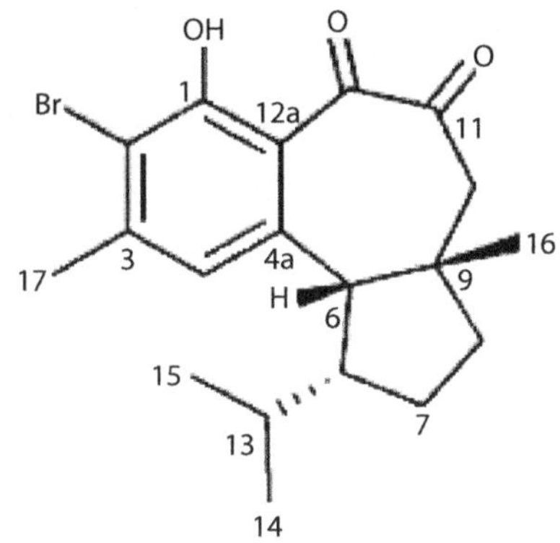

Hamigeran G

Antiviral and cytotoxic activities: Among seven new compounds isolated from this species, compounds 3–6 showed moderate in vitro cytotoxicity against P-388, while 3 showed 100% in vitro virus inhibition against both the herpes and polioviruses, with only slight cytotoxicity (Wellington et al., 2000).

Hemimycale arabica (Ilan, Gugel, and van Soest, 2004)

Order: Poecilosclerida Family: Hymedesmiidae

Common name: Arabian crust sponge

Distribution: Northern and Central Red Sea, Israel

Ecology: It is a thinly encrusting sponge on dead corals. It inhabits shallow water (2 m) and is associated with polychaete tubes and sand grains.

Biology: It is a thick cushion-like sponge with a maximum thickness of 2.5 cm. The ectosome is dark blue-black-green and the choanosome is yellow green. Openings of the aquiferous system are organized in very numerous areolated pore fields that form slight depressions. Compared with other species, these sponges are rather large and rounded or angular. The pore fields result in a reticulated appearance of the surface. The oscules are raised into small tubes (1–5 mm in dia.) irregularly distributed over the surface. This sponge mostly builds thin crusts of several millimeters thick and covers areas of up to 15 cm. The sponge is usually very soft.

Compound(s) and Activities

Antiproliferative and antimicrobial activities: The compounds (Z)-5-(4-hydroxybenzylidene)imidazolidine-2,4-dione, hydantoin alkaloids, and hemimycalins A and B (1–3) have been isolated from this species. Among these compounds, compound 1 showed moderate antiproliferative activity against HeLa cells with an IC50 value of 28 μg/mL. The other compounds were weakly active with IC50 values of >50 μg/mL. In the antimicrobial screen, compounds 1–3 displayed good activity against *E. coli* with inhibition zones of 18, 10, and 20 mm, respectively, while these compounds showed inhibition zones of 22, 14, and 20 mm against *C. albicans* (Youssef et al., 2015b).

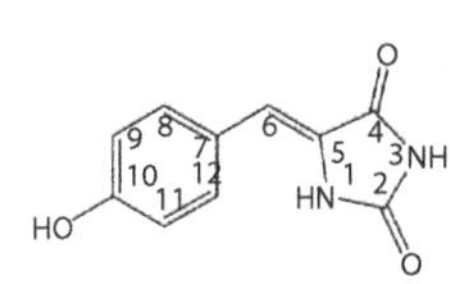

(Z)-5-(4-hydroxybenzylidene)imidazolidine-2,4-dione

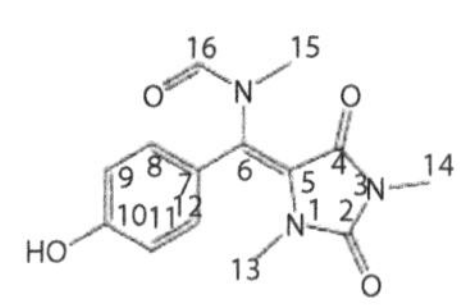

Hemimycalin A

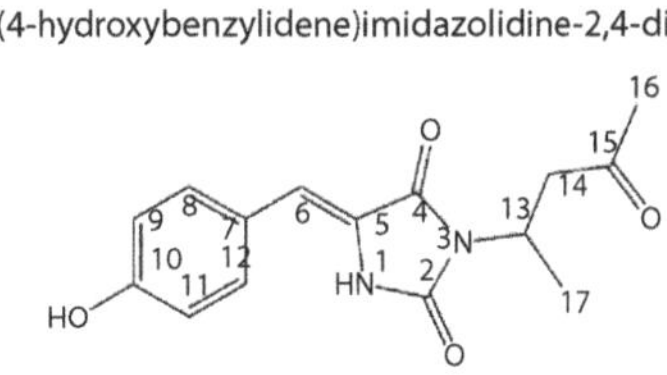

Hemimycalin B

The organic extracts of this species showed antiproliferation activity against the human colon adenocarcinoma cell line (Caco-2) (Rady, 2014).

Anticancer activity/cytotoxicity: The compound (Z)-5-(4-hydroxybenzylidene)-hydantin of this species showed anticancer activity *in vitro* against GSK3β cell line with an IC50 value of 13.7 μM (Brahmachari, 2017). The mesohyl of this sponge species exhibited a very potent cytotoxicity against Caco-2 cells with an IC50 value of 1 μl/mL (Rady, 2014).

Antidiabetic activity: The compound phenylmethylene hydantoin isolated from the extracts of this species showed antidiabetic activity by inhibiting GSK-3β and by increasing liver glycogen in rat (Lauritano and Lanora, 2016).

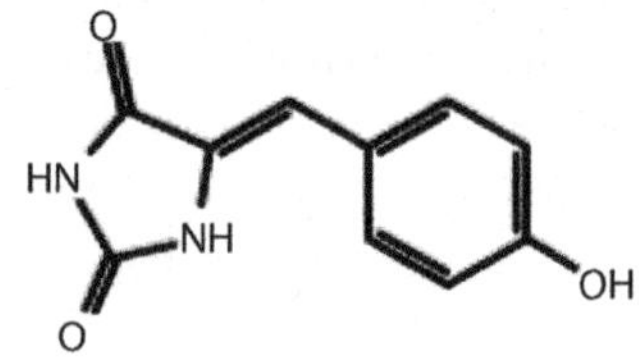

Phenylmethylene hydantoin

FAMILY: IOTROCHITIDAE

Iotrochota purpurea (Bowerbank, 1875)

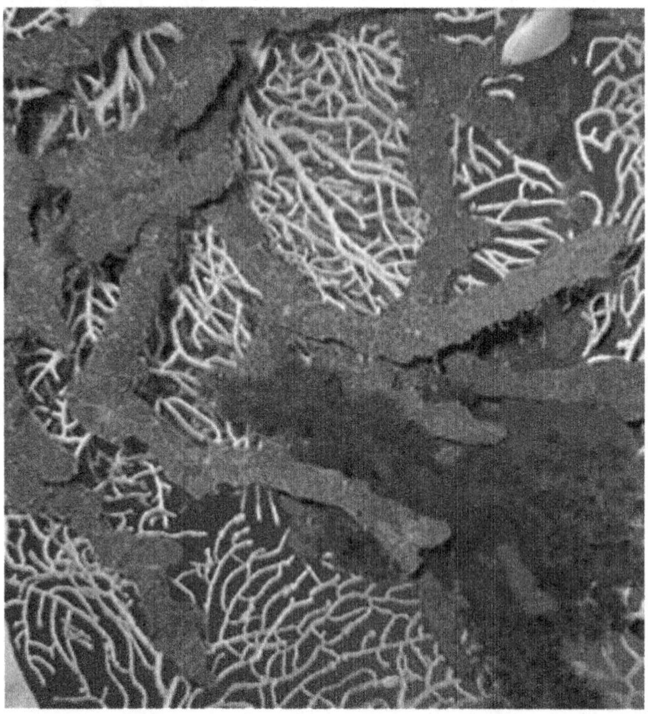

Order: Poecilosclerida Family: Iotrochitidae

Common name: Iodine finger sponge, Indian Ocean sponge

Distribution: Southeast Asia, Malacca Strait

Ecology: It is a reef-associated species.

Biology: It is sessile and massive sponge. Surface is rough and rugged. Oscula are simple and dispersed. Pores are inconspicuous. Dermal and interstitial membranes are spiculous. Color of the animal is dark purple in its dried state.

Compound(s) and Activities

Antimicrobial activity: The halogenated alkaloids purpuroines A–J isolated from this species showed antimicrobial activities (Perdicaris et al., 2013).

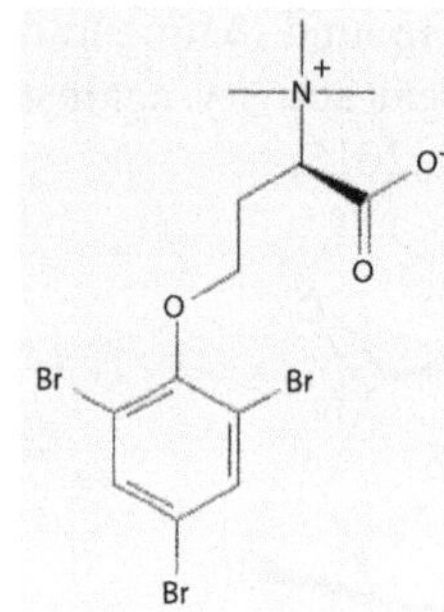

Purpuroine A

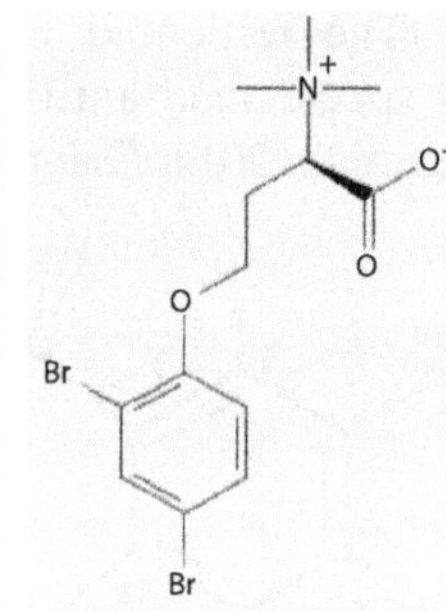

Purpuroine B

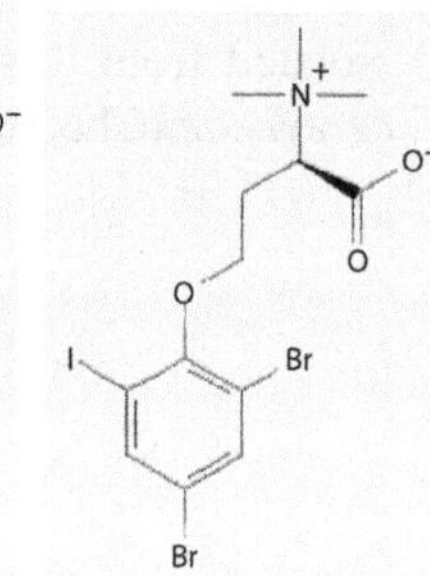

Purpuroine D

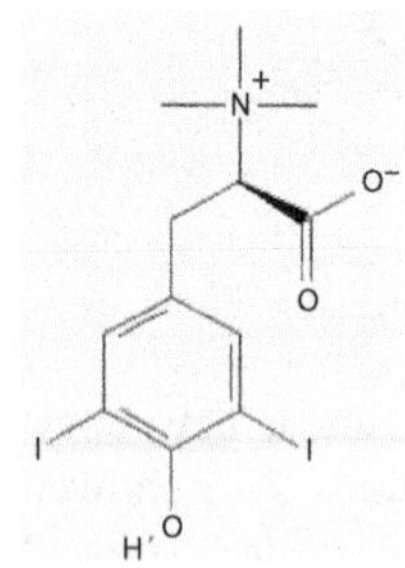

Purpuroine G

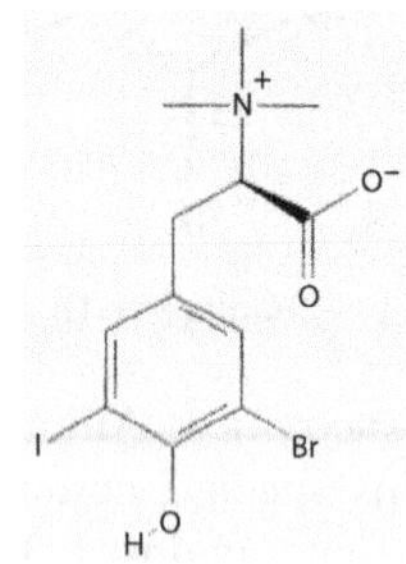

Purpuroine I

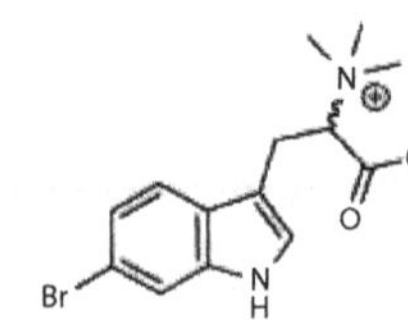

Purpuroine J

Ten new halogenated alkaloids, namely, purpuroines A–J, have been isolated from this species. Among them the compounds A and D possessed activity against human pathogenic fungi by inhibiting the enzyme kinase LCK (Shen et al., 2012).

The compound purpuroine I of this species showed selective inhibition of the human pathogen *S. pneumonia* with an IC50 value of 18.1 μg/mL (Gribble, 2015).

Antibacterial and cytotoxic activities: A new bioactive bromine-containing oxindole alkaloid, matemone (1), together with the known 6-bromoindole-3-carbaldehyde have been isolated from this species. Compound 1 showed mild cytotoxicity against three cancer cell lines and marginal antibacterial activity against *Staphyloccocus aureus* (Carletti et al., 2000).

Iotrochota sp.

Order: Poecilosclerida Family: Iotrochitidae

Compound(s) and Activities

Antioxidant activity: The ethanolic extract of this species yielded an aromatic purple pigment alkaloid compound, namely, purpurone, which showed with an IC50 value of 19 μM (Liu et al., 2008).

The ethanolic extract of this sponge reduced the radical 2,2-diphenyl-1-picrylhydrazyl (DPPH) (IC50 = 183.97 µg/mL). On the other hand, the purpurone-like compound isolated from this extract had potent antioxidant activity with an IC50 value of 19 µm (Tovar-Barrios et al., 2016).

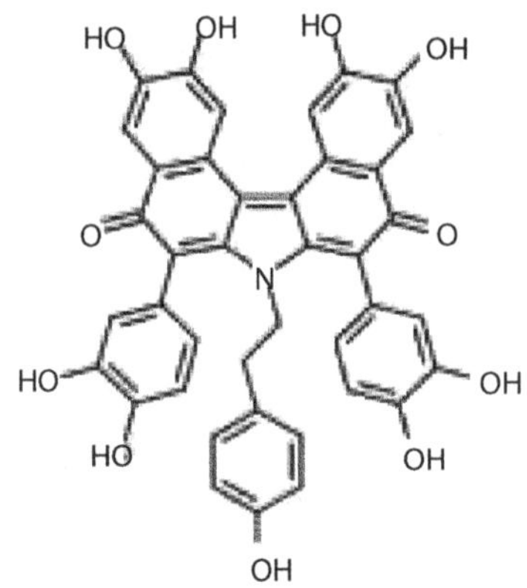

Purpurone

Cytotoxicity: A new glycosphingolipid compound iotroridoside-A (1-O-β-D-glucopyranosyl-2-[(4′Z)-2′-hydroxytetracosene amido]-4-tetradecyl-1, 3, 4-butantriol) isolated from this unidentified species exhibited strong cytotoxicity against L1210 murine leukemia cells *in vitro* (ED50 = 0.08 µg/mL) (Deng et al., 2001).

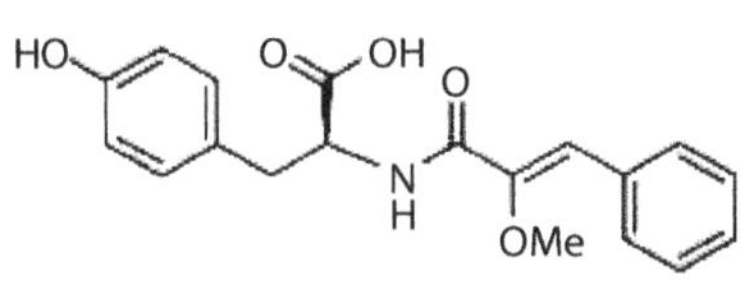

Iotroridoside A

Antiparasitic activity: The compounds iotrochamides A and B of this species exhibited moderate activity against *Trypanosoma brucei brucei* (IC50 3.4 and 4.7 µM, respectively) (Gribble, 2015).

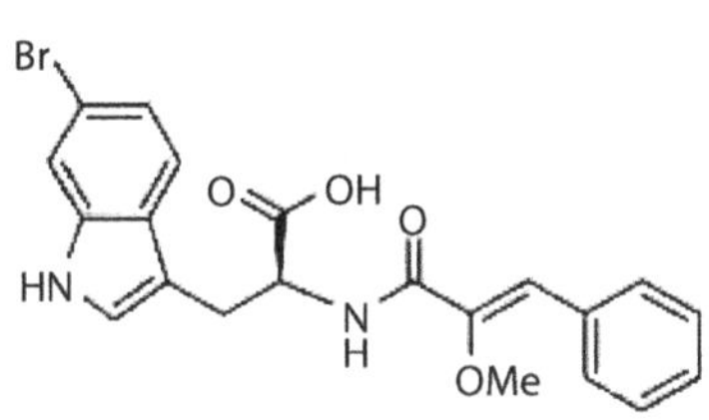

Iotrochamide B

FAMILY: ISODICTYIDAE

Isodictya erinacea (Topsent, 1916)

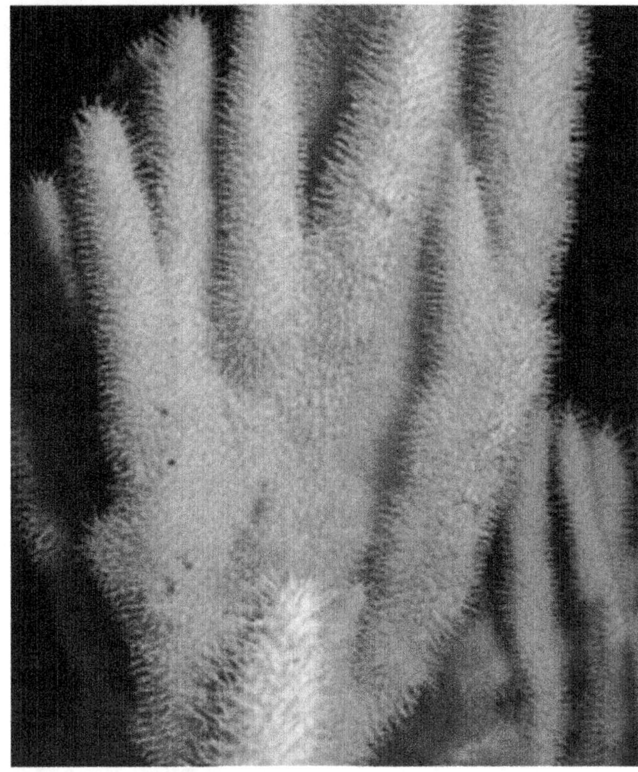

Order: Poecilosclerida Family: Isodictyidae

Common name: Antarctic sponge, polychaete sponge

Distribution: Throughout Antarctica, the Antarctic Peninsula, South Shetland Islands, and the Falkland Islands

Ecology: It inhabits rocky coral reefs at 30 to 900 m depth.

Biology: The body of this sponge is globular, cylindrical, and stem- or treelike. It is bright yellow with firm consistency. The ectoosme is very thin. Size of the specimen is 26 × 5 cm. The surface of this sponge has numerous conules in two morphs: short conules or long conules up to 9 cm long. Oscules are abundant with 2 mm diameter.

Association: On the surface of this sponge, small (3 mm long) bivalve mollusks are often seen. This species may also host diatoms within its food-capturing cells that line the passages through which the sponge circulates water. These endobiont diatoms live by consuming carbohydrates produced by the sponge and also by photosynthesis. Diatoms produce large amounts of polysaccharides, thus giving the sponge an alternative food source during food-scarce periods. This symbiotic adaptation enhances the survival of these sponges in the low light levels found down deep under the ice (as well as the dark months of winter).

Compound(s) and Activities

Neuroprotective activity: The extracts of this species have shown neuroprotective activity by regulating the activities of cholinesterase enzyme (it breaks down acetylcholine, the accumulation of which may impair the nervous system and muscles) (Berne et al., 2016).

Anti-inflammatory activity: The ethanol extract of this species showed anti-inflammatory activity by inhibiting significantly and dose-dependently the release of IL-1β and PGE2 (Moles et al., 2014).

Isodictya setifera (Topsent, 1901)

Order: Poecilosclerida Family: Isodictyidae

Common name: Stringy sponge

Distribution: Throughout Antarctica and the Antarctic Peninsula, South Orkney Islands, South Georgia Island, Falkland Islands, southern Argentina, Tierra del Fuego, and New Zealand

Ecology: It inhabits both shallow and deeper water from 0 to 710 m depth. It attaches to rocks or *Limatula hodgsoni* valves. Isodictya setifera is uncommonly seen, being 0.02% of the benthic surface cover at a Cape Armitage site. Isodictya setifera has been observed to settle and grow to sizes of 5–8 cm in diameter in a 3-year period.

Biology: The body of this sponge may be globular, spherical, cylindrical, finger shaped, or gradually thickening near the distal end and shaped like a club (clavate). It may have a stalk. The color of this sponge is cream, pink, pale yellow red, or different shades of yellow. Size of the specimen is 10 × 3 cm. Consistency is soft and is not very flexible. It has crumbly and hispid surface. Predators: Sea star *Odontaster meridionalis* and the nudibranch *Doris kerguelenensis*.

Compound(s) and Activities

Antibacterial activity: The symbiotic bacterium *Pseudomonas aeruginosa* of this species produced a compound, cyclo-(L-proline-L-methionine), which showed antibacterial activity against *Bacillus subtilis*, *Staphylococcus aureus*, and *Micrococcus luteus* (Thomas et al., 2010).

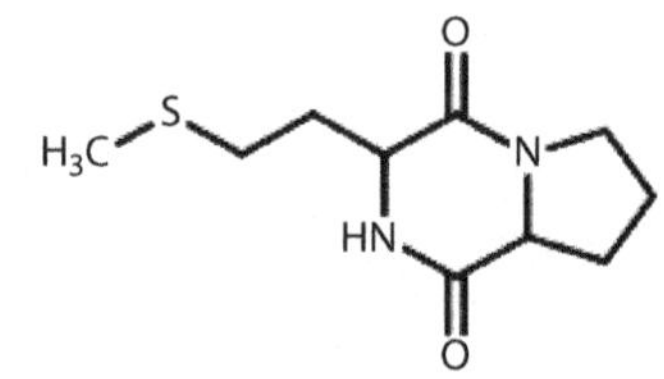

Cyclo-(L-proline-L-methionine)

The extracts of this species showed antibacterial activity against the environmental bacterial strains *Exignuobacterium* sp. and *Vibrio ruber* with the MIC values of 70 and 65 μg/mL, respectively (Turk et al., 2013).

Isodictya toxophila (Burton, 1932)

Order: Poecilosclerida Family: Isodictyidae

Common name: Antarctic sponge

Distribution: South Georgia

Ecology: It attaches to rock fragments.

Biology: It is a conical, erect sponge with 4 cm height, 2 cm width and 1 cm thickness. Surface is hispid. Oscules are abundant and are of 0.4 cm diameter. Preserved specimens possess firm consistency and grayish beige color. Ectosome is with abundant spongin and exogenous material. Megascleres are of oxeas, which are slightly curved and microscleres are also with curved contour.

Compound(s) and Activities

Anti-inflammatory activity: The EtOH extract of this species showed anti-inflammatory activity by reducing the release of IL-1β and LTB4 at two concentrations (50 and 125 μg/mL) and also diminished PGE2 release at the intermediate concentration (125 μg/mL) (Moles et al., 2014).

Cytotxicity: The EtOH extracts of this species showed cytotoxicity, only at two higher concentrations, namely, 125 and 250 μg/mL (Moles et al., 2014). The organic dried extract of this species exhibited cytotoxicity against lung fibroblast (V-79) and human colon epithelial (Caco-2) cancer cell lines at a concentration of 100 μg/mL. The percentage inhibition values recorded for these cell lines were found to be 85 and 98, respectively (Turk et al., 2013).

Antibacterial activity: The extracts of this species showed antibacterial activity against the laboratory, commensal, and clinically relevant bacterial strains with the MIC values.

Bacterial Strain	MIC (μg/mL)
Staphylococcus aureus (MRSA) S-943 A	600
Staphylococcus epidermidis EXB-V55	200
Macrococcus 1F	100
Micrococcus 2F	150

Source: Turk, T. et al., 2013, *Mar. Drugs*, 11: 1126–1139.

FAMILY: LATRUNCULIIDAE

Latrunculia (Latrunculia) austini Samaai (Gibbons and Kelly, 2006)

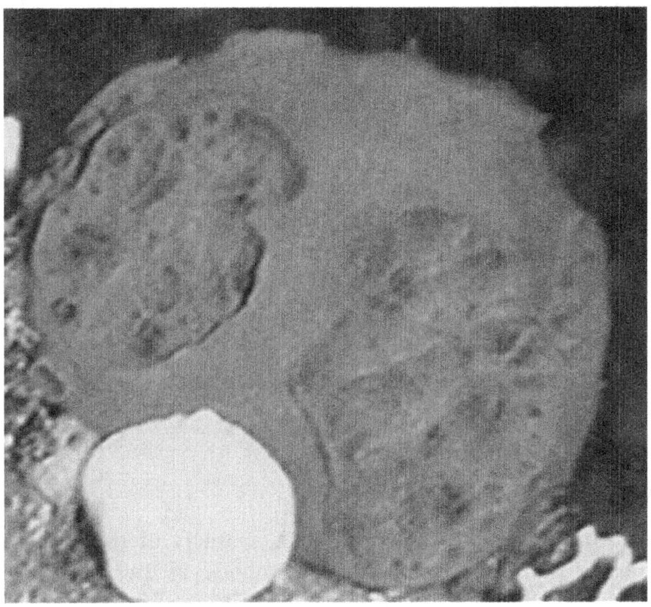

Order: Poecilosclerida Family: Latrunculiidae

Common name: Green deep-sea sponge, Alaska sponge

Distribution: Canada, Gulf of Alaska, North American Pacific Fijordland, Oregon, Washington, Vancouver Coast and Shelf, Puget Trough/Georgia Basin

Ecology: This sponge lives on rocks in patches at depths of about 70 to 200 m and it sticks very firmly to rocky surfaces.

Biology: It is a small, golf-ball sized, spherical to block-shaped sponge, typically about 5–8 cm diameter, but can reach up to 12 cm diameter. It is found often in large groups. Surface is distinctive with small, circular, crater-shaped pore fields, or pore fields that cover large sections of the surface. There are one to five large oscules that appear prominently on the apex of specimens. Texture is very soft and compressible, somewhat slimy in life. Color in life is khaki to emerald green in deeper waters. Ectosome is a palisade of vertically arranged anisodiscorhabds, below which is a thin paratangential layer of megascleres. Choanosome is a polygonal reticulation of wispy tracts of anisostyles.

Compound(s) and Activities

Anticancer activity: The discorhabdin compounds isolated from the organic extracts of this species have been reported to be the most active molecule against pancreatic cancer as these have the ability to kill pancreatic cancer cells. These compounds may therefore be potential drugs in the treatments for pancreatic cancer (AFSC News, 2017).

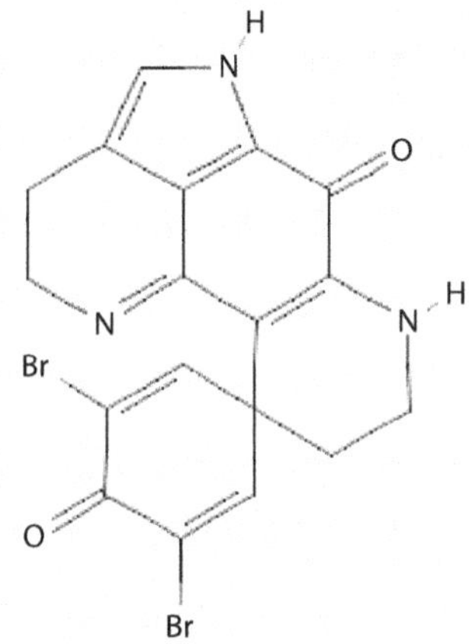

Discorhabdin

Latrunculia (Latrunculia) basalis (Kirkpatrick, 1908) (= *Latrunculia lendenfeldi*)

Order: Poecilosclerida Family: Latrunculiidae

Common name: Not designated

Distribution: Western Central Pacific, South Atlantic Ocean, Uruguay

Ecology: No reported

Biology: Not available

Compound(s) and Activities

Neuroprotective activity/cholinesterases (ChEs) sctivity: The ethyl acetate extracts of this species rich in terpenoid compounds showed AChE inhibition with an IC50 value of 1.3 ng/mL (Grosso et al., 2014; Turk et al., 2013).

Cytotoxic activity: The organic dried extract of this species exhibited cytotoxicity against lung fibroblast (V-79) and human colon epithelial (Caco-2) cancer cell lines at a concentration of 100 μg/mL. The percentage inhibition values recorded for these cell lines were found to be nil and 2, respectively (Turk et al., 2013).

Antibacterial activity: The extracts of this species showed antibacterial activity against the laboratory, commensal, and clinically relevant bacterial strains, and the MIC values are given next.

Bacterial Strain	MIC (µg/mL)
Staphylococcus aureus (MRSA) S-943 A	15
S. pseudintermedius (MRSP) S-053 A	80
S. pseudintermedius (MRSP) S-043 A	80
Listeria monocytogenes	90
Staphylococcus epidermidis EXB-V55	100
Staphylococcus aureus 10F	100
Bacillus subtilis EXB-V68	8
Macrococcus 1F	10
Micrococcus 2F	70
Escherichia coli HB101	700
Acinetobacter 1C	70

Source: Turk, T. et al., 2013, *Mar. Drugs*, 11: 1126–1139.

The extracts of this species showed antibacterial activity against the environmental bacterial strains *Exignuobacterium* sp., *Pseudoalteromonas* sp., *Alteromonas* sp., *Vibrio ruber*, and *Janthinobacterium svalbardensis* with the MIC values of 1.8, 0.4, 90, 0.9, and 0.1 µg/mL, respectively (Turk et al., 2013).

Latrunculia (Latrunculia) biformis (Kirkpatrick, 1908)

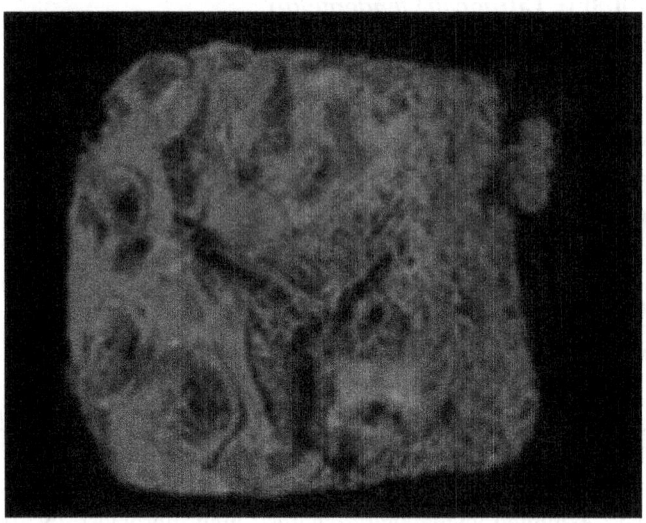

Order: Poecilosclerida Family: Latrunculiidae
Common name: Not designated
Distribution: Antarctica, Rio de la Plata, Ross Sea
Ecology: It is found on rocky substrata sheltered by algal cover and other invertebrates at about 20 m depth.
Biology: It is a hemispherical to globular sponge with a size of 6.5 × 4 × 4 cm. Its ectosomal layer is easily detachable from the underlying choanosome. Surface is microscopically smooth and is crowded with conical volcano-shaped oscules that are 1 × 0.2 mm. There are numerous flattened mammiform areolate porefields of 1 × 0.2 mm. Texture is firm, resilient, and slightly fleshy. External color in life is mottled chocolate brown and interior is dark chocolate brown.

Compound(s) and Activities

Antioxidant activity: The SC-CO2 extracts of this species showed the highest antioxidant activities, as 30.6 µg Trolox equivalent mg^{-1} extract (Botic et al., 2015).

Latrunculia (Latrunculia) bocagei (Ridley and Dendy, 1886)

Order: Poecilosclerida Family: Latrunculiidae
Common name: Not designated
Distribution: Antarctic Indian Ocean, Kerguelen Islands
Ecology: This sessile species occurs at a depth of about 300 m.
Biology: It is a massive, hemispherical sponge 3 × 3 × 4 cm in diameter. Surface is smooth with volcano-shaped or cylindrical oscules that are 5 mm high. Ectosome is thick and leathery and is not separable from the underlying choanosome. Color in life is dark brown and in preservative it is pale yellow. The choanosomal skeleton is a very irregular, polygonal meshed reticulation formed by wispy tracts of smooth styles.

Compound(s) and Activities

Antioxidant activity: The SC-CO_2 extracts of this species showed the highest antioxidant activities, as 29.7 µg Trolox equivalent mg^{-1} extract (Botic et al., 2015).
Neuroprotective activity/cholinesterases (ChEs) activity: The ethyl acetate extracts of this species rich in terpenoid compounds showed AChE inhibition with an IC50 value of 9 ng/mL (Grosso et al., 2014; Turk et al., 2013).

Latrunculia (Latrunculia) brevis (Ridley and Dendy, 1886)

Order: Poecilosclerida Family: Latrunculiidae
Common name: Not designated
Distribution: New Caledonia, New Zealand
Ecology: Not reported
Biology: It is a massive and amorphous sponge measuring 7 cm in length and 3.5 cm in width. Surface is densely areolated. Porefields measure about 1 cm diameter. Some areas are with smooth surface with a slight spicule protraction. Oscular openings are visible inside porefields.

Compound(s) and Activities

Antitumor activity: The pyrrolophenanthroline alkaloid compound discorhabdin D isolated from this species has shown antitumor activity (Anjum et al., 2016; Sipkema et al., 2005).

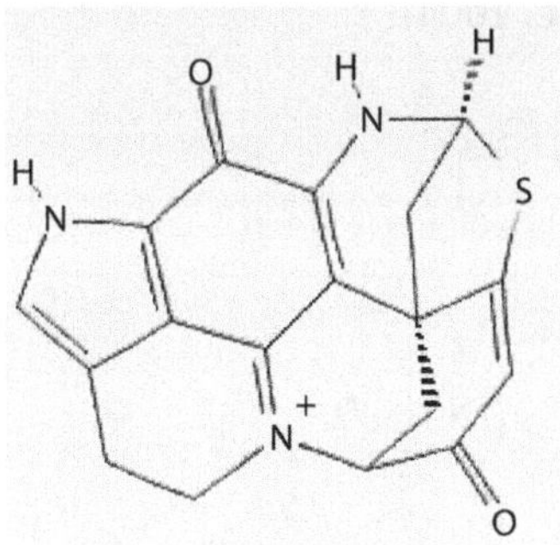

Discorhabdin D

Cytotoxicity: The alkaloids discorhabdins L and I isolated from this species showed strong *in vitro* cytotoxicity against a panel of 14 tumor cell lines (Reyes et al., 2004).

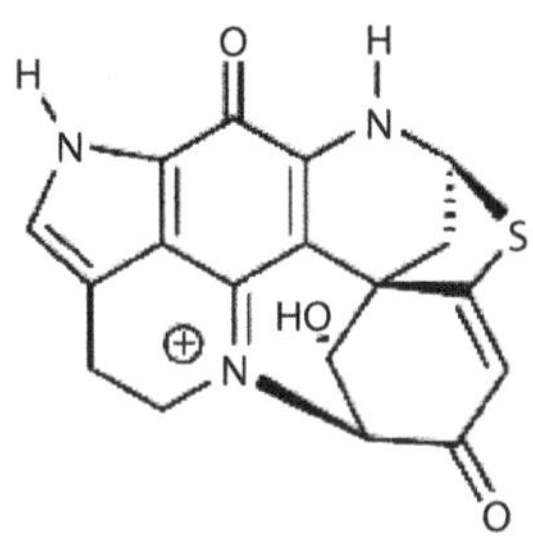

Discorhabdin L

Latrunculia (Latrunculia) hamanni (Kelly, Reiswig, and Samaai, 2016)

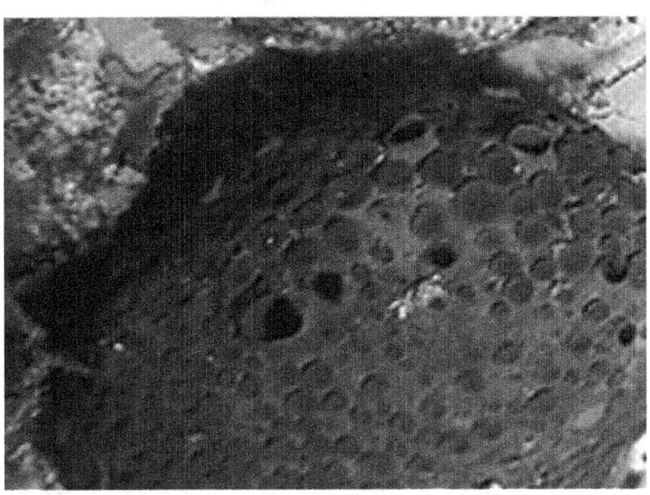

Order: Poecilosclerida Family: Latrunculiidae
Common name: Not designated
Distribution: Aleutian Islands
Ecology: It grows on bedrock, pebbles, cobbles, and boulders; on shelf and upper slope habitats; and at depths between about 80 and 300 m in areas with relatively high currents.
Biology: It is a hemispherical to spherical sponge with a 16 cm diameter and 5 cm thickness. There are short cylindrical oscules of 4–5 mm wide. These oscules are elevated 2 mm high. Numerous, small, raised, sucker-shaped areolate porefields cover the surface in life. Texture in life is firm, and surface is leathery and is difficult to tear. The interior is cavernous, fibrous, and stringy. Color in life is dark chocolate to purplish brown.

Compound(s) and Activities

Antiviral, antimicrobial, and antimalarial activities: The brominated pyrroloiminoquinones dihydrodiscorhabdin B and discorhabdin Y isolated from this species have shown significant antiviral activity against hepatitis virus C, antimalarial activity against *Plasmodium falciparum*, and antimicrobial effects against several AIDS opportunistic pathogens (Kelly et al., 2016).

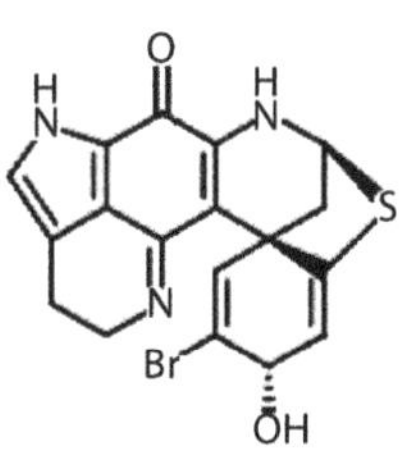

Dihydrodiscorhabdin B

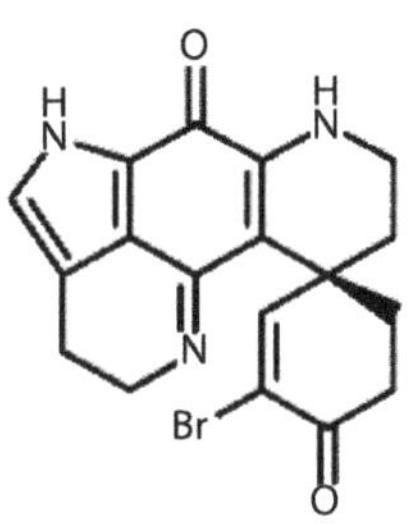

Discorhabdin Y

Latrunculia sp.

Order: Poecilosclerida Family: Latrunculiidae

Compound(s) and Activities

Cytotoxic, antimicrobial, antiviral, antimalarial, and immunomodulatory activities: The discorhabdins and discorhabdin-related alkaloids of *Latrunculia* spp. showed a plethora of biological effects, which include strong cytotoxic, antimicrobial, antiviral, antimalarial, and immunomodulatory activities (Botic et al., 2017).

Antifungal activity: The peptides callipeltins F–I of this species inhibit the growth of *Candida albicans* (ATCC24433) with an MIC value of 10^{-4} M. Further, the peptides, callipeltins J and K of this species inhibited the growth of *C. albicans* with MIC values of ca. 4^{-10} M (Agrawal et al., 2016).

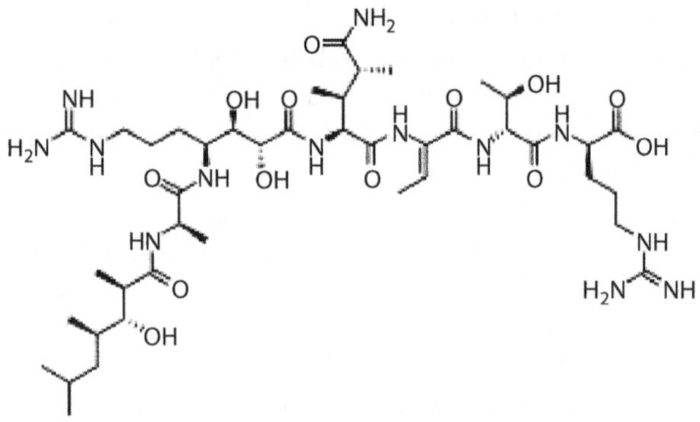

Callipeltin I

Acetylcholinesterase inhibition activity: The compounds discorhabdin G (1), 3-dihydro-7,8-dehydrodiscorhabdin C (2), discorhabdin B (3), and discorhabdin L (4) of this species have shown inhibitory activities against electric eel acetylcholinesterase (eeAChE), human recombinant acetylcholinesterase (hAChE), and equine serum butyrylcholinesterase (BChE) (Botic et al., 2017).

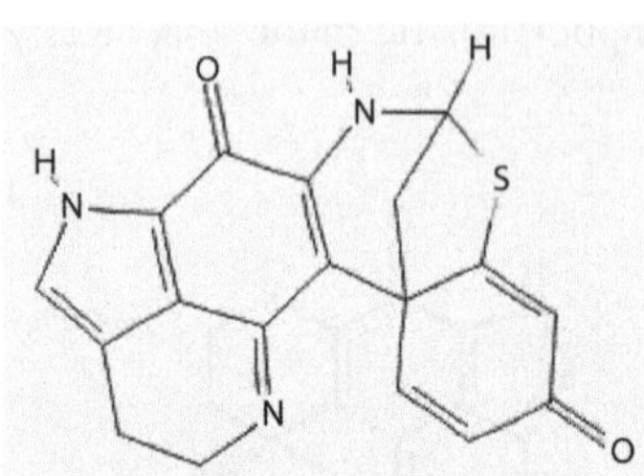

Discorhabdin G

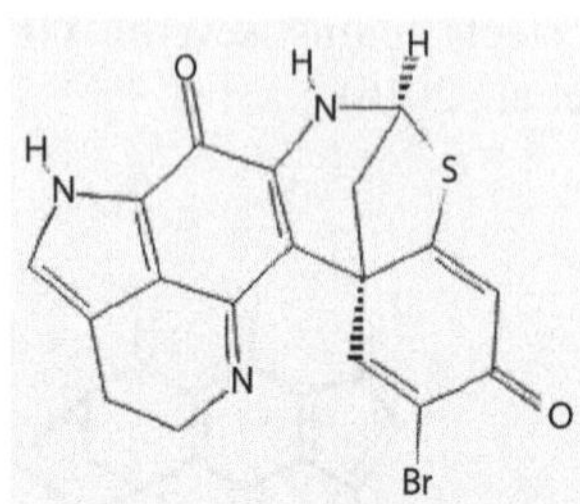

Discorhabdin B

The IC50 values recorded for these enzymes are given next.

Compound	eeAChE (IC50 mM)	hAChE (IC50 mM)	BChE (IC50 mM)
1	1.3	116	7.0
2	14.5	152	15.8
3	5.7	49.4	137
4	25.7	158	531

Source: Botic, T., 2017, *Eur. J. Med. Chem.*, 136: 294–304.

FAMILY: MICROCIONIDAE

Clathria (Thalysias) abietina (Lamarck, 1814)

Order: Poecilosclerida Family: Microcionidae
Common name: Philippine marine sponge
Distribution: Australia
Ecology: Not reported
Biology: No information

Compound(s) and Activities

Cytotoxicity and anti-TB activity: The compounds microcionamides A (1) and B (2) isolated from this species displayed significant cytotoxicity toward human breast tumor cell lines MCF-7 and SKBR-3 with IC50 values of 125 and 98 nM for compound (1) and 177 and 172 nM for compound (2), respectively. These compounds also displayed inhibitory activity against *Mycobacterium tuberculosis* H37Ra (Agrawal et al., 2016; Davis et al., 2004).

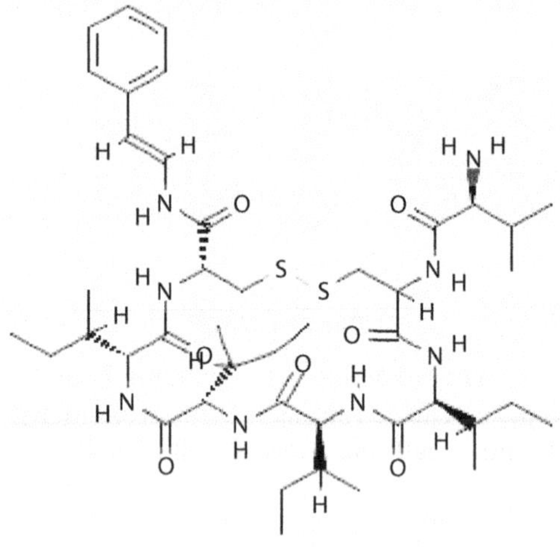

Microcionamide A

Clathria (Clathria) basilana (Lévi, 1961)

Order: Poecilosclerida Family: Microcionidae

Common name: Red vase sponge, Philippine marine sponge

Distribution: Indo-West Pacific, Philippines

Ecology: It inhabits caves, crevices, and slope areas where current is swift.

Biology: Not reported

Compound(s) and Activities

Antibacterial activity: The extract of this sponge species showed antibacterial activity against the following values of zone of inhibition

Zone Inhibition (mm) of Extract of Sponge

S. aureus	*E. coli*	*K. pneumoniae*	*S. typhi*	*P. aeruginosa* MDR	*S. aureus* MRSA
8.10	11.0	12.50	–	7.0	15.0

Source: Cita, Y.P. et al., 2017, *J. Marine Sci. Res. Dev.*, 7:237.

Antioxidant and anticancer activities: The fraction of methanol–water extract of this species, at concentration of 500 ppm, showed DPPH antioxidant activity of more than 50%, while the fraction of n-hexane at concentration of 500 ppm showed antioxidant activity only 10%. Further, the fractions of methanol–water and n-hexane had anticancer activity of more than 50% at concentration of 125 ppm (Karyawati et al., 2015).

Clathrina clathrus (Schmidt, 1864)

Order: Poecilosclerida Family: Microcionidae

Common name: Yellow clathrina, Mediterranean calcareous sponge

Distribution: British Isles, west coasts of France and Spain, Mediterranean, New Zealand and Indonesia

Ecology: It occurs in the *Laminaria* zone down to 50 m.

Biology: This species forms irregular cushions made up of anastomosed tubes. There are many contort and erect tubes, but these are all blindly ending. There are no visible oscules. Size of the sponge is 2–10 cm high and wide and the individual tubes are of 2 mm in diameter. Surface of the sponge is smooth and its consistency is soft. Color of the animal is white or yellow but brownish or reddish tinges occur also. There are calcareous (calcium carbonate) spicules.

Compound(s) and Activities

Antimycotic activity: The compound clathridine of this species showed significant antimycotic activity against *C. albicans* and *S. cerevisiae* at 40 µg/disk (Putra and Jaswir, 2014).

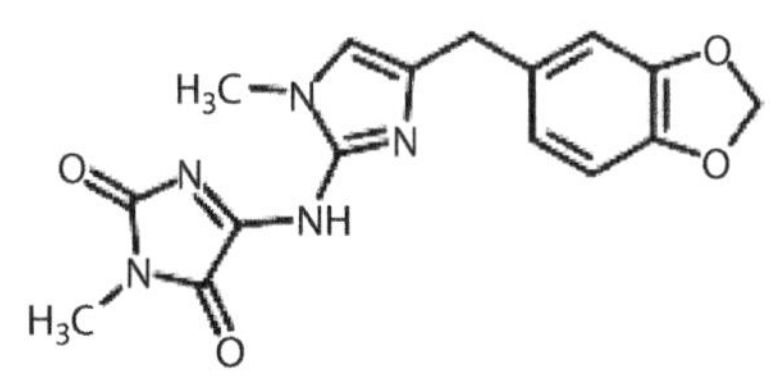

Clathridine

Antimicrobial activity: The CH2C12/MeOH 1:1 crude extract of this species displayed a broad spectrum of antimicrobial activities against Gram-positive *S. aureus* and Gram-negative *Escherichia coli* strains as well as against the yeast *Candida albicans* with the values of inhibition zone (diameter in mm), 11, 15, and 20, respectivley (Quévrain et al., 2014).

Clathria (Clathria) compressa (Schmidt, 1862)

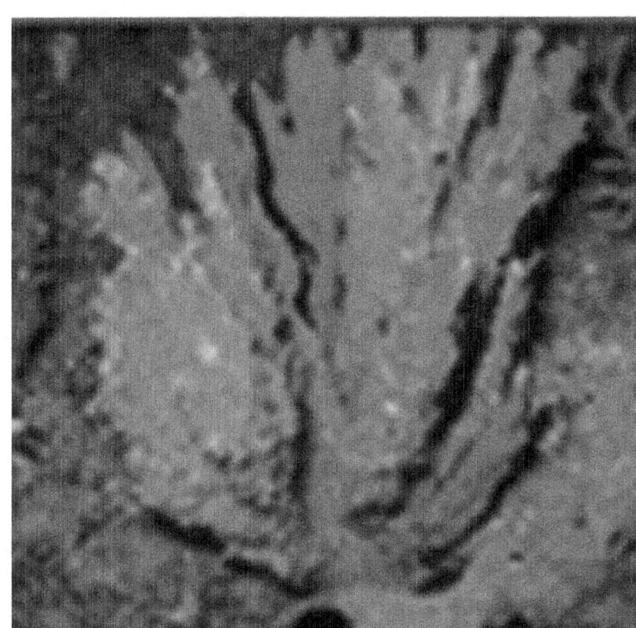

Order: Poecilosclerida Family: Microcionidae

Common name: Not designated

Distribution: Western Mediterranean

Ecology: Coral habitats

Biology: This sponge is erect, arborescent, thinly lamellate and branching. Surface is even and not hispid. The ectosomal skeleton is with tangential layer of subectodermal auxillary subtylostyles.

Compound(s) and Activities

Antibacterial activity: The bicyclic terpenoid clathric acid of this species showed mild antibacterial activity against *Staphylococcus aureus* (ATTC 6538P) (MIC value of 32 μg/mL); and against methicillin-resistant (ATTC 33591) and vancomycin-resistant *Staphylococcus aureus* (VRSA) (MIC value of 64 μg/mL) (Gupta et al., 2012).

Clathria (Microciona) echinata (Alcolado, 1984)

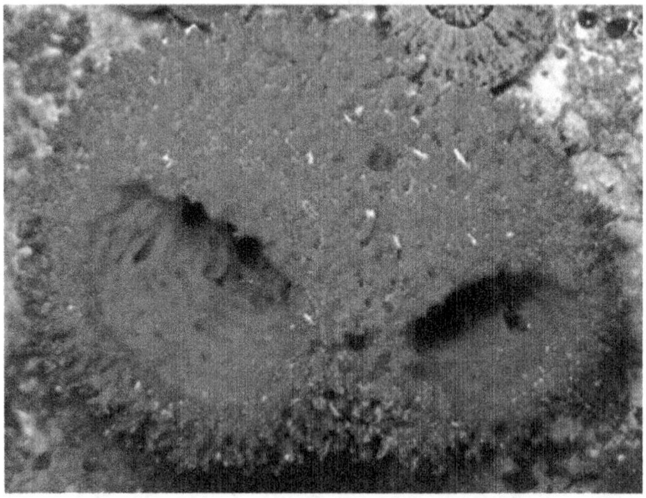

Order: Poecilosclerida Family: Microcionidae

Common name: Not designated

Distribution: Bermuda, Southwestern Caribbean, Western Caribbean, Cuba, Panamá, Belize

Ecology: This sponge grows on deeper edges of reefs (20 m deep) covered by mud; also found on reefs under 10 m deep.

Biology: This sponge forms clusters or single vase-like tubes (5–15 cm high). Color of the animal is bright red to orange externally and internally. Surface is very spiky. It is compressible but resilient to the touch.

Compound(s) and Activities

Antimalarial activity: The organic extracts of this species exhibited promising antimalarial activities in the infection model of C57BL/6 mice with *Plasmodium berghei* ANKA (Martínez et al., 2012; Perdicaris et al., 2013).

Clathria (Clathria) gombawuiensis (Kim and Sim, 2005)

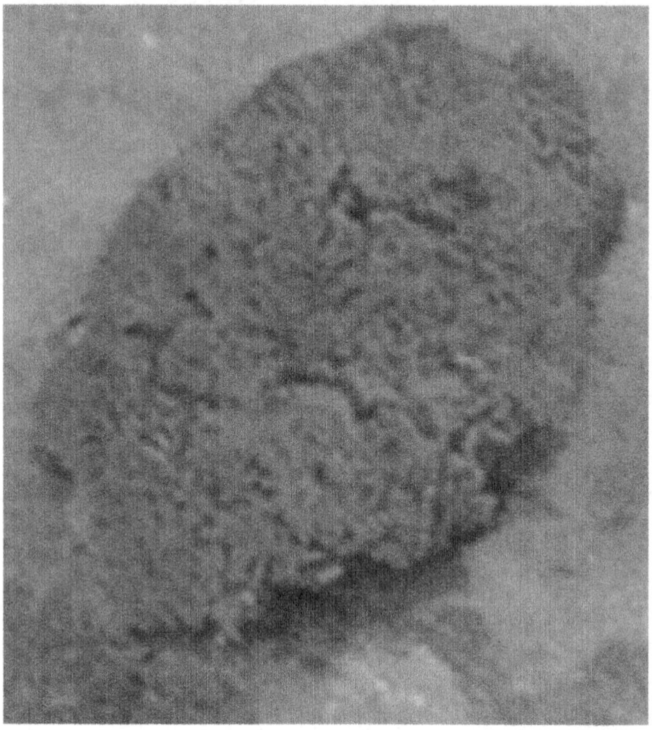

Order: Poecilosclerida Family: Microcionidae

Common name: Not designated

Distribution: Korea

Ecology: Not reported

Biology: Not available

Compound(s) and Activities

Cytotoxicity: The hexapeptide gombamide A isolated from this sponge showed weak cytotoxic activity against the K562 and A549 cell lines with LC50 values of 6.9 and 7.1 μM, respectively, as well as moderate inhibitory activity against Na+/K+-ATPase with an LC50 value of 17.8 μM (Agrawal et al., 2016).

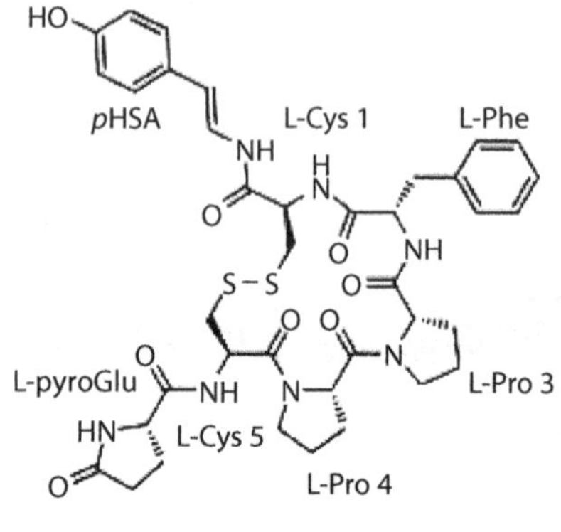

Gombamide A

Cytotoxicity and antibacterial activity: The sesterterpene metabolites gombaspiroketals A–C (1–3), phorone B (4), ansellone C (5), and saponin gombaside A (6) isolated from this species exhibited moderate cytotoxicity and antibacterial activity (Woo et al., 2015).

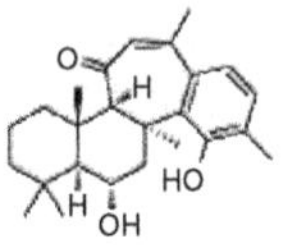
Phorone B

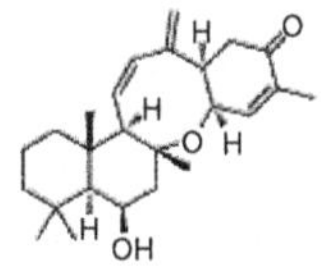
Ansellone C

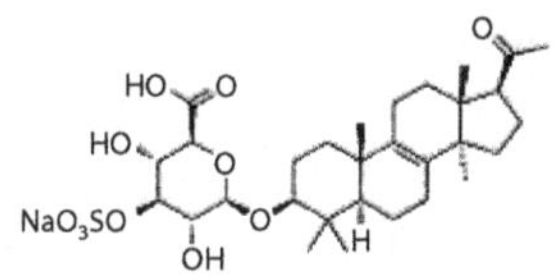
Gombaside A

Clathria (Clathria) gorgonioides (Dendy, 1916) (= *Echinodictyum gorgonoides*)

Order: Poecilosclerida Family: Microcionidae

Common name: Not designated

Distribution: Indian Ocean

Ecology: It grows on many marine habitats including chank beds.

Biology: It is a pretty sponge consisting of a number of slender stems springing close together from a somewhat spreading, encrusting base. The stems rise almost vertically from the points of attachment and bifurcate repeatedly. All the branches are of same diameter, about 3 mm. They mostly lie in approximately the same plane and occasionally anastomose with one another. The surfaces of the branches are subrugose and minutely conulose. The branches terminate in rounded extremities. The total height and width of the fully grown sponge is about 11 cm, and the greatest width nearly as much. The texture is fairly tough, very compressible, and resilient.

Compound(s) and Activities

Antibacterial activity: *Streptomyces collinus* ICN1, an endosymbiotic actinomycete, isolated from this species showed antibacterial activity against the clinical pathogen methicillin-resistant *Staphylococcus aureus* (MRSA) with minimum inhibitory concentration of 2 μg/mL (Nandhagopal et al., 2017).

Antibacterial activity: The associated bacterial strain MB2 (*Staphyloccous* sp) isolated from this sponge species showed antibacterial activity against human pathogenic bacteria and the values of zone inhibition (dia. mm) in different concentration of the bacterial extract (100 μl and 150 μl) are given next.

Zone Inhibition Diameter (mm)

Bacterial pathogen	100 μl	150 μl
E. coli	7.2	8
Proteus spp.	10	11.5
S. aureus	2.2	1
Pseudomonas sp.	5.1	7
B. subtilis	11	15.6

Source: Dhinakaran, D.I. et al., 2012, *Afr. J. Biotechnol.*, 11: 15469–15476.

Antifungal activity: The crude extracts of this species showed antifungal activity against *Aspergillus niger, Pencillium notatum*, and *Candida albicans*, and the values of zone of inhibition at different concnentrations of the extract are given next.

Zone of Inhibition (Diameter in mm)

Aspergillus niger		*Pencillium notatum*		*Candida albicans*	
50 μl	**100 μl**	**50 μl**	**100 μl**	**50 μl**	**100 μl**
1.1	1.3	1.0	1.2	0.9	1.4

Source: Dhinakaran, D.I., 2012, *J. Appl. Pharm. Sci.*, 2: 52–55.

Mosquito larvicidal activity: The methanol extract of this species showed mosquito larvicidal activity against the larvae of *Culex* sp. (the mosquito vector of filariasis) with an LC50 value of 0.6 mg/mL (Sonia and Lipton, 2012).

Clathria (Clathria) indica (Dendy, 1889)

Order: Poecilosclerida Family: Microcionidae

Common name: Not designated

Distribution: Indian Ocean, Mozambique, India, Sri Lanka

Ecology: It is attached to solid substrates in benthic habitats.

Biology: This sponge is erect and flabellate with a number of flattened branching and anastomosing trabeculae, fused together so as to form a more or less continuous frond. Texture in the dry state is coarse and hard, and fibrous and tough. Color of the sponge is bright red. Oscula are apparently represented by numerous minute openings. Maximum size of the sponge is 18 cm in height and 28 cm in breadth.

Compound(s) and Activities

Antimicrobial activity: The peptides of this species possessed antibacterial activity (Joseph et al., 2012). The crude extracts of this species showed antimicrobial activity against *S. aureus* and *C. albicans* at a concentration of 100 μg/mL. The chloroform fraction of the extracts exhibited selective activity on *S. flexineri* and *Vibrio cholerae*. Further, the methanolic extract and chloroform fraction of this species inhibited the growth *of S. typhi*, *C. albicans*, and *C. neoformans* (Ravichandran et al., 2011).

Antibacterial activity: The crude extracts of this species showed antibacterial activity. Among the various strains, a 12 mm maximum diameter zone of inhibition was recorded in the *E. coli* strain and a 6 mm minimum zone of inhibition was observed in S. *typhi*, *P. mirabilis*, and *L. vulgaris* strains (Duraikannu et al., 2009).

Antifungal activity: The n-butanol fraction of this species showed activity against different strains of fungi including *Aspergillus niger, Aspergillus fumigatus, Candida albicans, Fusarium solani*, and *Fusarium* sp. Maximum zone of inhibition was observed in *Fusarium* sp. and the minimum zone of inhibition was recorded in *Aspergillus fumigatus* (Duraikannu et al., 2009).

The aqueous, n-butanol, and chloroform fractions of the extracts of this species displayed antifungal activity against *C. albicans* and *C. neoformans* (Cheung et al., 2014).

Antiplasmodial activity: The ethyl acetate extracts of the bacteria isolated from this sponge species showed *in vitro* antiplasmodial activity. The activity of an isolate THB23 (IC50, 28.8 μg.mL^{-1}) extract was highly comparable with the positive control chloroquine (IC50, 19.6 μg.mL^{-1}) and 17 bacterial extracts showed IC50 values of more than 100 μg.mL^{-1}. The *in vitro* antiplasmodial activity might be due to the presence of carbohydrates and alkaloids in the ethyl acetate extracts of bacterial isolates. It is suggested that the novel compounds of the ethyl acetate extracts of THB23 may be of great use for the development of antiplasmodial drugs (Inbaneson and Ravikumar, 2012).

Others: The crude extract of this species showed antagonistic activity against urinary tract pathogens, namely, *Pseudomonas* sp., *Streptococcus* sp., and *Vibrio* sp., isolated from UTI-infected patients (Mauti et al., 2014).

Clathria (Clathria) lissosclera (Bergquist and Fromont, 1988)

Order: Poecilosclerida Family: Microcionidae

Common name: Not reported

Distribution: New Zealand

Ecology: Not reported

Biology: Not studied

Compound(s) and Activities

Anti-inflammatory activity: The steroids clathriol A and B isolated from this species were found to possess anti-inflammatory activity and inhibit the production of superoxide from human peripheral blood neutrophils stimulated with either fMLP (54: IC50, 33 μM; 55; IC50, 27 μM) or PMA (54: 140 μM; 55: 130 μM) (Keyzers, 2005).

Clathria (Microciona) prolifera (Ellis and Solander, 1786)

Order: Poecilosclerida Family: Microcionidae

Common name: Red beard sponge

Distribution: Western Atlantic Ocean from Prince Edward Island southward to Florida and Mexico; Brazil; west coast of the United States

Ecology: It grows in shallow water (subtidal) habitats and encrusts shells and pilings. It commonly occurs in bays and harbors in dock fouling and on the undersides of low intertidal and subtidal rocks, down to salinities of around 15 ppt.

Biology: This species has an encrusting base that may be up to 20 cm across but less than 3 mm thick. It forms bushy masses up to 20 cm tall of orange or red branches up to 6 mm thick. These branches resemble fingers, or may join together to form fanlike or drapery-like folded sheets. The spongy surface is covered with minute pores through which water is drawn into the sponge. It has spinose styles.

Food and feeding: It draws in water through its pores and filters out planktonic particles on which it feeds.

Reproduction: The larvae are released in the summer and autumn. At first they rise to the surface but after a day or so they sink to the bottom and crawl across the substrate until they find suitable locations for settlement. The juvenile sponges are encrusting at first.

Compound(s) and Activities

Antimicrobial activity: The associated Gammaproteobacteria with the dominant genus *Pseudoalteromonas* derived from this species have been reported to produce a variety of compounds possessing antimicrobial activity (Blasiak and Clardy, 2010; Nguyen, 2015).

Clathria (Thalysias) reinwardti (Vosmaer, 1880)

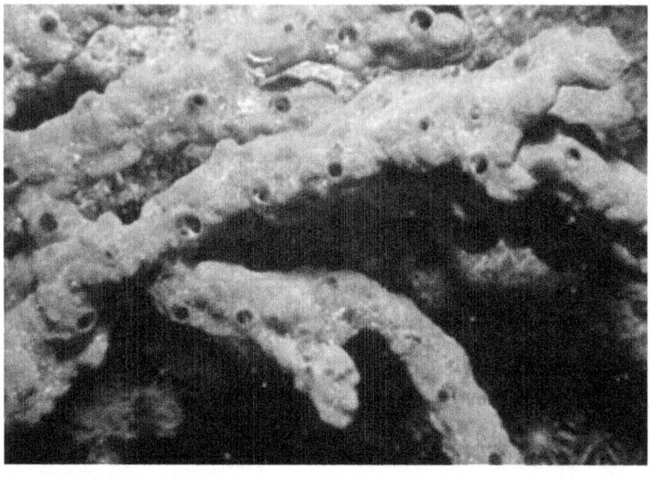

Order: Poecilosclerida Family: Microcionidae

Common name: Orange sprawling sponge

Distribution: Australia, Caroline Islands, Vietnam, Indonesia, Motupore Island, Papua New Guinea, Solomon Island, Zanzibar, Philippines

Ecology: It is found attached to some dead coral (5–10 m) at an area with sandy substrate and patches of corals.

Biology: It is massive, elongate, and ramose (ramose: 9 cm long and 5 cm wide; elongate: 13 cm long and 1 cm wide) with primary branch giving rise to many cylindrical fingers growing or rising just above the ground clinging into branching corals. Some may have as many as five branches growing at different directions forming a mass branching network attached to the substrate through several points, whereas others have only a single elongate branch. Surface is rough and hispid due to protruding spicules with oscula (1–2 mm in dia.) that are irregularly dispersed throughout the "body." Texture is semielastic and difficult to tear. In life, ectosome is bright orange and endosome is brick brown. Ectosomal skeleton is made up of a thin layer of smaller microspined styles that form discrete brushes erect on the surface in a continuous palisade.

Compound(s) and Activities

Antibacterial activity: The extract of this sponge species showed antibacterial activity against *E. coli* with a 10.8 mm of zone of inhibition (Cita et al., 2017).

Mosquito larvicidal activity: The extract of this species displayed larvicidal activity against the fourth instar larvae of *Aedes aegypti* (yellow fever mosquito vector) with an LC50 value of 110.8 ppm (Rao et al., 2008).

Clathria (Thalysias) vulpina (Lamarck, 1814)

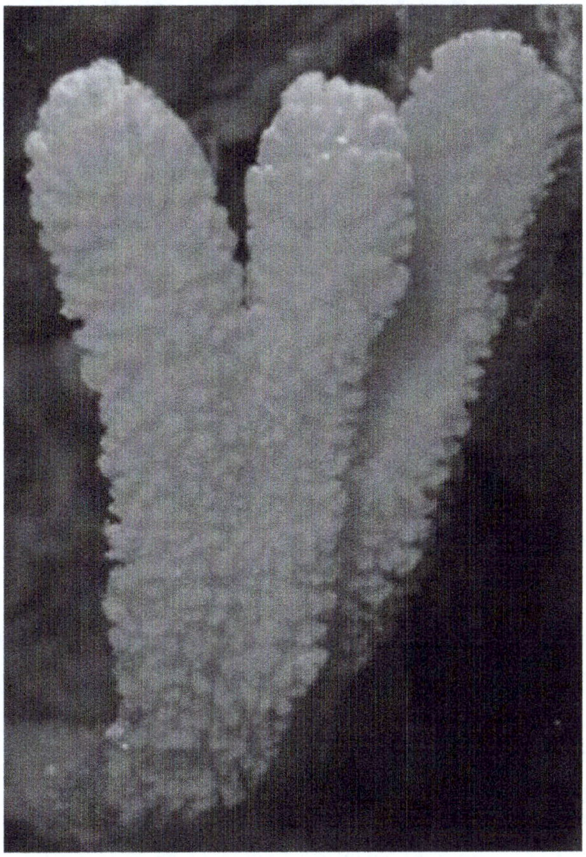

Order: Poecilosclerida Family: Microcionidae

Common name: Not designated

Distribution: Indian Ocean, Malaysia, Mozambique, Australia, Philippines, Singapore, Kenya

Ecology: It lives on the slope of fringing reefs.

Biology: Its gross morphology is more or less tubulo-digitate, variable in size, ranging from small single digits (8 cm high) to massive multiple lobes attached to a common base (45 cm high). The body of the sponge is composed of lamellate, cavernous, insubstantial, tight, or loosely anastomosing branches. Growth form is relatively consistent, but its surface may or may not have small digitate processes arising from free branches. Where present, surface processes range from tapering digitate to spiny protrusions. Oscules (pores) are usually flush with the surface, or raised only slightly above it, with a distinct membranous lip. These pores are found scattered between surface conules and not confined to any particular region. The surface is moderately dense to very dense, ornamented by minute grooves and striations and irregularly scattered microconulose processes. The whole surface is irregularly folded.

Compound(s) and Activities

Antiplasmodial activity: The crude extract of this species showed excellent antiplasmodial activity (IC50 = 14.8 μg/mL), which was highly comparable to the positive control chloroquine (IC50 = 7 μg/mL) (Kumar and Ravikumar, 2014).

Antibacterial and antifungal properties: The ethanol extracts of this species showed antimicrobial activity against *V. alginolyticus*, *V. fischerii*, *Micrococcus* spp., and *R. solani*. The values of inhibition zone recorded for *V. alginolyticus* were more than 2 mm (Rajendran et al., 2011).

Antifungal activity: The crude extracts of this species showed antifungal activity against *Aspergillus niger*, *Pencillium notatum*, and *Candida albicans*, and the values of zone of inhibition at different concentrations of the extract are given next.

Zone of Inhibition (Diameter in mm)

Aspergillus niger		*Pencillium notatum*		*Candida albicans*	
50 μl	100 μl	50 μl	100 μl	50 μl	100 μl
0.8	0.9	0.5	2.0	1.3	2.0

Source: Dhinakaran, D.I. et al., 2012, *J. Appl. Pharm. Sci.*, 2: 52–55.

Antiviral activity: The MeOH–EtOAc extract of this species yielded a novel sterol sulfate, clathsterol, which showed antiviral activity against HIV-1 at a concentration of 10 μM (Rudi et al., 2001).

Clathria sp.

Order: Poecilosclerida Family: Microcionidae

Compound(s) and Activities

Anti-HIV activity: The depsipeptide compound clathsterol of this species inhibited HIV-1 RT at 10 μM concentrations (Senthilkumar and Kim, 2013).

Echinochalina (*Protophlitaspongia*) *bargibanti* (Hooper and Lévi, 1993)

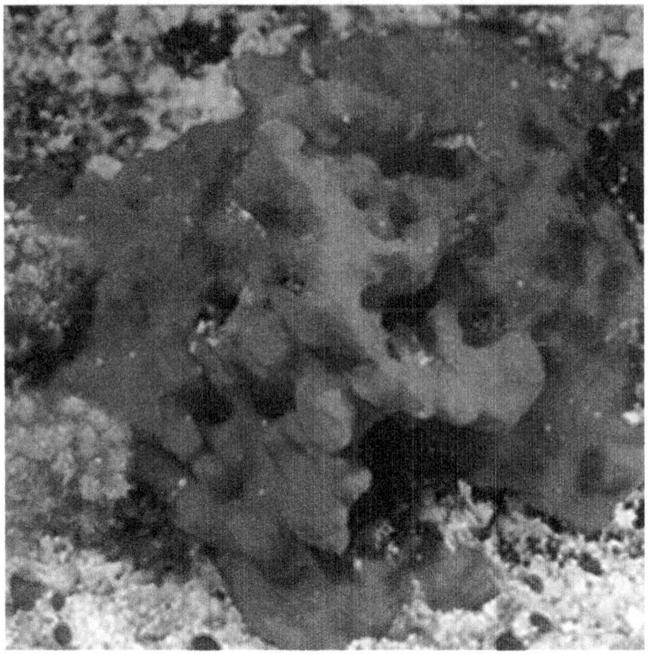

Order: Poecilosclerida Family: Microcionidae

Common name: Not designated

Distribution: Southwest lagoon of New Caledonia, northeastern coast of Grande Terre

Ecology: It is a free-growing sponge on sand substrate at 20 to 30 m depths. It encrusts on bivalve shells.

Biology: This sponge is a lobate-digitate mass composed of clumps of partially fused, small, thick digits, joined to a common base encrusting on molluscan shells. The digits are bulbous, cylindrical, and thicker at the apex than at the point of attachment to the base. These digits have even, rounded tips, either unornamented or with few, large conules or secondary digits) near the margins. Oscules (1–2 mm dia.) are small and are located mainly on the ends of, or between, the surface conules, and they hove slight subdermal sculpturing. Texture is compressible and is difficult to tear. The sponge is 65 mm high and 57 mm wide; digits are 18–35 mm long and 8–15 mm in diameter. Color of the sponge is dark red in life. Ectosomal skeleton is a sparse, tangential or paratangential layer of auxiliary subtylostyles dispersed over the surface. Choanosoma skeletal architecture is regularly reticulate.

Compound(s) and Activities

Antimicrobial and antiproliferative activities: A polyarsenic compound, arsenicin A, isolated from this species demonstrated potent bactericidal and fungicidal activities against human pathogenic strains such as *Staphylococcus aureus*, *Escherichia coli*, and *Candida albicans* with inhibition circles of 24, 28 and 26 mm, respectively, at 10 μg per disk. The synthesized (±)-arsenicin A showed potent antiproliferative activity on acute promyelocytic leukemia cell lines, and pancreatic adenocarcinomas and glioblastomas (Motuhi et al., 2016).

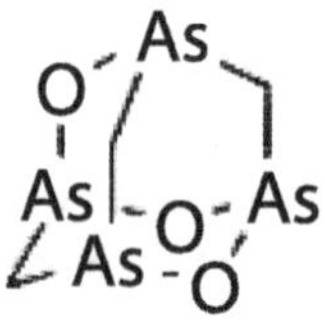

Arsenicin A

Pandaros acanthifolium (Duchassaing and Michelotti, 1864)

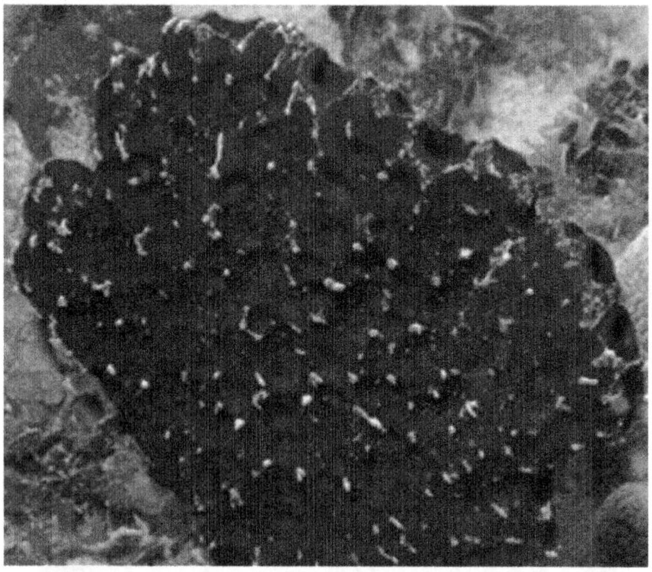

Order: Poecilosclerida Family: Microcionidae

Common name: Azure vase sponge, branching vase sponge, Caribbean marine sponge

Distribution: Western Central Atlantic, Florida and the Caribbean

Ecology: This sessile sponge is a reef dweller at depths of 2 to 35 m.

Biology: These sponges are erect, dark bushes with flattened branches. Surface is pronounced conulose, with blunt truncate conules of 0.5 to 2.0 mm high. Oscules are inconspicuous. Branches are up to 25 cm long and 4 cm wide. Consistency is soft, almost limp, and spongy compressible. It is purplish black in color. It gives off a purple exudate when squeezed.

Compound(s) and Activities

Antioxidant and cytoprotective activities: Among the steroidal glycosides acanthifoliosides G–J isolated from this species, the compound acanthifolioside G showed antioxidant and cytoprotective activities (Berrué et al., 2012; Perdicaris et al., 2013).

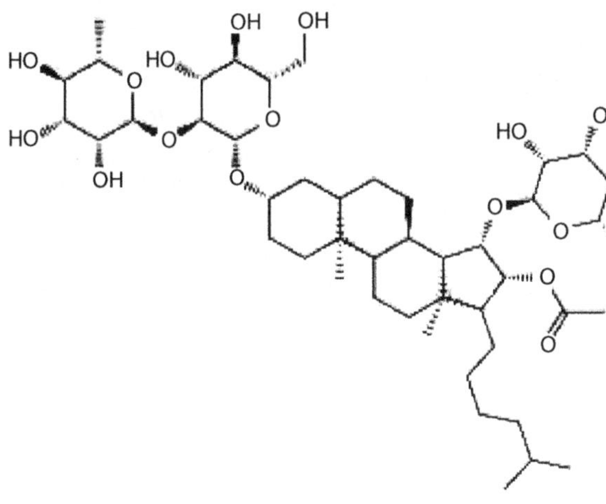

Acanthifolioside G

Antiparasitic activity: Steroidal glycosides, namely, pandarosides A–J and their methyl esters, have been isolated from this species. The majority of these metabolites showed *in vitro* activity against three or four parasitic protozoans. However, the compounds pandaroside G and its methyl ester potently inhibited the growth of *Trypanosoma brucei rhodesiense* (IC50 values, 0.8 and 0.04 µM, respectively) and *Leishmania donovani* (IC50s 1.3 and 0.05 µM, respectively) (Regalado et al., 2010).

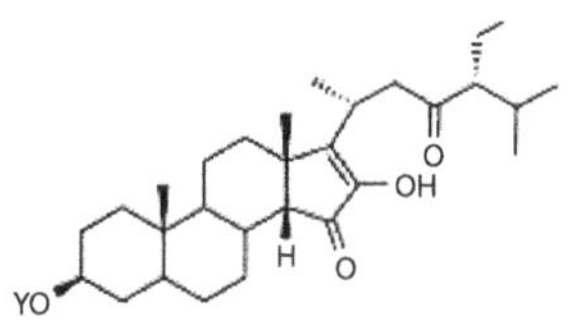

Pandaroside A

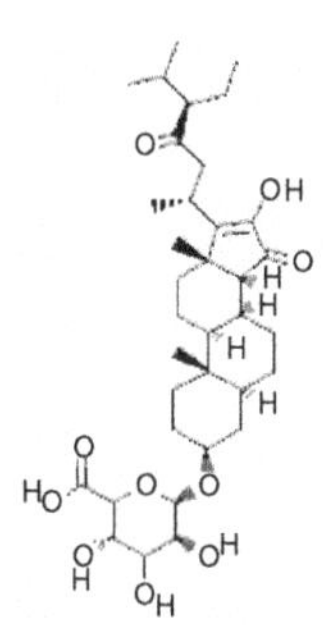

Pandaroside B

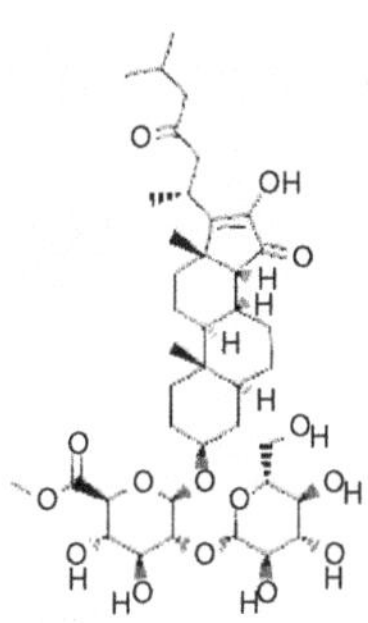

Pandaroside C

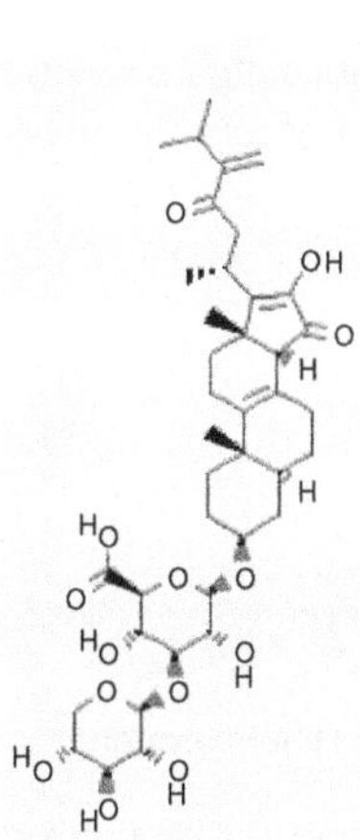

Pandaroside E

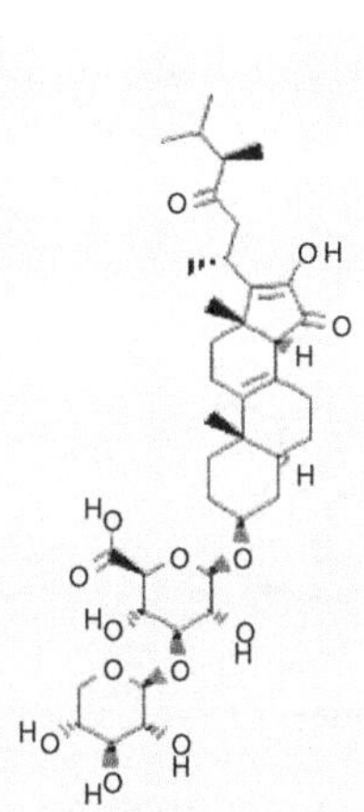

Pandaroside F

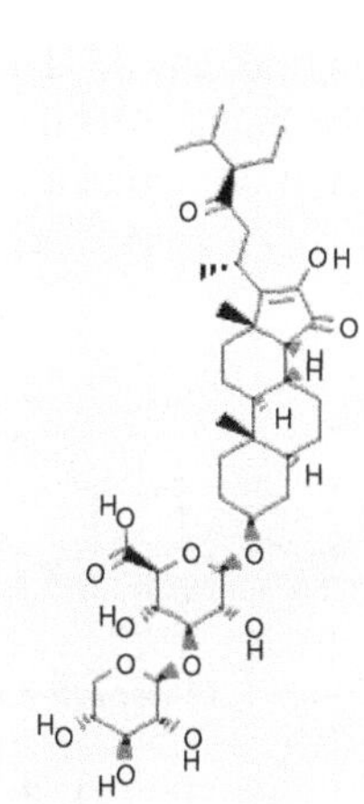

Pandaroside H

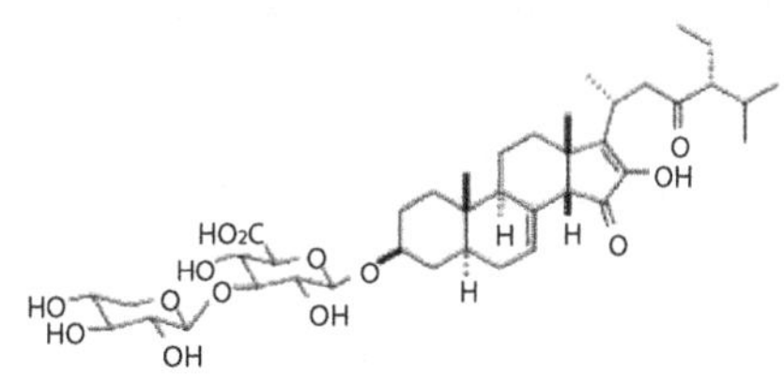

Pandaroside G

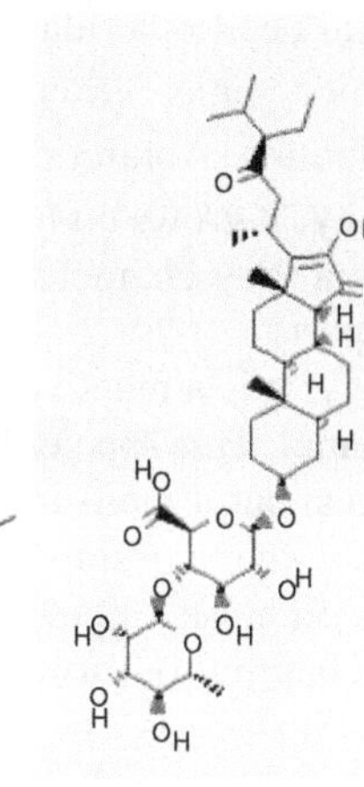

Pandaroside I

Antiparasitic and cytotoxic activities: Six new steroidal saponins, pandarosides K–M, and their methyl esters have been isolated from this species. All the new compounds (pandarosides K–M) showed moderate to weak activity against four parasitic protozoans. Additionally, these compounds exhibited moderate to strong cytotoxic effect against three human tumor cell lines (Regalado et al., 2011).

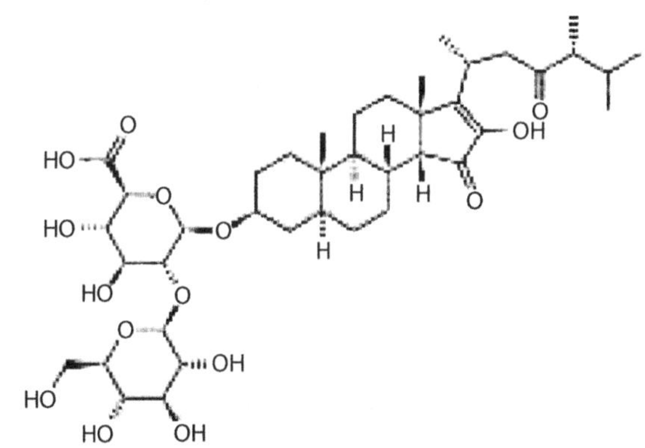

Pandaroside K

FAMILY: MYCALIDAE

Mycale (Oxymycale) acerata (Kirkpatrick, 1907)

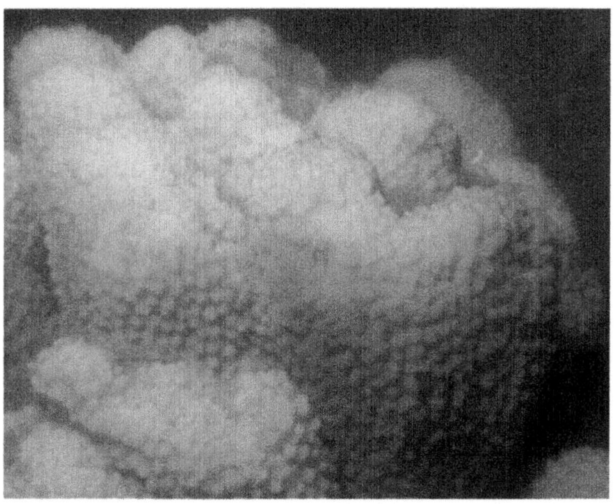

Order: Poecilosclerida Family: Mycalidae

Common name: Slimy sponge

Distribution: Antarctic

Ecology: It grows on hard rock ledges and cliffs. This sponge shows a very characteristic habit as a mixture of living and dead parts.

Biology: It is a massive, soft sponge with whitish appearance in ethanol. This species has no programmed shape (like many sponges) but it tends to form mounds like those made by termites, with many smooth, rounded lobes protruding. Skeleton is of a plumose reticulation of anastomosing spicular-spongin fibers fanning out toward the surface. Basal fibers are very thick. Surface is covered by a distinct reticular ectosome. Oxeas of the spicules are 0.5 to 1 mm long. This sponge is particularly notable as it grows quickly. Mounds of *Mycale acerata* and the starfish that eat them are common features of hard bottom communities especially in Antarctic waters. This species has proven to be toxic for goldfish.

Compound(s) and Activities

Cytotoxicity: The EtOH extracts of this species exhibited cytotoxicity at two higher concentrations, 125 and 250 μg/mL (Moles et al., 2014).

Anti-inflammatory activity: The EtOH extracts of this species showed anti-inflammatory activity by reducing the release of IL-1β, PGE2, and LTB4 at the lower concentration of 50 μg/mL (Moles et al., 2014).

Mycale (Aegogropila) adhaerens (Lambe, 1893)

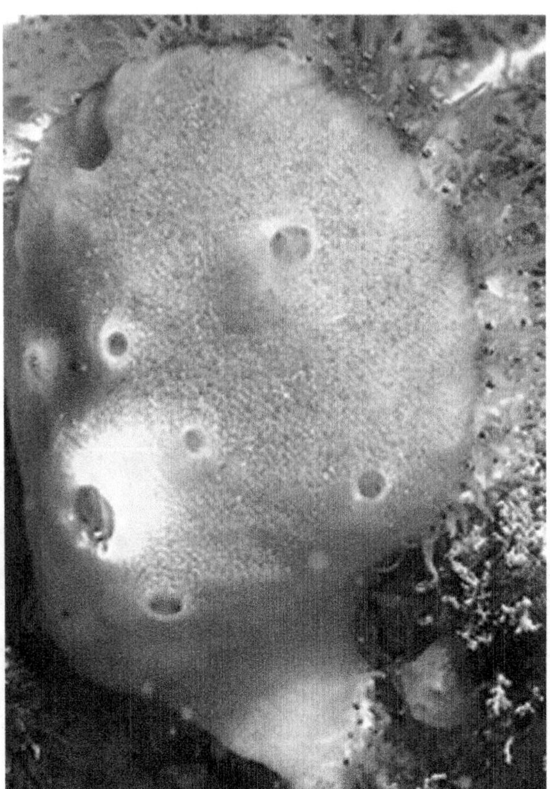

Order: Poecilosclerida Family: Mycalidae

Common name: Purple scallop sponge

Distribution: North Pacific Ocean, Canada; North-Eastern Pacific boreal species

Ecology: It lives symbiotically on the valves of the scallops *Chlamys hastata* or *Chlamys rubida*, and *Pecten hastatus*.

Biology: This sponge encrusts the upper layer of bivalve shells to a thicknes of 9 mm in the center. It ranges in color from yellowish brown through purple to pink. Texture is soft and elastic. Surface is even and minutely hispid. Dermal membrane is thin. Pores are in the form of circular openings. Oscula (0.3 mm wide) are small and very slightly raised above the surface.

Association: The symbiosis of this sponge with the scallops is likely mutualistic. If the predator of the scallop, the seastar *Evasterias troschelii,* encounters the scallop, the seastar often turns away if it touches the sponge. This may be the response to some secretion or to the spicules from the sponge. If the sponge is removed from the scallop and the scallop is prevented from swimming, it is readily captured by the seastar. The scallop will also swim from predators of the sponge, such as *Archidoris* spp, so the sponge is benefited as well. The swimming scallop may also help carry the sponge into areas with good currents and help prevent fouling of the sponge.

Predators: Nudibranches that graze on this sponge include *Archidoris odhneri* and *Anisodoris nobilis.*

Compound(s) and Activities

Antibacterial activity: The associated fungal strain, *Exophiala pisciphila* NI10102, of this species yielded a compound, exophilin A, which showed antibacterial activity against Gram-positive bacteria (Thomas et al., 2010).

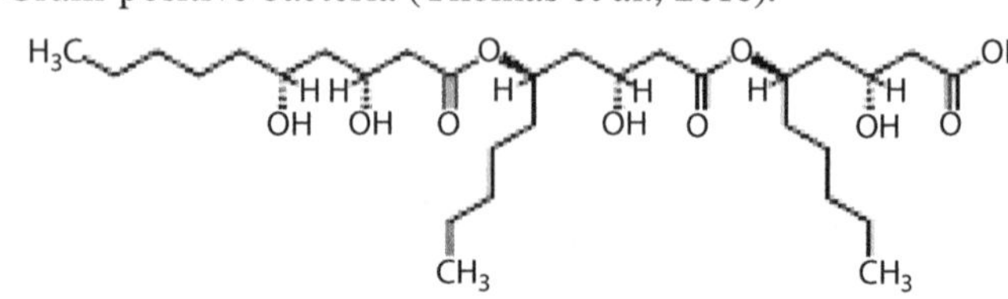

Exophilin A

Cytotoxicity: The lipophilic extract of this species showed the compounds 13-deoxytedanolide and a brominatd isocoumarin, namely, hiburipyronone, which showed potent cytotoxicity (Fusetani et al., 1991).

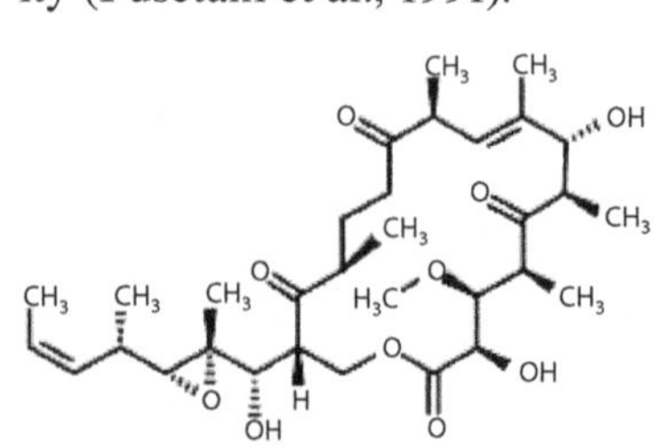

13-deoxytedanolide

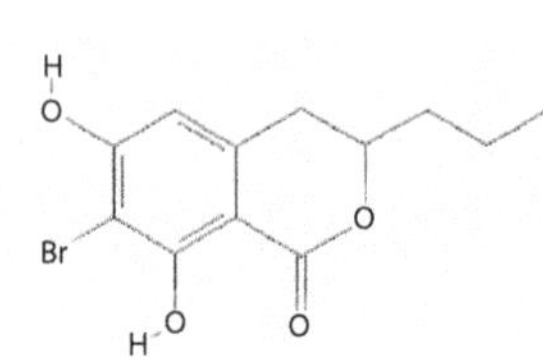

Hiburipyronone

Uchida et al. (1997) reported that the compound hiburipyranone of this species was cytotoxic against P388 murine leukemia cells (IC50, 0.19 μg/mL).

Mycale (Carmia) cecilia (de Laubenfels, 1936)

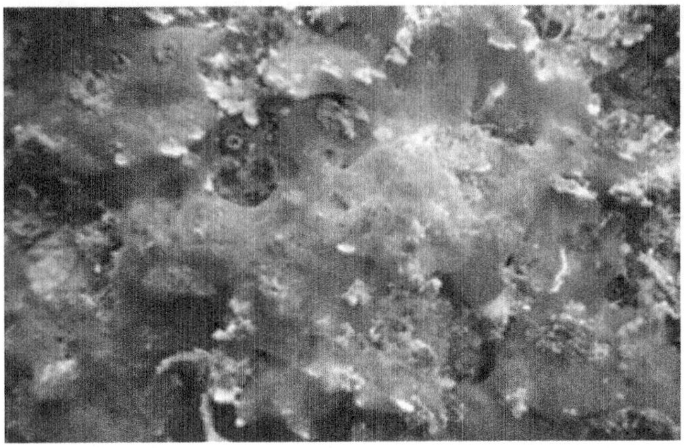

Order: Poecilosclerida Family: Mycalidae
Common name: Fouling sponge
Distribution: East Pacific Ocean, Panama
Ecology: It is found on corals as on other types of substrata.
Biology: It is an encrusting to cushion-shaped sponge with 1–9 mm thickness. It is red to reddish orange, or green with yellow, or almost blue, but always with distinctive small orange patches Consistency is very soft and fragile after preservation. Surface is smooth. Subectosomal canals converge to oscula. Small ectosomal pores are seen on the surface. Oscules are circular to oval, 0.5 to 4.0 mm in diameter, and are elevated from the surface by a translucent dermal membrane. Megascleres are mycalostyles, and the microscleres are anisochelae.

Compound(s) and Activities

Antitumor activity: A total of 14 new pyrrole-containing metabolites, namely, mycalazals, have been isolated from this species. All these mycalazals have shown activity as growth inhibitors of several tumor cell lines, in particular the LNcaP cell line (Ortega et al., 2004).

Mycalazal

Mycale izuensis (Tanita and Hoshino, 1989)

Order: Poecilosclerida Family: Mycalidae
Common name: Not designated
Distribution: Japan
Ecology: Not reported
Biology: Not available

Compound(s) and Activities

Cytotoxicity: The mycalolide, 30,32-dihydroxymycalolide A isolated from the lipophilic extract of this species was found to be cytotoxic against HeLa cells with an IC50 value of 2.6 ng/m (Phuwapraisirisan et al., 2002).

Anticancer activity: The compound azumamide E of this species displayed human histone deacetylase inhibitory activity (Agrawal et al., 2016).

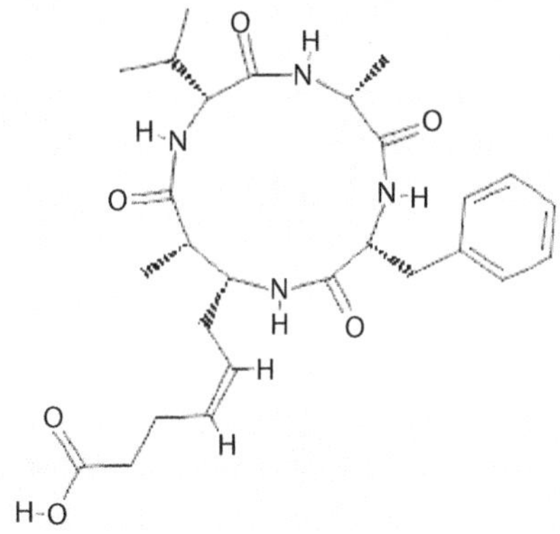

Azumamide E

Mycale (Arenochalina) laxissima (Duchassaing and Michelotti, 1864)

Order: Poecilosclerida Family: Mycalidae
Common name: Strawberry vase sponge
Distribution: Gulf of Mexico, South Florida and throughout the Caribbean, East and West Africa
Ecology: It inhabits chiefly on coral reefs at a depth range of 2 to 60 m. It occasionally occurs in mangrove peat and harbors and tolerates brackish water conditions.

Biology: It is a massively encrusting to vase- or cuplike sponge. It may also be tubular to globular and solitary or in clusters. Surface is strongly conulose due to flimsy fleshy parts. A large pseudoscule (3–6 cm wide) with transparent membrane is seen. Oscules are indistinct. Consistency of the sponge is soft. The species is notable for its copious slime production. Color of the body is reddish purple with bluish tinges.

Compound(s) and Activities

Antimalarial activity: The organic extract of this species has shown antimalarial activity against *Plasmodium berghei* ANKA in the infection model of C57BL/6 mice (Perdicaris et al., 2013; Martínez et al., 2012).

Mycale (Carmia) lissochela (Bergquist, 1965)

Order: Poecilosclerida Family: Mycalidae

Common name: China Sea sponge

Distribution: West Caoline Islands, Palau Archipelago

Ecology: It occurs on coral fragments and other hydroids.

Biology: This thickly encrusting sponge may contain small zoanthids, and around each of these the sponge dermis may be raised in a membranous chimney. In alcohol, this sponge is pinkish gray to pale creamy yellow. Texture is extremely soft and very slightly elastic. Surface is smooth, and the dermis is variable in structure. The fleshy tissues of the dermis may be arranged so as to form a reticulate pattern. For the most part, the fleshy tissues are uniformly dispersed, and no pores or oscules are visible. The skeleton is composed of thin, curving, sparsely branched tracts composed of subtylostyles.

Diabetes and obesity inhibitory activities: Two new 5-alkylpyrrole-2-carboxaldehyde derivatives, mycalenitrile-15 (1) and mycalenitrile-16 (2), along with five known related ones (3–7), have been isolated from this species. Of these compounds, compounds 1 and 7 exhibited significant PTP1B (protein-tyrosine phosphatase 1B, a recognized target for diabetes and obesity) inhibitory activities with IC50 values of 8.6 and 3.1 mmol/L, respectively (Xuea et al., 2017).

Mycale (Carmia) microsigmatosa (Arndt, 1927)

Order: Poecilosclerida Family: Mycalidae

Common name: Not designated

Distribution: Southwestern Atlantic, South Caribbean, Brazil

Ecology: It occurs on sublittoral rocky shores.

Biology: It is an orange-dotted to reddish, crustose sponge. Larger specimens are encrusting or cushion-shaped (5 cm thick), and very soft and fragile. Subectosomal channels of the aquiferous system converge to scattered oscula (5 mm). Live color is highly variable, but it is often reddish orange or grayish greenish. Skeleton is composed of scattered megascleres on the thin surface membrane. Spicules are slender subtylostyles that are mostly straight.

Association: This sponge was in association with foraminiferans (26 species) and other invertebtates such as crustaceans (83%, 31 spp.), polychetes (10%, 18 spp.), and mollusks (3.7%, 15 spp.). The Foraminifera species inside the sponges were found inside the sponges in various ontogenetic stages, suggesting the occurrence of commensalism between them, as the sponge uses this microhabitat for nutrition and reproduction purposes. Similarly, the sponge was important for the reproduction and survivorship of the above said invertebrates (Mazzoli-Dias et al., 2007; Ribeiro et al., 2003).

Compound(s) and Activities

Antioxidant activity: The methanol extracts of this species showed percent inhibition of radical ABTS with 67% (Tovar-Barrios et al., 2016).

Mycale (Aegogropila) plumosa sensu (Hoshino, 1981)

Order: Poecilosclerida Family: Mycalidae

Common name: Not designated

Distribution: Western Japan, Mauritius

Ecology: It occurs in the intertidal, low tide subzone.

Biology: It is an amorphous, massive sponge with a size of 4 × 3 × 6 (height) cm. Few or many irregular projections or anastomosing branches are seen. Color is olive ocher. Consitency is vey soft, slightly compressible and fragile. Surface is uneven and coarsely cornulose with cornules of 2–3 mm height. Oscules open at the summit of each projection. Pores are invisible.

Compound(s) and Activities

Cytotoxic activity: The associated actinobacterium strain *Saccharopolyspora* of this species showed cytotoxic activities against temperature-sensitive mutant cell lines of mouse (tsFT210) (Thomas et al., 2010).

Antitumor activity and cytotoxic effects: The associated fungus *Penicillium auratiogriseum* of this species produced the compounds (S)-2,4-dihydroxy-1-butyl-(4-hydroxy) benzoate and fructigenine A. These compounds showed both antitumor activity and potent cytotoxic effects. Further, the two new quinazoline alkaloids, aurantiomides B and C, isolated from the associated *Penicillium auratiogriseum* showed moderate cytotoxic activities (Thomas et al., 2010).

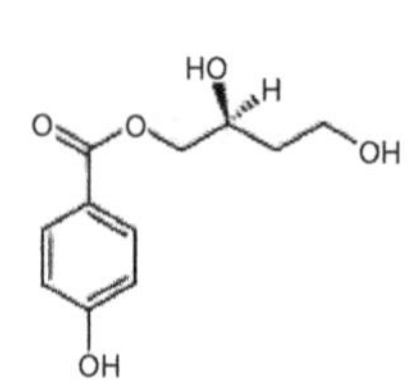

(S)-2,4-dihydroxy-1-butyl-(4-hydroxy) benzoate

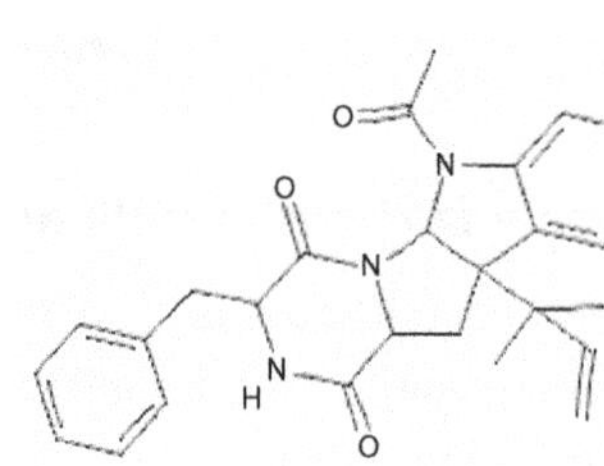

Ffructigenine A

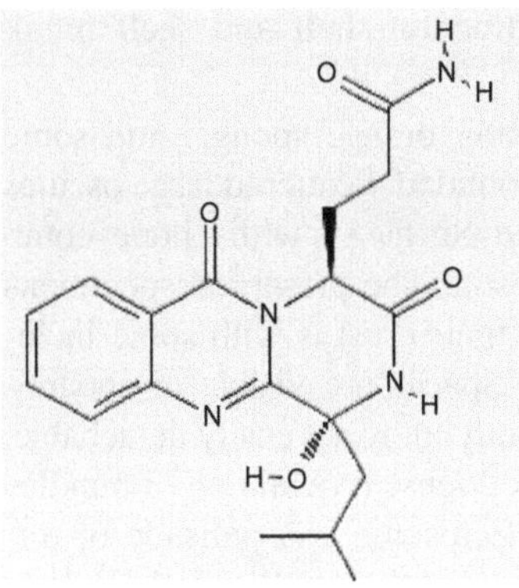
Aurantiomide B

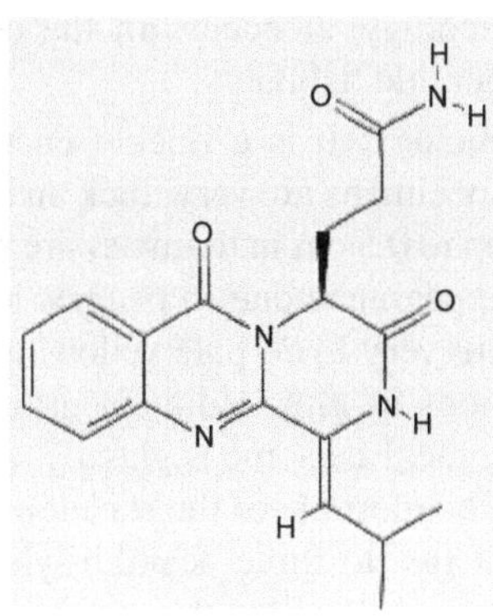
Aurantiomide C

Mycale (Arenochalina) trincomalensis (Rao, 1941)

Order: Poecilosclerida Family: Mycalidae
Common name: Not designated
Distribution: Japan; Gulf of Mannar, India
Ecology: It is a deep-sea sponge.
Biology: Not available

Compound(s) and Activities

Anticancer activity: The extract of this species showed anticancer activity against A549, HeLa, and HT-29 cancer cell lines with GI50 values of >100 µg ("Sponge crude extraction," n.d.).

Mycale sp.

Order: Poecilosclerida Family: Mycalidae

Compound(s) and Activities

Antitumor activity: The oxazole macrolide mycalolide B isolated from this unidentified sponge showed antitumor activity through actin depolymerization (Sipkema et al., 2005).

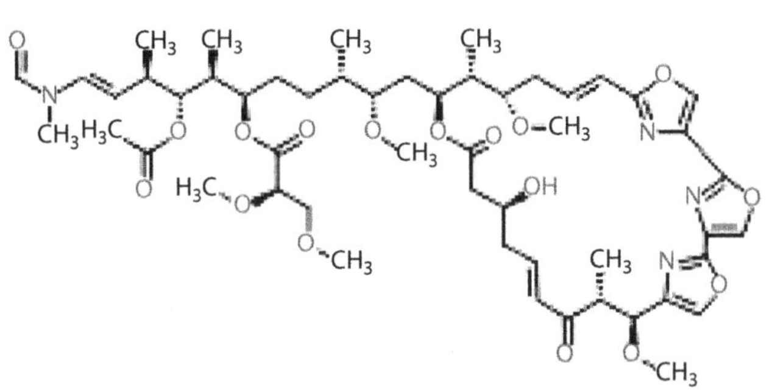
Mycalolide B

Immunosuppressive activity: The thiazole macrolide pateamine A isolated from this sponge showed immunosuppressive activity by inhibiting IL-2 (Anjum et al., 2016; Sipkema et al., 2005).

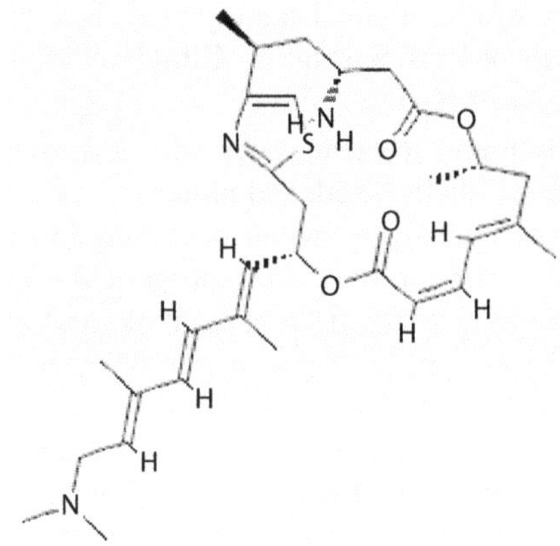
Pateamine A

Antiviral activity: The nucleosides, mycalamides A and B of this species showed antiviral activity against A59 coronavirus (HSV-1) (Anjum et al., 2016).
Cytotoxicity: A new cyclic diamine, 1,5-diazacyclohenicosane, isolated from this species displayed cytotoxicity at the micromolar level against three human tumor cell lines: lung (A549), colon (HT29), and breast (MDA-MB-231) tissues. The compound exhibited moderate activity with GI50 values in the micromolar range (5.4 µM (A549), 5.1 µM (HT29) and 5.7 µM (MDA-MB-231) (Coello et al., 2009).
Cytotoxicity: The compound mycaperoxide H isolated from this sponge showed cytotoxicity against HeLa cells with an IC50 value of 0.8 µg/mL (Muraleedharan and Avery, n.d.).
Northcote et al. (1991) reported that the thiazole-containing macrolide, pateamine 1, was an exceedingly potent, but possibly selective, cytotoxin (P388: IC50 0.15 ng/mL).

FAMILY: MYXILLIDAE

Myxilla (Myxilla) incrustans (Johnston, 1842)

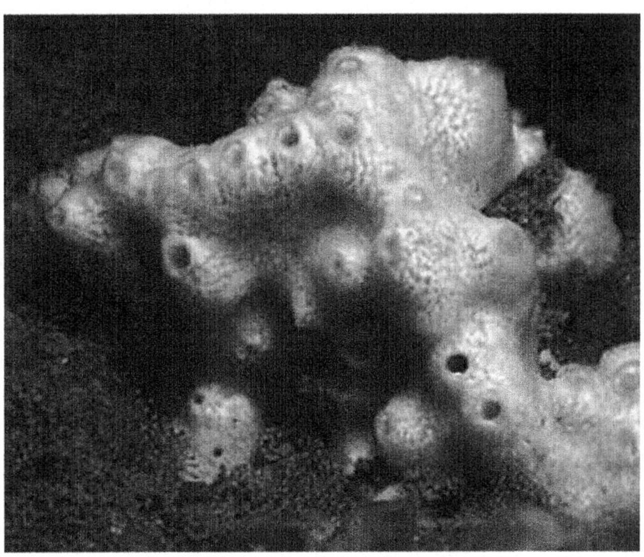

Order: Poecilosclerida Family: Myxillidae
Common name: Rough scallop sponge

Distribution: Arctic, Bear Island, Faroe Islands, Norway, Atlantic coasts of Europe south to Gibraltar; Mediterranean, British Isles, North Sea

Ecology: It is found in clear water with low water mark to 400 m on stones, shells, sand, and mud.

Biology: It is an encrusting sponge occurring in patches up to 20 cm across and 5 cm high. Coloration of this sponge may vary from shade of yellow to orange, pink, and white. It has a bubbly-looking appearance with internal channels visible through the surface and large, raised oscules. The consistency of the sponge is fairly soft and elastic but the surface is crisp. The skeleton is built out of megascleres (tornotes) with spear-shaped ends with tiny spines on them. The microscleres are a mixture of curved, shovel-like chelae and *C*-shaped sigmas.

Reproduction: It is viviparous and the developing embryos are are found inside the tissues in the months of August and September. These embryos are spherical, measuring 0.5 mm across, before becoming detached. Initiation of the sexual processes involved in reproduction largely depends on the water temperature. The male tissue releases sperm into the water column. These may get sucked into another sponge, and then fertilization takes place.

Compound(s) and Activities

Antifungal activity: The associated fungal strain *Microsphaeropsis* sp. H5-50 of this species yielded the metabolites microsphaeropsisin, (R)-mellein, (3R,4S)-hydroxymellein, (3R,4R)-hydroxymellein, and 4,8-dihydroxy-3,4-dihydro-2H-naphthalen-1-one. Among these compounds, microsphaeropsin and the eremophilane derivative showed antifungal activity at the concentration of 50 µg/mL (Thomas et al., 2010).

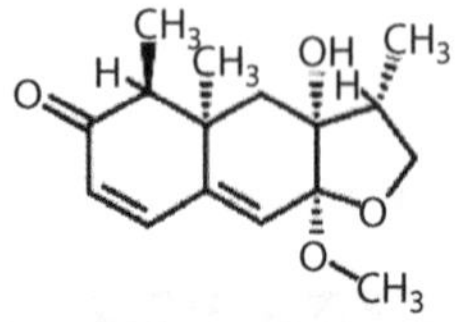

Microsphaeropsin

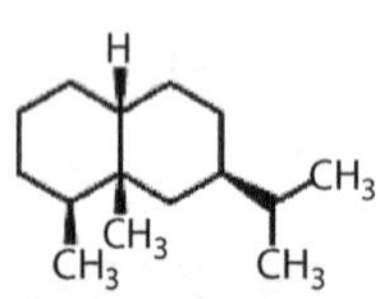

Eeremophilane

Immunomodulating/antitumor/anti-inflammatory activity: The compounds myxillin A, myxillin B, and myxillin C isolated from this species were tested for immunomodulating activity in an *in vitro* dendritic cell model. Dendritic cells matured and stimulated with myxillin A (1) secreted lower levels of IL-12p40, whereas dendritic cells matured and stimulated with myxillin C (3) secreted lower levels of IL-10. These molecules may therefore be important in preventing inflammatory diseases besides inducing a response to fight tumors and/or pathogens (Einarsdottir et al., 2016).

Myxilla (Ectyomyxilla) kerguelensis (Hentschel, 1914)

Order: Poecilosclerida Family: Myxillidae

Common name: Not designated

Distribution: Indian Ocean, Kerguelen Islands, New Zealand, South Georgia

Ecology: It occurs in the continental shelf and shelf break benthic habitats.

Biology: It is a thickly encrusting orange sponge, and some specimens are very thick and mounded. Scattered large oscules up to 0.5 cm in diameter are seen. Surface is with a honeycomb appearance due to patches of ostia. The preserved specimens are very firm, pale yellow, and their crust is with some inclusions of gray sediment grains. Spaces are visible in choanosome. Ectosome is glassily smooth but is not easily detachable. Choanosome of the skeleton is of dense reticulation of bundles of two to three acanthostyles. Ectosome is a palisade of tornotes covered with tangential crust of the small acanthostyles.

Compound(s) and Activities

Cytotoxic activity: The organic extracts of this species showed cytotoxicity against the HeLa cervical cell line with an EC50 value of 11.7 µg/mL (Olsen et al., 2016).

Myxilla sp.

Order: Poecilosclerida Family: Myxillidae

Compound(s) and Activities

Antibacterial activity: The extracts of this species showed antibacterial activity against the laboratory, commensal, and clinically relevant bacterial strains *Bacillus subtilis* EXB-V68 and *Macrococcus* 1F with the MIC values of 60 and 400 µg/mL, respectively (Turk et al., 2013).

The extracts of this species showed antibacterial activity against the environmental bacterial strains *Exignuobacterium* sp., *Pseudoalteromonas* sp., *Alteromonas* sp., *Vibrio ruber*, and *Janthinobacterium svalbardensis* with the MIC values of 110, 90, 150, 90, and 95 µg/mL, respectively (Turk et al., 2013).

FAMILY: PODOSPONGIIDAE

Diacarnus erythraeanus (Kelly-Borges and Vacelet, 1995)

Order: Poecilosclerida Family: Podospongiidae

Common name: Not known

Distribution: Saudi Arabia, Northern and Central Red Sea

Ecology: It is common on coral substrate or rocks. Large specimens are free living or attached to debris or small coral fragments in sand. The depth range is 2 to 30 m.

Biology: It is an anastomosing sponge from a sprawling mass up to 1 m diameter. Its repent or erect branches are of 2 to 3 cm in diameter. The slightly enlarged ends of branches may be of 5 cm diameter. Texture is tough or just compressible and elastic. Surface is irregular with broad, blunt cornules. Ostia are localized in groups of about 10 in small surface depressions. Oscules, which are of 2 to 3 mm diameter, are present on the upper surface of branches or at tips of enlarged branch ends. Color of the sponge is rose pink.

Compound(s) and Activities

Antimalarial activity: The compound sigmosceptrellin B isolated from this species has shown antimalarial activity against *P. falciparum* (clone, D10) with an IC50 value of 1200 ng/mL (Anjum et al., 2016).

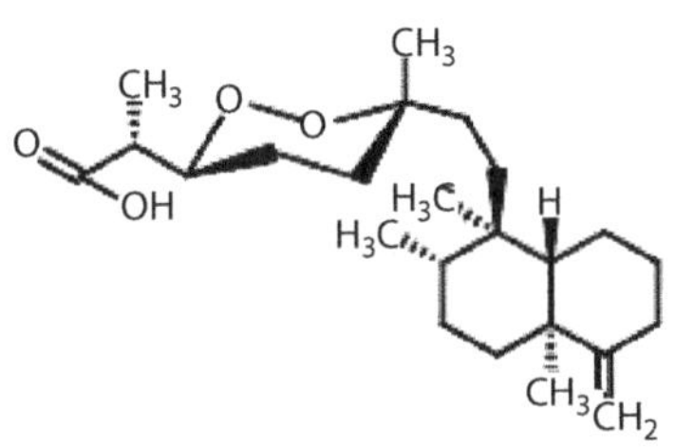

Sigmosceptrellin B

Antimalarial and aniviral activities: A new norsesterterpene acid, named muqubilone (1), along with the known sigmosceptrellin B and muqubilin have been isolated from this species. Sigmosceptrellin B exhibited significant *in vitro* antimalarial activity against *Plasmodium falciparum* (D6 and W2 clones) with IC50 values of 1200 and 3400 ng/mL, respectively. Muqubilin and compound 1 showed *in vitro* antiviral activity against herpes simplex type 1 (HSV-1) with ED50 values of 7.5 and 30 μg/mL, respectively. Muqubilin and sigmosceptrellin B also displayed significant *in vitro* activity against *Toxoplasma gondii* at a concentration of 0.1 μM (El Sayed et al., 2001).

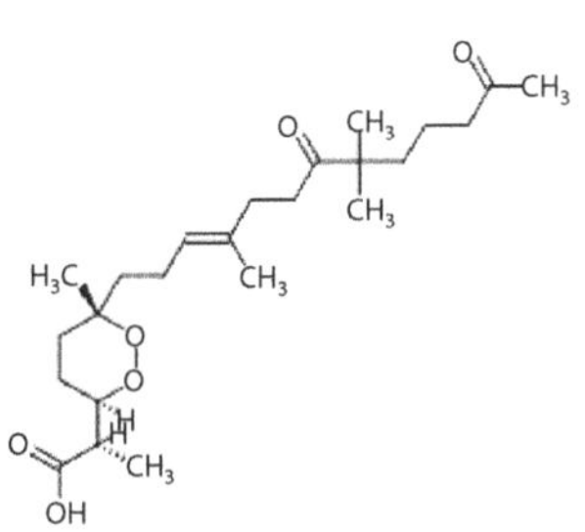

Muqubilone

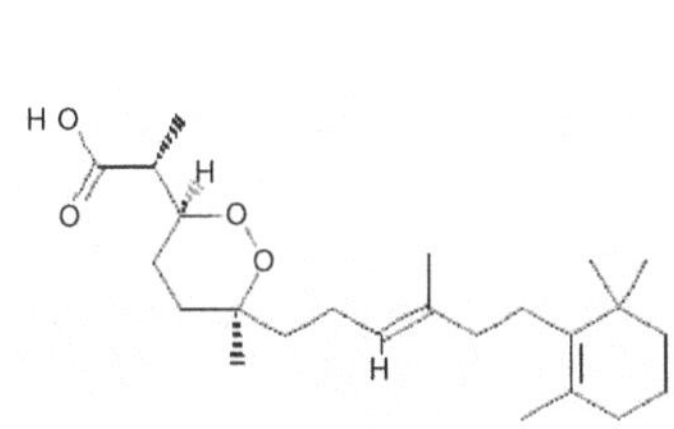

Muqubilin

The compound muqubilone of this species exhibited *in vitro* antiviral activity against HSV-1 with ED50 = 30 μg/mL (Muraleedharan and Avery, n.d.).

Anticancer activity: Eight cyclic peroxide norterpenoids, compounds 1–8, have been isolated from this species. Among these compounds, (–)-muqubilin A (5) and the new compounds 3 and 4 exhibited growth inhibitory concentrations in cancer cell lines with IC50 values of <10 μM (Lefranc et al., 2001).

Diacarnus levii (Kelly-Borges and Vacelet, 1995)

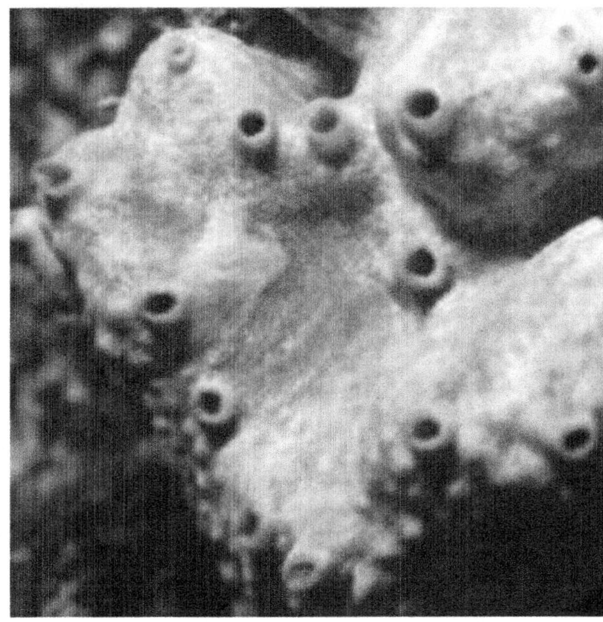

Order: Poecilosclerida Family: Podospongiidae

Common name: Papua New Guinea marine sponge

Distribution: New Caledonia

Ecology: Not reported

Biology: Not known

Compound(s) and Activities

Antimalarial activity: The alkaloid manzamine A isolated from this species showed antimalarial activity against the chloroquine-resistant *P. falciparum* with an IC50 value of 4.5 ng/mL (Anjum et al., 2016).

Antitumor and antiviral activities: Four new orsesterterpene peroxides, diacarnoxides A–D, have been isolated from this species. Among these compounds, diacarnoxide B exhibited a significantly enhanced ability to suppress the growth, proliferation/viability (97%) of tumor cells, and hypoxia-induced HIF-1 activation (95%) following a 16 h exposure. However, all the compounds (isolated from two pooled chromatographic fractions) were found to inhibit hypoxia-activated HIF-1 by 80% and 91% (at 5 μg mL^{-1}), and suppressed tumor cell proliferation/viability by 38% and 34%, respectively (Dai et al., 2007).

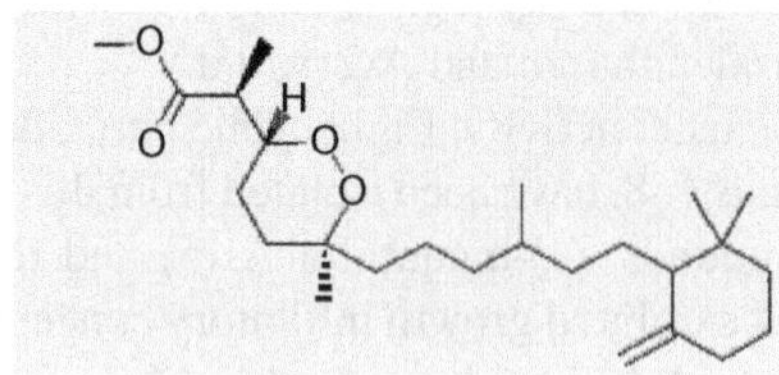

Diacarnoxide A

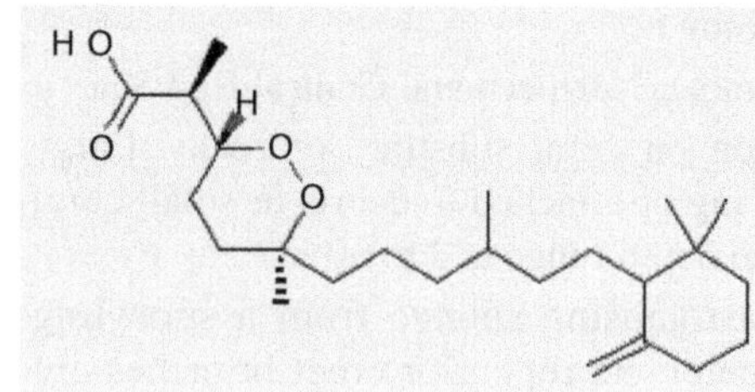

Diacarnoxide B

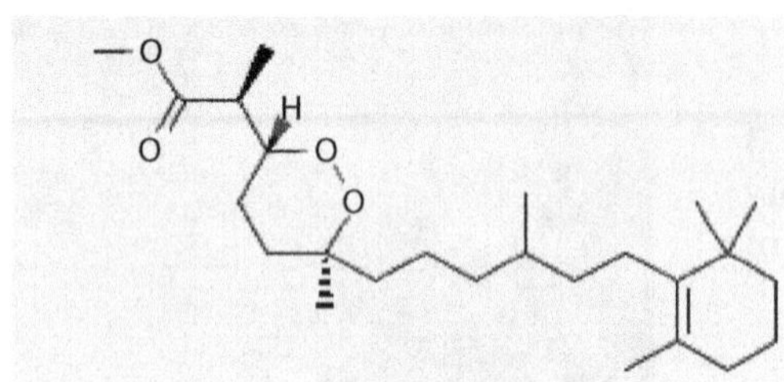

Diacarnoxide C

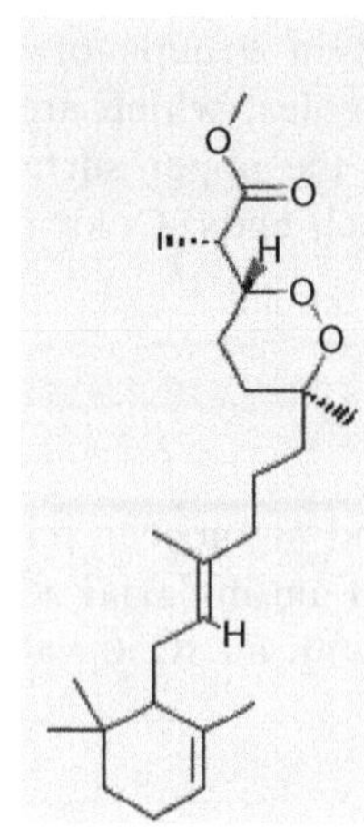

Diacarnoxide D

Antimalarial activity: The new cyclic norditerpene peroxides, which were esterified from the carboxylic acids isolated from this species, showed antimalarial activity. The epidioxides 1–3, 6, and 10 proved active *in vitro* against *Plasmodium falciparum*. Further, its methyl 3-epinuapapuanoate was found active against a chloroquine-resistant strain with a good security index (D'Ambrosio et al., 1998).

FAMILY: TEDANIIDAE

Tedania (Tedania) anhelans (Vio in Olivi, 1792) (= *Tedania digitata*)

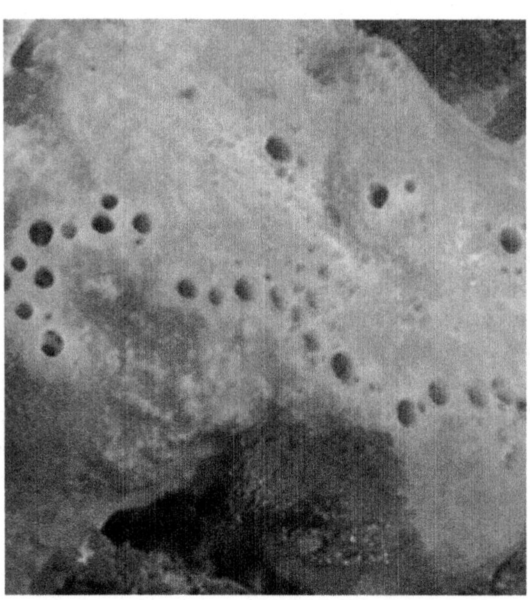

Order: Poecilosclerida Family: Tedaniidae

Common name: Not designated

Distribution: China, Bretagne, Galicia

Ecology: It inhabits the intertidal zone and is from subtidal localities (reefs) down to 40 m.

Biology: It is a massive species, up to 8 cm long, 4 cm wide, and 6 cm high. It has conical oscular elevations. Oscules are of 4 mm in diameter. Surface is smooth and wrinkled in preserved specimens. Consistency is soft and compressible. Color of the sponge is reddish brown, orangegreen, or bluish. The megascleres of the spicules are microspined tylote. Microscleres occur in a single but variable size category. Ectosomal skeleton consists of tangential and partly erect bundles of tylotes strewn rather randomly at the surface. The choanosomal skeleton consists of an irregular ladder-like reticulation of multispicular bundles of styles.

Compound(s) and Activities

Antimicrobial activities: The associated bacterial isolates belonging to α-proteobacteria, γ-proteobacteria, cytophaga/flavobacterium/bacteroidetes exhibited antimicrobial activities against Gram-positive bacteria, Gram-negative bacteria, and fungi; and antitumor indicators *Escherichia coli* 343/591 (with DNA repair deficiency) and regular *E. coli* 343/636 (Zeng et al., 2013).

Antioxidant activity: The associated *Bacillus* sp. (S2, S4, S6 strain) of this species showed free radical scavenging activity the values of which was ranging from 40% to 46% (Balakrishnan et al., 2014).

Immunomodulatory and immunostimulator activities: The organic extract of this species has been reported to stimulate the chemotatic, phagocytic, and intracellular killing of human neutrophils at a dose of 50 µg/mL. Further this extract has also shown moderate immunostimulator activity (Chandraraj et al., 2010).

Neurosuppressive, muscle relaxant, and antiallergic activities: The compound 1-methylisoguanosine, a nucleoside analog of this species, showed neurosuppressive, muscle relaxant, and antiallergic activities (Sipkema et al., 2005).

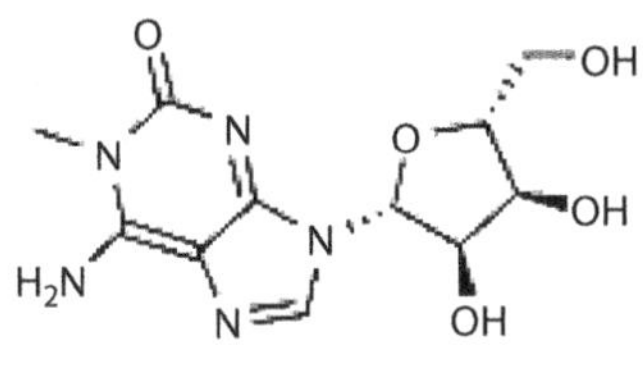

1-Methylisoguanosine

Tedania (Tedania) ignis (Duchassaing and Michelotti, 1864)

Order: Poecilosclerida Family: Tedaniidae

Common name: Fire sponge

Distribution: Southern Pacific; Florida; Caribbean to Brazil

Ecology: It is mostly known from sheltered bays and lagoons, on mangroves and among eel grasses or algae. It is a typical member of the fouling community.

Biology: It is a massively encrusting to lobate sponge with a smooth, tuberculate, or villous surface. The oscules are of 0.5 cm or more in diameter and are located on top of volcano-like projections. Size of the sponge is up to 20 cm in horizontal expansion and is up to 15 cm thick. Consistency is soft and compressible. Color of the sponge is a bright, warm red; occasionally it is rosy or even orange. This sponge exudes large quantities of mucus and produces toxic substances that may cause inflammation of the skin.

Compound(s) and Activities

Antibacterial activity: The aqueous and ethyl acetate extracts of this species showed antibacterial activity. While both the aqueous and ethyl acetate extracts showed activity against reference strains *Corynebacterium fimi* NCTC 7547 and *Micrococcus luteus* ATCC 4698, the ethyl acetate extract displayed activity against *Staphylococcus aureus* ATCC 29213 and *Staphylococcus epidermidis* ATCC 12228. Similarly, the ethyl aacetate extract showed activity against the clinical strain, *Enterococcus* spp. A 256 and *Staphylococcus aureus* 42AE. The aqueous extract showed activity against the clinical strain *Enterococcus faecalis* V583 (Muricy et al., 1993).

Tedania sp.

Order: Poecilosclerida Family: Tedaniidae

Compound(s) and Activities

Antiprotozoal activity: The actinobacterial strain *Streptomyces* sp. 11 derived from this species showed antiprotozoal activity against *L. major* and *T. brucei brucei* with IC50 values of 1.2 µg/mL and 0.005 µg/mL, respectively (Indraningrat et al., 2016).

Antitumor activity: The macrocyclic lactone compound tedanolid isolated from this species showed antitumor activity (Sipkema et al., 2005).

FAMILY: POLYMASTIIDAE

Polymastia tenax (Pulitzer-Finali, 1986)

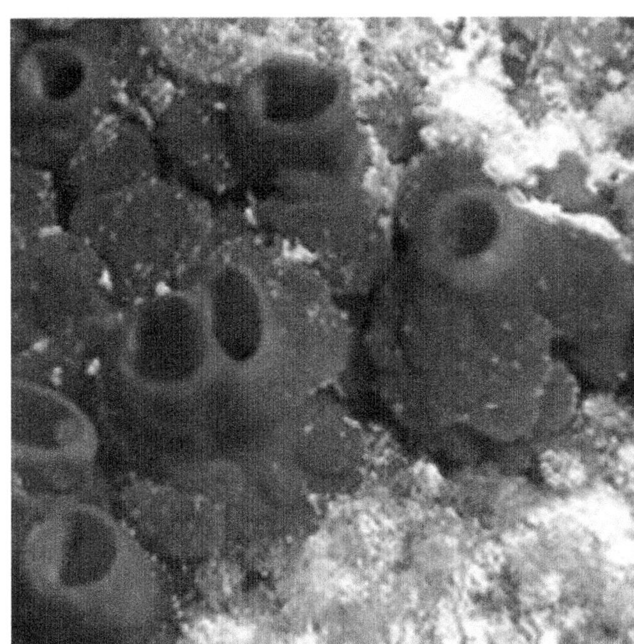

Order: Polymastiida Family: Polymastiidae
Common name: Not designated
Distribution: Western Central Atlantic, Belize, Caribbean
Ecology: It is a reef-associated species.
Biology: It is an encrusting species, and its upper surface is covered by numerous mammiform processes, the appearance of which is very similar to the papillae of *Cliona viridis.* There is one exhalant papilla for 15 to 20 inhalant papillae. The size of the inhalant papillae is quite variable as they are often fused together. These papillae contract upon contact. Reddish groups of large papillae surround the tall membranous oscules. The inhalant apertures seem limited to the surface of the inhalant and exhalant papillae. The color *in vivo* is dull yellow for the choanosame and whitish to light brown for the cortex. Consistency is tough.

Compound(s) and Activities

Antiproliferative activity: Three new sterols, 5alpha,6alpha-epoxy-24R-ethylcholest-8(14)-en-3beta,7alpha-diol (1), 5alpha,6alpha-epoxy-24R-ethylcholest-8-en-3beta,7alpha-diol (2), and 3beta-hydroxy-24R-ethylcholesta-5,8-dien-7-one (3), have been isolated from this species. Among the compounds, 1 and 2 showed antiproliferative activity of A-549, HT-29, H-116, MS-1, and PC-3 tumor cells in the range 0.5 to 10 μg/mL (Jha and Zi-rong, 2004; Santafé et al., 2002).

FAMILY: SCOPALINIDAE

Stylissa caribica (Lehnert and van Soest, 1998)

Order: Scopalinida Family: Scopalinidae
Common name: Caribbean sponge
Distribution: Greater Antilles, Jamaica
Ecology: It occurs on coral reefs.
Biology: It is a dark-orange sponge in life and dark brown in dry state. This sponge starts from a narrow base and broadens to the top. It is flattened in one dimension and is up to 15 cm high. Surface is undulating and corrugate, and even lamellate in certain places. Oscules are circular and they flush with the surface. The skeleton consists of an organic ectosome. Spicules are curved styles, which sometimes have blunt ends.

Compound(s) and Activities

Antimicrobial activities: The cyclic heptapeptides stylisin 1 (1) and stylisin 2 (2) and phakellistatin 13 (3); and bromopyrrole alkaloids sceptrin (4), stevensine (5), and oroidin (6) have been isolated from this species. Among them the compound 2 showed antimicrobial activities against *Bacillus subtilis, Staphylococcus aureus, Pseudomonas aeruginosa, Klebsiella pneumoniae, Candida albicans, Microsporum audouinii,* and *Trichophyton mentagrophytes* with MIC values of 25, 13, 28, 6, 6, 6, and 6 μg/mL, respectivly. The corresponding values of the zone of inhibition diameter (mm) were 13, 16, 26, 28, 27, 15, and 18 mm, respectively (Dahiya and Gautam, 2010; Mohammed et al., 2006).

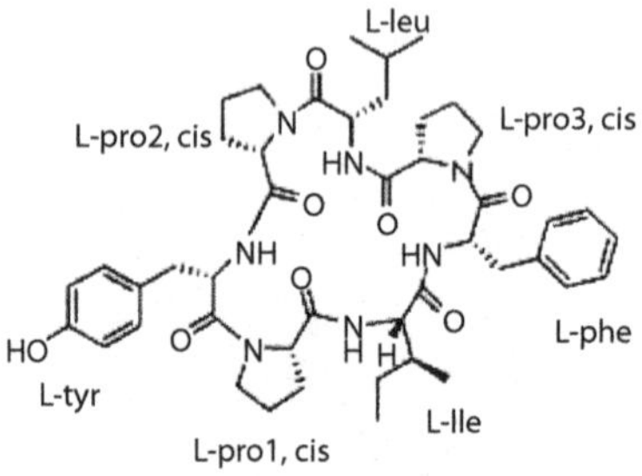

Stylisin 1

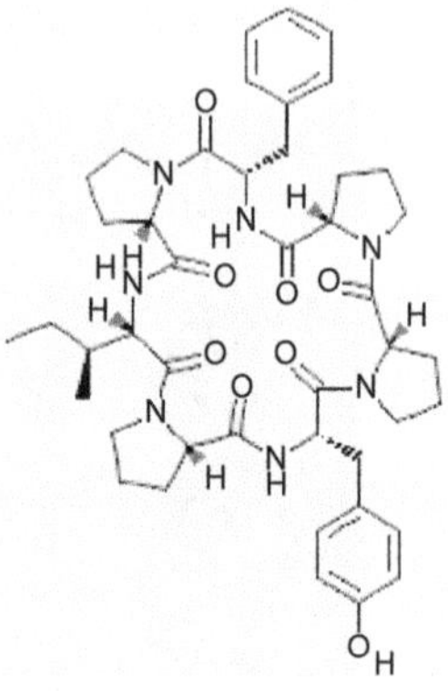

Stylisin 2

Cytotoxic activity: The compound phakellistatin 13 of this species showed moderate cytotoxicity against the HeLa cell line with an IC50 value of 15 μM (Afifi, 2017).

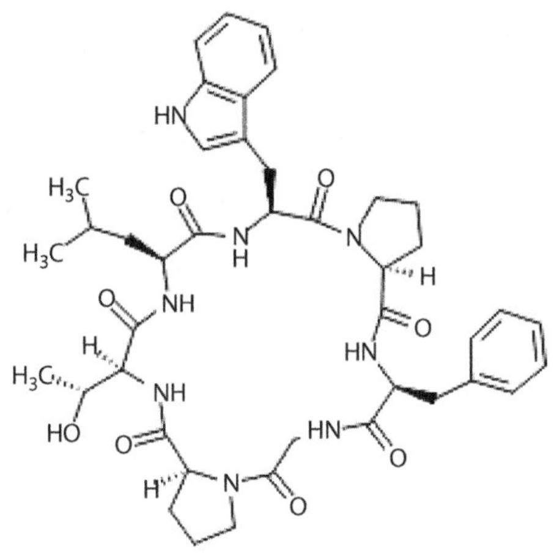

Phakellistatin 13

Stylissa carteri (Dendy, 1889) (= *Phakellia carteri; Acanthella carteri; Axinella carteri; Acanthella aurantiaca*)

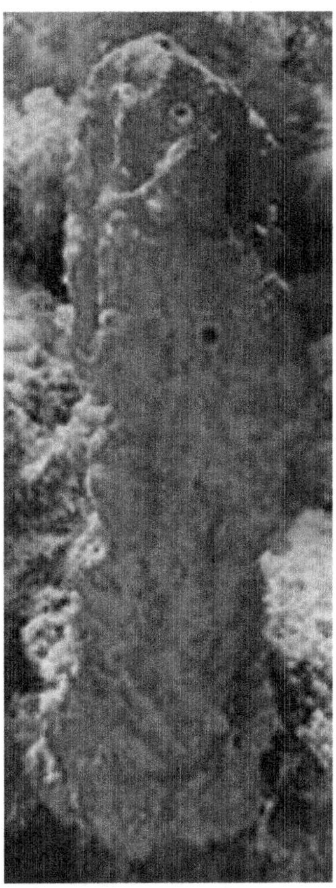

Order: Scopalinida Family: Scopalinidae

Common name: Elephant ear sponge, orange sponge

Distribution: Indian and Pacific Oceans; the Red Sea to Australia; Indonesia, Papua New Guinea, and the Great Barrier Reef, New Caledonia; Red Sea, Saudi Arabia, Zanzibar; Cebu, Philippines

Ecology: It is attached to coral stones of sandy fringing reefs and lagoons at 15–18 m.

Biology: It is a flabellate sponge (94 mm long, 42 mm wide, and 12 mm thick) with relatively thick-buttressed lamellae having irregular margin. It is attached to the substrate by a small basal stalk. Surface is hispid and rugged with ridges (5 mm high) and conules throughout. There is only one osculum (1 mm in dia.). Texture is rubbery, compressible, and easy to tear. The color is bright orange brown in life and pale orange brown in spirit. Megascleres are styles that are relatively long, either slender or robust, and slightly curved. Microscleres are absent. The ectosome is membranous with 13 sparsely protruding extra-axial spicules. Choanosome is composed of multispicular bundles fully cored. Fiber reticulation formed is relatively close-meshed.

Compound(s) and Activities

Antiproliferation activity: The organic extracts of this species showed antiproliferation activity against the human colon adenocarcinoma cell line (Caco-2) (Rady, 2014).

Antitumor activity: The cyclic peptide compounds axinellins A and B of this species displayed antitumor activity (Sipkema et al., 2005).

The two new bioactive cyclopeptides axinellins A and B isolated from this sponge exhibited moderate *in vitro* antitumor activity against human broncopulmonary non-small cell lung carcinoma lines (NSCLC-N6) with IC50 values of 3.0 and 7.3 μg/mL, respectively (Randazzo et al., 1998).

Cytotoxicity: The debromohymenialdisine of this species was found to be the most active cytotoxic compound against L5178y mouse lymphoma cells (ED50, 1.8 μg/mL), followed by its other compounds hymenialdisine and 3-bromohymenialdisine, which were essentially equitoxic and exhibited ED50 of 3.9 μg/mL in both cases (Supriyono et al., 1995).

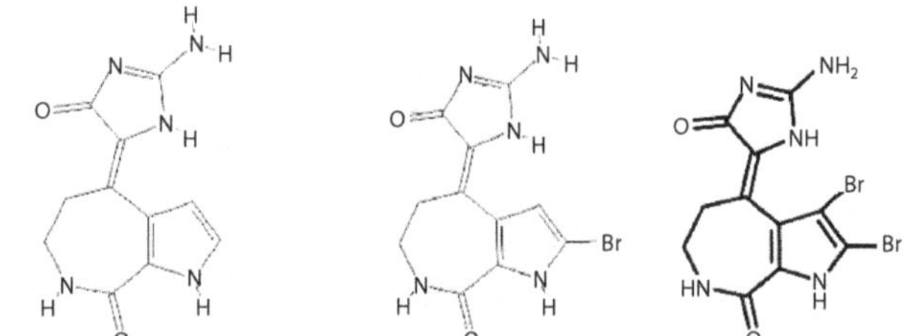

Debromohymenialdisine Hymenialdisine 3-bromohymenialdisine

The cycloheptapeptide named axinastatin 4 of this species showed activity against the P388 lymphocytic leukemia cell line with an ED50 value of 0.06 μg/mL. This compound also showed comparable activity against a series of human cancer cell lines (Pettit et al., 1993b).

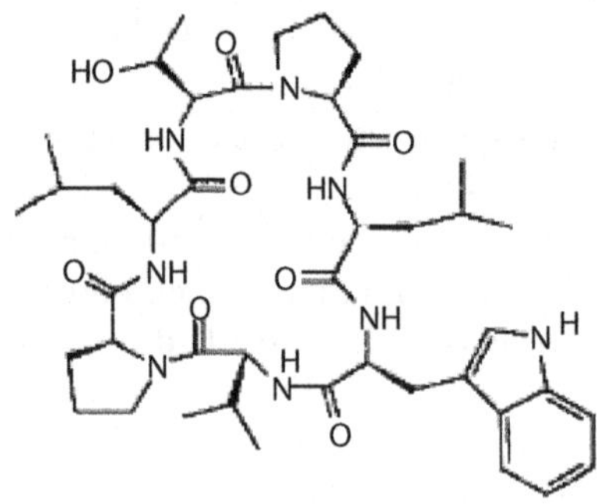

Axinastatin 4

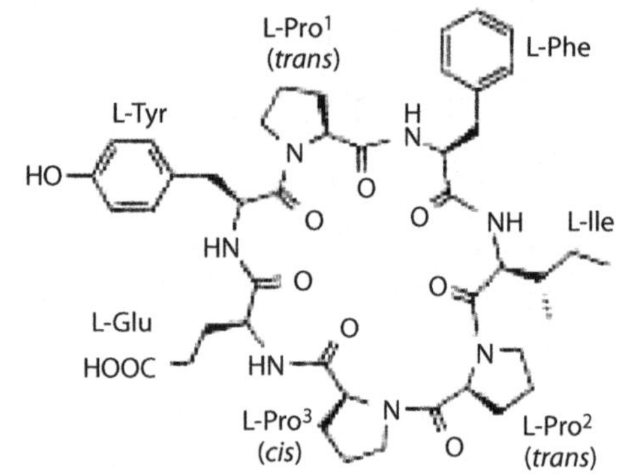

Carteritin A

The phakellistatins isolated from this species inhibited leukemia cell growth. Another related compound, phakellistatin 13, of this species was found to be cytotoxic against the human hepatoma BEL-7404 cell line with an ED50 $<10^{-2}$ µg/mL (Suarez-Jimenez et al., 2012).

The compounds halistatin 2, halistatin 1, halichondrin B, and homohalichondrin B isolated from this species showed anticancer activity. Halistatin 2, like halichondrin B and homohalichondrin B, also caused the accumulation of cells arrested in mitosis, inhibited tubulin polymerization, and inhibited the binding of radio-labeled vinblastine and GTP to tubulin (Pettit et al., 1993c).

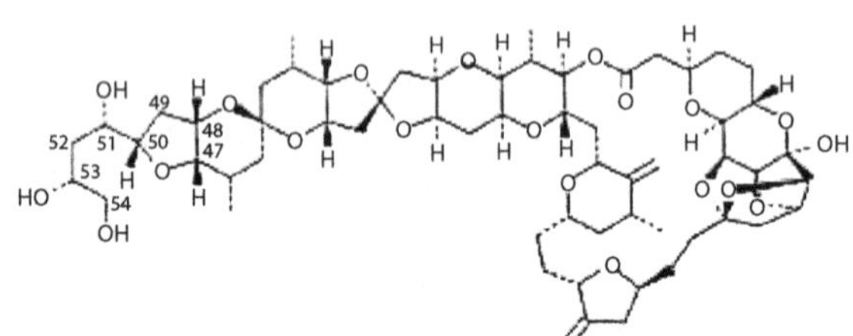

Halistatin 1

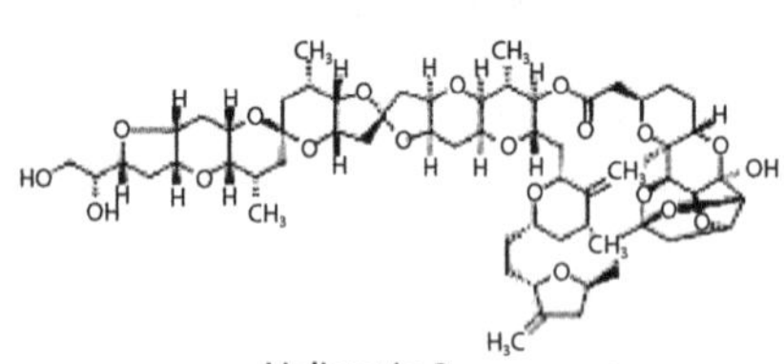

Halistatin 2

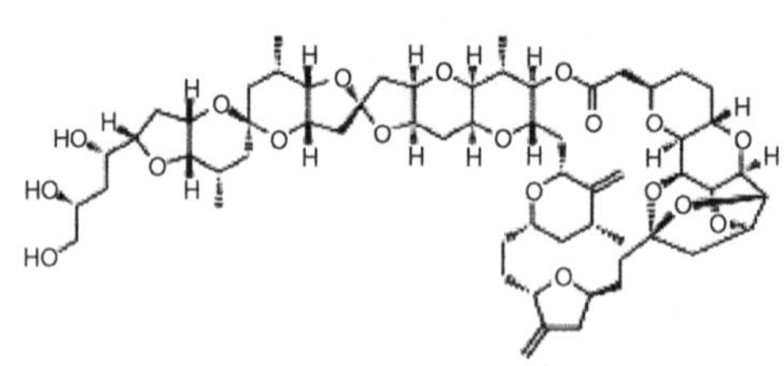

Halichondrin B

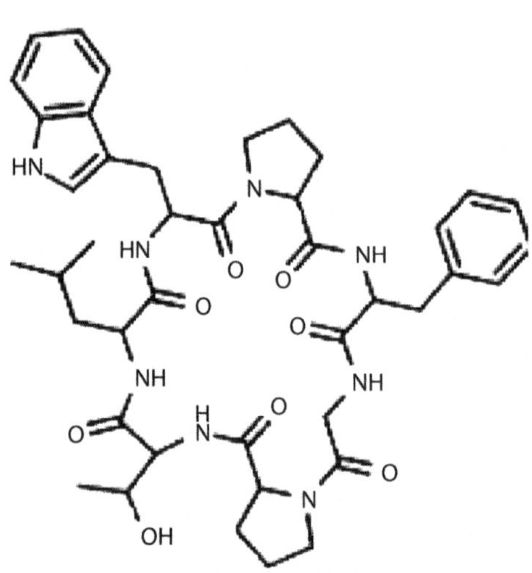

Phakellistatin 13

The cyclo-heptapeptide compound phakellistatin 2 of this species was found to show cell growth inhibitory activity against the P388 cell line (ED50 0.34 µg/mL) (Pettit et al., 1993d).

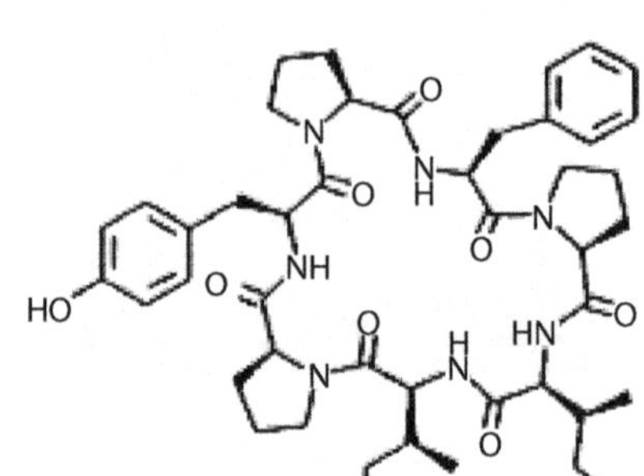

Phakellistatin 2

The cyclic heptapeptide, carteritin A, of this species showed cytotoxicities against HeLa, HCT116, and RAW264 cells with IC50 values of 0.70 to 1.5 µM (Afifi et al., 2016).

The cyclo-octapeptide, axinastatin 5, isolated from this species showed inhibitory activity against human and murine tumor cell lines with GI50 values ranging from 0.3 to 3.3 µg/mL. Other peptide cancer cell growth inhibitory constituents of this orange sponge were found to be the cyclic hepta- and octapeptide compounds axinastatin 1 and hymenistatin 1 (1) (Pettit et al., 1994a).

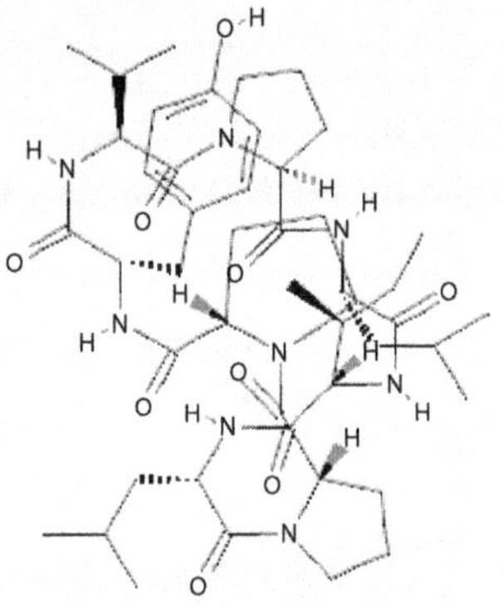

Axinastatin 5

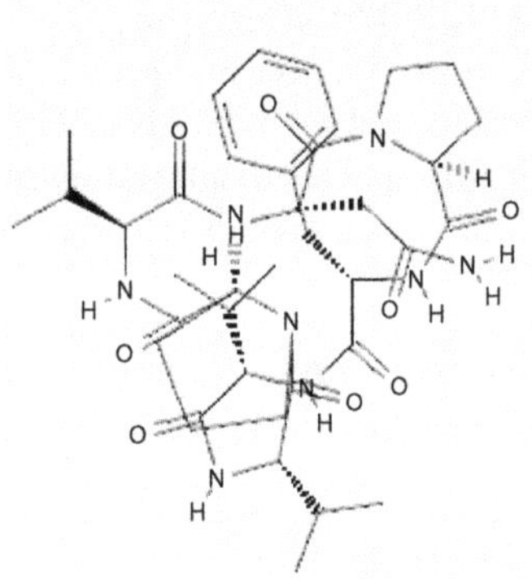

Axinastatin 1

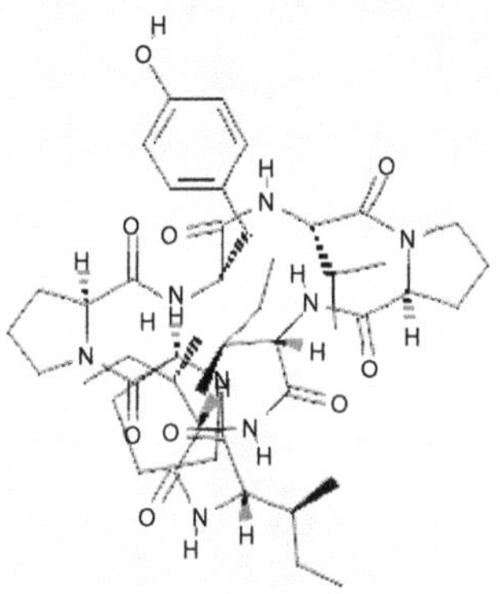

Hymenistatin 1

The compounds lantonduine B ethyl ester and stevensine of this species showed cytotoxicity with IC50 values of 15 and 13 μM, respectively (Afifi, 2017).

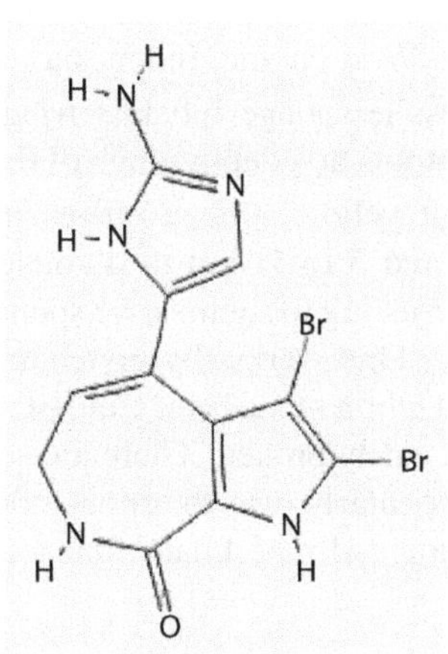

Stevensine

Antiplasmodial activity: The unidentified bacterial isolate THB17 derived from this species displayed antiplasmodial activity against *P. falciparum* with an IC50 value of 20.6 μg/mL (Indraningrat et al., 2016).

Chemotherapeutic agent: The halichondrin B isolated from this species is a potential chemotherapeutic agent (Sepčić et al., 2010).

Antibacterial activity: The extract of this sponge species showed antibacterial activity with the following values of zone of inhibition.

Zone Inhibition (mm)

S. aureus	*E. coli*	*K. pneumoniae*	*S. typhi*	*P. aeruginosa* MDR	*S. aureus* MRSA
–	11.7	–	–	10.0	8.0

Source: Cita, Y.P. et al., 2017, *J. Marine Sci. Res. Dev.*, 7:237.

Anti-inflammatory activity: The alkaloid compound carteramine A of this species showed anti-inflammatory activity by inhibiting neutrophil chemotaxis (IC50 = 5 μM) (Senthilkumar and Kim, 2013).

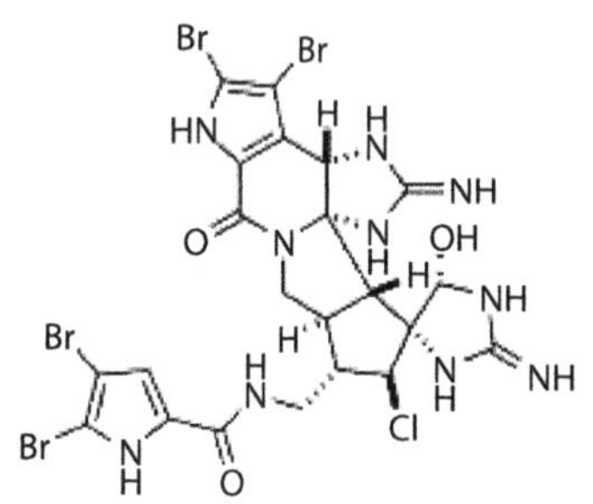

Carteramine A

Neuroprotective activity: The alkaloid hymenialdisine of this species inhibits phosphorylation of the protein tau (which is hyperphosphorylated in Alzheimer's disease) with promising potential against human neurodegenerative diseases (Senthilkumar and Kim, 2013).

Neutrophil inhibitory activity: The alkaloid carteramine A isolated from this species hs been reported to inhibit neutrophil chemotaxis. Neutrophils are the initial responders to bacterial infection or other inflammatory stimuli (Senthilkum and Kim, 2013).

Stylissa flabelliformis (Hentschel, 1912)

Order: Scopalinida Family: Scopalinidae

Common name: Orange fan sponge, orange paddle sponge, Indonesian sponge

Distribution: Indo-West Pacific, Caribbean, Arafura Sea

Ecology: It prefers to grow on rocky shelves.

Biology: It is an erect, flabellate, or compressed-lobate sponge with irregularly conulose and/or ridged surface. Conules are blunt. Surface is smooth. It is usually red, orange, or yellowish in color. Skeleton is confused, but some plumose reticulation is usually recognizable. In the interior and in the stem of erect forms, there is axial condensation. Styles are curved, stout, and relatively short, and are of a single size category.

Aquarium use: It is a suitable species for a reef- or fish-only aquarium. The following parameters have to be maintained while keeping this species (Goemans, 2012).

Minimum tank size: 50 gallons
Temperature range: 72–83°F (22–28°C)
Specific gravity: 1.023–1.025
pH: 8.0–8.4

Acclimation: It is a species sensitive to specific gravity/salinity changes. Therefore, the collection bag's water has to be adjusted to that of the aquarium in which it will be placed by slowly adding small amounts of aquarium water to the bag every few minutes. This process may take at least 15 minutes.

Compound(s) and Activities

Enzyme inhibiting activity: The compound dispacamide E was active against protein kinases GSK-3, DYRK1A, and CK-1 (IC50 2.1, 6.2, and 4.9 μM, respectively) (Gribble, 2015).

Antibacterial and cytotoxic activity: The organic extracts of this species showed fungicidal activity against *Candida albicans*. Further, these extracts showed cytotoxicity against myeloma cell, with an LC50 value of 0.08 mg/mL (Setyowati et al., 2004).

Antiproliferative and protein kinase inhibitory activities: The bromopyrrole alkaloids isolated from the methanolic extracts of this species displayed antiproliferative and protein kinase inhibitory activities (Ebada et al., 2015).

Stylissa massa (Carter, 1887) (= *Stylotella aurantium*; *Hymeniacidon aldis*)

Order: Scopalinida Family: Scopalinidae

Common name: Lumpy orange sponge, mango sponge

Distribution: Indo-West Pacific, Indonesia

Ecology: This sessile sponge inhabits fringing reef or rock wall on sandy bottoms at a depth range of 0 to 100 m.

Biology: This soft yellow sponge ranges in size from 7 to 20 cm in length and 5 to 11 cm in diameter. It is an amorphous, thickly encrusting to submassive sponge with irregular-shaped colonies. This vertically extending growth form narrows to the attachment surface. Consistency is soft, compressible, and not easily broken. There are several oscula that are distributed irregularly over the sponge body. It has a dark maroon outer color and pale khaki inner. It is white brown in spirit.

Compound(s) and Activities

Anticancer/cytotoxic activity: The cycloheptapeptide compound phakellistatin 1 isolated from this species showed cell growth inhibitory property against P-388 murine leukemia with an ED50 value of 7.5 μg/mL (Pettit et al., 1993a).

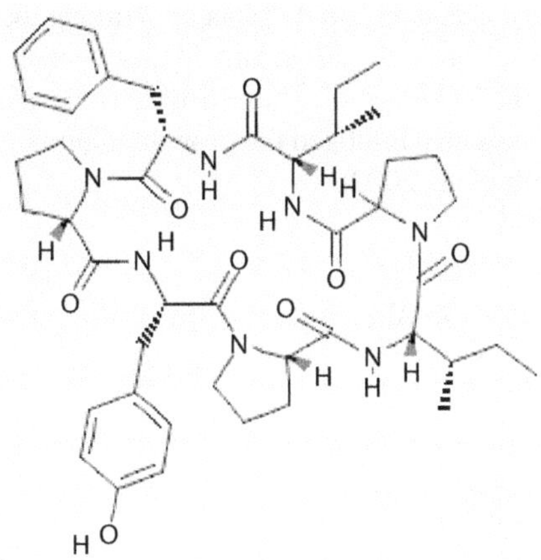

Phakellistatin 1

The cyclic hexapeptide wainunuamide isolated from this species displayed anticancer activity against A2780 ovarian and K562 leukemia cancer cells with ID50 values of 19.2 and 18.4 µg/mL, respectively (Agrawal et al., 2016).

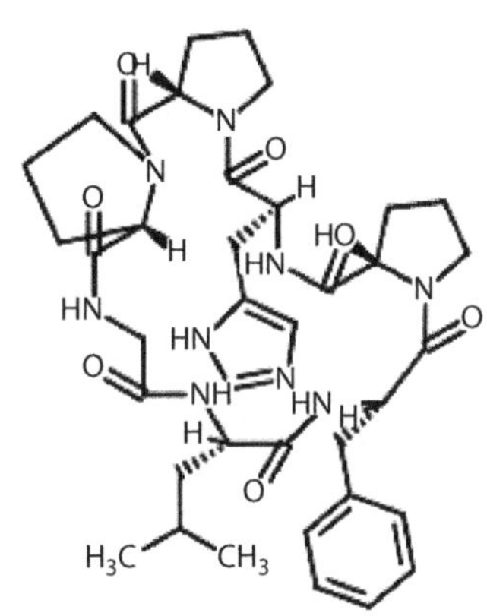

Wainunuamide

The cyclic octapeptide axinellin C isolated from this species displayed anticancer activity against A2780 ovarian and K562 leukemia cancer cells with ID50 values of 13.2 and 4.5 mg/mL, respectively (Agrawal et al., 2016).

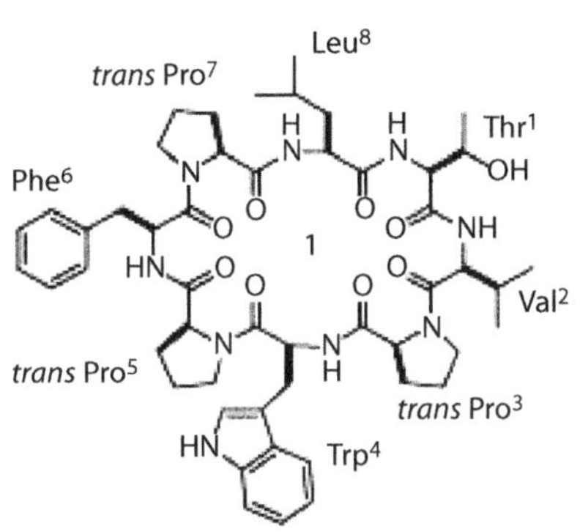

Axinellin C

Three new cycloheptapeptides, stylissatins B–D (1–3), together with 10 known cyclopeptides have been isolated from this species. Of these, the compound stylissatin B exhibited the inhibitory effects against a panel of human tumor cell lines including HCT-116, HepG2, BGC-823, NCI-H1650, A2780, and MCF7 with the ID50 values ranging from 2.4 to 9.8 µM (Sun et al., 2016).

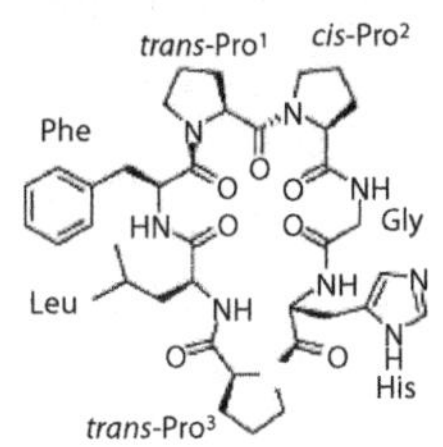

Stylissatins B

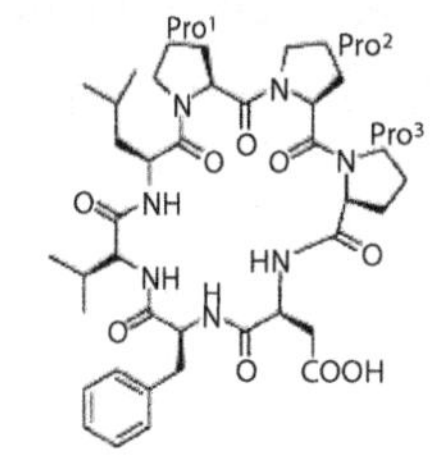

Stylissatins C

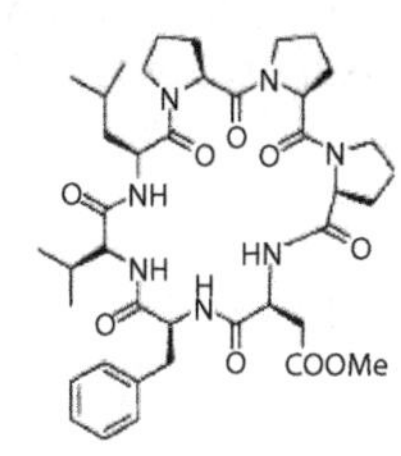

Stylissatins D

The compounds hymenamide C, 4,5 dibromopyrolle-2-carbomide, and lantonduine B ethyl ester of this secies showed cytotoxicity against the human cell line HeLa, wih IC50 values of 25, 18, and 15, µM respectivley (Afifi, 2017).

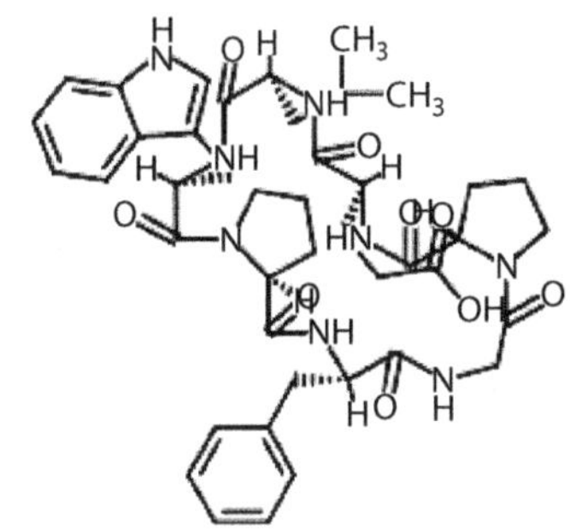

Hymenamide C

Antitumor activity: The uanidine alkaloid compound pyrrole of this species showed antitumor activity by serving as a protein kinase C inhibitor (Anjum et al., 2016; Sipkema et al., 2005).

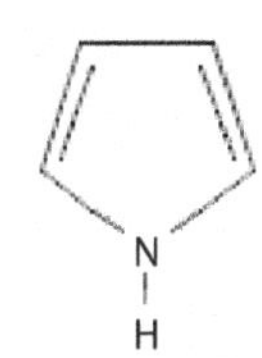

Pyrrole

Antimalarial activity: The compounds 8-isocyanato-15-formamidoamphilect-11(20)-ene and 8-isothiocyanato-15-formamidoamphilect-11(20)-ene, along with two derivatives, namely, 8-isocyano-15-formamidoamphilect-11(20)-ene and 7-formamidoamphilect-11(20),15-diene, from this species showed antimalrial activity (Perdicaris et al., 2013).

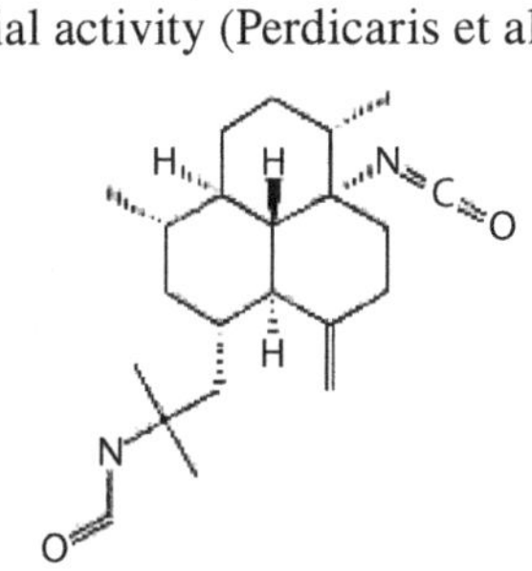

8-isocyanato-15-formamidoamphilect-11(20)-ene

Anti-inflammatory activity: The cyclic peptide stylissatin A of this species showed anti-inflammatory activity by inhibiting the nitric oxide production in LPS-stimulated murine macrophage RAW264.7 cells with an IC50 value of 87 μM (Agrawal et al., 2016).

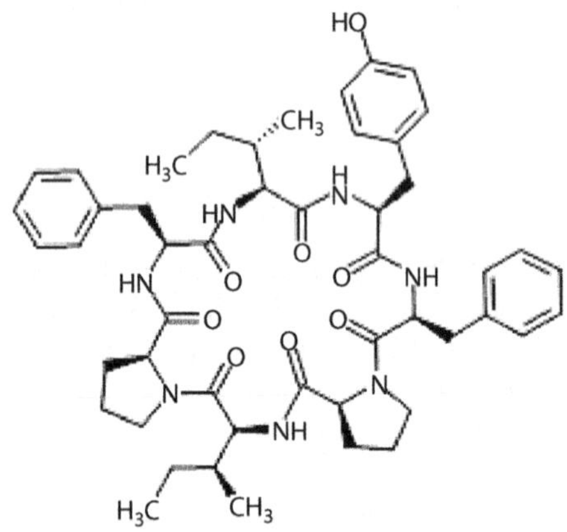

Stylissatin A

Neuroprotective activity: The alkaloid hymenialdisine of this species inhibits phosphorylation of the protein tau (which is hyperphosphorylated in Alzheimer's disease) with promising potential against human neurodegenerative diseases (Senthilkumar and Kim, 2013).

Arthritis-preventing activity: The alkaloid hymenialdisine isolated from this species has been reported to prevent arthritis (Senthilkumar and Kim, 2013).

Enzyme-inhibiting activity: The compound spongiacidin C isolated from this species most strongly inhibited the enzyme ubiquitin-specific-processing protease (USP7) with an IC50 of 3.8 μM (Yamaguchi et al., 2013).

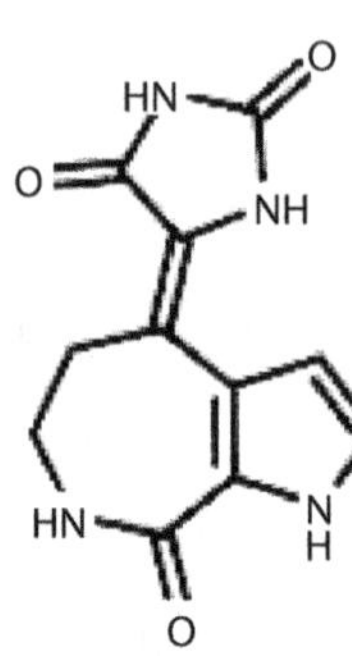

Spongiacidin C

Eight alkaloids—aldisine (1), 2-bromoaldisine (2), 10Z-debromo hymenialdisine (3), 10E-hymenialdisine (4), 10Z-hymenialdisine (5), hymenin (6), oroidin (7), and 4,5-dibromopyrrole-2-carbonamide (8)—have been isolated from this species. Among them, the compounds 4 and 5 showed significant MEK-1 enzyme inhibitory activity with IC50 values of 3 and 6 nM, respectively. Further, these compounds also inhibited the growth of human tumor LoVo cells (Tasdemir et al., 2002).

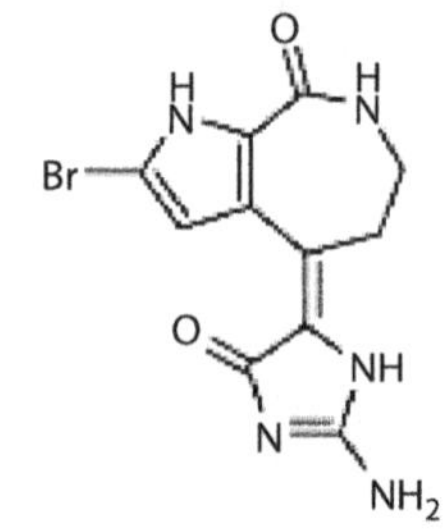

10E-hymenialdisine

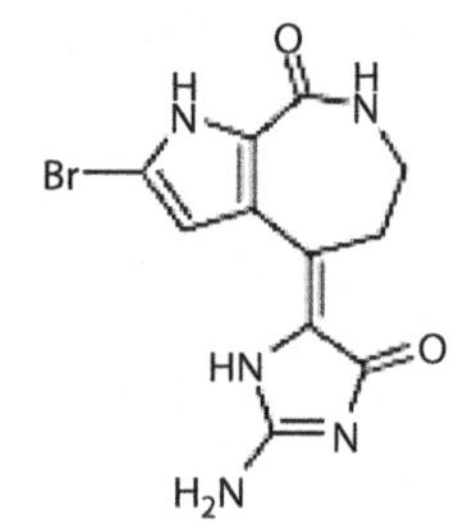

10Z-hymenialdisine

The compound, dispacamide E was active against protein kinases GSK-3, DYRK1A, and CK-1 (IC50 2.1, 6.2, and 4.9 μM, respectively) (Gribble, 2015).

Stylissa sp.

Order: Scopalinida Family: Scopalinidae

Compound(s) and Activities

Antibacterial activity: The hexane-soluble fraction of this sponge showed antibacterial activity with an MIC value of 0.13 mg/mL against *S. aureus* and *E. coli* and 0.25 mg/mL against *E. faecalis* (Govinden-Soulange et al., 2014).

Cytotoxic activity: The compounds 8-isothiocyanatoamphilect-10-ene (1), 8-isocyanatocycloamphilect-10-ene (2), and amphilectanes (3,4) isolated from this species showed weak cytotoxicity toward HeLa cells with IC50 values of 89, 38, 11, and 20 μg/mL, respectively (Emsermann et al., 2016).

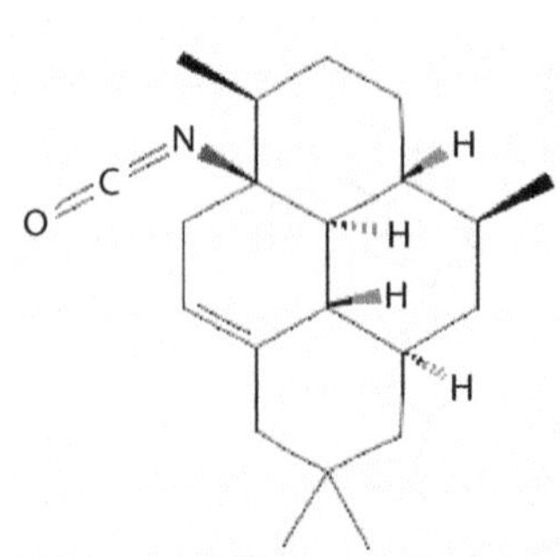

8-isocyanatocycloamphilect-10-ene

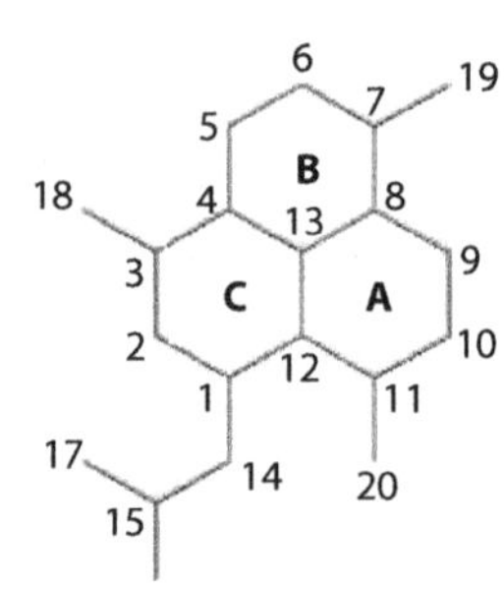

Amphilectane

The compound 12-N-methyl stevensine of this species showed significant *in vitro* activity for cytotoxicity against the mouse lymphoma cell line L5187Y with an EC50 value of 3.5 μg/mL (Putra and Jaswir, 2014).

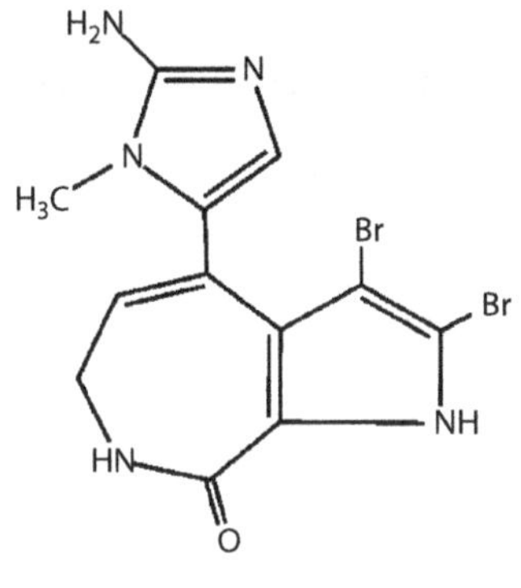

12-N-methyl stevensine

Anticancer activity: A proline-rich octapeptide stylissamide X isolated from this species inhibited HeLa cells in the concentration range 0.1 to 10 μM through wound-healing assay (Agrawal et al., 2016).

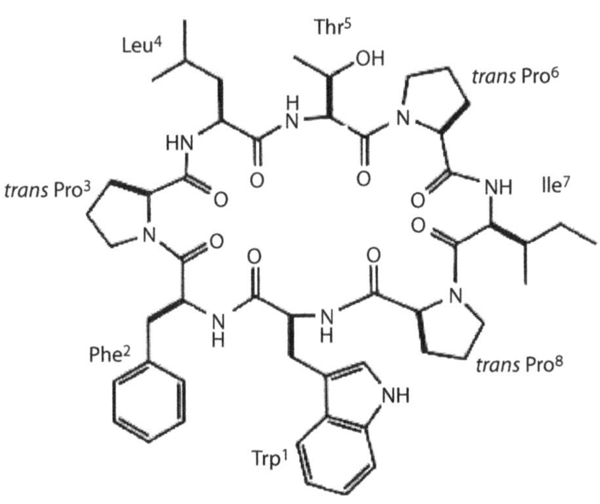

Stylissamide X

Antitumor activity: The 12 unidentified brominated alkaloids of this species showed cytotoxicity against mouse lymphoma cells L5187Y with EC50 values ranging from 1.8 to 9.0 μg/mL at an extract concentration of 10 μg/mL (Gribble, 2015).

Svenzea flava (Lehnert and van Soest, 1999)

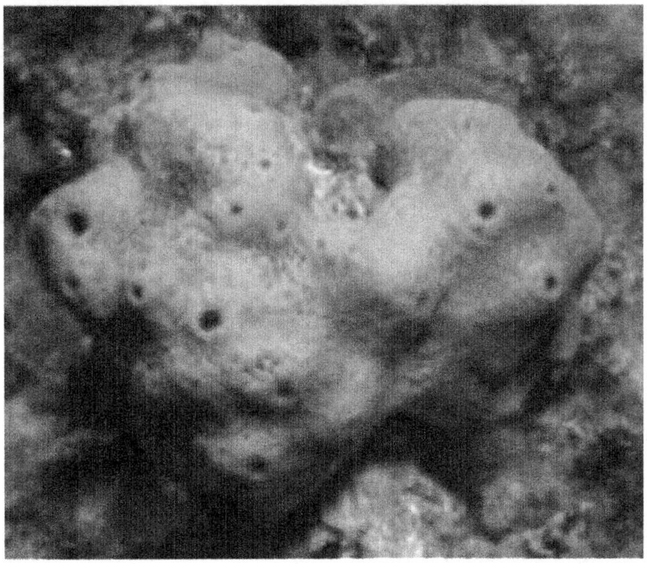

Order: Scopalinida Family: Scopalinidae
Common name: Caribbean sponge
Distribution: Discovery Bay, Jamaica
Ecology: It inhabits deep fore reefs.
Biology: It is a yellow sponge with a smooth suface of the size 8 × 6 × 4 cm. In the dry state, numerous openings are visible, and these are covered by a skinny, organic veneer. One large osculum (1.2 × 0.8 cm) is seen and it is divided into two canals. Ectosome is an organic veneer into which spicule brushes protrude. Choanosome consists of long, ascending polyspicular tracts, which are embedded in sponging. Spicules are styloids that are tapering at one end but are mostly blunt. True styles are rare.

Compound(s) and Activities

Anti-TB activity: The tricyclic diterpene 7-isocyanoneoamphilecta-1(14),15-diene isolated from this sponge demonstrated strong growth inhibition potency against *Mycobacterium tuberculosis* H37Rv with an MIC value of 8 μg/mL (Emsermann et al., 2016).

FAMILY: HALICHONDRIIDAE

Amorphinopsis atlantica (Carvalho, Hajdu, Mothes and van Soest, 2004)

Order: Suberitida Family: Halichondriidae
Common name: Not designated, as it is a new species
Distribution: Atlantic Ocean, Caribbean Sea, Southeastern Brazil

Ecology: It is found in rocky coasts (5–16 m depths), on vertical surfaces or in overhangs. It is rarely exposed to direct sunlight. It is often associated with ophiuroids.

Biology: It is a massive, cushion-shaped lobate, occasionally encrusting. Live color of the animal is white (rarely), or yellow to dark green. Largest specimen measures 10 cm long, 4 cm wide, and 4 cm high. Surface is smooth and the oscules are 2 to 7 mm across, and are scattered. This sponge is only slightly compressible, and becomes friable when dried.

Compound(s) and Activities

Antioxidant activity: The methanol extracts of this spcies showed percent inhibition of radical ABTS with 53% (Tovar-Barrios et al., 2016).

Amorphinopsis fenestrata (Ridley, 1884) (= *Leucophloeus fenestratus*)

Order: Suberitida Family: Halichondriidae

Common name: Not designated

Distribution: Indo-West Pacific, Malay Archipelago, Vietnam

Ecology: Known from a depth of 320 m

Biology: This species has a porous body and a hard surface like a stone. It can absorb oxygen from water through the diffusion process.

Compound(s) and Activities

Anthelmintic/antiparasitic activity: Amorphane sesquiterpenes, namely, (1R, 6S, 7S, 10S)-10-isothiocyanato-4-amorphene, (1R*, 4S*, 6R*, 7S*)-4-isothiocyanato-9-amorphene, 10-isothio cyanato-4,6-amorphadiene, and (4S*, 10S*)-10-isothiocyanato-5-amorphen-4-ol, isolated from this species have shown potent antihelmintic/antiparasite activity (Alvi et al., 1991).

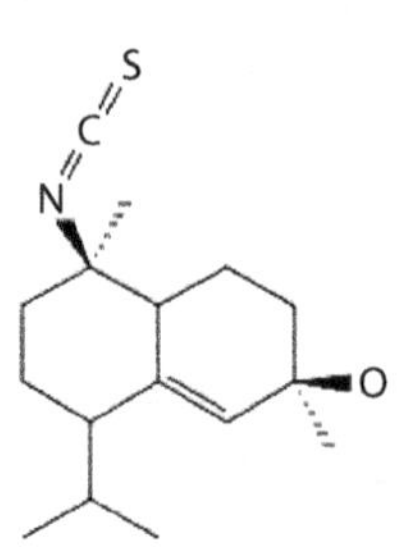

(4S*, 10S*)-10-isothiocyanato-5-amorphen-4-ol

Amorphinopsis sp.

Order: Suberitida Family: Halichondriidae

Compound(s) and Activities

Anticancer, anti-inflammatory, antimicrobial, and analgesic activities: The organic extracts of this species showed anticancer, anti-inflammatory, antimicrobial, and analgesic activities (Breitinger et al., n.d.)

Axinyssa ambrosia (de Laubenfels, 1936)

Order: Suberitida Family: Halichondriidae

Common name: Caribbean marine sponge

Distribution: Widespread in the Caribbean

Ecology: It is found at depths of 10 to 176 m in deeper areas of reefs.

Biology: It is a massive-amorphous, fleshy species with lobate protuberances. The surface is smooth, from finely to more obviously conulose. Consistency is hard but compressible. Color of the animal is cream to yellow.

Compound(s) and Activities

Cytotoxic activity: The acetone–methanol (1:1) extract of this species yielded three new sesquiterpenes: (4R*,5R*,7S*,10R*)-eudesm-11-en-4-ylamine hydrochloride (1), axinyssamine hydrochloride, (4R*,5R*,7S*,10R*)-4-isocyanatoeudesm-11-ene (2), and (4R*,5R*,7S*,10R*)-formamidoeudesm-11-ene (3). Among these compounds, compound 1 exhibited significant cytotoxic activity against cancer cells (Petrichtcheva et al., 2002).

Axinyssa isabela (Carballo and Cruz-Barraza, 2008)

Order: Suberitida Family: Halichondriidae

Common name: Not available

Distribution: Gulf of California

Ecology: It grows over rocky and coral reef regions.

Biology: It is an encrusting cushion-shaped to massive amorphous form with a thickness of about 5 cm. Sometimes it is lobated with rounded borders and apical oscules. The surface can be smooth, although generally it is irregular, conulose–microconulose, and minutely hispid, due to the terminal part of the choanosomal spicule and tracts that form brushes on the surface. Sometimes small, raised ridges may be observed running over the surface. Circular or oval-shaped oscula (1 to 6 mm in dia.) are found scattered irregularly on the surface. The consistency is firm, fleshy, compressible, and resilient in life. The color in life is light yellow orange.

Compound(s) and Activities

Cytotoxicity: Nine new sesquiterpenes, the eudesmanes, axinisothiocyanates M and N (1, 2), the bisabolane axinythiocyanate A (3), and the aristolane derivatives axinysones A–E (4–8) and axinynitrile A (9), together with four known sesquiterpenoids (10–13), isolated from this species have shown cytotoxic activity against three human tumor cell lines (Zubía et al., 2008).

Antimalarial activity: The compound 4-isothiocyanatoeudesm-11-ene isolated from this species has shown moderate antimalarial activity against *Plasmodium falciparum* (chloroquine-sensitive strain D6: IC50 = 4000 ng/mL, chloroquine-resistant strain W2: IC50 = 550 ng/mL) (Emsermann et al., 2016).

Axinyssa sp.

Order: Suberitida Family: Halichondriidae

Compound(s) and Activities

Cytotoxic activity: New bicyclic sesquiterpenes axinisothiocyanates A–L (1–12) together with the known compounds (1R,6S,7S,10S)-10-isothiocyanato-4-amorphene (13), (4R*,5R*,7S*,10R*)-4-isocyanoeudesm-11-ene, (–)-epipolasin A, and (+)-aristolone isolated from this unidentified species have shown cytotoxic activity against three human tumor cell lines (Zubía et al., 2008).

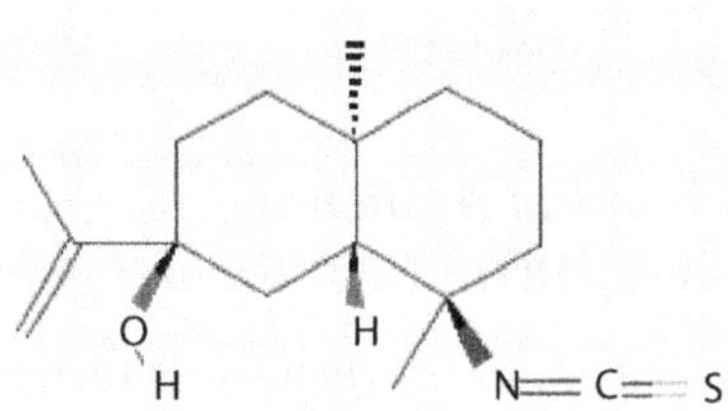

Axinisothiocyanate M

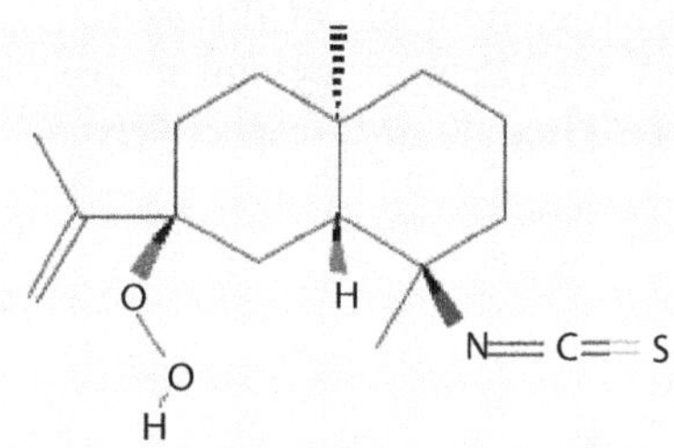

Axinisothiocyanate N

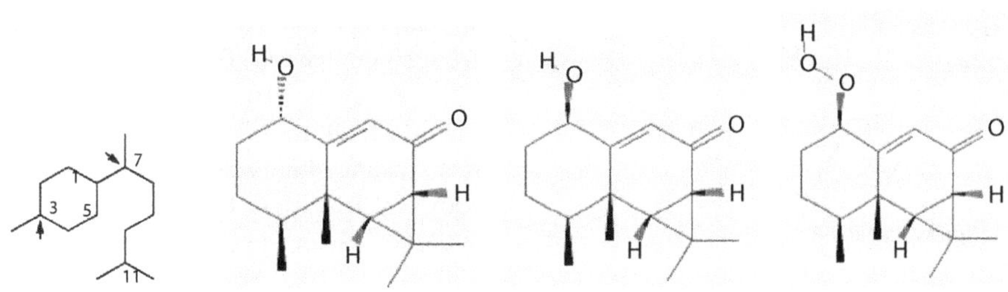

1 2 3 4

1: Bisabolane axinythiocyanate A ; 2-4: Axinysones A-C

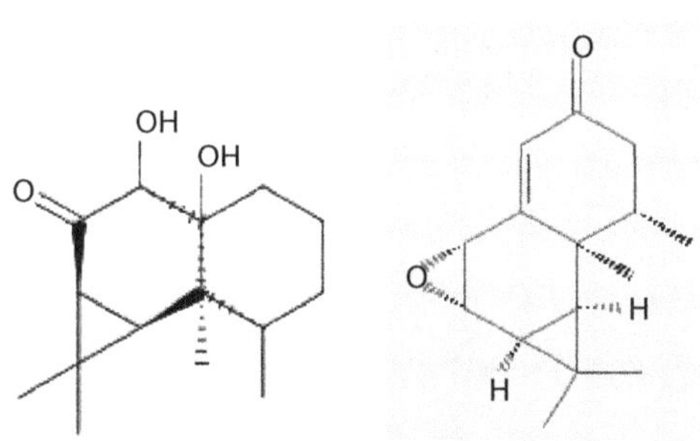

Axinysones D-E

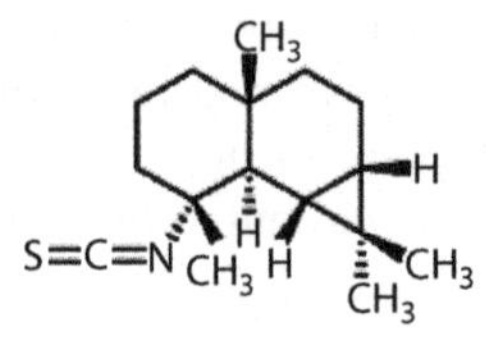

Epipolasin A

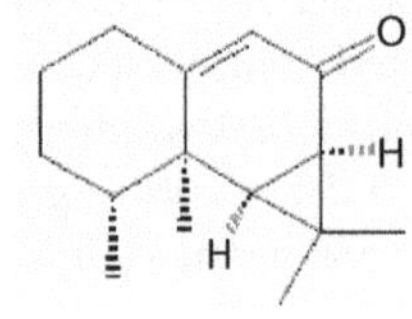

Aristolone

Emsermann et al. (2016) reported that the compound 4-formamidoeudesm-7-ene isolated from this unidentified species exhibited significant cytotoxic activity against the human cancer cell lines CNE-2 (IC50 = 13.8 μg/mL), HeLa (IC50 = 7.5 μg/mL), and LO2 (IC50 = 38.0 μg/mL).

Epipolasis sp.

Order: Suberitida Family: Halichondriidae

Compound(s) and Activities

Antitumor activity: The macrolide compounds of this species, namely, spirastrellolides A and B, were cytotoxic to HeLa cells, with IC50 values of 20 and 40 nM, respectively (Gribble, 2015).

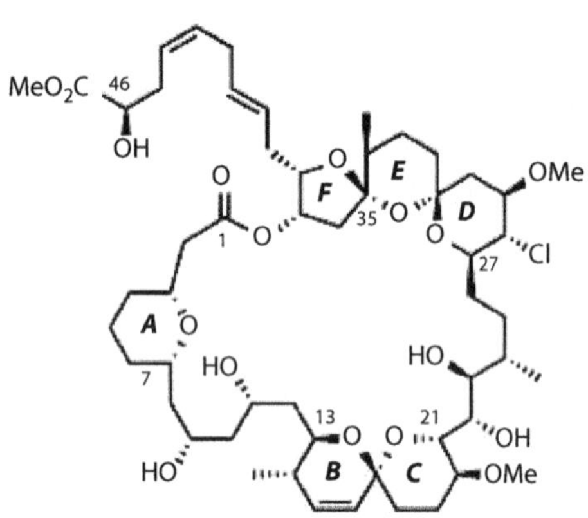

Spirastrellolide A

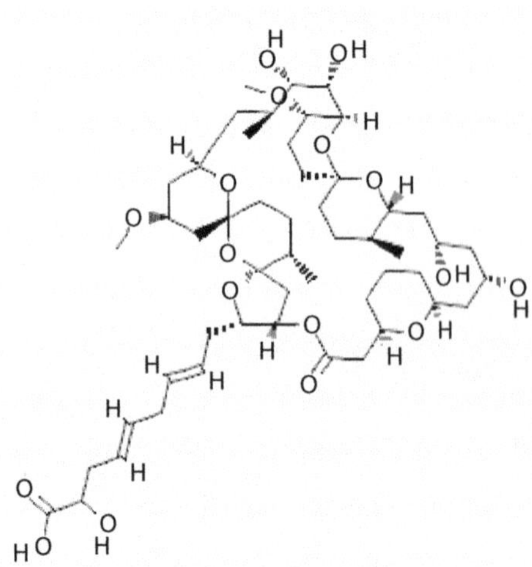

Spirastrellolide B

Cytotoxic activity: The compounds spirastrellolides A and B isolated from this species showed cytotoxic activity (Suzuki et al., 2012).

Enzyme inhibitory activity: The sesquiterpenes, namely, (+)-curcuphenol and dehydrocurcuphenol, isolated from this species showed inhibitory activities of the enzyme H,K-ATPase. Some ATPase inhibitors have entered clinical trials, which could be new cancer therapies (Fusetani et al., 1987).

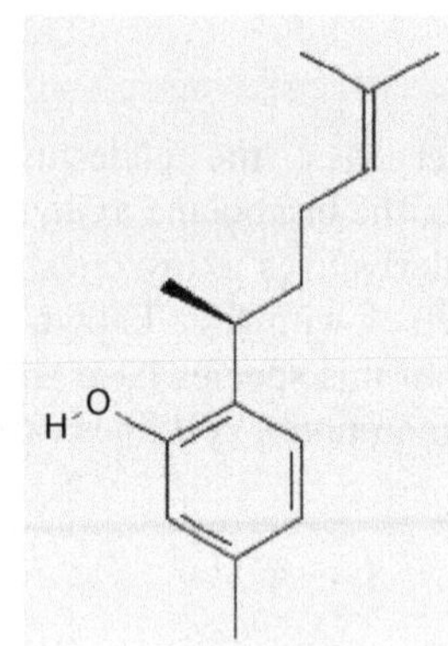

Curcuphenol

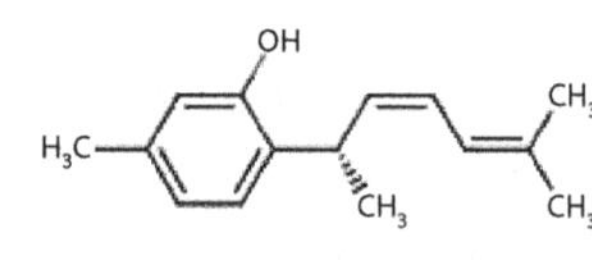

Dehydrocurcuphenol

Halichondria (Halichondria) genitrix (Schmidt, 1870) (= *Topsentia genitrix*)

Order: Suberitida Family: Halichondriidae

Common name: Not known

Distribution: Greenland, Roscoff, west coast of Ireland, Norway

Ecology: It occurs on stones at greater depths, that is, below 16 m.

Biology: Shape, size, surface and consistency: It is a massive, flattened sponge wih a size of 10 to 15 cm diameter. Oscules are seen in grooves. The interior is riddled by holes and canals. Consistency is firm and crumbly. Spicules are in the form oxeas, which are often irregular in shape and are of variable size. A tangential ectosomal skeleton is present. Color of the body is yellowish or brownish green, exhibiting a color change to darker, violet tinges when brought above the surface.

Compound(s) and Activities

Anti-inflammatory activity: The bis-indole alkaloid topsentin isolated from this species showed anti-inflammatory activity by serving as an phospholipase A2 inhibitor (Sipkema et al., 2005).

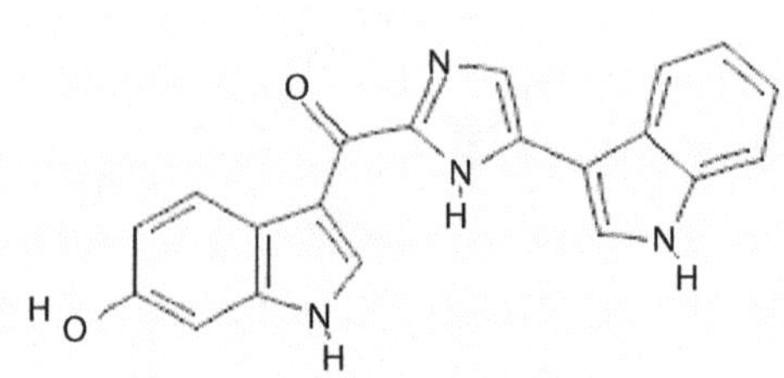

Topsentin

Keyzers (2005) reported that the bis(indoyl)imidazole alkaloids topsentin and bromotopsentin of this species have shown to inactivate PLA2 (IC50, 0.5 and 6.0 μM, respectively) and to inhibit PMA-induced mouse ear edema (ED50, 15 and 30 μg per ear, respectively).

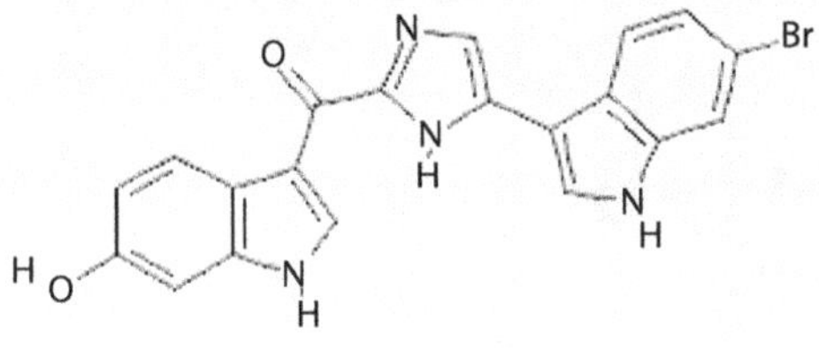

Bromotopsentin

Halichondria (Halichondria) glabrata (Keller, 1891)

Order: Suberitida Family: Halichondriidae
Common name: Not designated
Distribution: Eritrean Exclusive Economic Zone
Ecology: Not available
Biology: No report

Compound(s) and Activities

Anticancer activity: The extract of this species showed anticancer activity against A549, HeLa, and HT-29 cancer cell lines with GI50 values of >100 μg ("Sponge crude extraction," n.d.).

Halichondria (Halichondria) japonica (Kadota, 1922)

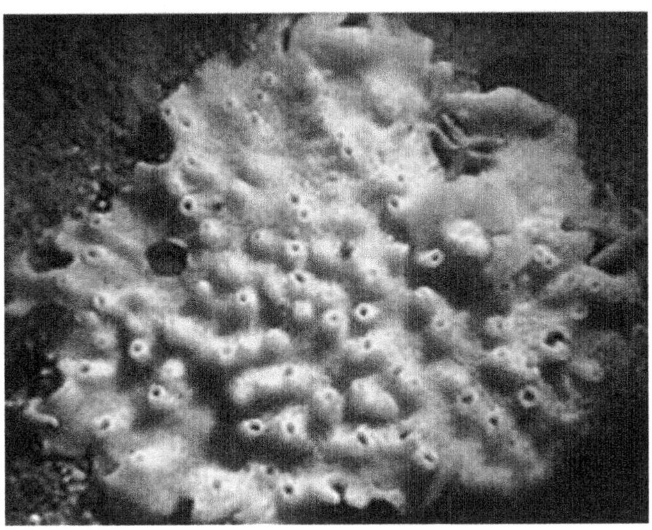

Order: Suberitida Family: Halichondriidae
Common name: Not designated
Distribution: Pacific coast of central Japan, East China Sea
Ecology: It occurs on rocky shores, exposed areas, tide pools, or narrow crevices between rocks.
Biology: Growth form of this sponge is thickly and broadly encrusting. There are two morphs, namely, orange and yellow types. The orange type has abright reddish body. Oscula (1–1.5 mm in dia.) are opened on each apex of conical or cylindrical processes (0.5–2 cm in height) that are irregularly scattered on the surface. The spiculation consists of only styles that are straight or slightly curved. Individuals of this species are able to distinguish an allogeneic individual from an autogeneic one, and showed rejection reactions against allogeneic individuals. For example, it showed rejection reaction against individuals of *Halichondria okadai*, but the intensity of rejection was less than that of allogeneic rejection.

Compound(s) and Activities

Antibacterial activity: The compounds cyclic thiopeptides of its associated bacterial species, *B. cereus*, exhibited potent antibacterial activities (Joseph et al., 2012).
Antimicrobial activity: The associated fungal strain *Phoma* sp. Q60596 isolated from this species showed potent antifungal activity against *Candida albicans*, *Cryptococcus neoformans*, and *Aspergillus fumigatus*. Similarly, *Bacillus cereus* QN03323 of this sponge species exhibited potent antibacterial activities against staphylococci and enterococci including multiple drug-resistant strains (Thomas et al., 2010).
Cytotoxicity: Novel compounds, namely, gymnastatins A–H, Q and R; gymnasterone A, B, C, and D; and steroid dankasterones A and B, and dankastatins A and B, isolated from an ascomycete fungal strain *Gymnascella dankaliensis* OUPS-N134, derived from this sponge species showed cytotoxic activities. Gymnastatins A, B, C, F, G, Q, and R, and dankastatins A and B exhibited potent cytotoxicity and growth inhibition in a P388 lymphocytic leukemia. Gymnastatin Q was equally active against breast and human cancer cell lines. Gymnasterones B, C, and D, and dankasterone A showed significant cytotoxic activity in P388 lymphocytic leukemia. Dankasterone A was also active against human cancer cell lines (Amagata et al., 1999; Thomas et al., 2010).

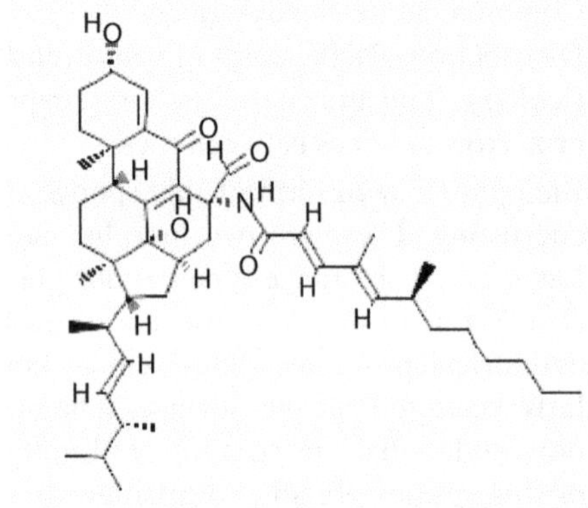

Gymnasterone A

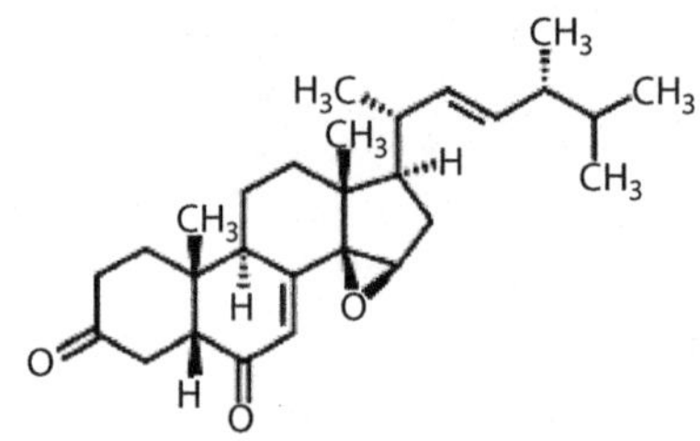

Gymnasterone B

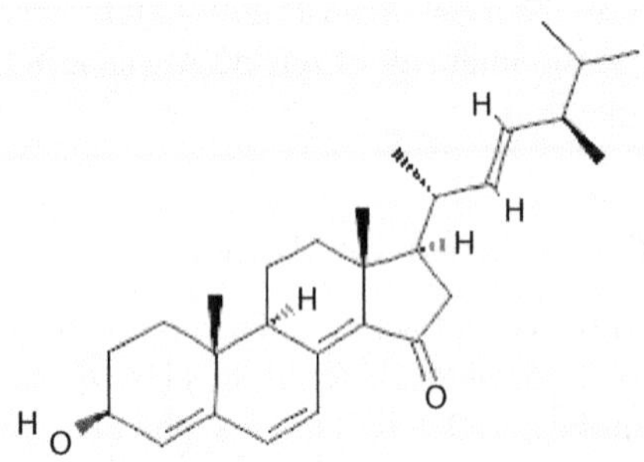

Gymnasterone C

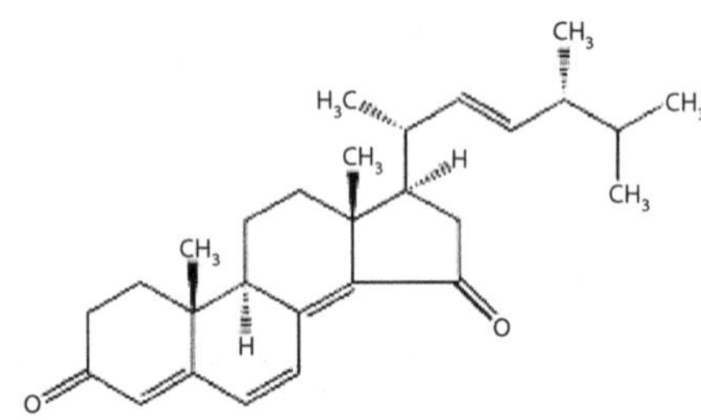

Gymnasterone D

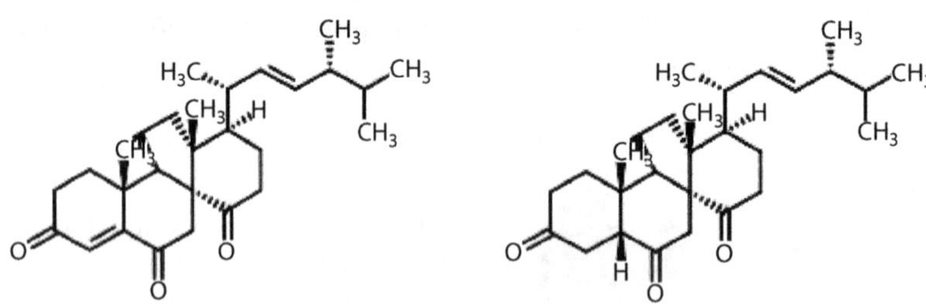

Dankasterones A and B

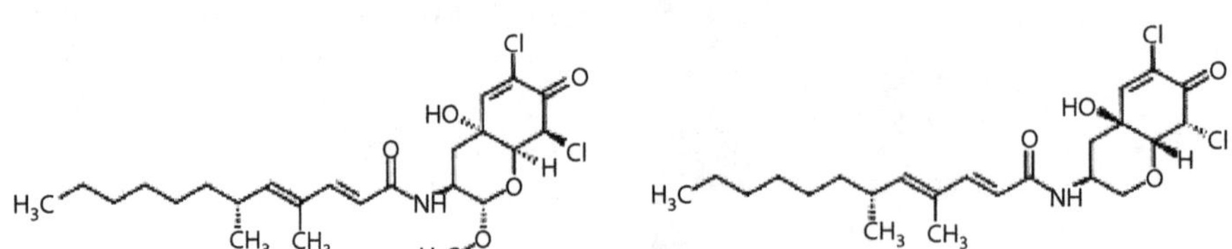

Dankastatins A and B

Halichondria (Halichondria) melanadocia (de Laubenfels, 1936)

Order: Suberitida Family: Halichondriidae

Common name: Not designated

Distribution: Bahamas, Eastern Caribbean, Greater Antilles, Northern Gulf of Mexico, Southern Caribbean, Southwestern Caribbean

Ecology: It commonly occurs in shallow water with little circulation, often in brackish waters or areas with broad salinity fluctuations.

Biology: This sponge is amorphous, thickly encrusting, or massive. Specimens are often lobate, with digitate processes and oscular chimneys that are of one to several centimeters thick. Live individuals are black with olive tinges externally, and greenish yellow internally. Living tissue is softly spongy, limp, and easily torn. Dried specimens are, however, black and somewhat shiny, hard, and tough. The surface of all individuals is smooth to the touch and is slightly tuberculate and rugose, with intermittent lobes of variable size. The oscules are flush with the surface, or raised on uneven conical projections with membranous collars 1–10 mm wide. The ectosome is a detachable skin containing haphazardly placed spicules. The spicules are fusiform oxeas, slightly bent or curved, with long, conical, constricted points. A single specimen of this species can cover an area of at least 225 cm^2.

Predators: Angelfish, filefish, parrotfish, trunkfish, and cowfish have been observed to feed on this sponge, consuming over 50% of the total tissue in some cases.

Associated species: It has been found to occur with several other sponges. In areas where competition occurs, this species often lives alongside or overgrows other sponges, but is rarely overgrown by its neighbors.

Compound(s) and Activities

Cardioprotective activity: The okadaic acid extracted from the organic extracts of this species has shown cardioprotective activity by serving as a protein phosphatase inhibitor, which is of therapeutic use in the treatment of heart failure and cardiac arrhythmia (Patella and Ehrlich, 2016; Thomas et al., 2010).

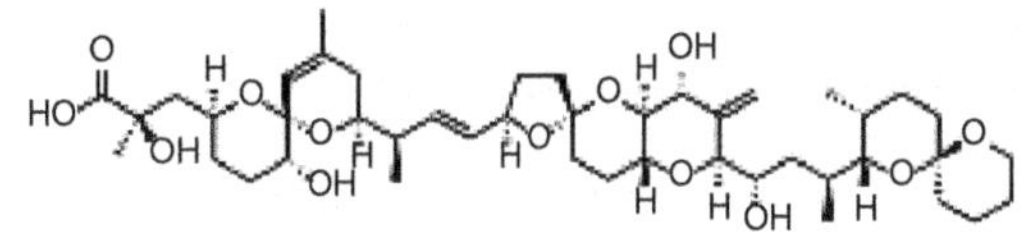

Okadaic acid

Halichondria (Halichondria) moorei (Bergquist, 1961)

Order: Suberitida Family: Halichondriidae

Common name: Not known

Distribution: Off the coast of New Zealand

Ecology: It prefers to settle on basalt and shell in 5 m deep water.

Biology: It is an irregularly folded sponge with a wrinkled surface. The color of the animal is pinkish orange. During the mating period, the adults show a change in color.

Compound(s) and Activities

Antimicrobial activity: A new C29 a steroid sulfate, 24ξ,25-dlmethylcholestane-2β,3α,6α-triyl trisodium sulfate, isolated from this species has shown antimicrobial properties (Fusetani et al., 1981).

Anti-inflammatory effect: The potassium fluorosilicate, a derivative of fluorine of this species, has shown anti-inflammatory effect. The fluorine constitutes about 10% of the dry weight of this sponge species (Gregson et al., 1979).

Halichondria (Halichondria) okadai (Kadota, 1922)

Order: Suberitida Family: Halichondriidae

Common name: Not designated

Distribution: Japan, Taiwan

Ecology: It inhabits intertidal zone as an encrusting sponge.

Biology: These are massive, amorphous sponges with clearly separated inner and outer skeletons, which consist of bundles of spicules arranged in a seemingly random pattern. These sponges are known to accommodate infaunal crustacean organisms, especially gnathiid isopods.

Compound(s) and Activities

Cytotoxic and antimicrobial properties: The *Alteromonas* spp., found commonly associated with this species, produce macrolactam and amide ester compounds with cytotoxic and antimicrobial properties. The tetracyclic alkaloid alteramide A exhibited cytotoxic activity against leukemia P-388, lymphoma L-1210, and epidermal carcinoma KB cells (Thomas et al., 2010).

Cytotoxicity: The fungal strain *Trichoderma harzianum* OUPSN115 isolated from this sponge yielded novel cytotoxic compounds such as trichodenone A, B, and C. They exhibited significant cytotoxicity against the leukemia P388 cell line (Thomas et al., 2010).

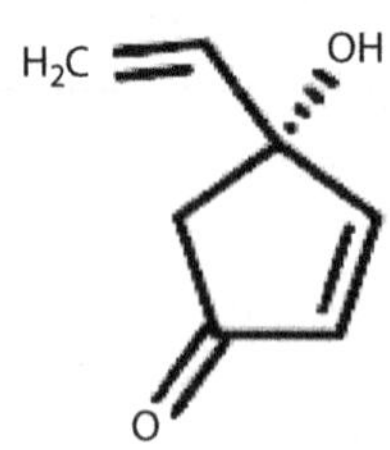

Trichodenone A

Anticancer activity: The novel sesquiterpene alkaloids halichonines A (1), B (2), and C (3) have been isolated from this species. Among these compounds, compound 2 induced apoptosis in HL60 human leukemia cells (Ohno et al., 2011).

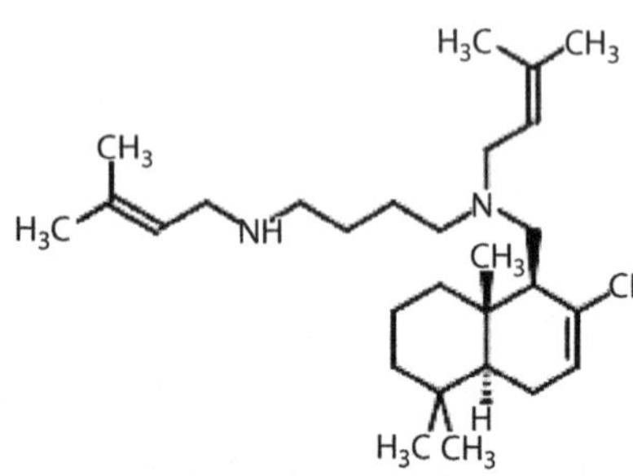

Halichonine B

Antitumor activity: The polyether macrolide compound halicondrin B isolated from this species showed antitumor activity by serving as a tubulin polymerization inhibitor (Braga, 2014; Sipkema et al., 2005).

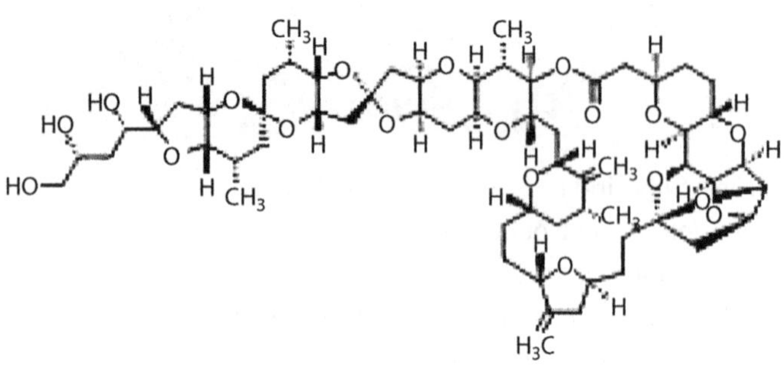

Halicondrin B

Antioxidant activity: The Gram-bacterial strain *Rubritalea squalenifasciens* HOact23T obtained from this species yielded potent red pigmented antioxidants acyl glycol-carotenoic acids such as diapolycopenedioic acid xylosyl esters A, B, and C (Thomas et al., 2010).

Cardioprotective activity: The cyclic aza-polyketide compound halichlorine isolated from this species showed cardiovascular protective activity by serving as a VCAM-1 (vascular cell adhesion molecule) inhibitor (Anjum et al., 2016; Senthilkumar and Kim, 2013; Sipkema et al., 2005).

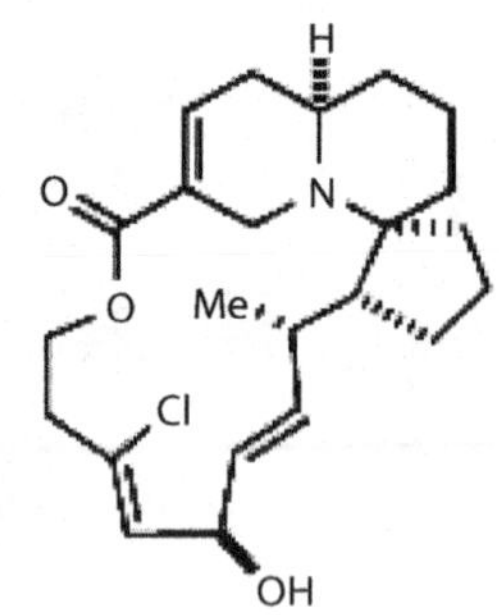

Halichlorine

The okadaic acid extracted from the organic extracts of this species has shown cardioprotective activity by serving as a protein phosphatase inhibitor that is of therapeutic use in the treatment of heart failure and cardiac arrhythmia (Patella and Ehrlich, 2016; Thomas et al., 2010).

Chang et al. (2013) reported that the okadaic acid, a 38-carbon fatty acid isolated from this species, is useful in the treatment of cardiovascular diseases. This compound induces the phosphorylation and detachment of lipid droplet–associated perilipin A and B from the lipid droplet surface and thereby leads to accelerated lipolysis.

Thomas et al. (2010) reported that the bacterial strain *Pseudomonas* sp. KK10206C of this sponge species produced a novel C50-carotenoid, which is a well-known source of okadaic acid.

Chemotherapeutic agent: The halichondrin B isolated from this species is a potential chemotherapeutic agent (Sepčić et al., 2010).

Anti-inflammatory activity: The macrolide, eribulin mesylate (E7389) isolated from this species displayed anti-inflammatory activity by serving as a microtubule inhibitor. This compound is presently used in clinical trials (Senthilkumar and Kim, 2013).

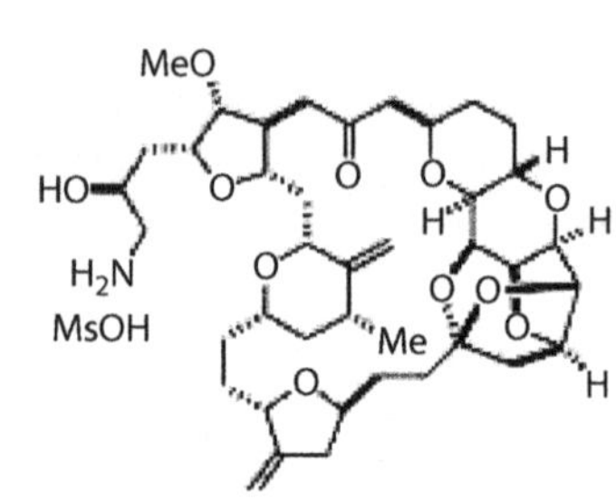

Eribulin mesylate

Halichondria (Halichondria) osculum (Lundbeck, 1902)

Order: Suberitida Family: Halichondriidae

Common name: Not designated

Distribution: Micronesia

Ecology: Not reported

Biology: This sponge has an erect, more or less compressed form, which may, however, be varying. The height of the sponge is about 3 cm. The color of the preserved specimen is light brown, gray, or yellowish gray. The surface is smooth, but due to its dermal skeleton it may have wrinkled or reticular appearance. It has only one osculum, which may be situated close to the summit of the sponge or farther down the side. This osculum has a form as a conical spout formed by the dermal membrane. The oscular opening leads into a kind of oscular tube, stretching as a hollow through the whole sponge or the greater part of it. The skeleton consists of a rather dense but irregular network of loose or not sharply marked fibers or bundles of spicules. The spiaila are slender oxea, which are slightly curved in the middle.

Compound(s) and Activities

Cytotoxic activity: The organic dried extract of this species exhibited cytotoxicity against lung fibroblast (V-79) and human colon epithelial (Caco-2) cancer cell lines at a concentration of 100 µg/mL. The percentage inhibition values recorded for these cell lines were found to be 63 and 57, respectively (Turk et al., 2013).

Antibacterial activity: The extracts of this species showed antibacterial activity against the laboratory, commensal, and clinically relevant bacterial strains with the following MIC values.

Bacterial Strain	MIC (µg/mL)
Staphylococcus aureus (MRSA) S-943 A	200
S. pseudintermedius (MRSP) S-053 A	200
S. pseudintermedius (MRSP) S-043 A	200
Listeria monocytogenes	250
Staphylococcus epidermidis EXB-V55	250
Staphylococcus aureus 10F	250
Bacillus subtilis EXB-V68	3
Macrococcus 1F	4
Micrococcus 2F	25
Escherichia coli HB101	250
Acinetobacter 1C	250

Source: Turk, T. et al., 2013, *Mar. Drugs*, 11: 1126–1139.

The extracts of this species showed antibacterial activity against the environmental bacterial strains *Exignuobacterium* sp., *Pseudoalteromonas* sp., *Alteromonas* sp., *Vibrio ruber*, and *Janthinobacterium svalbardensis* with the MIC values of 80, 65, 70, 85, and 75 µg/mL, respectively (Turk et al., 2013).

Halichondria (Halichondria) panicea (Pallas, 1766)

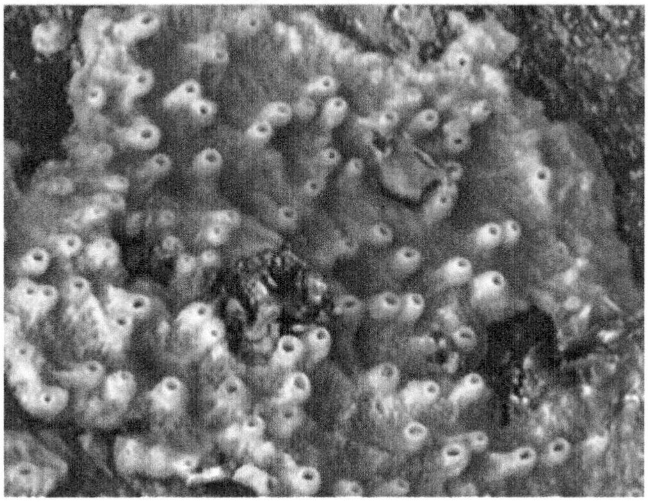

Order: Suberitida Family: Halichondriidae

Common name: Breadcrumb sponge, Okinawan marine sponge

Distribution: North Atlantic, Mediterranean Sea, English Channel, North Sea, Baltic Sea

Ecology: From the intertidal coastal areas to a depth of over 500 m. Its only requirement is a rocky substrate, which can include small cobbles. It is very tolerant of a wide range of coastal habitats, including strong currents, high salinity, and exposure to powerful wave action.

Biology: It is an encrusting sponge covering surfaces of several square centimeters. Depending on the habitat, it may form groups or even massive structures. Borders of the sponge are irregular and lobe shaped. Oscules are clearly visible and crater shaped. The surface is smooth and glossy. The color is variable from orange yellow to brownish green. This sponge may also get a green color or, more rarely, a reddish brown color due to its symbiotic algae. This sponge is also strongly odoriferous.

Food and feeding: It is a suspension feeder feeding mainly on phytoplankton.

Reproduction: Relatively little is known about its reproduction. It appears to be a hermaphrodite and oogenesis occurs in a very narrow timeband within a single population.

Compound(s) and Activities

Antiobesity and antidiabetic activities: A new polyacetylene compound, isopetrosynol (1), together with petrosynol (2), adociacetylene D (3), (5R)-3,15,27-triacontatriene-1,29-diyn-5-ol (4), and petrosterol (5) have been isolated from this species. Compound 1 inhibited protein tyrosine phosphatase 1B (PTP1B) activity with an IC50 value of 8.2 µM, whereas compound 2 showed only 29% inhibition at 22 µM. The IC50 values of compounds 3 and 4 were 7.8 and 12.2 µM, respectively. The inhibitory activity of 1 was stronger than that of compound 2. The enzyme PTP1B is a therapeutic target for obesity and diabetes (Abdjul et al., 2016).

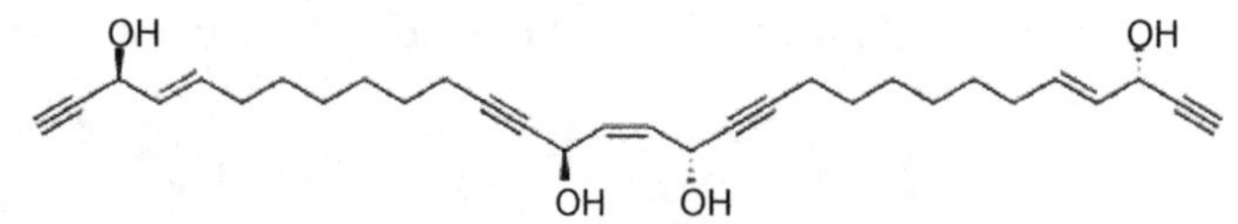

Isopetrosynol

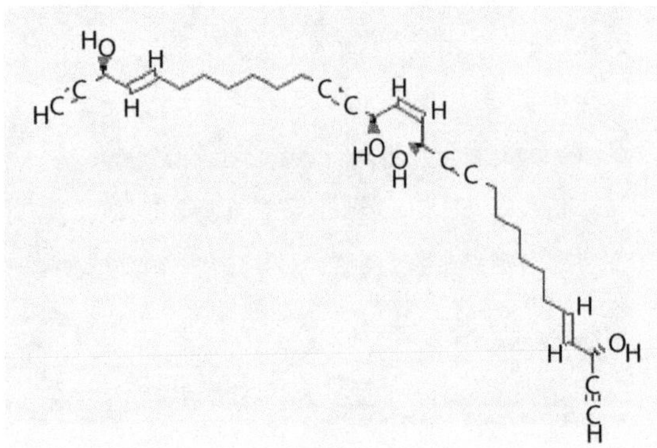

Petrosynol

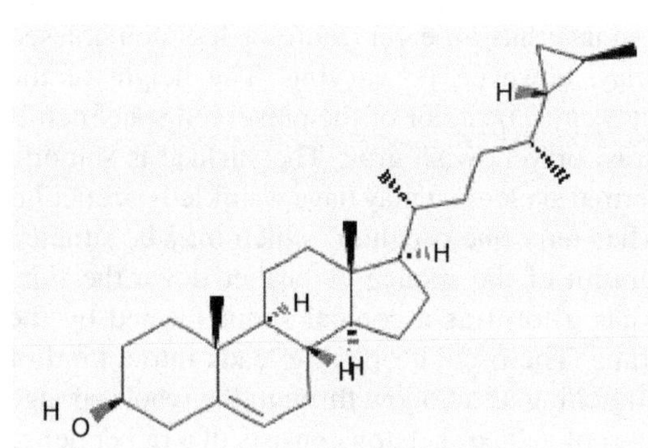

Petrosterol

Neuromodulatory activity: The methanolic, chloroform-methanol, and aqueous extracts of this species showed neuromodulatory activity by inhibing the enzyme Na+-K+ ATP-ase. The values of inhibition of this enzyme were found to be 4.1% to 354%, 243% to 320%, and 272%, respectively, in the extracts (Purushottama et al., 2009).

PTP1B inhibition activity: The EtOH extract of this sponge species containing the compounds isopetrosynol (1), petrosynol (2), adociacetylene D (3), (5R)-3,15,27-triacontatriene-1,29-diyn-5-ol (4), and petrosterol (5) inhibited PTP1B activity. Compound 1 inhibited the PTP1B activity with an IC50 value of 8.2 μM, whereas compound 2 showed only 29% inhibition at 21.6 μM. The IC50 values of compounds 2 and 3 were 7.8 and 12.2 μM, respectively. The inhibitory activity of compound 1 was stronger than that of compound 2 (Abdjul, 2016).

Antimicrobial, antiviral, cytotoxic, and other activities of the associated actinobacteria: The actinobacteria belonging to the genera *Actinoalloteichus*, *Micrococcus*, *Micromonospora*, *Nocardiopsis*, and *Streptomyces* associated with this sponge have yielded several secondary metabolites, which showed activities as given next.

Group	Compound(s)
	Antibacterial Activity
Aminoglycoside	6‴-O-α-d-Mannopyranosylmannosido-streptomycin
Aminoglycoside	Streptomycin, Tobramycin, Gentamicin C1 and A2
Dipeptide	Bovinocidin
Macrolide	Antibiotic M-2846 (against G+)
Macrolide	Lankamycin
Alkaloid	1-Epimanzamine D, Pyrocoll
Phenazine	Senacarcin A, Saphenyl ester D
Phenazine	Streptophenazin A–H and Hemipyocyanine (against G+)

(Continued)

Group	Compound(s)
	Antibacterial Activity
Sesquiterpene ketone	Albaflavenone
Aminocyclitol	Fortimicin KR1, Fortimicin AH
Lipopeptide	Massetolide K
Amino acid	4-Hydroxythreonine
Anthraquinone	Blanchaquinone
Nucleoside analog	4-Thiouracil
	Antifungal Activity
β-Lactam antibiotic	Clavamycin D
Aminoglycoside	N-Didemethylallosamidin, Validamycin D
Alkaloid	Pyrocoll
Macrolide	Venturicidin A, Galbonolide A, Blastamycin, Antimycin A0, Urauchimycin A or B
Lactone	Irumamycin
Hydroxamic acid	Neoenactin B1
Polyene	Polyene with 7 or 8/8 or 9/12 double bonds
Peptide	Histidinomycin, Albopeptin A
Amino acid	5-Hydroxy-4-oxonorvaline
Aminophosphonic acid	Fosfazinomycin A
Octadepsipeptide	Antibiotic SW 163B
Phenazine	Hemipyocyanine
Macrolactam	Dihydromaltophilin, Xanthobaccin A
Polyketide	Polyene
	Cytotoxic Activity
Nucleoside analog	4-Thiouracil
Peptide	Antibiotic NA 22598A1
Alkaloid	1-Epimanzamine D, Pyrocoll, Streptazone B1 or B2
Phenazine	Senacarcin A, Phencomycin ethyl ester
Depsipeptide	Detoxin E1
Lactone	5-(6′-Methyl-7′-oxo-octyl)-SH-furan-2-one
Aromatic polyketide	Fredericamycin A
Macrolide	Antimycin A3
Thiazolylindole	Antibiotic BE 10988

(Continued)

Group	Compound(s)
Antibacterial Activity	
Polyketide	Daryamide C
Aromatic polyketide	Fredericamycin A
Nucleoside	Antibiotic A 58365B, Futalosine
Pyrrol, amide	PD 125375
Lipopeptide	Rakicidin B
Enzyme-Inhibitory Activity	
Peptide	Pepstatin AC, BU and CU (Pepsin inhibitor)
Tripeptide	Valistatin (Aminopeptidase M inhibitor)
β-Lactam	N-Acetylglycylclavaminic acid (β-Lactamase inhibitor)
Lactone	5-Hydroxy-3-(1-hydroxy-2-methylpropyl)-4-methyl-2(5H)-furanone(chitinase and phosphatase inhibitor)
Macrolide	Antimycin A1, A2, A3, and A4; Blastamycin (inhibitor of ATP-citrate lyase)
Macrolide	Antimycin A1 (inhibitor of ATP-citrate lyase and ubiquinol-cytochrome C oxidoreductase); Blastamycin (inhibitor of ATP-citrate lyase)
Macrolide	Bafilomycin D (Inhibitor of membrane ATPases)
Thiazolylindole	Antibiotic BE 10988 (topoisomerase II inhibitor)
Alkaloid	Sulfostin (inhibitor of dipeptidyl peptidase IV)
Pyrone	Germicidin (inhibitor of porcine Na+/K+-activated ATPase)
Antiparasitic Activity	
Aminoglycoside	Gualamycin (acaricidal)
Polyether	Antibiotic X 14931A (antiprotozoal)
Alkaloid	Pyrocoll (antiprotozoan, antiparasitic)
Antiviral Activity	
Macrolide	Antimycin A1, A3
Phenazine	Hemipyocyanine
Neuroprotective and Antioxidant Activities	
Phenazine	Aestivophoenin C

Source: Schneemann, I. et al., 2010, *J. Nat. Prod.*, 73: 1309–1312; Schneemann, I. et al., 2010, *Appl. Envtl. Microb.*, 76: 3702–3714.

Halichondria sp.

Order: Suberitida Family: Halichondriidae

Compound(s) and Activities

Antiplasmodial activity: The compound 10-isothiocyanato-4-amorphene isolated from this unidentified species showed weak antiplasmodial activity against *Plasmodium falciparum* (strains K1 and NF54 with an IC50 value of 5.7 μM (Emsermann et al., 2016).

Antimicrobial and antimalarial activities: The compound halochonadin F isolated from this sponge displayed antimicrobial activity against *Micrococcus luteus* (MIC, 4 μg/mL), *Trichophyton mentagrophytes* (MIC, 8 μg/mL), and *Cryptococcus neoformans* (MIC, 16 μg/mL). Further, the compound thiocyanatoneopupukeanane of this sponge showed antimalarial activity against *Plasmodium falciparum* (strain D6: IC50 = 4700 ng/mL; strain W2: IC50 = 890 ng/mL) (Emsermann et al., 2016).

Antifungal and antimalarial activities: The macrolid compound halichondramide showed *in vitro* antimalarial activity and antifungal activity (Thomas et al., 2010).

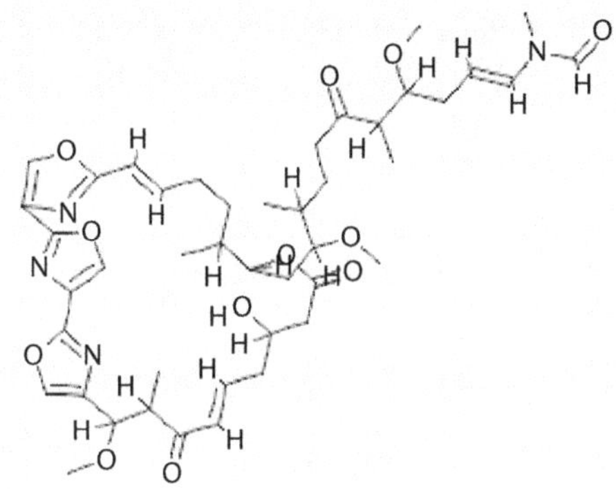

Halichondramide

Platelet aggregation inhibitory activity: The compound 24-methylscalarane, 12α-Acetoxy-24-methyl-24-oxoscalar-16-en-22,25-dial of the unidentified sponge displayed significant inhibitory activity (IC50 = 0.5 μg/mL) on the platelet aggregations caused by adenosine 5′-diphosphate, collagen, or arachidonic acid (González, 2010).

Anticancer activity: The mycelial MeOH extract of this unidentified sponge yielded the compounds gymnastatins F–H (1–3) and gymnamide (4). Among these compounds, gymnastatins F (2) and G (2) were found to exhibit potent growth inhibition against the P388 cancer cell line (Amagata et al., 2006).

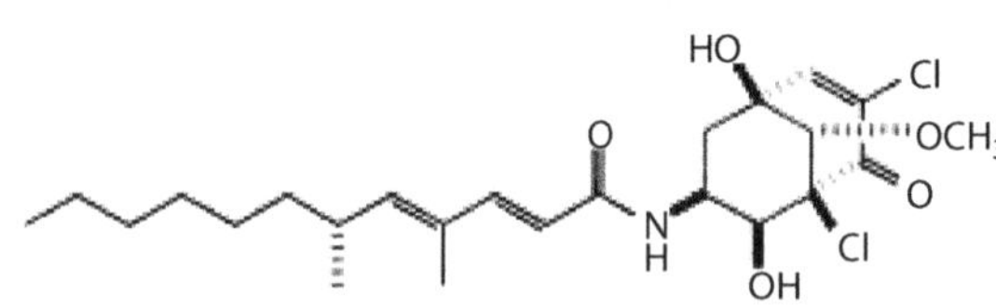

Gymnastatin F

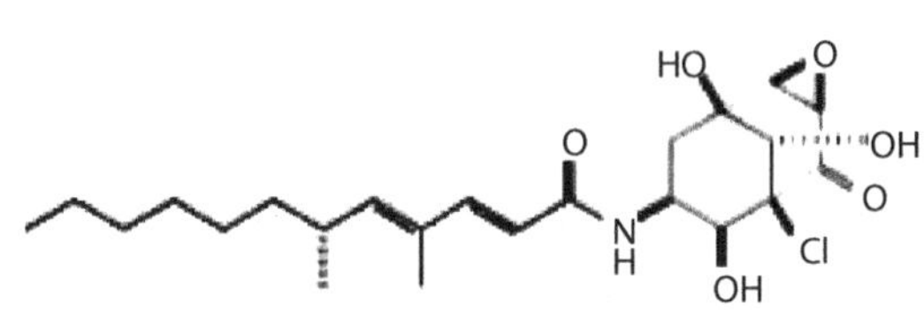

Gymnastatin G

Antitumor/cytotoxic activity: The compound 16β,22-dihydroxy-24-methyl-24-oxoscalaran-25,12β-olactone of this species inhibited tumor cell growth with very little specificity for individual cell lines (average GI50 value of 20.4 μM), as well as it exhibited cytotoxicity against lung carcinoma

A549 cells (GI50 = 0.5 μg/mL, 1.27 μM) (González et al., 2010).

The 22-acetoxy-16β-hydroxy-24-methyl-24-oxoscalaran-25,12β-olactone of this species showed cytotoxicity against lung carcinoma A549 cells (GI50 = 1.0 μg/mL, 2.13 μM) (González et al., 2010).

The compounds (1*S**,3*S**,4*R**,7*S**,8*R**,13*R**)-7-isocyano-11-cycloamphilectene and (3*S**,4*R**,7*S**,8*R**,11*S**,12*R**,13*S**)-7-isocyano-1-cycloamphilectene isolated from the Palauan *Halichondria* species showed cytotoxicity (Emsermann et al., 2016).

Hymeniacidon heliophila (Wilson, 1911)

Order: Suberitida Family: Halichondriidae

Common name: Sun sponge

Distribution: Western Central Atlantic and Western Central Pacific

Ecology: This is an epibiotic species found in bays; shallow, intertidal, sheltered, euryhaline areas at a depth range of 1 to 6 m; and attached to exposed oyster clumps, anchored to buried shells on sand flats, on concrete pilings and rock jetties. It is uniquely tolerant of exposure to air and direct sunlight during low tides.

Biology: The sun sponge is amorphous-lobulate in shape. The base is 1 to 3 cm thick, with crowded, erect chimneys, fistules, and uneven digitations, which rise from a few millimeters to 4 cm high. Sponges occurring on partially shaded pilings and jetties have a size of 40 cm in diameter. Live specimens are generally yellow to yellow orange, with a soft, limp, and fragile consistency. Specimens growing under high sunlight are lighter yellow to olive, while those in shady areas are often a more intense orange to reddish orange. The underside and interstitial surfaces are smooth, and the sides and projections are finely conulose, denticulate, or lacinulate. Fasiculate or denticulate ridges appear normally on the chimneys. Algae, sediment, shells, and rock fragments often cover the surface. The oscules are inconspicuous and are hidden in the interstices or on top of the chimneys. Tangentially placed spicules are found in the ectosome. The spicules are variable in size, straight, or slightly curved.

Associated species: This sponge provides habitat for a variety of smaller animals, including the spiny brittle star *Ophiothrix angulata* and the nematode *Leptosomatum acephalatum*.

Compound(s) and Activities

Antioxidant activity: The amino acid l-5-hydroxytryptophan (l-5-HTP) content of this species has shown antioxidant activity (Lysek et al., 2003).

Hymeniacidon perlevis (Montagu, 1814) (= *Hymeniacidon sanguinea*)

Order: Suberitida Family: Halichondriidae

Common name: Crumb-of-bread sponge

Distribution: North Atlantic Ocean, North Sea, Brazil

Ecology: This sponge occurs on stones, rocks, shells, Laminaria holdfasts; and intertidal and shallow sublittoral areas. It may be buried deep into the sediment with long projection sticking out to the surface.

Distribution: British Isles, Netherlands, Belgium, France, Spain, Mediterranean, Canary Islands, Azores

Biology: The shape of this sponge varies from thin sheets, cushions, to massive flanged, and rarely branching erect. Lateral expansion of all forms may be 20 cm or more.

The thickness may vary from a few millimeters to 10 cm or more. Flanges and irregular projections may be up to 3 or more centimeters. Color of the animal varies from yellow to orange to blood red. Shore forms are blood red, whereas sublittoral forms are pinkish red. The surface of the sponge is variable; smooth, tuberculate, and thrown into irregular folds; or else covered with branching processes. Oscules are found scattered. Consistency is firm, compact, and compressible. Megascleres are styles only of simple, smooth form. Ectosomal skeleton is a tangential layer of spicules intercrossing in all directions, or of brushes of spicules set vertically to surface.

Association: Many invertebrates such as nematodes, annelids, and crustaceans are found associated with this species.

Compound(s) and Activities

Antibacterial and antifungal activities: The extracts of epiphytic fungus *Fusarium oxysporum* DLFP2008005 of this sponge species exhibited effective antibacterial and antifungal activities against *Staphylococcus epidermidis*, *Bacillus subtilis*, *Pseudomonas fluorescens*, *Pseudomonas aeruginosa*, and the yeast *Candida albicans* (Thomas et al., 2010).

Hymeniacidon sp. (= *Stylotella* sp.)

Order: Suberitida Family: Halichondriidae

Compound(s) and Activities

Antiplasmodial activity: The diterpenoid β-lactam compound monamphilectine A isolated from this species showed antiplasmodial activity against chloroquine-resistant *P. falciparum* with an IC50 value of 0.6 μM (Anjum et al., 2016; Perdicaris et al., 2013).

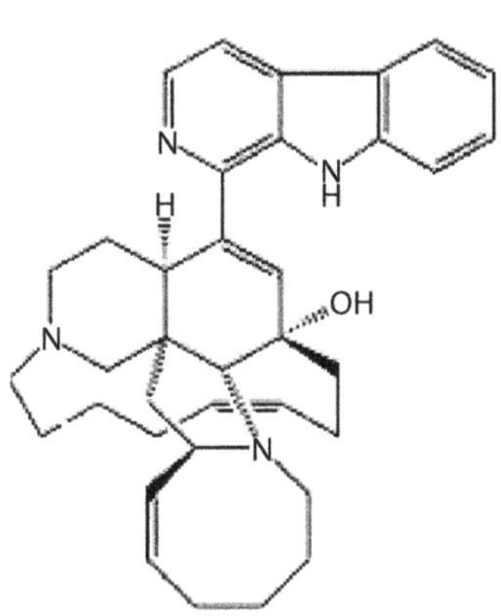

Monamphilectine A

Anticancer activity: The proline-rich cyclodecapeptide, stylopeptide 2 of this species inhibited the growth of BT-549 and HS 578T 2 breast cancer cell lines by 77% and 56%, respectively (Agrawal et al., 2016).

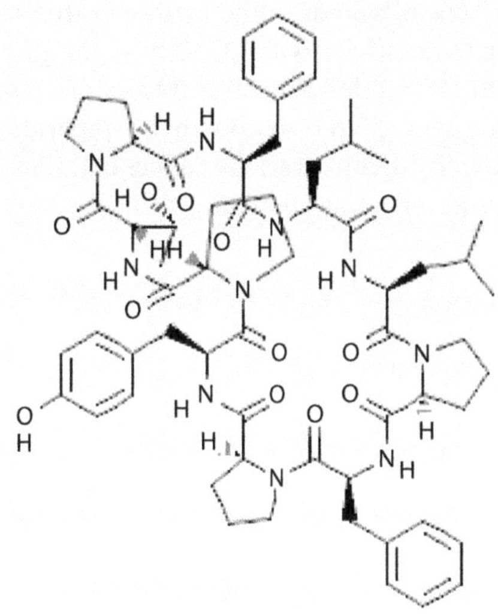

Stylopeptide 2

Spongosorites halichondriodes (Dendy, 1905)

Order: Suberitida Family: Halichondriidae

No other information is available for this species.

Compound(s) and Activities

Anti-inflammatory and antioxidant properties: The ethyl acetate extract of this species showed negligible antioxidant activity and the values of activity at 10 to 100 μg/mL concentrations were varying from 4.4% to 12.3%. Further, the ethyl acetate extract of this species containing glycyrrhetinic acid (pentacyclic triterpenoid derivative of the beta-amyrin type), 5α-choletan-3β-ol, dihydrocholesterol, cholesterol, and 3-β,6α-dihydroxy-5α-cholan-24-oic acid methyl ester have shown strong anti-inflammatory activity in rats by significantly reducing the hind paw swelling with increasing dose at 200 and 500 mg/kg (Kumar and Pal, 2016; Kumar et al., 2014).

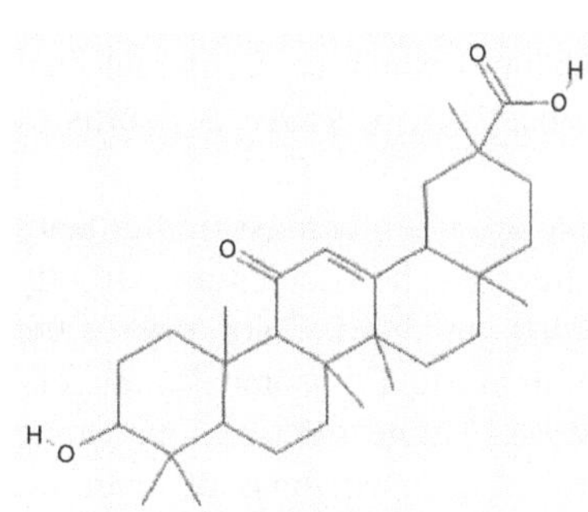

Glycyrrhetinic acid

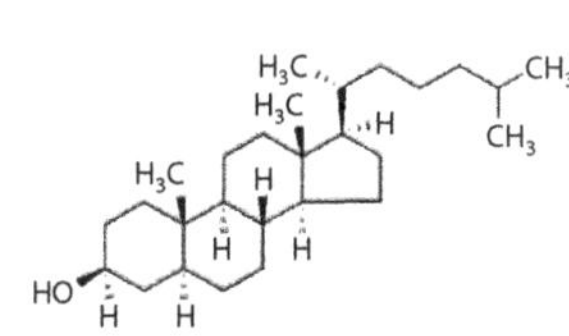

5α-Choletan-3β-ol

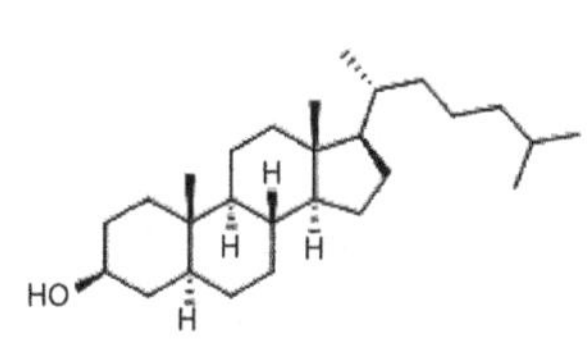

Dihydrocholesterol

Cytotoxicity: The alkaloids of the extract of this species have shown strong cytotoxicity with LC50 dose ranging from 5.24 to 1000 μg/mL. The ethyl acetate extract containing the compound inosine and 2,2,6,6-tetramethyl-4-piperidone showed strong cytotoxicity against various cancer cell lines with GI50 value ≤20 μg/mL (Kumar and Pal, 2016).

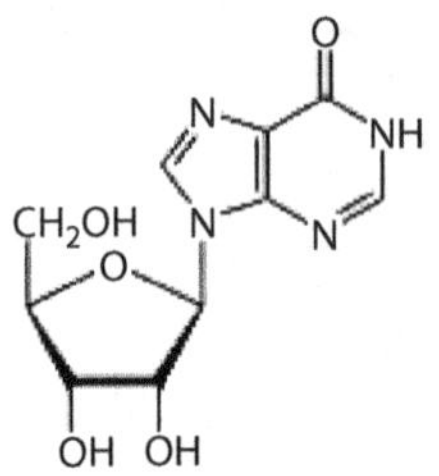

Inosine

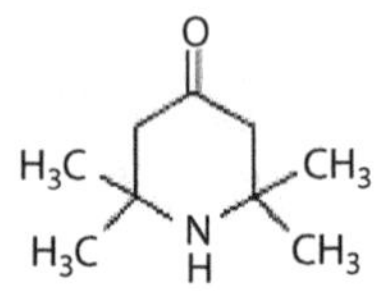

2,2,6,6-Tetramethyl-4-piperidone

Antimicrobial activity: The nonpolar and polar extracts from this species have shown promising antimicrobial activity against both Gram-negative and Gram-positive bacteria (Kumar and Pal, 2016).

Immunosuppressive activity: The methanolic extract of this species has been reported to suppress both humoral and cellular immune responses. The extract not only suppressed nonspecific immune response, but also effectively decreased humoral as well as cell-mediated immunity (Kumar and Pal, 2016).

Immunomodulatory activity: The methanolic extract of this species showed a significant decrease in total white blood cell (WBC) count. Further, it prevented myelosuppression in cyclophosphamide drug–treated rats (Kumar et al., 2012).

Spongosorites ruetzleri (van Soest and Stentoft, 1988) (= *Halichondria ruetzleri*)

Order: Suberitida Family: Halichondriidae

Common name: Caribbean deep-sea sponge

Distribution: Eastern Caribbean region

Ecology: It occurs in the shallow water (1–5 m) reef flat, where coral rubble is most abundant. It is also at depths of 100 to 150 m.

Biology: This sponge is an oblong, irregular mass riddled with vennetid tubes. Surface is smooth. The oscules are slightly raised. Consistency (dry) is corky and hard. Color (dry) of the specimen is black and yellowish in life. Ectosome is a tangential crust of confusedly arranged single spicules and vague tracts. Choanosome is a dense confused mass of spicules. Spicules are oxea, with stairstepped or ragged apices.

Compound(s) and Activities

Cytotoxicity: The indole alkaloid, topsentin of this species exhibited *in vitro* activity against P-388 (IC50, 3 μg/mL) and human tumor cells (HCT-8, A-549, T47D) (IC50, 20 μg/mL) and *in vivo* activity against P-388 and B16 melanoma (Gul and Hamann, 2005).

The indole alkaloid compounds nortopsentins A, B, and C isolated from this species inhibited activity against P-388 cells with IC50 values of 7.6, 7.8, and 1.7 μg/mL, respectively (Gul et al., 2005).

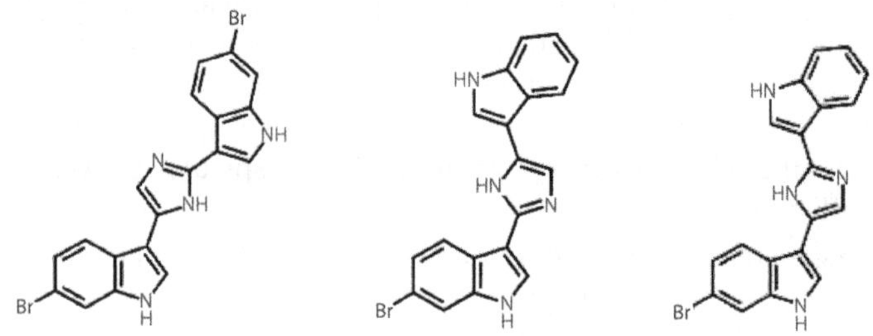

Nortopsentins A,B, and C

Antiviral activity: The indole alkaloids topsentin and bromotopsentin of this species exhibited antiviral activities *in vitro* against HSV-1, vesivular stomatitis virus (VSV), and the corona virus A-59 (Gul and Hamann, 2005).

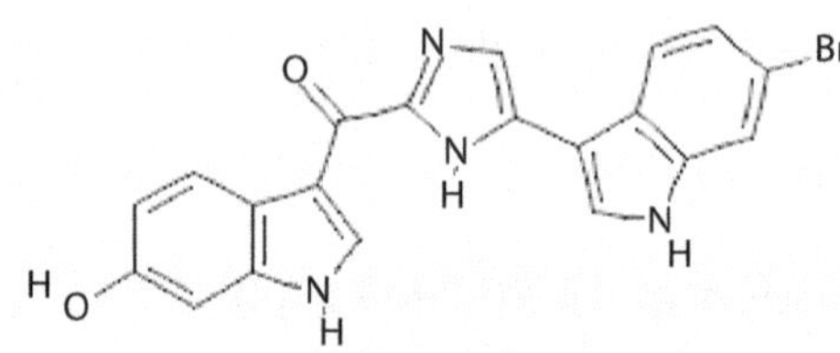

Bromotopsentin

Antifungal activity: The compounds nortopsentins A, B, and C exhibited antifungal activity against *C. albicans* (Gul and Hamann, 2005).

Spongosorites sp.

Order: Suberitida Family: Halichondriidae

Compound(s) and Activities

Neurosuppressives and muscle relaxant activity: The bisindole alkaloid bromotopsentin isolated from this unidentified species neurosuppressives and muscle relaxant activity by serving as a adrenergic receptor antagonist (Sipkema et al., 2005).

Antimicrobial activity: The indole alkaloid hamacanthin A showed antimicrobial activity against *Candida albicans* with MIC values of 6.25 μg/mL. The compounds macanthins A and B showed activity against *C. albicans* and *C. neoformans*, respectively, with MIC values of 1.6 and 6.2 μg/mL, respectively (Anjum et al., 2016).

The indole alkaloid dragmacidin D of this sponge inhibited the growth of several microbes, including *E. coli* (MIC, 15.6 μg/mL), *B. subtilis* (MIC, 3.1 μg/mL), *Candida aeruginosa* (MIC, 62.5 μg/mL), *C. albicans* (MIC, 15.6 μg/mL), and *C. neoformans* (MIC, 3.9 μg/mL) (Gul and Hamann, 2005).

Oh et al. (2006) reported that the compounds deoxytopsentin and hamacanthin A exhibited significant antibacterial activity against methicillin-resistant *Staphylococcus aureus*, with MIC values of less than 12.5 μg/mL. Further, hamacanthin A also showed potent inhibitory activity against medically important pathogenic fungi.

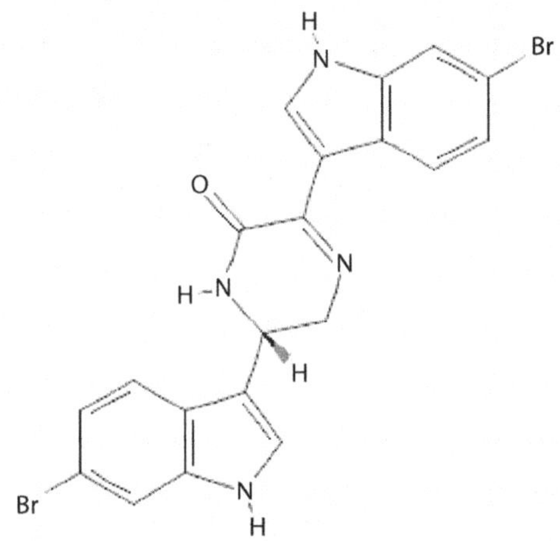

Hamacanthin A

Anticancer activity: The new bis(indole)-derived sponge metabolite, dragmacidin D isolated from this sponge exhibited anticancer activity *in vitro* against P-388 and A-549 with IC50 values of 1.4 and 4.4 μg/mL, respectively (Gul and Hamann, 2005).

Cytotoxicity: The compounds of the topsentin class (1–4) and hamacanthin class (5–9) isolated from this species exhibited moderate cytotoxicity against cancer cell lines at concentrations between 1.1 and >20 μg/mL (Oh et al., 2006).

Antiviral activity: The indole alkaloid dragmacidin D, a bis(indole)-derived sponge metabolite, isolated from this sponge inhibited *in vitro* replication of feline leukemia virus (FeLV) with an IC50 of 6.25 μg/mL (Gul and Hamann, 2005).

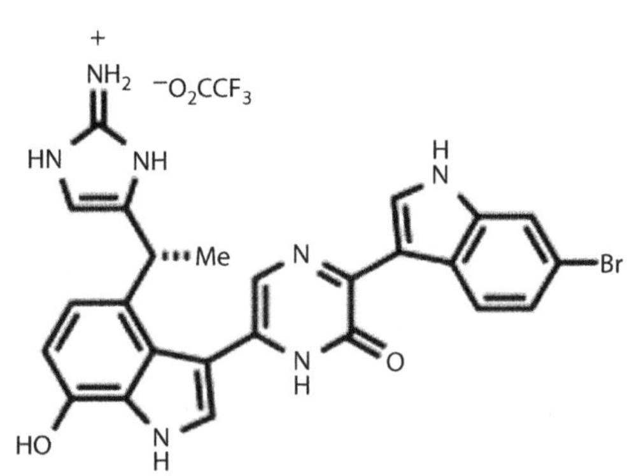

Dragmacidin D

Topsentia kushimotoensis **(Hoshino, 1977) (= *Epipolasis kushimotoensis*)**

Order: Suberitida Family: Halichondriidae
Distribution: Japan
Habitat: It inhabits rocky substrate of the subtida1 zone. The depth range is 20 to 30 m.
Biology: It is an irregualr funnel-like or massive sponge of the size 7 × 8 × 2 cm. Color of the specimen is pansy purple. Consistency is hard like stone and is incompressible. Surface is minutely hispid and uneven. Pores are invisible. Oscula (2–3 mm dia.) open inside in funnel-like specimens. Ectosome and endosome are made of oxea with a mixture of small and larger spicules. Each oxeon is smooth and is slightly arched or curved at both ends.

Compound(s) and Activities

Cytotoxic and antiplasmodial activities: The compound epipolasinthiourea-B of this species showed moderate cytotoxic activities against the murine L1210 leukemia cell line (ED50 = 3.7 μg/mL). On the other hand, its (+)-epipolasin A exhibited antiplasmodial activity against *Plasmodium falciparum* strains D6 (IC50 = 5600 ng/mL) and W2 (IC50 = 5550 ng/mL) (Emsermann et al., 2016).

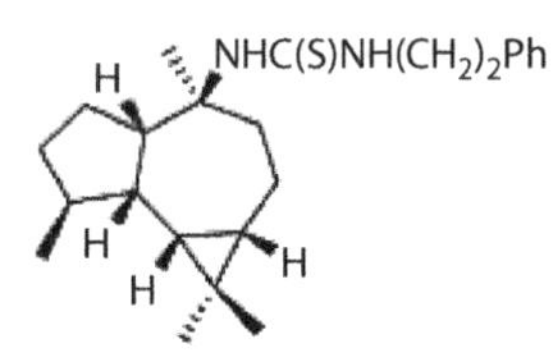

Epipolasinthiourea-B

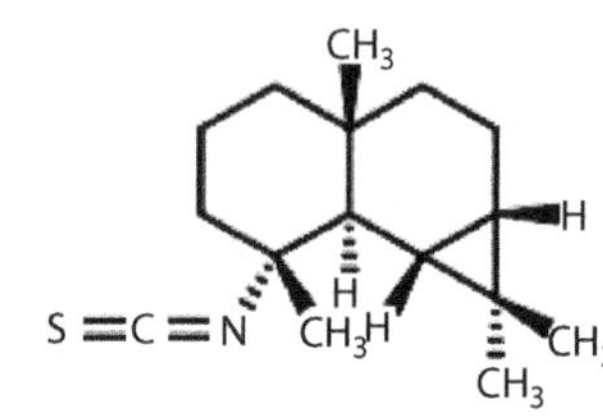

Epipolasin A

Cytotoxic activity: Two isothiocyanates, epipolasin A and B (1,2), and two corresponding thiourea derivatives, epipolasinthiourea-A (3) and -B (4), have been isolated from *Epipolasis kushimotoensis*. Their structures were elucidated on the basis of chemical and spectral evidence. Among these compounds, thioureas 3 and 4 showed cancer cell growth inhibition activity of L1210 cells *in vitro* with ED50 values of 4.1 μg/mL and 3.7 μg/mL, respectively (Tada and Yasuda, 1985).

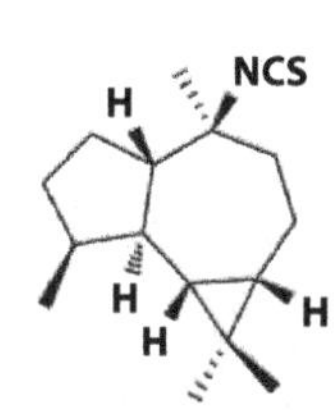

Epipolasin B

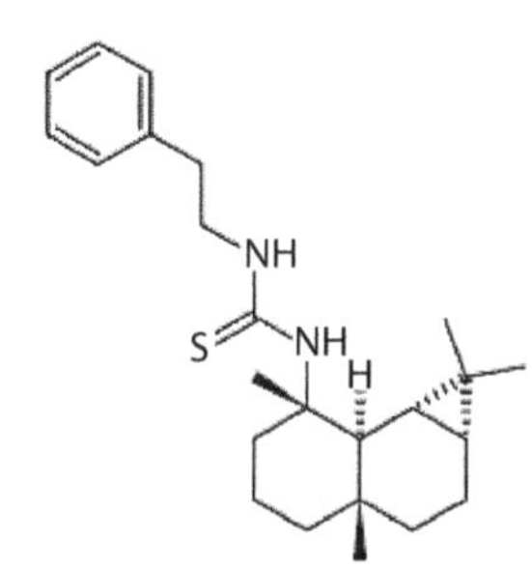

Epipolasinthiourea A

Topsentia ophiraphidites (de Laubenfels, 1934)

Order: Suberitida Family: Halichondriidae

Common name: Colombian Caribbean sponge, pink or brown sponge

Distribution: Japan

Ecology: It is found in shallow and deep reef areas. It is common at depths between 10 and 20 m.

Biology: It is a massive amorphous sponge and sometimes it is subspherical to lobate. Some specimens have sharp projections of 2 to 8 cm high. Surface is usually uneven and lumpy. Color of the sponge is gray, brown, pink, or purple externally and tan internally. The hydrozoan *Parazoanthus swiftii* is growing symbiotically in this sponge species.

Compound(s) and Activities

Antibacterial activity: The organic extracts of this species showed considerable activity toward Gram-negative bacterial strain. The activity of the unheated aqueous extract (MIC = 48 μg/mL) was about 100-fold more potent than the heated extracts, indicating the active compound of this extract may be a peptide or a protein (Sepčić et al., 2010).

AChE inhibitory activity: The butanolic extract (34 μg) from this species showed 50% AChE inhibitory activity (Sepčić et al., 2010).

Antiprolifertive activity: The steroid trisulfate ophirapstanol trisulfate isolated from the ethanol extracts of this species exhibited siginificant inhibition of guanosine diphosphate/G protein. The latter has been reported to lead unregulated cell proliferation of human colon and pancreatic carcinomas (Gunasekera et al., 1994).

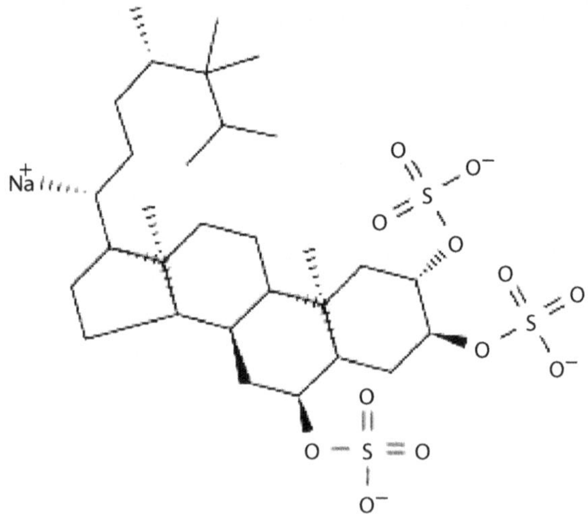

Ophirapstanol trisulfate

Topsentia rugosa (Ridley and Dendy, 1886) (= *Halichondria rugosa*)

Order: Suberitida Family: Halichondriidae

Common name: South China Sea sponge

Distribution: Micronesia, New Hebrides

Ecology: Not reported

Biology: It is a massive and irregular sponge with the size of 10 × 4 × 6 (height) cm. Surface is raised into jagged projections. Color in alcohol is white. Oscules are normally single, but they may also be in groups of two or three. They are slightly sunken below the surface of the sponge. Dermal membrane is well developed with tangentially placed oxea. Sometimes the larger oxea are aggregated in bundles of eight to ten resembling short fibers.

Compound(s) and Activities

Anti-HIV and antitumor activities: A new sulfur-containing guanidino derivative, halichondria sulfonic acid (1), showing anti-HIV-1 activity and halistanol trisulfate (2) with antitumor activity have been isolated from this species. Compound 1 exhibited activity against HIV-1 with an EC50 value of 29 μg/mL and IC50 value of more than 200 μg/mL. Halistanol trisulfate (2) was also active against HIV-1 and possessed additional activity against tumor cell lines, the human hepatoma QGY-7701, and chronic myelogenous leukemia K562 with IC50 values of 26 and 52 μg/mL, respectively (Jin et al., 2006).

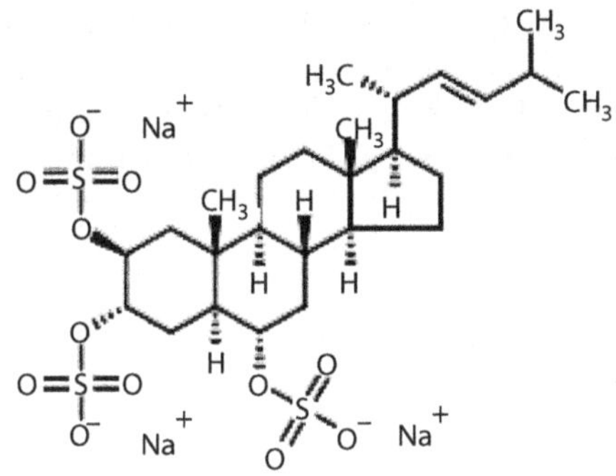

Halistanol trisulfate

Antimicrobial activity: The bacterial isolated derived from this species have shown broad-spectrum antimicrobial activities against fungi and Gram-positive and Gram-negative bacteria (Zhang et al., 2009).

Cytotoxicity: The steroids, 24ξ,25-dimethyl-3α-hydroxylcholest-5-ene-2β-ol sodium sulfate and 24ξ,25-dimethylcholest-5-ene-2β,3α-diol disodium sulfate isolated from the ethanol extract of this species showed cytotoxicity to four human cancer cell lines (BEL-7402, HT-29, SPC-A1, and U-251) with IC50 values between 6.5 and 23.1 μM (Zhang et al., 2007).

Topsentia sp.

Order: Suberitida Family: Halichondriidae

Compound(s) and Activities

Antimicrobial activity: The sulfated sterols topsentiasterol sulfates A–E of this species have shown antimicrobial activity (Sipkema et al., 2005).

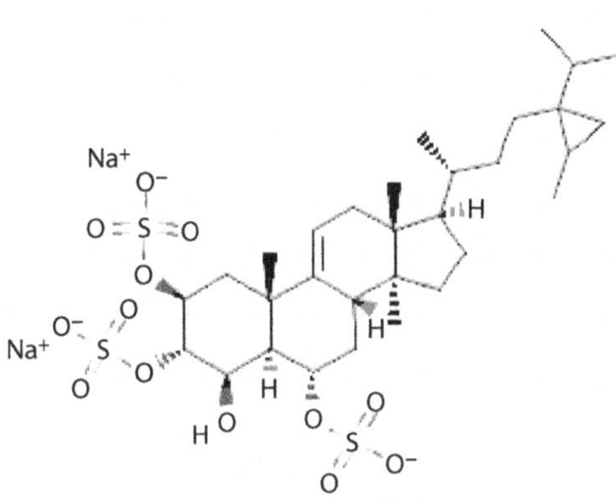

Topsentiasterol sulphate E

Anthelmintic/antiparasitic activity: The sesquiterpene axisonitrile 3 isolated from this unidentified species has shown anthelmintic/antiparasitic activity (Alvi et al., 1991).

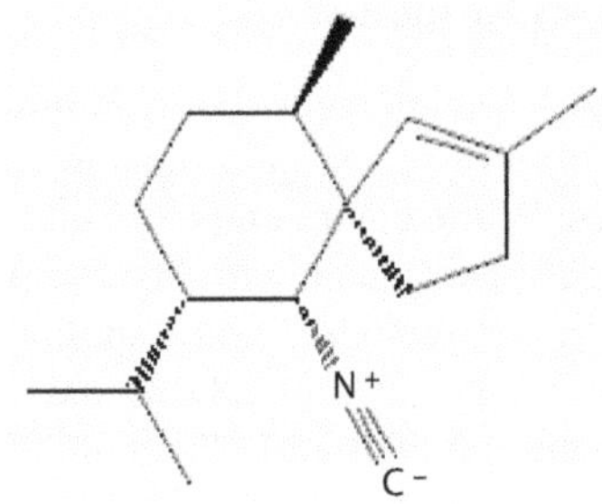

Axisonitrile 3

FAMILY: SUBERITIDAE

Aaptos aaptos (Schmidt, 1864)

Order: Suberitida Family: Suberitidae

Common name: Okinawan sponge

Distribution: Northwest Spain and Portugal, Mediterranean, Azores, Madeira, and Canary Islands

Ecology: It is mostly known from deeper water, but occasionally occurring in the shallow sublittoral areas.

Biology: It is a massive, lumpy, and blunt lobate form that is brown yellow, pale yellow or grayish, or reddish dark brown. Interior body is yellow. It typically grows to a size of 3 to 5 cm in diameter, occasionally larger, fist-size. Surface of the sponge is irregular, hispid, occasionally smooth, and is seldom elevated in distinct papillae. The oscules are up to 3 mm in diameter, few in number, contractile, and are occasionally elevated. Consistency of the sponge is compact, firm, and hard.

Owing to its pharmaceutical compounds, high growth potential, and high survival rate, it is considered to be a promising candidate for mariculture development (de Voogd, 2007).

Compound(s) and Activities

Antiviral activity: A total of eight alkaloid (unidentified) compounds isolated from this species have shown antifungal and antiviral activities. While compounds 3, 7, and 8 showed antifungal activity against six fungi, with MIC values in the range of 4 to 64 μg/mL, compounds 7 and 8 exhibited anti-HIV-1 activity, with inhibitory rates of 88.0% and 72.3%, respectively, at a concentration of 10 μM (Yu et al., 2014a).

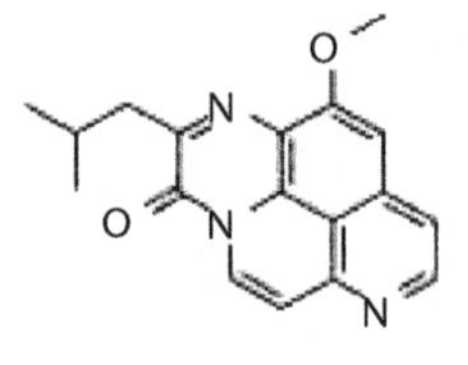

1

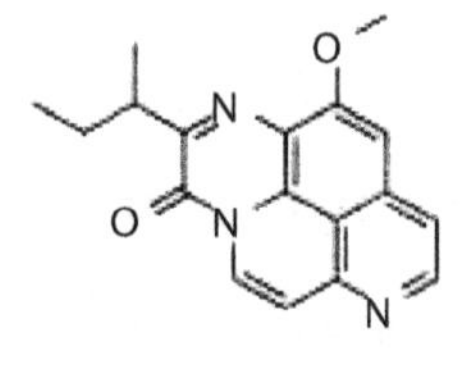

2

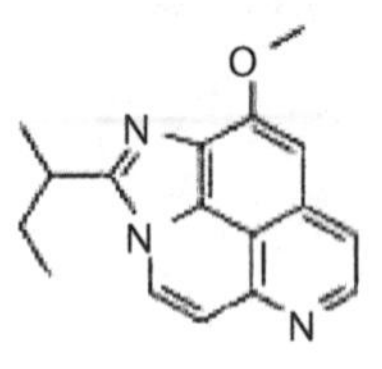

4

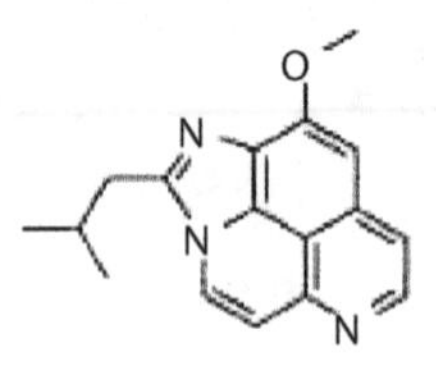

5

Unidentified compounds (1, 2, 4, 5).

Anjum et al. (2016) also reported that the alkaloid compound 4-methylaaptamine of this species has shown antiviral activity against HSV-1.

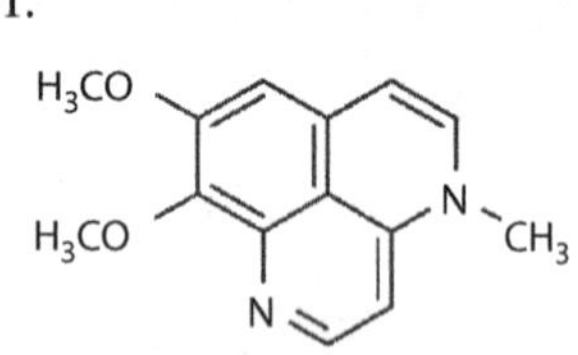

4-methylaaptamine

Souza et al. (2007) that that the alkaloid 4-methylaaptamine isolated from this species inhibited HSV-1 replication in Vero cells in a dose-dependent manner with an EC50 value of 2.4 μM. Further, this compound sustained antiherpetic activity even when added to HSV-1–infected Vero cells at 4 h after infection, suggesting that this compound inhibits initial events during HSV-1 replication. These findings suggest that this alkaloid could be considered as a potential compound for HSV-1 therapy.

Cytotoxicity: The crude methanolic extract of this species showed cytotoxicity on the HL-60 (promyelocytic leukemia), CEM-SS (T-lymphoblastic leukemia), MCF-7 (breast cancer), HeLa (cervical cancer), HT-29 (colon cancer), and L929 (murine fibrosarcoma from mouse) cell lines with IC50 values of 9.5, 5.4, 8.0, 2.8, 24.1, and 3.2 μg/mL, respectively (Shaari et al., 2009).

A total of 12 aaptamine derivatives (1–12) with cytotoxic and anticancer activity have been isolated from this species. Among them the compounds 2, 8, 11, and 12 showed potent cytotoxicities against human cancer cell lines, namely, HL60, K562, MCF-7, KB, HepG2, and HT-29 cells, with IC50 values in the range of 0.03 to 8.5 μM (Yu et al., 2014b).

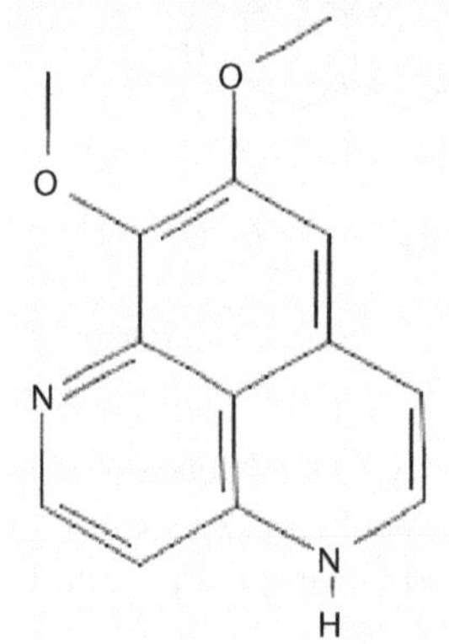

Aaptamine

Antitumor and antiinflammatory activities: The isoaaptamine compound benzonaphthyridine of this species has shown antitumor and anti-inflammatory activities by inhibiting protein kinase C (Anjum et al., 2016; Sipkema et al., 2005).

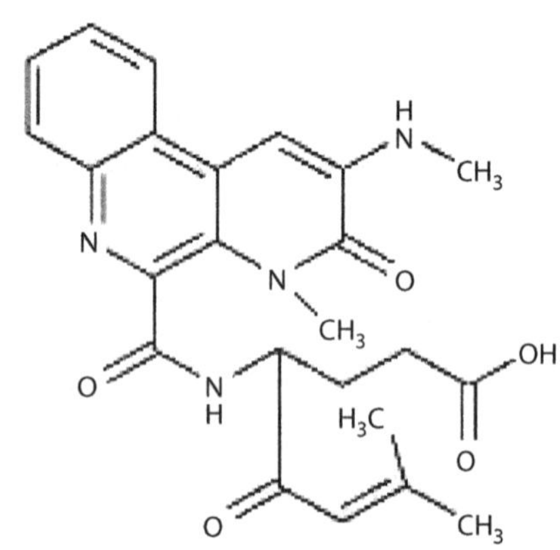

Benzonaphthyridine

Antibacterial activity: The alkaloid isoaaptamine of this species has shown antibacterial activity against *S. aureus* with an MIC value of 3.7 μg/mL (Anjum et al., 2016; Braga, 2014).

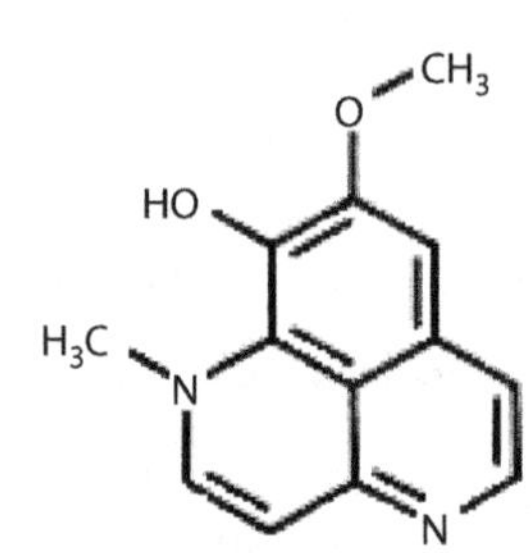

Isoaaptamine

Aaptos ciliata (Wilson, 1925)

Order: Suberitida Family: Suberitidae
Common name: Japanese sponge
Distribution: Philippines
Ecology: It is a deep-sea sponge.
Biology: Not available

Compound(s) and Activities

Anticancer and antileishmanial activities: The lipopeptide compounds ciliatamides A–C have been isolated from this sponge. Of them ciliatamides A–B have shown antileishmanial activity at 10 µg/mL with 50% and 45.5% growth inhibition, respectively. On the other hand, ciliatamides A–C displayed anticancer activity by inhibiting the growth of HeLa cells with IC50 values of 50, 4.5, and 50 µg/mL, respectively (Agrawal et al., 2016).

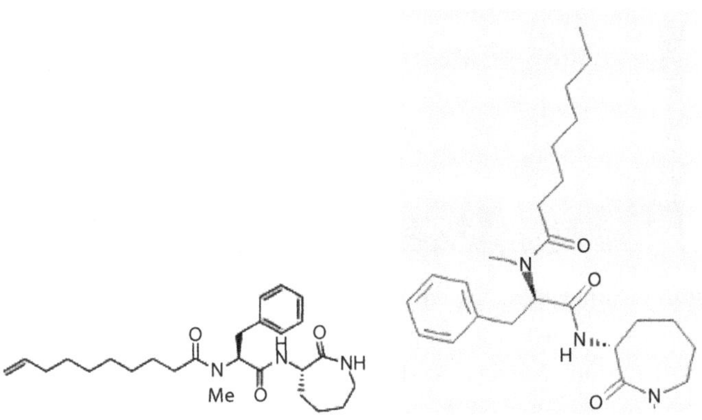

Ciliatamide A Ciliatamide B

Anti-leishmanial activity: The organic extract of this species has shown activity against *Leishmania major* with 86% growth inhibition at the concentration of 10 lg/mL (Oliveira et al., 2016).

Aaptos suberitoides (Brøndsted, 1934)

Order: Suberitida Family: Suberitidae
Common name: "Century egg" sponge
Distribution: Indo-West Pacific Ocean, Western Central Pacific, Singapore, Vietnam and Indonesia
Ecology: This sessile species inhabits lagoons where it encrusts/bores into stony corals. It is often seen partially buried in the sand.
Biology: It is an oval or spherical sponge 10 to 15 cm in diameter. Surface of the sponge is lumpy and looks smooth but is rough to the touch. The holes are tiny on low humps. Color where exposed to sunlight is yellow black and it is bright yellow on the inside. It is hermaphroditic and the zygote develops into parenchymella larva (free-swimming) before settling down on a substrate where it grows into a young sponge.

Compound(s) and Activities

Antibacterial activity: The extract of this sponge species showed antibacterial activity with the following values of zone of inhibition.

Zone Inhibition (mm) of Extract of Sponge

S. aureus	*E. coli*	*K. pneumoniae*	*S. typhi*	*P. aeruginosa* MDR	*S. aureus* MRSA
6.25	11.15	–	–	–	–

Source: Cita, Y.P. et al., 2017, *J. Marine Sci. Res. Dev.*, 7:237.

The secondary metabolites isolated from the ethyl acetate and methanol extracts of the associated bacteria of this species showed antibacterial activity against *Staphylococcus aureus*, *Escherichia coli*, *Candida albicans*, and *Aspergillus niger*. The values of inhibition zone recorded for these extracts for various bacterial species are given next.

Diameter of Inhibition Area (mm)

Extract	*S. aureus*	*E. coli*	*C. albicans*	*A. niger*
Ethyl acetate	9.8	9.6	8.4	6.3
Methanol	9.8	9.2	7.9	6.7

Source: Abdillah, S. et al., 2014, *IOSR J. Pharm. Biol. Sci.*, 9: 34–38.

Ly et al. (2017) also reported that the crude organic extracts of this species showed antibacterial activity, and the values are given next.

Zone of Inhibition (mm, dia)

Staphylococcus aureus	12
Escherichia coli	7
Bacilus cereus	9

Source: Ly, B.T.K. et al., 2017, *J. Sci., Ho Chi Minh City Open Univ.*, 7: 66–71.

The ethyl acetate extract of this species displayed significant activities toward *Staphylococcus aureus* and *Escherichia coli* (Quiao and Uly, 2013).

Cytotoxicity: Of the aaptamines derivatives, namely, 11-methoxy-3H-[1,6] naphthyridino [6,5,4-def] quinoxalin-3-one, 2,11-dimethoxy-3H-[1,6] naphthyridino[6,5,4-def]-quinoxalin-3-one, 5-benzoyldemethyaaptamine, and 3-amino demethyl (oxy)-aaptamine, isolated from this species, the compound 5-benzoyldemethyaaptamine inhibited the growth of mouse lymphoma L5178Y cells, with an IC50 value of 5.5 μM (Pham et al., 2013; Putra and Jaswir, 2014).

Tsukamoto et al. (2010) also reported that the compounds aaptamine, isoaaptamine, and demethylaaptamine isolated from this species showed cytotoxic activities against HeLa cells.

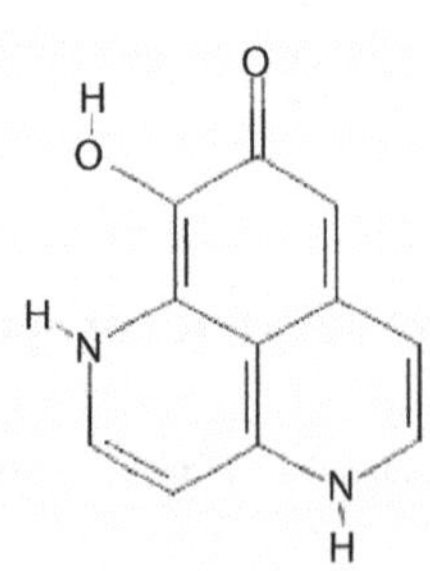

Demethylaaptamine

Among the bis-aaptamine alkaloid compounds suberitines A–D isolated from this species, suberitines B and D showed potent cytotoxicity against P388 cell lines, with IC50 values of 1.8 and 3.5 μM, respectively (Liu et al., 2012).

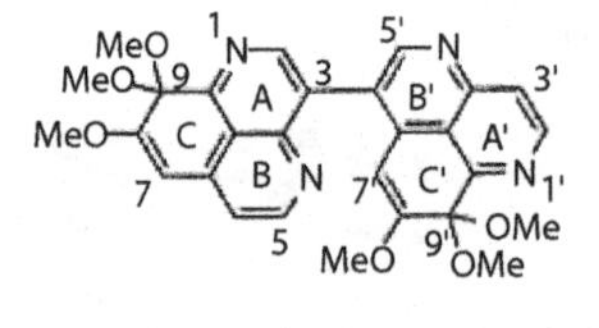

Suberitine A

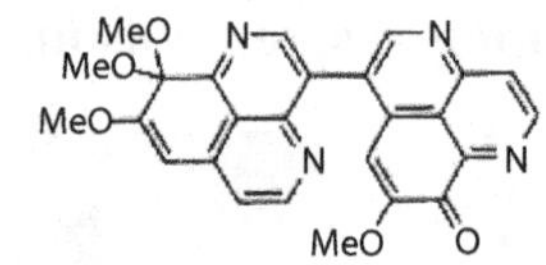

Suberitine B

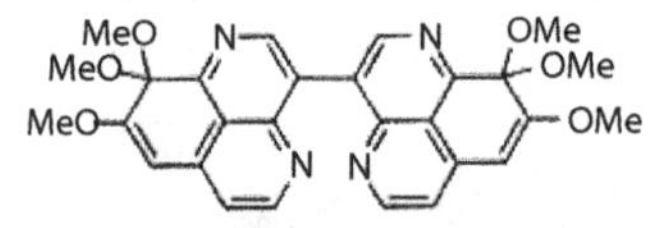

Suberitine C

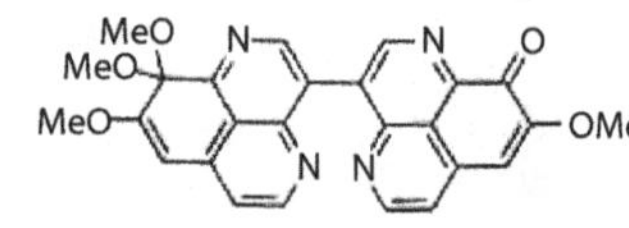

Suberitine D

Anticancer activity: The ethanolic extract of this species showed anticancer activity in the benzo(a)pyrene induced mice. The extract at a concentration of 1000 and 1500 mg/weight repressed the expression of protein 115 kDa in the blood plasma of fibrosarcoma mice (*Mus musculus*) (Sa'adah et al., 2016).

Aaptos sp.

Order: Suberitida Family: Suberitidae

Compound(s) and Activities

Antimycobacterial activity: The alkaloid 2-methoxy-3-oxoaaptamine isolated from this species was antimycobacterial against *Mycobacterium smegmatis* in both growing and dormancy-inducing hypoxic conditions with an MIC value of 6.3 and 5 μg/mL, respectively (Putra and Jaswir, 2014).

Caulospongia biflabellata (Fromont, 1998)

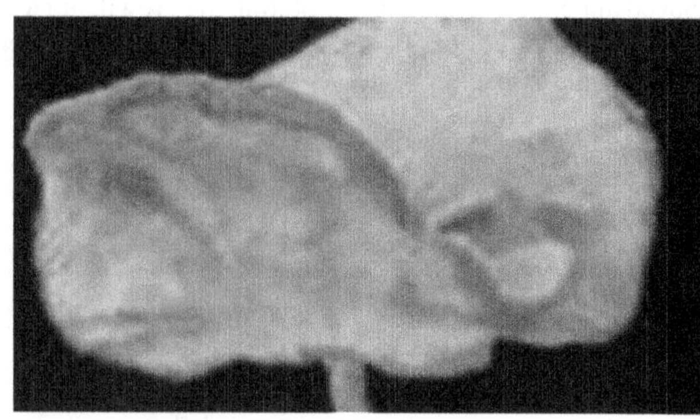

Order: Suberitida Family: Suberitidae

Common name: Not designated

Distribution: Western Australia

Ecology: It grows on the vertical or horizontal limestone reef with a component of either coral or kelp.

Biology: These sponges are erect fans and are commonly bilobed, with basal stalks. Specimens may have additional lobes on the outer surfaces, as irregular raised ridges. Specimens are 9–40 cm tall with a stalk of 5–12 cm long. Fans are 9–24 cm wide and are split into two lobes approximately 6–8 cm above the stalk. This sponge harbors barnacles within its tissue and ophuiroids on the surface.

Compound(s) and Activities

Cytotoxic activity: Tubercidin, a known nucleoside antibiotic, was isolated as the major metabolite from this species. This compound exhibited potent cytotoxic activity against P388 and A549 tumor cells (Biabani et al., 2002).

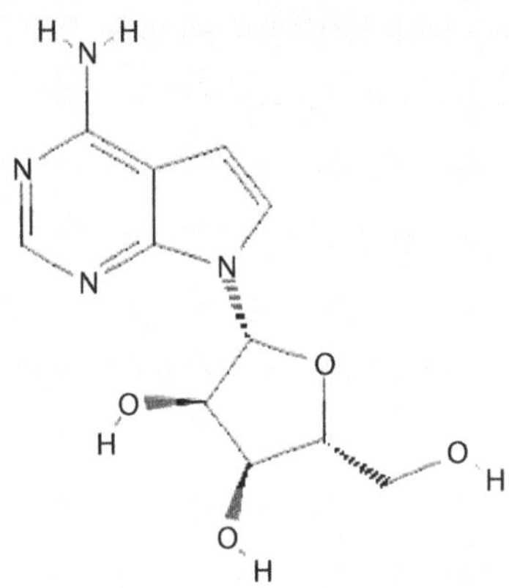

Tubercidin

Homaxinella balfourensis (Ridley and Dendy, 1886)

Order: Suberitida Family: Suberitidae

Common name: Bush sponge

Distribution: Kerguelen Islands, Antarctica

Ecology: It often grows on the upper valve of the Antarctic scallop (*Adamussium colbecki*) when these shells reach a length of about 7 cm. It is usually attached near the shell margin where the sponge may benefit from an increased flow of water. It is found at depths down to about 500 m.

Biology: Many predators find the spicules of *Homaxinella balfourensis* distasteful but the seastars *Odontaster meridionalis* and *Odontaster validus* feed on it, leaving behind the sponging, which provides fibrous support to the sponge. Small copepods in the family Arcturidae are often to be found among the sponge's branches.

This sponge grows in two forms, either branching out in one plane like a fan or forming an upright, clublike structure. This sponge is attached to the substrate by roots and has creeping stolons. It is usually arborescent with a main trunk and a dichotomous branching habit of growth but can also be clavate, thickening upward like a club. The lower branches are usually stout and cylindrical, while the upper branches are soft and spongy and are sometimes clavate themselves. It can grow to a length of about 50 cm with flattened branches up to 10 cm long and 0.8 cm wide. The surface of the sponge is smooth with a range of color from white, cream, or orange to light gray. It has an axially condensed, choanosomal skeleton of spicules with an extra-axial skeleton of bundles of larger spicules (megascleres). There are no microscleres.

Compound(s) and Activities

Anti-inflammatory activity: The EtOH extract of this species displayed anti-inflammatory activity by significantly reducing the release of IL-1β and PGE2 only at the highest (Moles et al., 2014).

Others: This species has been reported to produce antifreeze proteins (AFPs), which have potential applications in pharmaceutical industries (Wilkins et al., 2002).

Pseudosuberites andrewsi (Kirkpatrick, 1900)

Order: Suberitida Family: Suberitidae

Common name: Not designated

Distribution: Christmas Island

Ecology: It occurs in the shaded intertidal overhangs and under coral boulders.

Biology: This sponge is loosely encrusting or forms free thick lamellae. It is pale yellow (fawn) in color. Surfce is smooth and with canalicular markings below the dermis. Consistency is soft and is easily torn. Oscules, if present, are small and circular (7.5 cm dia.) and are guarded by a silvery fringe. Dermal skeleton is very distinct and is formed of tangentially arranged bundles of spicules. Spicules are tylotes that are slightly curved basally. Their head is round and are slightly knobbed at the summit or swollen laterally.

Compound(s) and Activities

Antioxidant activity: The n-hexane and ethanol extracts of this species displayed antioxidant activity with IC50 values of 779 μg/mL and 1043 μg/mL, respectively (Azizah et al., 2017).

Pseudosuberites sp.

Order: Suberitida Family: Suberitidae

Compound(s) and Activities

Antioxidant activity: Seradj et al. (2012) reportd on the hydroxyl radical scavenging activities of the sponge's extracts.

Percent of OH Radical Inhibition in CH_2Cl_2 Extracts

Concentrations in milligrams of dried sponge extract

0.5	0.25	0.125	0.062
20.3	14.9	39.0	43.5

Percent of OH Radical Inhibition in MeOH Extracts

Concentrations in milligrams of dried sponge extract

0.5	0.25	0.125	0.062
81	74	52	23

Source: Seradj, H. et al., 2012, *Iranian J. Pharm. Sci.*, 8: 249–255.

Suberites aurantiacus (Duchassaing and Michelotti, 1864)

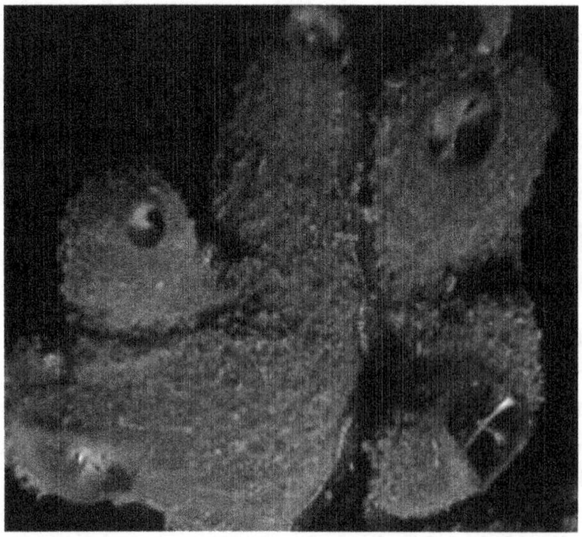

Order: Suberitida Family: Suberitidae
Common name: Lobate sponge
Distribution: Western Central Atlantic and Central Pacific
Ecology: It cccurs in mangrove ponds.
Biology: It is an encrusting to massive growth form, usually with large lobate or digitate projections. External color is usually red, red orange, yellow, dark green, greenish purple, or turquoise. Internal color is yellow ocher. Texture is dense, flexible, and rubbery. Surface is generally smooth and some with occasional goosebumps. Spicules are straight sharply pointed tylostyles with large variation in size. Some smaller ones have irregularly shaped or lumpy heads. Ectosome consists of brushes of smaller tylostyles.

Compound(s) and Activities

Antioxidant activity: Both the 70% and 100% fractions of the extracts of this species showed low antiradical efficiency, the values of which which ranged between 21.27% and 48.47% (Tovar-Barrios et al., 2016).

Suberites clavatus (Keller, 1891) (= *Pseudosuberites clavatus*)

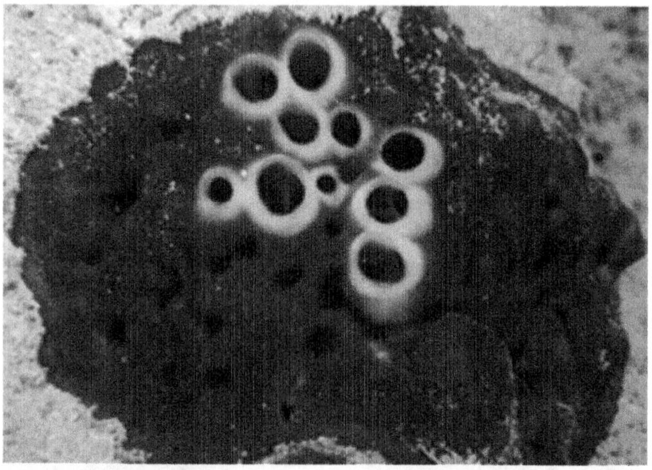

Order: Suberitida Family: Suberitidae
Common name: Red Sea sponge
Distribution: Northern and Central Red Sea, Southern Red Sea
Ecology: It inhabits coral reefs in depth range of 5 to 30 m.
Biology: Description not available.
Association: This species is in symbiotic association with the barnacle *Membranobalanus longirostrum*, which is found on the same side as the sponge's ostia. This adaptation allows the barnacles to catch more suspended particles from the water, provides them with more oxygen and prevents their exposure to discharged sponge waste.
Heavy metal removal: *Suberites clavatus* was found to contain a significantly higher concentration of Cd, Co, Ni, and Zn. It also has the second highest Fe concentration. This indicates the active uptake and accumulation of Fe by this species. This species may therefore be of great use in the treatment of wastewater (Mayzel et al., 2014).

Compound(s) and Activities

Quorum-quenching activity: The bacterial extract from this species has been reported to possess quorum-quenching, bioactive lead molecules for antivirulence therapy (Saurav et al., 2016).

Suberites domuncula (Olivi, 1792)

Order: Suberitida Family: Suberitidae

Common name: Large encrusting sponge

Distribution: Adriatic Sea

Ecology: It inhabits sandy, muddy, and detritic bottoms at depths of 26 to 36 m.

Biology: It is a massive sponge with smooth surface and brown color (in spirit). Spicules are tylostyles with well-formed heads variable in length and thickness. Microstrongyles are abundant. This species contains suberitine, a neurotoxin that can cause fatal hemolytic hemorrhaging in various animals. While it is highly toxic to fish, it is known to be preyed upon by the hawksbill turtle, *Eretmochelys imbricate*.

Association: A kind of mutualistic association exists between this sponge and hermit crabs, which live in the shells of this species. The advantage of this association to the sponge is that it is carried by the mollusk; the hermit crab gains protection not only by living in the shell of the mollusk but also through the unfavorable smell and taste of the sponge, which discourages attack by predators such as fishes and other animals.

Compound(s) and Activities

Antibacterial activity: The cyclic peptides cyclo-(glycyl-Lseryl-L-prolyl-L-glutamyl) and cyclo- (glycyl-L-prolyl-Lglutamyl) isolated from the cell extract of a *Ruegeria* strain associated with cell cultures of this sponge showed moderate activity against *B. subtilis* (Joseph et al., 2012).

The extracts of the associated microorganisms, namely, α-Proteobacterium MBIC3368, *Idiomarina loihiensis*, and *Pseudomonas* sp. of this sponge exhibited antiangiogenic, antimicrobial, hemolytic, and cytotoxic properties. These organisms were found to be strongly active against multidrug-resistant clinical strains of *Staphylococcus aureus* and *Staphylococcus epidermidis*, isolated from hospital patients (Thomas et al., 2010).

Cytotoxic activity: The drimane sesquiterpenoids extracted from the associated marine fungus *Aspergillus ustus* of this species exhibited a potent inhibitory effect on tumor cell lines L5178Y, HeLa, and PC12, with half maximal effective concentration (EC50) values ranging from 0.6 to 5.3 μg/mL (Zhang et al., 2017b).

Suberites japonicus (Thiele, 1898)

Order: Suberitida Family: Suberitidae

Common name: Not designated

Distribution: Western Central Pacific, Indonesia and Singapore, Japan

Ecology: It is a coral-reef associated species.

Biology: Not available

Compound(s) and Activities

Antioxidant activity: The two methyl esters of this species strongly inhibited nitric oxide (NO) production from RAW 264.7 murine macrophase cells, with IC50 3.9 and 7.0 μM, respectively. These compounds also inhibited NO production in BV2 microglia cells with IC50 values of 3.1 and 1.8 M, respectively (Gribble, 2015).

Anticancer activity: Six depsipeptides, seragamides A–F (1–5), isolated from this sponge showed multinuclei formation in NBT-T2 cells at 0.01, 0.02, 0.01, 0.01, and 0.04 mg/mL, respectively. Compound (1) also promoted the polymerization of G-actin and stabilizes F-actin filaments (Agrawal et al., 2016).

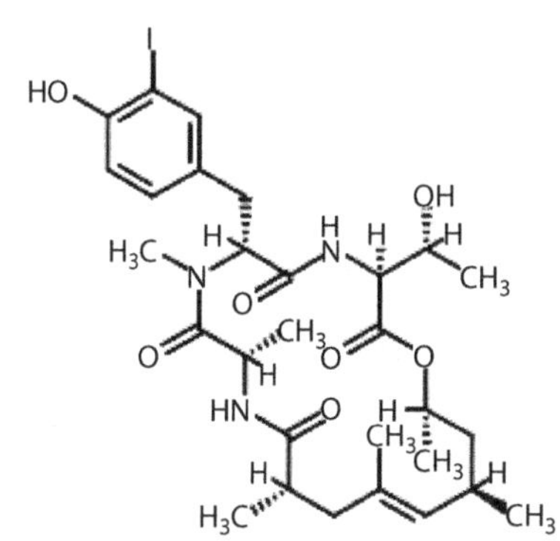

Seragamide A

Cytotoxicity: One novel sterol, bendigole, produced by the associated fungal strain *Actinomadura* sp. SBMs009 of this species displayed a moderate cytotoxic effect on the L929 cells with an IC50 value of 30 μM (Zhang et al., 2017b).

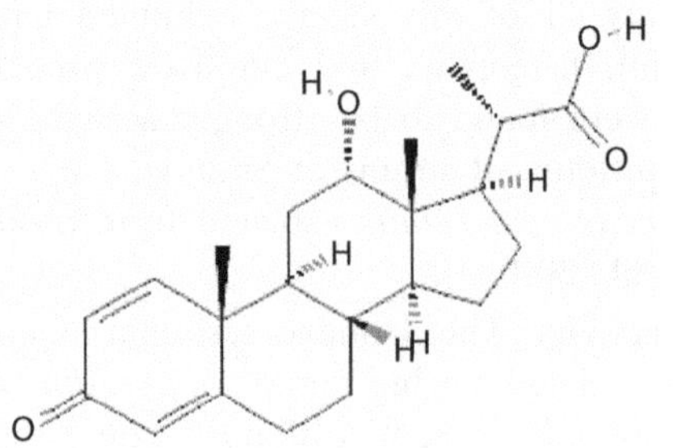

Bendigole

Anti-inflammatory effects: The unknown iodinated acetylenic acids isolated from this species possessed anti-inflammatory effects. While two of these compounds exhibited a strong NO inhibitory effect on RAW264.7 cells, another two of these compounds were found to be highly active in BV2 microglia cells (Hwang et al., 2013).

Suberites waedoensis n. sp.

Order: Suberitida Family: Suberitidae

Common name: Not designated

Distribution: Korea

Ecology: Not reported, as it is a new species.

Biology: It is a cushion-shaped sponged with the size of 12 × 6.5 × 4 cm (thickness). Texture is soft and surface is covered with wrinkles. Oscula, which are 2 to 4 cm in diameter, are rarely seen. Ectosome skeleton consists of small tylostyles that point outward. Choanosome skeleton is densely packed with larger tylostyles. Color of the sponge is orange both in life and alcohol. Megascleres are in two types of tylostyles. Microscleres are absent.

Compound(s) and Activities

Cytoxicity and enzyme inhibitory activity: The new peptides chujamides A (1) and B (2) isolated from this species exhibited weak cytotoxicities against A549 and K562 cell lines. The LC50 values were 37 and 10 μM for compound 1 and were 56 and 26 μM for compound 2, respectively. Further, the compound chujamide B moderately inhibited the action of Na+/K+-ATPase with an IC50 value of 17.2 μM (Song et al., 2014).

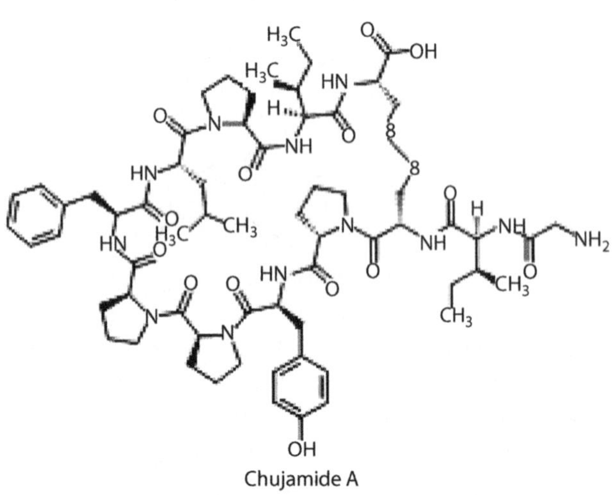

Chujamide A

Suberites sp.

Order: Suberitida Family: Suberitidae

Compound(s) and Activities

Antibacterial activity: The alkaloid nakijinamine A showed antibacterial activity against *S. aureus* (MIC 16 μg/mL), *B. subtilis* (MIC 16 μg/mL), and *Micrococcus luteus* (MIC 2 μg/mL) (Gribble, 2015).

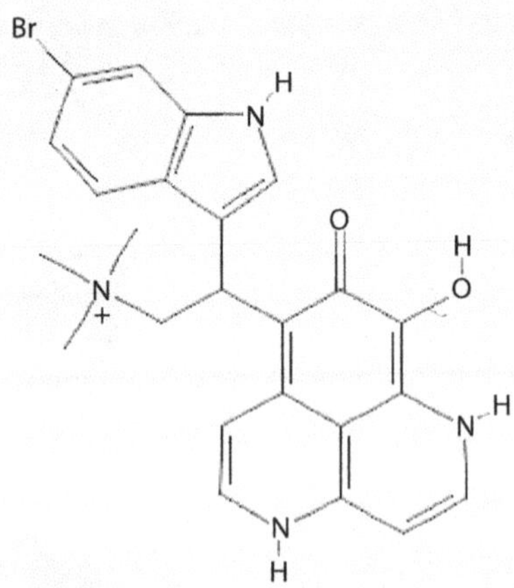

Nakijinamine A

FAMILY: HEMIASTERELLIDAE

Hemiasterella vasiformis var. *minor* (Kirkpatrick, 1903) (= *Hemiasterella minor*)

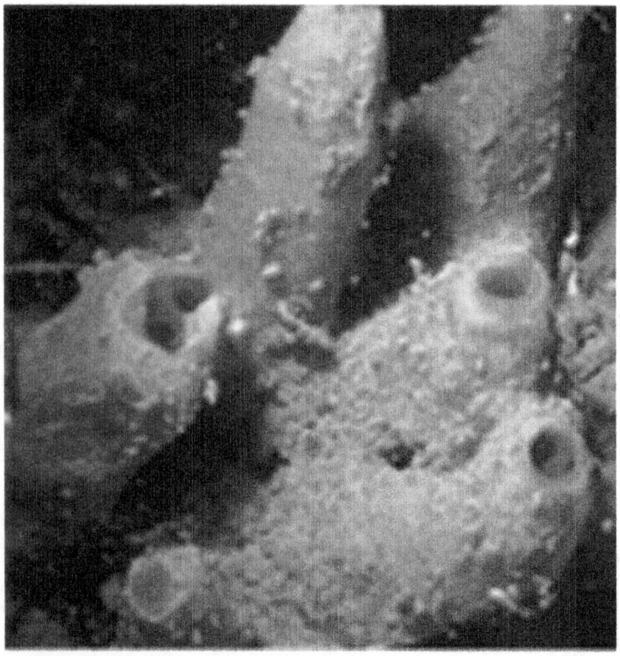

Order: Tethyida Family: Hemiasterellidae

Common name: Not designated

Distribution: South Africa

Ecology: Not reported.

Biology: It is a vase shaped sponge.

Compound(s) and Activities

Antitumor activity: The peptide taltobulin (HTI-286) isolated from this species showed antitumor activity (Joseph et al., 2012).

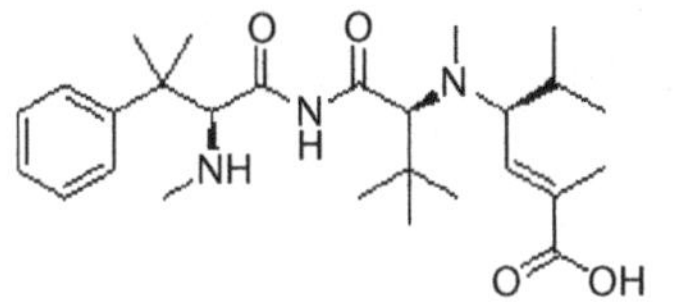

Taltobulin (HTI-286)

The macrocyclic lactam/lactone, jaspamide showed antitumor activity inducing apoptosis in the HL-60 human promyelocytic leukemia cell (Sipkema et al., 2005; Suarez-Jimenez et al., 2012).

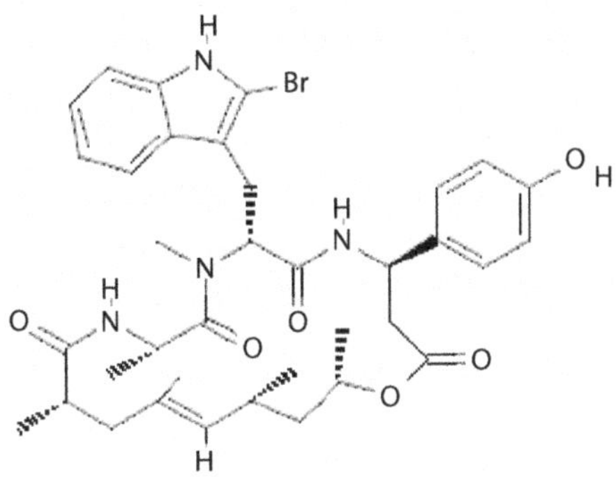

Jaspamide

FAMILY: TETHYIDAE

Tectitethya crypta (de Laubenfels, 1949) (= *Cryptotethya crypta; Tethya crypta*)

Order: Tethyida Family: Tethyidae

Common name: Caribbean sponge

Distribution: Eastern Caribbean, Atlantic Ocean, Gulf of Mexico, North America

Ecology: It grows in lagoonal environments on sediment-covered bottoms at a depth range of 1 to 20 m.

Biology: This species (4 × 7 × 12 cm) has a porous body and a hard surface like a stone. Body is conical, hemispherical, or stubby cylindrical. It is covered on much of its surface by low, irregularly rounded, and flattened tubercles of 3 to 5 mm in diameter and by abundant sediment. Color of surface of sponge (without sediment) is gray olive. Megascleres are seen in radiating bundles and are frequently branched and fused.

Compound(s) and Activities

Antiviral activity: The nucleoside Ara-A (vidarabine) isolated from this species inhibited the viral DNA polymerase and DNA synthesis of herpes, vaccinica, and varicella zoster viruses (Perdicaris et al., 2013).

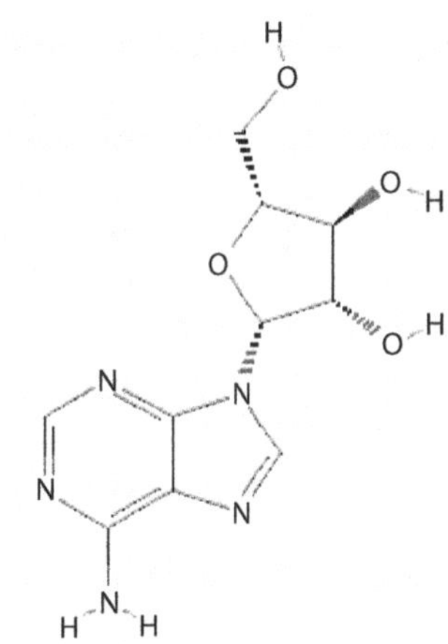

Vidarabine

The nucleosides spongothymidine and spongouridine were the basis for the compound Ara-A (vidarabine), which was active against the herpes simplex virus. These compounds have also shown tumor-inhibiting activity and are of great use in the production of antiviral and anticancer drugs (Indraningrat et al., 2016; Jha and Zi-rong, 2004; *Wikipedia*).

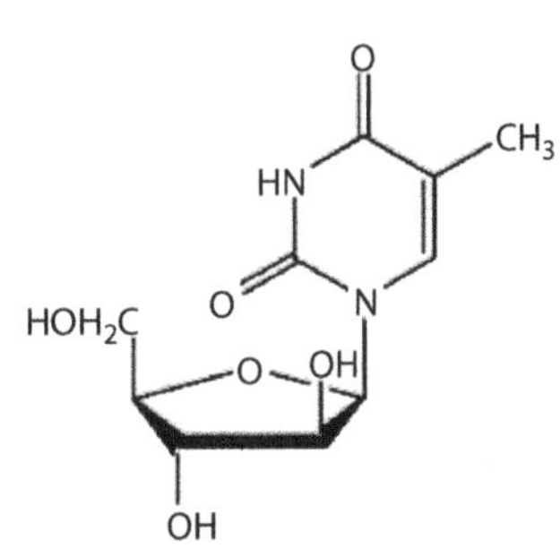

Spongothymidine

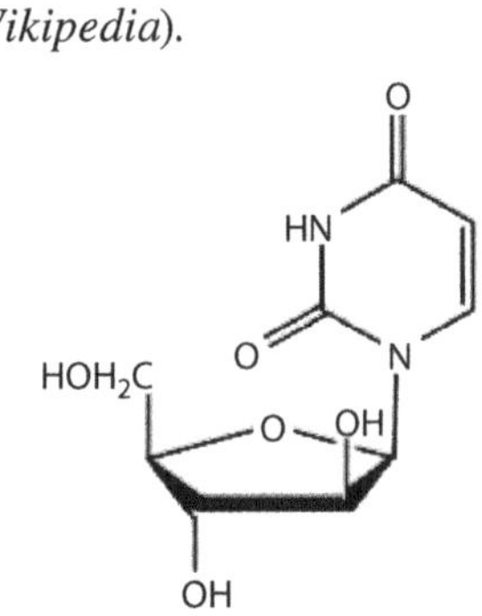

Spongouridine

The compound vidarabine of this species possessed antiviral activity by inhibiting viral DNA polymerase and DNA synthesis of herpes, vaccinica, and varicella zoster viruses

(Sagar et al., 2010). This compound is of great use in the production of antiviral drugs for the treatment of viral infections caused by herpes simplex and herpes zoster (Anjum et al., 2016; Sagar et al., 2010).

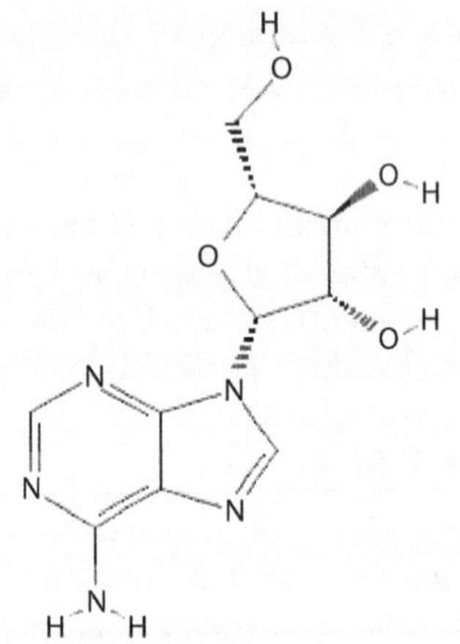

Vidarabine

Anticancer activity: The compound cytarabine, a synthetic derivative of the spongothymidine, has anticancer activity against nonlymphocytic leukemia, the leukemia meninges, and chronic myelocytic leukemia (Moura et al., 2012). This compound, besides serving as an anticancer drug, is also considered as a liposomal drug known as Liposomal ara-C, which is classified as a antimetabolite, or a cell replication inhibitor, and is used in most the chemotherapy treatments ("*Cryptotethya crypta*," 2007).

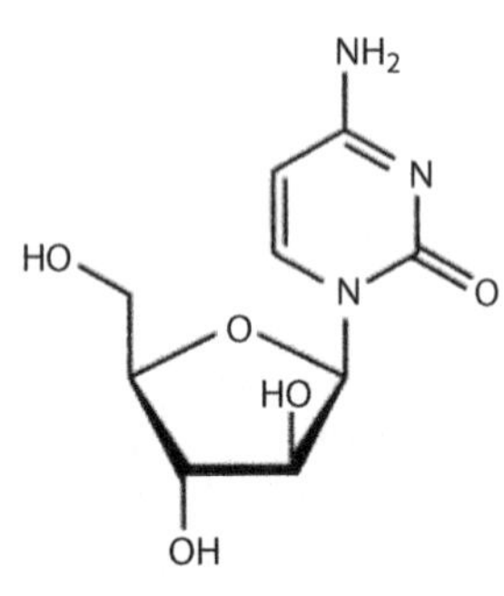

Cytarabine

The compound gemcitabine, a fluorinated derivative of cytarabine of this species possesses anticancer activity and is used in the treatment of pancreatic, breast, bladder, and non-small cell lung cancer (Schwartsmann et al., 2001).

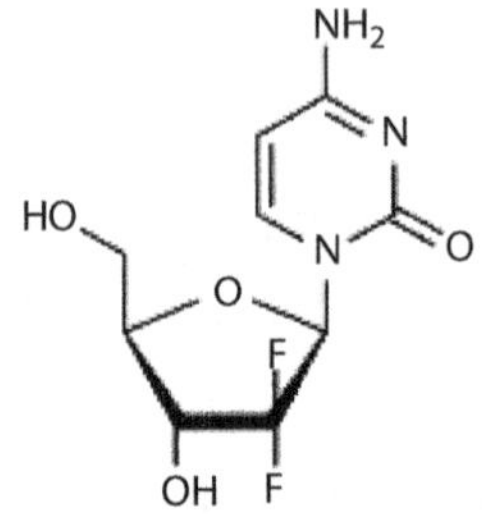

Gemcitabine

Anti-inflammatory activity and analgesic and vasodilation properties: The compound spongosine, a methoxyadenosine derivative of this species, has shown anti-inflammatory activity and analgesic and vasodilation properties (Bertin et al., 2015).

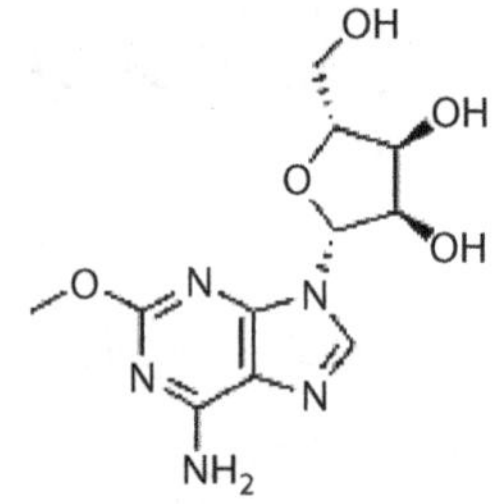

Spongosine

Antimicrobial, anticancer, antimetastatic, and immunomodulatory properties: The compound deoxyspongosine of this species exhibited antimicrobial, anticancer, antimetastatic, immunomodulatory, and insecticidal properties (Bertin et al., 2015).

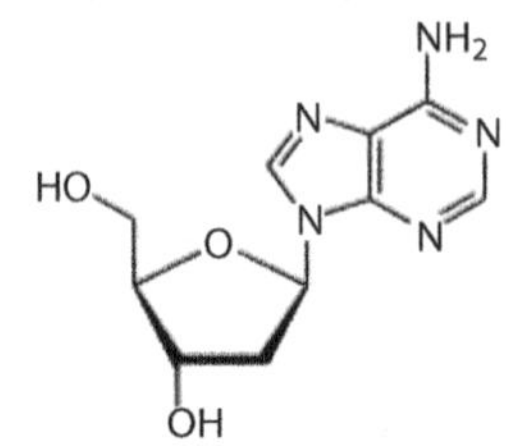

Deoxyspongosine

Tethya aurantium **(Pallas, 1766)**

Order: Tethyida Family: Tethyidae

Common name: Orange puffball sponge, sea orange, ball sponge

Distribution: Northeast Atlantic and Indo-West Pacific; English Channel, the North Sea and the Mediterrannean Sea

Ecology: It dwells usually on shallow reefs and rock pools at a depth of about 400 m. It is attached to rocks by means of its base and some extensions resembling roots.

Biology: It is a ball-shaped, spherical sponge 2 to 10 cm in diameter. It is yellow or pale orange. It has a single, large oscule, generally at the top. Its surface is granulous sometimes with small, round, and spiny outgrowths. Its consistency is hard. Megascleres are rounded at the base and their apex is usually blunt but sometimes sharp. Styli are faintly tylote and there are no oxea or strongyla. Microscleres are very small chiasters.

Compound(s) and Activities

Antibacterial and antitumor activites: Two novel cyclodepsipeptides, scopularides A and B [cyclo-(3-hydroxy-4-methyldecanoyl-Gly-L-Val-D-Leu-L-Ala-L-Phe) and cyclo-(3-hydroxy-4-methyloctanoyl-Gly-L-Val-D-Leu-L-Ala-L-Phe) for scopularide A and B, respectively], have been recorded in the fungus *Scopulariopsis brevicaulis*, which was isolated from this species. These compounds showed a weak antibiotic activity against Gram-positive bacteria. But the activity of these compounds against several tumor cell lines was significant at 10 μg/mL (Yu et al., 2008).

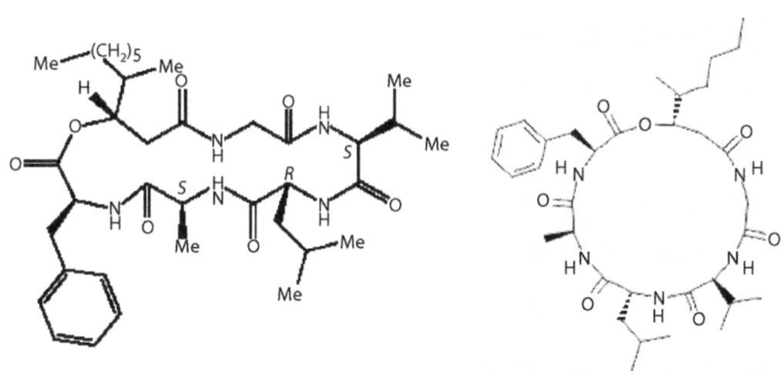

Scopularides A and B

Antibacterial activity: The fungus *Bartalinia robillardoides* (strain LF550) derived from this species yielded the chloroazaphilone compound isochromophilone XI, which showed antibacterial activity against *B. subtilis* (IC50 55.6 μM) and *Staphylococcus lentus* (IC50 78.4 μM) (Gribble, 2015).

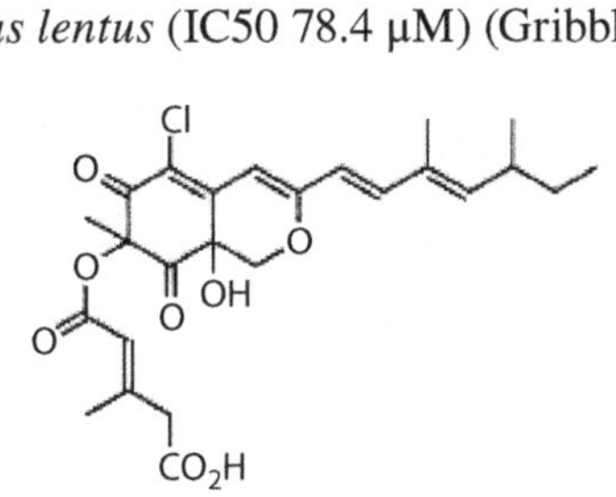

Isochromophilone XI

Tethya sp.

Order: Tethyida Family: Tethyidae

Compound(s) and Activities

Antiprotozoal activity: The actinobacterial strain *Streptomyces* sp. T03 derived from this species yielded a compound, butenolide, which showed antiprotozoal activity against *T. brucei brucei* with an IC50 value of 7.92 μg/mL (Indraningrat et al., 2016).

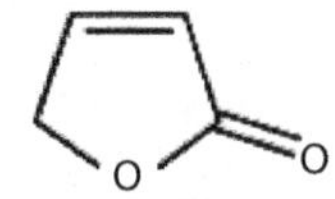

Butenolide

FAMILY: ANCORINIDAE

Asteropus sp.

Order: Tetractinellida (Astrophorida) Family: Ancorinidae

Compound(s) and Activities

Antibacterial activity: The compound indolo[3,2-a]carbazole of this species showed some antibacterial activity against methicillin-resistant *S. aureus* (MRSA) with an MIC of 50 μg/mL (Gribble, 2015).

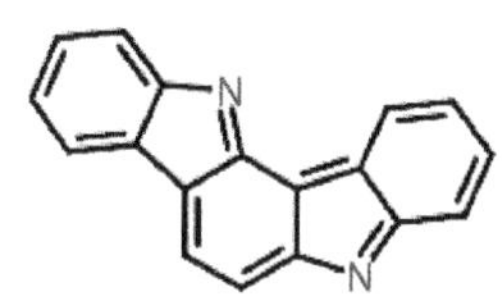

Indolo[3,2-a]carbazole

Antifungal activity: The compound indolo[3,2-a]carbazole from this species was antifungal toward *Candida albicans* (MIC 25 μg/mL). Further, its purpuroine D was active against *C. albicans* (IC50, 19 μg/mL), and purpuroines A, C, and D inhibited the human disease-causing *Aspergillus fumigatus* (IC50, 29, 26, and 26 μg/mL, respectively). Furthermore, another compound, nakijinamine A, showed antifungal activity toward *C. albicans* (IC50, 0.3 μg/mL), *Cryptococcus neoformans* (IC50, 0.5 μg/mL), and *Trichophyton mentagrophytes* (IC50, 0.3 μg/mL). However, less activity was observed against *C. albicans* with nakijinamines B and F (IC50, 8 μg/mL each) (Gribble, 2015).

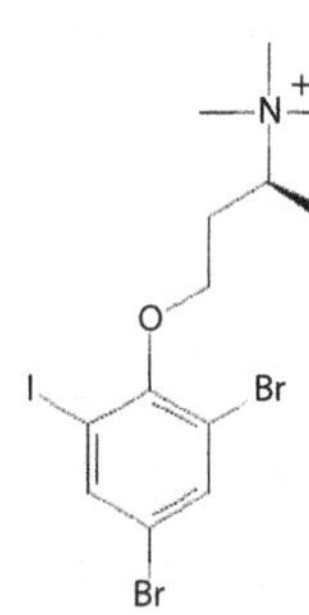

Purpuroine D

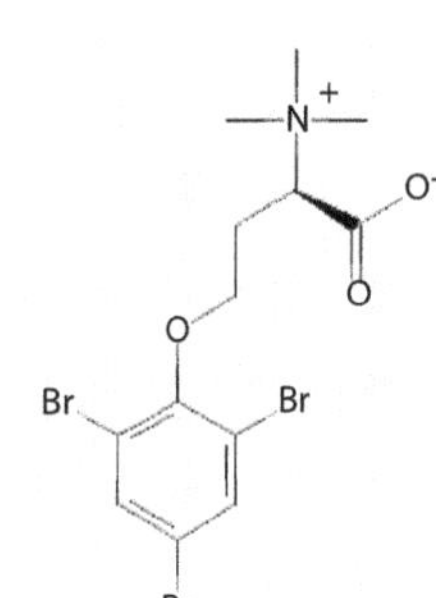

Purpuroine A

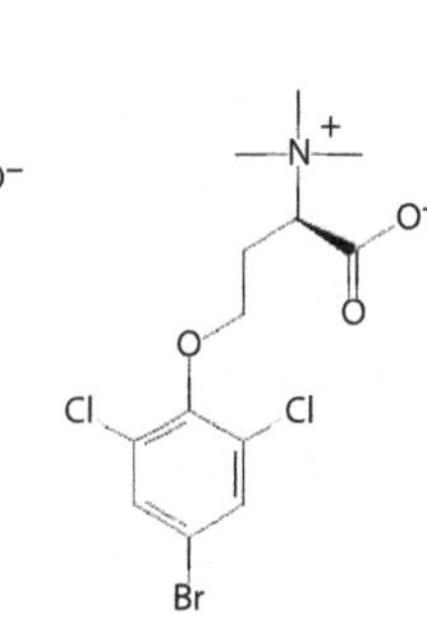

Purpuroine C

Dercitus (Halinastra) japonensis (van Soest, Beglinger, and de Voogd, 2010)

Order: Tetractinellida Family: Ancorinidae

Common name: Not known

Distribution: Central Kuroshio

Ecology: It is a newly identified species so there is no report.

Biology: This species possesses the characteristic feature, namely, toxa-like spicules.

Compound(s) and Activities

Anti-HIV activity: The compound ancorinolate B isolated form the MeOH soluble extract of this species showed moderate anti-HIV activity with an IC50 value of 102 μM (Hirade et al., 2017).

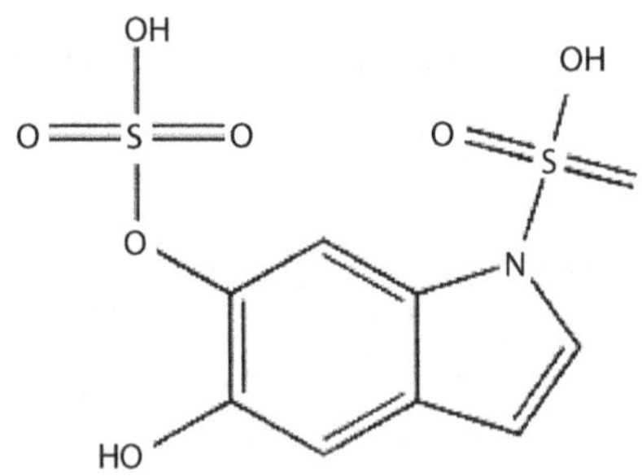

Ancorinolate B

Dercitus (Stoeba) sp.

Order: Tetractinellida Family: Ancorinidae

Compound(s) and Activities

Cytotoxic activity: An aminoacridine alkaloid, dercitin, isolated from this deepwater sponge, displayed cytotoxic activities in the low nanomolar concentration range. Further this compound prolonged the life of mice-bearing ascitic P388 tumors, and was also active against B16 melanoma cells and small-cell Lewis lung carcinoma (Jha and Zi-rong, 2004).

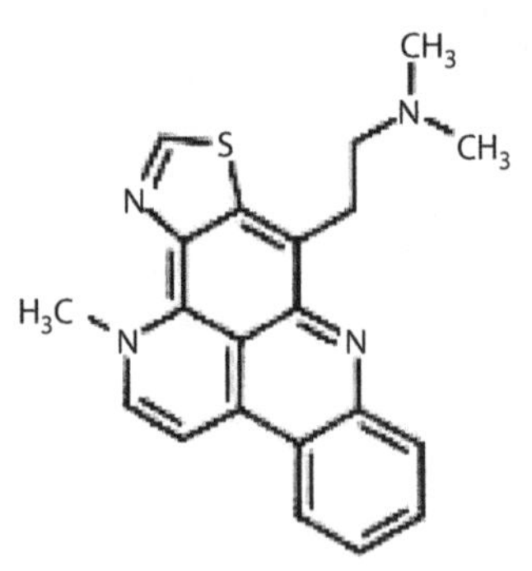

Dercitin

Antiprotozoal activity: The fungl strain, *Penicillium citrinum* F40 derived from this species showed antiprotozoal activity against *T. vaginalis* ATCC 30236 with an MIC value of 250 μg/mL (Indraningrat et al., 2016).

Antiviral activity: The polysaccharides extracted from this species were all highly active against HIV-1 (90% to 95% inhibition) (Esteves et al., 2011).

Jaspis digonoxea (de Laubenfels, 1950)

Order: Tetractinellida Family: Ancorinidae
Common name: Not designated
Distribution: Hawaii, Indian Ocean, Mahe Island
Ecology: Not reported
Biology: Not available

Compound(s) and Activities

Antitumoral activity: The compound, digonazole triacetate, an oxazole derivative of this species has shown antitumoral activity (Gravalos et al., 1995).

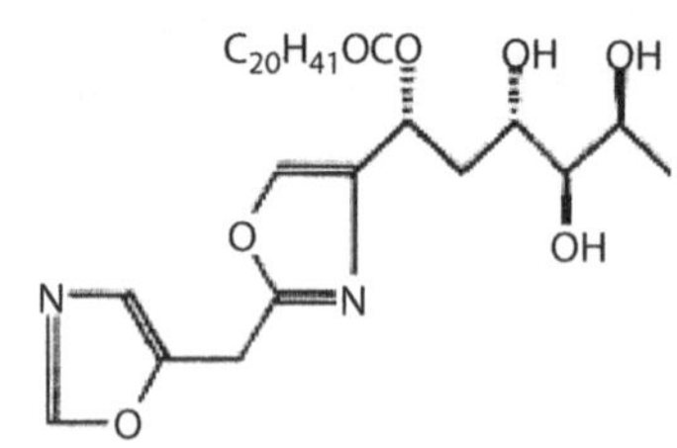

Digonazole

Jaspis johnstonii (Schmidt, 1862)

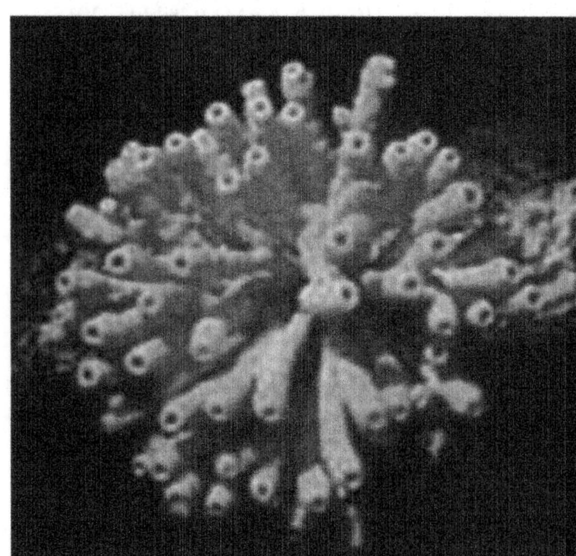

Order: Tetractinellida Family: Ancorinidae
Common name: Bright orange sponge

Distribution: Adriatic, Mediterranean (coasts of France and Spain), Atlantic (Azores and Cape Verde), Rema Island

Ecology: It is a thinly encrusting sponge inhabiting rock crevices or covering bivalve shells.

Biology: Surface of this sponge is smooth and is perforated by sparse oscules, of 0.5 mm in diameter. Cortex is undifferentiated. Ectosome is reinforced by paratangential small oxeas and asters. The megascleres are oxeas that are fusiform and slightly curved. The microscleres are oxyasters without a marked centrum. Regarding the skeletal arrangement, the large oxeas are radially arranged at the sponge periphery and confused toward the interior. The small oxeas are arranged tangentially to the sponge surface. Oxyasters are dispersed thoroughout the sponge but more abundant in the peripheral region.

Compound(s) and Activities

Cytotoxicity: The tricyclic sesquiterpenes, namely, coriolin B and dihydrocoriolin C, and chloriolines A–C have been isolated from the associated hypomycete fungus of this sponge species. Among these compounds, coriolin B exhibited strong inhibition of human breast and CNS cell lines with IC50 values of 0.7 µg (breast) and 0.5 µg (neuroblastoma) (Thomas et al., 2010).

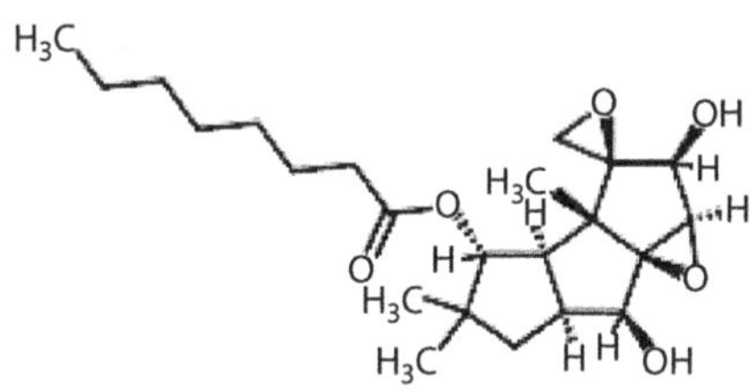

Coriolin B

The pyrrolo [2,3-d]pyrimidine nucleosides, namely, toyocamycin and 5-(methoxycarbonyl)tubercidin isolated from this species have shown cytotoxicity against tumor cell lines (Zabriskie and Ireland, 1989).

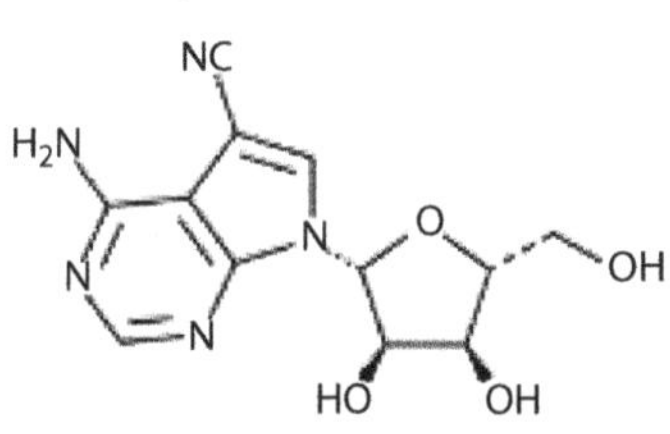

Totocamycin

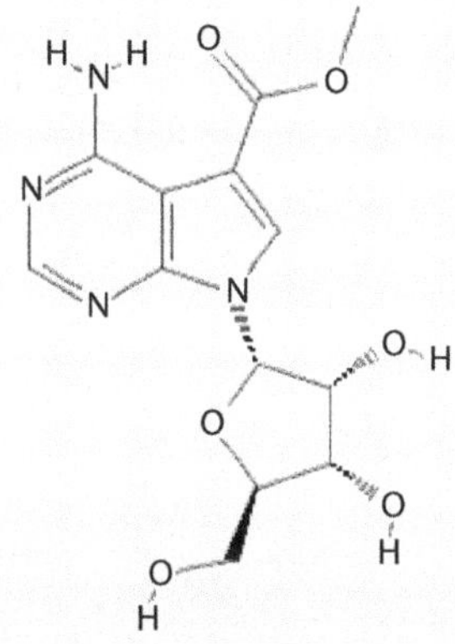

5-(methoxycarbonyl)tubercidin

The cyclodepsipeptide compound, jasplakinolide, of this species has been reported to enhance apoptosis induced by interleukin 2 (IL-2) deprivation. This compound has also displayed fungicidal, and anticancer properties ("Jasplakinolide, *Jaspis johnstonii*," n.d.).

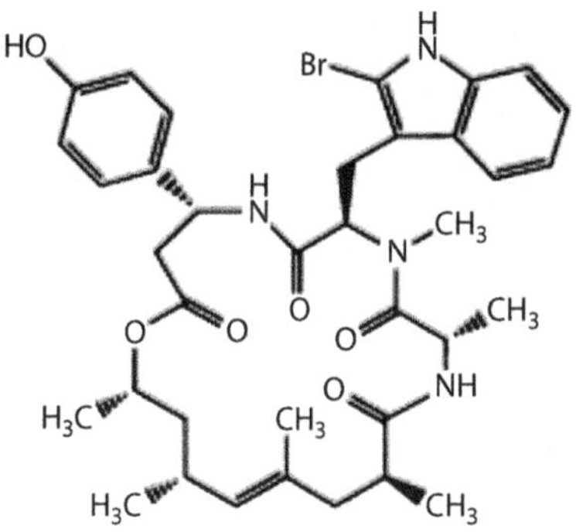

Jasplakinolide

Jaspis splendens (de Laubenfels, 1954)

Order: Tetractinellida Family: Ancorinidae
Common name: Not designated

Distribution: Western Central Pacific, New Caledonia and Micronesia

Ecology: This sessile sponge occurs at a depth range of 0 to 100 m.

Biology: It is an encrusting or massive sponge without triaenes. Choanosomal skeleton is composed of oxeas that are irregularly interlaced. The ectosomal skeleton is formed by a layer of paratangential oxeas, which are generally smaller than those in the choanosome. Microscleres are euasters without a centrum.

Compound(s) and Activities

Anticancer activity: Two new jaspamide derivatives, 2 and 3, together with the parent compound jaspamide (1) have been isolated from this species. The new derivatives were found to inhibit the growth of mouse lymphoma (L5178Y) cell line *in vitro* with IC50 values of <0.1 μg/mL (Ebada et al., 2009).

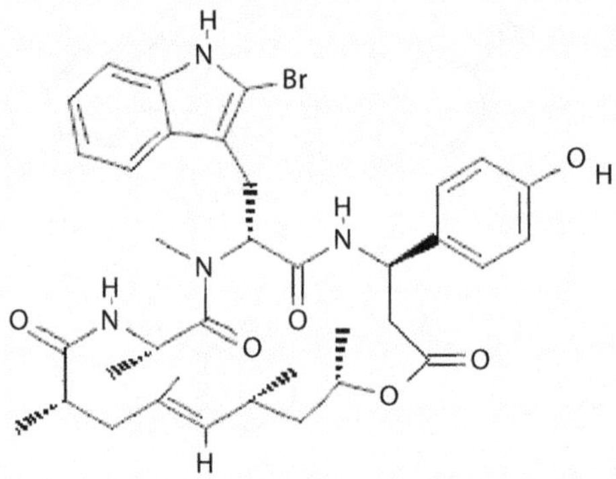

Jaspamide

Anti-inflammatory activity: The diterpene benzenoid jaspaquinol isolated from this species has shown anti-inflammatory activity by inhibiting the enzyme lipoxygenase (Sipkema et al., 2005).

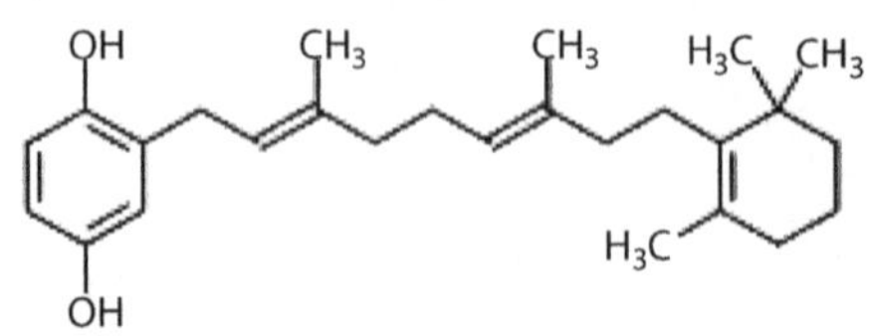

Jaspaquinol

Jaspis stellifera (Carter, 1879)

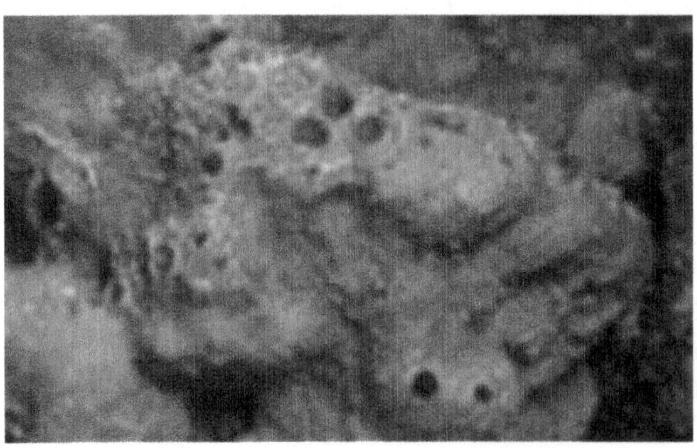

Order: Tetractinellida Family: Ancorinidae

Common name: Okinawan marine sponge

Distribution: Indo-West Pacific Ocean, Australia, Bassian, Palau

Ecology: It inhabits shallow lagoons where it encrusts dead corals, coral rubble, and other hard substrates. The depth range is subtidal to 40 m depth.

Biology: This sponge is irregularly lobate-massive with dull pinkish purple-brown color. Surface is optically smooth, even, and is unornamented. The ectosomal skeleton consists of a densely packed tangential arrangement of oxeas.

Compound(s) and Activities

Cytotoxicity: The isomalabaricane triterpene compounds, stelliferins A–F, isolated from this species have shown antineoplastic activity (Tsuda et al., 1991).

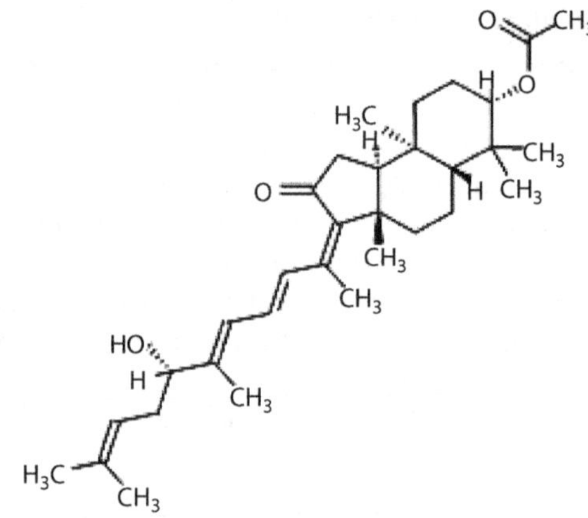

Stelliferin A

The isomalabaricane-type nortriterpenoids, jaspiferals A–G, isolated from this species have shown cytotoxicity against tumor cell lines (Kobayashi et al., 1996).

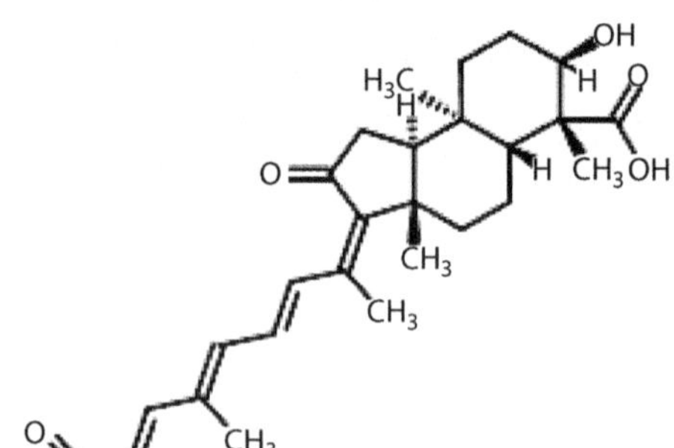

Jaspiferal A

Antitumor activity: The compound stellettin B of this species exhibited highly potent inhibition against the growth of a human glioblastoma cell line SF295, with a GI50 of 0.01 μM. Further, stellettin B increased the production of ROS, the activity of caspase 3/7, as well as the cleavage of PARP, each of which is known to be involved in apoptosis (Tang et al., 2014).

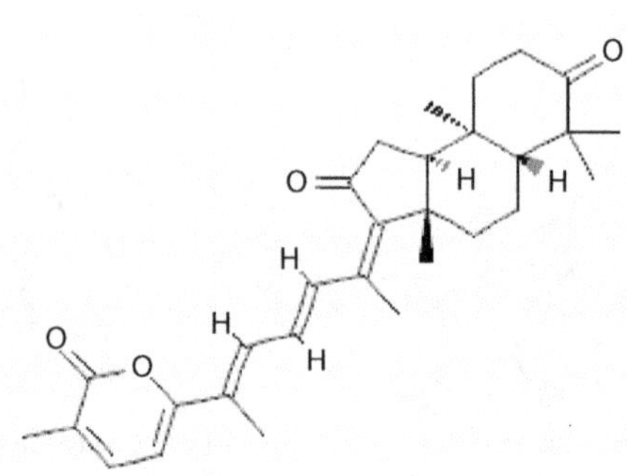

Stelletin B

Jaspis sp.

Order: Tetractinellida Family: Ancorinidae

Compound(s) and Activities

Cytotoxicity: The isomalabaricane triterpenes stelliferin G (1), 29-hydroxystelliferin A (2), and 29-hydroxystelliferin E (3), together with the known triterpenes 3-epi-29-hydroxystelliferin E (4), 13(E)-29-hydroxystelliferin E (5), 29-hydroxystelliferin B (6), 13(E)-stelliferin G (7), and 13(E)-3-epi-29-hydroxystelliferin E (8), have been isolated from the organic extract of this unidentified species. Among these compounds, the mixtures of 29-hydroxystelliferin B (6) and 13E-stelliferin G (7) have shown highest growth-inhibitory (IC50 0.11 and 0.23, μg/mL, respectively) activities against melanoma (MALME)-3M (Li et al., 2013).

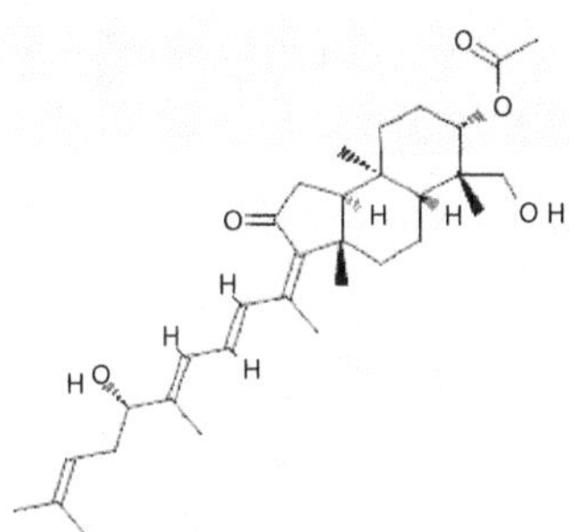

29-hydroxystelliferin B

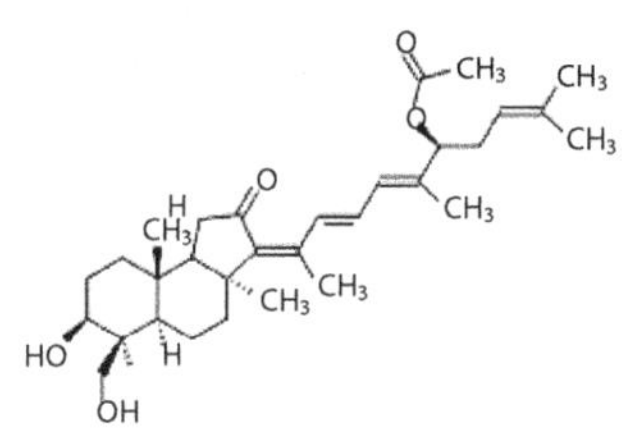

Stelliferin G

The isomalabaricane-type triterpenoids and nortriterpenoids with a 3α-acetoxy group have been isolated along with the known globostellatic acids B and C from this sponge. All these compounds have shown cytotoxicity against tumor cell lines (Zampella et al., 2000).

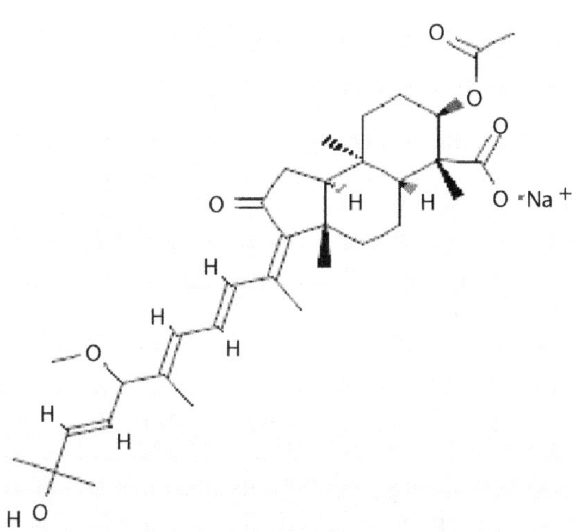

Globostellatic acid B

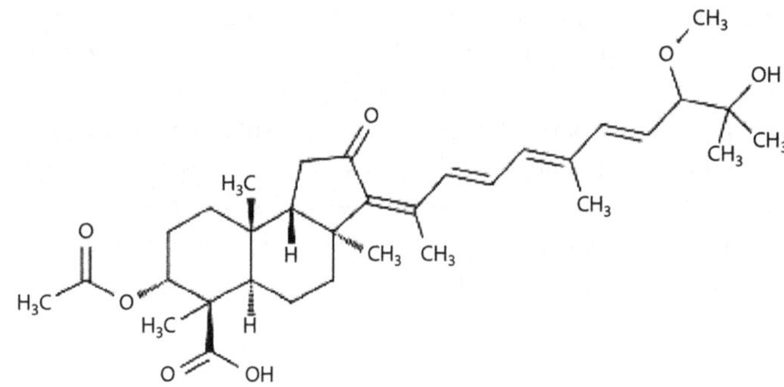

Globostellatic acid C

Antifungal activity: The macrocyclic depsipeptide jaspamide isolated from this species has shown activity against *Candida albicans* with an MIC value of 25 μg/mL (Crews et al., 1986).

Antiproliferative activity: The cyclic depsipeptide compound jaspamide isolated from this unidentified species has shown antiproliferative activity and induced apoptosis in the HL-60 human promyelocytic leukemia cell line (Suarez-Jimenez et al., 2012).

The bengamides isolated from this sponge showed *in vitro* antiproliferative activity on MDA-MB-435 human mammary carcinoma cells and the IC50 (μM) recorded for various compounds are given next.

In vitro Antiproliferative Activity of Bengamides (IC50 [μM]) on MDA-MB-435

Compound	IC50 (μM)
A	0.001
B	0.012
E	3.3
F	2.9
M	0.0101
O	0.0003
P	1.2
Z	2.9

Source: García-Ruiz, C., and F. Sarabia, 2014, *Mar. Drugs*, 12: 1580–1622.

Rhabdastrella distincta (Thiele, 1900)

Order: Tetractinellida Family: Ancorinidae

Common name: Not known

Distribution: Halmahera (Moluccas); Indonesia, North Sulawesi Island

Ecology: It lives on rocky reefs in 2 to 90 m of water.

Biology: This sponge has a massive and irregular shape with a large size, up to 50 cm in diameter. Color of the animal is lemon yellow or dark green and turns black when cut or preserved. Wide oscular areas are often seen. Skeleton is of spherasters, which are located in the outer part of the sponge, but do not form a real cortex. The choanosomal skeleton consists of scattered oxeas that tend to form radial tracts toward the peripheral part. Oxyasters and oxyspherasters are dispersed in the choanosome. Spicules are of megascleres, which are fusiform oxeas with rather sharp tips. Microscleres are spherasters of variable size and are with a large center and thick rays with sharp or bifurcated tips.

Compound(s) and Activities

Cytotoxicity: This sponge speices yielded six new isomalabaricane triterpenes, rhabdastrellins A–F (1–6) (minor components), along with stellettins L and M. Among these compounds, compounds 1–6 showed cytotoxicity against a small panel of human tumor cell lines (Lv et al., 2008).

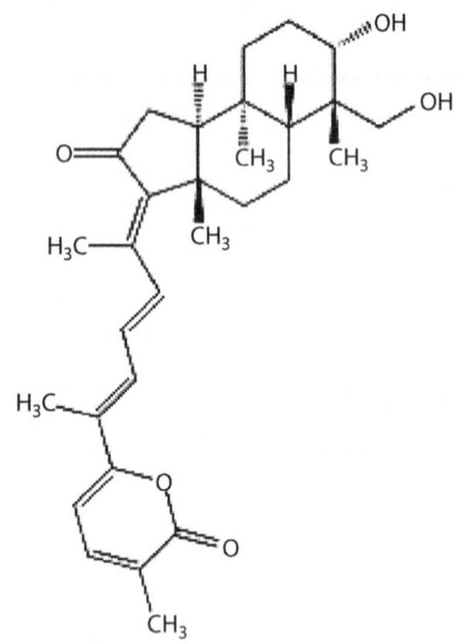

Rhabdastrellin A

Four isomalabaricane triterpenes—geoditin A, geoditin B, isogeoditin A, and isogeoditin B—have been isolated from this species. Among them, geoditin A was the most cytotoxic to HL60 cells (IC 50 Z3 mg/mL [<6.6 mM]), and geoditin B exhibited relatively weak cytotoxicity (Y. Li et al., 2013).

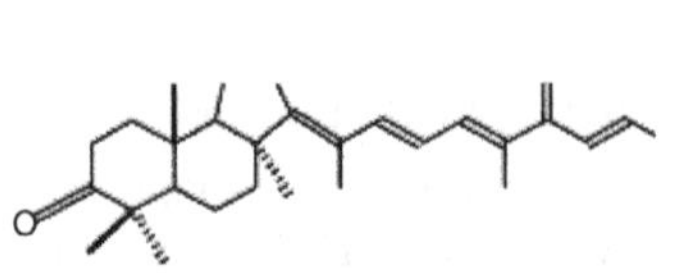

Geoditin A

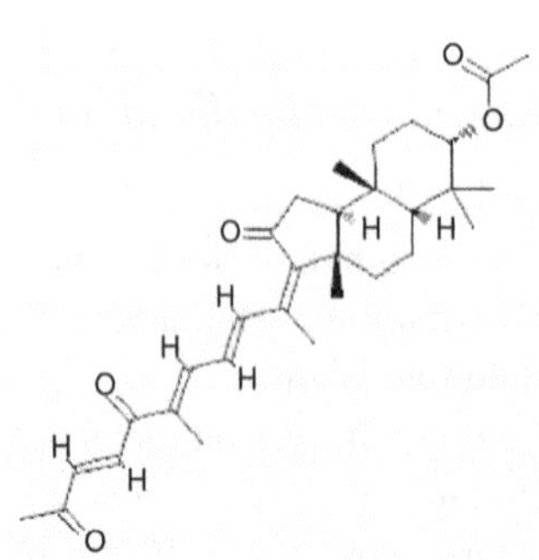

Geoditin B

Rhabdastrella globostellata (Carter, 1883) (= *Aurora globostellata; Stelletta globostellata; Jaspis carteri*)

Order: Tetractinellida Family: Ancorinidae

Common name: Yellow pot sponge

Distribution: Indian Ocean including the shores of Madagascar, the Seychelles, and Australia; Malayan Peninsula and Singapore; off South Africa

Ecology: It occurs along the shores, among coral rubble.

Biology: It is a smooth, soft, cheeselike sponge in life with an orange-brown outside and orange-yellow inside. Dry samples, however, display a dull grayish beige-brown ectosome, with dirty yellow or beige-brown choanosome. It is about 15 cm wide and is looking like a squat sphere with one or several shallow depressions at the top, so that the entire sponge resembles a shallow pot. The depression is smooth and often has many closely spaced holes. There are bumps on the outside of the "pot," and sometimes ridges are seen.

Compound(s) and Activities

Immunomodulatory effects: The organic extract of this secies has shown immunomodulatory effects (Perdicaris et al., 2013).

Cytotoxicity: Among the compounds, stellettins A–E, H, and I, and rhabdastrellic acid-A isolated from this sponge, compounds stellettin B and (-)-stellettin E showed selective cytotoxicity toward p21(WAF1/Cip1)-deficient human colon tumor (HCT-116) cells with IC50 values of 0.043 and 0.039 μM, respectively. Further, the compounds stelletin J and K of this species displayed varying levels of activity toward the A2780 ovarian cancer cell line. In addition, rhabdastrellic acid-A (11) showed potent inhibition against HL-60 cells, and it induced the apoptosis of HL-60 cells in M/G2 phase (Li et al., 2010; Y. Li et al., 2013; Tasdemir et al., 2002).

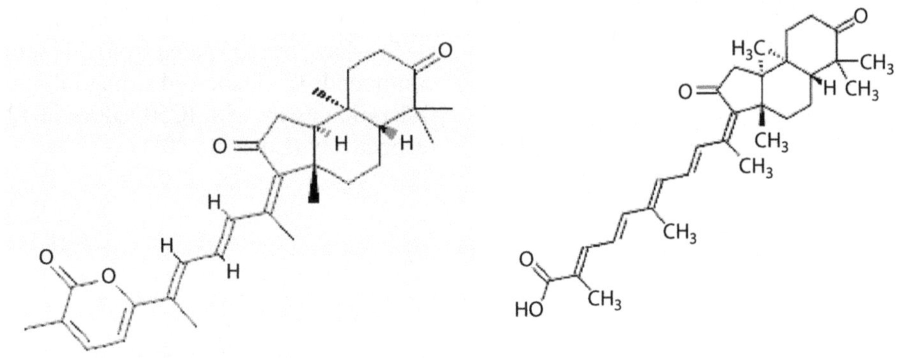

Stellettins B,E

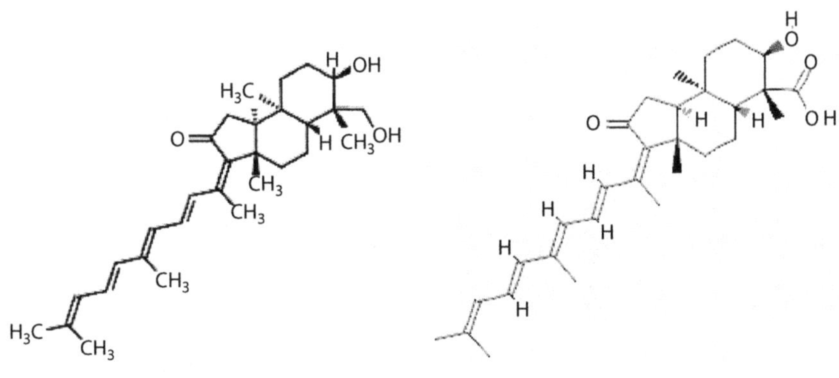

Stellettins J,K

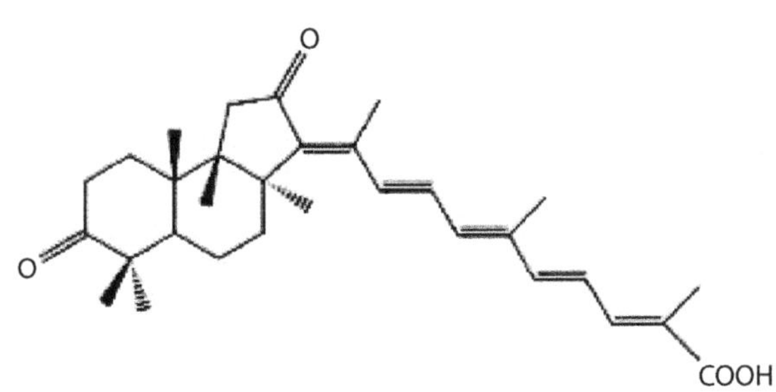

Rhabdastrellic acid-A

The compounds globostellatic acids A and D, stelliferin ribosides, and globostelletin isolated from this species showed cytotoxicity against three different cancer cell lines: L5178Y (mouse lymphoma), HeLa (human cervix carcinoma), and PC-12 (rat pheochromocytoma) (Fouad et al., 2006).

Anticandidal activity: The compounds bengamides A and B, G–J. and K derived from this species showed a remarkable anticandidal activity (García-Ruiz and Sarabia, 2014).

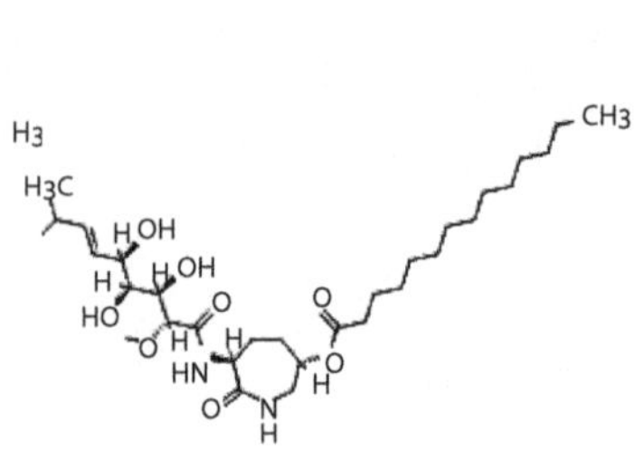

Bengamide A

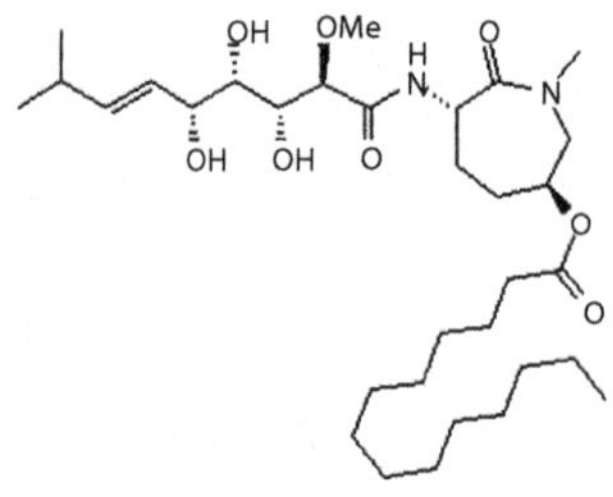

Bengamide B

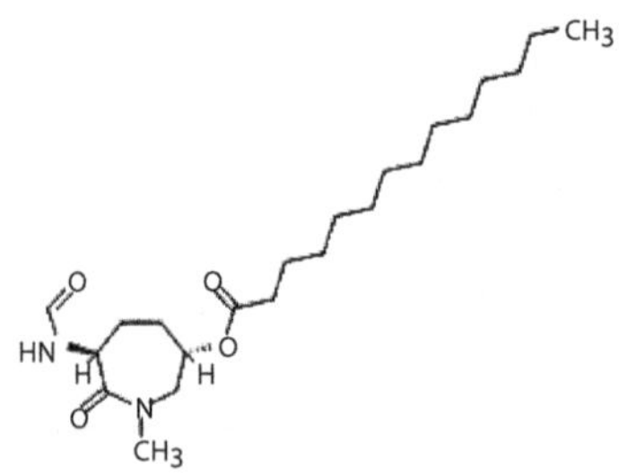

Bengamide K

Antimicrobial activity: The extracts of this species yielded the compounds bengamides A, B, and G–K. All these compounds have shown a remarkable anticandidal activity (García-Ruiz and Sarabia, 2014).

Stelletta clavosa (Ridley, 1884) (= *Myriastra clavosa*)

Order: Tetractinellida Family: Ancorinidae

Common name: Thick white prickly sponge

Distribution: From Ceylon to the waters North of Australia

Ecology: It occupies diverse marine habitats.

Biology: It is a kidney-shaped sponge of the size 1.5 × 0.8 cm. Surface is smooth. Osculum is present at the concave surface and is 1.5 mm in diameter. Consistency is compressible and color is light gray.

Compound(s) and Activities

Cytotoxicity: The new isotactic polymethoxydienes 1–5 isolated from this species exhibited moderate cytotoxicity (Rao and Faulkner, 2002).

The compounds clavosines A and B isolated from this species were found to be highly potent cytotoxins and inhibitors of type 1 and 2A serine/threonine protein phosphatases (Xiong et al., 1998).

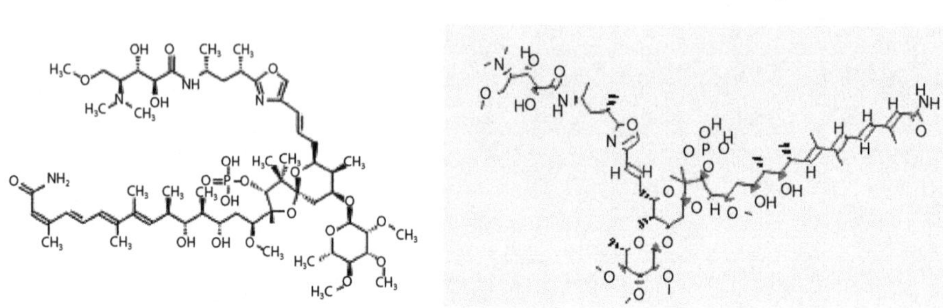

Clavosines A, B

Anti-HIV activity: Five depsipeptides—mirabamides C, E–H (1–5)—have been isolated from this sponge. Among them, the compounds (2–5) showed strong inhibition of HIV-1 in a neutralization assay with IC50 values of 121, 62, 68, and 41 nM, respectively (Lu et al., 2011).

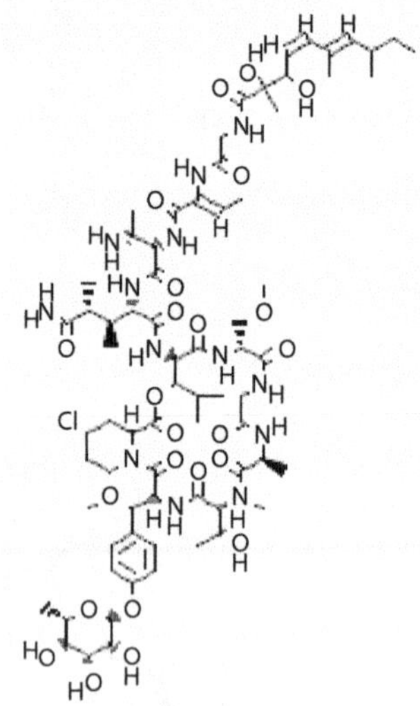

Mirabamide E

Stelletta tenuis (Lindgren, 1897)

Order: Tetractinellida Family: Ancorinidae

Common name: Not reported

Distribution: South China Sea, Southeast Asia, Java

Ecology: It is known from seamounts and knolls.

Biology: Not reported

Compound(s) and Activities

Antimicrobial activity: The associated bacterium, *Alcaligenes faecalis*, of this species produced L,L-diketopiperazine known as cyclo-(L-Pro-L-Phe), which showed moderate antimicrobial activity (Thomas et al., 2010).

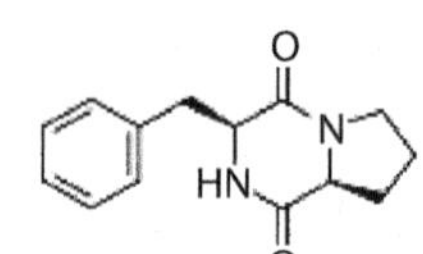

Cyclo-(L-Pro-L-Phe)

Cytotoxicity: The compounds stellettin A (1) and B (2) have been isolated from this species. Stellettin A was significantly toxic to P388 leukemia cells, exhibiting an ED50 value of 0.001 μg/mL. On the other hand, stellettin A and stellettin B induced cytotoxicity in HL-60 cells treated for 24 h at 3 μM concentration of the extract of this species (Y. Li et al., 2013). Three new isomalabaricane-type triterpenes named stellettins N, O, and P isolated from this species showed cytotoxicity inhibitory activities against the human cancer cell line AGS with IC50 values of 4.52, 9.61, and 7.44 μM, respectively (Li et al., 2015).

The isomalabaricane-type triterpenes stellettins L and M exhibited significant cytotoxic activity against stomach cancer (AGS) *in vitro* (Lin et al., 2007).

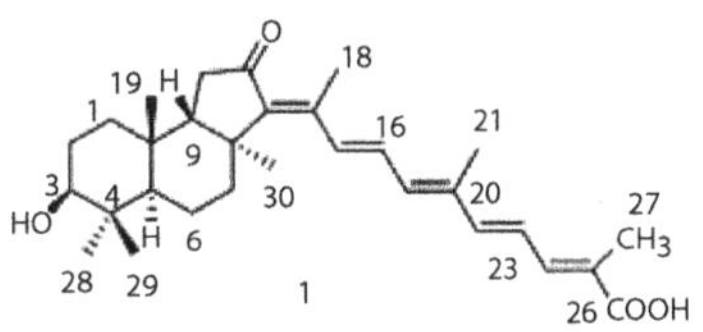

1. stellettin L

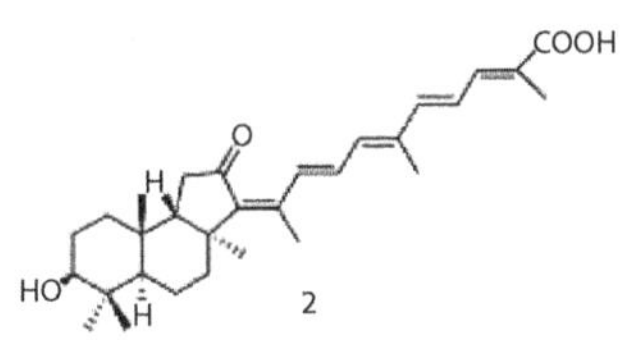

2. stellettin M

Stelletta sp.

Order: Tetractinellida Family: Ancorinidae

Compound(s) and Activities

Cytotoxicity: The acetylenic metabolites (1–3) isolated from this species displayed weak cytotoxicity against the human leukemia cell line K562 with LC50 values of 44, 51, and 63 μg/mL, respectively. (Dembitsky et al., 2006).

Stryphnus fortis (Vosmaer, 1885)

Order: Tetractinellida Family: Ancorinidae

Common name: Arctic sponge, boreal sponge

Distribution: Norway, Arctic

Ecology: It is a deepwater species inhabiting depths of 60 to 1400 m.

Biology: It is a globular-massive or lumpy size growing up to 15 cm or more in diameter and height. Surface is extremely rough and hispid, with spicules sticking out far beyond the surface, but otherwise even. Consistency is hard and crumbly. Color of the animal is brownish red on the outside and ocher yellow on the inside. Megascleres of spicules are oxeas. Microscleres are amphiasters that are short irregular rays and almost spiraster-like.

Compound(s) and Activities

Cytotoxic activity: The compound ianthelline isolated from its organic extract showed cytotoxity (through inhibition of mitotic spindle formation and inhibition of protein kinase activity) against the human melanoma cell line A2058 in a dose- and time-dependent manner. The IC50 values recorded from 24, 48, and 72 h exposure were 125, 108, and 91.6 μM, respectively (Hanssen et al., 2012).

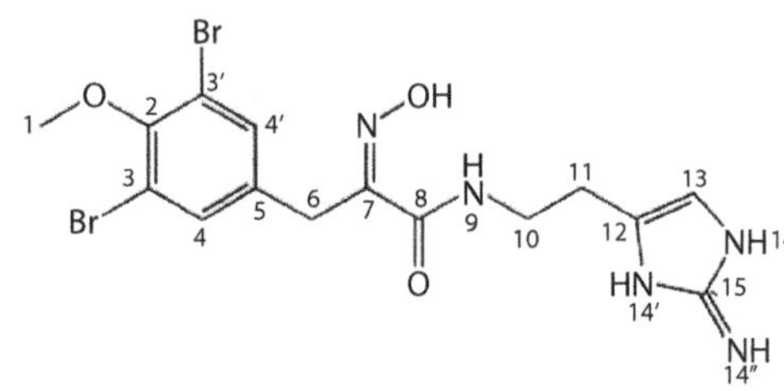

Ianthelline

Enzyme inhibitory activity: The secondary metabolite stryphnusin isolated from this sponge displayed acetylcholinesterase inhibitory activity. Similarly, its synthetic phenethylamine analog was the strongest competitive inhibitor of both acetylcholinesterase and butyrylcholinesterase with IC50 values of 57 and 20 μM, respectively (Moodie et al., 2016).

FAMILY: CORALLISTIDAE

Neoschrammeniella fulvodesmus (Lévi and Lévi, 1983) (= *Corallistes fulvodesmus*)

Order: Tetractinellida Family: Corallistidae

Common name: Rock sponge

Distribution: Western Central Pacific, New Caledonia, and New Zealand

Ecology: This sessile species inhabits deep seamounts; dredged from hard rocky substrata; depth range 180 to 1680 m; temperature 5°C to 15°C.

Biology: This sponge mostly had coral into their bases. No other information is available.

Compound(s) and Activities

Cytotoxic activity: The chloromethylenic extract of this species was cytotoxic againt KI3 and P388 cells (DI50, 10 pg) (Debitus et al., 1989).

Neoschrammeniella sp. (= *Corallistes* sp.)

Order: Tetractinellida Family: Corallistidae

Compound(s) and Activities

Cytotoxicity: The crude dichloromethanol extract shows cytotoxic activities against the KB nasopharyngeal carcinoma cell line with an IC50 of 10 μg/mL (Motuhi et al., 2016).

FAMILY: GEODIIDAE

Caminus sphaeroconia (Sollas, 1886)

Order: Tetractinellida Family: Geodiidae
Common name: Not designated
Distribution: Eastern Brazil
Ecology: It is a shallow water species.
Biology: It is a massive sponge with massive vertical lobes. The oscules are locatd on the summits of lobes. Pores are in sieves. Color of the animal is purplish when exposed to the light and yellowish below.

Compound(s) and Activities

Antibacterial activity: The glycolipid compound caminoside-A isolated from this species showed antibacterial activity against *S. aureus* with an MIC value of 16 μg/mL (Anjum et al., 2016; Jha and Zi-rong, 2004).

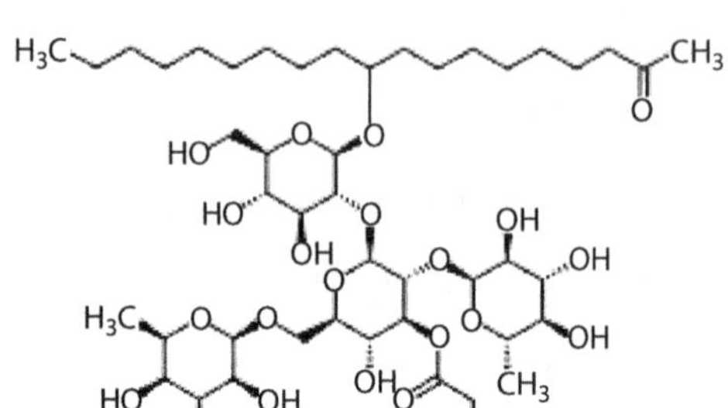

Caminoside-A

Antimicrobial activity: The antimicrobial glycolipid caminoside B isolated from this species serves as a small molecule inhibitor (IC50 = 20 μM) targeting the type III secretory pathway of bacterial pathogenesis (Zhang et al., 2010).
Linington et al. (2006) reported on the isolation of antimicrobial glycolipids caminosides B–D.

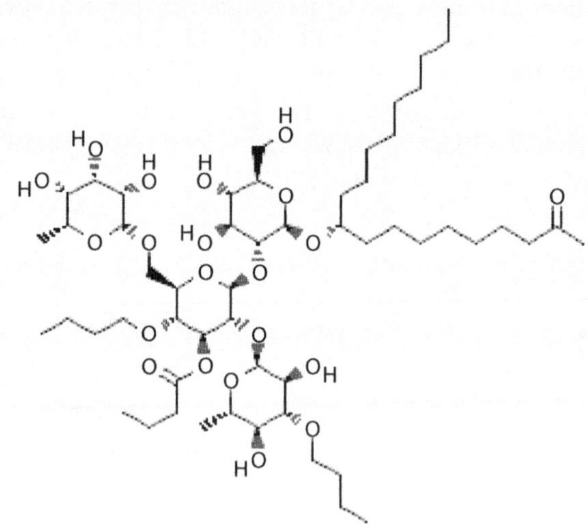

Caminoside D

Erylus deficiens (Topsent, 1927)

Order: Tetractinellida Family: Geodiidae
Common name: Not designated
Distribution: Azores, Canaries, Madeira, North Atlantic Ocean, Western Mediterranean
Ecology: It is associated with deep coral bank and maerl habitats.
Biology: Not reported

Compound(s) and Activities

Antibacterial activity: The organic extract of the bacterial isolates associated with this sponge species showed antibacterial activity against *Bacillus subtilis* (inhibition, 87.9%), *Staphylococcus aureus* MRSA (10.6%), *Aliivibrio fischeri* (9.1%), and *Vibrio harveyi* (6.1%). Among the different groups of associated bacteria, Firmicutes were found to be the most bioactive (61.5%) followed by Gammaproteobacteria (40%), Alphaproteobacteria (26.7%), and Actinobacteria (23.5%) (Graca et al., 2015).
Enzyme-inhibiting activity: The dichloromethane extract of this species led to the isolation of four new glycolipids, erylusamides A–D, which showed inhibitory activity against of indoleamine 2,3-dioxygenase (IDO), which is an immunosuppressive enzyme capable of inhibiting a destructive maternal T cell response against allogeneic fetuses (Gaspar et al., 2016).

Erylus discophorus (Schmidt, 1862)

Order: Tetractinellida Family: Geodiidae
Common name: Not known

Distribution: Adriatic Sea, Aegean Sea, Azores, Canaries, Madeira, Ionian Sea, Levantine Sea, South European Atlantic Shelf, Western Mediterranean

Ecology: It occurs on rocks at the low water mark and in shallow sublittoral areas.

Biology: This sponge is a very irregular, flattened, tuberose mass. It surface is smooth. Small oscules are seen. Consistency is cheesy, incompressible, and is easily damaged. Color of the sponge is black or dark brown externally and lighter inside.

Compound(s) and Activities

Antimicrobial activity: A total of 212 heterotrophic bacteria belonging to Alpha- and Gammaproteobacteria, Actinobacteria, and Firmicutes were isolated from this sponge species. The three most bioactive genera were *Pseudovibrio* (47.0%), *Vibrio* (22.7%), and *Bacillus* (7.6%). Other less bioactive genera were *Labrenzia*, *Acinetobacter*, *Microbulbifer*, *Pseudomonas*, *Gordonia*, *Microbacterium*, *Micrococcus*, *Mycobacterium*, *Paenibacillus*, and *Staphylococcus*. Of the total isolates, 31% produced antimicrobial metabolites, which showed antibacterial activity against *B. subtilis* and some bioactivity against *S. aureus* MRSA and *V. harveyi* (Graça et al., 2013).

Antiviral activity: The polysaccharides extracted from this species showed high activity against HIV-1 (90% to 95% inhibition) (Esteves et al., 2011).

Erylus formosus (Sollas, 1886)

Order: Tetractinellida Family: Geodiidae

Common name: Not designated

Distribution: Bahamas, Belize, Venezuela

Ecology: It grows exclusively in reef habitats, under 5 m deep.

Biology: It is massive, amorphous to regularly, and lobate. Color of the sponge is black to purplish gray externally and tan internally. The sponge skin is very smooth and it wrinkles when lifted out of the water. Consistency is compressible, dense, and is easily torn. Oscules are seen on the top of the lobes (diameter 0.5 cm). A tangential section at the surface shows the presence of numerous ostia (inhalant pores), which are regularly scattered on the surface. The surface is formed by a well-differentiated cortex containing strictly aspidasters, micro-oxeas, tylasters, and tangential orthotrianes.

Compound(s) and Activities

Cytotoxic activity: The triterpene glycosides erylosides F1–F4 and eryloside F have been isolated from this species. All these compounds showed cytotoxicity and induced apoptosis of Ehrlich carcinoma cells. The compound eryloside F3, however, has shown the highest activity at a concentration of 100 μg/mL (Y. Li et al., 2013).

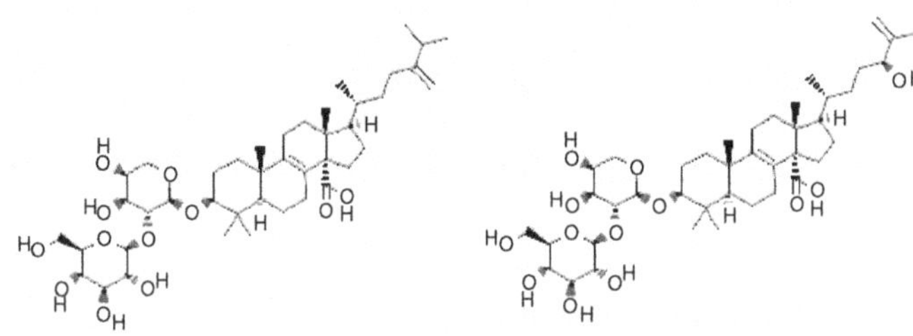

Erylosides F1, F3

Thrombin receptor antagonist: The compound eryloside F of this speices is a potent thrombin receptor antagonist. Thrombin receptor activation plays a key role in arterial thrombosis and atherosclerosis. Atherosclerosis and associated deposition of fats, cholesterol, and other substances in the artery wall would lead to shrinkage of the arterial diameter leading to heart diseases (Anjum et al., 2016; Senthilkumar and Kim, 2013; Sipkema et al., 2005).

Erylus lendenfeldi (Sollas, 1888)

Order: Tetractinellida Family: Geodiidae

Common name: Not designated

Distribution: Red Sea, Indian Ocean, Australia, Kenya, Madagascar

Ecology: Not reported

Biology: Sponge of this species is thickly encrusting on molluscan shell mass and is irregular in shape and hard. It is 4 cm long, 3.5 cm wide, and 2 cm high. Surface is even. Color of the animal is pale yellow in spirit. Oscules are found scattered (1–1.5 mm dia.), and evenly distributed over surface.

Compound(s) and Activities

Antitumor, antibacterial, and antifungal activities: Fouad et al. (2004) reported on the isolation of erylosides A, G–J, K, and L from this sponge species. Among these, eryloside A displayed mild antibiotic and antifungal activity against *E. coli*,

B. subtilis, and *C. albicans*. The zones of inhibition at 10 μg per disc for these species were 6, 7, and 7 mm, respectively. Further, this compound showed antitumor activity against P-388 cells with an IC50 = 4.2 μg/mL. Erylosides G–J, however, displayed moderate cytotoxicity against the human leukemia cell line K562.

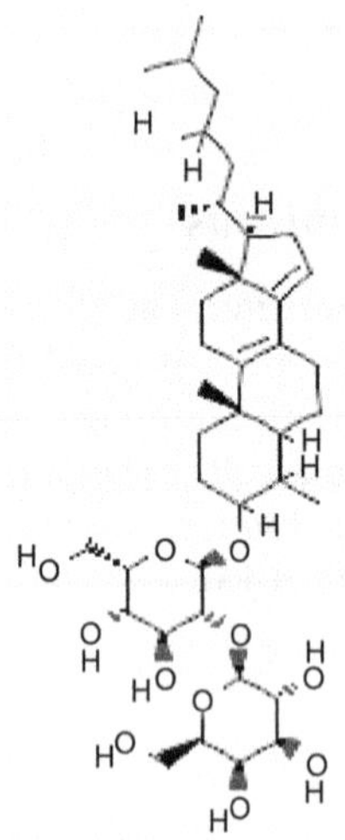

Eryloside A

Erylus nobilis (Thiele, 1900)

Order: Tetractinellida Family: Geodiidae
Common name: Not available
Distribution: Indonesia
Ecology: Not kown
Biology: Not reported

Compound(s) and Activities

Cytotoxicty: The compounds erylosides G–J isolated from this species exhibited moderate cytotoxicty against a human leukemia cell line (Shin et al., 2001).

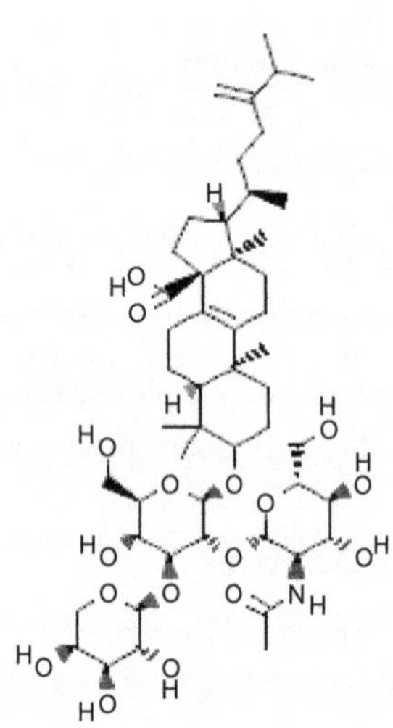

Eryloside G

Antiviral activity: A neuraminidase inhibitor, nobiloside, has been isolated from this species. This compound inhibited neuraminidase from the bacterium *Clostridium perfringens* with an IC50 value of 0.5 μg/mL (Takada et al., 2002). Neuraminidase inhibitors are drugs that block the function of the viral neuraminidase protein. By blocking this protein enzyme, it stops the release of viruses from the infected host cell and prevents new host cells from being infected. These antiviral agents inhibit all subtypes of neuraminidase enzymes and are effective against influenza viruses A and B.

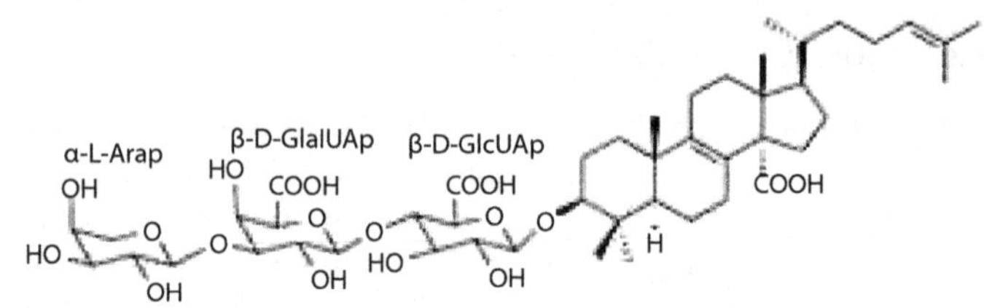

Nobiloside

Anticancer activity: The dichloromethane extract of this species led to the isolation of four new glycolipids, namely, erylusamides A–D, which serve as IDO inhibitors. The IDO inhibition can delay tumor growth, enhance dendritic cell vaccines, and synergize with chemotherapy through an immune-mediated mechanism (Gaspar et al., 2016).

Erylus placenta (Thiele, 1898)

Order: Tetractinellida Family: Geodiidae
Common name: Not known
Distribution: Japan
Ecology: Not available
Biology: Not reported

Compound(s) and Activities

Anti-inflammatory activity: The compound erylusamine B isolated from this species serves as an IL-6 antagonist. Anti-interleukin-6 has inhibitory activity against inflammatory diseases and many cancers (Fusetani et al., 1993).
Keyzers (2005) reported on the isolation of alkaloids, erylusamines B–E which showed anti-inflammatory activity by inhibiting the binding of IL-6 to its receptor with IC50 values of 66, 33, 37, and 17 μg/mL, respectively.
Anticancer activity: The steroid glycosides, sokodosides A and B, isolated from this species have shown inhibitory activities against several strains of yeast and a cancer cell line (Okada et al., 2006).

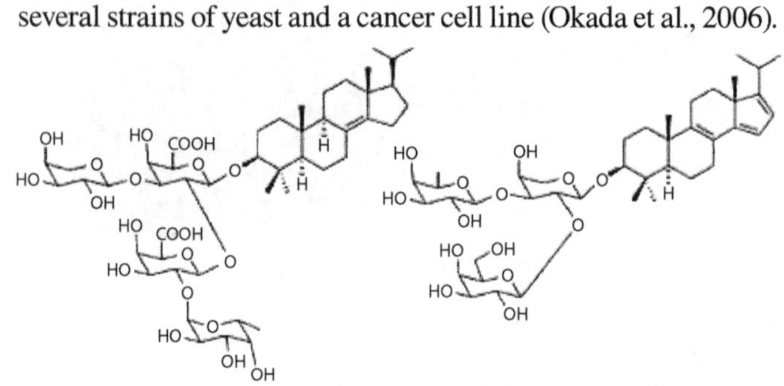

Sokodosides A and B

Geodia baretti (Bowerbank, 1858)

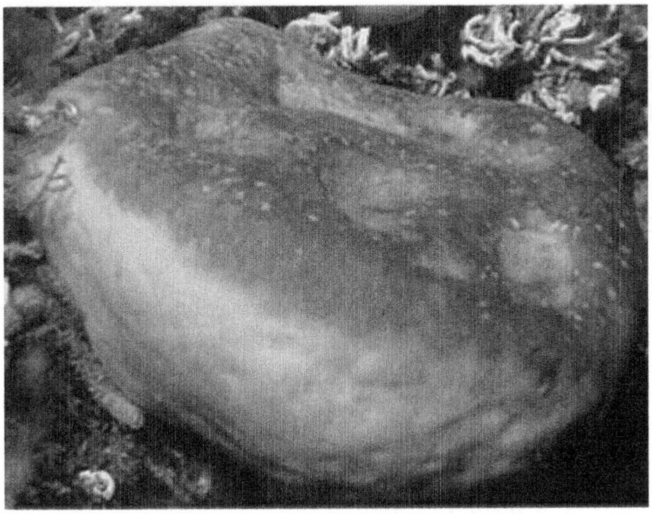

Order: Tetractinellida Family: Geodiidae

Common name: Not known

Distribution: Northeast Atlantic, Mediterranean Sea, North Atlantic Ocean, Norway, Sweden

Ecology: It inhabits deeper water, 15 to 330 m.

Biology: This sponge is globular and larger specimens become irregular with one or more "sunken" oscules, which are only cavities in which grouped oscules are found. Size may be up to 50 cm or more. Surface is smooth and slightly undulate. Consistency is hard. Color of the sponge is cream yellow or pale yellow white.

Reproduction: It is oviparous and a simultaneous spawner. Oogenesis normally lasts for 4 months and spermatogenesis lasts for 2.5 months. During spawning, both oocyte and spermatozoa leave the sponge through an excurrent canal. Spawning occurs in May/June and October.

Compound(s) and Activities

Anticancer activity: The two bioactive low molecular weight compounds isolated from the organic extract of this species showed anticancer activity against melanoma cells (A2058) and normal lung fibroblasts (MRC-5) (Olsen et al., n.d.).

Antibacterial activity: The two bioactive low molecular weight compounds isolated from the organic extract of this species showed antibacterial activity against *E. faecalis, E. coli, P. aeruginosa, S. aureus, and Streptococcus* sp. (Olsen et al., n.d.).

Antioxidant activity: The two bioactive low molecular weight compounds isolated from the organic extract of this species showed antioxidant activity at concentrations of 1 to 100 μM of the extract (Olsen et al., n.d.).

Antidiabetic activity: The two bioactive low molecular weight compounds isolated from the organic extract of this species showed antidiabetic activity at concentrations of 10 to 100 μM of the extract (Olsen et al., n.d.).

Acetylcholinesterase inhibitory activity: The compounds barettin, 8,9-dihydrobarettin, and bromoconicamin have been isolated from this species. Among them, barettin and 8,9-dihydrobarettin displayed significant inhibition of the acetylcholinesterase enzyme, with inhibition constants (Ki) of 29 and 19 μM, respectively (Olsen et al., 2016b).

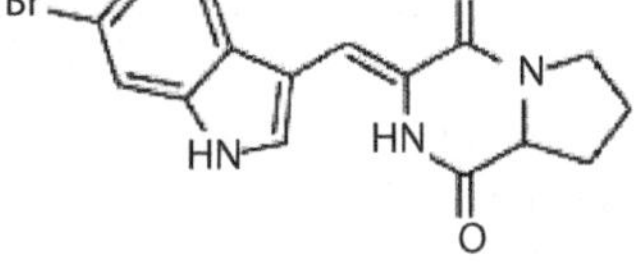

Barettin

Geodia corticostylifera (Hajdu, Muricy, Custodio, Russo, and Peixinho, 1992)

Order: Tetractinellida Family: Geodiidae

Common name: Not designated

Distribution: Southwestern Atlantic; Southeastern Brazilian coast and Venezuela

Ecology: These sponges can be almost totally buried in the sediment at photophilous habitats or be hanging from the roofs of small caves. Sometimes great numbers of ophiuroids (brittle stars), can be found in the calyciform grooves of the surface of this species. It is known to occur from 3 to 40 m depths.

Biology: These sponges are massive-globose, cerebriform and are normally bearing deep grooves at the surface. The largest specimen may cover an area of more than 60 cm^2 and 6 cm high. Consistency is firm. Color of the sponge is orange externally and beige internally. The surface is minutely hispid, with or without abundant wrinkles. Oscula are approximately 1 mm in diameter and are abundant inside some of the calyciform-like grooves of the surface. The spicules of this species are extremely slender or variably robust. They are bent near the middle.

Compound(s) and Activities

Cytotoxic activity: The hydroethanolic extracts of this species showed cytotoxicity against tumor cell lines. The IC50 value recorded for the all the tested cell lines, namely, HL-60, HCT-8, MDA-MB435, and SF-295 was 1.6 μg/mL (Ferreira et al., 2007).

Antiproliferative activity: The compound geodiamolide H isolated from this species showed demonstrated antiproliferative activity against breast cancer cells (Suarez-Jimenez et al., 2012).

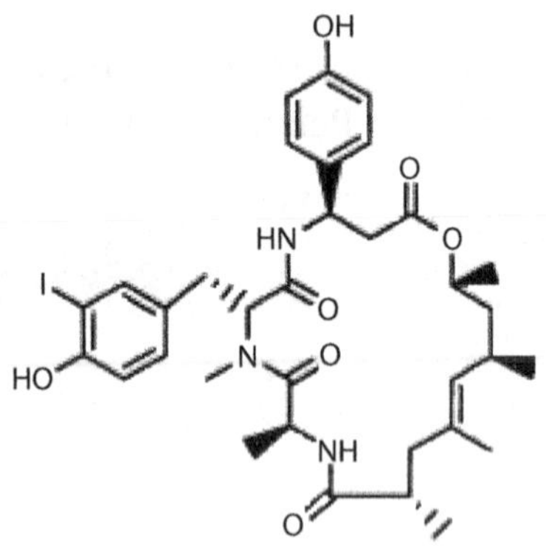

Geodiamolide H

Geodia cydonium (Linnaeus, 1767)

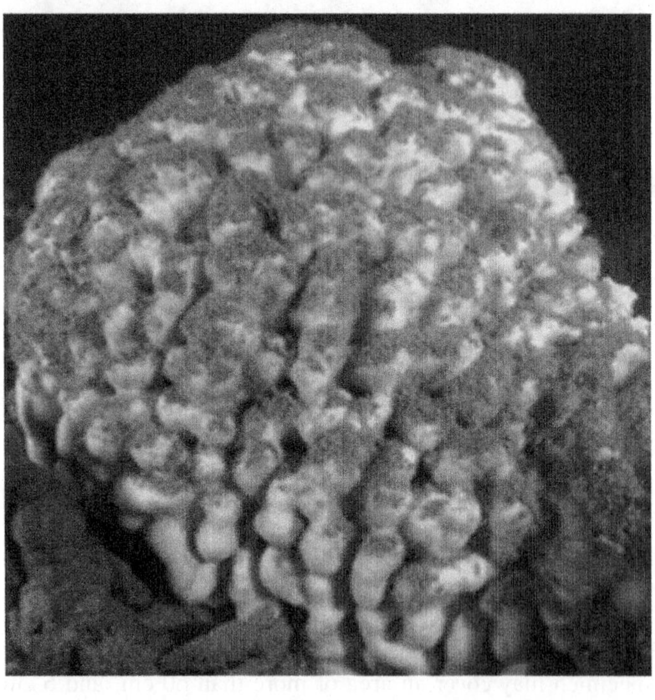

Order: Tetractinellida Family: Geodiidae

Common name: Orange ball sponge

Distribution: North Atlantic Ocean, British Isles, Roscoff, Norway, Mediterranean

Ecology: It inhabits pebble bottoms at depths of 80 to 90 m

Biology: This sponge is subspherical and is somewhat flattened with a diameter 8 to 15 cm. Surface is irregular and is generally hispid; but here and there it may be smooth and encrusted by shells and pebbles. Oscules (1.5–2 mm in dia.) and ostia are grouped in cribral areas. Consistency is hard but fragile. Color of the sponge is white gray (exterior) and orange (interior). Polychaete worms may be associated to this sponge.

Compound(s) and Activities

Anti-inflammatory effect: The methanol extract of this species showed anti-inflammatory effect against human breast cancer cells (MCF-7) by inducing a decrease in levels of VEGF and five proinflammatory cytokines (CCL2, CXCL8, CXCL10, IFN-γ, and TNF-α) in a dose-dependent manner (Costantini et al., 2015).

Geodia japonica (Sollas, 1888)

Order: Tetractinellida Family: Geodiidae

Common name: Not designated

Distribution: Philippines, Japan

Ecology: Not reported

Biology: Not studied

Compound(s) and Activities

Cytotoxicity: The isomalabaricane triterpenes, geoditin A and geoditin B, isolated from this species showed cytotoxicity of tumor cell lines. While geoditin A was the most cytotoxic to HL60 cells (IC 50, <6.6 μM), geoditin B exhibited relatively weak cytotoxicity (Y. Li et al., 2013).

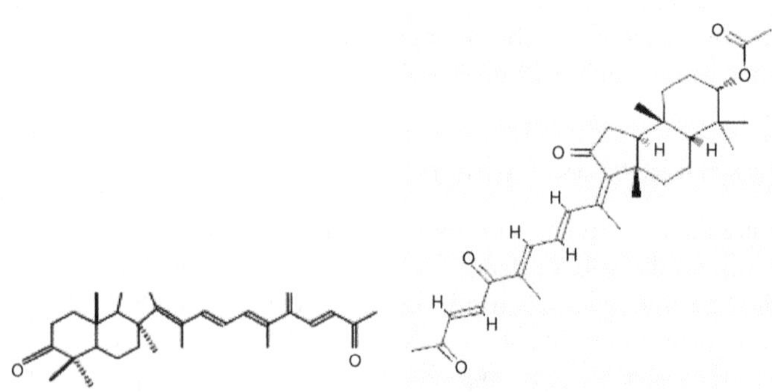

Geoditins A, B

Four isomalabaricane triterpenes, namely geoditin A, geoditin B, stellettin A, and stellettin B, were isolated from this species. Among these compounds, geoditin A was the most cytotoxic to HL60 cells (IC50 =3 μg/mL [<6.6 μM]), followed by stellettins A and B. Geoditin B, however, exhibited relatively weak cytotoxicity. At extract concentrations of 1.25 to 25 μg/mL, when the HL60 cells were exposed to geoditin A, a dose-dependent increase of reactive oxygen species, a progressive dissipation of mitochondrial membrane potential, and an increase in annexin V-FITC binding were noted (Liu et al., 2005; Zhang et al., 2001).

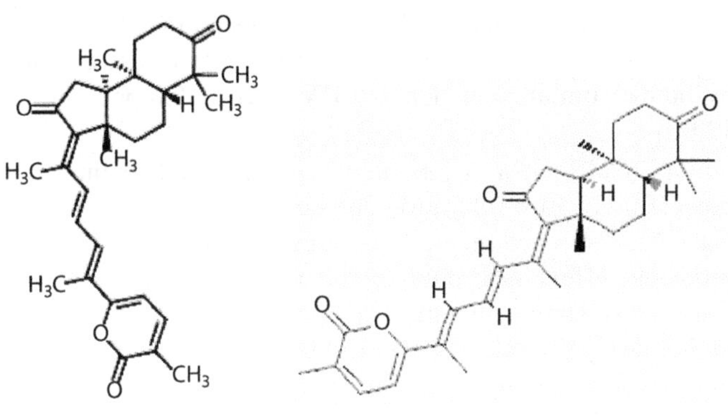

Stellettins A, B

Anticancer/cytotoxic activity: The compound stellettin A an isomalabaricane triterpene isolated from this species was found to inhibit the growth of melanoma cells by induction of abnormal protein glycosylation, endoplasmic reticulum (ER) stress, and cell death. Stellettin A was also cytotoxic to mouse melanoma B16F10 cells, with IC50 = 0.15 μg/mL (Liu et al., 2013).

Geodia microspinosa (Wilson, 1925) (= *Sidonops microspinosa*)

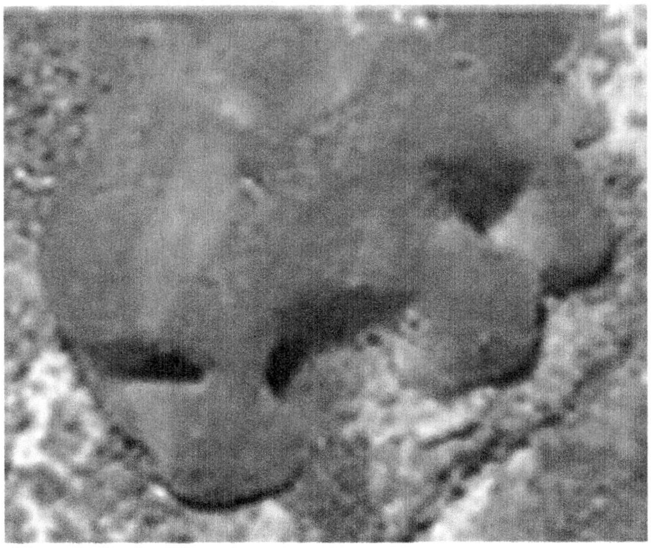

Order: Tetractinellida Family: Geodiidae
Common name: Not known
Distribution: Eastern Philippines
Ecology: Not reported
Biology: Not available

Compound(s) and Activities

Antiviral activity: The cyclic depsipeptide compound microspinosamide isolated from the extracts of this species inhibited the cytopathic effect of HIV-1 infection with an EC50 value of approximately 0.2 μg/mL (Joseph et al., 2012; Rashid et al., 2001).

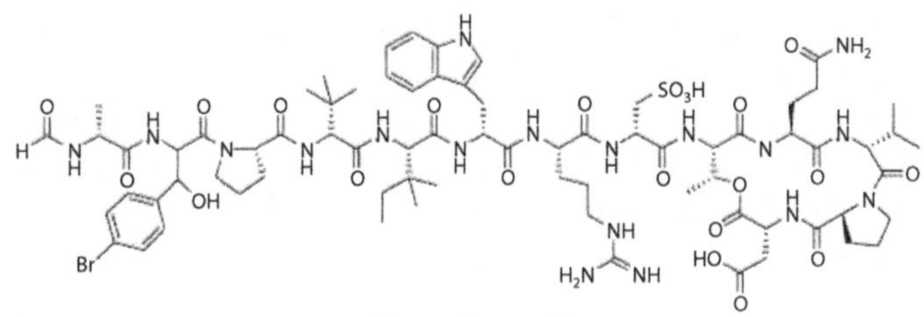

Microspinosamide

Melophlus sp.

Order: Tetractinellida Family: Geodiidae

Compound(s) and Activities

Antifungal activity: The compound aurantoside K of this species displayed broad antifungal activity toward *Candida albicans* (wild type ATCC 32354 and amphotericin-resistant ATCC 90873; MIC 31.25 and 1.95 μg/mL, respectively) along with *Cryptococcus neoformans*, *Aspergillus niger*, *Penicillium sp.*, *Rhizopus sporangia*, and *Sordaria* sp. (Cheung et al., 2014; Gribble, 2015).
Anticancer activity: The compounds papuamides E and F of this species showed anticancer activity in brine shrimp with IC50 values of 92 and 106 μg/mL, respectively (Agrawal et al., 2016).

Penares sp.

Ordr: Tetractinellida Family: Geodiidae

Compound(s) and Activities

Neurosuppressives and muscle relaxant activity: The azetidine alkaloid penaresidin A isolated from this species showed neurosuppressive and muscle relaxant activity by inhibiting the enzyme actomyosin ATPase (Sipkema et al., 2005).

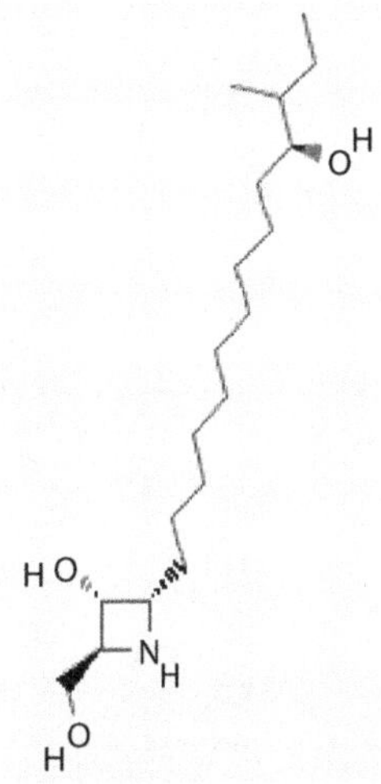

Penaresidin A

Enzyme inhibitory activity: The compounds azetadins isolated from this species have shown protein kinase C (PKC) inhibitory activities. The PKC has a greater role in diabetic nephropathy (Sepčić et al., 2010).

Antitumor activity: An unidentified alkaloid isolated from this species was moderately cytotoxic to the human tumor cell lines HL-60 (lung) and HeLa (cervix), with IC50 values of 16 and 33 μM, respectively (Gribble, 2015).

FAMILY: NEOPELTIDAE

Callipelta sp.

Order: Tetractinellida Family: Neopeltidae

Compound(s) and Activities

Cytotoxicity: The compound callipeltoside A isolated from this sponge exhibited cytotoxic activity against NSCLC-N6 human bronchopulmonary non-small cell lung carcinoma and P388 cell lines. On the other hand, the callipeltosides A, B, and C were found to be moderately cytotoxic against NSCLC-N6 cells with IC50 values of 10, 15, and 30 μg/mL, respectively (Dembitsky et al., 2006).

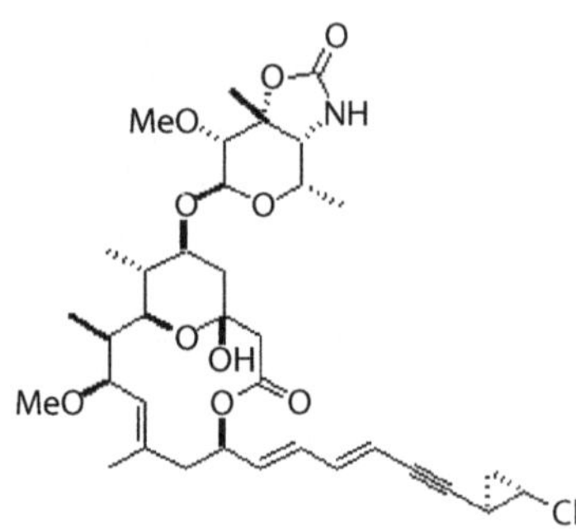

Callipeltoside A

Callipeltins A–C isolated from this species have been found to be cytotoxic against various human carcinoma cells (D'Auria et al., 1996).

Antiviral activity: The cyclic depsidecapeptide callipeltin A isolated from this speices was found to protect cells infected by human immunodeficiency (HIV) virus (Zampella et al., 1996).

Others: Callipeltin A has also been reported to induce a powerful (IC50 = 0.85 μM) and selective inhibition of the Na(+)/Ca(2+) exchanger. It is suggested that the positive inotropic effect is linked to the inhibition of the Na(+)/Ca(2+) exchanger and that callipeltin A may be a useful tool to study the role of the cardiac Na(+)/Ca(2+) exchanger in physiological and pathological conditions (Trevisi et al., 2000).

Homophymia lamellosa (Vacelet and Vasseur, 1971)

Order: Tetractinellida Family: Neopeltidae

Common name: Not known

Distribution: Western and Northern Madagascar

Ecology: It inhabits the continental shelf regions.

Biology: It is a stipitate, clavate, or lamellate. The ectosomal megascleres are pseudotriaenes, modified desmas that have the appearance and disposition of a triaene. Monaxonal megascleres are strongyloxea-like with a central canal that dilates in the distal third of each end rendering the spicule syringelike. Microscleres are spined amphiasters.

Compound(s) and Activities

Cytotoxic activity: Two new cyclic depsipeptides, pipecolidepsins A and B, isolated from this species displayed cytotoxic activity against a panel of different human tumor cell lines (Coello et al., 2014).

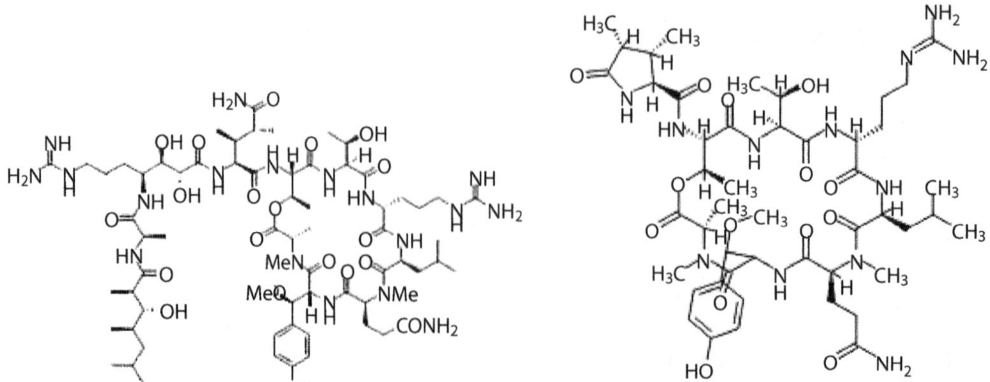

Callipeltins A,B

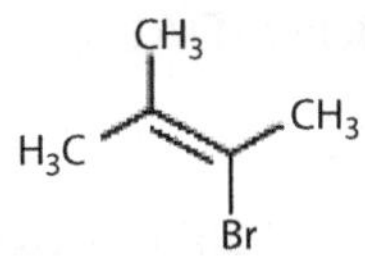

Pipecolidepsin A

Homophymia sp.

Order: Tetractinellida Family: Neopeltidae

Compound(s) and Activities

Antimicrobial, antiviral, antimalarial, and cytotoxic activities: The associated bacterium *Pseudomonas* species of this unidentified sponge species produced three compounds, namely, 2-undecyl-4-quinolone (1) and 2-undecen-1′-yl-4-quinolone (2) and 2-nonyl-4-hydroxyquinoline N-oxide (3). Compound 1 was found active against the malarial parasite *Plasmodium falciparum* and HIV-1 with IC50 values of 1 µg/mL and 0.01 µg, respectively. The compound 2 showed mild toxicity (Indraningrat et al., 2016; Thomas et al., 2010).

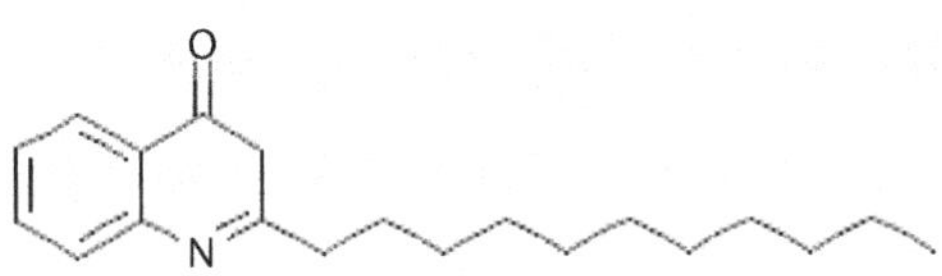

2-undecyl-4-quinolone

Anticancer activity: Nine new cyclodepsipeptides, homophymines B–E and A1–E1, isolated from this sponge have shown very potent cytotoxic activity against several human cancer cell lines with IC50 values in the nanomolar range and moderate selectivity against human prostate (PC3) and ovarian (OV3) carcinoma. Homophymines A1–E1 exerted stronger potency than the corresponding A–E compounds (Suarez-Jimenez et al., 2012).

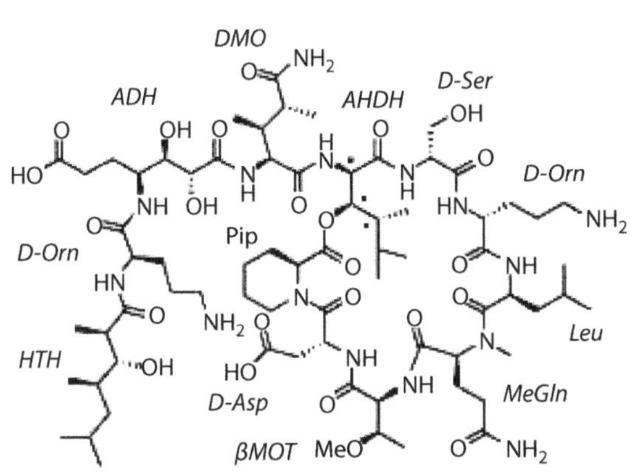

Homophymine B

Antiviral activity: The cyclodepsipeptide compound homophymine A of this sponge also exhibited cytoprotective activity against HIV-1 infection with an IC50 value of 75 nM (Zampella et al., 2008).

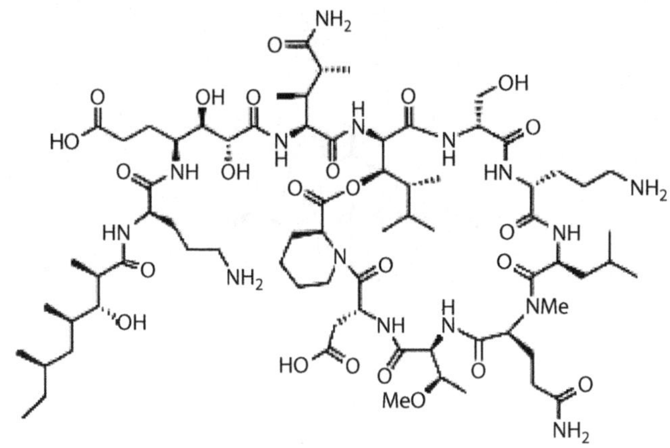

Homophymine A

Indraningrat et al. (2016) reported that the compound 2-undecyl-4-quinolone produced by its associated protobacterium *Pseudomonas* sp. 1531-E7 displayed activity against HIV-1 with an IC50 value of 10^{-3} µg/mL.

Antiprotozoal activity: The associated bacterial strain *Pseudomonas* sp. 1531-E7 of this species yielded the compound 2-undecyl-4-quinolone, which displayed antiprotozoal activity against *Plasmodium falciparum* with an IC50 value of 1 µg/mL (Indraningrat et al., 2016).

FAMILY: PACHASTRELLIDAE

Brachiaster sp.

Order: Astrophorida Family: Pachastrellidae

Compound(s) and Activities

Antitubercular and cytotoxic activities: The compound 25-dehydroxy-12-epi-scalarin isolated from this unidentified species has shown both antitubercular (MIC = 117 µM) and cytotoxic activities (human breast adenocarcinoma MCF-7, IC50 = 5.2 µM; colorectal carcinoma HT-29, IC50 = 2.3 µM; human cervical carcinoma HeLa, IC50 = 5.5 µM); and human oral epidermoid carcinoma KB, IC50 = 3.0 µM) (González, 2010).

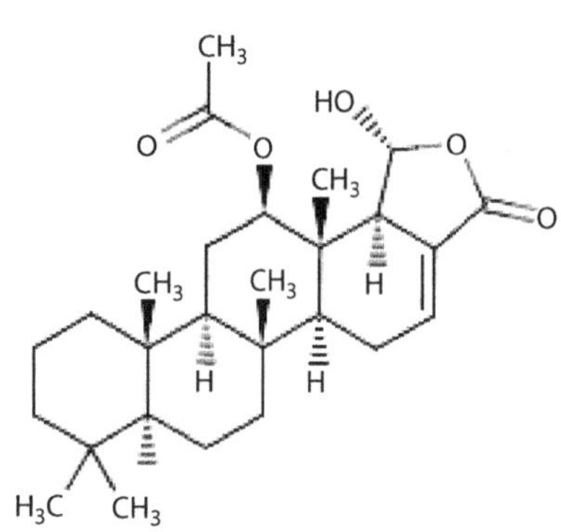

12-epi-scalarin

The compound heteronemin acetate of this species showed cytotoxicity against human breast adenocarcinoma MCF-7 cells (IC50 = 6.45 µM) and antitubercular activity (MIC = 6.0 µM) (González, 2010).

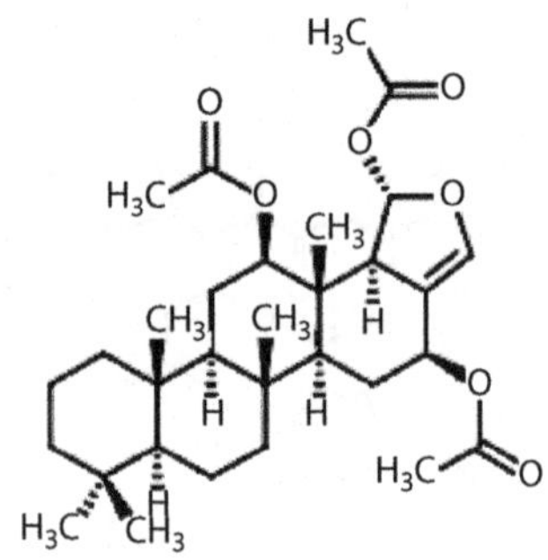

Heteronemin acetate

Antitubercular activity: The sesterpene compounds 12-deacetoxyscalarin 19-acetate and manoalide 25-acetate isolated from this sponge showed antitubercular activity against *M. tuberculosis* (H37Rv) with MIC values of 4 and 7 µg/mL, respectively (González, 2010; Jyothi et al., 2016).

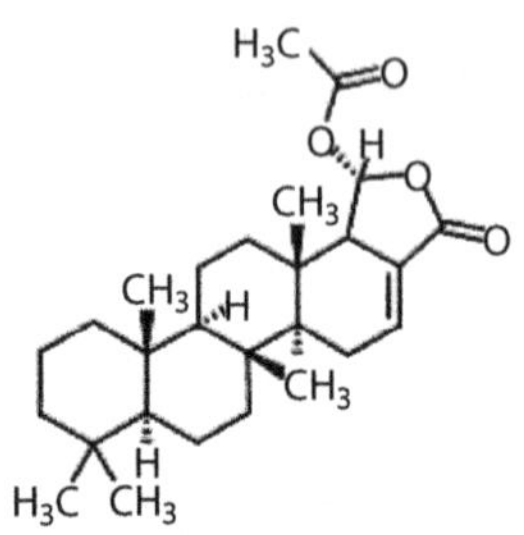

12-deacetoxyscalarin 19-acetate

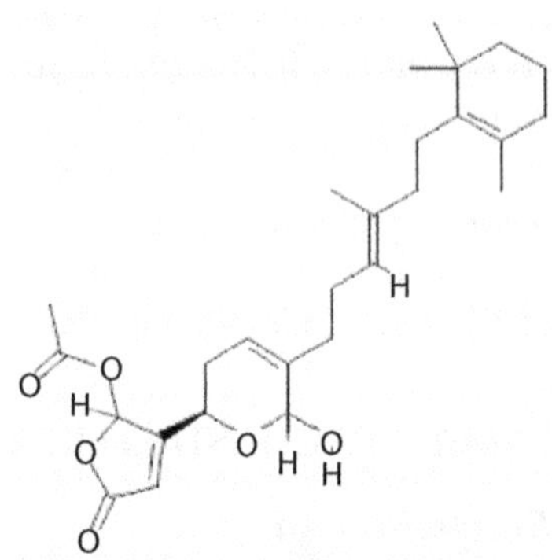

Manoalide 25-acetate

Asif (2017) reported that the scalarane-type sesterterpenes of this unidentified species, namely, massetolide A, heteronemin acetate, 12-deacetyl-12-epi-19-deoxyscalarin, 12-epi-19-deoxyscalarin, and 12- deacetoxyscalarin acetate, have shown anti-TB activity with MIC values in the range of 100 to 102 µM.

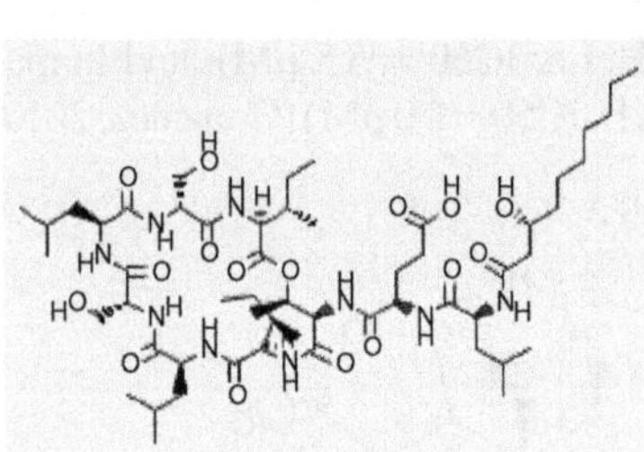

Massetolide A

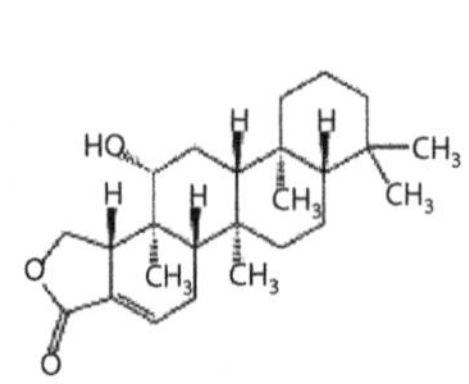

12-deacetyl-12epi-19-deoxyscalarin

Pachastrella sp.

Order: Tetractinellida Family: Pachastrellidae

Compound(s) and Activities

Antibacterial activity: The extract of this South African sponge species showed antibacterial activity against *E. coli* and *S. aureus* MRSA with a zone of inhibition 12.7 and 10.0 mm, respectively (Cita et al., 2017).

FAMILY: PACHASTRELLIDAE

Microscleroderma sp.

Order: Tetractinellida Family: Scleritodermidae

Compound(s) and Activities

Anticancer activity: The cyclic peptide compounds microsclerodermins F–I of this species showed anticancer activity against the HCT-116 cell line with IC50 values of 1.8, 2.4, 1.0, and 1.1 µg/mL, respectively (Agrawal et al., 2016).

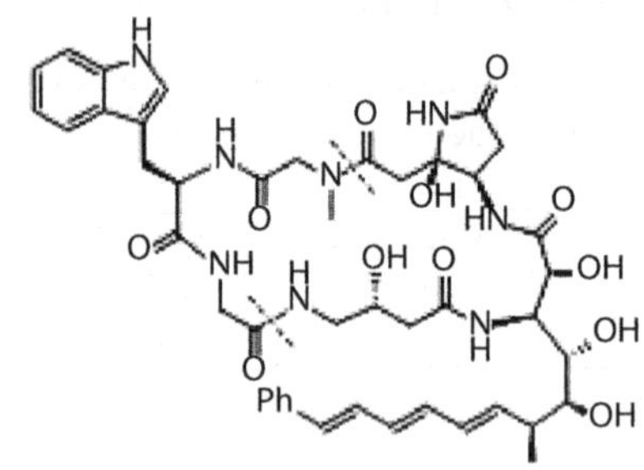

Microsclerodermin F

Scleritoderma nodosum (Thiele, 1900)

Order: Tetractinellida Family: Scleritodermidae
Common name: Not designated
Distribution: Indo-West Pacific Region, Indonesia, Halmahera (Moluccas)
Ecology: It inhabits seamounts and knolls; depth range 107 to 80000 m.
Biology: Not available

Compound(s) and Activities

Antitumor activity: The cyclic peptide compound scleritodermin A isolated from this sponge inhibited tubulin polymerization and demonstrated significant *in vitro* cytotoxicity against a panel of human tumor cell lines (IC50 <2 µM), including colon (HCT116), ovarian (A2780), and breast carcinoma cell lines SKBR3 (Agrawal et al., 2016).

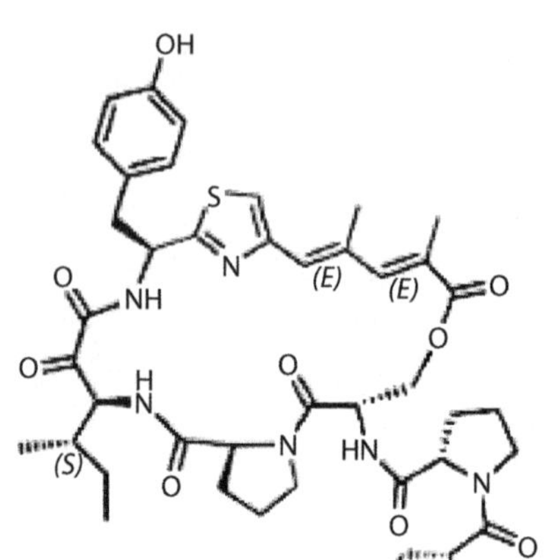

Scleritodermin A

FAMILY: TETILLIDAE

Antarctotetilla leptoderma (Sollas, 1886) (= *Tetilla leptoderma*)

Order: Tetractinellida Family: Tetillidae
Common name: Not designated
Distribution: Uruguay-Buenos Aires Shelf
Ecology: It is found in the benthic zone of 15 to 33 m depths.
Biology: The surface of this animal is smooth or bristly with a spicule felt. It has small crests and conules. Size of the animal can be large, up to 35 cm, and it may have rootlike spicules entangled at its base. The color of the body is grayish beige, reddish gray, dirty gray, light yellow, or light brown.
Predators: The predators of this species include the seastars *Perknaster fuscus antarcticus*, *Odontaster validus*, and *Acodontaster conspicuus* and the nudibranch *Doris kerguelenensis*.

Compound(s) and Activities

Cytotoxic activity: The organic dried extract of this species exhibited cytotoxicity against lung fibroblast (V-79) and human colon epithelial (Caco-2) cancer cell lines at a concentration of 100 μg/mL. The percentage inhibition values recorded for these cell lines were found to be 92 and 97, respectively (Turk et al., 2013).

Cinachyra barbata (Sollas, 1886)

Order: Tetractinellida Family: Tetillidae
Common name: Not known
Distribution: Weddell Sea, Kerguelen Islands
Ecology: It occurs at depths of 20 to 300 m.
Biology: Sponge of this species is subspherical or subcylindrical and is seated on a dense mass of its own anchoring filaments. Basal mass, which was absent in the smallest collected specimen, is apparently a fused root system that gradually expands during the growth of the individual. Surface is hispid-bristly, with megascleres protruding far beyond the ectosome. Size of the main body is up to 7 × 8.5 × 7 cm, and the size of basal mass is up to 5 × 10 × 7 cm. The ectosome of this species is a thick cortex, which consists chiefly of a dense fibrous felt of fusiform cells, only passing into a thin layer of collenchyma close to the outer surface. The cortex is solid throughout, without intercortical cavities. The excurrent and incurrent apertures are confined to special flask-shaped recesses.

Compound(s) and Activities

Cytotoxic activity: The organic dried extract of this species exhibited cytotoxicity against lung fibroblast (V-79) and human colon epithelial (CaCo-2) cancer cell lines at a concentration of 100 μg/mL. The percentage inhibition values recorded for these cell lines were found to be 98 and 95, respectively (Turk et al., 2013).

Cinachyrella apion (Uliczka, 1929)

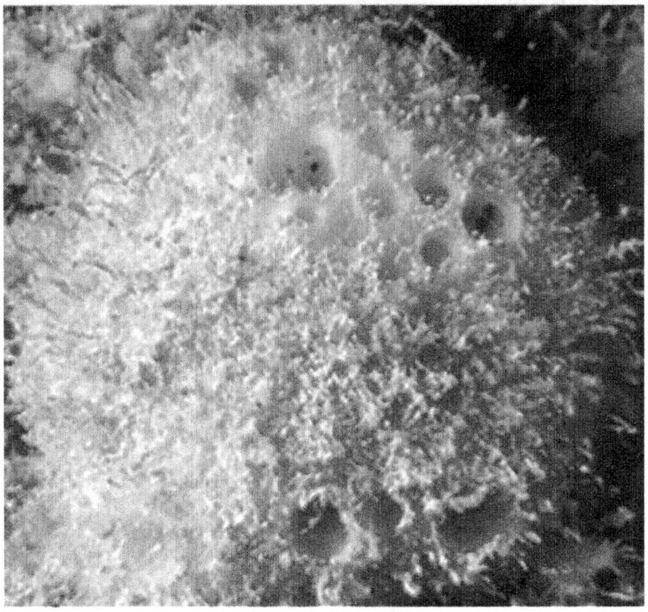

Order: Tetractinellida Family: Tetillidae
Common name: Marine ball sponge
Distribution: North and South Carolina to Bermuda, southwestern Florida, Bahamas, Virgin Islands, Belize, and Panamá
Ecology: It grows on mangrove roots and peat (>0.3 m deep).
Biology: This sponge is spherical to subspherical (up to 10 cm in dia.), yellow to gray externally and lighter internally. Surface is strongly hispid or furry. Porocalices are abundant (up to 3 mm wide). Oscules are rare and are 2 to 3 mm wide when available. Texture of the sponge is soft and compressible.

Compound(s) and Activities

Cytotoxic and antiproliferative properties: The lectin obtained from the extracts of this species showed antiproliferative properties, besides the ability to induce cell death in tumor cells. With regard to antiproliferative activity, the highest growth inhibition was recorded for HeLa, reducing cell growth at a dose-dependent manner (0.5–10 μg/mL). Further, the lectin-induced cell death by apoptosis activation by proapoptotic protein Bax, promoting mitochondrial membrane permeabilization, cell cycle arrest in S phase, and acting as both dependent and/or independent of caspases pathway. These results indicate the potential of the lectin of this species in studies of medicine for treating cancer (Perdicaris et al., 2013; Rabelo et al., 2012).

The lection of this species also showed significant cytotoxicity of CaL to HeLa, PC3 and 3T3 cells. HeLa and PC3 cell proliferation was inhibited in a dose-dependent manner in response to increasing concentrations of lectin (0.5–10 μg/mL). The 50% inhibition (IC50) was obtained with a concentration of 10 μg/mL of lectin. Further, the lectin also presented toxicity against 3T3 cells (Rabelo et al., 2012).

Cinachyrella australiensis (Carter, 1886) (= *Craniella australiensis*)

Order: Tetractinellida Family: Tetillidae

Common name: Golf ball sponge, Chinese marine sponge

Distribution: Indo-West Pacific

Ecology: This species is sometimes seen on coral rubble. It is one of the most common and abundant sponges on intertidal and shallow coral reefs. It also thrives in silty areas at a depth range of 0 to 5 m.

Biology: This furry, ball-shaped, spherical sponge resembles a golf ball. It has a size of 6 to 8 cm in diameter and is densely covered with short, fine spines. These are needle-shaped spicules that stick out of the surface. There may be circular pits or indentations on the surface that are bare of spines. These are specialized pore-bearing pits called porocalices. Young sponges often lack the pits on the surface and often resemble furry balls. Color of the animal is usually yellow, although sediment accumulated on the spines may hide the color.

Compound(s) and Activities

Antibacterial activity: The MeOH extract of this species displayed antibacterial activity against *E. coli*, *P. aeruginosa* MDR, and *S. aureus* MRSA with the inhibition zone (mm, dia.) values of 10.5, 13.0 and 14.5, respectivley (Cita et al., 2017)

Antifungal activity: The chitinase obtained from the microbial symbiont, namely, *Streptomyces* sp. DA11, of this species showed activity against *Aspergillus niger* and *Candida albicans* (Han et al., 2009; Thomas et al., 2010).

Antimicrobial activity: The actinomycetes isolated from the associated bacterial community of this sponge species were found to possess broad-spectrum antimicrobial potentials (Li and Liu, 2006).

Cytotoxicity: The fatty acid metabolites, such as the 2-methoxy acetylenic acid derivative cinachylenic acid A, from this species showed strong cytotoxicity against L5178Y mouse lymphoma cells with an IC50 value of 0.3 μM (Mokhlesi et al., 2017).

Cinachyrella sp.

Order: Tetractinellida Family: Tetillidae

Antitumor activity: The oximated steroid compound 6-hydroximino-4-en-3-one of this unidentified speices showed antitumor activity by inhibiting the enzyme aromatase (Sipkema et al., 2005).

Cytotoxicity: The ethanolic extract of this species showed cytotoxicity against Hela, T47D, and W1DR cell lines (Nurhayati et al., 2014).

Antibacterial activity: The aqueous extracts of this unidentified species showed a large action spectrum over resistant-bacteria such as *Staphylococcus aureus*, coagulase-negative staphylococci, and *Enterococcus faecalis* (Marinho et al., 2010).

The extract of this sponge species showed antibacterial activity with the following values of zone of inhibition.

Zone Inhibition (mm, dia) of Extract of Sponge

S. aureus	*E. coli*	*K. pneumoniae*	*S. typhi*	*P. aeruginosa* MDR	*S. aureus* MRSA
–	10.5	–	–	13.0	14.5

Source: Cita, Y.P. et al., 2017, *J. Marine Sci. Res. Dev.*, 7:237.

Estrogen activity: Cinanthrenol A, a phenathrene-containing steroid, isolated from this species showed estrogen (female sex hormone) activity (Machida et al., 2014).

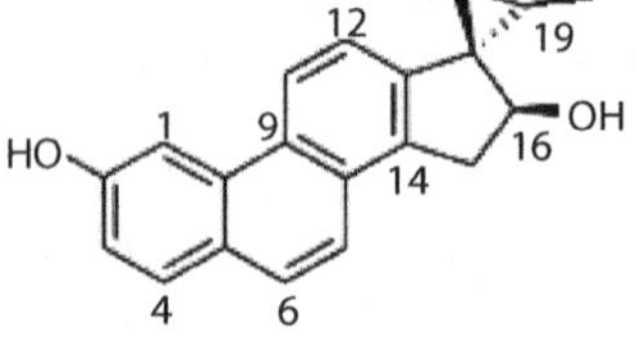

Cinanthrenol A

Others: The aqueous sponge extract of this speices yielded *Cinachyrella* galectins (CchGs) (lectin protein) possessed biochemical properties as well as an unexpected activity on mammalian excitatory amino acid receptors. These findings suggest that further analysis of these new members of the galectin family will yield further glycobiological and neurophysiological insights (Ueda et al., 2013).

FAMILY: THENEIDAE

Tisiphonia penetrans (Carter, 1880) (= *Jaspis penetrans*)

Order: Tetractinellida Family: Theneidae
Common name: Not reported
Distribution: India: Gulf of Mannar, Mahe Island
Ecology: It is a boring sponge living on corals.
Biology: This sponge has a dark-brown outer layer and soft light-yellow color body. Encrustation of this species is with a firm consistency. Chambers are irregular. Skeleton consists of large oxeas, which are scattered irregularly intermingled with microxeas.

Compound(s) and Activities

Anticancer activity: The extract of this species showed anticancer activity against A549, HeLa, and HT-29 cancer cell lines with the following GI50 values.

GI50 (μg) Values of Cancer Cell Lines

A549	HeLa	HT-29
1.0	0.6	1.3

Source: Sponge crude extraction, n.d., http://shodhganga.inflibnet.ac.in/bitstream/10603/64131/10/10_chapter%205.pdf.

FAMILY: THEONELLIDAE

Discodermia calyx (Döderlein, 1884)

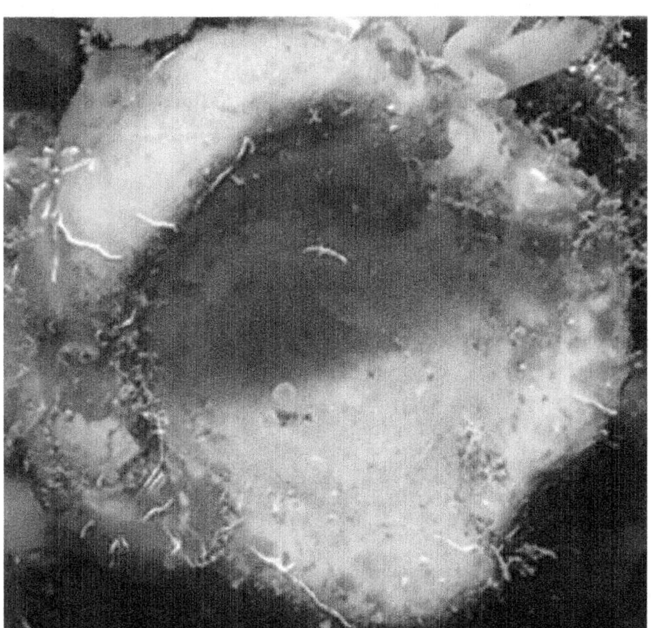

Order: Tetractinellida Family: Theonellidae
Common name: Japanese marine sponge
Distribution: Eastern Philippines
Ecology: Not reported
Biology: Not known

Compound(s) and Activities

Cytotoxicity: The major compound, calyculin A, isolated from this species displayed cytotoxicity by inhibiting protein phosphatases 1 and 2A (Wakimoto et al., 2014).

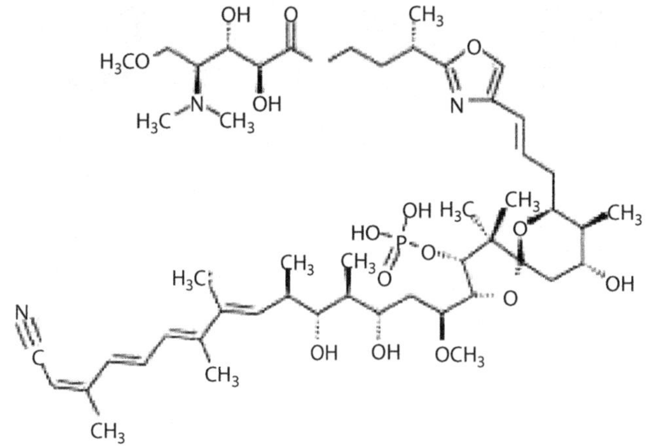

Calyculin A

The organic extract of this species showed cytotoxicity by inhibiting protein phosphatase 2A (IC50 = 0.5–1 nM) and protein phosphatase 1 (IC50 = 2 nM) (Millipore Sigma, 2018). Further, this extract is known to prevent γ-radiation–induced apoptosis in Burkitt's lymphoma cell line BM13674.
Antiprolific and antitumoral activities: The compounds calyculins isolated from this species were found to be cytotoxic and inhibited the cancer cell proliferation with antitumoral activity (Sepčić et al., 2010).
Antitumor activity: The extraordinary metabolites isolated from this species showed marked cytotoxicity against murine leukaemia cells. Further, the major compound calyculin A also displayed potent antitumor activity (Kato et al., 1988).
Anticancer activity: The compounds debromohamacanthin B, tetrahydrofurospongin, and deoxytopsentin of this species showed good interaction with the epidermal growth factor receptor (EGFR) forming hydrogen bonds with the tyrosine and arginine residue preventing the binding of EGF to the EGFR receptor by inhibiting dimerization. Overexpression of EGFR has resulted in triple negative breast cancer (Raj and Inbakandan, 2013).

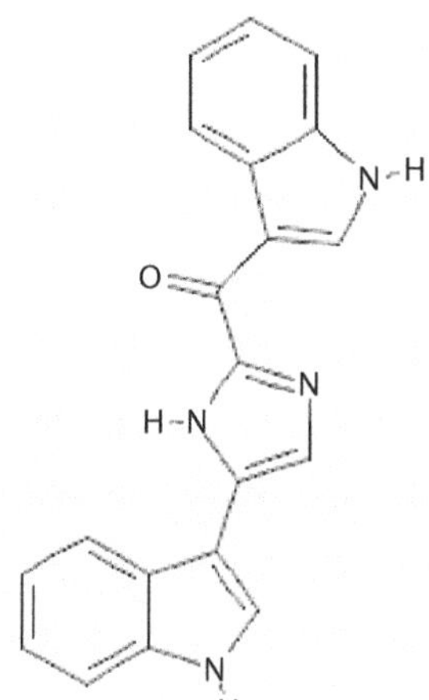

Deoxytopsentin

Discodermia dissoluta (Schmidt, 1880)

Order: Tetractinellida Family: Theonellidae

Common name: Not designated

Distribution: Eastern Caribbean, Southwestern Caribbean, Guyana, Brazil

Ecology: It inhabits depths greater than 30 m. It was recorded in greater proportion in hard substrate (60%) than in gravel (38%) and other substrates (2%). The hard substrate comprised metamorphic rock, coral rock, and dead coral. The other substrates were artificial, octocoral, and mollusk shells.

Biology: Not studied in detail

Compound(s) and Activities

Antitumor/cytotoxic, antifungal, and immunosuppressive properties: The polyhydroxylated lactone, discodermolide of this species has shown antitumor activity by stabilizing microtubules. Further, this compound has also displayed immunosuppressive activity (Anjum et al., 2016). Discodermolide has also been reported to inhibit *in vitro* proliferation of P388 murine leukemia cells and inhibit the growth of *Candida albicans* (Gunasekera et al., 1990).

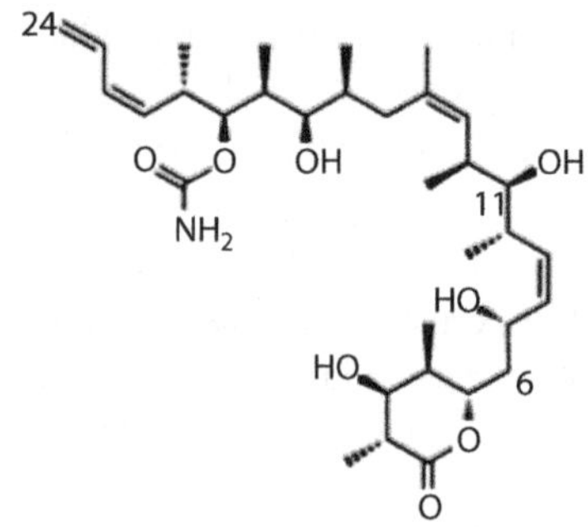

Discodermolide

Antiprolific and antitumoral activities: The compound discodermolide isolated from this species was found to be cytotoxic and inhibited the cancer cell proliferation with antitumoral activity (Sepčić et al., 2010).

Discodermia kiiensis (Hoshino, 1977)

Order: Tetractinellida Family: Theonellidae

Common name: Not designated

Distribution: Japan

Ecology: Not reported

Biology: Not studied

Compound(s) and Activities

Antibacterial activity: The cyclic peptide compounds discodermins A–D of this species have shown antibacterial activity (Joseph et al., 2012; Sipkema et al., 2005). Anjum et al. (2016) reported that the compounds B, C, and D have shown activity against *B. subtilis* with an MIC value of 3 μg/mL.

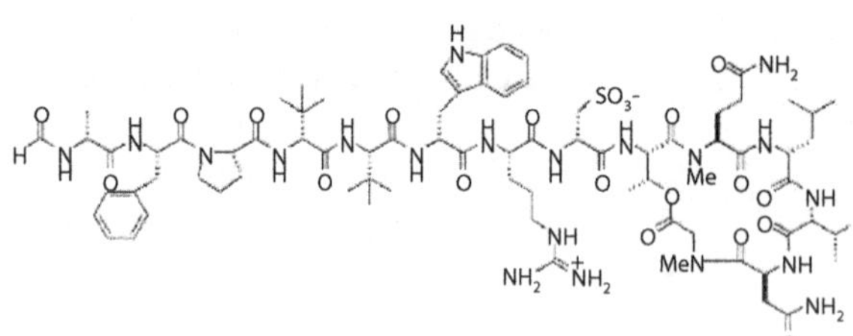

1

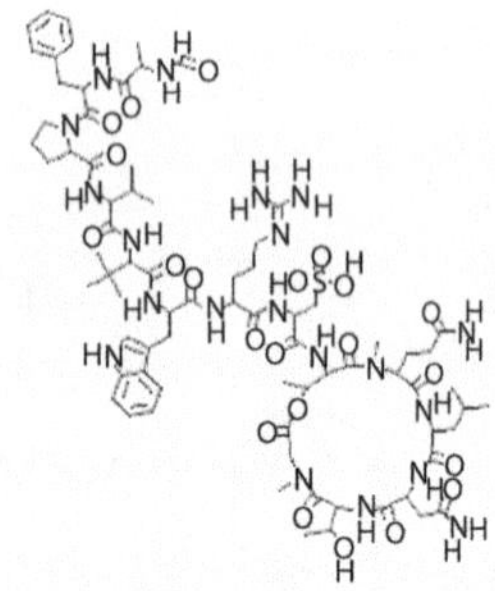

2

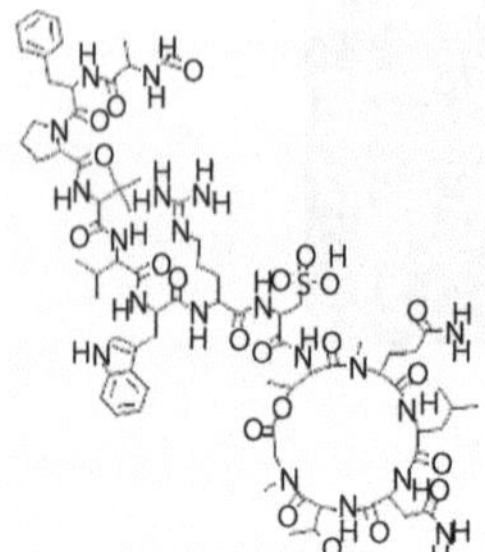

3

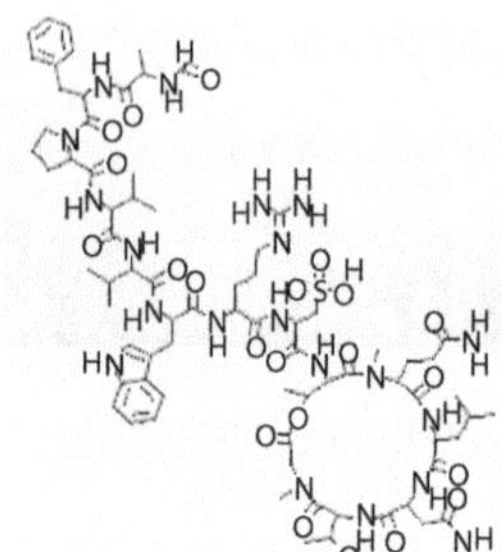

4

Discodermins A-D

Antimicrobial and cytotoxic activity: The peptides discodermins E–H isolated from this species have shown both cytotoxicity and antimicrobial activity (Ryu et al., 1994a, 1994b).

Cytotoxicity: Three new lipodepsipeptides, lipodiscamides A–C, isolated from this species have shown cytotoxicity against P388 and HeLa cells (Tan et al., 2014).

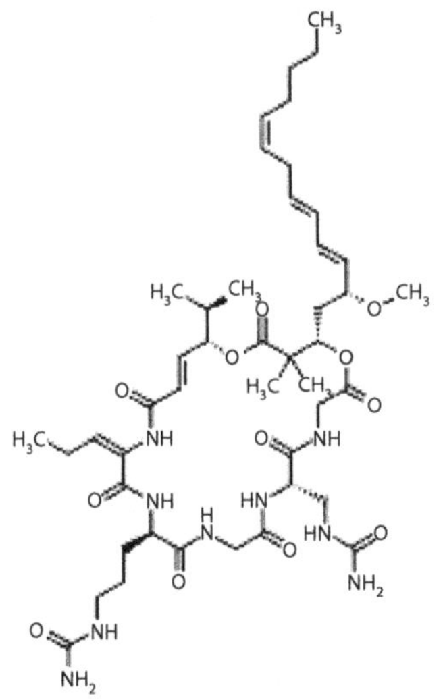

Lipodiscamide B

Tan et al. (2016) also reported that the new N-sulfoureidylated lipopeptides, sulfolipodiscamides A–C, isolated from this species have shown 2.3-fold increase in cytotoxicity against murine leukemia (P388) cells, compared to the unconjugated parent compound.

Discodermia sp.

Order: Tetractinellida Family: Theonellidae

Compound(s) and Activities

Antifungal activity: The alkaloids discobahamin A and B isolated from this unidentified Bahamas deepwater marine sponge showed weak activity against the yeast *C. albicans* (Gul et al., 2005).

Cytotoxicity: The cytotoxic peptides discodermins A–H isolated from this unidentified species showed cytotoxicity against the A549 human lung cell line and P388 murine leukemia cells (Suarez-Jimenez et al., 2012).

Siliquariaspongia japonica (Hoshino, 1981)

Order: Tetractinellida Family: Theonellidae

Common name: Not designated

Distribution: Japan

Ecology: It inhabits intertidal zones and midtide subzones.

Biology: It is an irregular massive sponge with a size of 3 × 3 × 4 cm. Color of the sponge is orange. Consistency is very slightly compressible and quite hard. Surface is smooth and even. Oscules and pores are invisible. Ectosome is of discotriaenes, which are fully packed in a jigsaw puzzle arrangement. Spicules are discotriaenes and microacanthostrongyles.

Association: This species is symbiotic with the gastropod *Siliquaria cumingi*.

Compounds and Activities

Antifungal activity: The tetramic acid glycosides—aurantosides D, E, and F—isolated from this species showed antifungal activity against *Aspergillus fumigatus* and *Candida albicans* (Kon and Rai, 2016).

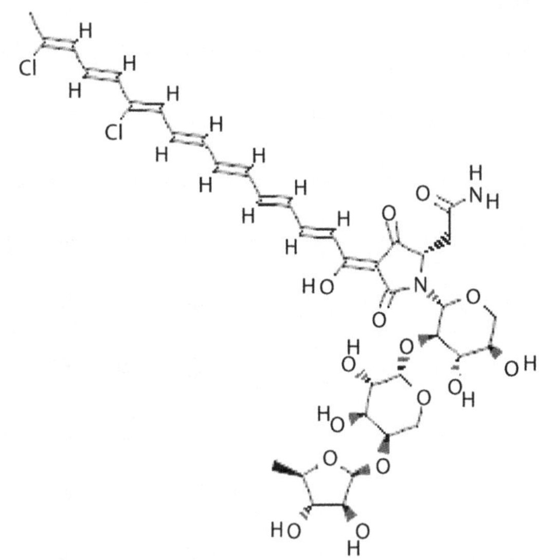

Aurantoside D

Cytotoxic and antifungal activities: Eight new tetramic acid glycosides named rubrosides A–H have been isolated from this species. Among them, rubrosides A, C, D, and E were cytotoxic against P388 murine leukemia cells with IC50 values of 0.046–0.21 μg/mL. Further, most rubrosides also showed antifungal activity against *Aspergillus fumigatus* and *Candida albicans* (Sata et al., 1999).

Siliquariaspongia mirabilis (de Laubenfels, 1954)

Order: Tetractinellida Family: Theonellidae

Common name: Not designated

Distribution: Palau; Papua New Guinea; Indonesia; Philippines—Davao, Panglao, Bohol

Ecology: It is attached to coral reefs at a depth of 10 m.

Biology: It is an irregularly encrusting (10 cm long and 4 cm wide) with short tubular projections (7–12 mm high, 10–11 mm wide) that are distributed 14–20 mm apart. Surface is wrinkly and uneven but the tubes are smooth. Ostia are not visible, but the oscula are terminally located at each tube (4 mm in dia.). Texture is spongy and crumbly. In life, ectosome is reddish brown and endosome is yellowish brown. In spirit, the color is orange brown. Spicules are strongyles that are long and smooth. Ectosome seems to be devoid of desmas or may be present sparsely, but it has high concentration of rhabds. In areas with conules, ascending tracts of strongyles are present terminating at its crest. Desmas are present in great number at the choanosome or at the area below the conules.

Compound(s) and Activities

Antiviral activity: The peptides mirabamides A, C, and D isolated from this species showed antiviral (HIV-1) activity (Anjum et al., 2016). Joseph et al. (2012) reported that the peptide compounds celebesides A–C and theopapuamides B–D showed anti-HIV effects.

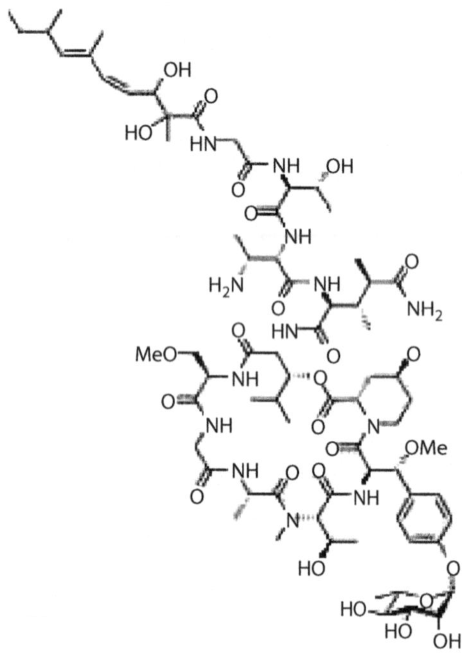

Mirabamide A

Agrawal et al. (2016) reported that cyclic depsipeptide neamphamide A of this species displayed potent cytoprotective activity against HIV-1 infection with an EC50 value of ~28 nM.

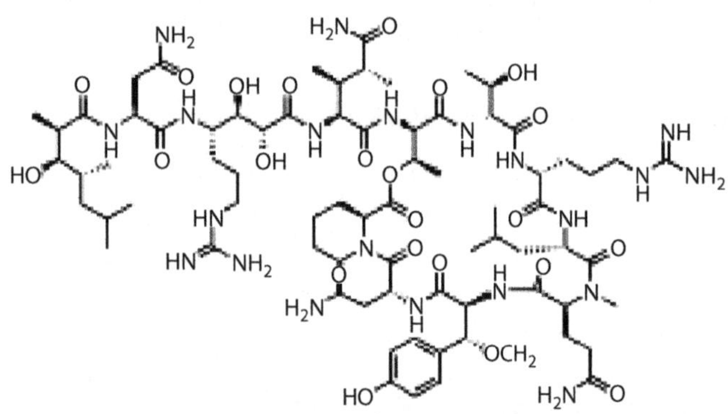

Neamphamide A

Antiviral and antimicrobial activities: The cyclic depsipeptides, namely, mirabamides A–D (1–4), isolated from this sponge potently inhibited HIV-1 fusion. Mirabamide A inhibited HIV-1 in neutralization and fusion assays with IC50 values between 40 and 140 nM, as did mirabamides C and D (IC50 values between 140 nM and 1.3 μM for 3, and 190 nM and 3.9 μM for 4), indicating that these peptides can act at the early stages of HIV-1 entry. Further, mirabamides A–C inhibited the growth of *Bacillus subtilis* and *Candida albicans* at 1–5 μg/disk in disk diffusion assays (Kon and Rai, 2016; Plaza et al., 2007).

Antifungal and enzyme-inhibitory activities: The compound namalide isolated from this species showed protease inhibition and selectivity. Further, the synthetic version of natural namalide containing d-Lys and l-Ile was the most potent inhibitor of CPA (carboxypeptidase A) with an IC50 value of 250 nM. Further, its aurantosides A–B displayed antifungal activity (Cheruku et al., 2012).

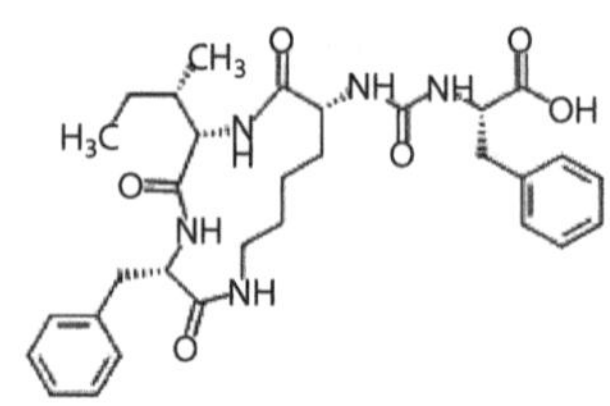

Namalide

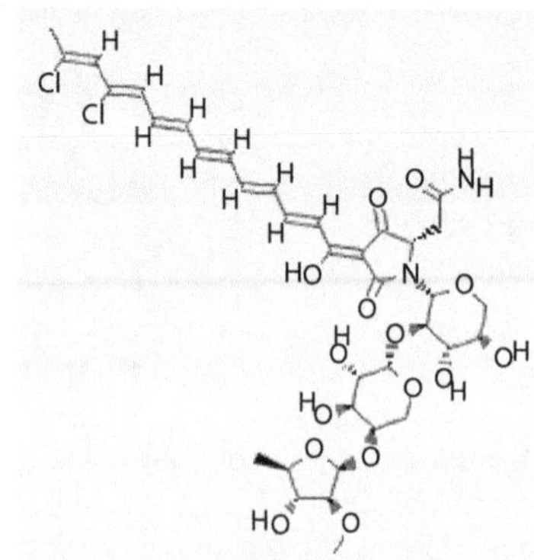

Aurantoside A

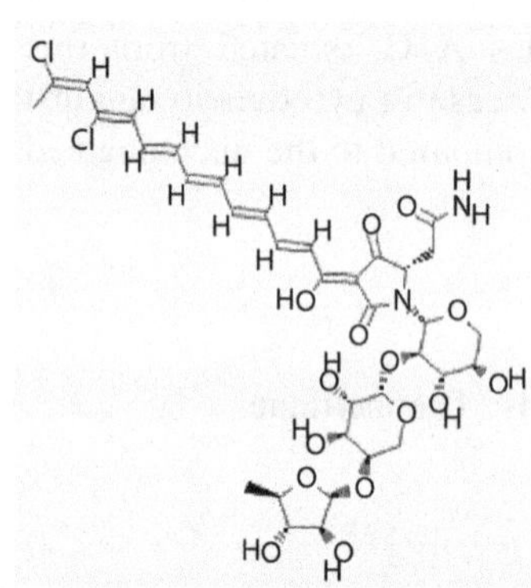

Aurantoside B

Antitumor activity: A highly unsaturated macrolide lactam, termed mirabilin, isolated from the aqueous extract of this sponge inhibited the growth of the tumor cell line HCT-116 with an IC50 value of 0.3 μM (Plaza et al., 2008).

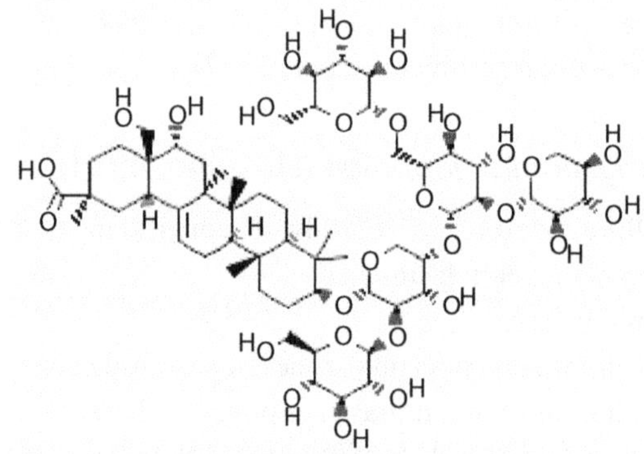

Mirabilin

Theonella cupola (Burton, 1928)

Order: Tetractinellida Family: Theonellidae
Common name: Not designated
Distribution: Palau
Ecology: It is a deepwater, reef-associated sponge.
Biology: It is a massive sponge with a convex upper surface that rests on a circular base of 2 cm in diameter. Larger specimens measure 8 mm high. Surface is smooth with a depression located excentrically on the surface. A single osculum (1 mm dia.) surrounded by a thin membrane is present on the above depression. Pores have a diameter of 0.05 mm. This sponge is grayish yellow in spirit. There are bundles of six slender monactinal megascleres that protrude beyond the ectosome. Spicules are oxeas that are straight or curved.

Compound(s) and Activities

Antiviral activity: The N-methylated peptides, koshikamides F and H of this species inhibited HIV-1 entry at IC50 values of 2.3 and 5.5 μM, respectively (Agrawal et al., 2016; Joseph et al., 2012; Plaza et al., 2010).
Cytotoxicity: A new cyclic heptapeptide cupolamide A isolated from this species showed cytotoxicity against P388 murine leukemia cells with an IC50 value of 7.5 μg/mL (Bonnington et al., 1997).

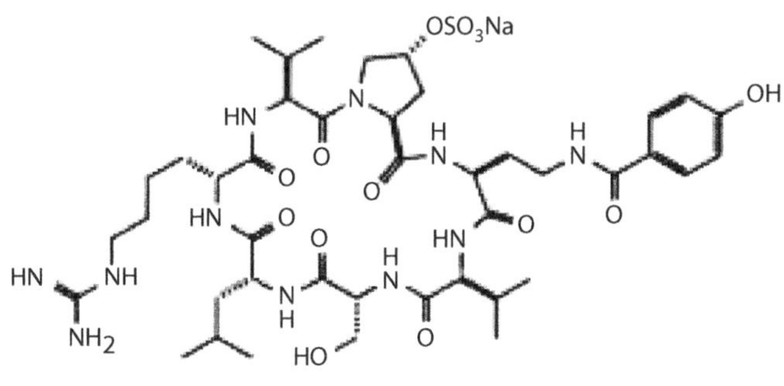

Cupolamide A

Theonella mirabilis (de Laubenfels, 1954)

Order: Tetractinellida Family: Theonellidae
Common name: Not designated
Distribution: Marshall Islands
Ecology: Not reported
Biology: Not available

Compound(s) and Activities

Antiviral activity: The cyclic peptides papuamides C and D isolated from this species have shown antiviral (HIV-1) activity (Sipkema et al., 2005). Nakao et al. (2000) reported that the compounds tokaramide A and miraziridine A of this species have shown antiviral activity.

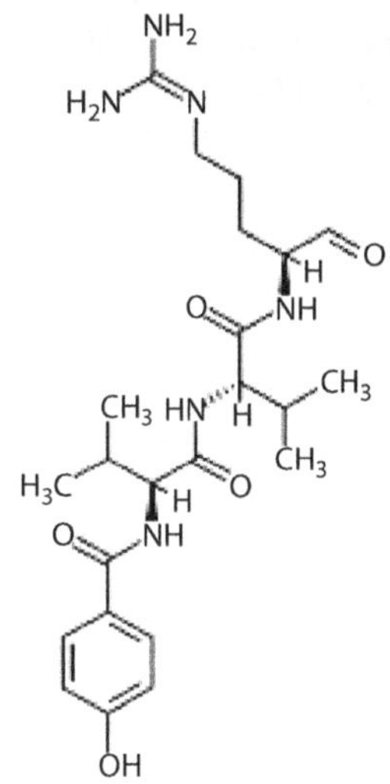

Tokaramide A

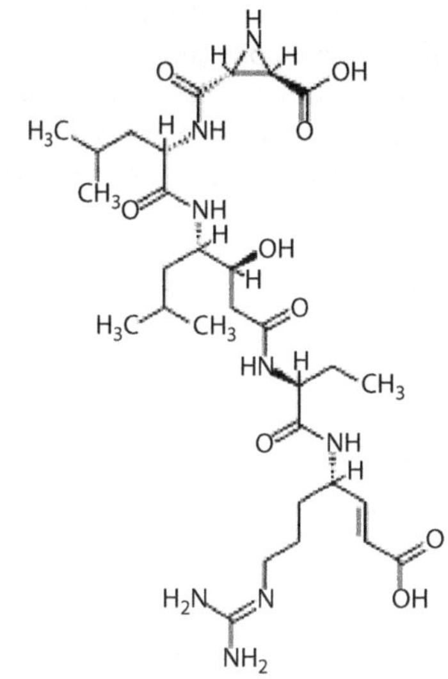

Miraziridine A

Anticancer and enzyme-inhibitory activities: The linear penta peptide compound miraziridine A isolated from this species showed anticancer and protease cathepsin B inhibitory activities with an IC50 value of 1.4 μg/mL. Cathepsin B has been implicated in tumor invasion and metastasis (Agrawal et al., 2016).

Theonella swinhoei (Gray, 1868)

Order: Tetractinellida Family: Theonellidae
Common name: Tube sponge, Red Sea sponge
Distribution: Indo-West Pacific; Andaman Island and Palau; Taiwan, Philippines, and Indonesia

Ecology: It inhabits depths of 20 to 180 m at coral reefs.

Biology: It is a massive, thick-walled tube or vase sponge with a single osculum at the top. Ectosomal spicules are strongly differentiated phyllotriaenes with a very short head. There are numerous microscleres on the dermal membrane as well as in subdermal lacunae. Surface of the sponge is smooth. There are slender slightly curved strongyles. It is an oviparous species.

Compound(s) and Activities

Cytotoxic activity: The polyketide compound swinholide A derived from the associated unicellular cyanobacteria, unicellular bacteria, and filamentous bacteria of this species has shown cytotoxic activity. Similarly, the compound onnamide A of these microorganisms showed antitumor activity (Thomas et al., 2010).

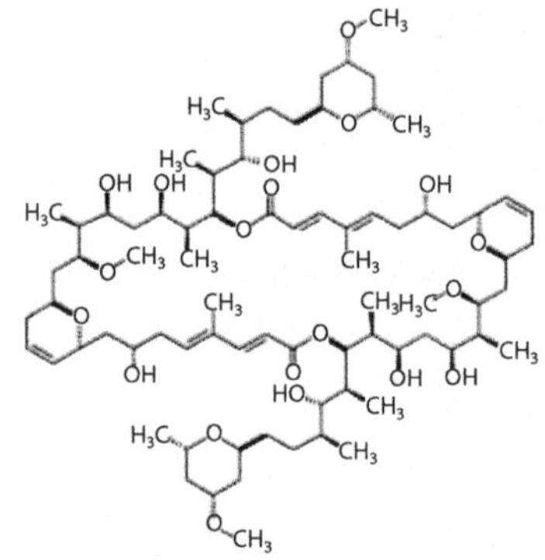

Swinholide A

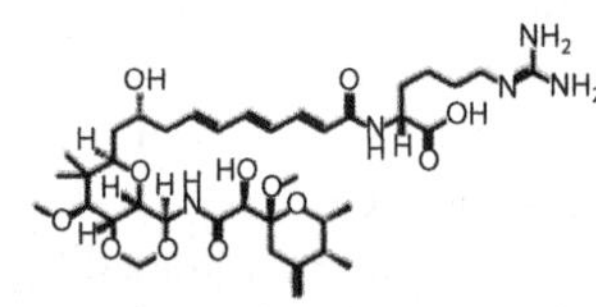

Onnamide A

The polypeptides polytheonamides A and B of this species were found to be cytotoxic against P388 murine leukemia cells with IC50 values of 78 and 68 pg/mL, respectively (Agrawal et al., 2016).

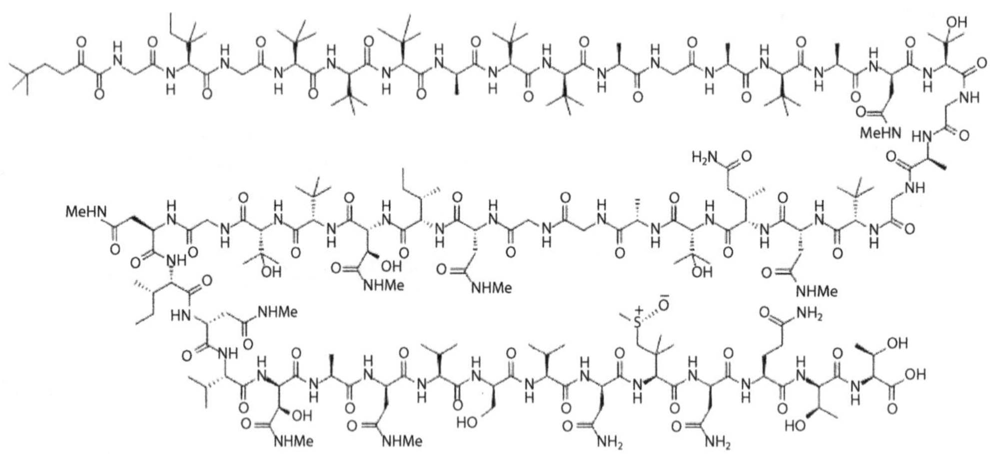

Polytheonamide A

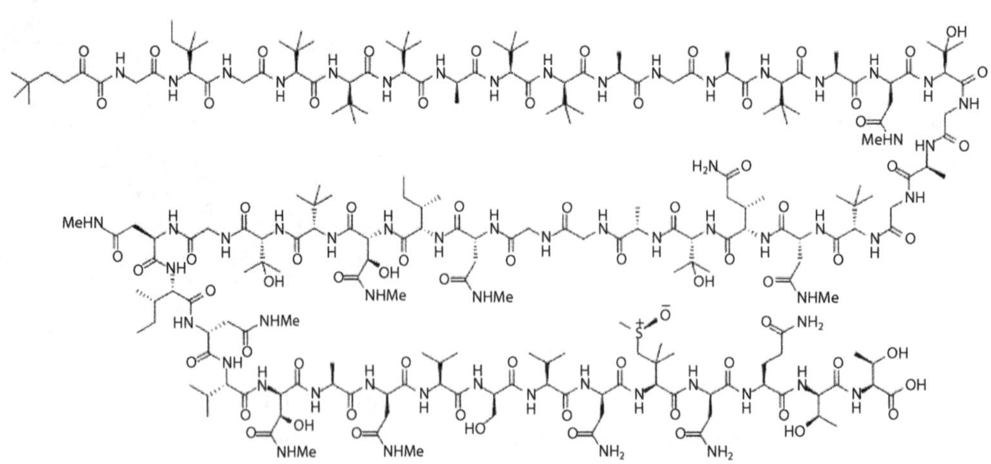

Polytheonamide B

The cyclic depsipeptide theopapuamide of this species showed cytotoxicity against CEM-TART (EC50 = 0.5 μM) and HCT-116 (EC50 = 0.9 μM) cell lines (Agrawal et al., 2016).

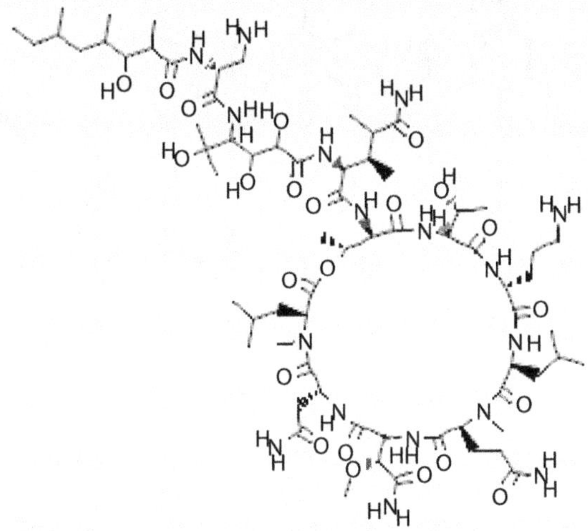

Theopapuamide

The peptide koshikamide H of this species has shown cytotoxicity against the tumor cell line HCT-116 with an IC50 value of 10 μM (Plaza et al., 2010).

Antitumor activity: The brominated C17 acetylenic acid of this species inhibited the cell proliferation of U937 and HL60 (human leukemia), A549 and H1299 (human lung), and HEK293 (human embryonic kidney) with IC50 values of 24, 27, 58, 72, and 40 mg/mL, respectively (Gribble, 2015).

The macrocyclic lactone compound swinholide A has shown antitumor activity thorugh actin-depolymerization (Sipkema et al., 2005).

The crude dichloromethane extract containing the compound miraziridine A has shown antiviral activity through protease inhibiting activity. The protease inhibitors (PIs) are a class of antiviral drugs that are widely used to treat HIV/AIDS and hepatitis C (Tabares, 2011).

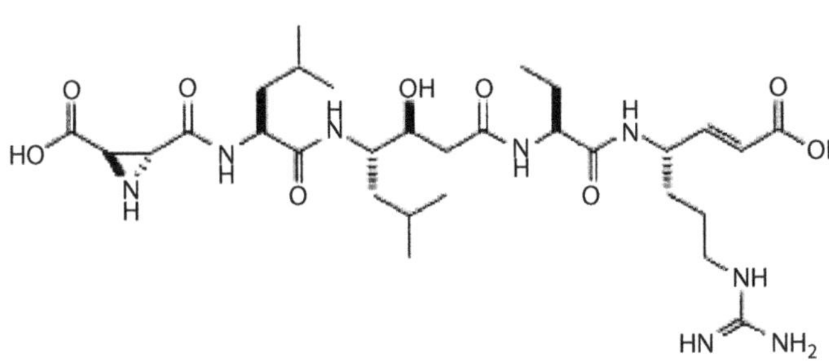

Miraziridine A

Antifungal and cytotoxic activities: The compound theonellamide G of this species showed potent antifungal activity toward wild and amphotericin B–resistant strains of *Candida albicans* with IC50 values of 4.49 and 2.0 μM, respectively. Further, it displayed cytotoxic activity against the human colon adenocarcinoma cell line (HCT-16) with an IC50 value of 6.0 μM (Youssef et al., 2014).

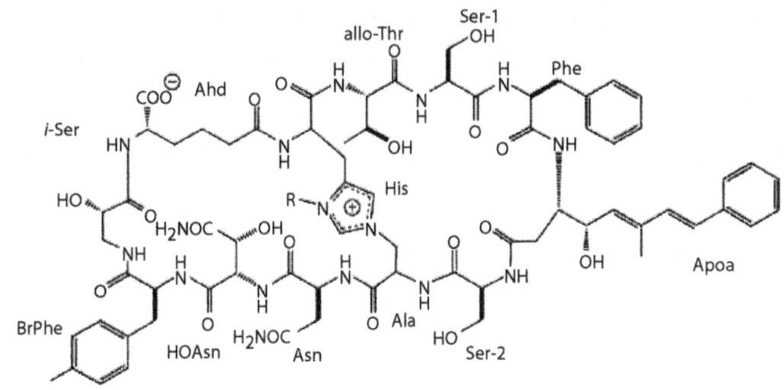

Theonellamide G

Antibacterial activity: The depsipeptide compound, nagahamide A of this species showed antibacterial activity (Agrawal et al., 2016).

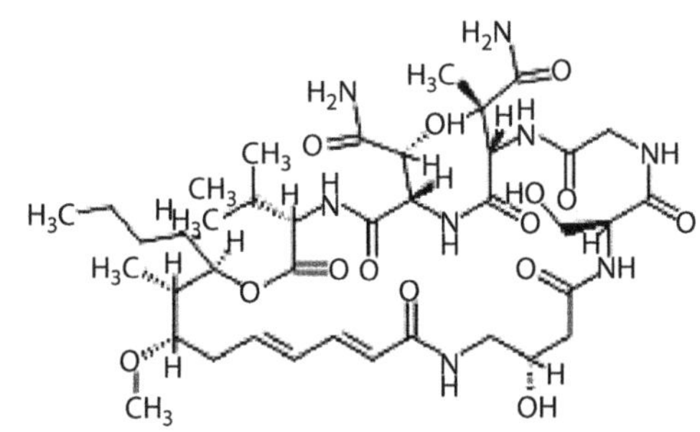

Nagahamide A

Antifungal activity: The compound aurantoside J of this species showed antifungal activity. Further, theonellamide G showed potent activity against both wild and amphotericin B–resistant strains of *Candida albicans* with IC50 values of 4.49 and 2.0 μM, respectively (Gribble, 2015).

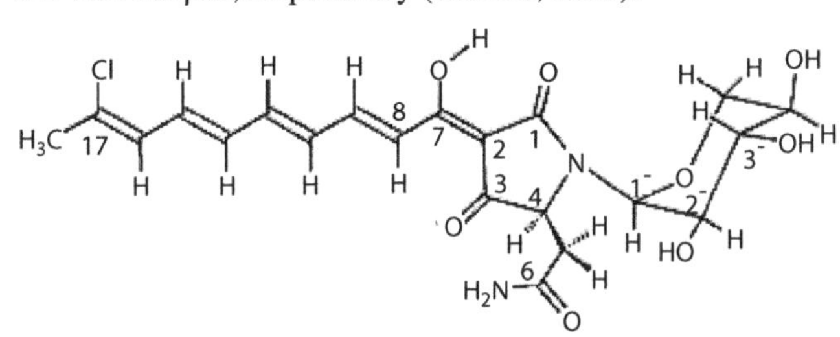

Auramtoside j

Anti-inflammatory activity: The cyclopeptides perthamides C and D of this species showed potent anti-inflammatory activity. Further, the cyclic peptide solomonamide B of this species also showed anti-inflammatory activity (Agrawal et al., 2016).

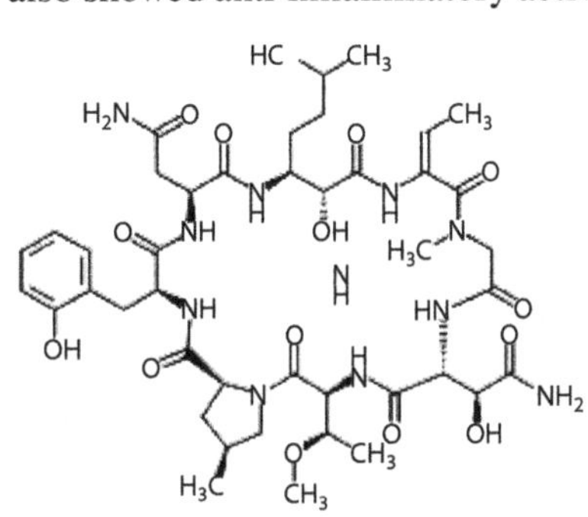

Perthamide C

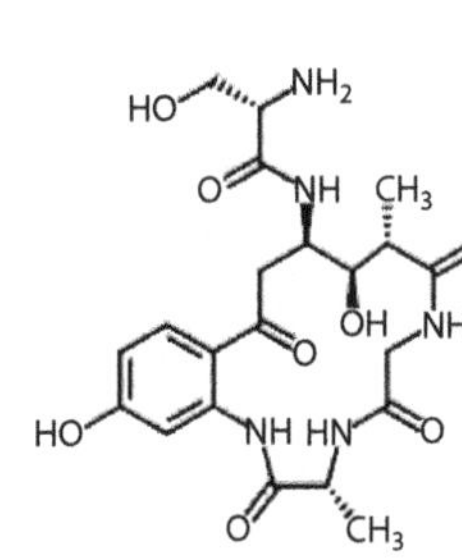

Solomonamide B

Enzyme inhibitory activity: The linear peptides from this species act as inhibitors of thrombin and serine proteases (Sepčić et al., 2010).

Theonella sp.

Order: Tetractinellida Family: Theonellidae

Compound(s) and Activities

Antiproliferative/anticancer and antiviral activities: Among the cyclic depsipeptides, namely, papuamides A–D, isolated from this sponge, the compounds papuamides A and B inhibited the infection of human T-lymphoblastoid cells by HIV-1 *in vitro*. Further these compounds have also shown antiviral (HIV-1) activity (Anjum et al., 2016; Suarez-Jimenez et al., 2012).

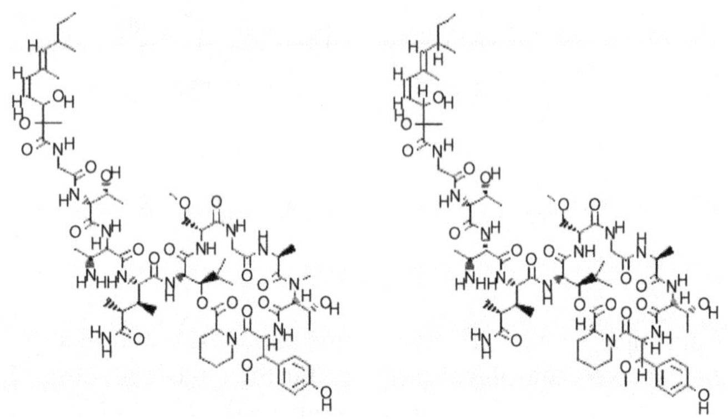

Papuamides A, B

FAMILY: THOOSIDAE

Neamphius huxleyi (Sollas, 1888)

Order: Tetractinellida Family: Thoosidae
Common name: Australian sponge
Distribution: Vanuatu
Ecology: It is a deepwater sponge inhabiting gravely hard-bottom habitats on the outer shelf and upper slope.
Biology: It is a massive sponge with large oscula of 4 to 10 mm in diameter and is elevated on mounds that are fused. Size of the sponge is 11 × 6 cm and 7 cm high. The ectosome (skin) is brown and wrinkled. Texture is soft and elastic. Pores are clustered and sievelike. Spicules are of two types: megascleres and microscleres. The megascleres are oxeas, partly scattered but also organized in radial strands. The microscleres are amphiasters that are delicate and dispersed throughout the choanosome.

Compound(s) and Activities

Anti-HIV activity: The cyclic depsipeptide neamphamide A isolated from this species displayed potent cytoprotective (cytopathic) activity against HIV-1 infection with an EC50 value of 28 nM (Agrawal et al., 2016; Senthilkumar and Kim, 2013).

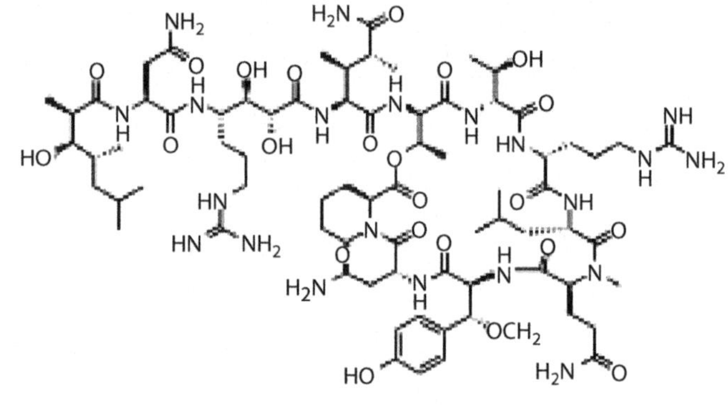

Neamphamide A

Cytotoxicity: The cyclic depsipeptides neamphamides B, C, and D isolated from this species inhibited the growth of human cell lines (A549, HeLa, LNCaP, PC3, and NFF) with IC50 values ranging from 88 to 370 nM. However, neamphamide D caused A549 cell proliferation at subcytotoxic doses (Tran et al., 2012).

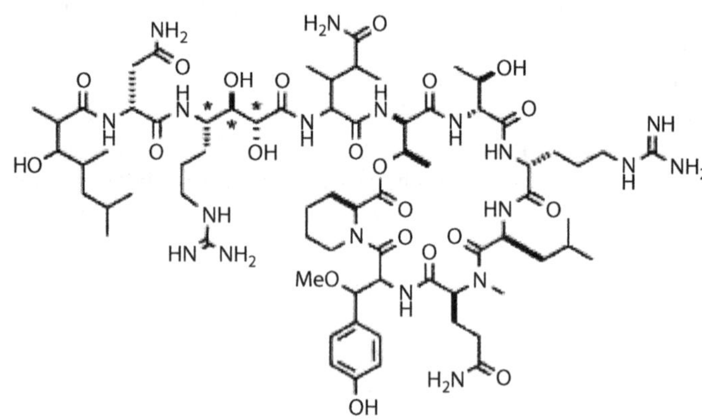

Neamphamide B

FAMILY: TRACHYCLADIDAE

Trachycladus spinispirulifer (Carter, 1879) (= *Spirastrella spinispirulifera*)

Order: Trachycladida Family: Trachycladidae
Common name: Southwest Indian Ocean sponge, South African marine sponge, wall sponge
Distribution: Southwest India
Ecology: It inhabits intertidal zones on rocky and sandy shore.
Biology: Not available

Compound(s) and Activities

Cytotoxicity: The compounds trachycladindoles A–G isolated from this species displayed promising selective cytotoxicity against a panel of human cancer cell lines (Capon et al., 2008).

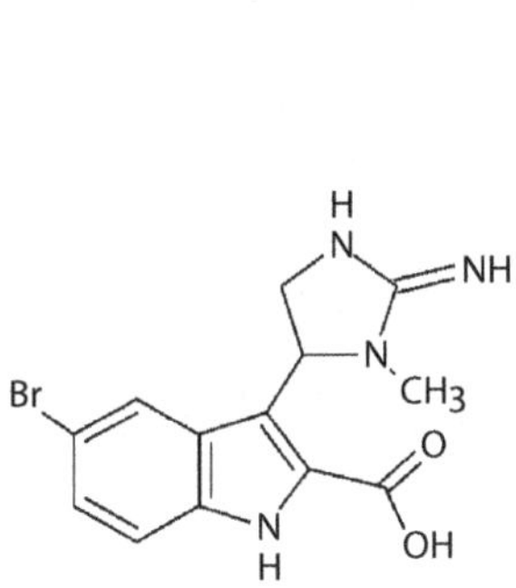

Trachycladindole A

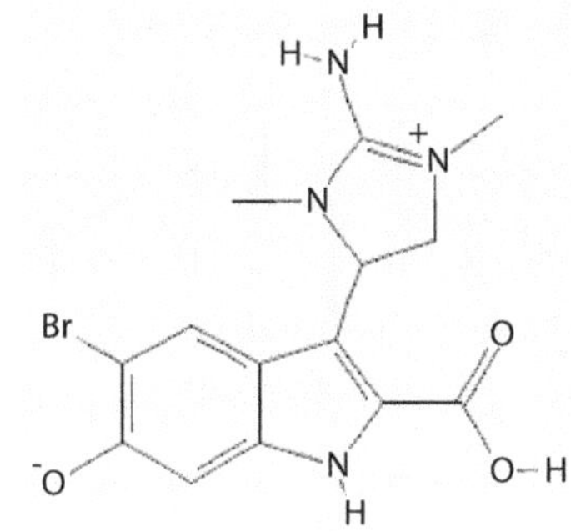

Trachycladindole D

Pettit et al. (1993c) reported on the isolation of two powerful human cancer cell growth inhibitors, spongistatins 4 and 5, that showed a GI50 value of about 10^{-10} µg/mL.

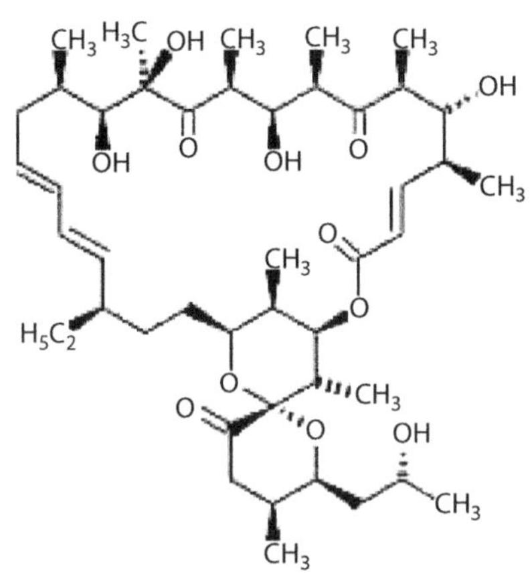

Spongistatin

Antitumor activity: The compound spongistatin 1 isolated from this species effectively caused cell death of leukemic tumor cells, especially in the clinically highly relevant situation of chemoresistance. Further, this compound has also been shown to potently kill several solid tumor cell lines, including melanoma, lung, and colon cancer cells. Recently, spongistatin 1 was shown to have an apoptosis-inducing effect in A549 cells by activating caspase-3 and by the cleavage of vimentin (Schyschka et al., 2008; Senthilkumar and Kim, 2013).

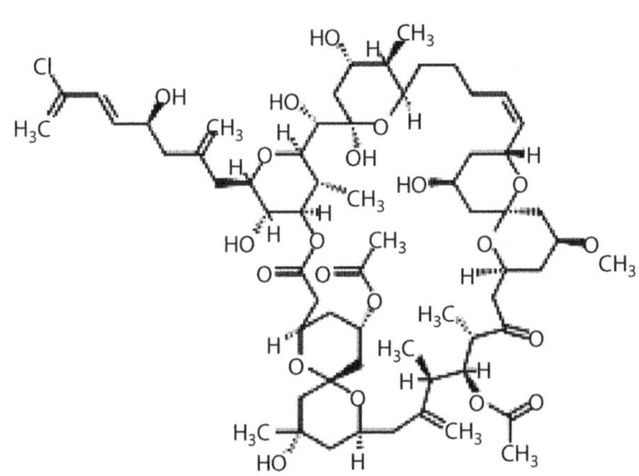

Spongistatin 1

Nematocidal activity: The compound onnamide F isolated from this species is a potent new nematocide with antifungal activity (Vuong et al., 2001).

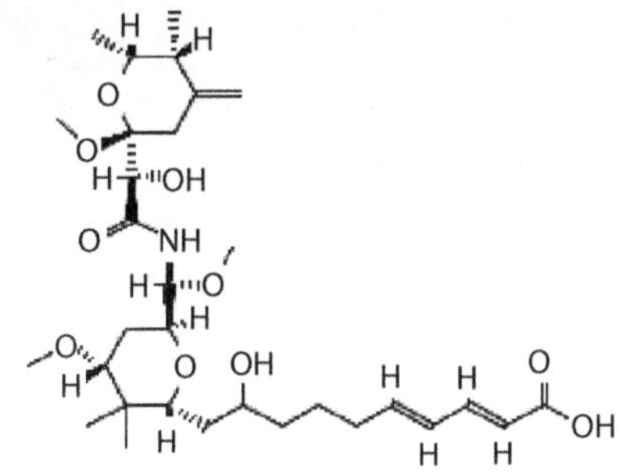

Onnamide F

Antibacterial, antifungal, and anticancer properties: The compound, 9-(5′-deoxy-5′-thio-β-D-xylofuranosyl)adenine disulfide isolated from this species possessed antibacterial, antifungal, and anticancer properties (Peng et al., 2010).

FAMILY: DARWINELLIDAE

Aplysilla glacialis (Merejkowski, 1878)

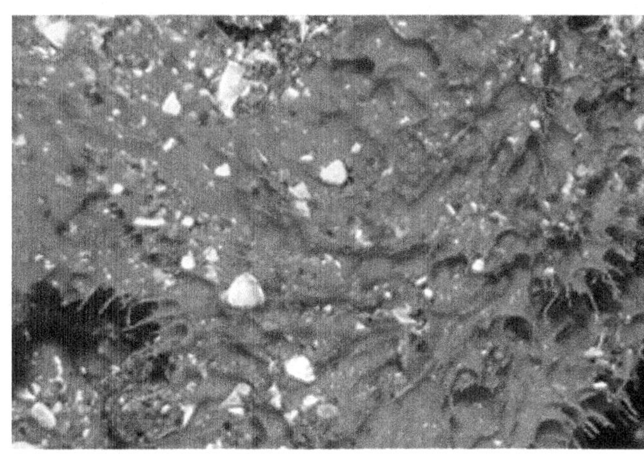

Subclass: Keratosa: Order: Dendroceratida
Family: Darwinellidae
Common name: Keratose sponge
Distribution: White Sea
Ecology: It inhabits coral reefs.
Biology: Not reported

Compound(s) and Activities

Antitumor activity: The glaciasterols A and B of this speices have shown antitumor activity (Anjum et al., 2016; Sipkema et al., 2005).

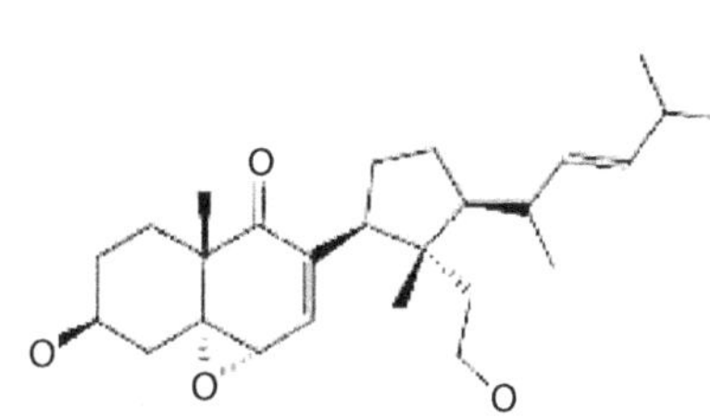

Glaciasterol A

Aplysilla rosea (Barrois, 1876)

Order: Dendroceratida Family: Darwinellidae

Common name: Rose sponge, Australian sponge

Distribution: Atlantic; North Sea; Norway; Sweden; Ireland; Channels of the Mediterranean Sea

Ecology: It inhabits undersurfaces of stones on the shore, and on rock and stones below low water from the littoral to 600 m depth.

Biology: Growth form is encrusting, thin (3–6 mm), and is in irregular patches of up to 20 cm in diameter. Surface is evidently conulose (1–3 mm) because of the dense dendritic "forest" of "small horny trees" forming the typical skeleton. Oscules (1–3 mm) are found scattered and are not evident. Inhalant apertures are rarely visible *in vivo*. Color of body is from rose to yellow. Skeleton of large ramified fibers are arising from a spongin basal plate strictly adhering to the substratum. Consistency is soft and compressible.

Reproduction: Larvae are yellow with a pigmented posterior pole and are scarcely ciliated but with a longer ring of cilia near the posterior pole.

Compound(s) and Activities

Anti-inflammatory activity: The diterpenes, 1, 5, and 6 aplyroseols, isolated from this species have shown anti-inflammatory activity (Keyzers, 2005).

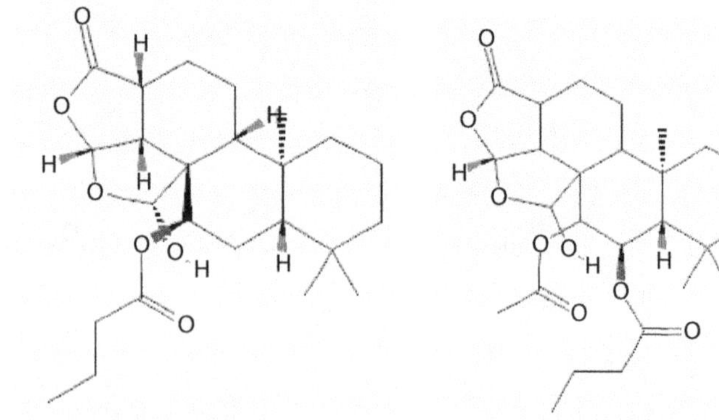

Aplyroseols 1,6

Antibacterial activity: The extracts of the associated actinobacterium, *Streptomyces* ACT-52A, of this sponge spcies have shown antibacterial activity (Mehbub et al., 2016).

Aplysilla sp.

Order: Dendroceratida Family: Darwinellidae

Compound(s) and Activities

Anti-inflammatory activity: The diterpenes gracilin A and 12-acetoxytetrahydroaplysulfurin-1 of this unidentified species have shown anti-inflammatory activity (Keyzers, 2005).

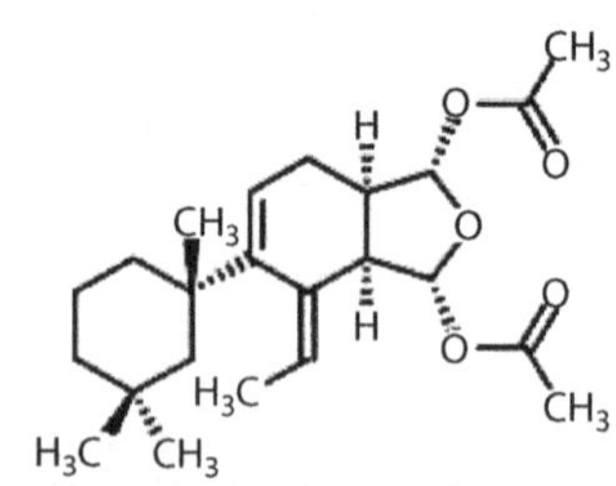

Gracilin A

Cytotoxicity and antihelmintic activity: The chromodorolides A (1) and B (2), and a new derivative, chromodorolide C (3), isolated from the ethyl acetate fraction of this unidentified species exhibited significant cytotoxicity against the P388 mouse leukemia cell line. At extract concentrations of 10 μg/mL, the percentage values of inhibition for the above compounds were 66, 70, and 42, respectively. The small-scale extract of the sponge, however, showed 94% inhibition at a concentration of 100 μg/mL. Chromodorolide A (1) also showed nematocidal activity against the larval stages of *Haemonchus contortus* and *Trichostrongylus colubriformis*, two important pathogens of sheep and other ruminants. A concentration of 100 μg/mL caused 94% and 95% inhibition of the development of *H. contortus* and *T. colubriformis* larvae, respectively (Rungprom et al., 2004).

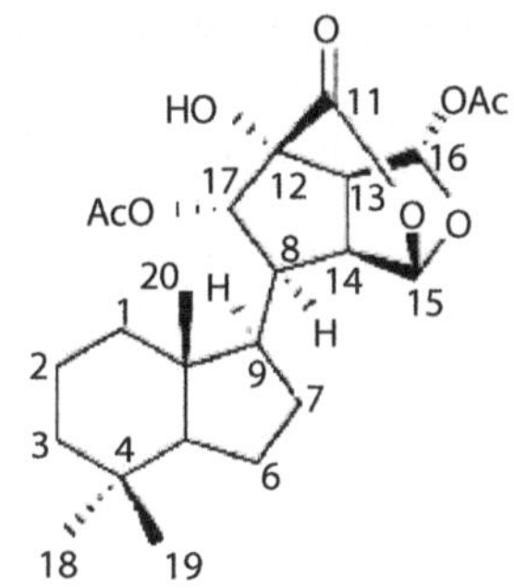

Chromodorolide A

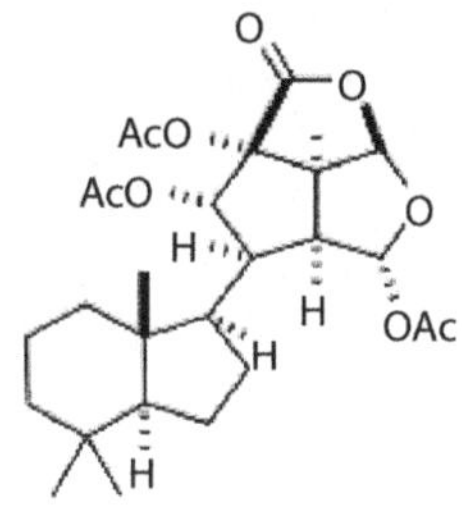

Chromodorolide B

Chelonaplysilla violacea (Lendenfeld, 1883) (= *Aplysilla violacea*)

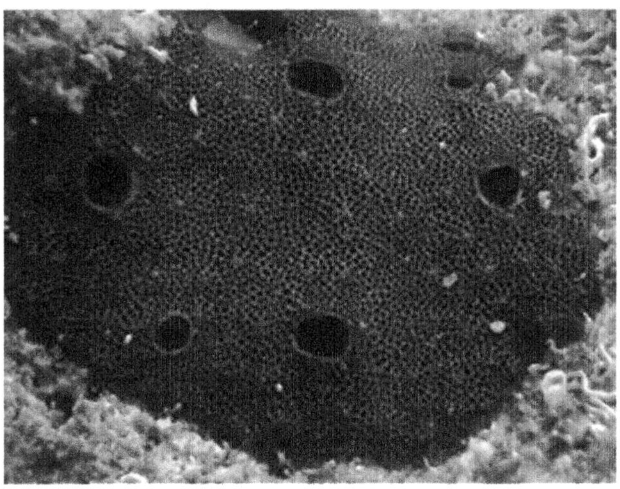

Order: Dendroceratida Family: Darwinellidae

Common name: Not known

Distribution: Central New Zealand, Hawaii, Northeastern New Zealand, Solomon Archipelago, South New Zealand

Ecology: It is a subtidal, shade-loving, encrusting sponge occurring in caves and archways, and boulders and solid reefs at all degrees of slopes.

Biology: This sponge is dark purple in color externally and internally with a size of 50 × 50 cm with 0.5 cm thickness. Texture is firm and is minutely corrugated. Surface is cornulose with sandy reticulation. It has a pin-prickled appearance.

Compound(s) and Activities

Anti-inflammatory activity: The diterpenes pourewic acid A, methylpourewate B, and cadlinolide C isolated from this species inhibited the production of superoxide in human peripheral blood neutrophils stimulated with either fMLP (IC50 74, 58, 13 μM, respectively) or PMA (IC50 77, 58, 13, μM respectively) (Keyzers, 2005).

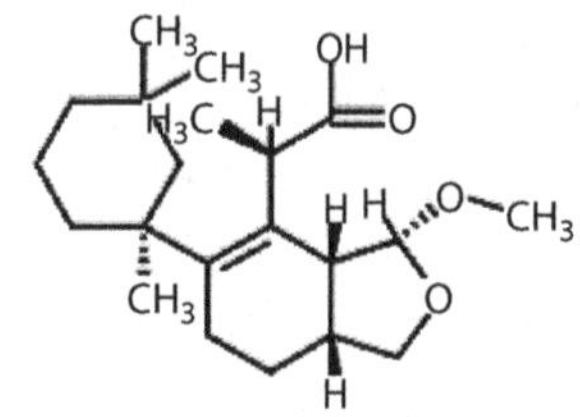

Pourewic acid A

Chelonaplysilla sp.

Order: Dendroceratida Family: Darwinellidae

Compound(s) and Activities

Anti-inflammatory and antimicrobial activities: The indole alkaloid chelonin A isolated from this unidentified species exhibited antiinflammatory activity in addition to the antimicrobial activity. Chelonin A showed a 60% inhibition against PMA-induced inflammation in a mouse ear model (50 μg/ear) (Gul et al., 2005).

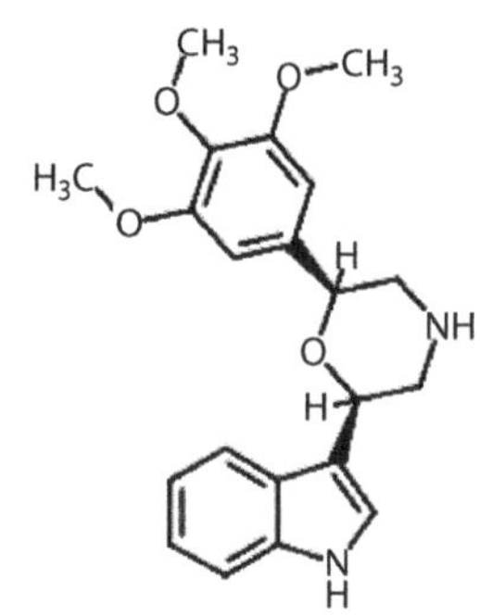

Chelonin A

Dendrilla cactos (Selenka, 1867)

Order: Dendroceratida Family: Darwinellidae

Common name: Antarctic sponge, prickly rose sponge

Distribution: Antarctica to the tropics; Indo-Pacific area, including Port Phillip and Victoria

Ecology: It occurs in coral reef and jetty areas, at depths of 0 to 50 m.

Biology: It is a dark pink sponge species with encrusting to branching form. Size of this sponge is about 30 cm.

Compound(s) and Activities

Anticancer activity: Two new bastadin alkaloids, bastadins-22 and -23, together with six known bastadins, namely, bastadins-6, -12 (formerly basatadin-9), -14, -15, -16, and -19, were isolated from this sponge. All these compounds have been shown to be active against Sup-T(1) cancer cells (T cell lymphoma) (Reddy et al., 2006).

This species has also yielded laminarins O, P, Q, and R. Among them, lamellarin O proved to be an effective cytotoxic agent in a number of leukemia and lymphoma screens. It was also active in human HeLa–S3 uterine and glioma screens. Further, this compound exhibited modest cytotoxic activity against both wild-type and MDR tumor cell lines, suggesting it may serve as new lead for the development of antitumor agents insensitive to MDR. Further, its antiproliferative activity was also much less efficient (IC50 >10 μM) (Bailly, 2015; Urban et al., 1994, 1995).

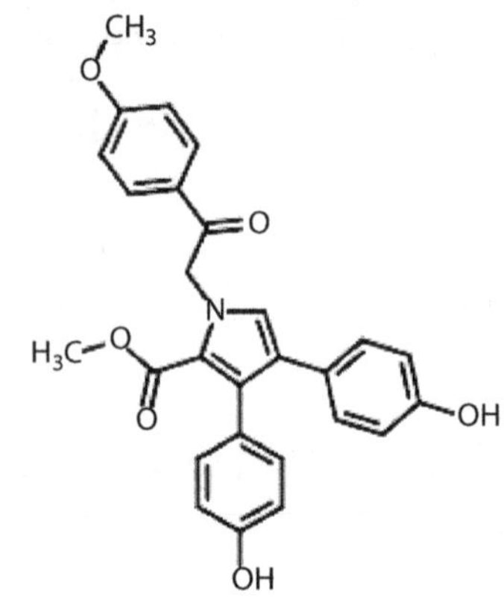

Lamellarin O

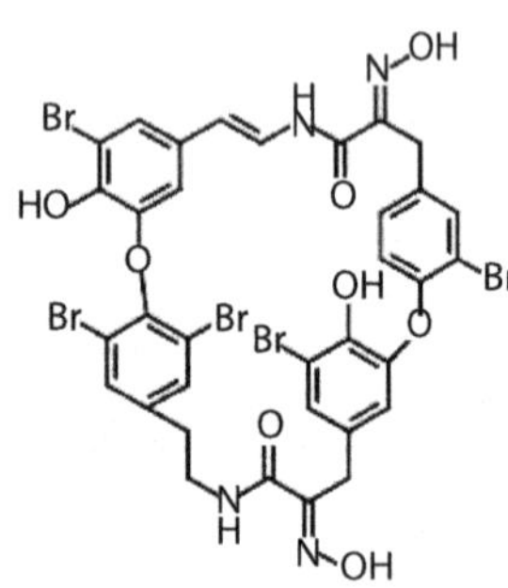

Bastadin 23

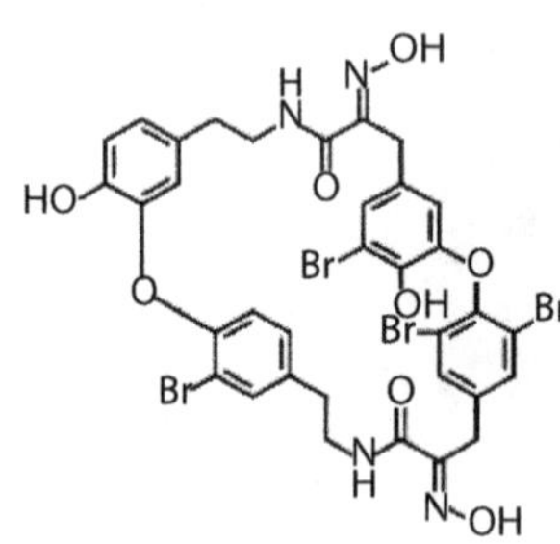

Bastadin 12

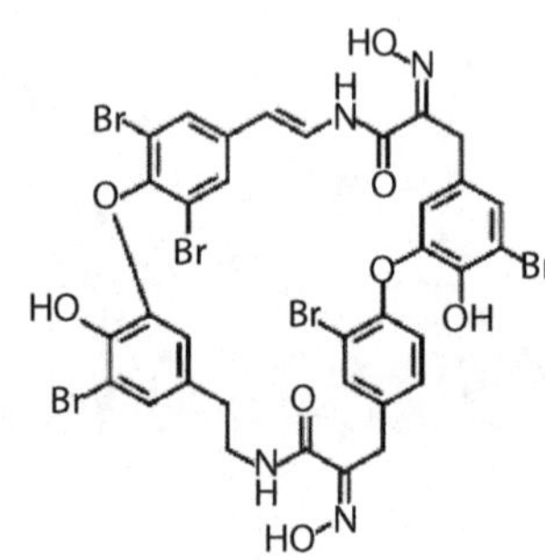

Bastadin 14

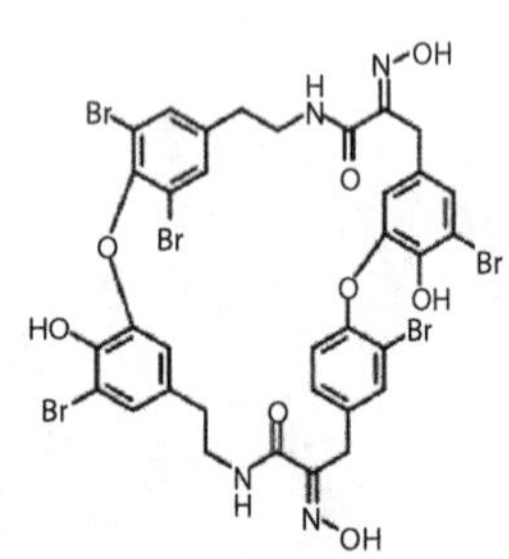

Bastadin 15

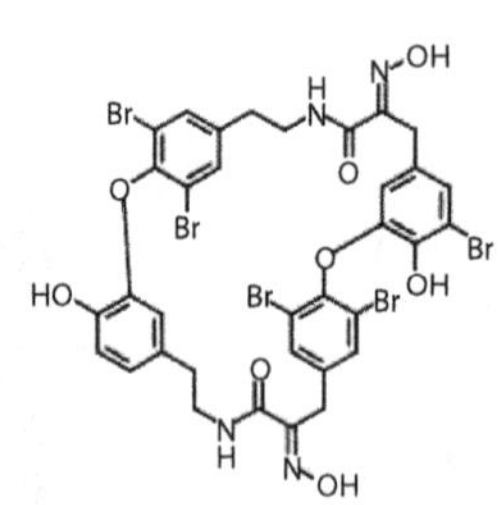

Bastadin 16

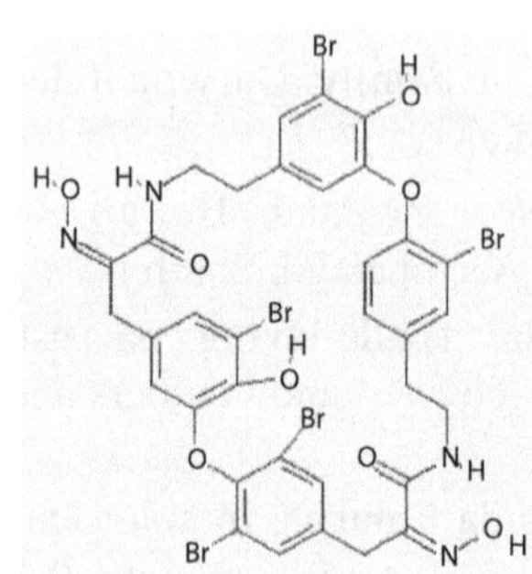

Bastadin 19

Antibacterial, anti-inflammatory, and enzyme-inhibitory properties: It contains cyclic bastadins, which are bromotyrosine-derived peptides endowed with Gram-positive antibacterial and anti-inflammatory properties; in addition they inhibit topoisomerase II, dehydrofolate reductase, and the endothelin A receptor (Motuhi et al., 2016).

Dendrilla membranosa (Pallas, 1766)

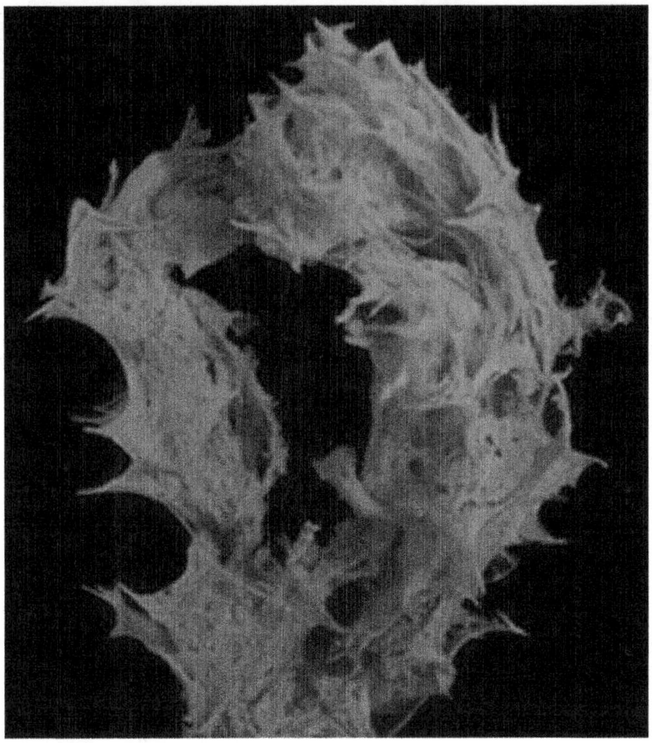

Order: Dendroceratida Family: Darwinellidae

Common name: Antarctic sponge

Distribution: Indian Ocean; Anvers Island, Antarctica

Ecology: It inhabits deeper waters.

Biology: It forms a cylindrical mass, with about 40 mm in diameter. The skeleton consists of a very open and irregular network of fibers. The dermis and the internal membranes consist of a tough membrane of a puce or dull purple color in spirit, almost black in the dry state.

Compound(s) and Activities

Antimicrobial activities: The diterpenes membranolides B–D of this sponge displayed Gram-negative antibiotic and antifungal activities (Ankisetty et al., 2004).

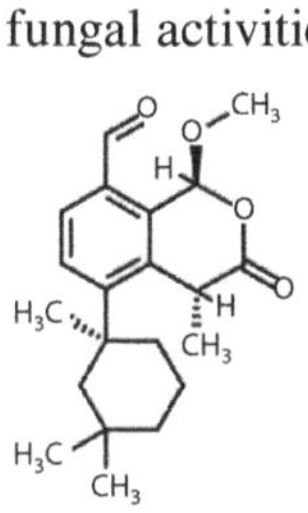

Membranolide B

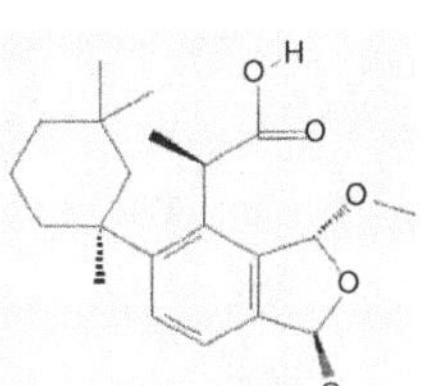

Membranolide C

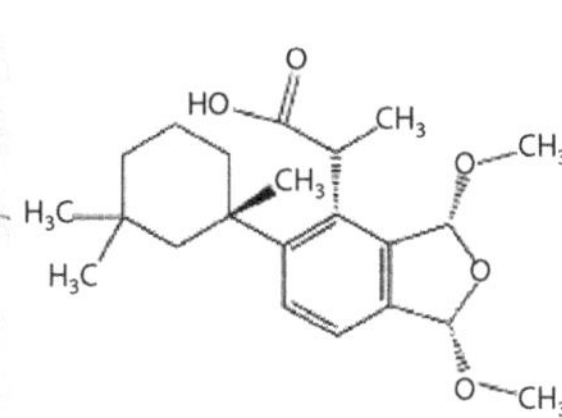

Membranolide D

The compound, 4,5,8-trihydroxyquinoline-2-carboxylic acid of this species has shown antimicrobial activities (Molinski and Faulkner, 1988).

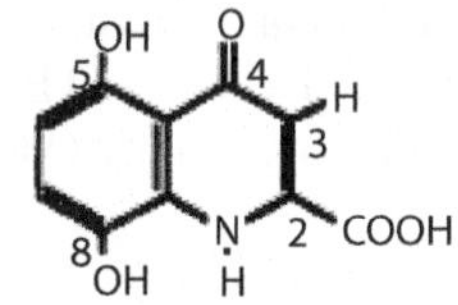

4,5,8-trihydroxyquinoline-2-carboxylic acid

Antibacterial activity: The diterpene compound, darwinolide of this species showed antibacterial activity against the highly methicillin-resistant *Staphylococcus aureus* (MRSA) with an MIC value of 132.9 μM. This compound has the potential to provide the backbone for new drug development in the fight against MRSA (von Salm et al., 2016).

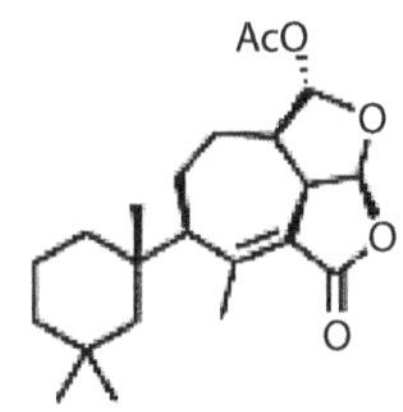

Darwinolide

Antileishmanial (antiprotozoal) activity: The compounds membranolides G and H of this species showed antileishmanial (antiprotozoal) activity against *Leishmania donovani* axenic amastigotes with IC50 values of 0.8 and 1.1 μM, respectively. The infection of this parasite manifests in humans in three disease forms: visceral leishmaniasis, cutaneous leishmaniasis, and mucosal leishmaniasis. The most common form (cutaneous leishmaniasis) causes skin sores, while the most deadly (visceral leishmaniasis) affects internal organs and can be fatal if left untreated (Witowski, 2015).

Dendrilla sp.

Order: Dendroceratida Family: Darwinellidae

Compound(s) and Activities

Anti-inflammatory activity: The diterpene compounds dendrillolide A and norrisolide showed anti-inflammatory activity by inhibiting PLA2 wih percentage inhibition values of 48 and 66, respectively, at an extract concentration of 2 μg/mL (Keyzers, 2005).

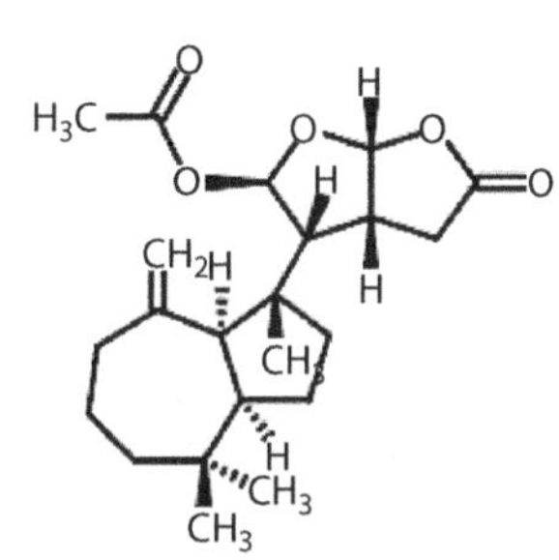

Dendrillolide A

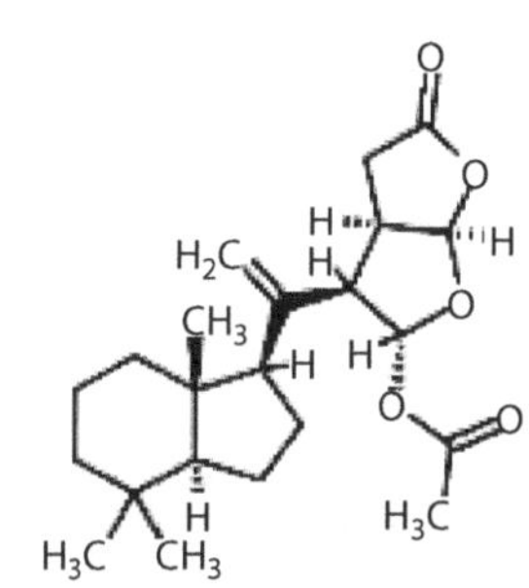

Norrisolide

Antitumoral (HIF-1 inhibitors) activity: The aromatic alkaloids (pyrrole derivatives) lamellarins A and B of this species showed antitumoral HIF-1 inhibiting activity (Motuhi et al., 2016).

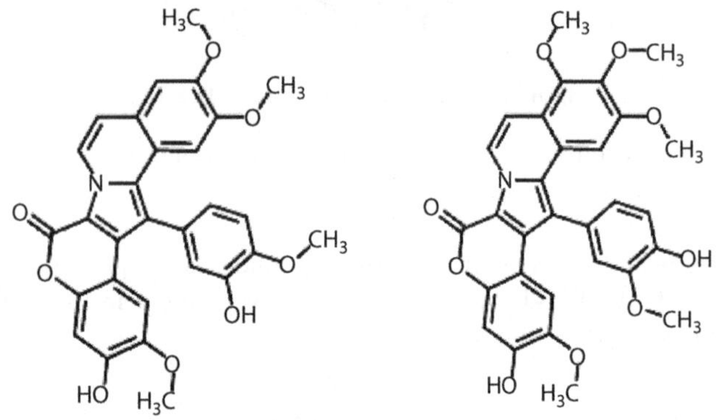

Lamellarins A,B

Enzyme inhibiting activity: The crude extracts of this unidentiied species (R171 stems) showed 100% inhibition on both bovine trypsin and porcine elastase activities (Motuhi et al., 2016).

FAMILY: DICTYODENDRILLIDAE

Dictyodendrilla verongiformis

Order: Dendroceratida Family: Dictyodendrillidae
Nothing is known about the biology and ecology of this species.

Compound(s) and Activities

Anticancer activity: The compounds dictyodendrins (A–E) isolated from this species were the potent inhibitors of the enzyme telomerase at 50 μg/mL level. Since telomerase is found nearly in 90% of cancer cells but not in any significant amount in normal cells, this finding could pave the way to the development of new kinds of anticancer drugs (Sugumaran and Robinson, 2010).

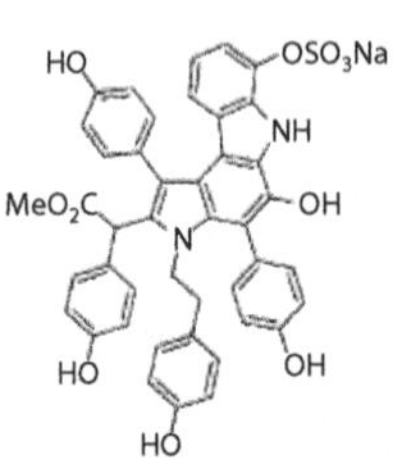

Dictyodendrin A

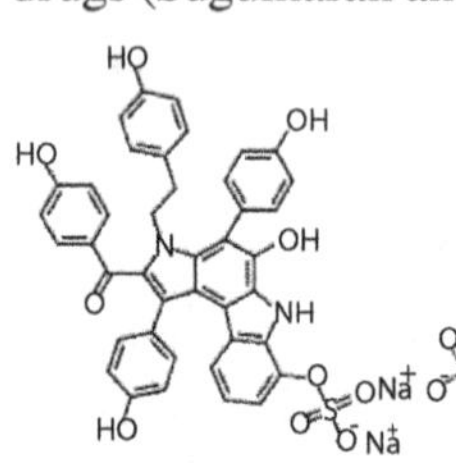

Dictyodendrin B

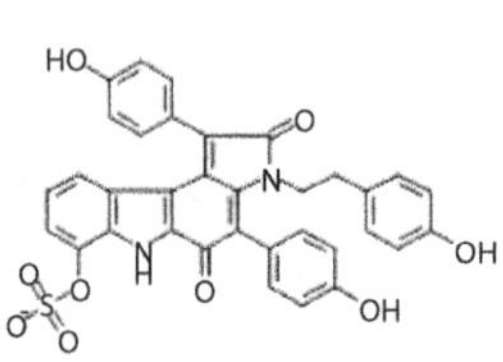

Dictyodendrin C

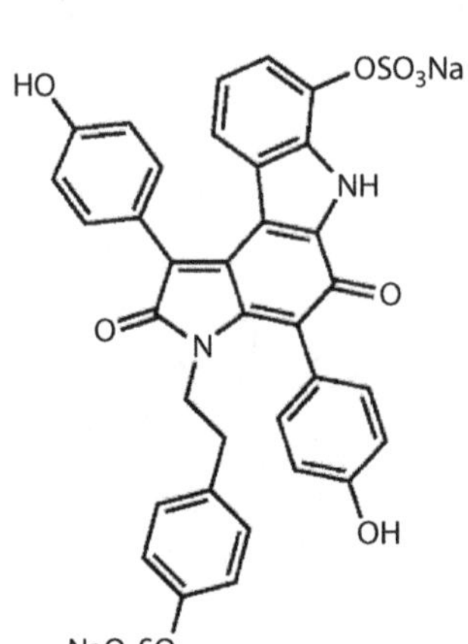

Dictyodendrin D

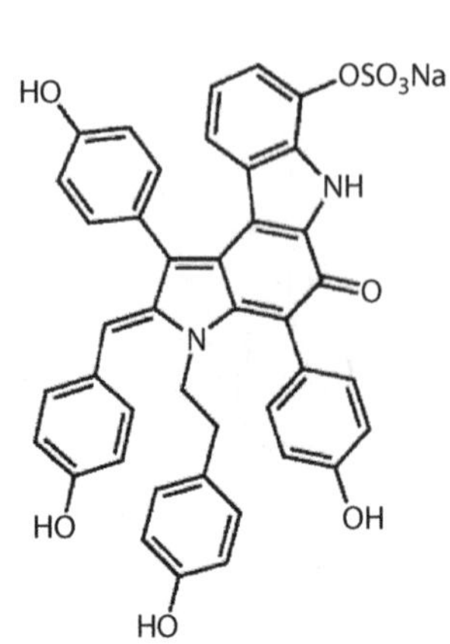

Dictyodendrin E

Telomerase inhibitory/anticancer activity: Five new alkaloids, dictyodendrins A–E, inhibited the enzyme telomerase completely at a concentration of 50 μg/mL. In normal cells, the activity of telomerase is suppressed, but it is activated in cancer cells. Telomerase continues to build telomeres in cancer cells (Warabi et al., 2003).

Acanthodendrilla sp.

Order: Dendroceratida Family: Dictyodendrillidae
Common name: Not designated, as it is an unidentified species.
Distribution: Western Mediterranean
Ecology: It dwells on detritic bottoms at a depth of 100 to 130 m.
Biology: The species of the genus possess reticulate fibrous skeleton with irregular mesh arrangement with all elements cored with detritus. Reticulation is more pronounced superficially, and ascending primary fibers project above the sponge surface. Spongin spicules are absent.

Compound(s) and Activities

Antifungal activity: The sulfated sterols isolated from this unidentified species, namely, acanthosterol I and J have shown antifungal activity (Sipkema et al., 2005).

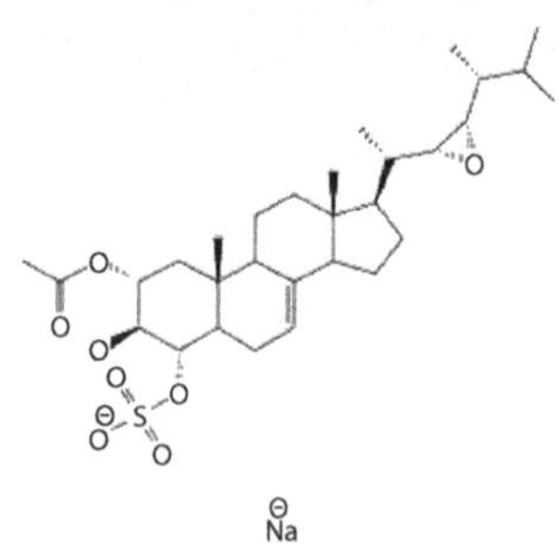

Acanthosterol sulphate I

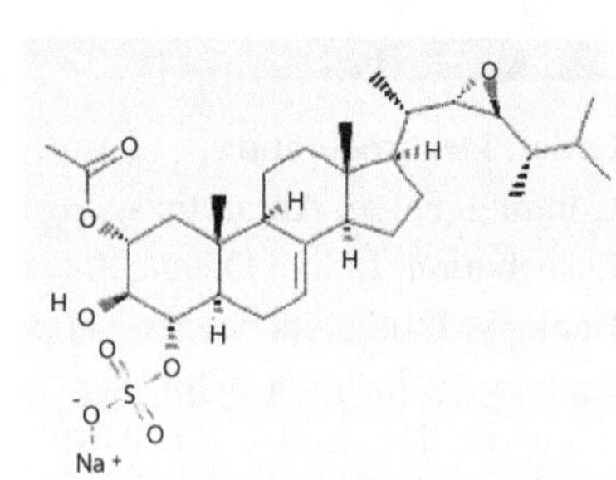

Acanthosterol sulphate J

Spongionella nigra (Dendy, 1889) (= *Dendrilla nigra*)

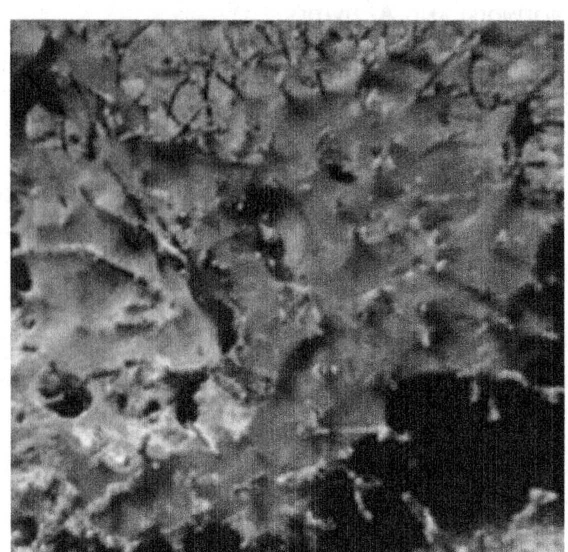

Order: Dendroceratida Family: Dictyodendrillidae
Common name: Horny sponge

Distribution: South India and Sri Lanka; Papua New Guinea; Bismark Sea

Ecology: It is a common shallow-water (lagoon) species living in a depth of about 20m.

Biology: This sponge is large and lamellate. Surface is very even and finely cornulose. Texture is compressible. There is a reticulate skeleton and the fiber reticulation is dense. It features an unusual "scouring pad" appearance with its spongin fibers protruding from an elastic tissue mass that can vary in color from dark green to reddish.

Compound(s) and Activities

Antimicrobial activity: The associated bacteria strains *Streptomyces dendra* and *Streptomyces* sp. BLT7 of this species showed antibacterial activity. The latter strain showed inhibitory interaction with *Micrococcus luteus*. The MIC of BTL7 against *M. luteus* was 44 µg protein/mL and the minimum bactericidal concentration was 88 µg protein/mL. Further, its other strain, *Nocardiopsis dassonvillei* MAD08), was prominent in its antibacterial and anticandidal activity against the multidrug-resistant pathogenic microbial strains (Selvin et al., 2004; Thomas et al., 2010).

Antibacterial activity: The ethanolic extract of this species exerted antibacterial activity against cerebrospinal fluid (CSF) disease (neurological disease)–causing pathogenic bacteria (Sujath et al., 2014). The inhibition zone diameter values for different species of bacteria are given next.

Inhibition Zone Diameter (in mm)

E. coli	*S. aureus*	*H. influenzae*	*N. meningitides*	*K. pneumoniae*	*Streptococcus agalactiae*	*E. faecalis*
4	11	9	5	17	2	6

Source: Sujatha, S. et al., 2014, *Pharma Innovation*, 3: 81–86.

Antitubercular activity: The 3,4-diaryl pyrrole alkaloid compounds, denigrins A–C, isolated from the ethyl acetate extract of this species showed potent antitubercular activity. Among these compounds, denigrin C showed the highest potency (MIC, 4 µg/mL) against Mycobacterium tuberculosis H37Rv (Kumar et al., 2014).

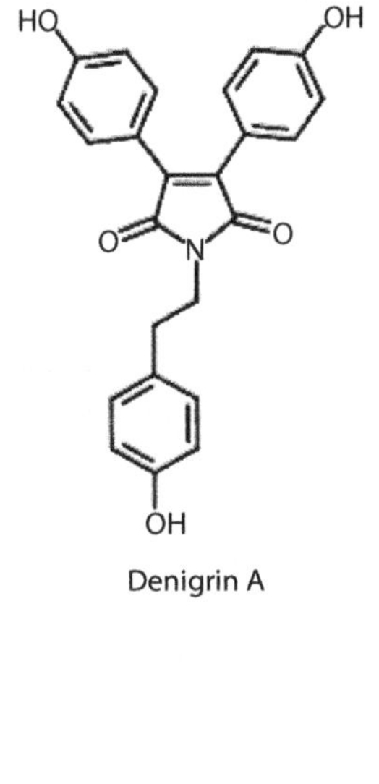

Denigrin A

The crude extracts of this were found to exhibit good antibacterial activity at concentration of 100 and 125 µL against most of the Gram-positive and Gram-negative organisms (Bhimba et al., 2011). The values of zone of inhibition for these bacterial species are given next.

Zone of Inhibition Diameter (in mm)

	Concentrations of Crude Extract (µL)			
	25	50	75	100
Proteus sp.	10	15	17	18
S. aureus	11	13	15	17
Pseudomonas aeroginosa	12	13	16	18
E. coli	10	15	17	18
Klebsiella sp.	12	15	17	18

Source: Bhimba, B.V. et al., 2011, *Asian Pac. J. Trop. Dis.*, 1: 299–303.

Antioxidant activity: The extract of this species showed DPPH antioxidant activity with a value of 50.8% (Bhimba et al., 2011).

Anti-inflammatory activity: The crude organic extracts of this species showed anti-inflammatory activity, the values of which were 58% at an extract concentration of 200 µL and 72% at 100 µL (Bhimba et al., 2011).

Antitumor activity: The crude organic extracts of this species showed antitumor properties (Motuhi et al., 2016).

Antitumor/HIF-1 inhibitory activity: The compound 7-hydroxyneolamellarin A (lamellarin-like phenolic pyrrole) isolated from this species inhibited hypoxia-induced HIF-1 activation (IC50 1.9 µM) in human breast tumor cell (T47D cells). Hypoxic induction of vascular endothelial growth factor (VEGF), a potent angiogenic factor and HIF-1 target gene, was also inhibited by this compound (Liu et al., 2007; Motuhi et al., 2016).

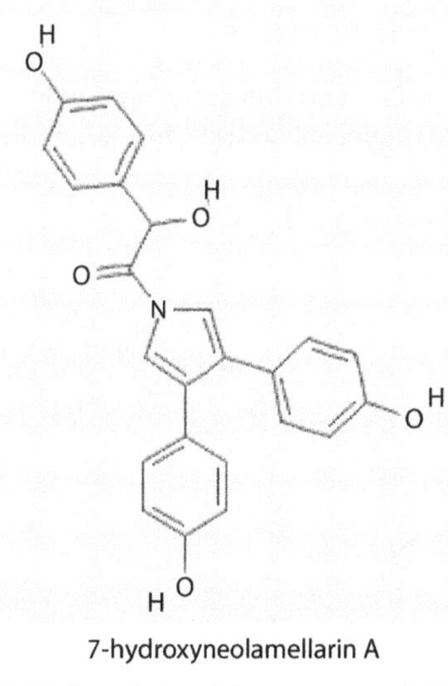

7-hydroxyneolamellarin A

Mosquito larvicidal activity: The extract of this species displayed larvicidal activity against the fourth instar larvae of *Aedes aegypti* (yellow fever mosquito vector) with an LC50 value of 64.0 ppm (Rao et al., 2008).

Spongionella sp.

Order: Dendroceratida Family: Dictyodendrillidae

Compound(s) and Activities

Neurosuppressives and muscle relaxant activity: The furanosesterterpenoid compound okinonellin B isolated from this unidentified sponge showed neurosuppressives and muscle relaxant activity (Sipkema et al., 2005).

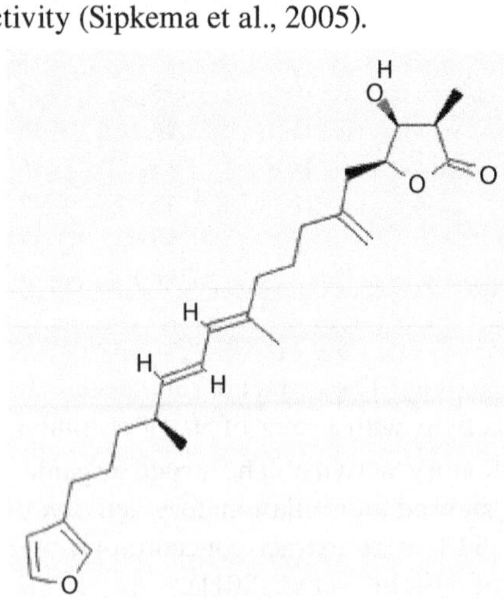

Okinonellin B

FAMILY: DYSIDEIDAE

Citronia sp.

Order: Dictyoceratida Family: Dysideidae
Common name: Australian sponge
Distribution: Australia
Ecology: Not available
Biology: Not reported

Compound(s) and Activities

Antifungal activity: The linear tetrapeptide citronamide A isolated from this species showed moderate antifungal activity against *Saccharomyces cerevisiae* at an MIC value of 8 μg/mL (Agrawal et al., 2016).

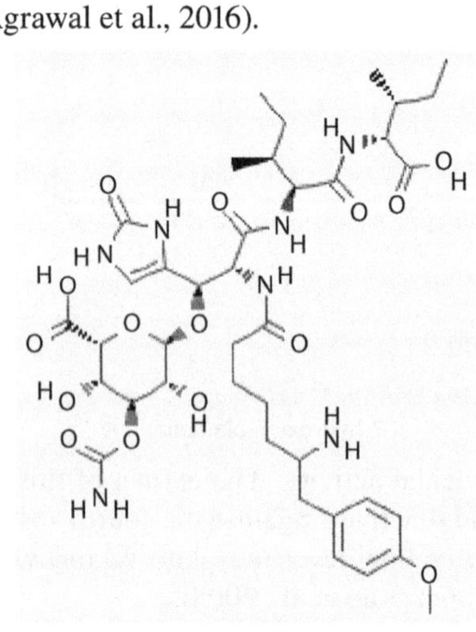

Citronamide A

Dysidea arenaria (Bergquist, 1965)

Order: Dictyoceratida Family: Dysideidae
Common name: Horny sponge
Distribution: Pacific Ocean, New Caledonia, Society Islands, Palau Islands
Ecology: It occurs in reef habitats and is found at a depth zone 15–18 m in an area with sandy substrate and coral patches.
Biology: In this species, a mineral skeleton is absent and a skeleton of organic fibers is present instead. Surface is conulose, with conules measuring 1–5 mm high and 20–50 mm apart. Texture is rubbery and less compressible. Its color in life is light brown. However, the color is apparently variable, including pale blue, cream, white, pink, and gray. Fibers are not differentiated into primary and secondary fibers and are arranged in a reticulate pattern forming irregular meshes. All fibers are fully cored with detritus.

Compound(s) and Activities

Antitumor activity: The macrocyclic lactan arenastatin A isolated from this species showed antitumor activity by serving as a tubulin polymerization inhibitor (Sipkema et al., 2005).

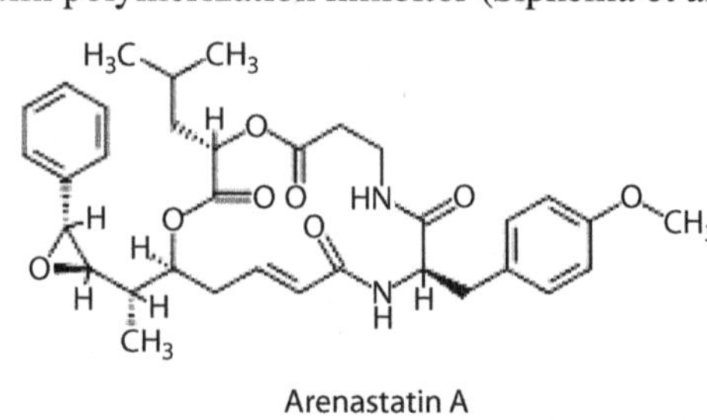

Arenastatin A

Cytotoxic activity: The compound arenastatin A, a cyclodepsipeptide, isolated from this species has demonstrated a potent cytotoxicity against KB cells with an IC50 value of 5 pg/mL (Suarez-Jimenez et al., 2012).

Protein kinase inhibitors: The novel sesquiterpene hydroquinone isoarenarol and the known compound arenarol isolated from the organic extracts of this species showed potent and selective protein kinase inhibition *in vitro* (Yoo et al., 2003).

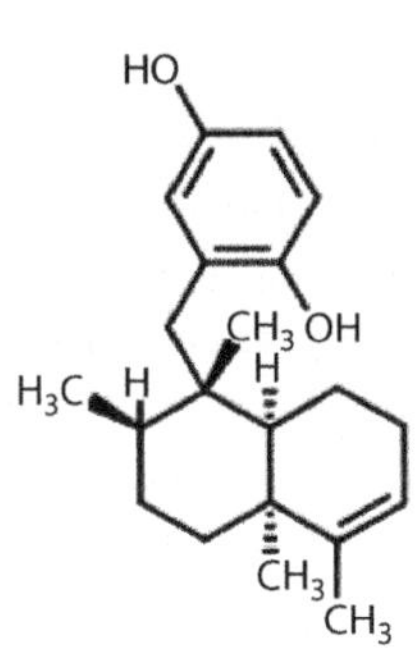

Isoarenarol

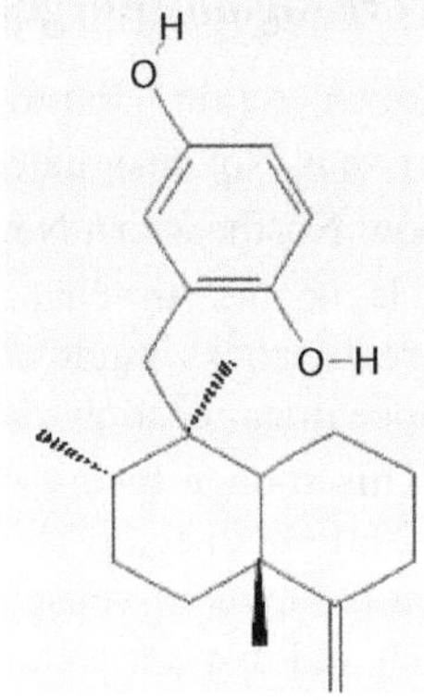

Arenarol

Antiviral activity: The new sesquiterpenoid hydroquinone, 19-hydroxy-polyfibrospongol B (1), along with five known compounds—polyfibrospongol B (2), isosemnonorthoquinone (3), ilimaquinone (4), smenospongine (5), and smenotronic acid (6)—have been isolated from this species. Among these, compounds 3–6 exhibited moderate *in vitro* anti-HIV activity on HIV-1 RT with IC50 values of 239.7, 16.4, 176.1, and 130.4 μM, respectively (Qiu and Wang, 2008).

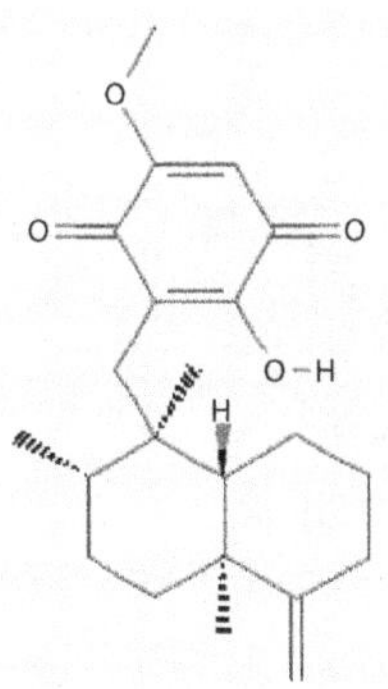

Ilimaquinone

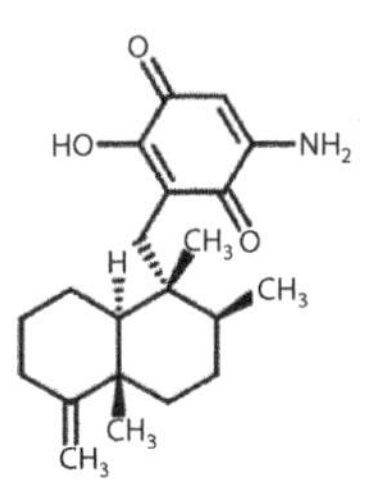

Smenospongine

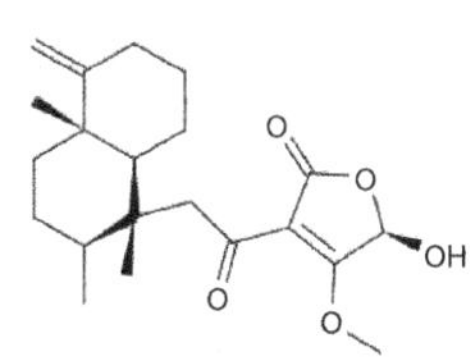

Smenotronic acid

Dysidea avara (Schmidt, 1862)

Order: Dictyoceratida Family: Dysideidae

Common name: Acquisitive sponge

Distribution: Galicia, Mediterranean

Ecology: It occurs in caves and on vertical cliffs, from 1 m downward (to at least 100 m).

Biology: It is massive and occasionally somewhat lobate. Its size is up to 25 × 15 × 5 cm. Surface is strongly conulose and the conules are up to 4 mm high and 2–5 mm apart. The conules have a strongly reticulate appearance. Oscules (1–3 mm in dia) are present at the summit lobes. Consistency of the sponge is soft. Spicules are absent. Color of the animal is soft rosy to dark maroon or violet.

Compound(s) and Activities

Antimicrobial activity: The organic extracts of this species have shown antimicrobial activity against Gram-positive bacteria and fungi. The MIC values recorded for the different species are given next.

MIC (μg mL^{-1})				
S. aureus	*E. faecalis*	*B. subtilis*	*C. albicans*	*C. parapsilosis*
128	128	64	16	16

Source: Orhan, I.E. et al., 2012, *Rec. Nat. Prod.*, 6: 356–367.

Cytotoxicity: The sponge contains the compounds avarol (1) and avarone (2). Avarol on acylation yielded 2′,5′-O-dibenzoylavarol (3); 2′,5′-O-(4-chlorobenzoyl)avarol (4); 2′,5′-O-dicinnamoylavarol (5); 2′,5′- O -(4-bromobenzoyl)avarol (6); 2′,5′-O-dioctanoylavarol (7); 2′,5′-O-(4-fluorobenzoyl)avarol (8) and diacetylavarol (9) have been isolated from this species. Among these compounds, compounds 6 and 9 showed selective cytotoxicity against Hepa (human hepatoma) and KB cell lines, respectively (Shen et al., 2003).

Four new sesquiterpene quinols, dysiquinols A–D (1–4), and four new sesquiterpene quinones, (5S,8S,9R,10S)-18-ethoxyneoavarone (5), (5S,8S,9R,10S)-19-ethoxyneoavarone (6), (5R,8R,9S,10R)-18-ethoxyavarone (7), and (5R,8R,9S,10R)-19-ethoxyavarone (8), have been isolated from this species. Among these compounds, the compound dysiquinol D (4) showed cytotoxicity against the human myeloma cancer cell line NCI-H929 (IC50 value, 2.8 μM) and NF-κB inhibitory activity (IC50 value, 0.8 μM) (Jiao et al., 2015).

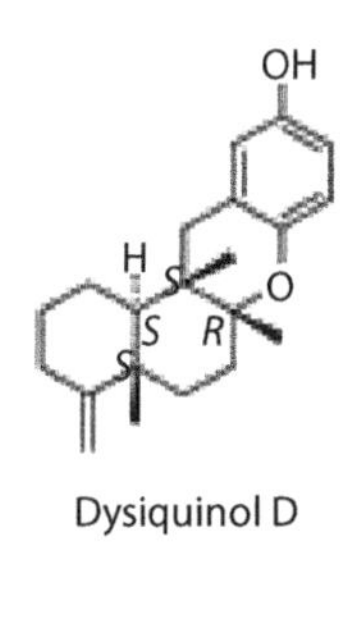

Dysiquinol D

AChE inhibitory activity: The organic extracts of this species have shown acetylcholinesterase (AChE) inhibitory activity (67%) at a concentration of 200 μg mL^{-1} (Orhan et al., 2012).

Antioxidant activity: The extracts of this species have shown DPPH radical scavenging effect (with 92% inhibition) at a concentration of 800 μg/mL (Aktas et al., 2013). Orhan et al. (2012) however reported that the extract of this species showed DPPH radical scavenging activity (with 15% inhibition) at a concentration of 2000 μg mL^{-1}.

Anti-inflammatory activity: The hydroquinone avarol isolated from this species showed anti-inflammatory activity by inhibiting mouse paw with an ED50 value of 9.2 mg/kg (Keyzers, 2005).

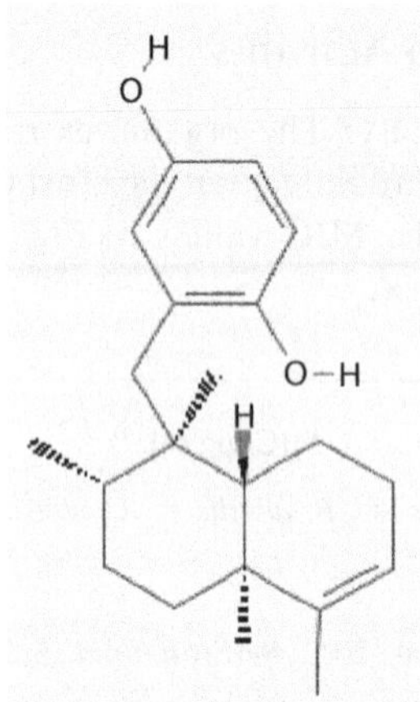

Avarol

Antiangiogenic and antimicrobial properties: The compound, 2-methylthio-1,4-naphthoquinone isolated from the associated, unidentified species of bacterium of this sponge species showed strong antiangiogenic and antimicrobial properties (Thomas et al., 2010).

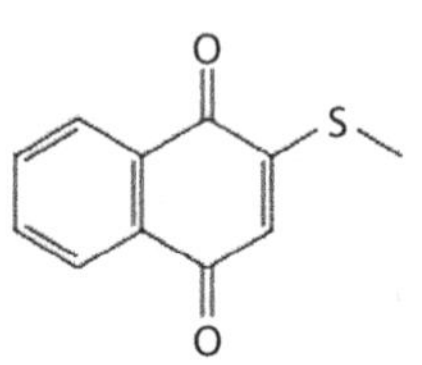

2-methylthio-1,4-naphthoquinone

Antiprotozoal activity: The associated actinobacterial strain *Nocardiopsis* sp. RV163 of this speices has yielded the compound, 1,6-dihydroxyphenazine, which showed antiprotozoal activity against *T. brucei* with an IC50 value of 4.03 μg/mL (Indraningrat et al., 2016).

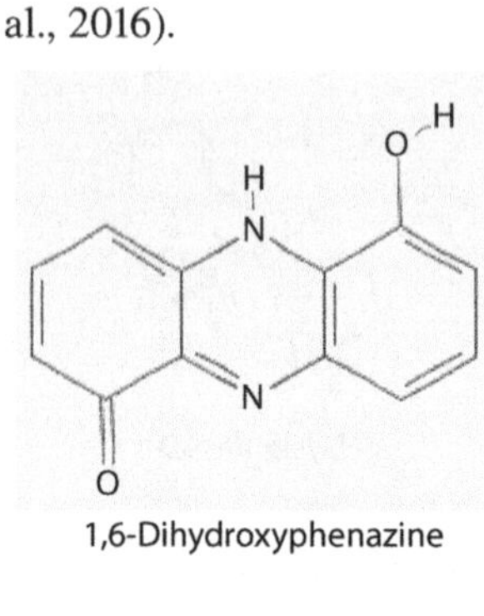

1,6-Dihydroxyphenazine

Dysidea cristagalli (Bergquist, 1961)

Order: Dictyoceratida Family: Dysideidae

Common name: Not designated

Distribution: Northeastern New Zealand

Ecology: It occurs in caves, coralligenous communities, artificial reefs, rocky/muddy/detritic bottoms, lagoons, and Posidoniaoceanica meadows at a depth range of 1 to 100 m.

Biology: This sponge forms a spherical mass of meandering ridges and interconnected short blunt branches. Oscules are situated on the ends of short branches. The surface is sandy to the touch and distinctly conulose, raised by the subsurface sandy fibers. Texture in life is somewhat compressible and is elastic. The color of the animal in life is grayish white.

Compound(s) and Activities

Anti-inflammatory activity: The compounds sesquiterpene, quinone, and acetylated hydroquinone isolated from this species showed anti-inflammatory activity by inhibiting superoxide production *in vitro* with IC50 values of 11 μM and 3 μM, respectively (McNamara et al., 2005).

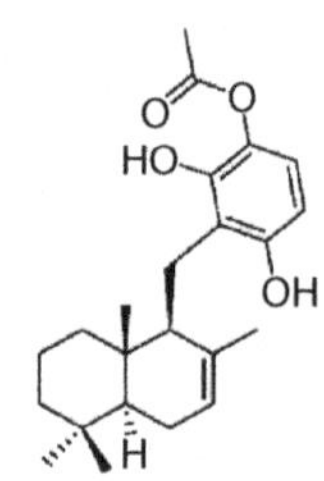

Acetylated hydroquinone

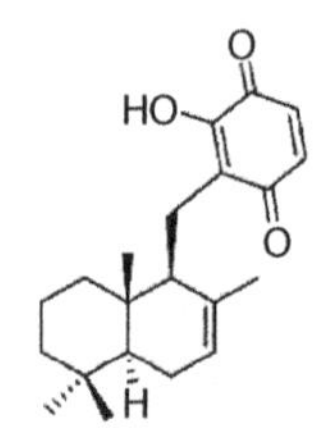

Quinone

Dysidea frondosa (Berquist, 1995)

Order: Dictyoceratida Family: Dysideidae

Common name: Not designated

Distribution: Central Pacific, French Polynesia and New Caledonia; Society Islands

Ecology: It is inhabiting crevices on open reef slopes and channels between 8 and 30 m depth.

Biology: It is a repent, multiple flattened lobate sponge with projections arising from a spreading base. It has a size of 12 × 12 × 8 cm. Color of the animal is pink-purple in life. Scules are large and are scattered. Texture is soft, flexible, and is easily torn. Surface is with irregularly distributed conules that are 1 to 1.5 mm (height).

Compound(s) and Activities

Anti-inflammatory activity: The hydroquinone sesquiterepene compounds frondosins A–E isolated from this species showed anti-inflammatory activity by inhibiting two different IL-8 receptors (IL-8Ra and 8Rb) and the protein kinase C enzyme (Keyzers, 2005; Patil et al., 1997).

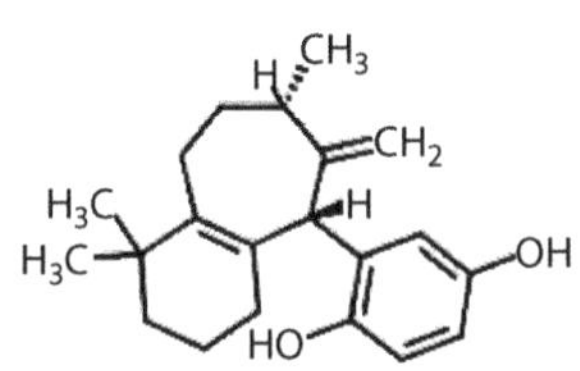

Frondosin A

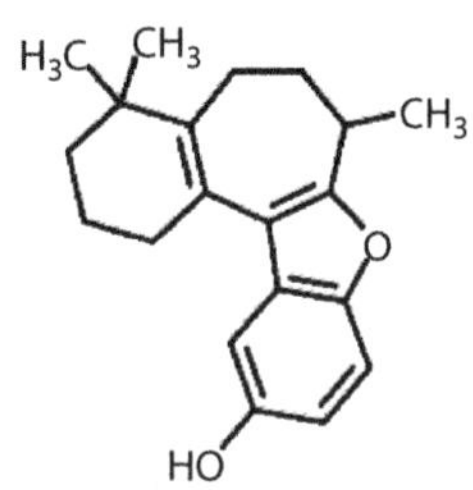

Frondosin B

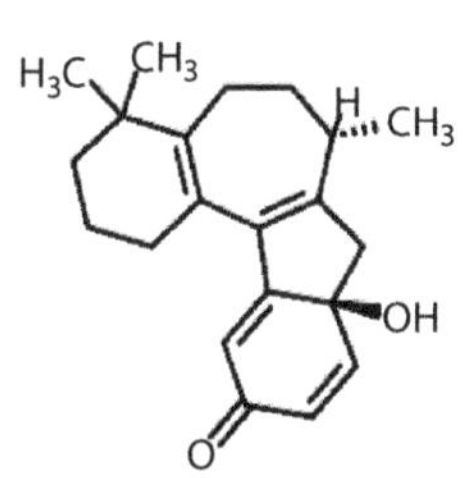

Frondosin C

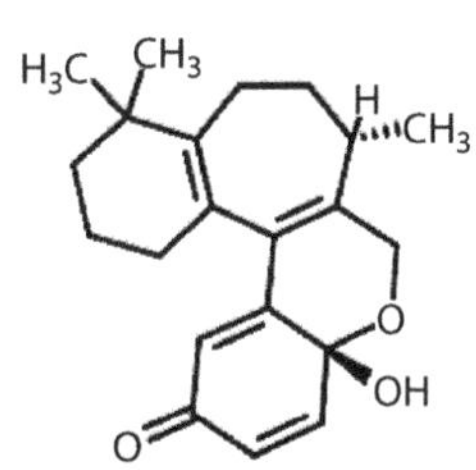

Frondosin E

The IC50 values recorded for these compounds are given next.

Interleukin Inhibition by Frondosins A–E (IC50 [mM])

Compound	IL-8 Rα	IL-8 Rβ	PKC-α
A	3.4	3.2	1.8
B	9.6	10.8	4.8
C	84	26.3	20.9
D	98	10.8	26
E	64	37.1	30.6

Source: Keyzers, R.A., and M.T. Davies-Coleman, 2005, *Chem. Soc. Rev.*, 34: 355–365.

Dysidea incrustans (Schmidt, 1862)

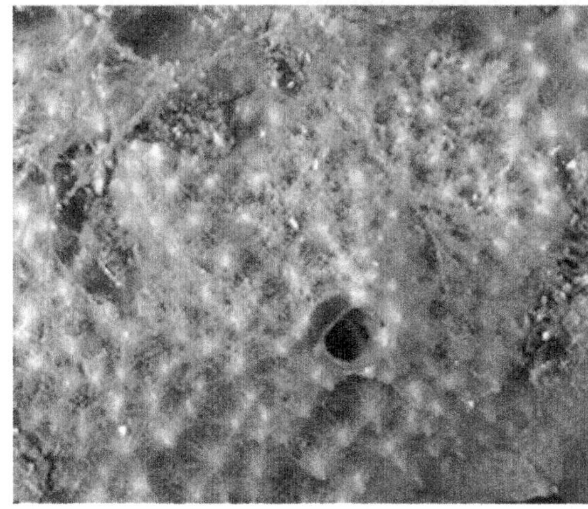

Order: Dictyoceratida Family: Dysideidae

Common name: Horny sponge

Distribution: Adriatic Sea, Azores Canaries Madeira, Ionian Sea, North Atlantic Ocean, eastern Mediterranean

Ecology: It inhabits caves.

Biology: It is an encrusting species with scattered small oscules and visible inhalant pores. The size of the animal is about 20 cm. There is a reticulate skeleton with a secondary network of slimmer fibers, which are almost free of inclusions.

Compound(s) and Activities

Antitumor/cytotoxic activity: The steroids incrustasterols A and B isolated from this species have shown antitumor activity (Sipkema et al., 2005). Casapullo et al. (1995) reported that these steroids were cytotoxic against human non-small cell lung, renal, and melanoma carcinoma cell lines.

Dysidea pallescens (Schmidt, 1862)

Order: Dictyoceratida Family: Dysideidae
Common name: Not designated
Distribution: North Atlantic Ocean, Black Sea, Adriatic Sea
Ecology: It inhabits depths of 14 m.
Biology: It is an encrusting sponge. No other information is available.

Compound(s) and Activities

Antimicrobial activity: The diethyl ether extract of this species showed bactericidal activity against *Staphylococcus aureus aureus* (ATCC 1764) (MBC = 10 mg/mL) and *Bacillus subtilis spizizenii* (ATCC 6633) (MBC = 20 mg/mL). On the other hand, the methanol extract showed fungicidal activity against *Aspergillus fumigatus* (PTCC5009) (MFC = 5mg/mL) and *Candida albicans* (ATCC10231) (MFC = 1.5 mg/mL) (Nazemi et al., 2017b).

Dysidea tupha (Martens, 1824)

Order: Dictyoceratida Family: Dysideidae
Common name: Mediterranean sponge
Distribution: North Atlantic Ocean
Ecology: It inhabits caves, rocky/detritic/muddy bottoms, coralligenous communities, and lagoons at a depth range of 1 to 450 m.
Biology: The growth form of this species is as a meshed irregular network of cylindrical processes (8–10 cm in length, 0.5–1 cm in dia) lying on the substratum. Color of the animal is whitish to pale light brown. Surface is finely and irregularly conulose (0.3–1 mm high and apart). Oscules are small (1 mm) and are irregularly scattered. Skeleton network is with irregular or quadrangular meshes (ca. 0.5 mm).

Compound(s) and Activities

Antiprotozoal activity: The associated actionobacterial strain, *Kocuria* sp. RV89, of this species has shown antiprotozoal activity against *T. brucei* and the percentage of growth inhibition was found to be 19% (Indraningrat et al., 2016).
Antihelmintic activity: The naphthacene glycoside SF2446A2 isolated from the associated *Streptomyces* sp. strain RV15 of this sponge species showed inhibitory activity against *Chlamydia trachomatis* and was able to inhibit the primary infection in a dose-dependent manner, as well as progeny formation. Further, it caused disruptive effects on the surface area of *Schistosoma mansoni* and affected the gonads by impairing oogenesis and spermatogenesis. The species of chlamydia or schistosomes (trematode blood flukes) cause infectious diseases that are a major health problem worldwide, particularly in developing countries (Reimer et al., 2015).

Dysidea villosa (Lendenfeld, 1886)

Order: Dictyoceratida Family: Dysideidae
Common name: Not designated
Distribution: Bassian
Ecology: Not reported
Biology: Not studied

Compound(s) and Activities

Neuroprotective activity: The compound 12-O-deacetyl-scalarin isolated from this species exhibited nerve growth factor (NGF) synthesis and stimulated the activity in cultured astroglial cells at a concentration of 30 μg/mL. This has led to a patent for the treatment of Alzheimer's disease (González et al., 2010).

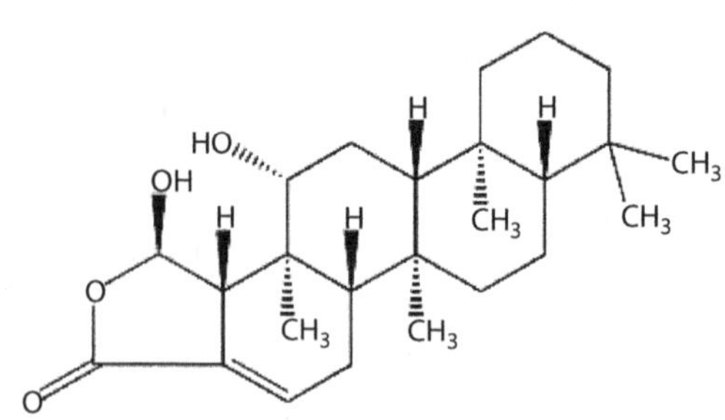

12-O-Deacetyl-scalarin

Anti-allergic activity: The meroterpenoids dysivillosins A–D (1–4) isolated from the organic extract of this species have shown anti-allergic activity. All these compounds potently inhibited the release of β-hexosaminidase in a dose-dependent manner with IC50 values ranging from 8.2 to 19.9 μM. Further, these compounds were found to downregulate the production of lipid mediator leukotrienes B4 (LTB4) and pro-inflammatory cytokine interleukin-4 (IL-4) in the antigen-stimulated RBL-2H3 mast cells (Jiao et al., 2017a).

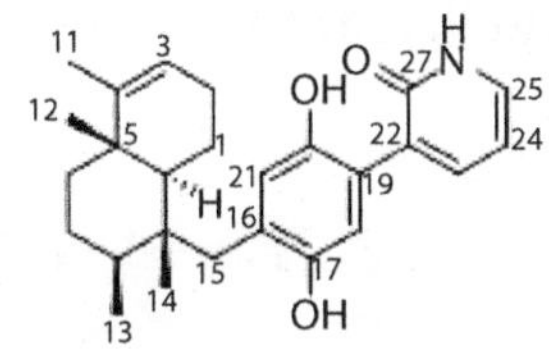

Dysivillosin B

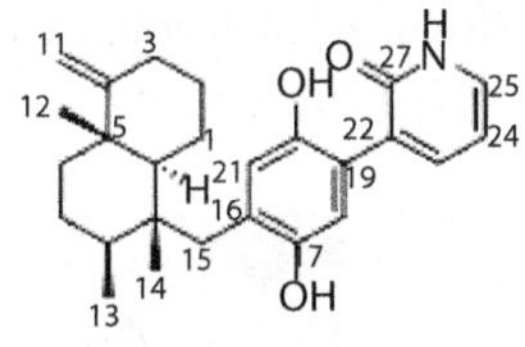

Dysivillosin A

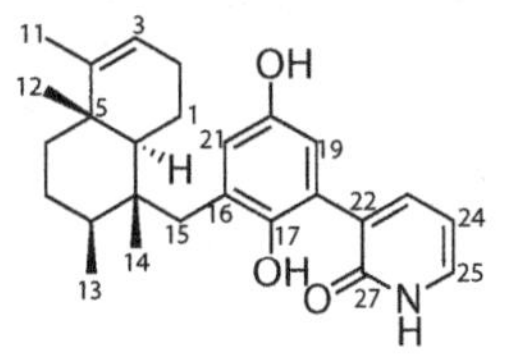

Dysivillosin C

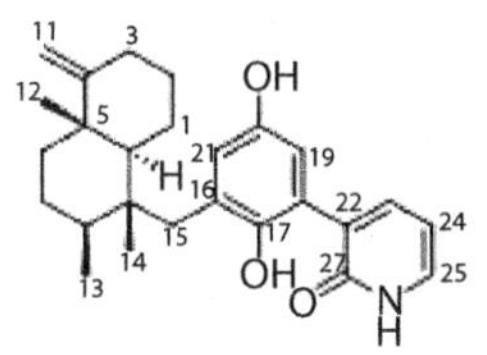

Dysivillosin D

Antidiabetic activity and cytotoxicity: A new sesquiterpene quinone, 21-dehydroxybolinaquinone (5), alongwith two known related analogs, namely, bolinaquinone (6) and dysidine (7), have been isolated from this species. Among these compounds, compound 7 showed the strongest hPTP1B inhibitory activity with an IC50 value of 6.7 μM, and compound 6 had significant cytotoxic activity against the HeLa cell line with an IC50 value of 5.5 μM. Compound 5, however, showed moderate PTP1B inhibitory activity and cytotoxicity with IC50 values of 39.5 and 19.5 μM, respectively (Li et al., 2009). Interestingly, the compound dysidine has now entered preclinical trials for the treatment of diabetes (Lauritano and Lanora, 2016).

Dysidea sp.

Order: Dictyoceratida Family: Dysideidae

Anti-inflammatory activity: The drimane sesquiterpenoid compound dysidotronic acid isolated from this unidentified species showed anti-inflammatory activity by inhibiting phospholipase A2 (Sipkema et al., 2005).

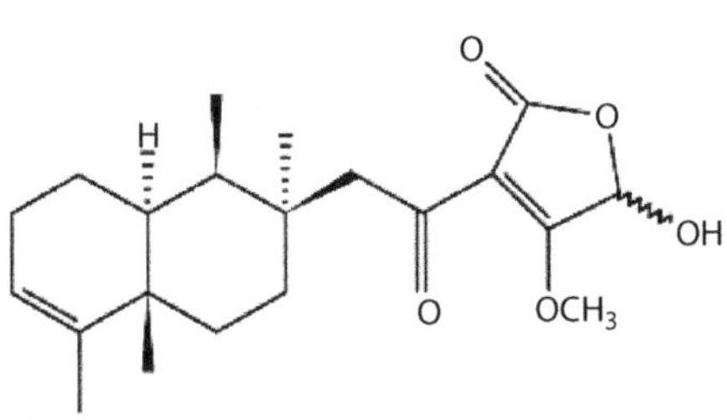

Dysidotronic acid

The metabolites bolinaquinone, dysidotronic acid, dysidenones A and B, and dysidine of this species possess anti-inflammatory activity against four different secretory PLA2 enzymes, belonging to groups I (Naja naja venom and porcupine pancreatic enzymes), II (human synovial recombinant enzyme), and III (bee venom enzyme). The compound bolinaquinone has also been shown to inhibit mouse ear edema induced by TPA via both topical and oral administration (ID50 76.7 μg per ear and 5.6 mg kg^{-1}, respectively). Further, the sterols 56–58 of this species have shown anti-inflammatory activity by inhibiting IL and 8Ra with IC50 values of 20, 5.5, and 4.5 μM, respectively (Keyzers, 2005).

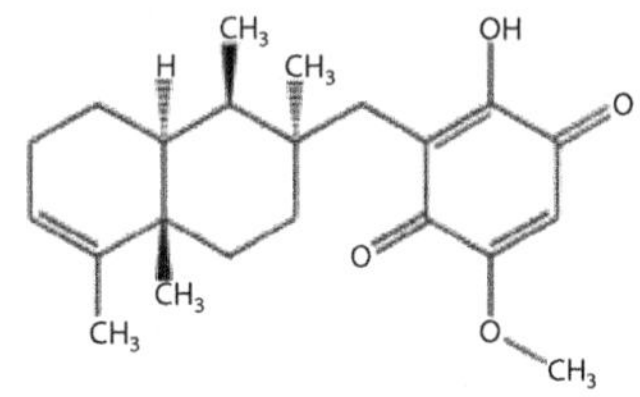

Bolinaquinone

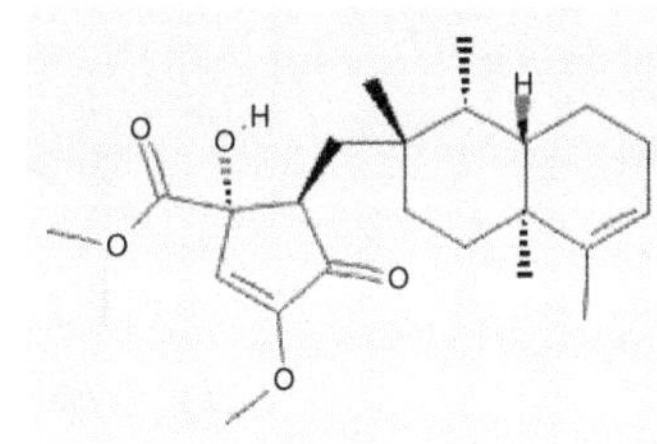

Dysidenone A

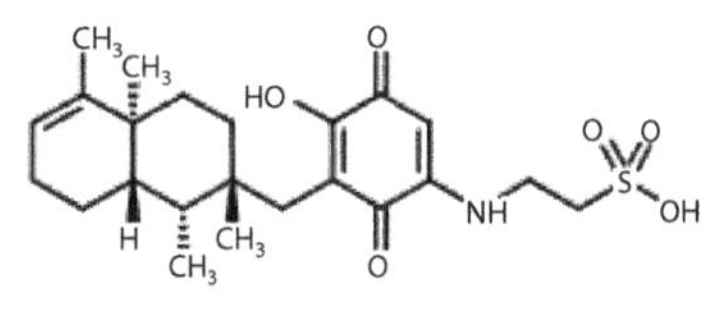

Dysidine

Cytotoxic and antibacterial activities: The associated bacterium, *Vibrio* sp. of this sponge species has been reported to synthesize cytotoxic and antibacterial tetrabromodiphenyl ethers (Thomas et al., 2010).

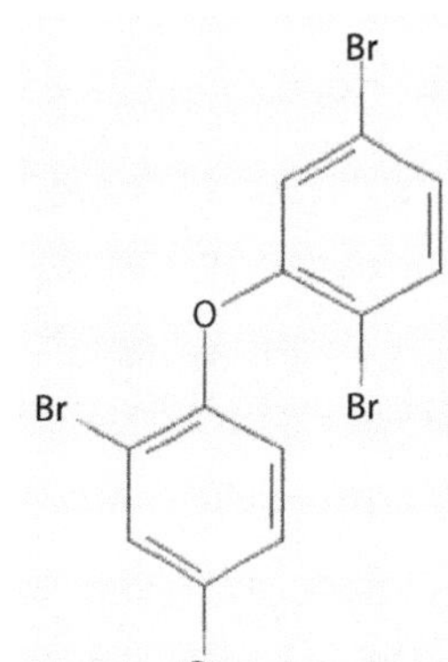

Tetrabromodiphenyl ether

The compounds 12α-acetoxy-20, 24-dimethyl-25-norscalar-16-en-24-one, 12α-acetoxy-13β, 18β-cyclobutan-20, 24-dimethyl-24-oxoscalar-16-en-25β-ol, 12α-acetoxy-24,25-epoxy-24-hydroxy-20, 24-dimethylscalarane, and scalardysin-A isolated from this unidentified sponge showed cytotoxicity and antibacterial activity against some Gram-positive and Gram-negative strains (Hassan et al., 2015).

Immunosuppressive activity: The polyoxygenated sterols isolated from this species have shown immunosuppressive activity by inhibiting IL 8. Further, these sterols showed activities against blood-related diseases (Anjum et al., 2016).

Neurosuppressives and muscle relaxant activity: The benzothiazole derivative S1319 of this unidentified species has shown neurosuppressives and muscle relaxant activity (Sipkema et al., 2005).

Antibacterial activity: The compound dysidinoid A of this species showed potent antibacterial activity against two strains of pathogenic bacteria namely methicillin-resistant *Staphylococcus aureus* (MRSA) with an MIC90 values of 8.0 µg/mL (Jiao et al., 2014).

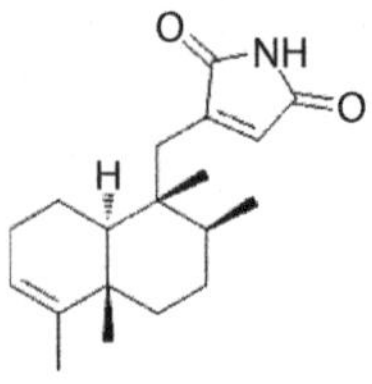

Dysidinoid A

Euryspongia sp.

Order: Dictyoceratida Family: Dysideidae

Compound(s) and Activities

Antifungal activity: The steroidal sulfates eurysterols A and B isolated from this unidentified species showed antifungal activity by inhibiting amphotericin B-resistant and wild-type strains of *C. albicans* with MIC values of 15.6 and 62.5 µg/mL (Anjum et al., 2016; Cheung et al., 2014).

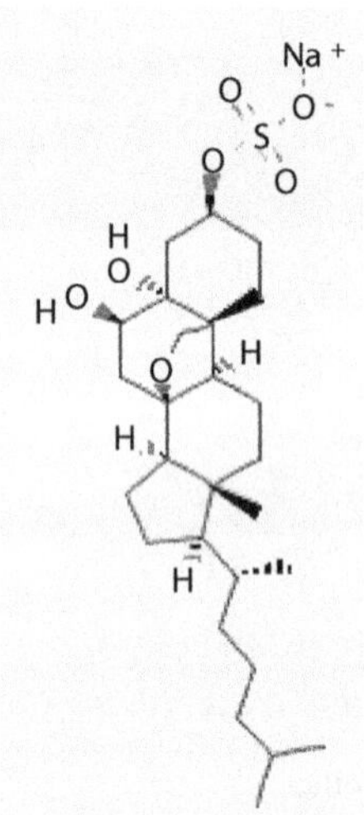

Eurysterol A

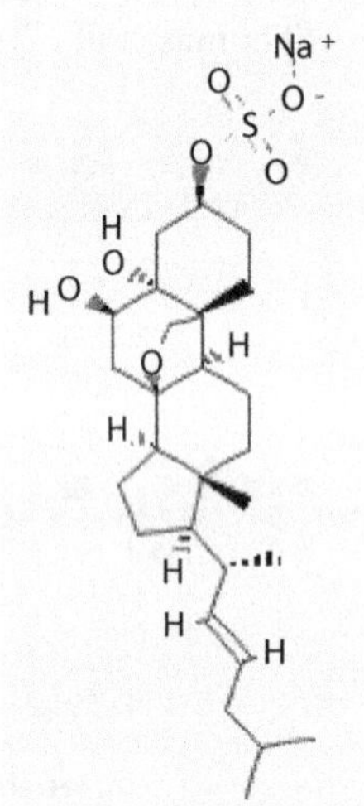

Eurysterol B

Antidiabetic activity: Three new unique sesquiterpenes, euryspongins A–C, isolated from this sponge possessed antidiabetic potential by inhibiting the protein tyrosine phosphatase 1B (PTP1B) (Barde et al., 2015).

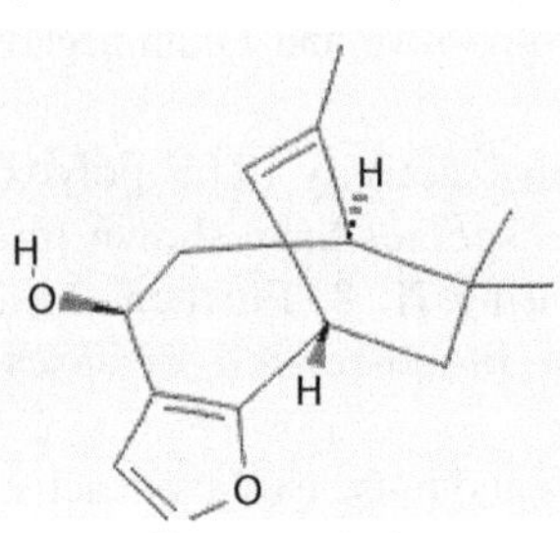

Euryspongin A

Lamellodysidea herbacea (Keller, 1889) (= *Dysidea herbacea*)

Order: Dictyoceratida Family: Dysideidae

Common name: Silvery blue sponge

Distribution: Red Sea, Indian Ocean, Indo-Pacific Ocean

Ecology: Inhabits rubble and rocky shelf areas in caves and under overhangs where there are strong currents. It can also be found growing over coral rubble.

Biology: It is a small sponge, covering an area of 10–15 cm. It can take various forms: flat spatula-like fingers, lettuce-like "leaves," spindly branches, and short thicker lobes, which all emerge from an encrusting base. Younger ones may look like cabbages with upright lobes arranged in a mazelike pattern. Surface texture is prickly. A few small holes are seen at the top edges. Color of the animal is from a pale grayish blue to greenish due to the cyanobacteria that inhabit the sponge.

Compound(s) and Activities

Antimicrobial activity: This sponge has yielded four new polybrominated diphenyl ether congeners and three known derivatives. All these compounds were found active against the Gram-positive bacteria *Bacillus subtilis* and the phytopathogenic fungus *Cladosporium cucumerinum* (Handayani et al., 1997).

Antibacterial and antimycobacterial activity: The bioactive compounds isolated from the hexane and dichloromethane extracts of this species have shown antibacterial activity against *Staphylococus aureus* and *Bacilus subtilis*, and antimycobacterial activity against *Mycobacterium avium* (Palma and Sumera, 1995).

Antibacterial activity: Polybrominated biphenyl ethers isolated from this species have shown antibacterial activity against both Gram-negative and Gram-postitive bacteria and unicellular marine cyanobacteria. The compound, 2-(2′,4′-dibromophenyl)-4,6-dibromophenol of this species showed antibacterial activity against *Staphylococcus aureus*, *Escherichia coli*, and *Bacillus subtilis* (Thomas et al., 2010).

The extracts of this species have displayed activity against two Gram-positive bacteria (*S. aureus* and *B. subtilis*) and the Gram-negative bacteria *V. cholerae*, with inhibition zones of 10.3, 9.2, and 9.5 mm, respectively (Putra and Hadi, 2017).

Antifungal activity: The compound 3,5-dibromo-2-(3,5-dibromo-2-methoxyphenoxy)phenol isolated from this species has shown antifungal activity against *C. albicans in vitro* (Sionov, 2005).

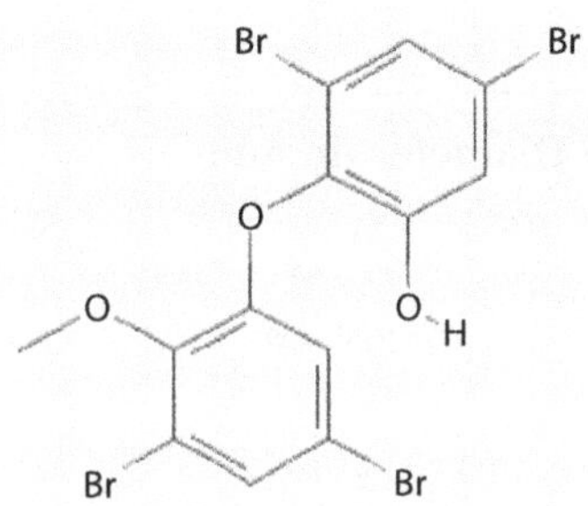

3,5-dibromo-2-(3,5-dibromo-2-methoxyphenoxy)phenol

Neurosuppressives and muscle relaxant activity: The unusual amino acid dysiherbaine isolated from this species has served as a glutamate receptor antagonist in the neurosuppressive and muscle relaxant activity (Sipkema et al., 2005).

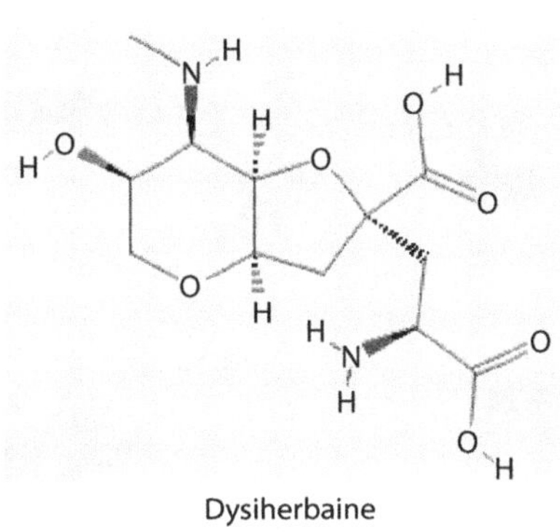

Dysiherbaine

Mosquito larvicidal activity: The extract of this species displayed larvicidal activity against the fourth instar larvae of *Aedes aegypti* (yellow fever mosquito vector) with an LC50 value of 199.5 ppm (Rao et al., 2008).

Antidiabetic activity: A polybromodiphenyl ether compound isolated from this species inhibited protein tyrosine phosphatase 1B, an important target for diabetes treatment (Lauritano and Ianora, 2016; Senthilkumar and Kim, 2013).

Others: The chlorinated diketopiperazines dihydrodysamide C and didechlorodihydrodysamide C isolated from the associated cyanobacterium *Oscillatoria spongeliae* of this sponge species have been reported to be be useful scaffolds for the rational design of receptor probes and therapeutic agents (Thomas et al., 2010).

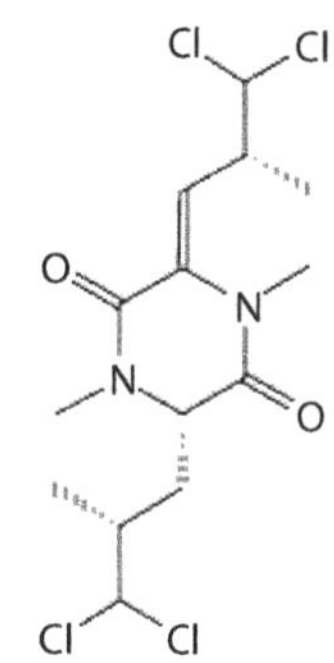

Didechlorodihydrodysamide C

FAMILY: IRCINIIDAE

Ircinia campana (Lamarck, 1814)

Order: Dictyoceratida Family: Irciniidae

Common name: Vase sponge, vase garlic sponge, Colombian marine sponge

Distribution: Bahama, Brazil, Caribbean, India, Mexico

Ecology: It inhabits shallow to midrange reefs, and prefers brightly lit areas.

Biology: It is a cup-shaped sponge, although asymmetrical specimens with a lower and a higher cup wall are common. It grows up to 50 cm high and 40 cm in diameter. Thickness of the cup wall is 0.5 to 3 cm and the consistency is fleshy but tough. Surface is strongly conulose. The conules are about 4 mm high and mostly 5 to 6 mm apart. Conules on the inside of the cup are distinctly lower and closer together than those on the outside. Oscules are seen only on the inside and are 4 to 10 mm in diameter. Color of the animal is reddish brown.

Compound(s) and Activities

Antileishmanial activity: Two sterol fractions of this species showed *in vitro* cytotoxicity and antileishmanial activity against amastigotes of the *Leishmania (Viannia) panamensis* parasite (Márquez et al., 2007). The LC50 and EC50 values for these fractions are given next.

Fraction	CL 50 (μg/mL)	CE 50 (μg/mL)
Fraction A	25.5	30.0
Fraction B	39.7	25.7

Source: Márquez, D.M. et al., 2007, *Porifera Res.: Biodiversity, Innovation Sustainability,* 433–437.

Antibacterial activity: The crude extract of this species showed antibacterial activity in different extracts as given next.

Organic Solvent	*Escherichia coli*	*Staphylococcus aureus*	*Salmonella typhi*	*B. subtilis*
Methanol	–	++	++	+
Chloroform	–	+	+	–
Acetone			+	–
	–			
	–			

Source: Subugade, S. et al., 2015, *Indian J. Pharm. Biol. Res.,* 3:12–14.
Note: no activity; +, weak activity (inhibition zone, 1–3 mm dia.); ++, good activity (inhibition zone, 4–10 mm dia.).

Antifungal activity: The crude extract of this species showed antifungal activity as given next.

Marine Sponge Ircinia fusca

Organic Solvent	*Aspergillus* sp.	*Penicillium* sp.	*Alternaria* sp.	*Fugarium* sp.
Methanol	++	++	–	–
Chloroform	–	++	–	+
Acetone	++	++	–	+

Source: Subugade, S. et al., 2015, *Indian J. Pharm. Biol. Res.,* 3:12–14.
Note: no activity; +, weak activity (inhibition zone, 1–3 mm dia.); ++, good activity (inhibition zone, 4–10 mm dia.).

Antibacterial activity: The ethanolic extract of this species exerted antibacterial activity against cerebrospinal fluid (CSF) disease (neurological disease)–causing pathogenic bacteria (Sujath et al., 2014). The values of inhibition zone diameter for different species of bacteria are given next.

Inhibition Zone Diameter (in mm)

E. coli	*S. aureus*	*H. influenzae*	*N. meningitides*	*K. pneumoniae*	*Streptococcus agalactiae*	*E. faecalis*
No inhibition	5	7	No inhibition	10	9	No inhibition

Source: Sujatha, S. et al., 2014, *Pharma Innovation,* 3: 81–86.

Antioxidant activity: While the aqueous extract if this species showed superior antioxidant activity with 146.73%, the fractions 30% and 50% exhibited free radical scavenging activity higher than 50% (Tovar-Barrios et al., 2016).

Ircinia dendroides (Schmidt, 1862)

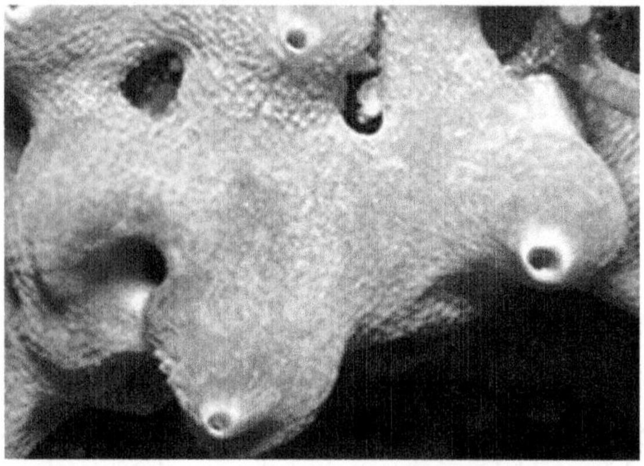

Order: Dictyoceratida Family: Irciniidae

Common name: Not designated

Distribution: Adriaitc Sea, Portugal, Northwest Spain, Mediterranean

Ecology: It inhabits on rocks and coralline algae in the *Laminaria* zone.

Biology: This sponge has creeping cylindrical or flattened branches of 1 to 2 cm in diameter. Surface is conulose with conules of 1 mm height. There are no apparent oscules. Consistency of the sponge is firm and very solid. Spicules are absent. Color of the body is gray to grayish white.

Compound(s) and Activities

Antidiabetic activity: The sesquiterpene compound palinurin isolated from the extracts of this species showed antidiabetic activity by inhibiting GSK-3β (Lauritano and Ianora, 2016).

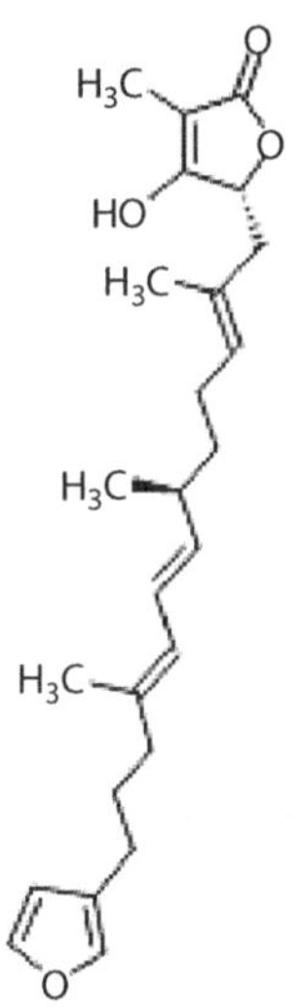

Palinurin

Others: The isopropanolic extracts of this species containing palinurin have the significant human recombinant GSK-3β inhibitory effect *in vitro*. The glycogen synthase kinase 3β (GSK-3β) enzyme has been reported to play a crucial role in several disorders such as cancer, Alzheimer's disease, diabetes, and mood disorders. The selective binding mechanism of palinurin to GSK-3β may therefore be used for the treatment of GSK-3b-mediated diseases (Barde et al., 2015).

Ircinia felix (Duchassaing and Michelotti, 1864)

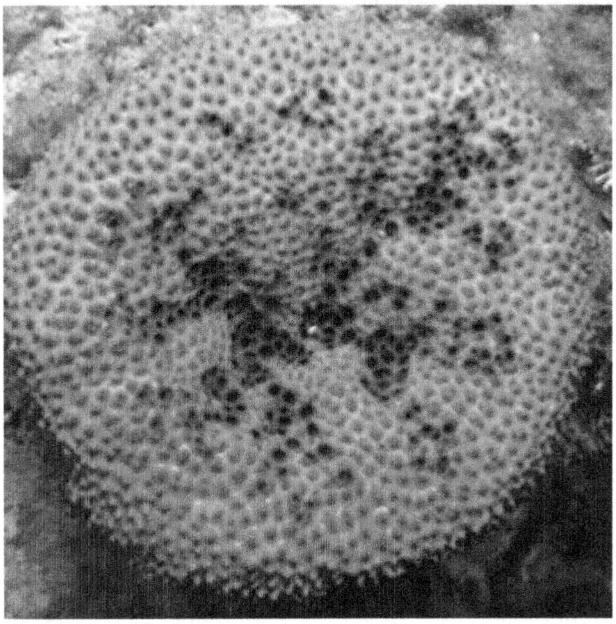

Order: Dictyoceratida Family: Irciniidae

Common name: Stinker sponge, Formosan sponge

Distribution: From Florida to Northeast Venezuela

Ecology: It inhabits shallow to midrange reefs and often grows near living corals. It is also found in lagoons.

Biology: It is a globular or angularly cake-shaped sponges 12 to 14 cm in diameter. Flatly encrusting, lamellar, or ramose specimens are also found. The surface is finely conulose. The conules are about 1–2 mm high and about 2–3 mm apart. The oscules are provided with a dark, blackish purple rim. They are irregularly scattered over the surface and have a diametere of 1–8 mm. The consistency of the sponge is tough, compressible, and is difficult to cut or tear. Color of the sponge is brown.

Compound(s) and Activities

Antioxidant activity: The methanol extracts of this species showed inhibition of radical ABTS with 61% (Tovar-Barrios et al., 2016).

Antitumor activity: Five new scalarane sesterterpenoids, felixins A–E (1–5), isolated from this species showed cytotoxicity against MOLT-4 (human acute lymphoblastic leukemia), SUP-T1 (human T-cell lymphoblastic lymphoma),

DLD-1 (human colorectal adenocarcinoma), LNCaP (human prostatic carcinoma), T-47D (human ductal carcinoma), and MCF7 (human breast adenocarcinoma) tumor cells. The IC50 values recorded for the different compounds of the cell lines are given next.

Cell Lines IC50 (μM)

Compound	MOLT-4	SUP-T1	DLD-1	LNCaP	T-47D	MCF7
1	NA	NA	10.9	24.3	NA	NA
2	14.9	27.1	8.5	NA	32.2	23.0
3	18.5	NA	15.0	NA	NA	NA
4	12.8	31.6	7.9	21.5	20.2	NA
5	14.0	31.1	7.2	NA	22.7	24.3

Source: Lai, Y. et al., 2015, *Mar. Drugs*, 13: 4296–4309.

Antimicrobial, PP1, and AChE inhibitory activities: The organic extracts from this species showed considerable antimicrobial, AChE, and PP1 inhibitory activities. The highest antimicrobial activity was, however, observed in the aqueous, unheated extract (Sepčić et al., 2010). Zea et al. (1999) reported that the linear furanosesterpene tetronic acids isolated from this species showed antimicrobial, cytotoxic, and antitumoral properties. During wound healing, increased FTA levels may also act as internal antibiotic protection.

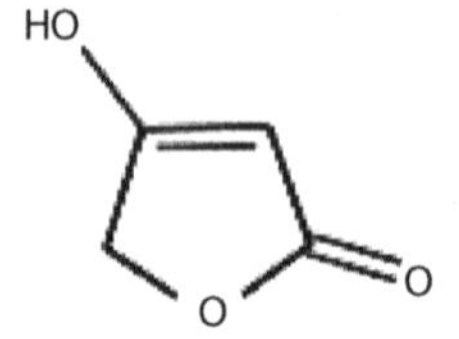

Tetronic acid

Ircinia fusa (Carter, 1880)

Order: Dictyoceratida Family: Irciniidae

Common name: Not designated

Distribution: Indo-Pacific region, South India and Sri Lanka

Ecology: It inhabits rocky shores of intertidal areas.

Biology: This sponge surface is conulose. Color of the sponge is dark brown and pale yellow internally. Texture is leathery. Dermal region contains sand particles. This species having long, spherical filaments with oblong heads. This sponge is often associated with the brittle star *Ophiactis savignyi*.

Compound(s) and Activities

Antibacterial activity: The ethanolic extract of this species exerted antibacterial activity against cerebrospinal fluid (CSF) disease (neurological disease)–causing pathogenic bacteria (Sujath et al., 2014). The inhibition zone diameter values for different species of bacteria are given next.

Inhibition Zone Diameter (in mm)

E. coli	*S. aureus*	*H. influenzae*	*N. meningitides*	*K. pneumoniae*	*Streptococcus agalactiae*	*E. faecalis*
No inhibition	4	No inhibition	No inhibition	21	10	12

Source: Sujatha, S. et al., 2014, *Pharma Innovation*, 3: 81–86.

Antibacterial and cytotoxic activities: The benzofuran sesquiterpene derivative, 7-methyl-7-((E)-4-methylpenta-1,3-dien-1-yl)-5,6,7,7a-tetrahydrobenzofuran-3a(4H)-ol (Iricinol A), isolated from this sponge showed broad spectrum activity against bacteria and fungi at 2 μg/μL. Also, it exhibited *in vitro* cytotoxic activity against RAW264.7 cells (mouse monotype macrophage leukemia) and N2A cells (neuro-blastoma) with IC50 values 55 ± 4.21 μg/mL (Meesala et al., 2016).

Mosquito larvicidal activity: The extract of this species displayed larvicidal activity against the fourth instar larvae of *Aedes aegypti* (yellow fever mosquito vector) with an LC50 value of 89.7 ppm (Rao et al., 2008).

Ircinia oros (Schmidt, 1864)

Order: Dictyoceratida Family: Irciniidae

Common name: Not designated

Distribution: Mediterranean

Ecology: This coralligenous species inhabits cave, detritic, and rocky bottoms at depths of 1 to 150 m. Sponges of this species are often covered by large specimens of *Haliclona (Reniera) cratera.*

Biology: Growth form of this species is massive and lobate, with large size (20–30 cm in dia. and 10–15 in height). Each lobe usually bears a large oscule (30–60 mm in dia.), sometimes at the end of a short funnel (1 cm high). Color of the sponge is medium to dark gray *in vivo.* Surface is covered by a slim layer of very fine and regular mineral sediment engulfed in a slender regular network showing a lighter color. Conules (1–2 mm in height) are regularly distributed, 24 mm apart. Choanosomal skeleton is rust colored and rich in fibers and filaments. Skeleton network is of cored primary fibers and free secondary fibers.

Compound(s) and Activities

Anti-inflammatory activity: The acyclic sesterterpenoid compounds ircinin 1 and 2 showed anti-inflammatory activity by inhibiting phospholipase A2 (Sipkema et al., 2005).

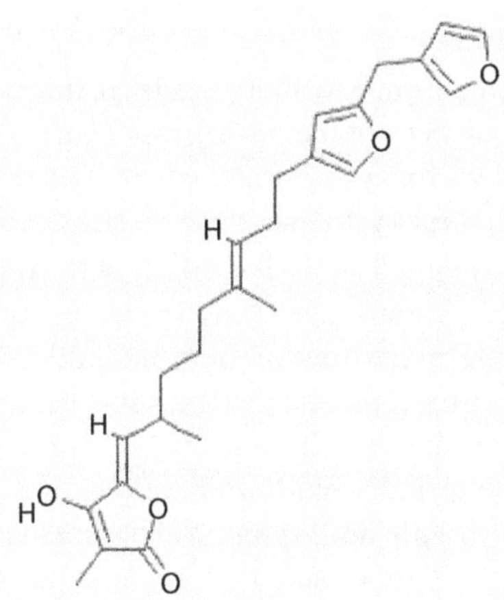

Ircinin

Leishmanicidal, trypanocidal, and antiplasmodial activities: The linear furanosesterterpenoids, namely, ircinin-1, ircinin-2, ircinialactam E (3), and ircinialactam F (4) isolated from this species showed showed moderate leishmanicidal, trypanocidal, and antiplasmodial activities (IC50 values 28–130 μM) (Chianese et al., 2017).

Ircinia ramosa (Keller, 1889)

Order: Dictyoceratida Family: Irciniidae

Common name: Indo-Pacific marine sponge

Distribution: Indo-West Pacific, Brazil, Kenya, Palau, Mexico, Arafura Sea, Bermuda, Mozambique Channel, Madagascar

Ecology: This sessile sponge inhabits waters of depth range 15 to 70 m.

Biology: In this species, the distal branches arise from a massive base but remain coalesced throughout most of their length. Only the terminal 1 or 2 cm of the branches are free; these vary in diameter from 1.5 to 2.0 cm and are rounded. The total height of the colony is 12.5 cm; at the base it measures 5 cm wide and 4 cm thick. The surface is conulose with conules varying from 1 to 2 mm high and are placed 1 to 4 mm apart. The dermis is covered with abundant foreign spicules and sand. The pores are sparsely distributed. The oscules occur at random over the surface and vary from 0.2 to 2.5 mm in dia. The skeleton consists of a loose network of fasciculated fibers cored with foreign spicules and some sand. The high growth potential and high survival rate of this sponge species suggest that it is a promising candidate for mariculture development (de Voogd, 2007).

Compound(s) and Activities

Anticancer activity: The compounds irciniastatins A (1) and B (2) isolated from this species showed anticancer activity (GI50 from 0.001 to <0.0001 μg/mL) by inhibiting the growth of murine and human cancer cells (Pettit et al., 2004).

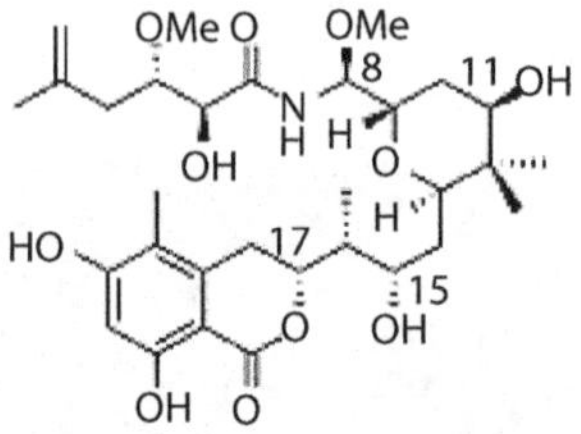

Irciniastatin A

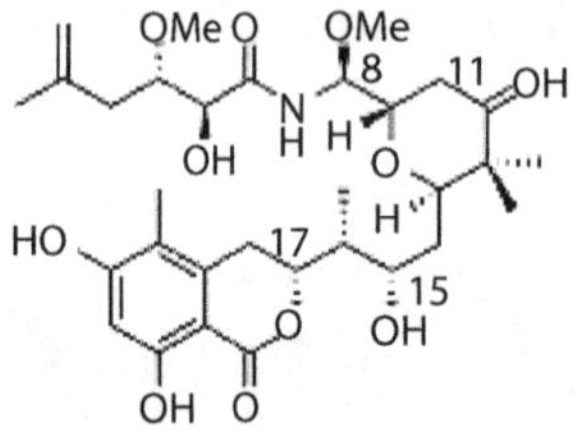

Irciniastatin B

Antitumor activity: The compounds irciniastatin A (also known as psymberin) and its analogs showed antitumor activity (Hirano et al., 2015).

Ircinia spiculosa (Hentschel, 1912)

Order: Dictyoceratida Family: Irciniidae

Common name: Not designated

Distribution: Western Central Pacific, Indonesia, Arnhem Coast to Gulf of Carpenteria

Ecology: It lives on rocky reefs subtidally down to 180 m.

Biology: It is a massive, spherical sponge with 7 cm thickness. Upper surface is black, and interior and lateral areas are brick red. Texture is extremely tough but easily compressible. Surface is flattened and covered with small, evenly spaced cornules of 1.5 mm height. Skeleton is composed of fasciculate columns of spongin fibers.

Compound(s) and Activities

DPPH radical scavenging activity: The extracts of this species showed DPPH radical scavenging activity (inhibition, 11.7%) at a concnentration of 2000 μg mL^{-1} (Orhan et al., 2012).

AChE inhibitory activity: The extracts of this species showed AChE inhibitory activity with a percentage inhibition value of 68 at an extract concentration of 200 μg mL^{-1} (Orhan et al., 2012).

Antimicrobial activity: The extracts of this species showed antibacterial activity against Gram-positive bacteria and antifungal activity, and the MIC values are given next.

MIC (μg mL)				
S. aureus	*E. faecalis*	*B. subtilis*	*C. albicans*	*C. parapsilosis*
128	64	32	16	16

Source: Orhan, I.E. et al., 2012, *Rec. Nat. Prod.*, 6: 356–367.

Ircinia strobilina (Lamarck, 1816)

Order: Dictyoceratida Family: Irciniidae

Common name: Black-ball sponge, bumpy ball sponge, cake sponge, pillow stinking sponge

Distribution: Florida to the Bahamas and throughout the Caribbean to Brazil

Ecology: It occurs from 3 to 30 m, mostly in inner-reef areas and in lagoons among beds of the turtle grass *Thalassia testudinum.* It is also found on rocky substrate, where it grows successfully on the bottoms of rocky reef channels in strong currents.

Biology: This sponge is massive, often with stubby cylinders wider than high. Sponges growing in strong currents usually become elongate in a direction perpendicular to the current, with a longitudinal apical crest bearing the oscules. Most specimens are cake or pillow shaped. Live individuals are gray to black externally, with drab yellow at the base. Their consistency is tough and spongy. The surface is coarsely conulose. Thick, blunt conules are seen on the surface of the sponge. The oscules are 2–5 mm in diameter and scattered in groups on the top plane, and are often very crowded.

Reproduction: This sponge exhibits reproductive activity year round on the reefs. Spermatic cysts are produced to varying degrees throughout the year, while oocytes and larvae have been observed from September to April, for an annual breeding season of 8 months. The sponge is viviparous with separate sexes, although the protandry cannot be ruled out.

Predators: Its main predator is the queen angelfish *Holocanthus ciliaris.*

Associated species: Like many other sponges, a variety of organisms can be found living in association with this sponge. The coral *Porites astreoides* and the sponge *Haliclona viridis* have been found along withthis species. The tusked goby *Risor ruber* often lives beside or within the sponge.

Compound(s) and Activities

Antimicrobial activity: The butanol extracts of this species were found to be most active with an MIC value of 30 ng/mL (Sepčić et al., 2010).

Cytotoxic activity: The hydroethanolic extracts of this species showed 100% cytotoxicity at an extract concentration of 500 μg/mL (Ferreira et al., 2007).

Ircinia variabilis (Schmidt, 1862)

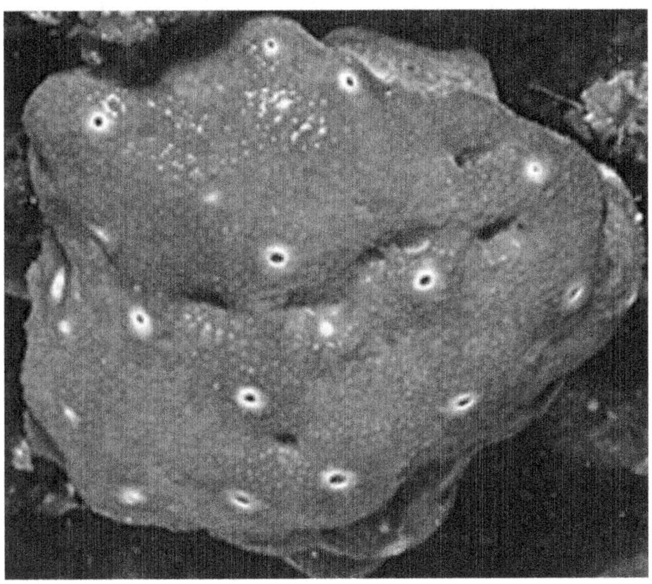

Order: Dictyoceratida Family: Irciniidae

Common name: Not designated

Distribution: Portugal, Galicia, Mediterranean

Ecology: It inhabits cave, coralligenous communities, detritic and rocky bottoms, Posidonia oceanica meadows, and lagoons. It is epibiotic on *Pinna nobilis*. It lives from the low water mark downward to at least 150 m.

Biology: This sponge is encrusting, massive, and it grows up to 20–25 cm in height and diameter. Color of the sponge is notably variable: from light or dark gray, to light or dark brown, and light or dark violet. Consistency is elastic and strong. Dimension and density of conules are variable. Oscules, which are up to 8 mm in diameter, are arranged in disorder. Skeleton network is of primary fibers cored by opaque foreign materials. Spicules are absent. The ectosome is strengthened by sand grains, which form a coat over the surface.

Compound(s) and Activities

DPPH radical scavenging activity: The extracts of this species showed DPPH radical scavenging activity (inhibition, 66%) at a concentration of 2000 μg mL^{-1} (Orhan et al., 2012).

Anti-inflammatory activity: The sesterterpenes of this species showed anti-inflammatory activity by inhibiting human synovial secretory PLA2 and U937 cytosolic PL with IC50 values of 6.9 μM and 7.9 μM, respectively (Keyzers, 2005).

Antiprotozoal activity: The actinobacterial strain *Geodermatophilus* sp. SBT381 derived from this species showed activity against *T. brucei brucei* with an IC50 value of 18.60 μg/mL (Indraningrat et al., 2016).

Antifungal activity: The fungus *Microascus* sp. K14 isolated from this species produced a compound, fungerin, which showed antifungal activity (Hardoim and Costa, 2014).

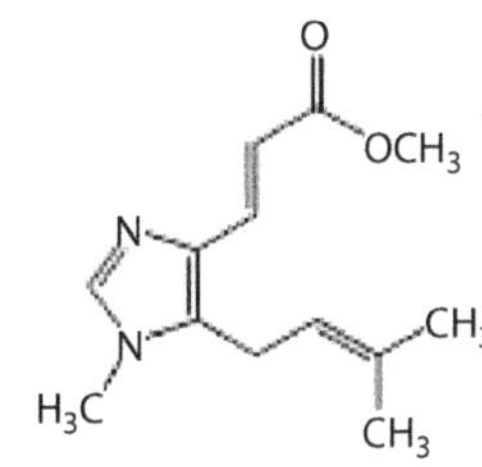

Fungerin

Ircinia sp.

Order: Dictyoceratida Family: Irciniidae

Compound(s) and Activities

Cytotoxicity: An indole alkaloid konbamidin isolated from this unidentified sponge species showed cytotoxicity against HeLa cells *in vitro* with an IC50 value of 5.4 μg/mL (Gul et al., 2005).

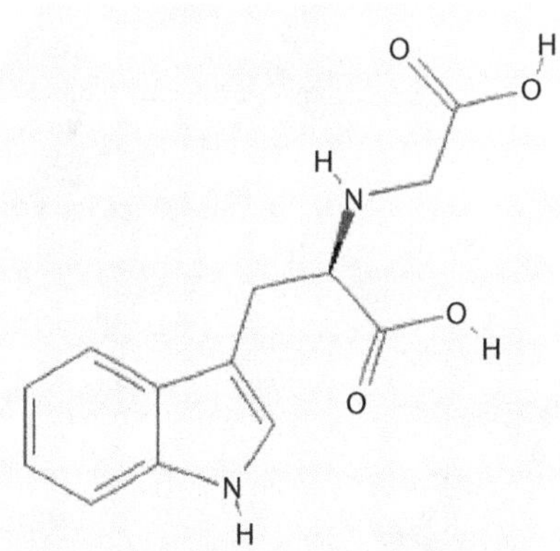

Konbamidin

Compounds ircinolin A (1), 15-acetylirciformonin B (2), 10-acetylirciformonin B (3), irciformonin B (4), and irciformonin F (5) have been isolated from this sponge. Among these metabolites, compounds 1 and 3–5 exhibited significant cytotoxic activity against K562, DLD-1, HepG2, and Hep3B cancer cell lines (Su et al., 2011).

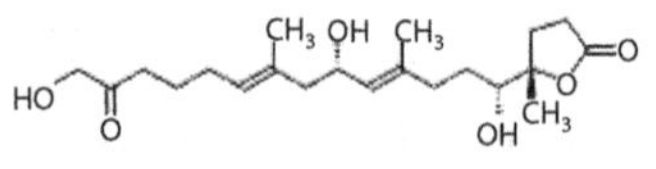

Ircinolin A

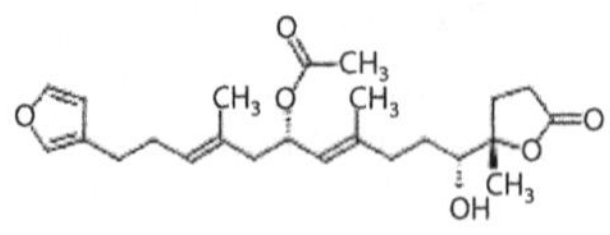

10-acetylirciformonin B

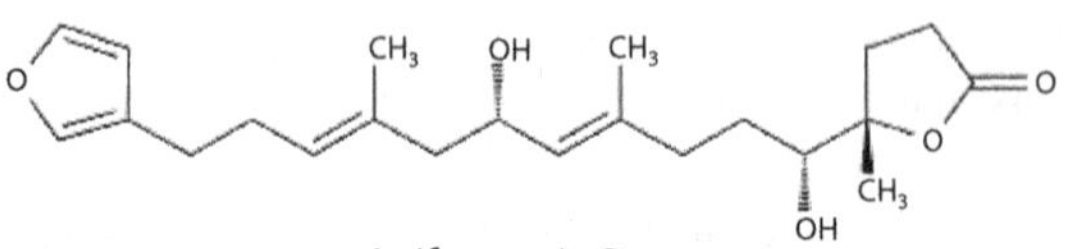

Irciformonin B

Antitumor and antiviral activities: The compound bacircine produced by the associated *Bacillus pumilus* strain of this sponge exhibited antitumor activity against Ehrlich ascites carcinoma and anti-HIV activity (Hardoim and Costa, 2014).

Antioxidant activity: In the DPPH assay, the antioxidant activity results showed 83.0 mg/mL and reduced the ferric ion with 52.0 mg AA/g (Bianco et al., 2015; Tovar-Barrios et al., 2016).

Cytotoxic, antitumor, and anti-HIV activities: *Bacillus pumilus* strain isolated from this sponge was found to produce five surfactin-like substances (cyclic acyldepsipeptides) called bacircines. These compounds showed cytotoxic effects besides exhibiting antitumor activity against Ehrlich ascites carcinoma and anti-HIV activity (Hardoim and Costa, 2014).

Sarcotragus fasciculatus (Pallas, 1766) (= *Ircinia fasciculata*)

Order: Dictyoceratida Family: Irciniidae

Common name: Stinking sponge, Mediterranean sponge

Distribution: Mediterranean, Western Bahamas

Ecology: This shallow water, coralligenous sponge inhabits caves, rocky bottoms, and Posidonia oceanica meadows at depths of 1 to 100 m.

Biology: These are massive, lobate sponges with occasionally branches. They grow up to 30 cm in width and 15 cm in height. Color of the sponge is yellowish brown or reddish brown, often with purplish or greenish tinges. The oscules have dark brown or black linings. Surface is regularly conulose (1 mm in height, 1–2 mm apart). Skeleton network is light brown, fragile, and reticulate with more or less square meshes from the sponge base to the surface. Almost parallel ascending primary fibers (200–300 μm in diameter) free from foreign inclusions, with apices supporting conules. Each primary fibre is a bundle of some (2–5) uncored secondary fibers joined by conspicuous spongin tracts and cribrose plates. Filaments are abundant and whitish.

Compound(s) and Activities

Antimicrobial activity: The extracts of this sponge species showed antibacterial activity against Gram-positive bacteria and antifungal activity, and the recorded MIC values are given next.

MIC (μg mL^{-1})				
S. aureus	*E. faecalis*	*B. subtilis*	*C. albicans*	*C. parapsilosis*
128	>128	32	16	16

Source: Orhan, I.E. et al., *Rec. Nat. Prod.*, 6: 356–367.

DPPH radical scavenging activity: The organic extract of this species showed DPPH radical scavenging activity (inhibition, 39%) at a concentration of 2000 μg mL^{-1} (Orhan et al., 2012).

AChE inhibitory activity: The organic extract of this species showed AChE inhibitory activity at a concentration of 200 μg mL^{-1} and the percentage of inhibition was 57.6 (Orhan et al., 2012).

Cytotoxicity and antiviral activity: The associated fungus *Penicillium chrysogenum* of this sponge species yielded a compound sorbicillactone A, which showed highly selective cytostatic activity against murine leukemic lymphoblasts (L5178y) and also showed the ability to protect human T cells against the cytopathic effects of HIV-1 (Thomas et al., 2010). The above compound also displayed cytoprotective effects on HIV-1–infected cells of the human cell line H9 at concentrations of 0.1–1 μg/mL. Further, this compound reduced the appearance of the HIV-1 protein up to 70% at a concentration of 0.3 μg/mL (Indraningrat et al., 2016).

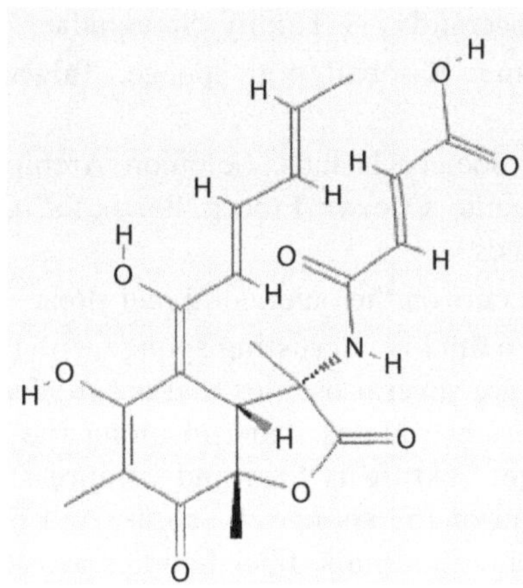

Sorbicillactone A

Anti-HIV and neuroprotect activities: The fungus *Penicillium chrysogenum* isolated from this species produced the sorbicillinoid alkaloids sorbicillactones A and B. Sorbicillactone A exhibited remarkable cytotoxic activity toward murine leukaemia lymphoblast (L5178y) cells and anti-HIV activity. This compound also showed promising neuroprotective metabolite (Hardoim and Costa, 2014).

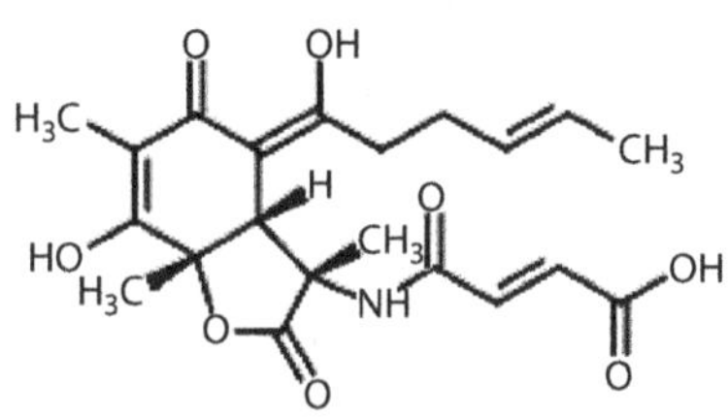

Sorbicillactone B

Indraningrat et al. (2016) reported that the compound sorbicillactone A produced by its associated fungus *Penicillium chrysogenum* showed antiviral activity by reducing protein expression and activity of reverse transcriptase (0.3–1 μg/mL) HIV-1.

Sarcotragus foetidus (Schmidt, 1862) (= *Ircinia foetida*)

Order: Dictyoceratida Family: Irciniidae

Common name: Dark stinging sponge, black sponge

Distribution: Northeast Atlantic and Mediterranean

Ecology: It inhabits rubble areas near algae growths. It develops on rocky bottoms or sometimes even debris, until 30 m deep.

Biology: It has a massive shape and it can reach a size of 1 m in diameter and 0.5 m in height. It shows numerous oscules (5–10 mm dia.), which are irregularly arranged on its surface and often appear grouped. The surface is irregular due to the presence of large conidia. The black sponge can also be gray or dark brown, with different shades, or even bluish or even purple. Color is often linked to the presence of symbiotic cyanobacteria.

Associations: The surface of the sponge is often covered by mostly small animals or algae, which colonize this sponge outside, making it appear partially covered with encrusting felts of various kinds. Arthropods were found to be highly abundant on this sponge, especially in the summer samples but other animals such as annelids and mollusks also showed very high abundances and dominated the winter assemblage.

Biosorption: This sponge is capable of removing uranium from the environment through biosorption. The optimum condition of the biosorption process was found to be at pH 4 and 35°C. At these conditions, the biosorption yield of uranium ions was found to be 93% (Celik et al., 2016).

Compound(s) and Activities

Antiprotozoal actrivity: The actinobacterial strains *Modestobacter* sp. SBT363 and *Nonomuraea* sp. SBT364 derived from this sponge species showed antiprotozoal activity against *T. brucei brucei* with IC50 values of <10 μg/mL (Indraningrat et al., 2016).

Sarcotragus spinosulus (Schmidt, 1862)

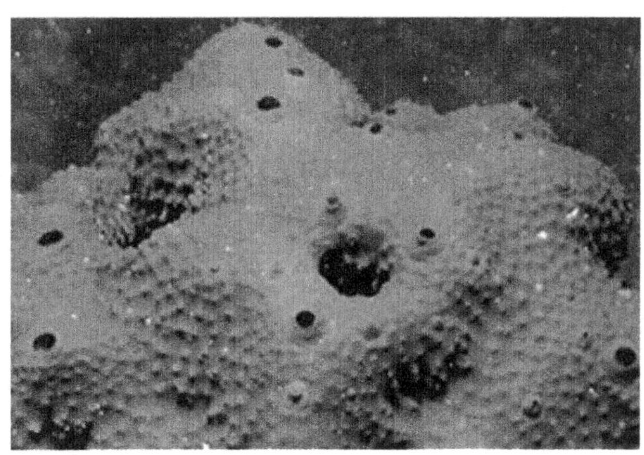

Order: Dictyoceratida Family: Irciniidae

Common name: Not designated

Distribution: Portugal, Galicia, Mediterranean

Ecology: It inhabits rocky substrate, between 1 and 300 m. It is also in caves, lagoons, and detritic and muddy bottoms. It is epibiotic on the bivalve *Pinna nobilis*.

Biology: It is a regular, massive sponge rarely exceeding 10 cm in diameter. Color of the body is black or dark gray *in vivo*. Consistency is strong and fairly elastic. Surface is finely conulose (1–2 mm in height and 2–3 mm apart). Oscules (up to 1 cm in dia.) are irregularly scattered. Skeleton network reticulation is of ascending primary fibers with a fibrous narrow core free of inclusions or bearing only rare spicules. Secondary fibers are uncored and laminated.

Compound(s) and Activities

DPPH radical scavenging activity: The extract of this sponge species showed DPPH radical scavenging activity (inhibition 11%) at a concentration of 2000 μg mL^{-1} (Orhan et al., 2012).

Sarcotragus sp.

Order: Dictyoceratida Family: Irciniidae

Compound(s) and Activities

Antitumor activity: The triterpenoid hydroquinones, namely, adociasulfates, isolated from this species showed antitumor activity by acting as A1, 3-fucosyltransferase, and kinesin motor protein inhibitors (Anjum et al., 2016; Sipkema et al., 2005).

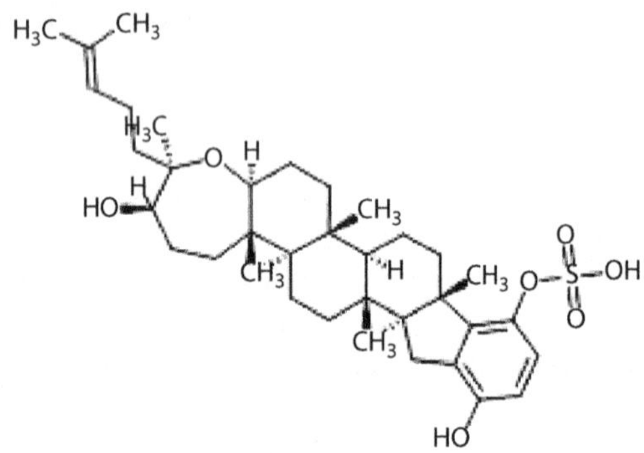

Adociasulfate

FAMILY: SPONGIIDAE

Dactylospongia metachromia (de Laubenfels, 1954) (= *Hippospongia metachromia*)

Order: Dictyoceratida Family: Spongiidae

Common name: Baseball mitt sponge, Taiwanese marine sponge

Distribution: Society Islands, Solomon Archipelago (western South Pacific Ocean); French Polynesia, Gambier and Tuamotu Islands

Ecology: It occurs on the outer coral reef slope.

Biology: It is a thickly encrusting sponge with bright yellow color. There are several oscules that are moderately small, grouped into sieve plates that run along the apical ridge of the sponge. Texture is firm and compressible. Surface ornamentation of the sponge is opaque and membranous. In unornamented animals, fiber bundles are seen protruding through the surface membrane especially in preserved specimens.

Compound(s) and Activities

Cytotoxicity: A total of five new sesquiterpene aminoquinones (1–5), two new sesquiterpene benzoxazoles (6 and 7), the known analog 18-hydroxy-5-epi-hyrtiophenol (8), and a known glycerolipid have been isolated from this species. Compounds 1–5 showed potent cytotoxicity against the mouse lymphoma cell line L5178Y with IC50 values ranging from 1.1 to 3.7 μM. When analyzed *in vitro* for their inhibitory potential against 16 different protein kinases, compounds 5, 6, and 8 showed the strongest inhibitory activity against ALK, FAK, IGF1-R, SRC, VEGF-R2, Aurora-B, MET wt, and NEK6 kinases (IC50 0.97–8.62 μM) (Daletos et al., 2014).

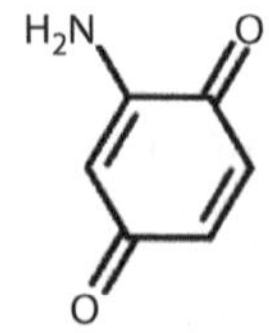

Aminoquinone

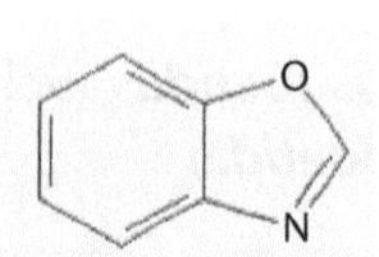

Benzoxazole

Cytotoxicity: Compounds metachromins A (1), B (2), and E (3); hippochromins A and B (4, 5); hippochromin A diacetate (6); and metachromin B monoacetate (7) have been isolated from this species. Among them, compounds, 1, 2, 6, and 7 exhibited potent cytotoxicities against human colon (COLO-205) and nasopharyngeal (KB) tumor cells, while compound 3 was inactive (Shen et al., 2001).

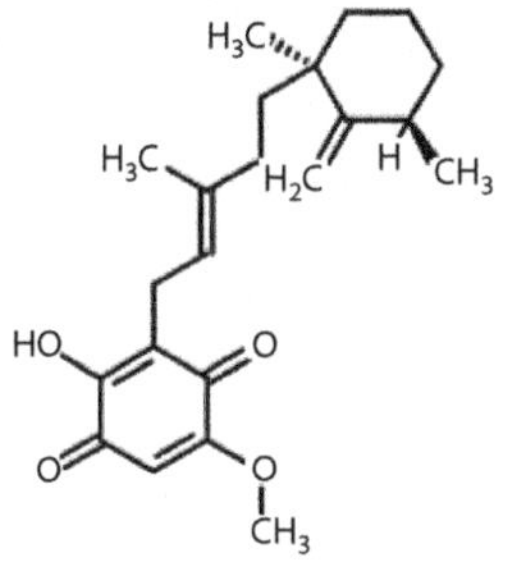

Metachromin A

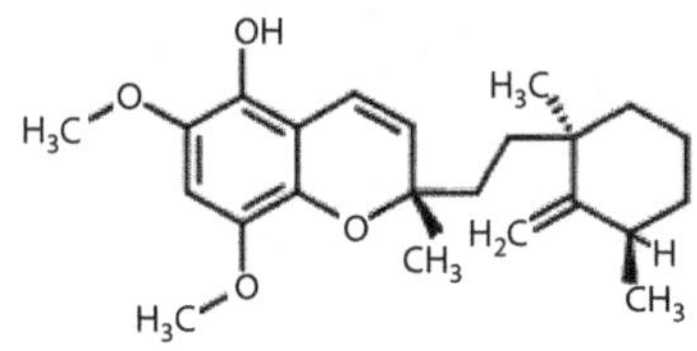

Metachromin B

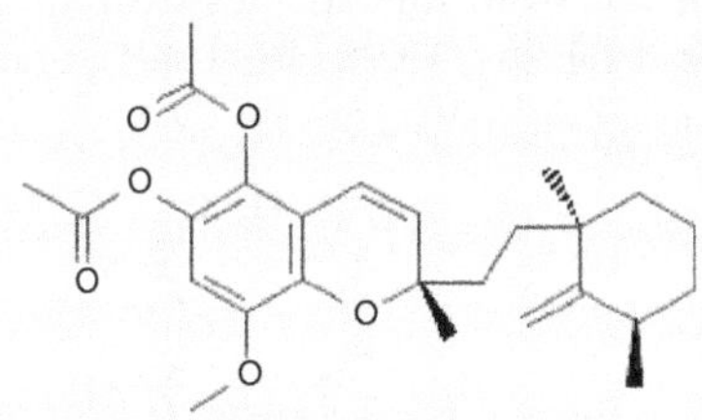

Hippochromin A diacetate

Among the sesterterpene sulfates, namely hipposulfates A and B, isolated from this species, hipposulfate A showed moderate cytotoxicity (Musman et al., 2001).

Antineoplastic activity: The ethyl acetate extract of this species showed antineoplastic activity (Ishibashi et al., 1988).

Anti-inflammatory activity: The compound quinone sesquiterpene ilimaquinone isolated from this species showed anti-inflammatory activity (Keyzers, 2005).

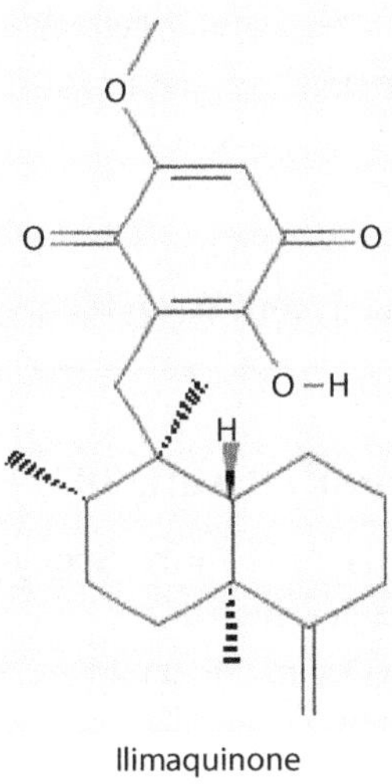

Ilimaquinone

Hippospongia lachne (de Laubenfels, 1936)

Order: Dictyoceratida Family: Spongiidae

Common name: Sheepwool sponge, commercial bath sponge, South China Sea sponge

Distribution: Western Central Atlantic; Bahamas, Greater Antilles, Northern Gulf of Mexico, Brazil, Southern Caribbean

Ecology: It is a reef-associated sponge at a depth range of 14 to 15 m.

Biology: This sponge is oblong to disc shaped, often with one large vent in the center. Its size is up to 18 cm in diameter. Surface grooves separate typical tufted knobs, which are 8–10 mm high and frayed. Consistency is soft. Color of the animal is blackish alive. Dermis is tough and is difficult to detach.

Compound(s) and Activities

Anti-inflammatory activity: The compounds hipposponlachnins A and B isolated from this species showed potent inhibitory activity on the release of β-hexosaminidase, a biomarker for degranulation, as well as the production of pro-inflammatory cytokine IL-4 and lipid mediator LTB4 in DNP-IgE-stimulated RBL-2H3 cells (Hong et al., 2017b).

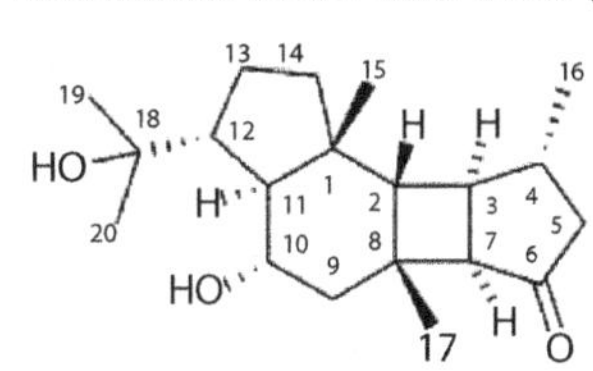

Hipposponlachnin A

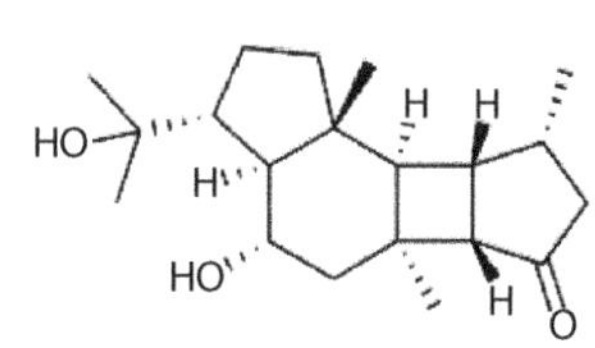

Hipposponlachnin B

Anti-inflammatory activity and cytotoxicity: Eight new acyclic manoalide-related sesterterpenes, hippolides A–H (1–8), along with two known manoalide derivatives, (6E)-neomanoalide (9) and (6Z)-neomanoalide (10), have been isolated from this species. Compound 1 (hippolide A) exhibited cytotoxicity against A549, HeLa, and HCT-116 cell lines with IC50 values of 5.2, 4.8, and 9.8 μM, respectively. Compound 1 also showed moderate PTP1B inhibitory activitiy with an IC50 value of 23.8 μM, and compound 2 showed moderate cytotoxicity against the HCT-116 cell line and PTP1B inhibitory activity with IC50 values of 35.1 and 39.7 μM, respectively. In addition, compounds 1 and 5 showed weak anti-inflammatory activity, with IC50 values of 62 and 40 μM for PKCγ and PKCα, respectively (Piao et al., 2011).

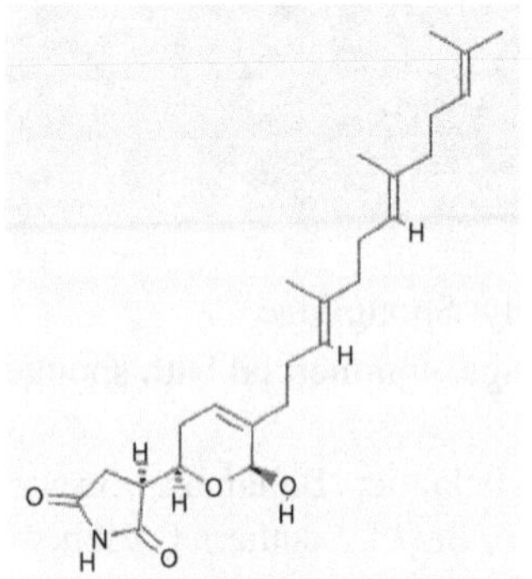
Hippolide A

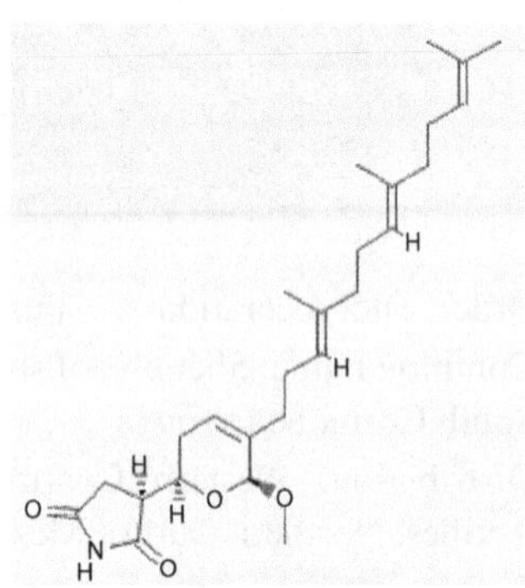
Hippolide B

Antifungal activity: The polyketide compound, hippolachnin A isolated from this species demonstrated potent antifungal activity against three pathogenic fungi, namely, *Cryptococcus neoformans*, *Trichophyton rubrum*, and *Microsporum gypseum*, with an MIC value of 0.4 μM for each fungal species (Cheung et al., 2014; Piao et al., 2013).

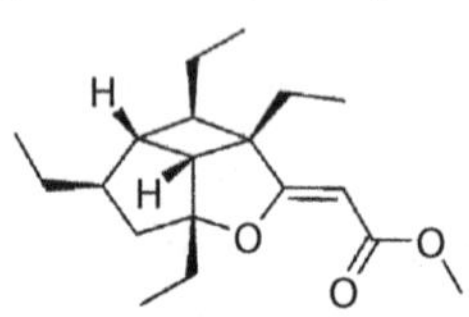
Hippolachnin A

The enantiomeric sesterterpenoids, (±)-hippolide J isolated from this species showed potent antifungal activity against three strains of hospital-acquired pathogenic fungi, namely, *Candida albicans* SC5314, *Candida glabrata* 537, and *Trichophyton rubrum* (Cmccftla), with MIC50 (minimum concentration to give 50% inhibition) values of 0.13–0.25 μg/mL (Jiao et al., 2017b).

Hippospongia sp.

Order: Dictyoceratida Family: Spongiidae

Compound(s) and Activities

Enzyme inhibitory activity: The compound taurospongin A isolated from this species showed inhibitory activity against c-erbB-2 kinase, which activates breast cancer (IC50, 28 μg/mL) (Dembitsky et al., 2006).

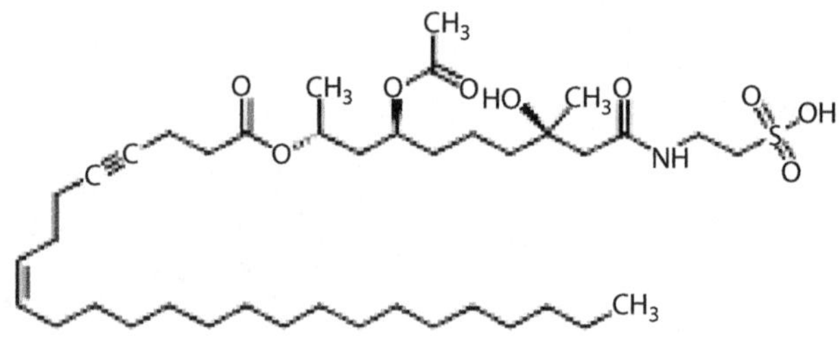
Taurospongin A

Cytotoxicity: The sesterterpenoids heteronemin, heteronemin acetate, and hyrtiosin E isolated from this species exhibited significant cytotoxicity against DLD-1, HCT-116, T-47D, and K562 cancer cell lines (Chang et al., 2012).

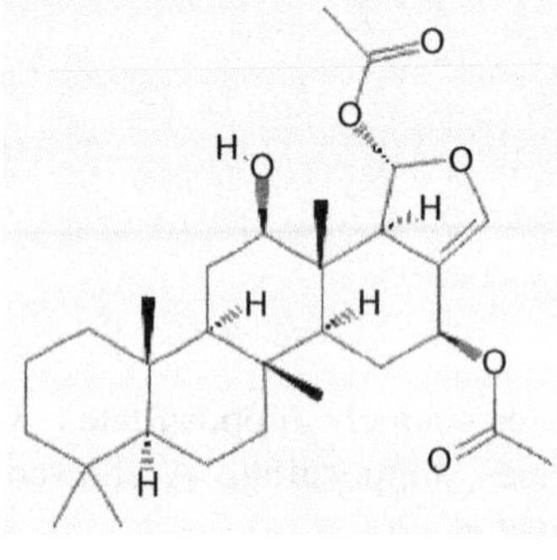
Heteronemin

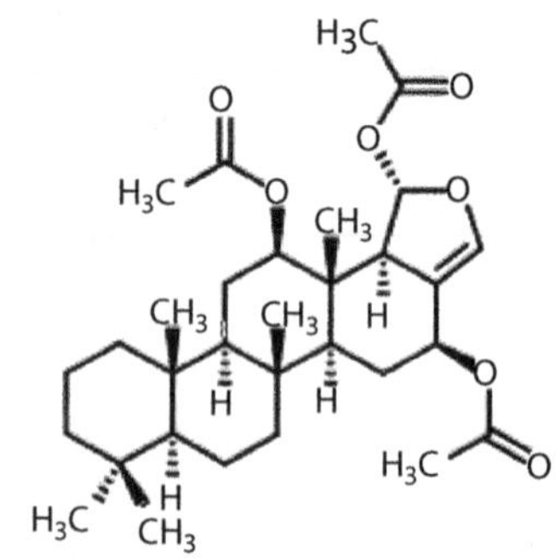
Heteronemin acetate

The IC50 values recorded for the different cell lines are given next.

Cytotoxicity (IC50 μM) of Compounds

	Cell Lines			
Compound	DLD-1	HCT-116	T-47D	K562
1	0.001	0.001	0.001	0.001
2	2.4	2.7	0.3	0.05
3	1.1	8.0	0.7	0.7

Source: Chang, Y. et al., 2012, *Mar. Drugs*, 10: 987–997.

Hyattella cribriformis (Hyatt, 1877)

Order: Dictyoceratida Family: Spongiidae

Common name: Not designated

Distribution: Indian Ocean, Mozambique, Red Sea, Caribbean region, Australia, New Zealand

Ecology: It inhabits tropical reefs and even the deep sea.

Biology: This sponge has coarse texture and abundant primary fibers. Body of this sponge is tuberous. Interior of the sponge is hollow and its walls are punctured by round openings. Thickness of wall is 1.5 mm. Color of the animal is pale yellow when dry. Consistency is hard with poor resiliency when dry. Surface is cornulose and the cornules are distinct toward the growing tip. Ectosome is semitransparent and detachable. Skeleton consists of a well-developed reticulation with main and connective fibers.

Compound(s) and Activities

Antibacterial activity: The organic extract of the associated bacterial strain *Bacillus cereus* SBS02 of this sponge species showed antibacterial activity against several pathogenic species. The values of the zone of for different bacterial species are given next.

Pathogen	Zone of Inhibition Diameter (mm)
Vibrio cholerae	12
Proteus mirabilis	8
Pseudomonas aeroginosa	10
Escherichia coli	5
Staphylococcus aureus	15
Lactobacillus bulgaricus	11
Klebsiella oxytoca	9
Klebsiella pneumoniae	10
Salmonella typhi	13
Salmonella paratyphi	11

Source: Brammavidhya, S., and G. Usharani, 2013, *IJEST*, 3: 61–64.

The associated bacterial communities derived from this species displayed antibacterial activity against the pathogenic bacteria *S. aureus*, *S. typhi*, *S. paratyphi*, *K. oxytoca*, *E. coli*, *P. mirabilis*, *L. vulgaris*, *K. pneumoniae*, *V. cholerae*, and *A. tumefaciens*. The values of the zone of clearance for different pathogens against each associated bacterial species are given next.

Cell Line	IC50 (μg/mL)
Murine leukemia L1210	1.6
Human cervix epithelioid carcinoma HeLa	16.5
Hhuman lung carcinoma A549	16.5
Human oral epidermoid carcinoma KB	17.1
Human breast adenocarcinoma MCF-7	0.3
Colorectal carcinoma HT-29	0.9
Human cholangiocarcinoma HuCCA-1	42

Source: González, M.A., 2010, *Curr. Bioact.Compd.*, 6: 178–206.

The ethyl acetate fraction of this species exhibited potent inhibition of cancer cell growth and resulted in 50% growth inhibition (GI50) of 0.27 μg/mL in A673 cell line; inhibition of sarcoma (MG-63, Saos-2) and ovarian (SK-OV-3 and OVCAR-3) cancer cell with GI50 of 1.0 μg/mL; and moderate activity (GI50 of 15.6 μg/mL) on colon and breast cancer cell lines and human lung fibroblast. Further, the above fraction promoted microtubule polymerization in tubulin polymerization assay and increased level of polymerized tubulin in the HeLa cells; induced the activation of caspase-3 and phosphorylation of Bcl-2 anti-apoptotic protein; and induced DNA fragmentation in HeLa cells as an evidence of apoptosis (Annamalai et al., 2015).

Hyattella intestinalis (Lamarck, 1814)

Order: Dictyoceratida Family: Spongiidae

Common name: Australian sponge

Pathogens Tested (Zone of Clearance in mm)

Associated Bacteria	*S. aureus*	*S. typhi*	*S. paratyphi*	*K. oxytoca*	*E. coli*	*P. mirabilis*	*L. vulgaris*	*K. pneumoniae*	*V. cholerae*	*A. tumefaciens*
B. cereus	–	–	4	5	3	3	3	2	2	6
B. megaterium	5	5	7	3	2	2	1	–	–	–
B. subtilis	4	–	6	6	5	4	–	–	2	5
C. freundii	2	2	2	1	11	6	6	–	4	–
N. mucosa	11	5	–	–	–	4	1	–	1	2
P. putida	10	4	3	3	7	8	5	6	7	9
L. plantarum	2	–	11	5	4	–	–	9	–	–

Source: Vasanthabharathi, V., and S. Jayalakshmi, 2012, *Afr. J. Biotechnol.*, 11: 7500–7511.

Anti-TB and anticancer/cytotoxic activities: The compound, 25-dehydroxy-12-epi-deacetylscalarin of this species showed anticancer activity aganst the murine lymphocytic leukemia P388 cell line with an ED50 value of 2.9 μg/mL. Further, this compound showed weak antitubercular activity (16% inhibition) against *Mycobacterium Tuberculosis* (H37Rv) with an MIC value of 16 μM at a concentration 6.25 μg/mL) (González, 2010).

The compound 25-dehydroxy-12-epi-deacetylscalarin also showed cytotoxicity against the following cell lines.

Distribution: Atlantic Ocean, Mediterranean Sea, Indian Ocean

Ecology: This sessile sponge grows at depths of 9 to 70 m.

Biology: Sponge of this species is minutely conulose. Oscules are terminal in position. Texture is hard with poor resiliency. This sponge harbors alpheid shrimps in its canals. It is hermaphroditic. The zygote develops into parenchymella larva (free-swimming) before settling down on a substrate where it grows into a young sponge.

Compound(s) and Activities

Anticancer activity: Norscalaral B and its deacetyl derivative, 12-deacetylnorscalaral B, of this species showed mild cytotoxic activity against three tumor cell lines (breast carcinoma MDA-MB-231, GI50 = 4.9 μg/mL; lung carcinoma A549, GI50 = 4.5 μg/mL; colon carcinoma HT29, GI50 = 4.2 μg/mL) (González, 2010).

The compound mooloolaldehyde showed cytotoxic activity toward the P388 mouse leukemia cell line with IC50 value in the range of 3 to 10 μg/mL (González, 2010).

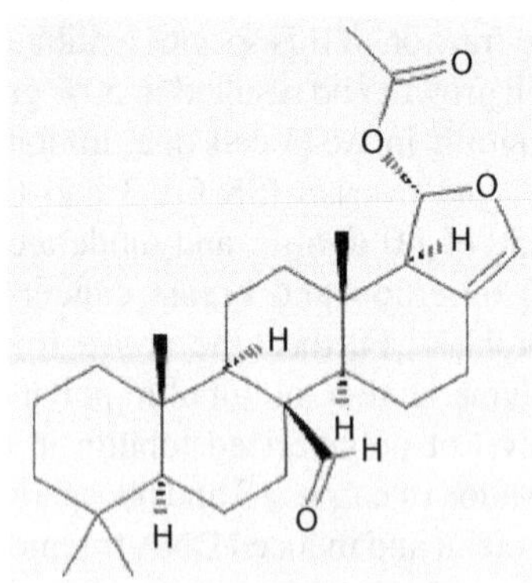

Mooloolaldehyde

The compound hyatelone A showed mild cytotoxic activity against the tumor cell lines MDA-MB-231 (breast carcinoma, GI50 = 5.4 μg/mL) and HT-29 (colon adenocarcinoma, GI50 = 9.2 μg/mL). Hyatolide A showed mild cytotoxic activity against the tumor cell lines MDA-MB-231 (breast carcinoma, GI50 = 4.8 μg/mL), A459 (lung adenocarcinoma, GI50 = 5.1 μg/mL), and HT-29 (colon adenocarcinoma, GI50 = 5.0 g/mL) (González, 2010).

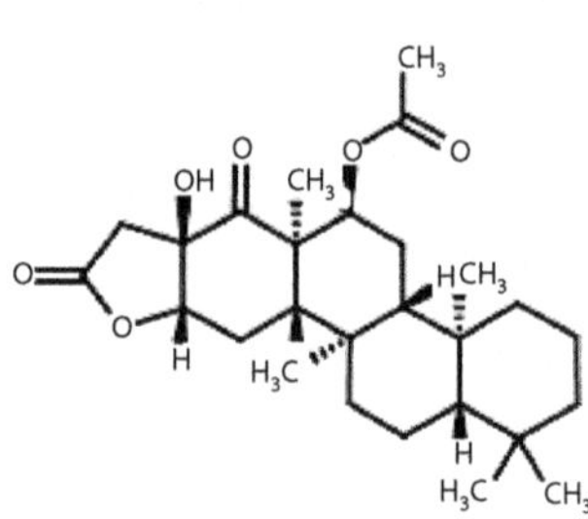

Hyatelone A

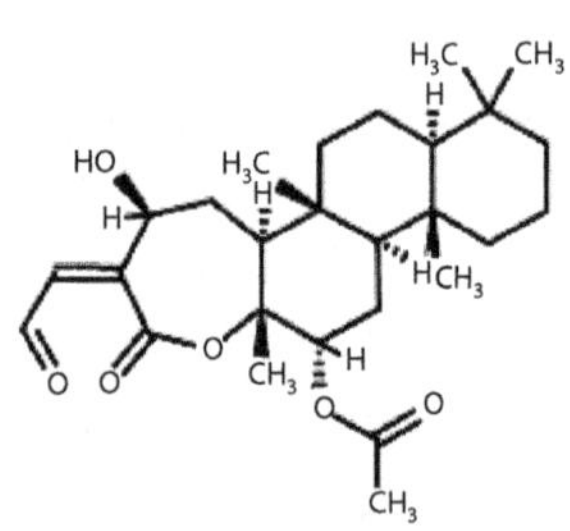

Hyatolide A

The compound hyatelactam was only cytotoxic against the HT-29 (colon adenocar cinoma, GI50 = 8.1 μg/mL) cell line (González, 2010).

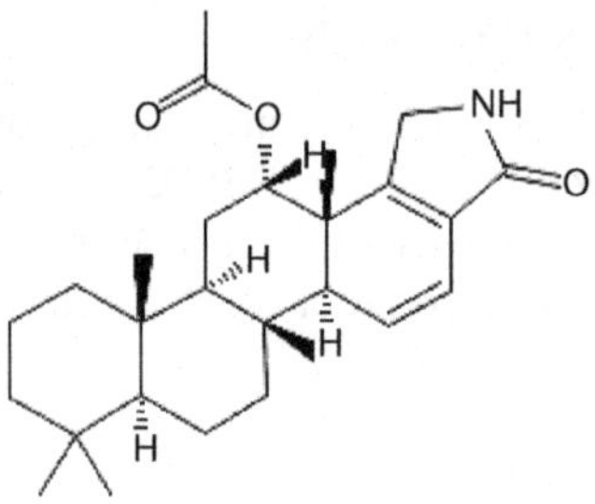

Hyatelactam

Ferreira et al. (2007) reported that the hydroethanolic extracts of this species showed cytotoxicity against the following tumor cell lines at a concentration of 20 mg/mL.

IC50 (μg/mL) Values of Tumor Cell Lines

HL-60	HCT-8	MDA-MB435	SF-295
17	19	38	23

Source: Ferreira, E.G. et al., 2007, *Porifera Res.: Biodiversity, Innovation Sustainability*, 313–318.

González (2010) reported that the compounds mooloolabenes A, B, and C were cytotoxic toward the P388 mouse leukemia cell line with IC50 values of 0.8, 1.2, and 3–10 μg/mL, respectively.

Somerville et al. (2006) and González (2010) reported that the new norsesterterpenes mooloolabenes A–E, and a new sesterterpene mooloolaldehyde related to the scalarane family of compounds isolated from the acetone extract of this species showed cytotoxic activity against the P388 mouse leukemia cell line. The compounds mooloolabene D and E showed with IC50 values 3–10 μg/mL.

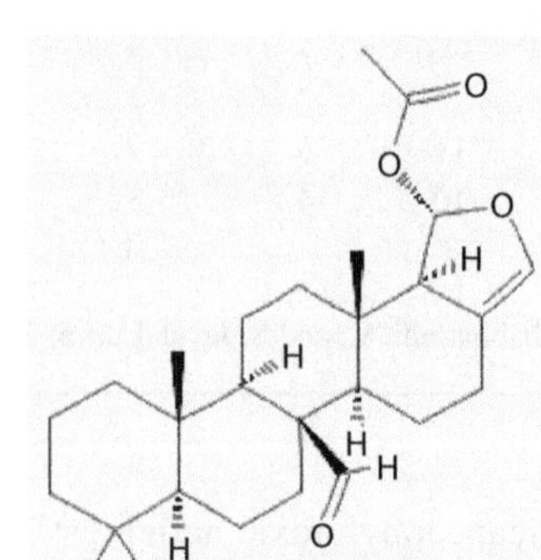

Mooloolaldehyde

Antiplasmodial activity: The unidentified bacterial isolate THB20 associated with this sponge species exerted antiplasmodial activity against *P. falciparum* with an IC50 value of 41.9 μg/mL (Indraningrat et al., 2016).

Hyatella sp.

Order: Dictyoceratida Family: Spongiidae

Compound(s) and Activities

Antibacterial properties: The associated bacterial strain, *Vibrio* M22-1 of this unidentified sponge species produced a peptide, namely, andrimid, which showed antibacterial properties against *Bacillus* sp. (Thomas et al., 2010).

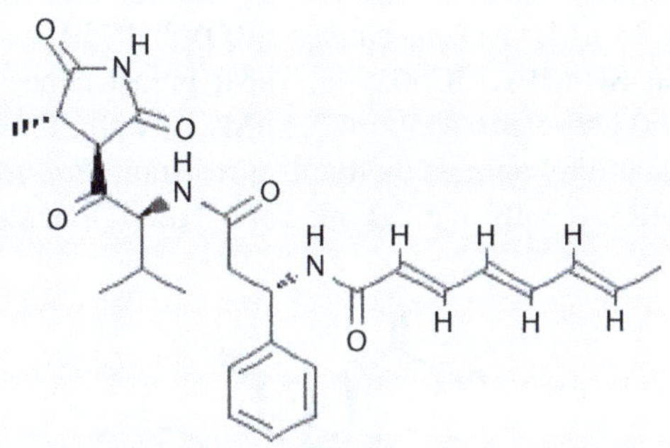

Andrimid

Leiosella idia (de Laubenfels, 1932) (= *Spongia idia*)

Order: Dictyoceratida Family: Spongiidae

Common name: Californian sponge

Distribution: Central California coast

Ecology: It occurs in intertidal or shallow subtidal habitats.

Biology: A structured sand cortex is present in this species. In the skeleton, its primary fibers are fasciculated. No other information is available.

Compound(s) and Activities

Antitubercular/cytotoxic/antimicrobial activity: The compound, 12-deacetyl-12,18-diepi-scalaradial extracted from this species displayed both antitubercular (MIC, 8 μM) and cytotoxic activities (ovarian cancer SK-OV-3, IC50, 19.4 μM; melanoma SK-MEL, IC50, 18.6 μM; breast cancer BT549, IC50, 12.9 μM; Vero, IC50, 20.7 μM). This compound also displayed potent antimicrobial activity against methicillin-resistant *Staphylococcus aureus* (MRSA) with an IC50 value of 3.8 μM (González, 2010).

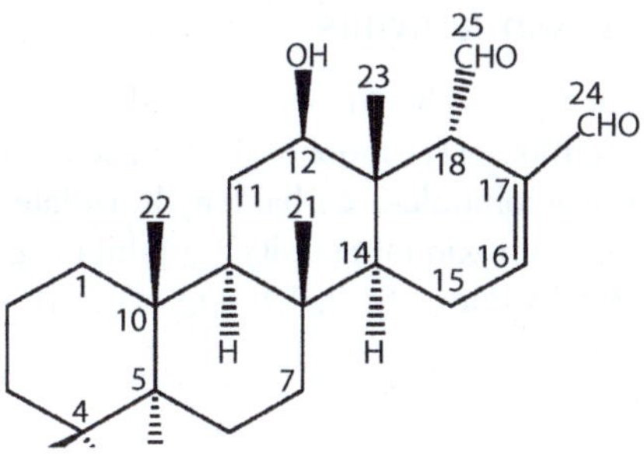

12-Deacetyl-12, 18-diepi-scalaradial

Rhopaloeides odorabile (Thompson, Murphy, Bergquist, and Evans, 1987)

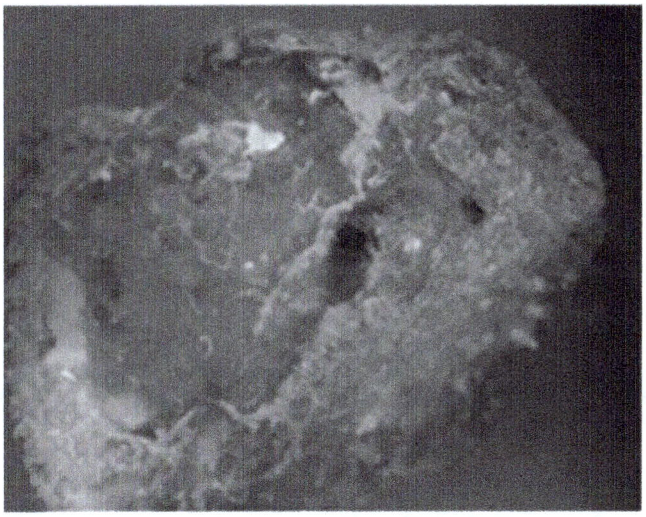

Order: Dictyoceratida Family: Spongiidae

Common name: Great Barrier Reef sponge

Distribution: South Pacific; Central Great Barrier Reef, Australia

Ecology: It is a coral reef–associated species. It occurs in fully illuminated and partially shaded forereef sites at a depth of 10 m from an inshore reef, midshelf reef, or an outershelf reef. It prefers medium to high current conditions.

Biology: It is an irregular, massive, or cylindrical sponge with colors of grayish beige or dark olive/green (in life), brown (exterior), and light brown (interior in ethanol). Oscules are several, moderately sized, and are scattered over the surface. Texture is firm and compressible. Surface is conulose; membranous between conules, or hispid; and conules are broad and multiple. Ectosomal skeleton consists of surface membrane with foreign spicules, which are scattered. Megascleres and microscleres are absent.

Compound(s) and Activities

Antioxidant activity: The monoindole alkaloids (1H-indol-3-yl) oxoacetamide, (1H-indol-3-yl) oxoacetic acid methyl ester, and 6-bromoindole-3-carbaldehyde isolated from this species showed antioxidant activity by inhibiting bee venom PLA2 with IC50 values of 1.2, 1.1, and 1.3 mM, respectively (Longeon et al., 2011).

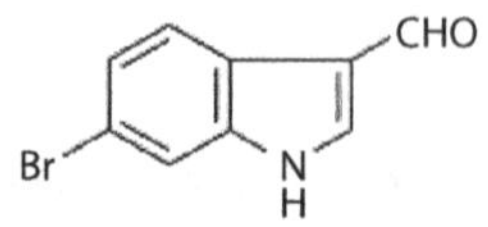

6-bromoindole-3-carbaldehyde

Antileukemic activity: This species contains the unusual group of C-20 terpene compounds. The norsester terpene rhopaleoic acid A isolated from this species showed antilekemic activity against human myeloid K-562 cells, MOT-4 leukemic cells, and murine L 1210 leukemic cells (Muller, 2003).

Spongia (Spongia) agaricina (Pallas, 1766)

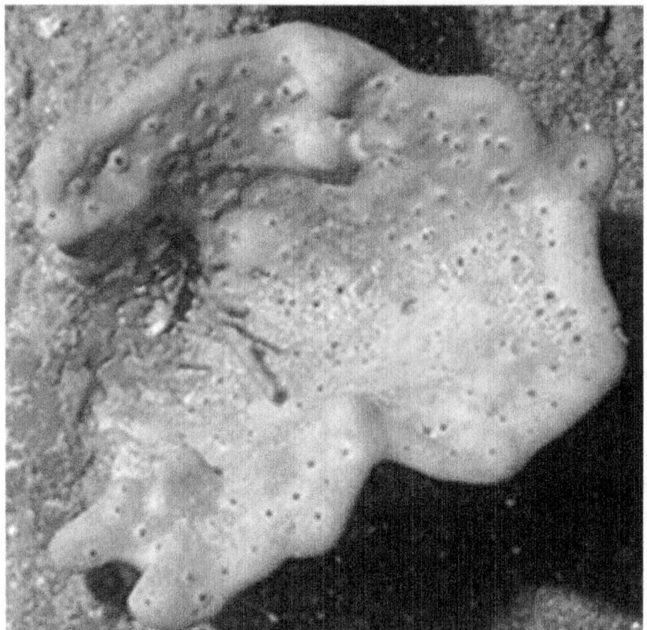

Order: Dictyoceratida Family: Spongiidae
Common name: Elephant ear sponge, Mediterranean horny sponge
Distribution: Atlantic Coast, Mid-Portugal, Mediterranean, Spain
Ecology: It inhabits caves at 5 to 15 m. It is also found on rock faces in deeper water, down to 300 m.
Biology: It is a flattened, platelike, or cuplike sponge, usually 15–25 cm wide and 1–4 cm thick. However, larger specimens up to 1 m wide are also seen. Surface is finely conulose. Oscules are only seen on the inside or upper side of the plate. Consistency is supple and elastic. Spicules are absent. Color of the sponge is beige or dark maroon. Skeleton (*Spongia agaricina* fibers) Ectosome is strengthened by a light reticulation of sand grains. Primary fibers are cored by sand grains. Secondary fibers are however, free from inclusions.

Compound(s) and Activities

Antimicrobial activity and cytotoxicity: The compound 12-epi-scalaradial isolated from this species showed antimicrobial activity against *Staphylococcus aureus* (10 µg/disk), *Bacillus subtilis* (10 µg/disk), and *Candida albicans* (10 µg/disk). Further, this compound was an inhibitor of human recombinant PLA2 (IC50 = 0.02 µM) and significantly reduced the induced mouse ear inflammation. Furthermore, it is was also an inhibitor of epidermal growth factor receptor (EGFR), which is implicated in many cancers. The semisynthetic 12-epi-scalaradial has displayed both antitubercular (MIC = 58 µM) and cytotoxic activity (human oral fibroblasts, IC50 = 1.26 µM; ovarian cancer SKOV3, IC50 = 23.3 µM; melanoma SK-MEL, IC50 >42.7 µM; breast cancer BT549, IC50 = 21.0 µM; Vero, IC50 = 22.2 µM), as well as potent antimicrobial activity against methicillin-resistant *Staphylococcus aureus* (MRSA) with an IC50 value of 6.3 µM (González, 2010).

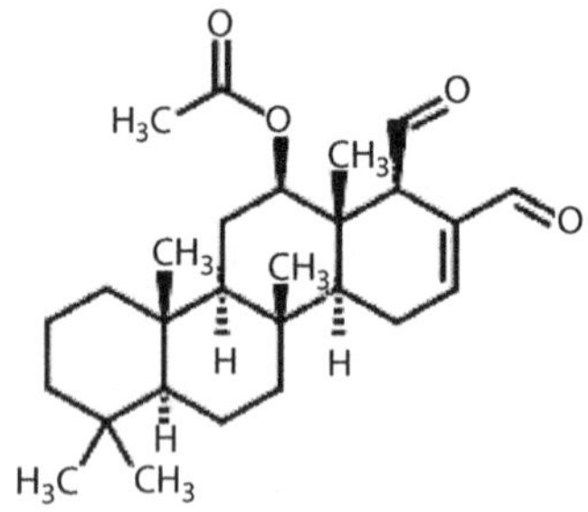

12-epi-scalaradial

Cytotoxicity: The compound, 12-epi-scalarin of this species exhibited moderate cytotoxicity toward four tumor cell lines (murine leukemia L1210, IC50 = 13.2 µg/mL; human cervix epithelioid carcinoma HeLa, IC50 = 26.0 µg/mL; human lung carcinoma A549, IC50 = 23.7 µg/mL; and human oral epidermoid carcinoma KB, IC50 = 18.5 µg/mL) as well as neurotrophic activity in pheochromocytoma (PC-12) cells at the extract concentration of 50 µg/mL (González, 2010).

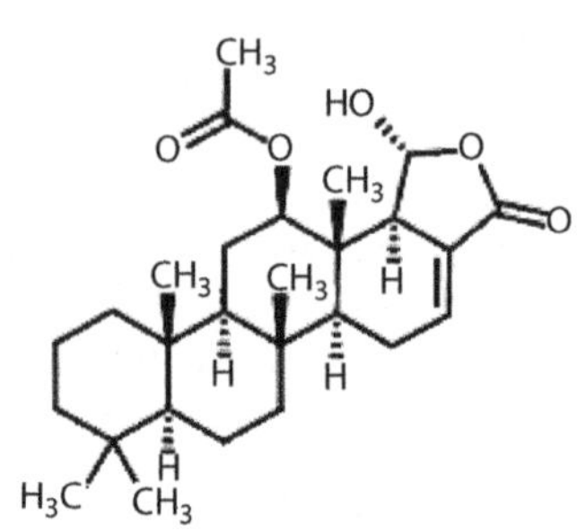

12-epi-scalarin

The compound 12,16-di-epi-12-deacetyl-16-acetylfuroscalarol showed significant cytotoxicity against four cancer cell lines (mouse lymphoma P388, human lung carcinoma A549, human colon carcinoma HT-29, and human melanoma MEL-28 with IC50 values of 1.0 μg/mL in all cases) (González, 2010).

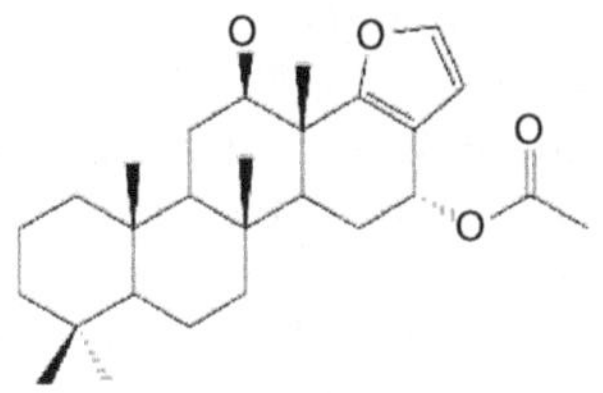

12, 16-di-epi-12-deacetyl-16-acetylfuroscalarol

Two new 9,11-secosterols, [3-O-deacetylluffasterol B (1) and 3-O-deacetyl-22,23-dihydro-24,28-dehydroluffasterol B (2)] and two new sesterterpenoids [12,16-di-epi-12-O-deacetyl-16-O-acetylfuroscalarol (3) and 16-epi-scalarolbutenolide (4)] have ben isolated from this species. Among them, the metabolites 1–3 showed significant cytotoxicity against four tumor cell lines (IC50, 1 μg/mL) (Rueda et al., 1998).

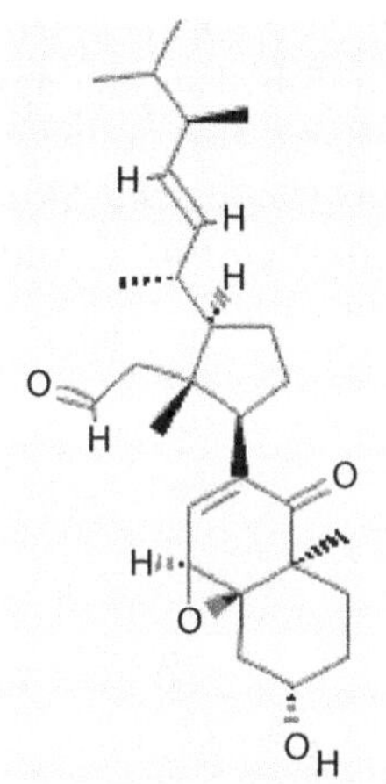

3-O-deacetylluffasterol B

Hypercholesterolemic activity: The compound, 12-epi-scalarin also showed inhibition of farnesoid X activated receptor (FXR) transactivation, which is related to potential drugs for hypercholesterolemia in humans, with an IC50 value of 60.4 μM (González, 2010).

Anti-TB activity: The compound, 16-epi-scalarolbutenolide of this species displayed 40% inhibitory activity against *Mycobacterium tuberculosis* (H37Rv) (González, 2010).

Spongia nitens (Schmidt, 1862)

Order: Dictyoceratida Family: Spongiidae

Common name: Shiny sponge, silky sponge

Distribution: Mid-Portugal, Mediterranean

Ecology: This sessile species is seen in caves and in coralligenous communities at a depth range of 0 to 60 m.

Biology: It is a massive, irregularly lobate sponge, with lobes often prolonged and anastomosing. Size is up to 7 × 6 × 4 cm. Specimes exceeding the size of 15–20 cm are also rarely seen. Surface is finely conulose. Oscules (1–4 mm) are seen on top of the lobes. Consistency is spongy, compressible, and elastic. Spicules are absent. Skeleton (spongia nitens fibers) is a reticulation of spongin fibers. At the surface there is a fine-meshed tangential reticulation of fibers with smaller meshes than those of the choanosome. Primary fibers of the choanosomal skeleton are devoid of any inclusions but have a distinct pith. Color of the animal is whitish to light brown.

Compound(s) and Activities

Antimicrobial and anti-inflammatroy activities: The compound 12-epi-scalaradial of this species showed antimicrobial activity against *Staphylococcus aureus* (10 μg/disk), *Bacillus subtilis* (10 μg/disk), and *Candida albicans* (10 μg/disk). This compound was also an inhibitor of human recombinant PLA2 (IC50 = 0.02 μM) and significantly reduced induced mouse ear inflammation (González, 2010).

Cytotoxic, antibacterial, and anti-TB activities: The 12-epi-scalaradial compound was also an inhibitor of epidermal growth factor receptor (EGFR), which is implicated in many cancers. Semisynthetic 12-epi-scalaradial displayed both antitubercular (MIC = 58 μM) and cytotoxic activity (human oral fibroblasts, IC50 = 1.26 μM; ovarian cancer SKOV3, IC50 = 23.3 μM; melanoma SK-MEL, IC50 >42.7 μM; breast cancer BT549, IC50 = 21.0 μM; Vero, IC50 = 22.2 μM). This semisynthetic compound also displayed potent antibacterial activity against methicillin-resistant *Staphylococcus aureus* (MRSA) with an IC50 value of 6.3 μM (González, 2010).

Cytotoxic, neurotrophic, and hypercholesterolemic activities: The compound 12-epi-scalarin of this species exhibited moderate cytotoxicity against four tumor cell lines (murine leukemia L1210, IC50 = 13.2 μg/mL; human cervix epithelioid carcinoma HeLa, IC50 = 26.0 μg/mL; human lung carcinoma A549, IC50 = 23.7 μg/mL; and human oral epidermoid carcinoma KB, IC50 = 18.5 μg/mL). Further, this compound showed neurotrophic activity in pheochromocytoma (PC-12) cells at the concentration of 50 μg/mL. Furthermore, it also showed inhibition of farnesoid X activated receptor (FXR) transactivation, which is related to potential drugs for hypercholesterolemia in humans, with an IC50 value of 60.4 μM (González, 2010).

Spongia officinalis (Linnaeus, 1759)

Order: Dictyoceratida Family: Spongiidae

Common name: Bath sponge

Distribution: Throughout the Mediterranean Sea

Ecology: It is seen in caves, coralligenous communities, rocky/detritic/muddy/sandy bottoms, lagoons, and *Posidonia oceanica* meadows at a depth range of 1 to 70 m.

Biology: It is a massive-lobate sponge. Surface is finely conulose. Single oscules are found scattered or at the apex of lobes. Pre-oscular cavities are clearly seen. Color *in vivo* is from light gray to black. Ectosomal skeleton consists of apices of primary fibers joining secondary fibers to form the conical reticulum, which supports the conules. Choanosomal skeleton is a dense network with irregular polygonal meshes of secondaries joining to form ascending primaries.

Compound(s) and Activities

Cytotoxicity: The compound deoxoscalarin isolated from this species displayed cytotoxic activity against murine lymphoma L1210 and human epidermoid carcinoma KB cell lines (IC50 = 1.4 and 6.4 μg/mL, respectively) (González, 2010).

Mosquito larvicidal activity: The extract of this species (var. ceylonensis) displayed larvicidal activity against the fourth instar larvae of *Aedes aegypti* (yellow fever mosquito vector) with an LC50 value of 172.1 ppm (Rao et al., 2008).

Anticonvulsant and analgesic activities: The crude extract and the semipurified fraction (F2) of this species exhibited interesting analgesic activity in a dose-dependent manner. The analgesic effect of the subcutaneous administration of the crude extract and its semipurified fractions has been studied through acetic acid (1%) and PBQ (phenylbenzoquinone) writihing tests in mice. The results showed that the values of nhibition of writhing were 62% (crude extract at 400 mg/kg) and 85% (F2 at 200 mg/kg) in the acetic acid test. The corresponding values were 75% (crude extract at 400 mg/kg) and 82% (F2 at 200 mg/kg) in the PBQ test (Dellai et al., 2012).

Antibacterial activity: The antibiotic compound furospongin-I isolated from this species showed activity againt *Diplococcus* sp. and *Streptococcus* sp. Similarly, its bioactive compounds 11β-hydroxyspongi-12-en-16-one and 11 β-acetoxyspongi-12-en-16-one (new diterpenoids) were active against *Bacillus sphaericus*, *Staphylococcus aureus*, and *Pseudomonas aeruginosa* ("Antibacterial activity," n.d.).

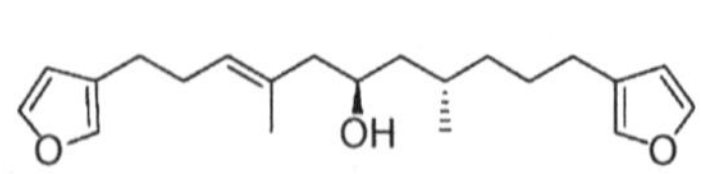

Furospongin-I

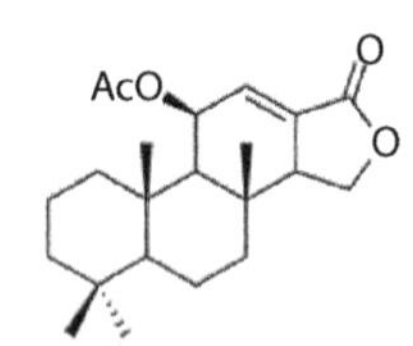

11β-hydroxyspongi-12-en-16-one

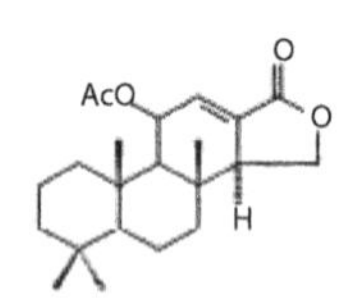

11β-acetoxyspongi-12-en-16-one

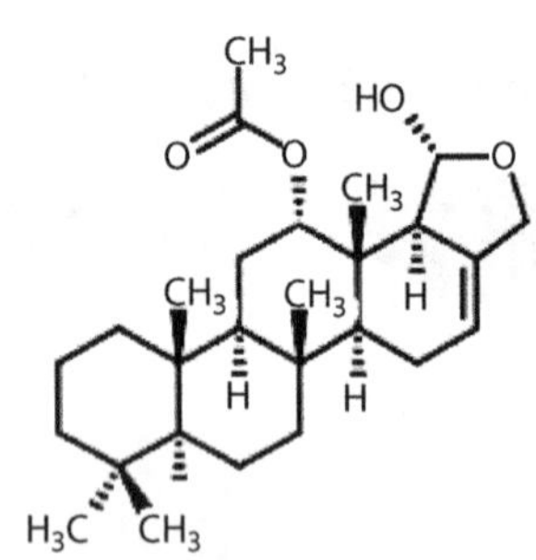

Deoxoscalarin

Spongia sp.

Order: Dictyoceratida Family: Spongiidae

Compound(s) and Activities

Anti-inflammatory activity: The pyridinium alkaloid compounds, spongidines A–D, of this sponge species showed anti-inflammatory activity by serving as a phospholipase A2 inhibitor (Sipkema et al., 2005).

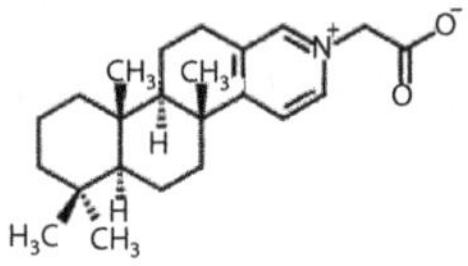

Spongidine A

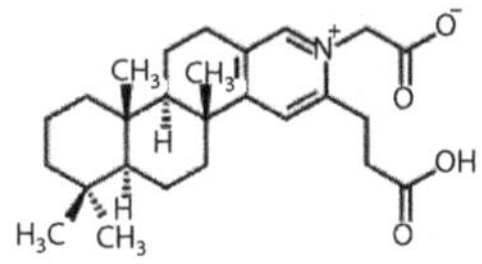

Spongidine C

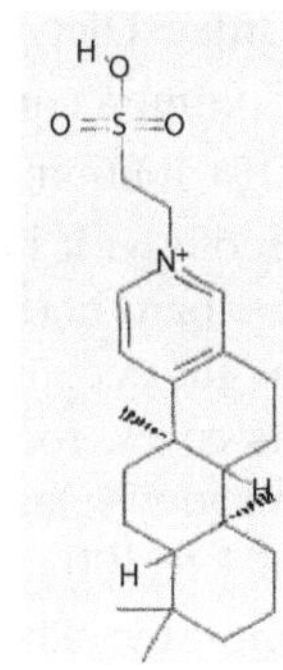

Spongidine D

The alkaloids 21-hydroxypetrosaspongiolide K and 21-hydroxypetrosaspongiolide P of this species anti-inflammatory activity by inhibiting the venom of cobra and bee (Keyzers, 2005).

The alkaloids spongidines A–D of this species showed anti-inflammatory activity by significantly inhibiting human synoviral PLA2 enzyme (Keyzers, 2005).

Anticancer/antitumor activity: The sterol agosterol A of this species reverses drug resistancy of cancer cells (Sipkema et al., 2005).

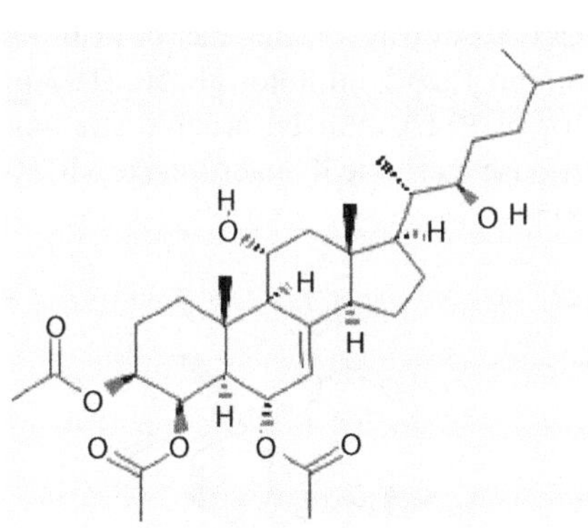

Agosterol A

The macrocyclic lactone polyether compound spongistatin 1 showed cytotoxic cell death in numerous cancer cell lines (Perdicaris et al., 2013).

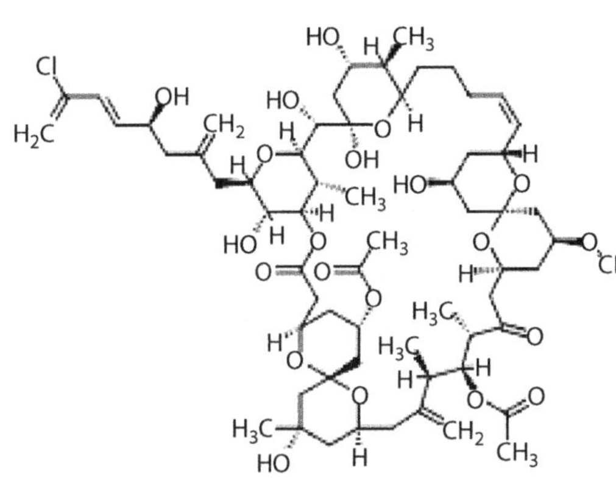

Spongistatin 1

The compounds spongistatins isolated from this species were found to be cytotoxic and inhibited the cancer cell proliferation with antitumoral activity (Sepčić et al., 2010).

Cytotoxicity: The extracts of the fungus *Myrothecium verrucaria* 973023 derived from this unidentified sponge species yielded trichothecene compounds, namely, 3-hydroxyroridin E; 13′-acetyltrichoverrin B; miophytocen C; roridins A, L, and M; isororidin A; epiroridin E; verrucarins A and M; and trichoverrins A and B. All the compounds except miophytocen C showed significant cytotoxicity against murine and human tumor cell lines (Thomas et al., 2010).

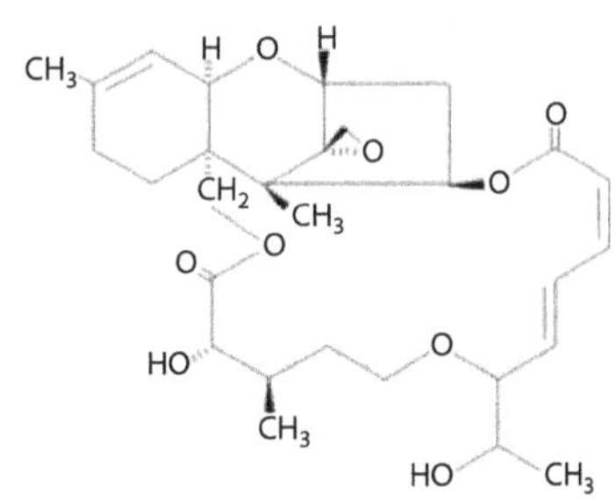

Roridin A

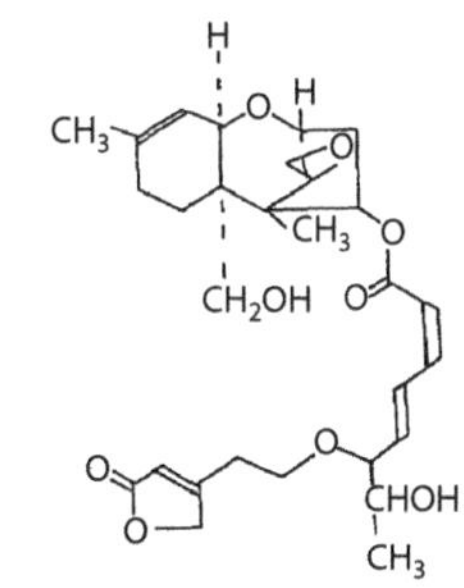

Roridin L

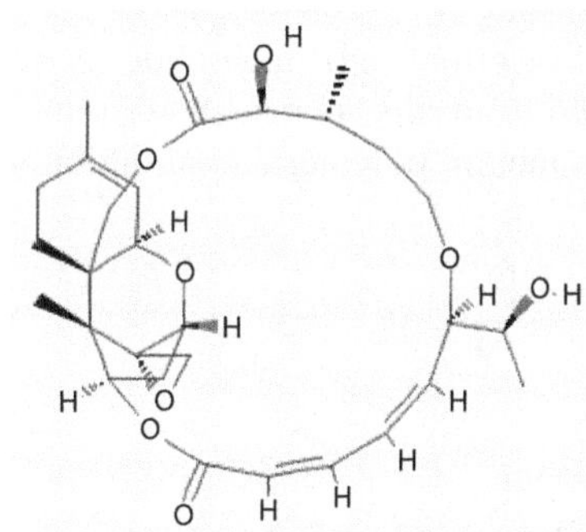

Isororidin A

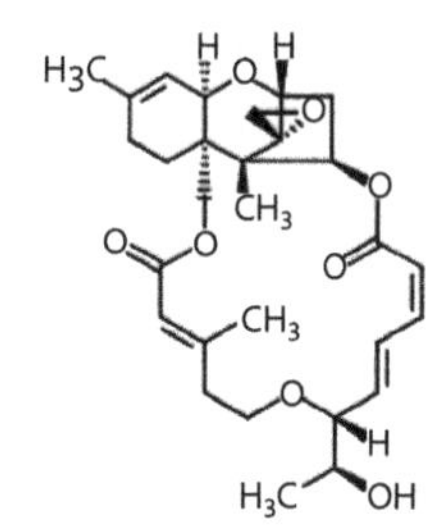

Epiroridin E

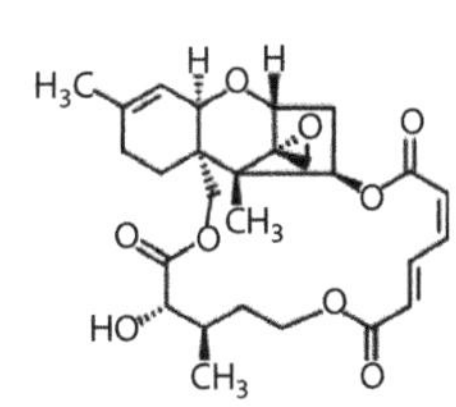

Verrucarin A

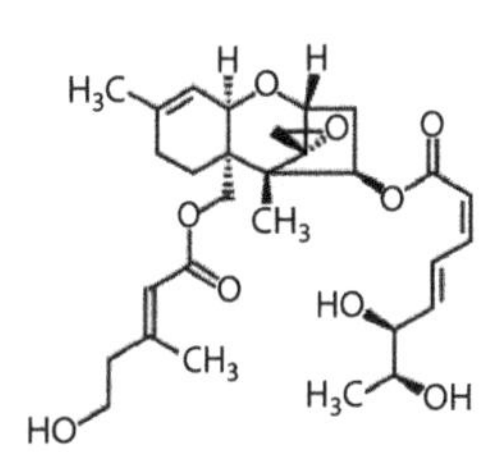

Trichoverrin B

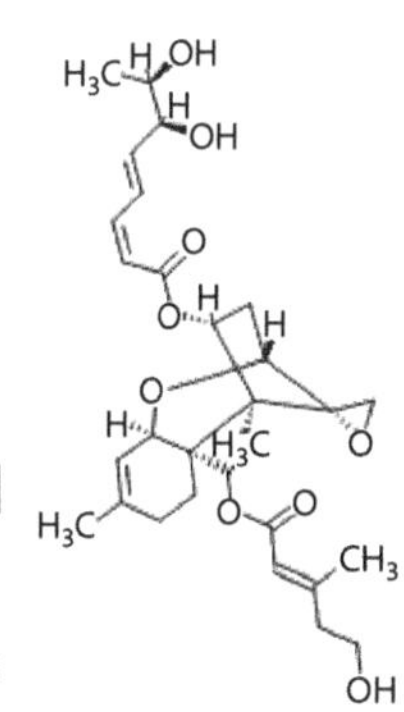

Trichoverrin B

The compound spongidepsin of this species showed cytotoxic activity against J774.A1, WEHI-164, and HEK-293 cancer cell lines, with an IC50 value in the submicromolar range (Dembitsky et al., 2006).

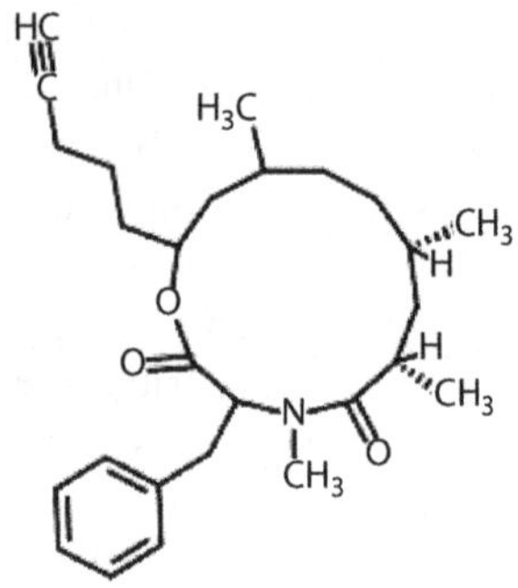

Spongidepsin

Cytotoxicity and neurotrophic activity: The compound 12-O-deacetyl-12-epi-scalarin of this species exhibited moderate cytotoxicity toward four tumor cell lines (murine leukemia L1210, IC50 = 2.3 μg/mL; human cervix epithelioid carcinoma HeLa, IC50 = 15.0 μg/mL; human lung carcinoma A549, IC50 =14.8 μg/mL; human oral epidermoid carcinoma KB, IC50 = 14.3 μg/mL). Further, this compound showed neurotrophic activity in pheochromocytoma (PC-12) cells at the extract concentration of 50 μg/mL (González, 2010).

The compound 16-deacetoxyscalara-furan of this species exhibited cytotoxicity against human cervix epitheloid carcinoma HeLa cells, IC50 = 19.5 μg/mL (González, 2010).

Antitumor activity: The bis(spiroacetal) macrolide compound spongistatin 1 isolated from this sponge showed antitumor activity by serving as tubulin polymerization inhibitor (Sipkema et al., 2005).

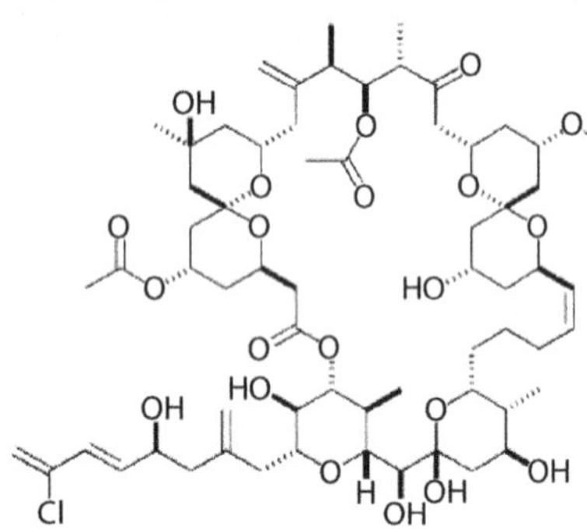

Spongistatin 1

FAMILY: THORECTIDAE

Cacospongia mollior (Schmidt, 1862)

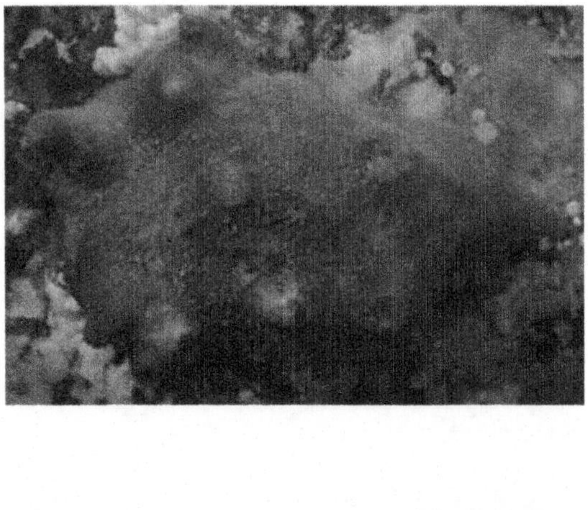

Order: Dictyoceratida Family: Thorectidae

Common name: Not designated

Distribution: North Atlantic Ocean

Ecology: It is a coralligenous species (coralligenous outcrops are hard bottoms of biogenic origin that are mainly produced by the accumulation of calcareous encrusting algae). It inhabits caves, rocky/detritic/muddy bottoms, *Posidonia oceanica* meadows, lagoons, and is epibiotic on *Pinna nobilis* at a depth of 1 to 100 m.

Biology: The growth form of this species is massive and lobate with 10–25 cm in diameter. Consistency is soft and spongy. Color of the body is dark gray with whitish, bluish, and magenta tinges. Surface is smooth and is regularly conulose (1–1.5 mm in height, 1–2 mm apart), forming regular characteristic "circular craters." Oscules are found scattered and are small and single, upward of 1 mm in diameter.

Compound(s) and Activities

Anti-inflammatory activity: The tetracyclic sesterterpene scalaradial isolated from this species has been reported to be a potent bee venom PLA2 inhibitor (IC50, 0.07 μM), as well as an inhibitor of PMA-induced arachidonic acid release *in vivo* in the mouse peritoneal macrophage (IC50, 0.05 μM) (Keyzers, 2005).

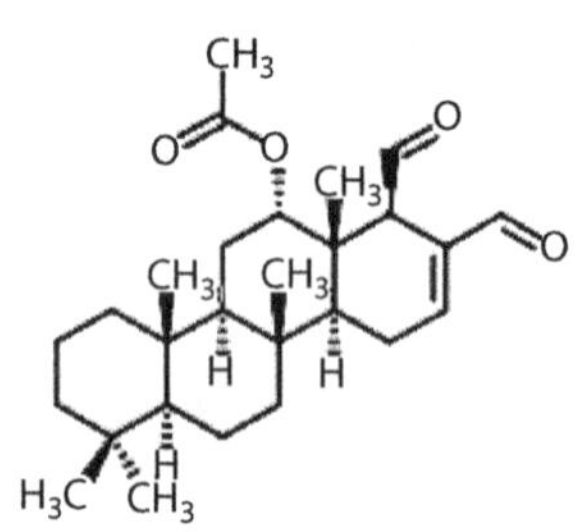

Scalaradial

Antitumor activity: The sesterterpene compound cacospongionolide isolated from this species has shown antitumoral properties (De Rosa et al., 1988).

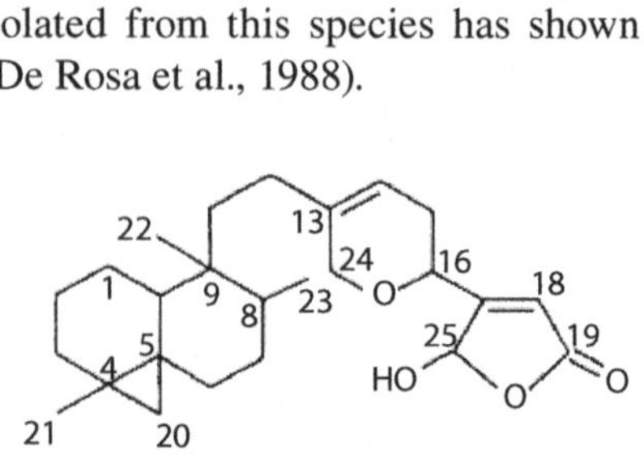

Cacospongionolide

Cacospongia mycofijiensis (Kakou, Crews, and Bakus, 1987) (= *Spongia mycofijiensis*)

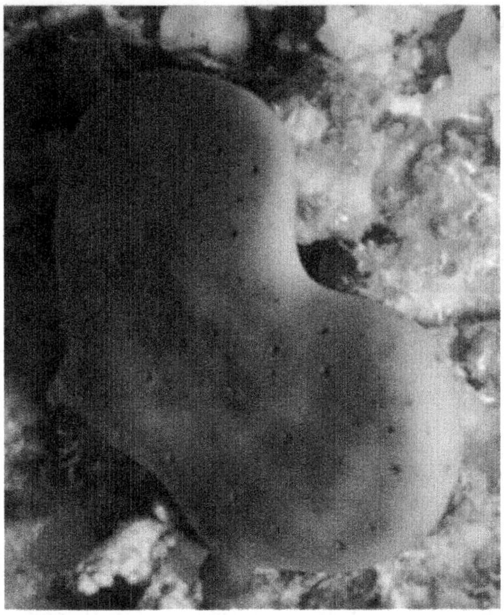

Order: Dictyoceratida Family: Thorectidae

Common name: Not designated

Distribution: Fiji Islands

Ecology: It grows in reef habitats, under ledges or in caves.

Biology: In skeletal structure, primary fibers of this species are of light color, visible inside and more heavily cored with spicules and sands. Further, the skeleton has large meshes. The color of the animal in this species does not change.

Compound(s) and Activities

Antitumor activity: The macrocyclic lactone, laulimalide isolated from this species showed antitumoral property through the stabilization of microtubules (Sipkema et al., 2005).

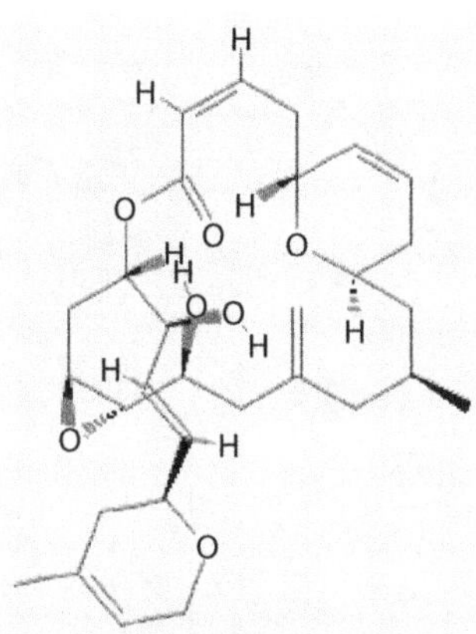

Laulimalide

Antitrypanosomial activity: Among the sesquiterpene compounds, namely, aignopsanoic acid A and B isolated from this species, compound A has been found to be active against *Trypanosoma brucei,* the parasite responsible for sleeping sickness (Hayes, 2012).

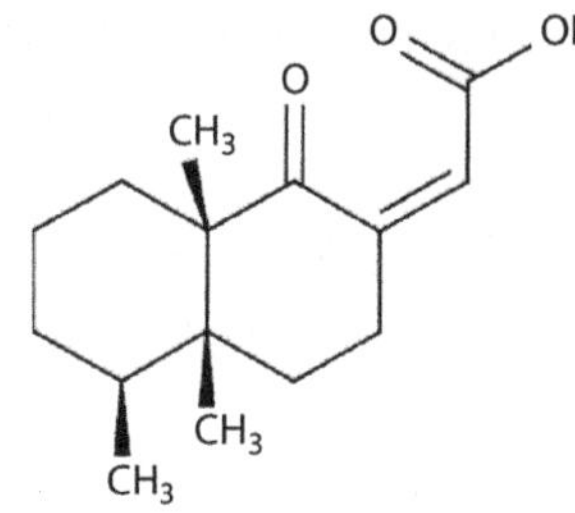

Aignopsanoic acid A

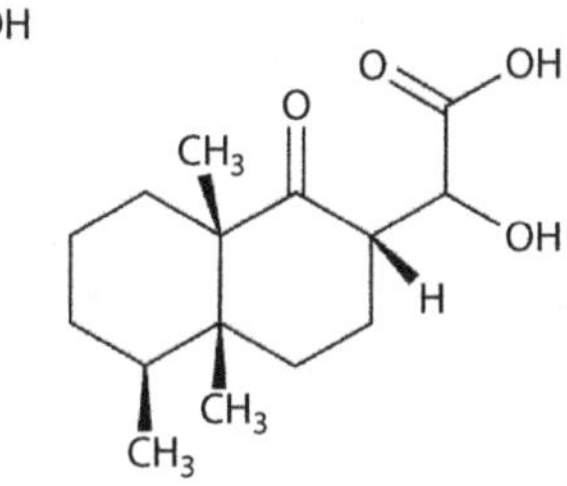

Aignopsanoic acid B

Johnson et al. (2009) reported that this species has yielded novel sesquiterpenes, namely, aignopsanoic acid A (1), methyl aignopsanoate A (2), and isoaignopsanoic acid A (3). Two of these metabolites, 1 and 2, were found to be moderately active against *Trypanosoma brucei*, the parasite responsible for sleeping sickness.

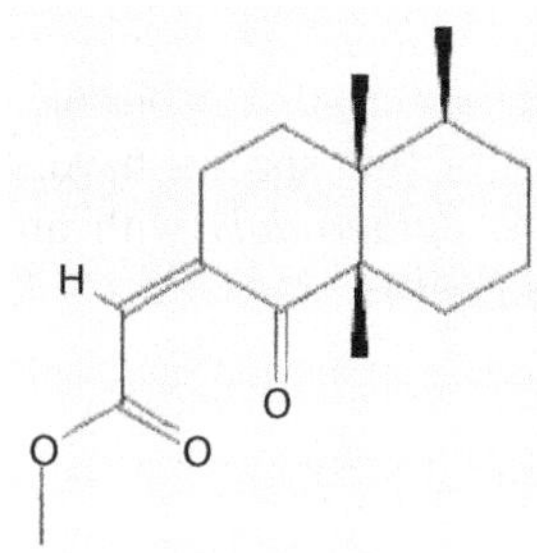

Methyl aignopsanoate A

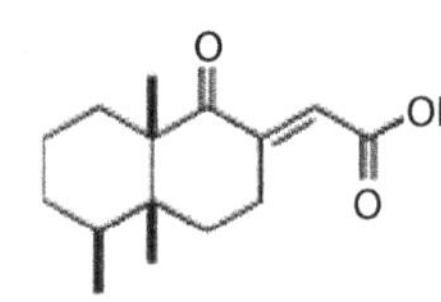

Isoaignopsanoic acid A

Cytotoxicity: The crude extract of this species was found to be extremely cytotoxic to epithelial carcinoma of the larynx (HEP-2) and monkey kidney (MA-104) cells (Kakou et al., 1987).

Amagata et al. (2008) reported that the compounds latrunculol A, latrunculol B, and 18-epi-latrunculol A showed cytotoxicities against both murine and human cancer cell lines with IC50 values ranging from 0.5 to 10 μM.

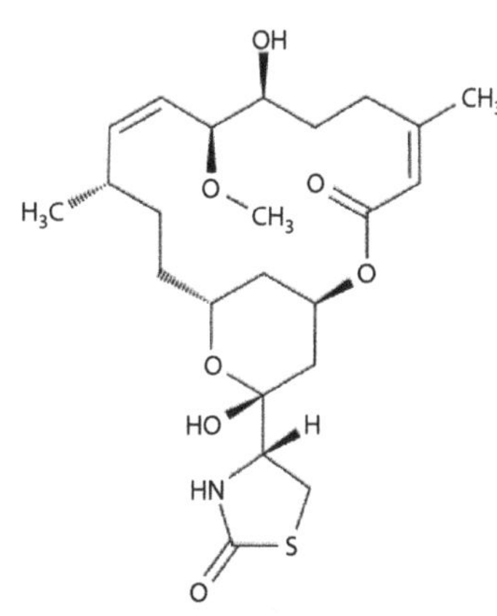

Latrunculol B

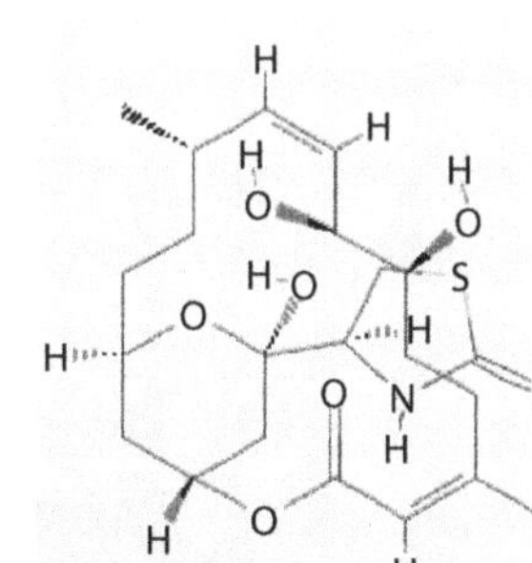

18-epi-latrunculol A

A 20-membered macrolide, zampanolide, isolated from this species has been reported to stabilize microtubules and block cells in G(2)/M of the cell cycle. Further this compound is cytotoxic by inducing microtubule bundles in cells (Field et al., 2009).

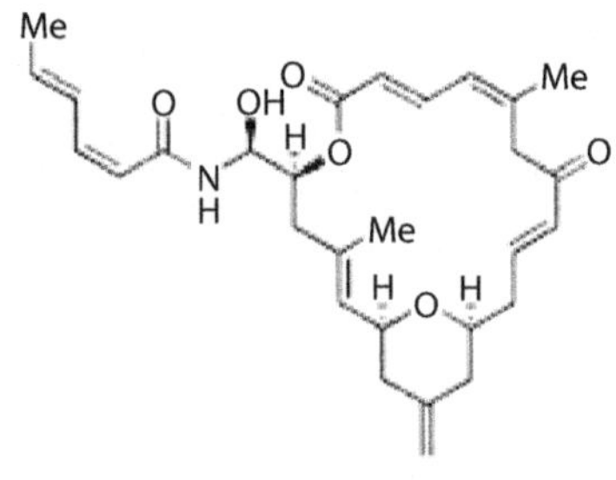

Zampanolide

Cacospongia sp.

Order: Dictyoceratida Family: Thorectidae

Compound(s) and Activities

Antimicroial activity: The meroditerpenes isojaspic acid, cacospongin D, and jaspaquinol of this species have shown antimicrobial activity against *S. epidermidis* with an MIC value of 20 µg/mL (Anjum et al., 2016).

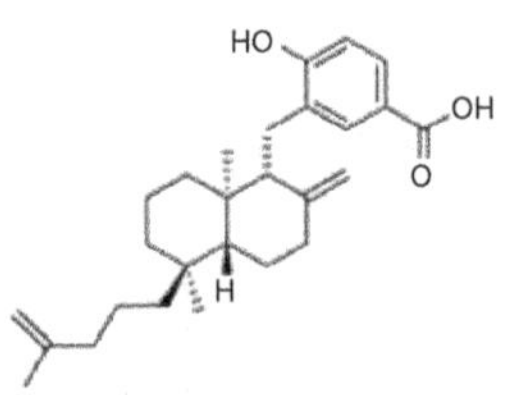

Isojaspic acid

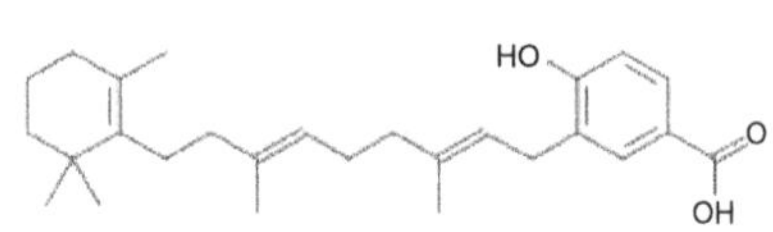

Cacospongin D

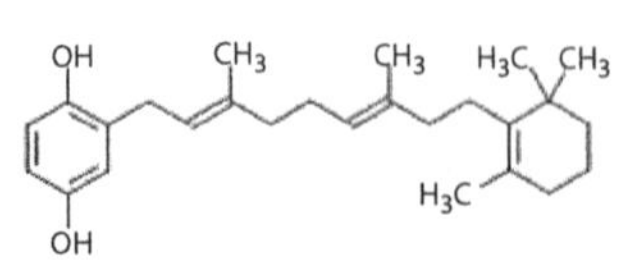

Jaspaquinol

Carteriospongia flabellifera (Bowerbank, 1877)

Order: Dictyoceratida Family: Thorectidae
Common name: Leaf sponge

Distribution: Central and Southern Great Barrier Reef, Eastern Philippines, Southeast Papua New Guinea

Ecology: It is found in the coral reefs of shallow, tropical coastal waters at 15–18 m deep. The foliose dictyoccratids are absent from inner-shelf reefs, which feature higher sediment loads and concentrations of organic nutrients.

Biology: This sponge is foliose (fragment: 13 cm wide, 6 cm long, and 0.2 cm thick) with a single and short attachment stalk. The surface displays a characteristic pattern of low, regularly aligned ridges and hispidous. The oscules are tiny. Texture is firm, flexible, and granular. In life, color is beige, and out of water it is brown. Siliceous spicules are absent.

Food and feeding: It is a sessile filter feeder with cyanobacterial symbionts.

Compound(s) and Activities

Antitumor activity: Two new sesterterpenoids named flabelliferins A and B were isolated from the lipophilic extract of this species. Among these, flabelliferin B exhibited inhibitory effect on the human colon tumor cell lines KM12 and COLO205 (Diyabalanage et al., 2012; Zhang et al., 2017b).

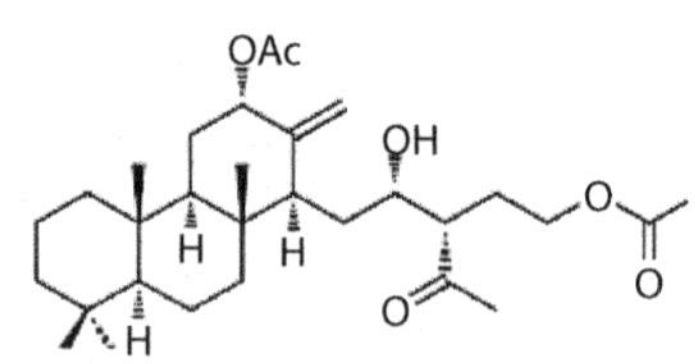

Flabelliferin A

Carteriospongia foliascens (Pallas, 1766) (= *Phyllospongia foliascens*)

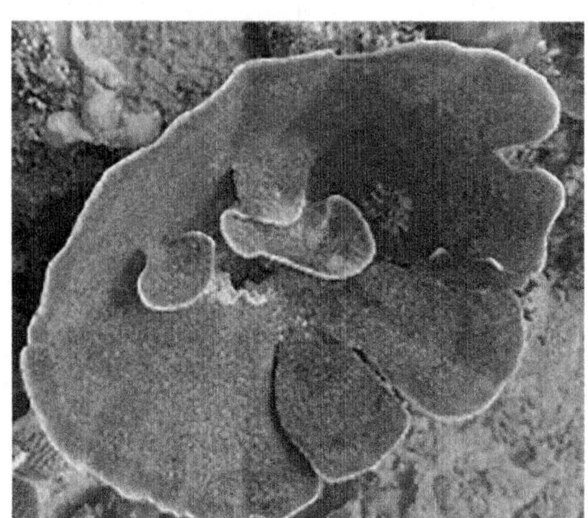

Order: Dictyoceratida Family: Thorectidae
Common name: Reef sponge

Distribution: Indo-West Pacific Ocean, Tanzania, Kenya, Mozambique Channel

Ecology: It inhabits reef flats and hard substrates. It is also in intertidal zones and shallow sandy lagoons.

Biology: Body of the sponge is flexible and fibrous. Surfce of the sponge is sandy and color of the animal is creamy gray brown.

Compound(s) and Activities

Cytotoxic, antithrombocyte, and vasodilative activities: The compounds dehydrofoliaspongin, phyllofoliaspongin, and foliaspongin isolated from this species exhibited a broad spectrum of pharmacological effects, such as cytotoxic, antithrombocyte, and vasodilative activities (González, 2010; Zhang et al., 2017a).

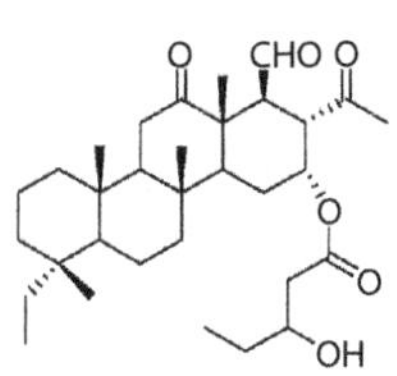

Dehydrofoliaspongin

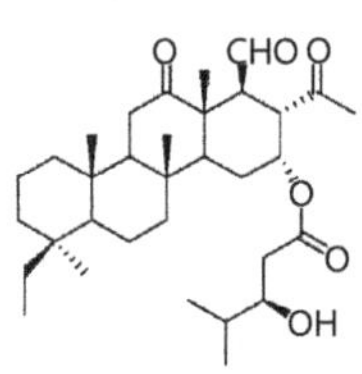

Phyllofoliaspongin

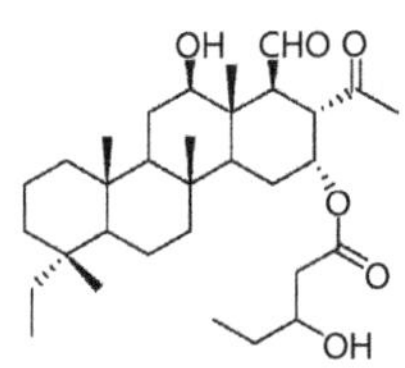

Foliaspongin

Cytotoxicity and antifungal activity: While the compound phyllofenone A of this species showed weak antifungal activity against *Candida pseudotropicalis*, phyllofenone B displayed cytotoxic activity against the P388 murine leukemia cell line with an IC50 value of 5 μg/mL (González, 2010; Zhang et al., 2017a).

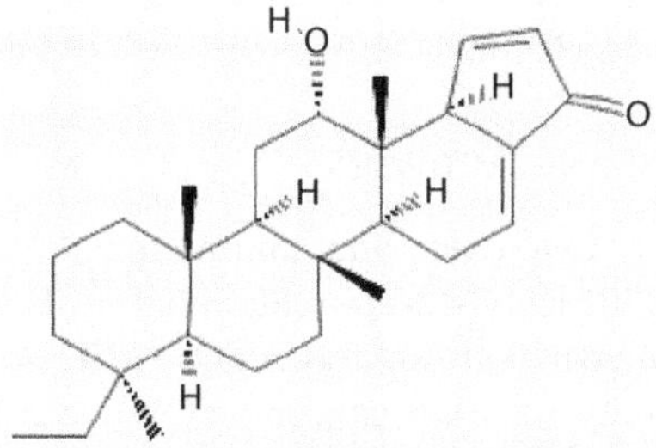

Phyllofenone B

Cytotoxicity: The compounds phyllactones A–C showed moderate cytotoxicity against KB cells (IC50 = 20 μg/mL). Phyllofenone D was cytotoxic against the P388 leukemia cell line with an IC50 value of 6.5 μg/mL, while phyllofenone B was cytotoxic against the same target with an IC50 value of 5.0 μg/mL (González, 2010; Zhang et al., 2017a).

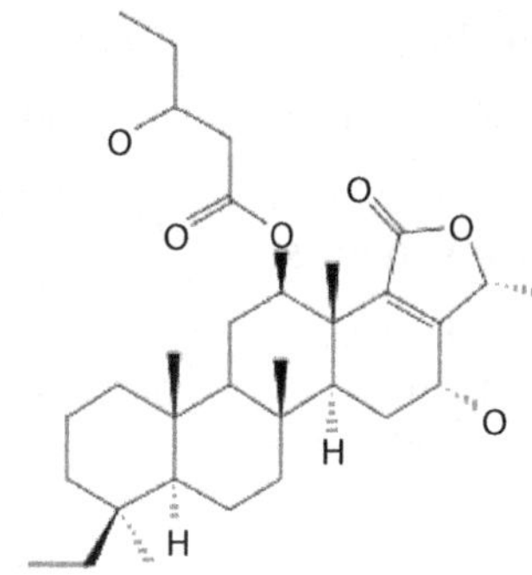

Phyllactone A

The compound 12α-acetoxy-16β-hydroxy-20,24-dimethyl-24-oxoscalaran-25-al exhibited potent cytotoxicity against 13 human tumor cell lines (IC50 = 1.13–1.99 μg/mL), 2 additional tumor cell lines (mouse lymphoma P388 and human lung carcinoma A549, IC50 = 0.1 μg/mL), and exhibited 30% to 95% inhibition of the growth of KB cells at 10 μg/mL (González, 2010).

Antitumor effect: The compound 20,24-bishomo-25-norscalarane, which is an inhibitor of Ras proteolytic enzyme (IC50 = 38 μg/mL), also inhibited the growth of four tumor cell lines (prostate PC3, IC50 = 3.8 μg/mL; colon LoVo, IC50 = 7.6 μg/mL; colon CAco-2, IC50 = 3.4 μg/mL; breast MDA 468, IC50 = 9.5 μg/mL) (González, 2010).

Eleven new scalarane sesterterpenoids, including three 20, 24-bishomo-25-norscalaranes [carteriofenones A–C (1–3)] and eight 20,24-bishomoscalaranes [carteriofenones D–K (4–11)], along with two known analogs (12 and 13), were isolated from this species. Among these compounds, the compound 4 (carteriofenone D) showed cytotoxicity against the mouse lymphocytic leukemia cell line (P388) with an IC50 value of 0.96 μM (Cao et al., 2015).

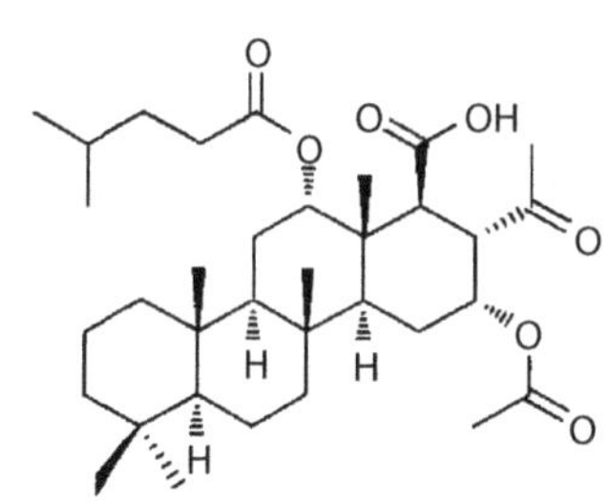

Carteriofenone D

Anti-inflammatory properties: The 20,24-bishomoscalarane member, foliaspongin, of this species showed anti-inflammatory properties (González et al., 2010). The bishomosesterterpene compound foliaspongin of this species showed anti-inflammatory activity (18.1% inhibition at 10 μg per disk utilizing the chorioallantoic chick embryo membrane assay) (Keyzers, 2005).

Carteriospongia sp.

Order: Dictyoceratida   Family: Thorectidae

Compound(s) and Activities

Cytotoxic and antibacterial activiites: The compounds 12α-acetoxy-20,24-dimethyl-25-norscalar-16-en-24-one, 12α-acetoxy-13β,18β-cyclobutan-20,24-dimethyl-24-oxoscalar-16-en-25β-ol, 12α-acetoxy-24,25-epoxy-24-hydroxy-20,24-dimethylscalarane, and scalardysin-A isolated from this unidentified sponge showed cytotoxic activity and antibacterial activity against some Gram-positive and Gram-negative strains (Hassan et al., 2015).

Antifungal and anti-inflammatory activities: The compound 24-methyl-24,25-dioxoscalar-16-en-12β-yl 3-hydroxybutanoate isolated from the organic extracts of this species showed antifungal and anti-inflammatory activities (González, 2010).

Anticancer activity: Three novel natural products—carteriosulfonic acids A, B, and C—of this species were found to nhibit GSK-3β, which inhibits Wnt signaling through phosphorylation of β-catenin. The Wnt signaling pathway plays major roles in controlling cell proliferation and differentiation. Therefore, misregulation of the Wnt pathway has been implicated in a number of human diseases including cancer and neurodegenerative diseases (McCulloch et al., 2009).

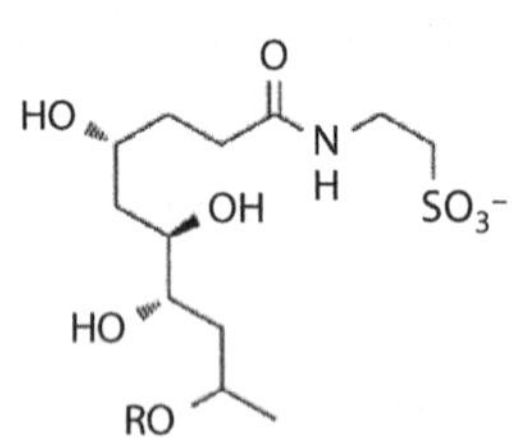

Carteriosulfonic acid

Antileukemic activity: The scalarane sesterterpenoid 12β-(3′β-hydroxybutanoyloxy)-20,24-dimethyl-24-oxo-scalara-16-en-25-al isolated from this species showed antileukemic activity of molt 4 cells at 0.0625 μg/mL (125 nM), by triggering mitochondrial membrane potential (MMP) disruption and apoptosis (Lai et al., 2016).

Antiviral activity: The compound hennoxazole A isolated from this sponge showed antiviral acitivty (Sipkema, et al., 2005).

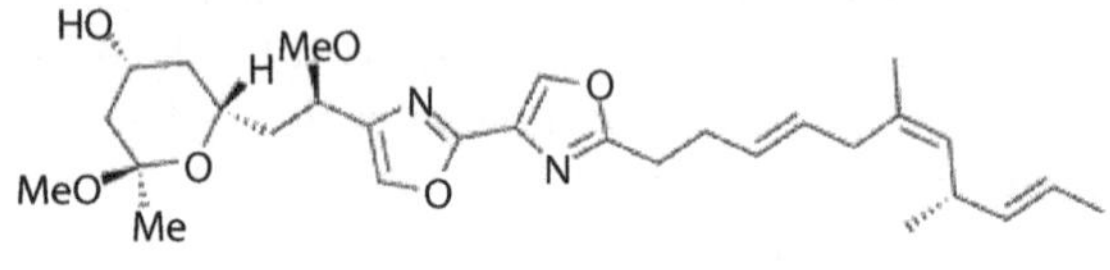

Hennoxazole A

Collospongia auris (Bergquist, Cambie, and Kernan, 1990)

Order: Dictyoceratida   Family: Thorectidae

Common name: Blue ear sponge

Distribution: Central and Southern Great Barrier Reef, Indo-West Pacific Ocean

Ecology: It inhabits fringing reefs where it seems to prefer calm but highly turbid waters.

Biology: It is a very beautiful sponge. Under dim light it is at least purple and in strong light the sponge gets iridescent blue. It is also an ideal species for saltwater aquariums. The aquarium parameters of this species are given next.

Aquarium Environment: Reef- or fish-only aquarium.

Temperature range: 72°F–83°F (22°C–28°C)
Minimum tank size: 50 gallons
Suitable specific gravity: 1.023–1.025
Optimal pH: 8.0–8.4

Acclimation time: This species is sensitive to specific gravity/salinity changes. Therefore, it is necessary to adjust the bag's water to that of the aquarium in which it will be placed by slowly adding small amounts of aquarium water to the bag every few minutes. In most cases, this process may take at least 15 minutes.

Compound(s) and Activities

Antitubercular, cytotoxic, and antimicrobial activities: The compound 12-deacetyl-12-epi-scalaradial isolated from this species has shown both antitubercular (MIC = 64 μM) and cytotoxic activity (ovarian cancer SKOV3, IC50 = 31.0 μM; melanoma SK-MEL, IC50 = 26.4 μM; breast cancer BT549, IC50 = 19.4 μM; Vero, IC50 = 19.4 μM), as well as antimicrobial activity against methicillin-resistant *Staphylococcus aureus* (MRSA) with an IC50 value of 22 μM (González, 2010).

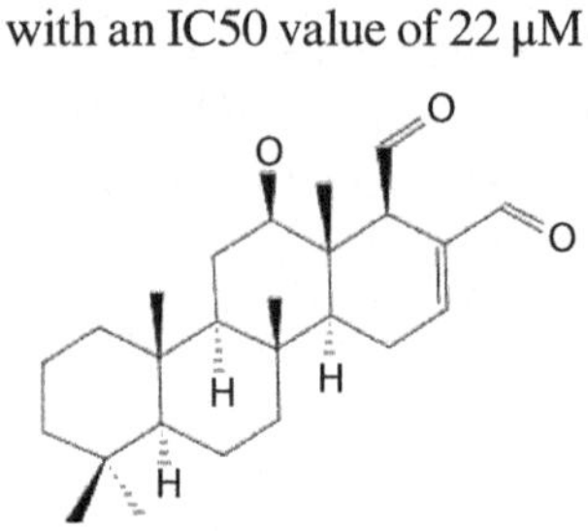

12-deacetyl-12-epi-scalaradial

Antimicrobial activity: The compound 12,18-di-epi-scalaradial of this species showed antimicrobial activity against *Staphylococcus aureus* (10 µg/disk), *Bacillus subtilis* (10 µg/disk), and *Candida albicans* (50 µg/disk) (González, 2010).

The compounds 12-deacetylhyrtial and methyl 18-hydroxy-25-norscalar-16-en-24-carboxylate of this species showed antimicrobial activity against *Staphylococcus aureus* (25 µg/disk) and *Bacillus subtilis* (50 µg/disk) (González, 2010).

The compound 24-hydroxy-scalarolide showed antimicrobial activity against *Staphylococcus aureus* (25 µg/disk), *Bacillus subtilis* (25 µg/disk), and *Candida albicans* (50 µg/disk) (González, 2010).

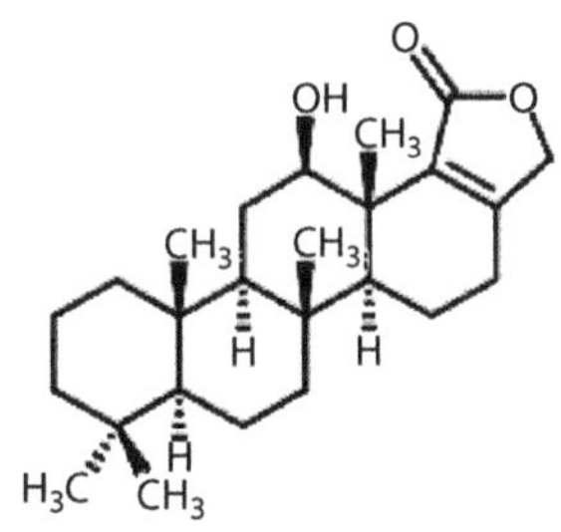

Scalarolide

Cytotoxic activity: The compound 12-epi-deoxoscalarin of this species showed cytotoxic activity against murine lymphoma L1210 and human epidermoid carcinoma KB cell lines (IC50 = 8.2 and >30.0 µg/mL, respectively). It is suggested that this compound may be a promising target to treat hypercholesterolemia in humans (González, 2010).

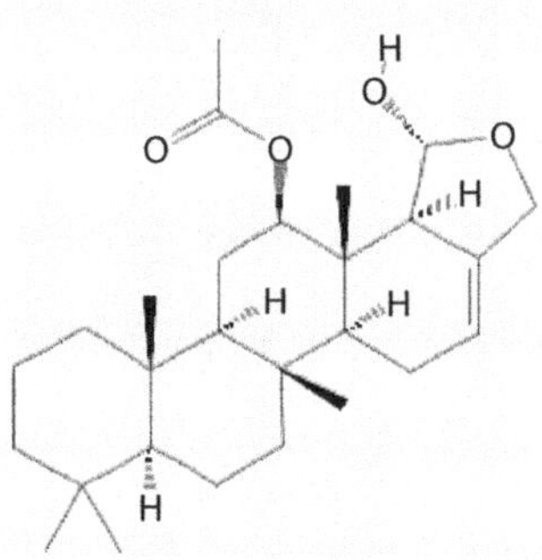

12-epi-deoxoscalarin

Others: The compound 12-epi-scalaradial of this species showed antimicrobial activity against *Staphylococcus aureus* (10 µg/disk), *Bacillus subtilis* (10 µg/disk), and *Candida albicans* (10 µg/disk). It is an inhibitor of human recombinant PLA2 (IC50 = 0.02 µM) and significantly reduces induced mouse ear inflammation. It is also an inhibitor of epidermal growth factor receptor (EGFR), which is implicated in many cancers. Semisynthetic 12-epi-scalaradial has displayed both antitubercular (MIC = 58 µM) and cytotoxic activity (human oral fibroblasts, IC50 = 1.26 µM; ovarian cancer SKOV3, IC50 = 23.3 µM; melanoma SK-MEL, IC50 >42.7 µM; breast cancer BT549, IC50 = 21.0 µM; Vero, IC50 = 22.2 µM), as well as potent antimicrobial activity against methicillinresistant *Staphylococcus aureus* (MRSA) with an IC50 value of 6.3 µM (González, 2010).

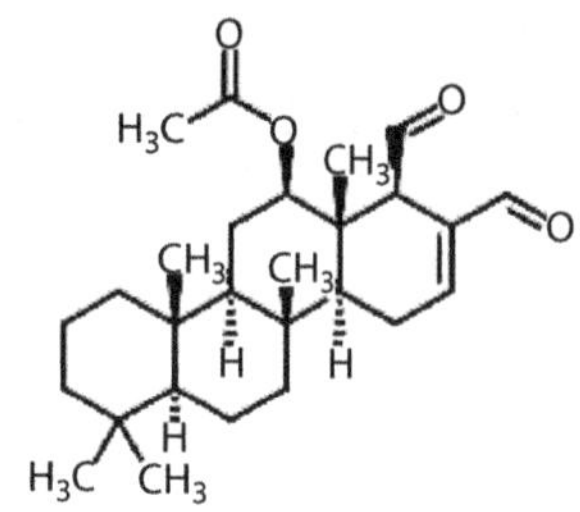

12-epi-scalaradial

Fascaplysinopsis reticulata (Hentschel, 1912)

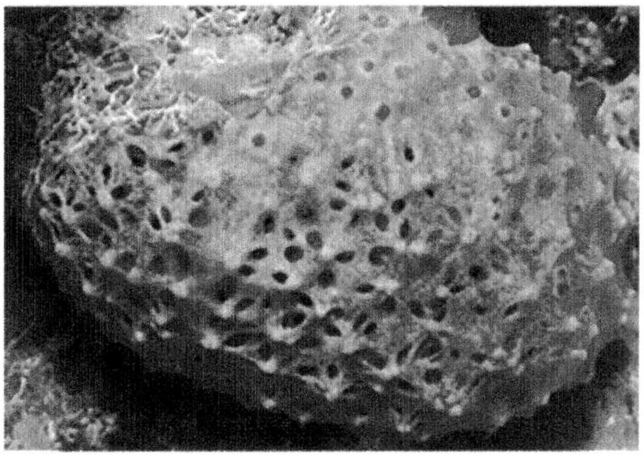

Order: Dictyoceratida Family: Thorectidae

Common name: Not designated

Distribution: Western Central Pacific, Indonesia, New Caledonia; Coral Sea, Indian Ocean, Palau

Ecology: This sessile sponge occurs in shallow water (depth range 0–100 m), on dead coral substrates, and in mangrove communities, especially on the pneumatophores of *Avicennia marina*.

Biology: Shape of this sponge is massive, irregular, and up to 15 cm high. The surface of this sponge is conulose. In life, it is characterized by a network of superficial hard fibers distributed over the whole surface. The surface is covered in some parts by a sand coat, which appears non-structural, induced by a high amount of suspended sediment. The texture is firm, hardly compressible, and difficult to tear. Color of the animal is black, with white fibers on surface.

Compound(s) and Activities

Cytotoxicity: The epibiotic bacterial strain *Pseudoalteromonas maricaloris* KMM 636T isolated from this sponge species yielded bromoalterochromide A, which showed moderate cytotoxicity (Thomas et al., 2010).

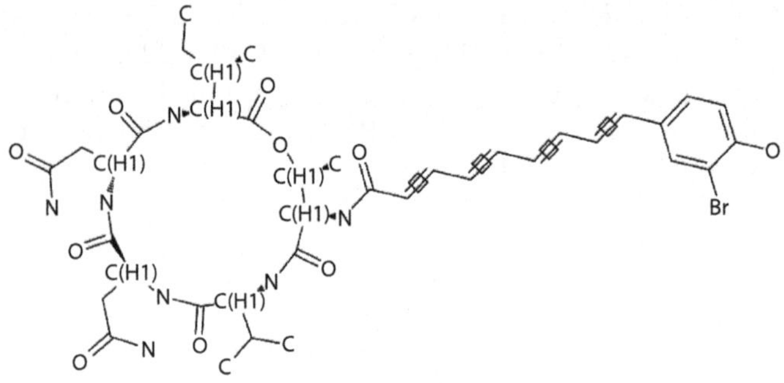

Bromoalterochromide A

Antiviral activity: The indole alkaloids fascaplysin and homofascaplysin A cation isolated from this species possessed antiviral properties, with the inhibitions against reverse transcriptase at a concentration of 1 mg/mL. The values of inhibition for the above alkaloids and reverse transcriptase were 81%, 58%, and 94%, respectively (Gul et al., 2005).

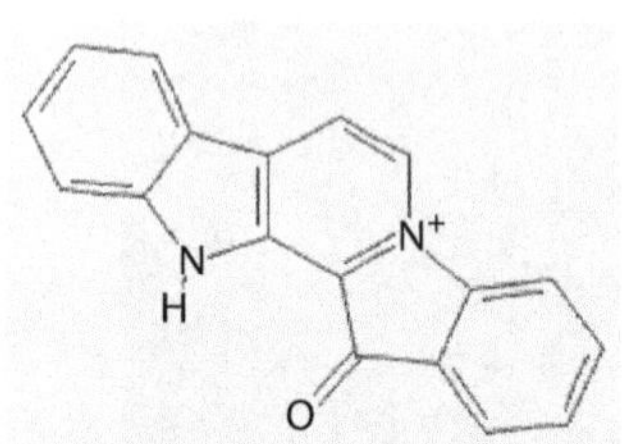

Fascaplysin

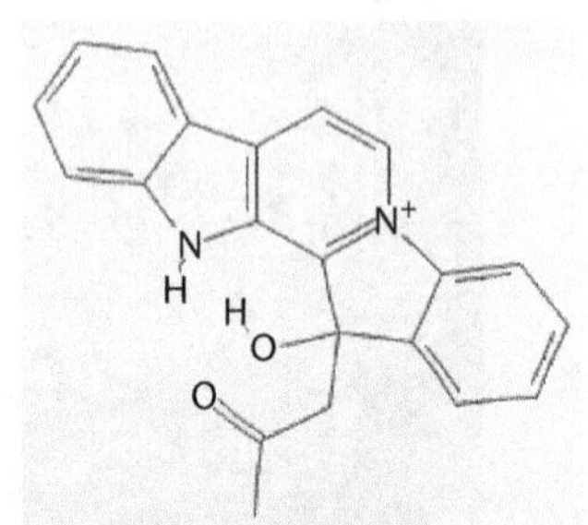

Homofascaplysin A

Antitumor activity (anti-interleukin-2 activity): The alkaloid compounds (+)- and (–)-spiroreticulatine A showed inhibitory activity on interleukin-2 (IL-2) production (http://www.scoop.it/t/natural-products-chemistry-burning-news).

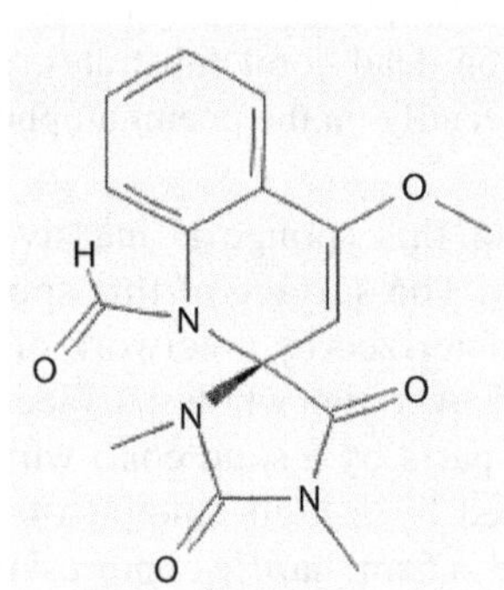

Spiroreticulatine

Antioxidant activity: The extract of this species showed moderate DPPH antioxidant activity with an IC50 <100 μg/mL (Abdillah et al., 2013).

Seradj et al. (2012) reportd on the hydroxyl radical scavenging activities of this sponge's extracts.

Percent of OH Radical Inhibition in CH_2Cl_2 Extracts

Concentrations in milligrams of dried sponge extract			
0.5	0.25	0.125	0.062
71.2	82.0	82.7	81.8

Percent of OH Radical Inhibition in MeOH Extracts

Concentrations in milligrams of dried sponge extract			
0.5	0.25	0.125	0.062
75.0	75.7	72.5	72.9

Source: Seradj, H. et al., 2012, *Iranian J. Pharm. Sci.*, 8: 249–255.

Fascaplysinopsis sp.

Order: Dictyoceratida Family: Thorectidae

Compound(s) and Activities

Anti-inflammatory activity: The sesterterpene palauolol isolated from this species showed anti-inflammatory activity by inactivating bee venom PLA2 with 82% inhibition at an extract concentration of 0.8 μg/mL (Keyzers, 2005).

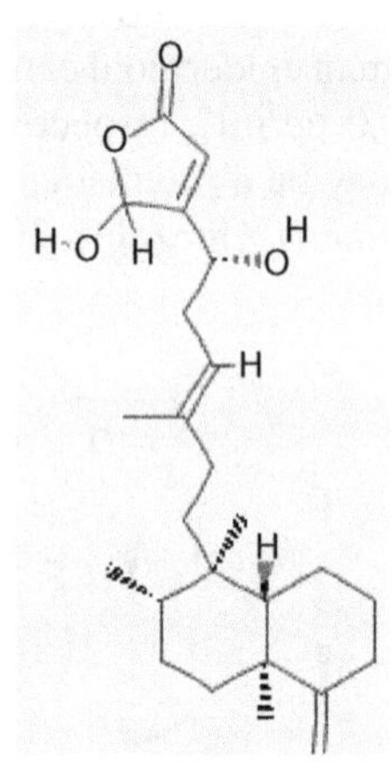

Palauolol

Antimicrobial activity: The organic extracts of this species have shown antimicrobial activity against *Staphylococcus aureus* (15 mm zone at 0.1 μg/disk), *Escherichia coli* (8 mm, at 5 μg/disk), and *Candida albicans* (11 mm, at 1 μg/disk) (Roll et al., 1988).

Gul et al. (2005) reported that the indole alkaloid fascaplysin showed activities against the growth of several microbes, with a 15 mm zone at 0.1 μg/disk against *S. aureus*, 8 mm zone at 5 μg/disk against *E. coli*, 11 mm zone at 1 μg/disk against *Candida albicans*, and 20 mm zone at 0.1 μg/disk against *S. cerevisiae*.

Fasciospongia cavernosa (Schmidt, 1862)

Order: Dictyoceratida Family: Thorectidae

Common name: Tyrrhenian sponge

Distribution: Mediterranean Sea

Ecology: It inhabits caves, coralligenous communities, rocky/detritic/muddy bottoms, and *Posidonia oceanica* meadows at a depth range of 1 to 350 m. Sometimes it presents a burrowing behavior.

Biology: Growth form of this species is tubular, massive, rounded, and is usually not larger than 10 cm, sometimes up to 25 cm in diameter. Color of the animal is dark brown at the surface. Consistency is strong and cartilaginous. The sponge surface is covered by very abundant conules (3–4 mm in height) giving a spiny aspect. External membrane is smooth, translucent, and resistant. Skeleton network is very strong with large (50–250 μm) rugose or granulated fibers.

Compound(s) and Activities

Anti-inflammatory activity: The sesterterpene lactone cacospongionolide B isolated from this species showed anti-inflammatory activity by serving as phospholipase A2 inhibitor (Sipkema et al., 2005). Keyzers (2005) reported that this compound showed activity against TPA-induced mouse ear edema after oral doses of 5, 10, or 20 mg/kg.

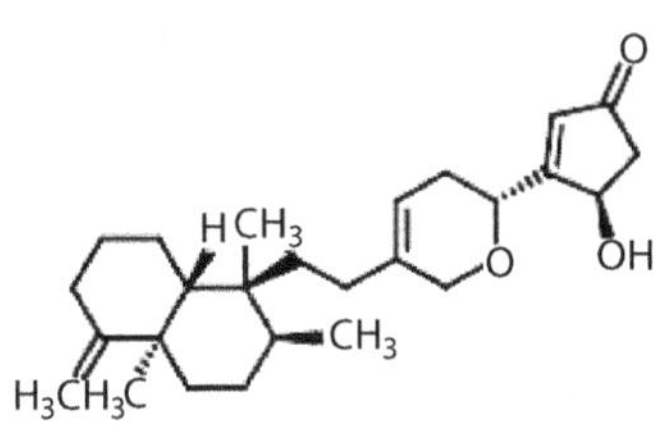

Cacospongionolide B

Antibacterial activity: The terpenoid microcionin-1 isolated from this species showed antibacterial activity against *M. luteus* with an MIC value of 6 μg/mL (Anjum et al., 2016).

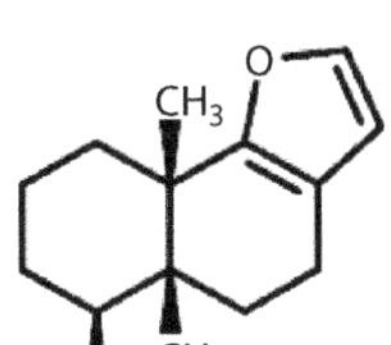

Microcionin-1

Mosquito larvicidal activity: The extract of this species displayed larvicidal activity against the fourth instar larvae of *Aedes aegypti* (yellow fever mosquito vector) with an LC50 value of 120.6 ppm (Rao et al., 2008).

Hyrtios altus (Poléjaeff, 1884)

Order: Dictyoceratida Family: Thorectidae

Common name: Not designated

Distribution: Not reported

Ecology: Not known

Biology: Not available

Compound(s) and Activities

Antibacterial activity: The trisindole derivative trisindoline obtained from the extracts of the associated *Vibrio* sp. of this sponge exhibited potential antibiotic activity against *Escherichia coli*, *Bacillus subtilis*, and *Staphylococcus aureus* (Thomas et al., 2010).

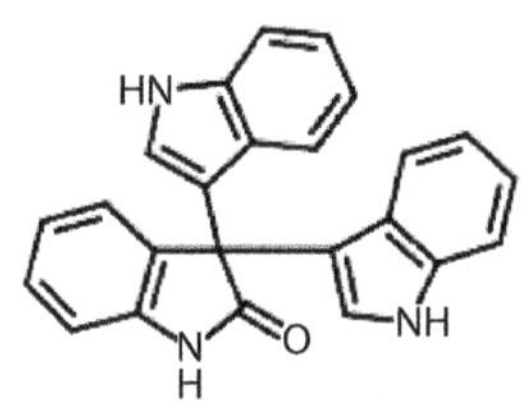

Trisindoline

Hyrtios communis (Carter, 1885) (*Hyrtios eubamma, Heteronema eubamma*)

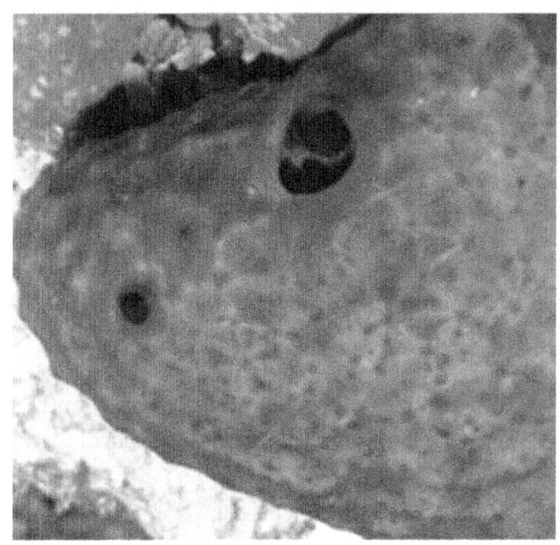

Order: Dictyoceratida Family: Thorectidae

Common name: Not designated

Distribution: Australia, Red Sea

Ecology: Not reported

Biology: Not reported

Compound(s) and Activities

Cytotoxicity: The extract of marine sponge *Hyrtios communis* was found to inhibit activation of the transcription factor hypoxia-inducible factor-1 (HIF-1) in T47D human breast tumor cells. Among the compounds isolated from the extract of this species, the sesterterpenes thorectidaeolide A, 4-acetoxythorectidaeolide A, and luffariellolide were the most potent inhibitors of hypoxia (1% O2)-induced HIF-1 activation (IC50 values of 3.2, 3.5, and 3.6 μM, respectively). Luffariellolide also exhibited a significant level of cytotoxicity that mirrored its HIF-1 inhibitory activity (Li et al., 2013).

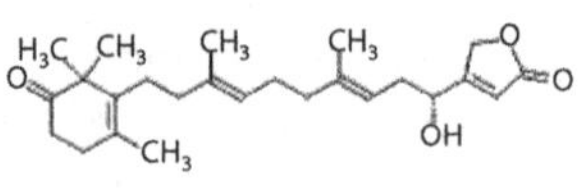

Thorectidaeolide A

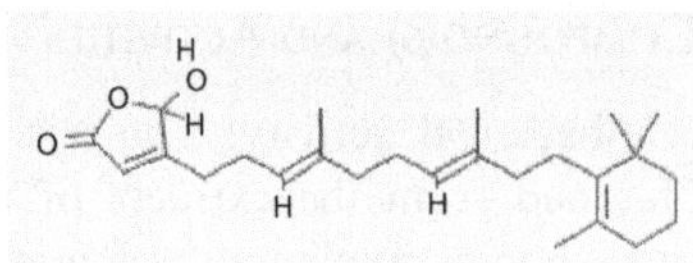

Luffariellolide

Anti-inflammatory activity: The extracts of this species inhibited the secretory phospholipase A2 isomers from bee venom and snake venom reducing inflammation with an IC50 value of 230 nM (Cayman Chemical, 2013).

Hyrtios erectus (Keller, 1889) (= *Hyrtios erecta; Heteronema erecta*)

Order: Dictyoceratida Family: Thorectidae

Common name: Tongan sponge, Red Sea sponge

Distribution: North Atlantic Ocean, Indian Ocean, Palau, Philippines

Ecology: It occurs in coral reef habitats and is found 9 to 13 m deep in areas with patches of hard and soft corals.

Biology: It is an elongate to massive (10 cm long, 6 m wide, and 2 cm thick) sponge. Surface is conulose (conules are 2 mm high and 3 mm wide). This sponge is compressible and may be difficult to tear. In life, the color of the ectosome ranges from brown to black while the endosome is light to dark brown. The siliceous spicules are absent. Skeleton is made up of primary and secondary fibers, fully cored with detritus. The surface is darkly pigmented.

Compound(s) and Activities

Chymotrypsin-like inhibitory activity: Among the β-carboline alkaloids, variabine A and variabine B, isolated from this species, variabine B inhibited chymotrypsin-like activity of the proteasome and Ubc13–Uev1A interaction with IC50 values of 4 and 5 μg/mL (16 and 20 dan μM), respectively. On the other hand, variabine A (a sulfonated derivative of variabine B) had little effect at 5 μg/mL (16 μM) (Putra and Jaswir 2014).

Antiparasitic activity: The compounds manadomanzamine A and B of this species showed antiparasitic activity against *Plasmodium falciparum* and *Mycobacterium tuberculosis*, respectively (Putra and Jaswir, 2014).

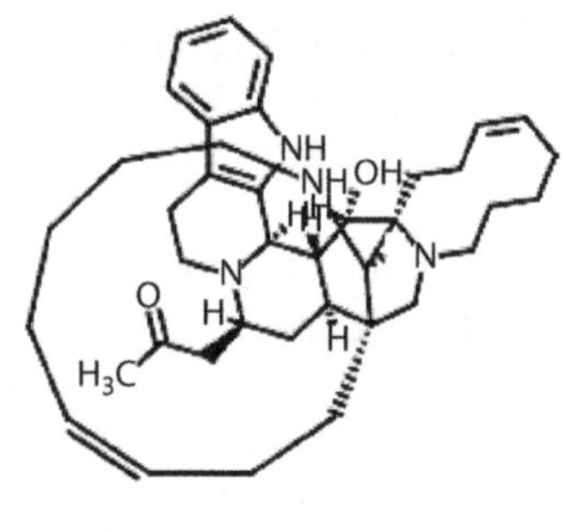

Manadomanzamine A

Antibacterial, antitumor, and anti-HSV-1 activities: The compound 8-hydroxymanzamine A of this species showed antibacterial and moderate antitumor and anti-HSV-1 activities (Putra and Jaswir, 2014).

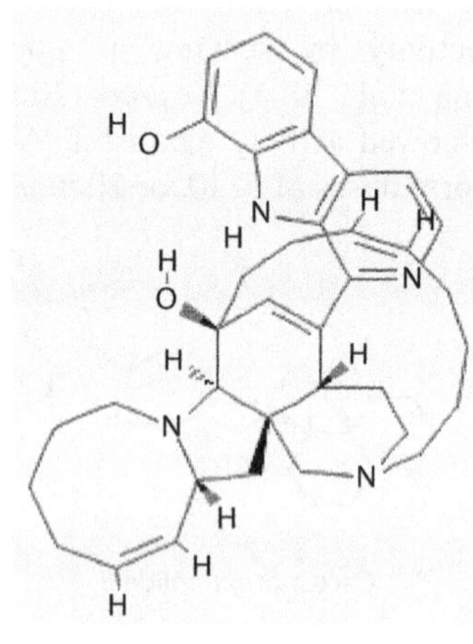

8-hydroxymanzamine A

Antimicrobial, anti-inflammatory, antitubercular, and cytotoxic activities: The compound 12-epi-scalaradial isolated from this species showed antimicrobial activity against *Staphylococcus aureus* (10 μg/disk), *Bacillus subtilis* (10 μg/disk), and *Candida albicans* (10 μg/disk). This compound was also an inhibitor of human recombinant PLA2 (IC50 = 0.02 μM) and significantly reduced the induced mouse ear inflammation. Further, it is also an inhibitor of epidermal growth factor receptor (EGFR), which is implicated in many cancers. Semisynthetic 12-epi-scalaradial has displayed both antitubercular (MIC = 58 μM) and cytotoxic activity (human oral fibroblasts, IC50 = 1.26 μM; ovarian cancer SKOV3, IC50 = 23.3 μM; melanoma SK-MEL, IC50 >42.7 μM; breast cancer BT549, IC50 = 21.0 μM; Vero, IC50 = 22.2 μM), as well as potent antimicrobial activity against methicillin-resistant *Staphylococcus aureus* (MRSA) with an IC50 value of 6.3 μM (González, 2010).

Anti-inflammatory activity: The 25-norscalarane compounds such as metabolites 19–26 and hyrtial showed anti-inflammatory activity at 50 μg/mL and a 43% decrease in ear weight of phorbolmyrstic-acetate-induced inflammation of mouse ear ca. 50 μg/ear (González, 2010; Keyzers, 2005).

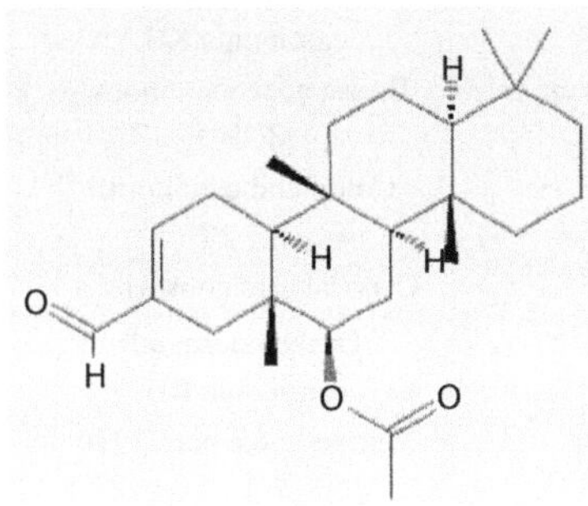

Hyrtial

The aplysinopsin-type compound, manzamine of this sspecies showed anti-inflammatory activity (Senthilkumar and Kim, 2013).

Antitubercular activity: The compound 25-dehydroxy-12-epi-scalarin isolated from this species showed anti-TB activity with an MIC value of 117 μM (González, 2010).

Heteronemin, a scalarin-type sesterterpene of this species, displayed a 99% inhibition of *M. tuberculosis* (H37Rv) with an MIC 6.25 mg/mL and IC50 1.3 mg/mL (González, 2010; Jyothi et al., 2016).

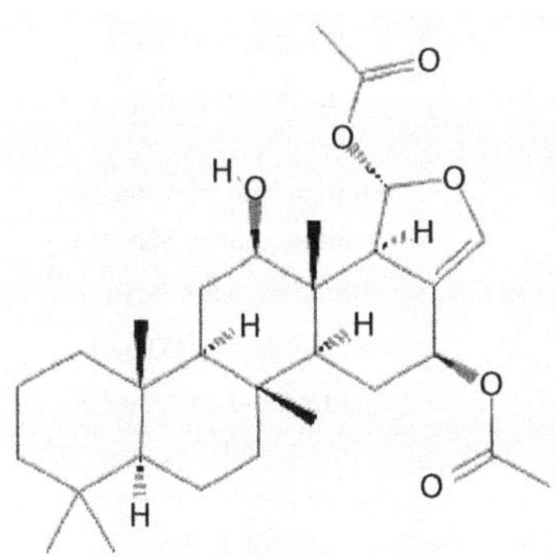

Heteronemin

3-acetylsesterstatin 1 and sesterstatin 7 of this species showed inhibition of *Mycobacterium tuberculosis* (H37Rv) (González et al., 2010).

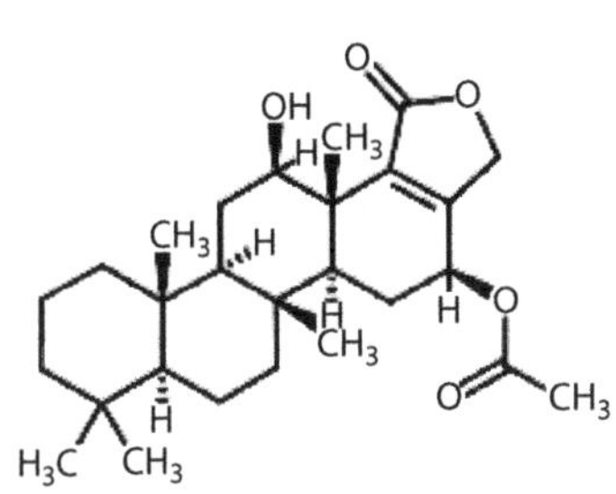

Sesterstatin 7

25-Dehydroxy-12-epi-deacetylscalarin of this species was a weak antitubercular agent (MIC = 16 μM with 16% inhibition against *Mycobacterium tuberculosis* (H37Rv) at 6.25 μg/mL (González, 2010).

The compound scalarafuran (semisynthesized) showed antitubercular activity with an MIC value of 14.0 μM (González, 2010).

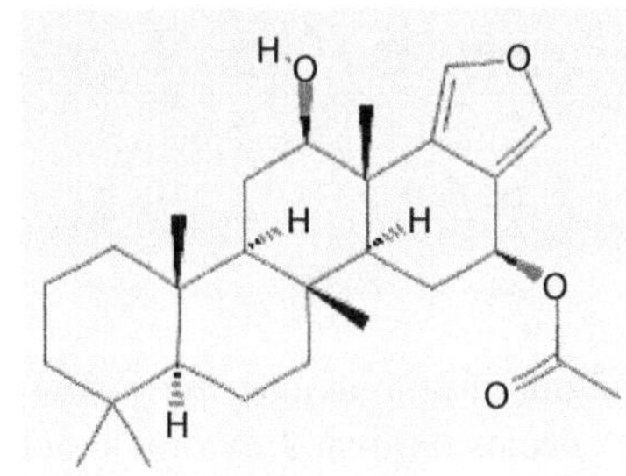

Scalarafuran

Antimicrobial/antimycobacterial activity: The compound heteronemin also showed antimicrobial activity against *Staphylococcus aureus* (25 μg/disk), *Bacillus subtilis* (25 μg/disk), and *Candida albicans* (50 μg/disk), as well as activity against *Mycobacterium intracellulare* (IC50 = 2.04 μM). (González et al., 2010).

Antibacterial activity: The compound sesterstatin 5 inhibited the growth of the Gram-positive bacterium *Micrococcus luteus* (MIC = 25–50 μg/disk) (González et al., 2010).

Antifungal activity: The macrocyclic lactone polyether compound spongistatin 1 isolated from this speices showed antifungal activity (Cheung et al., 2014; Sipkema et al., 2005).

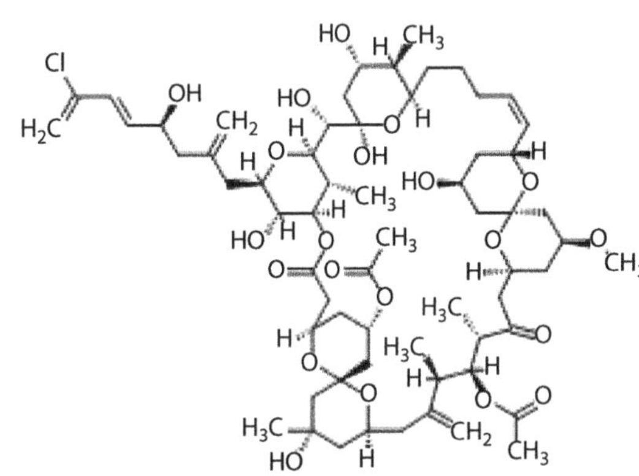

Spongistatin 1

Neurotrophic activity: The compound 12-epi-scalarin of this species showed neurotrophic activity in pheochromocytoma (PC-12) cells at the concentration of 50 μg/mL, and also showed inhibition of farnesoid X activated receptor (FXR) transactivation, which is related to potential drugs for hypercholesterolemia in humans, with an IC50 value of 60.4 μM (González et al., 2010).

Cytotoxicity: The lipophilic partition of a methanol extract of this species yielded a novel pentacyclic sesterterpene ester salmahyrtisol A (1), three new scalarane-type sesterterpenes [3-acetyl sesterstatin 1 (3), 19-acetyl sesterstatin 3 (4), and salmahyrtisol B (5)], together with the previously reported sesterterpenes hyrtiosal (2), scalarolide (6), and salmahyrtisol C (7). The new compounds showed significant cytotoxicity to murine leukemia (P-388), human lung carcinoma (A-549), and human colon carcinoma (HT-29) (Youssef et al., 2002).

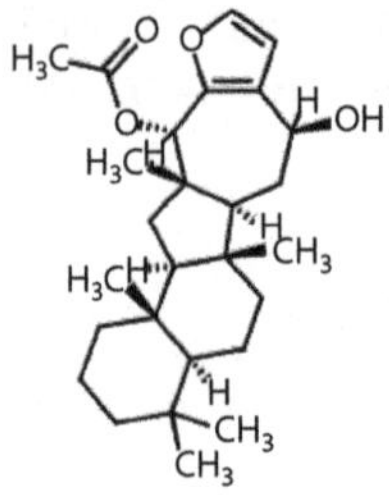

Salmahyrtisol A

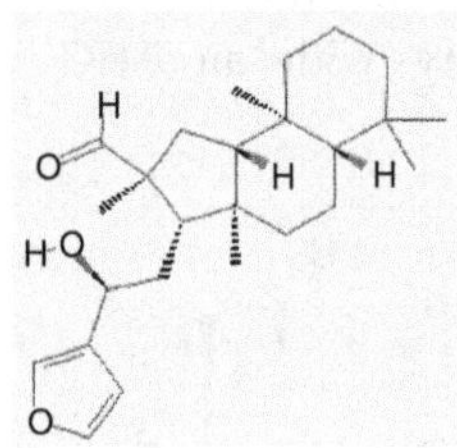

Hyrtiosal

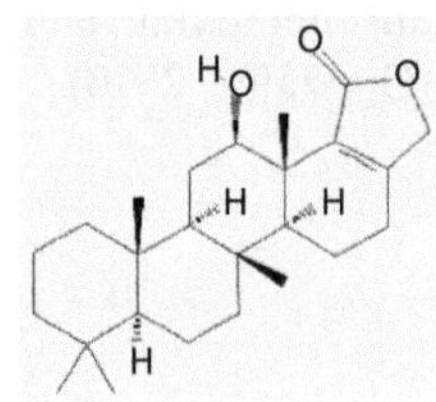

Scalarolide

The indole alkaloid compounds hyrtiosins A and B isolated from this species exhibited cytotoxic activities greater than 5-hydroxyindole-3-aldehyde, which was reported to be active *in vitro* against human epidermoid carcinoma KB cells (IC50 4.3 μg/mL) (Gul et al., 2005).

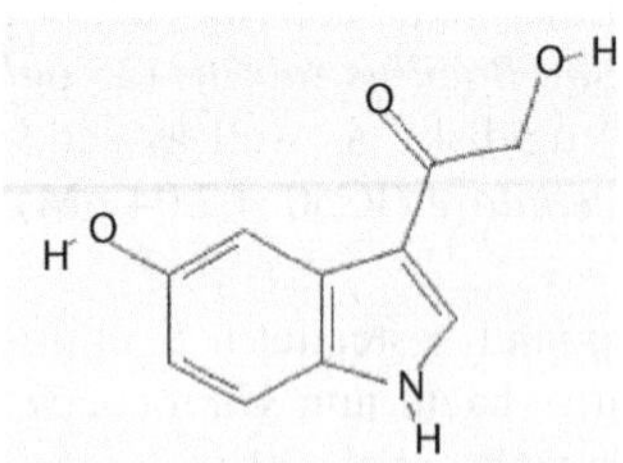

Hyrtiosin A

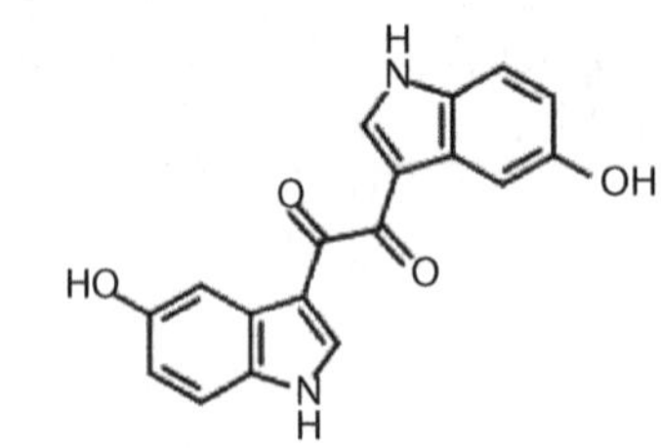

Hyrtiosin B

Youssef (2005) reported that the new alkaloids hyrtioerectines A–C isolated from this species were moderately cytotoxic.

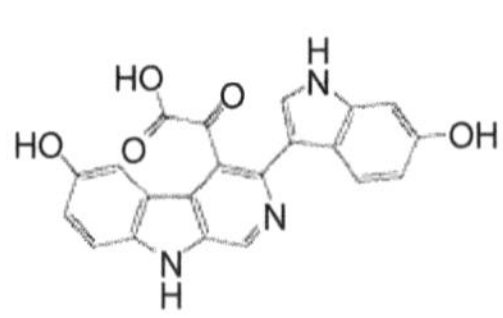

Hyrtioerectine A

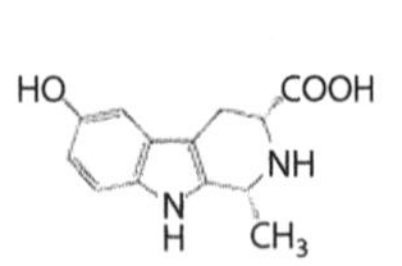

Hyrtioerectine B

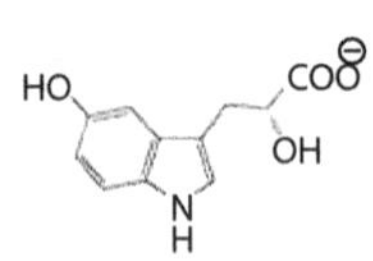

Hyrtioerectine C

González et al. (2010) reported on the isolation of several cytotoxic compounds from this species, and their activities along with the IC50 values for the different cell lines are given next.

Compound	Cell Line(s)	IC50 (μg/mL)
12-Deacetylhyrtial	KB cells	10.0
12-Deacetyl-17-hyrtial	KB cells	2.8
12-epi-Scalarin	Murine leukemia L1210	13.2
	Cervix epithelioid carcinoma HeLa	26
	Lung carcinoma A549	23.7
	Oral epidermoid carcinoma KB	18.5
12-O-Deacetyl-12-epi-scalarin	Murine leukemia L1210	2.3
	Cervix epithelioid carcinoma HeLa	15.0
	Lung carcinoma A549	14.8
	Oral epidermoid carcinoma KB	14.3
25-Dehydroxy-12-epi-scalarin	Breast adenocarcinoma MCF-7	5.2
	Colorectal carcinoma HT-29	2.3
	Cervical carcinoma HeLa	5.5
	Oral epidermoid carcinoma KB	3.0 (μM)
12-Oacetyl-16-O-methylhyrtiolide	Murine leukemia L1210	2.2
	Cervix epithelioid carcinoma HeLa	5.3
	Lung carcinoma A549	5.3
	Oral epidermoid carcinoma KB	15.6
Sesterstatin 6	Murine P388 lymphocytic leukemia	0.17 (ED50)
	Pancreas carcinoma BXPC-3	0.44 (GI 50)
	Thyroid KAT-4	0.4 (GI 50)
	Thyroid SW1736	0.87 (GI 50)
	Lung-NSC NCI-H460	0.18 (GI 50)
	Pharynx-sq FADU	0.89 (GI 50)
	Prostate DU-145	0.37(GI 50)
Sesterstatins 1–3	P-388 leukemia	0.46, 4.2 and 4.3 (GI 50)
Heteronemin	Human oral fibroblasts	5.3
	Ovarian cancer SKOV3	3.4
	Melanoma SK-MEL	15.3
	Breast cancer BT549	11.2
	Epidermoid carcinoma KB	1.2
	Breast adenocarcinoma MCF-7	0.3
	Colorectal carcinoma HT-29	0.4
	Cervical carcinoma HeLa	0.45
	Murine leukemia L1210	0.06
12-Epi-heteronemin acetate	Epidermoid carcinoma KB	2.7
3-Oxo-12-deacetyl-12-epi-deoxoscalarin	Mouse leukemia P388	14.5 (ng/mL)
3-Oxo-12-deacetyl-12-epi-deoxoscalarin	Gastric cancer (MKN-1)	57.7 (ng/mL)
	MKN-7	56.0 (ng/mL)
	MKN-74	36.8 (ng/mL)
19-Acetoxydeoxoscalarin	Murine lymphoma L1210	0.35
Scalarafuran	Epidermoid carcinoma KB	7.2
	Murine leukemia L1210	2.9
	Cholangiocarcinoma HuCCA-1	49.0 (μM)
	Cervical carcinoma HeLa	63.0 (μM)

(Continued)

Compound	Cell Line(s)	IC50 (μg/mL)
Salmahyrtisol B	Murine leukemia P388	1.0
	Lung carcinoma A549	1.0
	Colon carcinoma HT-29	1.0
12-Deacetoxy-19-acetoxyscalarin	P388 murine leukemia	0.9
Hyrtiolide	Cholangiocarcinoma HuCCA-1	57 (μM)
	Oral epidermoid carcinoma KB	12 (μM)
	Cervical carcinoma HeLa	22 (μM)
Sesterstatin 4	Murine leukemia P388	4.9 (GI 50)
	Pancreas BXPC-3	1.6 (GI 50)
	Thyroid KAT-4	2.0 (GI 50)
	Thyroid SW1736	2.1 (GI 50)
	Lung NSC NCI-H460	1.8 (GI 50)
	Pharynx FADU	2.0 (GI 50)
	Prostate DU-145	1.6 (GI 50)
Sesterstatin 5	Murine leukemia P388	>10 (GI 50)
	Pancreas BXPC-3	2.2 (GI 50)
	Melanoma RPMI-7951	2.1 (GI 50)
	CNS U251	1.9 (GI 50)
	Lung NSC NCI-H460	2.5 (GI 50)
	Pharynx FADU	1.9 (GI 50)
	Prostate DU-145	1.9 (GI 50)
12-Acetyl-16-epi-scalarolbutenolide	Breast cancer MCF-7	50.0
16-Deacetyl-16-episcalarolbutenolide	P388 leukemia	0.4
12-Acetyl-16-deacetyl-16-episcalarolbutenolide	P388 leukemia	2.1
25-Dehydroxy-12-epi-deacetylscalarin	Lymphocytic leukemia P388	2.9 (ED50)
	Murine leukemia L1210	1.6
	Cervix epithelioid carcinoma HeLa	16.5
	Lung carcinoma A549	16.5
	Oral epidermoid carcinoma KB	17.1
	Breast adenocarcinoma	0.3 (μM)
	Colorectal carcinoma HT-29	0.9 (μM)
	Cervical carcinoma HeLa	2.0 (μM)
	Oral epidermoid carcinoma KB	0.7 (μM)
	Cholangiocarcinoma HuCCA-1	42 (μM)
	Oral epidermoid carcinoma KB	7.0 (μM)
	Cervical carcinoma HeLa	23 (μM)

Source: González, M.A., 2010, *Bioact. Compd.*, 6: 178–206.
Note: ED50, median effective dose; GI50, growth inhibition.

The extract of this species showed anticancer activity against A549, HeLa and HT-29 cancer cell lines with the following GI50 values.

GI50 (μg) Values of Cancer Cell Lines

A549	HeLa	HT-29
79.8	50.5	80.2

Source: Sponge crude extraction, n.d., http://shodhganga.inflibnet.ac.in/bitstream/10603/64131/10/10_chapter%205.pdf.

Antileukemic activity: The macrocyclic lactone spongistatin 1 of this species inhibited leukemia (Jurkat) cancer cells by inducing apoptosis (Senthilkumar and Kim, 2013).

Hyrtios gumminae (Ridley, 1884)

Order: Dictyoceratida Family: Thorectidae
No ecological and biological information is available for this species.

Compound(s) and Activities

Antitubercular, antiviral, and cytotoxic activity: The semi-synthesized scalarafuran of this species showed antitubercular activity (MIC = 14.0 μM). This compolund also displayed cytotoxicity against several cancer cell lines (human epidermoid carcinoma KB, IC50 = 7.2 μg/mL; murine leukemia L1210, IC50 = 2.9 μμ/mL; human cholangiocarcinoma HuCCA-1, IC50 = 49.0 μM; human cervical carcinoma HeLa, IC50 = 63.0 μM). Further, it has been patented as an anti-HIV agent, since it inhibits the binding between HIV-1 integrase and its substrates (IC50 = 120.0 μM) (González et al., 2010).

Cytotoxicity: The compound 25-dehydroxy-12-epi-deace tylscalarin has shown cytotoxicity in the following cell lines:

Cell Line	ED50/IC50 (μg/mL)
Murine lymphocytic leukemia P388	2.9 (ED50)
Murine leukemia L1210	1.6 (IC50)
Cervix epithelioid carcinoma HeLa	16.5 (IC50)
Oral epidermoid carcinoma KB	17.1 (IC50)
Breast adenocarcinoma MCF	0.3 (IC50, μM)
Colorectal carcinoma HT-29	0.9 (IC50, μM)
Cervical carcinoma HeLa	2.0 (IC50, μM)
Oral epidermoid carcinoma KB	0.7 (IC50, μM)
Cholangiocarcinoma HuCCA-1	42.0 (IC50, μM)
Oral epidermoid carcinoma KB	7.0 (IC50, μM)
Cervical carcinoma HeLa	23.0 (IC50, μM)
Cholangiocarcinoma HuCCA-1	65.0 (IC50, μM)
Oral epidermoid carcinoma KB	14.0 (IC50, μM)
Cervical carcinoma HeLa	26.0 (IC50 (μM)

Source: González, M.A., 2010, *Curr. Bioact. Compd.*, 6: 178–206.

Antitubercular activity: The compound 25-dehydroxy-12-epi-deacetylscalarin of this species has shown weak antitubercular activity (MIC = 16 μM) with 16% inhibition against *Mycobacterium tuberculosis* (González et al., 2010).

Hyrtios proteus (Duchassaing and Michelotti, 1864)

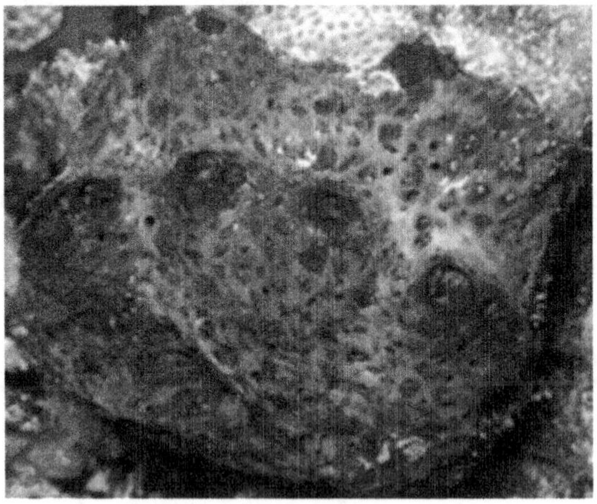

Order: Dictyoceratida Family: Thorectidae
Common name: Not designated
Distribution: Caribbean, Florida, Gulf of Mexico
Ecology: It is a shallow water sponge occurring in reef localities and mangroves.
Biology: It is a massive-amorphous to globular sponge. It is often cake-shaped, up to 10 × 9 × 7 cm, with protruding lobes. It has regular conules (spikes on the surface and are 1 mm high). Membrane-bearing ouscules (1–5 mm wide) are found scattered or on mound-like elevations. Color of the animal is black.

Compound(s) and Activities

Cytotoxicity: The culture extract of the fungus *Aspergillus niger* derived from this sponge species yielded compounds malformin C and asperazine, which displayed tumor and leukemia selective bioactivity (Thomas et al., 2010).

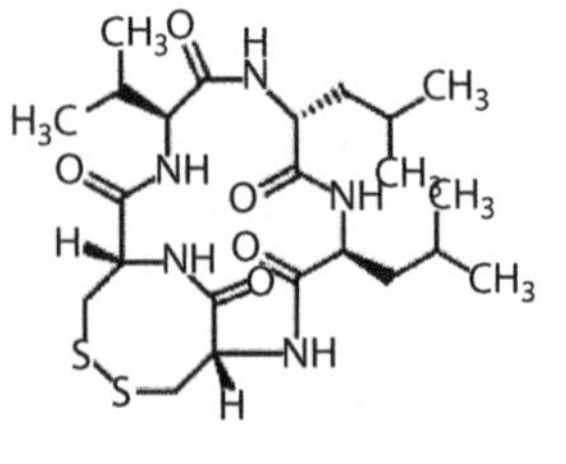

Malformin C

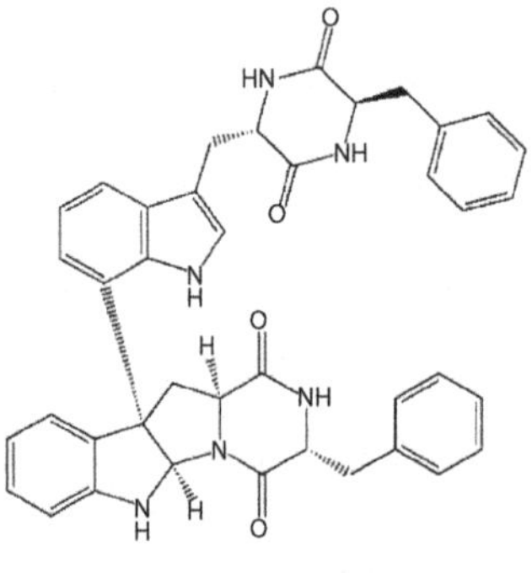

Asperazine

Hyrtios reticulatus (Thiele, 1899)

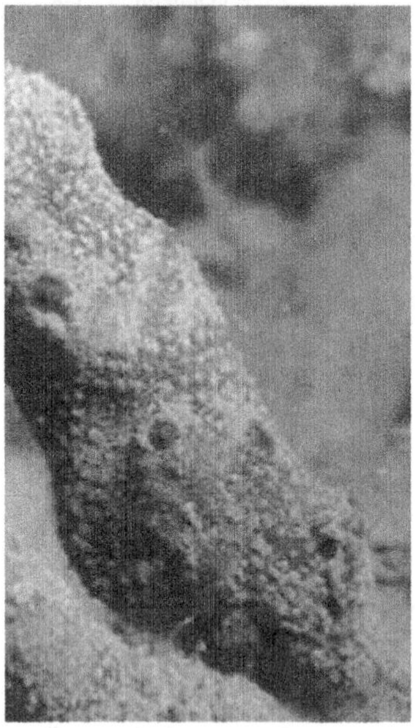

Order: Dictyoceratida Family: Thorectidae
Common name: Not designated
Distribution: New Caledonia, Sulawesi Sea/Makassar Strait
Ecology: It is a shallow water species.
Biology: The high growth potential and high survival rate of this sponge species suggest that it is a promising candidate for mariculture development (de Voogd, 2007).

Compound(s) and Activities

Antiproliferative activity: A new 1-imidazoyl-3-carboxy-6-hydroxy-beta-carboline alkaloid, named hyrtiocarboline, isolated from this species showed selective antiproliferative activity against H522-T1 non-small cell lung, MDA-MB-435 melanoma, and U937 lymphoma cancer cell lines (Inman et al., 2010).

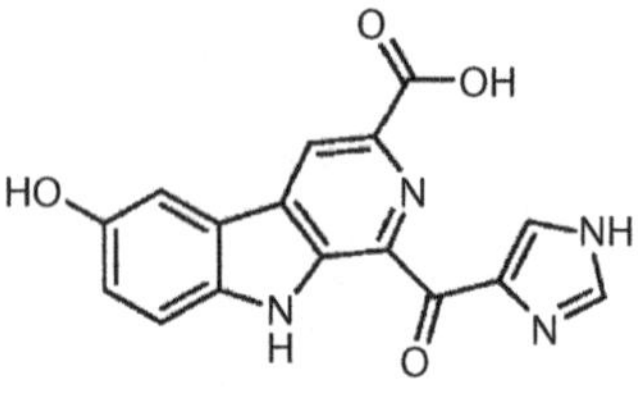

Hyrtiocarboline

Protein regulating activity: The compounds hyrtioreticulins A and B of this speices inhibited ubiquitin-activating enzyme (E1) with IC50 values of 0.75 and 11 μg/mL, respectively. This enzyme is responsible for protein degradation and regulation (Yamanokuchi et al., 2012).

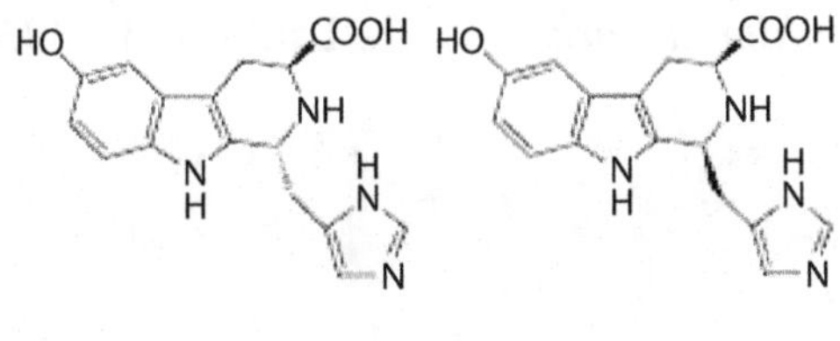

Hyrtioreticulins A, B

Hyrtios sp.

Order: Dictyoceratida Family: Thorectidae

Compound(s) and Activities

Antioxidant activity: The ethyl acetate extract of this sponge yielded three alkaloids called hyrtioerectines (D, E, and F). Whereas the compounds D and F showed free radical scavenging of 45% and 42%, respectively, by DPPH method, compound E only showed 31% of antioxidant capacity, which is considered moderately active (Tovar-Barrios et al., 2016).

The compound 5,6-dibromo-l-hypaphorine of this species showed significant antioxidant ability (Gribble, 2015).

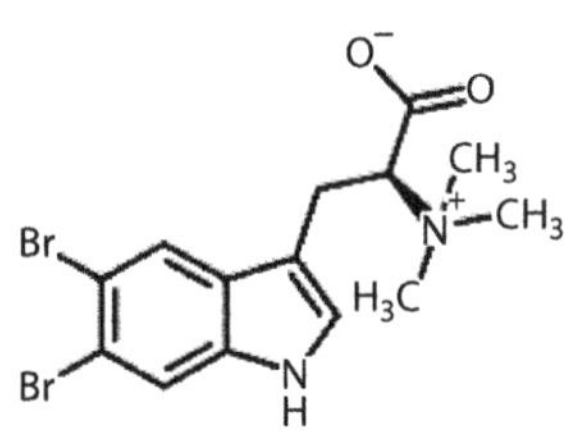

5, 6-dibromo-1-hypaphorine

Anti-inflammatory activity: The sesterterpene manoalide monoacetate isolated from this unidentified sponge species showed anti-inflammatory activity against bee venom PLA2 (IC50, 0.28 μM) (Keyzers, 2005).

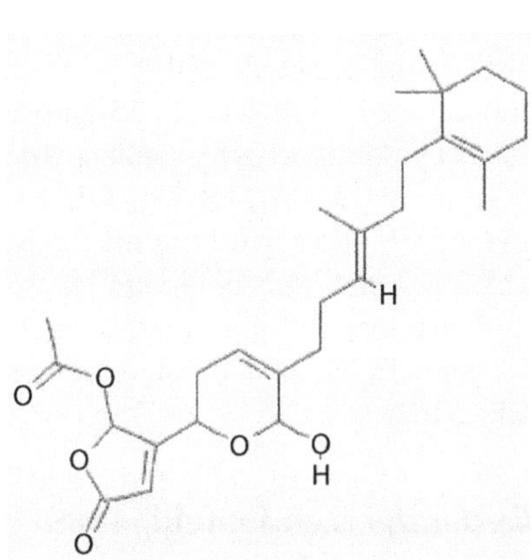

Manoalide monoacetate

The sesterterpene thorectolide monoacetate showed anti-inflammatory activity by inhibiting the cobra venom PLA2 up to concentration of 2 μM (Keyzers, 2005).

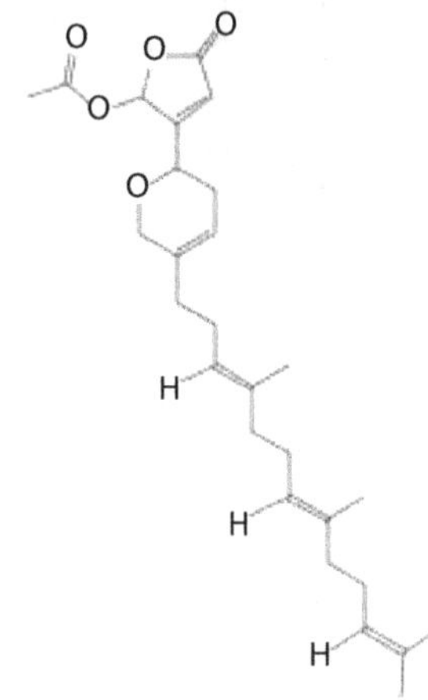

Thorectolide monoacetate

Cytotoxicity: The associated fungus *Aspergillus niger* of this sponge yielded a diketopiperazine alkaloid, asperazine, which displayed remarkable cytotoxicity and an interesting leukemia selectivity (Thomas et al., 2010).

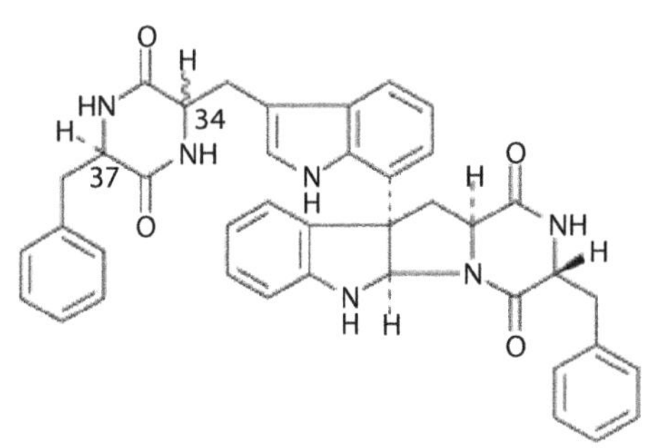

Asperazine

The compound 12-acetyl-12,16-diepi-16-deacetylscalarol butenolide showed cytotoxic activity against murine lymphoma L1210 (IC50 = 2.4 μg/mL) and human epidermoid carcinoma KB cells (IC50 = 7.6 μg/mL) (González et al., 2010).

The compound thorectolide monoacetate exhibited cytotoxic activity against KI3 cells with an IC50 value of 0.3 pg/mL. On the other hand, its compound thorectolide was much less active (IC50 = 5.3 pg/mL) (Bourguet-Kondracki et al., 1996).

The CH_2Cl_2 extract containing the compound aureol displayed cytotoxicity toward KB cells at 10^{-4} M (IC50 = 5 μM). It is also known for its antitumor activity (Longeon et al., 2011).

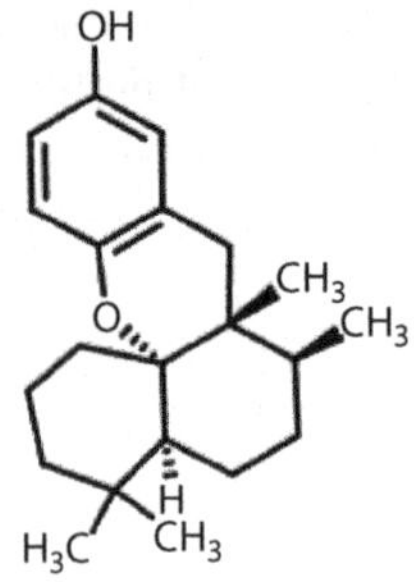

Aureol

Antifungal activity: The new alkaloid 1-carboxy-6-hydroxy-3,4-dihydro-b-carboline (1) together with the known metabolites 6- hydroxy-3,4-dihydro-1-oxo-b-carboline (2),5-hydroxy-1H-indole-3-carboxylic acid methyl ester (3),serotonin (4), hyrtiosin A (5), 5-hydroxyindole-3-carbaldehyde (6), and hyrtiosin B (7) have been isolated from this unidentified species. Of them, hyrtiosin B (7) displayed potent inhibitory activity against isocitrate lyase (ICL) of *Candida albicans* with an IC50 value of 89.0 lM (Lee et al., 2009).

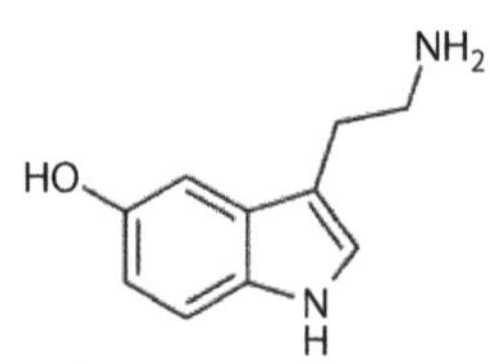

Serotonin

Anti-TB activity: The compound heteronemin showed cytotoxicity, protein function inhibition, and anti-TB activity. On the other hand, the compound pupehenone of this sponge showed only anti-TB activity (Asif, 2017).

The compound 12-epideacetyl-24α-methoxy-25α-acetoxy-deoxoscalarin of this species showed antitubercular activity (MIC = 54 μM) (González et al., 2010).

Enzyme inhibiting activity: The compound 5,6-dibromo-l-hypaphorine of this species was the weak inhibitor of bee venom, phospholipase A2 (PLA2) (Gribble, 2015).

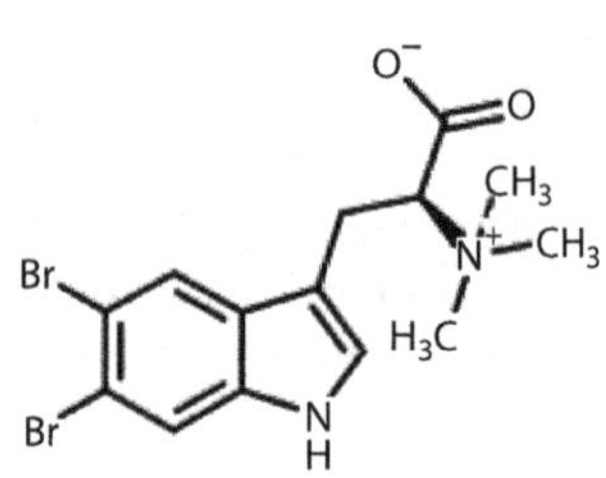

5, 6-dibromo-1-hypaphorine

Lendenfeldia chondrodes (de Laubenfels, 1954)

Order: Dictyoceratida Family: Thorectidae

Common name: Blue icing sponge

Distribution: Singapore

Ecology: This smooth sponge forms a layer over coral rubble.

Biology: The layer of this sponge encrusts coral rubble and may cover an area of 20 cm wide or more. Sometimes, it also encrusts living coral. Generally it has a smooth layer but is also often with irregular knobs, folds, and lobes. Tiny circular holes are seen at the top of folds and elsewhere. Texture is glossy, smooth, and rubbery. Color of the animal is pale to bright blue. Sometimes it is green.

Compound(s) and Activities

Antiutmor activity: The homoscalarane 24-methyl-12,24,25-trioxoscalar-16-en-22-oic acid was found to inhibit tumor cell growth in the National Cancer Institute's 60-cell line panel with some cell line selectivity (mean GI50 = 1.2 μM; range 0.26 to 3.55 μM) (González et al., 2010).

The compound 24-methyl-12,24,25-trioxoscalar-16-en-22-oate exhibited cytotoxic activity against three tumor cell lines (breast carcinoma MDA-MB-231, GI50 = 2.0 μg/mL; lung carcinoma A549, GI50 = 0.1 μg/mL; colon carcinoma HT29, GI50 = 1.0 μg/mL). This compound was also found to be cytotoxic to four tumor cell lines (breast carcinoma MCF-7, MDA-MB-231, and T47D, and prostate tumor PC-3 cells) (González et al., 2010).

Lendenfeldia dendyi (Lendenfeld, 1889) (= *Phyllospongia dendyi*)

Order: Dictyoceratida Family: Thorectidae

Common name: High mountain sponge, green sponge

Distribution: Indo-West Pacific Ocean, Palau, Western and Northern Madagascar, Madagascar, Mozambique Channel

Ecology: It inhabits reef flats and slopes where currents are strong.

Biology: In this species the body lamellae are arranged like petals of a partly opened flower. This sponge has a height of 5 cm and a thickness of about 2 mm.

Compound(s) and Activities

Antitumor activity: The bromides of this species showed inhibitory effects on the assembly of microtubule proteins (Zhang et al., 2017c).

Two new polybrominated diphenyl ethers (1 and 2) were isolated by bioassay-guided separations together with nine known compounds (3–11) have been iosolated from this species. Among them, the compounds 3, 7, and 10 showed inhibitory activities to the assembly of microtubule proteins (IC50 = 29.6, 33.5, and 20.9 μM, respectively) (Liu et al., 2004).

Cytotoxicity: The compound 22-hydroxy-25-methoxy-24-methyldeoxoscalarin of this species showed cytotoxicity against T-47D breast cancer cells (GI50 = 3.5 μM) (González et al., 2010).

Antimicrobial activity: Five polybrominated diphenyl ethers (1–5) have been isolated from this species. Among them, compound 2 (2a) displayed a significant and selective activity against methicillin-resistant *Staphylococcus aureus* (MRSa) with IC50 values of <0.04–1.18 μM. The antimalarial and antileishmanial activities of the isolated compounds were also evaluated (Radwan et al., 2015).

Lendenfeldia sp.

Order: Dictyoceratida Family: Thorectidae

Compound(s) and Activities

Cytotoxicity: The compound methyl 12β,16β-22-trihydroxy-24-methyl-24-oxoscalaran-25-carboxylate of this species exhibited cytotoxicity against the human lung carcinoma A549 cell line (IC50 = 3.7 μg/mL) (González et al., 2010).

The compounds 16β-hydroxy-24-methyl-12, and 24-dioxos calaran-25-al were found cytotoxic against 14 human tumor cell lines (IC50 = 1.5-3.9 μg/mL) (González et al., 2010).

Platelet aggregation inhibitory activity: The compound 24-methylscalarane,12α-acetoxy-24-methyl-24-oxoscalar-16-en-22,25-dial displayed significant inhibitory activity (IC50 = 0.5 μg/mL) on the platelet aggregations caused by adenosine 5′-diphosphate, collagen, or arachidonic acid (González et al., 2010).

Anti-HIV activity: The furanoterpene compound dehydrofurodendin of this species was found to be active against HIV-1 RT-associated RNA- and DNA-directed DNA polymerase (IC50 = 3.2–5.6 μM (Senthilkumar and Kim, 2013).

Luffariella variabilis (Polejaeff, 1884)

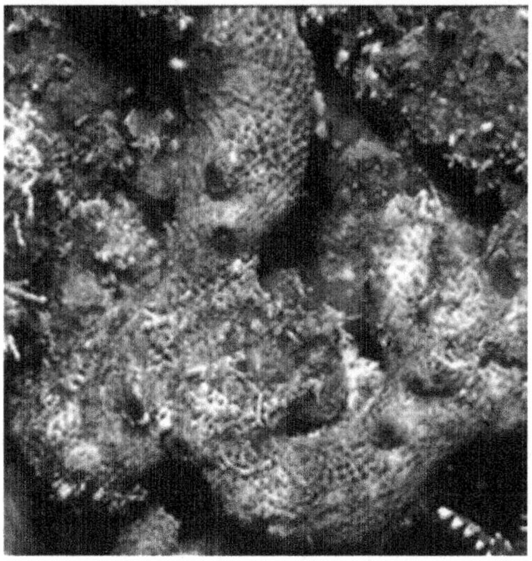

Order: Dictyoceratida Family: Thorectidae

Common name: Not designated

Distribution: Eastern Philippines, Malacca Strait, Singapore, Society Islands

Ecology: It is found at 10 to 18 m deep in coral reefs.

Biology: It is a massive species forming a shallow caliculate form (fragment: 130 mm wide, 80 mm and 5–10 mm thick). The surface has many depressions formed by four protruding tracts (4–8 mm in diameter and 9 mm deep) that are distributed close to each other and are connected by a membrane. This sponge is fairly firm and less compressible. Out of water, the sponge is yellowish brown to reddish brown, and in alcohol the external color is blackish brown while the internal is brown. Siliceous spicules are absent. Skeleton is formed by branching primary (cored with foreign debris and almost fasciculate near the surface) and uncored secondary and tertiary spongin fibers.

Compound(s) and Activities

Anti-inflammatory activity: The cyclohexane sesterterpenoid manoalide isolated from this species showed anti-inflammatory activity by inhibiting the enzyme phospholipase A2 (Braga, 2014; Sipkema et al., 2005). Keyzers (2005) reported on the isolation of sesterterpenes, namely, manoalide (1) and 4E,6E-dehydromanoalide (2). While compound 1 showed anti-inflammatory activity by inhibiting the bee venom (IC50, 0.1 μM) and cobra venom (IC50, 1.9 μM) PLA2 enzymes, compound 2 was against bee venom PLA2 (IC50, 0.28 μM).

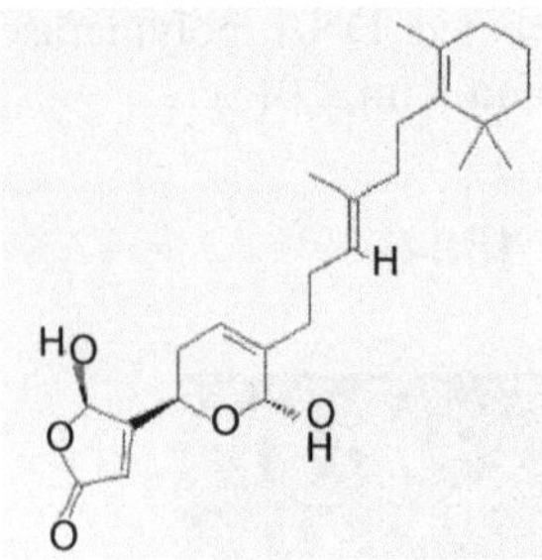

Manoalide

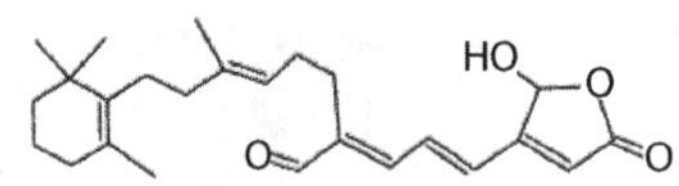

4E, 6E-dehydromanoalide

Antitumor and antiviral activities: Ten epidioxy sterols—5α,8α-epidioxycholesta-6,22-dien-3β-ol, 5α,8α-epidioxycholest-6-en-3β-ol,5α,8α-epidioxy-24(ξ)-methylcholesta-6,22-dien-3β-ol, 5α,8α-epidioxy-24(S)-methylcholest-6-en-3β-ol and its epimer (24R), 5α,8α-epidioxy-24(S)-ethylcholesta-6,22-dien-3β-ol and its epimer (24R), 5α,8α-epidioxy-24(S)-ethylcholest-6-en-3β-ol and its epimer (24R), and 5α,8α-epidioxy-24(ξ)-methylcholesta-6,9(11),22-trien-3β-ol—have been isolated from the lipid extract of this species. The mixture of all these steroids showed inhibitive activity against the human T-cell leukemia/lymphotropic virus type I (HTLV-I) and also displayed cytotoxic activity against the human breast cancer cell line (MCF7WT) (Gauvin et al., 2000).

Chymotrypsin-like inhibitory activity: Among the β-carboline alkaloids variabine A and variabine B, isolated from this species, variabine B inhibited chymotrypsin-like activity of the proteasome and Ubc13–Uev1A interaction with IC50 values of 4 and 5 μg/mL (16 and 20 dan μM), respectively. On the other hand, variabine A (a sulfonated derivative of variabine B) had little effect at 5 μg/mL (16 μM) (Putra and Jaswir, 2014).

Antiparasitic activity: The compounds manadomanzamine A and B of this species showed antiparasitic activity against *Plasmodium falciparum* and *Mycobacterium tuberculosis*, respectively (Putra and Jaswir, 2014).

Antibacterial, antitumor, and anti-HSV-1 activities: The compound 8-hydroxymanzamine A of this species showed antibacterial and moderate antitumor and anti-HSV-1 activities (Putra and Jaswir, 2014).

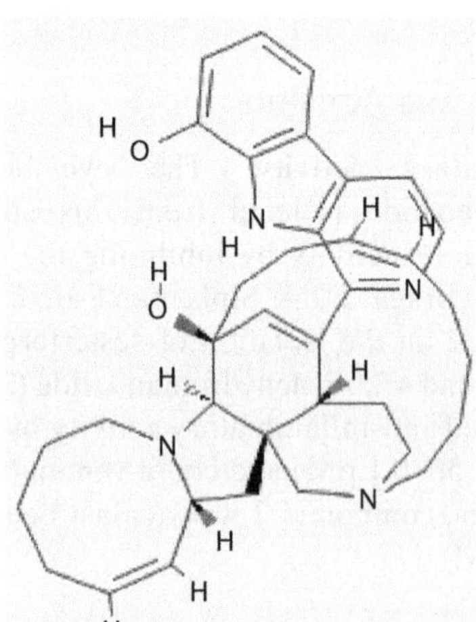

8-hydroxymanzamine A

Luffariella sp.

Order: Dictyoceratida Family: Thorectidae

Compound(s) and Activities

Anti-inflammatory activity: The sesterterpene luffolide isolated from this species showed anti-inflammatory activity by reducing the inflammation of mouse ear edema by 55% (50 μg per ear). Further, this compound completely inhibited the hydrolysis of phosphatidylcholine by bee venom PLA2 at a concentration of 3.5 μM (Keyzers, 2005).

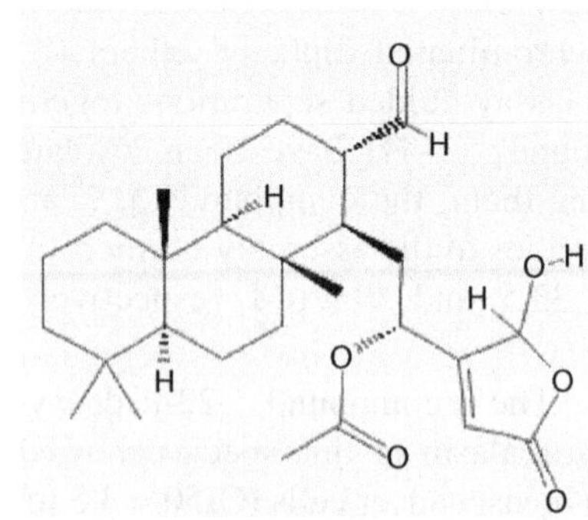

Luffolide

Keyzers (2005) also reported that the sesterterpene luffariellolide of this species also showed anti-inflammatory activity by inactivating the reversible bee venom PLA2 with an IC50 value of 0.23 μM.

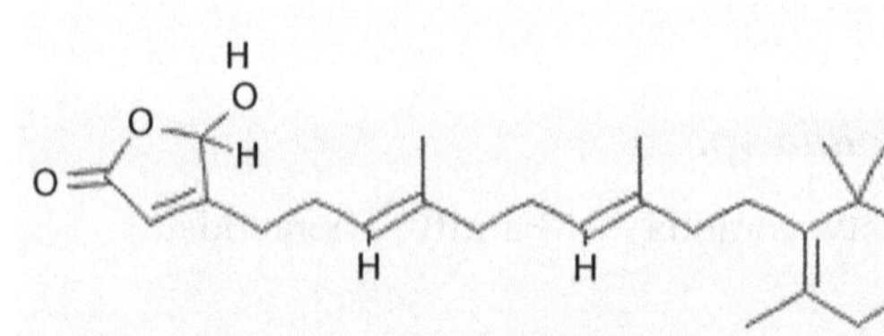

Luffariellolide

Petrospongia nigra (Bergquist, 1995)

Subclass: Keratosa Order: Dictyoceratida Family: Thorectidae

Common name: Not designated

Distribution: Throughout tropical and temperate oceans

Ecology: It occurs in coral reefs.

Biology: It is a large, massive and spreading sponge, with thick interlacing lobes arising from a spreading base. The surface is finely conulose and each conule has a fine projecting fiber. The dense skeletal reticulum is dominated by tightly interlocking, strongly laminated, uncored secondary fibers. The surface is encrusted by a fine, evenly dispersed sand layer that does not form an armored crust. The external pigmentation of the sponge is black, whereas the interior is beige to pale yellow. As a result of the density of the secondary network and the incorporation of moderate amounts of interstitial debris throughout the sponge, the texture is extremely hard and incompressible. There is also light deposition of collagen through the matrix, and choanocyte chambers are small, diplodal, and spherical.

Compound(s) and Activities

Anti-inflammatory activity: The cheilantane sesterterpenoid compounds petrosaspongiolides A–R isolated from this species showed anti-inflammatory activity by serving as a phospholipase A2 inhibitor (Keyzers, 2005; Sipkema et al., 2005).

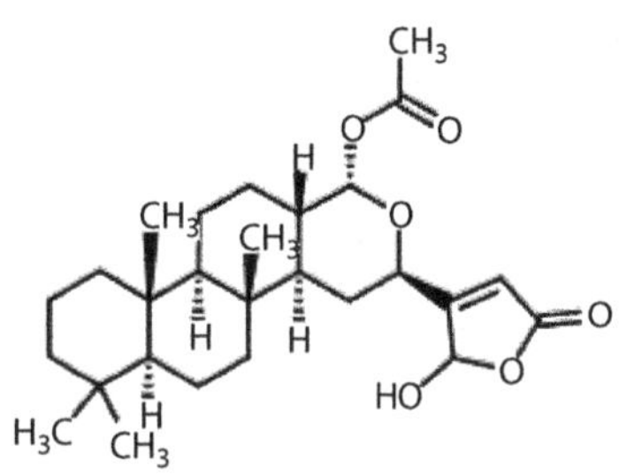

Petrosaspongiolide M

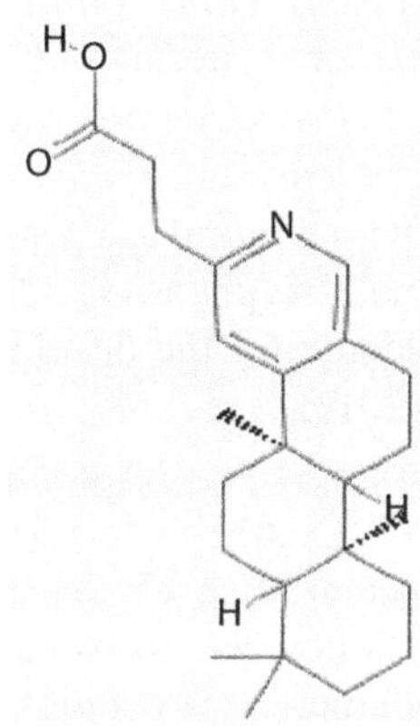

Petrosaspongiolide L

Phyllospongia lamellosa (Esper, 1794)

Order: Dictyoceratida Family: Thorectidae

Common name: Fan sponge

Distribution: Indo-West Pacific Ocean, Central and Southern Great Barrier Reef, East African Coral Coast, Palawan/North Borneo, Western and Northern Madagascar, Kenya

Ecology: It inhabits reef flats and slopes.

Biology: The body of this sponge consists of gravel seams and spongin fibers, and is able to absorb a lot of water.

Compound(s) and Activities

Antiviral activity: The sesterpenes phyllolactones A–E and phyllospongins A–E isolated from this species possessed an inhibitory effect on human immunodeficiency virus type 1 (HIV-1) envelope-mediated fusion *in vitro* with IC50 values of about 2 μM (Zhang et al., 2017a).

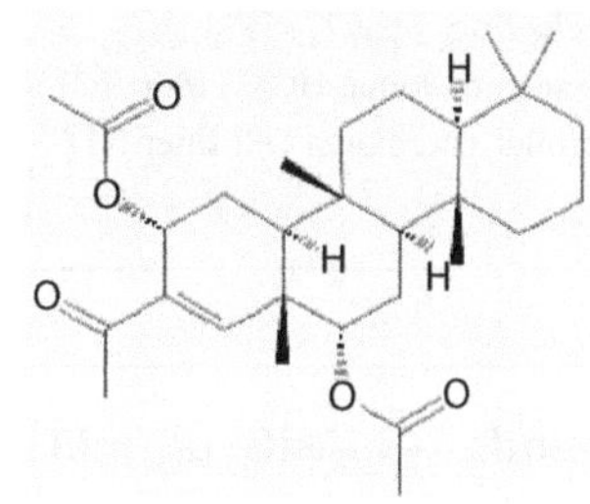

Phyllospongin A

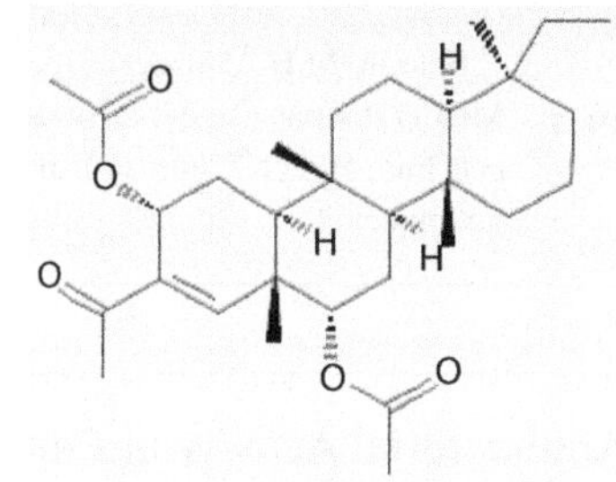

Phyllospongin B

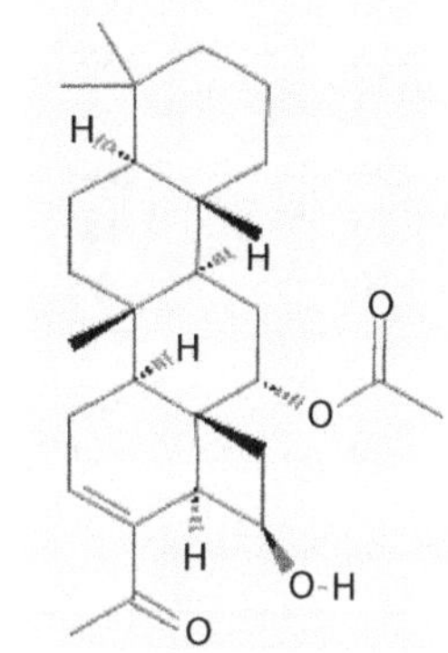

Phyllospongin D

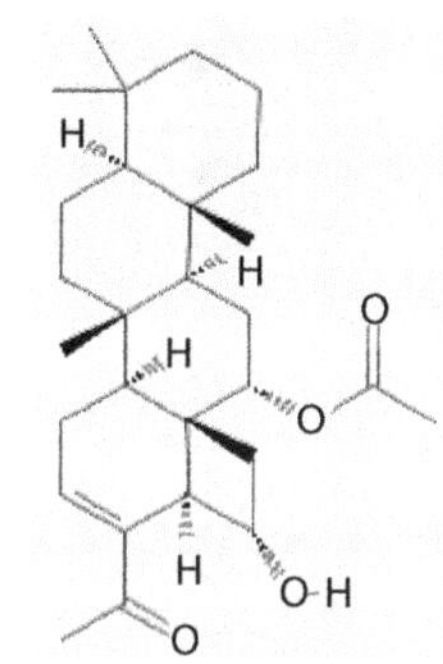

Phyllospongin E

Cytotoxic activity: The phyllospongins B–E and a derivative of this species had potent cytotoxic activity against HCT-116 as the positive control doxorubicin, while phyllospongin E showed cytotoxic activity against MCF-7 (Zhang et al., 2017a).

Cytotoxic and antibacterial activities: Five scalarane sesterpenes, namely, phyllospongins A–E (1–5), and the four derivatives, namely, 12α-acetoxy-20,24-dimethyl-25-norscalar-16-en-24-one (6),12α-acetoxy-13β,18β-cyclobutan-20,24-dimethyl-24-oxoscalar-16-en-25β-ol (7), 12α-acetoxy-24,25-epoxy-24-hydroxy-20, 24-dimethylscalarane (8), and scalardysin-A (9), isolated from this species showed *in vitro* cytotoxic activity against human cancer cell lines (Hep G-2, MCF-7, and HCT-116) and antibacterial activity against some Gram-positive and Gram-negative strains (Hassan et al., 2015).The IC50 values recorded for these activities are given next.

Cytotoxic Activity of Compounds 1–9 (IC50, mg/mL)

Compound	MCF-7	HCT-116	HePG-2
1	2.0	1.2	1.3
2	2.1	0.6	1.3
3	ND	0.8	1.8
4	1.1	0.3	2.1
5	0.6	0.8	2.1

(*Continued*)

Cytotoxic Activity of Compounds 1–9 (IC50, mg/mL)

Compound	MCF-7	HCT-116	HePG-2
6	ND	ND	1.8
7	1.3	1.4	1.7
8	1.9	0.8	1.5
9	ND	ND	ND

Source: Hassan, M.H.A. et al., 2015, *Tetrahedron*, 71: 577–583.
Note: MCF-7, human Caucasian breast adenocarcinoma; HCT-116, colon cell line; HePG-2, human hepatocellular carcinoma cell line; ND, not detected.

Antibacterial Activity of Compounds 1–9 (MIC, μg/mL)

Compound	*S. aureus*	*B. subtilis*	*E. coli*	*V. parahemolyticus*
1	6.8	10.3	>20	18.6
2	5.7	8.2	>20	15.5
3	16.6	18.9	>20	>20
4	1.9	1.7	>20	6.8
5	2.5	3.3	>20	9.8
6	15.8	17.5	>20	>20
7	1.8	1.7	>20	7.0
8	17.5	13.2	>20	>20
9	15.0	12.3	>20	>20

Source: Hassan, M.H.A. et al., 2015, *Tetrahedron*, 71: 577–583.

Anti-HIV activity: The phyllolactones A–D isolated from this species showed anti-HIV activity with an IC50 value of about 2 μM (González et al., 2010).

Phyllospongia papyracea (Esper, 1794)

Order: Dictyoceratida Family: Thorectidae
Common name: Not designated

Distribution: Indo-Pacific, Madagascar, Mozambique Channel, New Zealand

Ecology: This sessile sponge prefers shallow water and a depth range of 0 to 100 m.

Biology: This sponge is cup shaped and its branches have lobed margins. Walls of the branches are dark brown when alive. Texture is leathery. Skeleton consists of spongin fibers. Primaries contain sand grains.

Compound(s) and Activities

Antibacterial activity: The methanol extract of this species containing 2-(3′,5′-dibromo-2′-methoxyphenoxy)-3,5-dibromophenol showed siginificant antibacterial activity against *Bacillus subtilis* and *Staphylococcus aureus* (MIC value of 1 μg/mL); *Campylobacter jejuni* (MIC = 2 μg/mL); *Pseudomonas aeruginosa* (MIC = 4 μg/mL); and *Streptococcus pneumoniae* and *Listeria monocytogenes* (MIC = 8 μg/mL). These results suggest that 2-(3′,5′-dibromo-2′-methoxyphenoxy)-3,5-dibromophenol as a potential lead molecule for the development of antibacterial agents (Sun et al., 2017).

Cytotoxicity: The compounds phyllactones A–B (1–2) and D–G (3–6), and 12α, 24-dihydroxy-20, 24-dimethyl-15, 17-scalaradien-25, 24-olides (7–8); and an amorphous colorless powder (9) isolated from this species showed *in vitro* moderate cytotoxicities against cell lines A549, MCF-7, and HeLa with IC50 values of less than 25 μM. Among these 20, 24-bishomoscalarane sesterterpenes, phylactones F/G (5/6) exhibited the strongest inhibitory effects on A549, MCF-7, and HeLa with IC50 values of 7.8, 5.2, and 9.6 μM, respectively (Zhang et al., 2017a).

An idienitfied compound (a small molecule) of this species had an *in vitro* cytotoxic effect on the leukemia cancer cell line P388 with an IC50 value of 5 μg/mL. Further, the phyllactone H of this sponge was found to possess *in vitro* moderate cytotoxicities against cell lines A549, MCF-7, and HeLa with IC50 values of no more than 25 μM (Zhang et al., 2017a). González et al. (2010) reported that the compound, 12α-acetoxy-16β-hydroxy-20,24-dimethyl-25-norscalar-17-en-24-one of this species exhibited cytotoxic activity against the leukemia P388 cancer cell line with an IC50 value of 5.0 μg/mL.

Phyllospongia sp.

Order: Dictyoceratida Family: Thorectidae

Compound(s) and Activities

Cytotoxicity: Two 24-methylscalarane compounds of this species exhibited 30% to 95% inhibition of the growth of KB cells at an extract concentration of 10 μg/mL (González et al., 2010).

Two related homoscalaranes of 12α-acetoxy-16β-hydroxy-24-methyl-24-oxoscalaran-25-al exhibited 30% to 95% inhibition of the growth of KB cells at an extract concentrationof 10 µg/mL. Further, two dihydrodeoxoscalarins of this unidentified sponge exhibited 30% to 95% inhibition of the growth of KB cells at an extract concentration of 10 µg/mL (González et al., 2010).

Scalarispongia aqabaensis (Helmy, El Serehy, Mohamed, and van Soest, 2004)

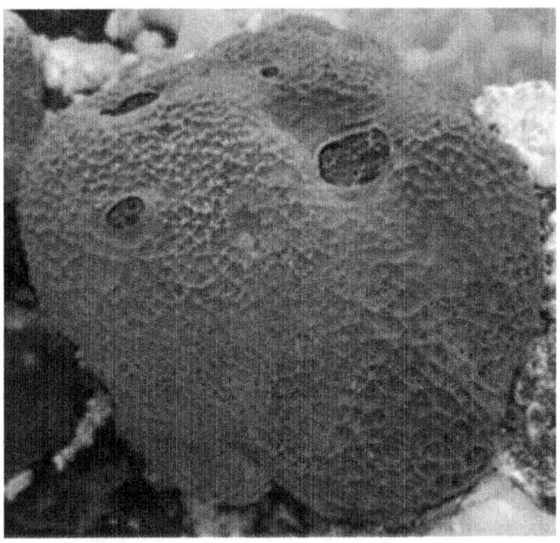

Order: Dictyoceratida Family: Thorectidae

Common name: Mediterranean sponge

Distribution: Northern and Central Red Sea, Gulf of Aqaba

Ecology: It is found attached to dead corals or hard rocks, mostly on the fore reef.

Biology: It is a regular or irregular massive pad or cushion sponge. Surface is unarmored and is microconulose with conules of 1 or 2 mm high and 2 to 4 mm apart. Oscules (3–5 mm dia.) are evenly distributed over the surface. Texture is tough and is difficult to tear and cut. It is compressible when in situ and incompressible when preserved in spirit. Skeleton is of primary spongin and secondary fibers, which are concentrically laminated. The primary fibers are cored with detritus, while the secondary fibers are uncored. Primary fibers and secondary fibers are arranged in a ladder-like pattern. Color of the sponge is black when alive and preserved.

Compound(s) and Activities

Anti-inflammatory activity: The chloroform fraction of the methanol extracts of this species yielded a compound, scalaristerol (5alpha,8alpha-dihydroxycholest-6-en-3beta-ol), which showed anti-inflammatory activity (Youssef et al., 2010).

Scalarispongia scalaris (Schmidt, 1862) (= *Cacospongia scalaris*)

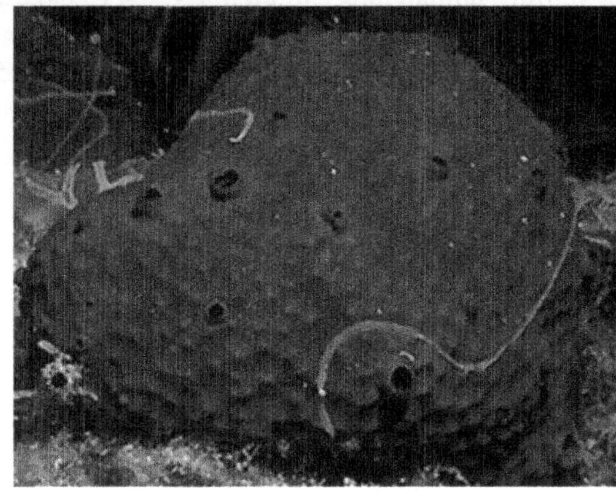

Order: Dictyoceratida Family: Thorectidae

Common name: Not designated

Distribution: Mid-Portugal, Mediterranean

Ecology: It grows in caves, rocky/detritic/muddy bottoms, coralligenous communities, *Posidonia oceanica* meadows, lagoons, and artificial reefs. It is epibiotic on the bivalve *Pinna nobilis*. The depth range is 1 to 250 m.

Biology: It is a massive, globose, lobate, and large sponge (up to 20–30 cm in dia). Color is dark gray with bluish shades. Surface is conulose (conules 1–2 mm high, 2–4 mm apart) with smooth, scattered circular depressions. Oscules (up to 1 cm in dia.), which are surrounded by a short collar, are abundant and are irregularly scattered on the sponge's upper part. Skeleton network lax is with hard, not elastic spongin fibers. Primary fibers are almost parallel and are interconnected by quite perpendicular secondary fibers. Primary fibers are cored by abundant inclusions and the secondary fibers are laminated with an evident fibrous core.

Association: This sponge harbors the nudibranch *Hypselodoris fontandraui* on its surface.

Compound(s) and Activities

Cytotoxic, antitumor, and anti-inflammatory effects: The compound scalaradial isolated from this species showed strong antitumor activity against several human cancer cell lines (leukemia HL60, IC50 = 3.4 µM; hepatocarcinoma BEL7402, IC50 = 5.8 µM; breast cancer MDA-MB435, IC50 = 4.8 µM). Further, this compound exhibits potent activity against phospholipase A2 (PLA2) enzymes, which are involved in the biosyntesis of eicosanoids responsible for the inflammatory response. This compound was also was found to inhibit PLA2 from bee venom (IC50 = 0.07 µM) and human recombinant PLA2 (IC50 = 0.07 µM) (González, 2010).

Cytotoxicity: The compounds 25-dihydroscalaradial and 18-epi-scalaradial showed significant cytotoxicity toward five tumor cell lines (mouse lymphoma P388 and Schabel, ED50 = 2 and 0.2 µg/mL; human lung carcinoma A549, ED50 = 2 and 0.2 µg/mL; human colon carcinoma HT29, ED50 = 2 and 0.2 µg/mL; human melanoma MEL28 ED50 = 2.5 and 0.5 µg/mL) (González, 2010).

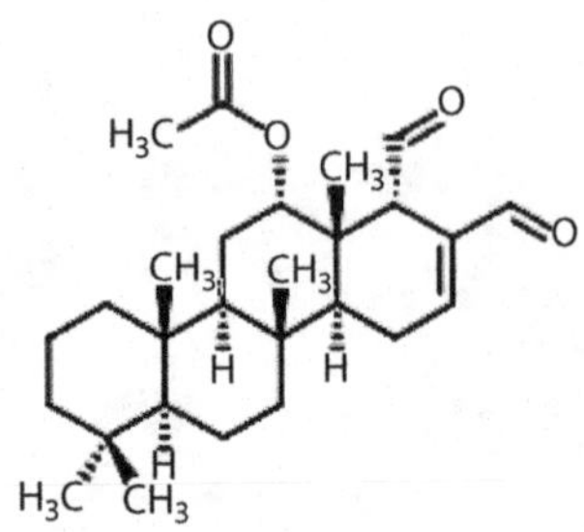

18-epi-scalaradial

Desacetylscalaradial of this compound showed interesting cytotoxic activity against L1210 cancer cells (IC50 = 0.58 µg/mL) (González, 2010).

The new compounds norscalarals A–C of this species showed significant cytotoxicity toward five tumor cell lines (mouse lymphoma P388, ED50 = 1–2 µg/mL; Schabel, ED50 = 1–2 µg/mL; human lung carcinoma A549, ED50 = 1–2.5 µg/mL; human colon carcinoma HT29, ED50 = 1–5 µg/mL; human melanoma MEL28 ED50 = 2–2.5 µg/mL) (González et al., 2010).

The compound acetylfuroscalarol showed significant cytotoxicity toward five cancer cell lines (mouse lymphoma P388 and Schabel, ED50 = 2.5 µg/mL; human lung carcinoma A549, ED50 = 5.0 µg/mL; human colon carcinoma HT29, ED50 = 2.5 µg/mL; human melanoma MEL28 ED50 = 10.0 µg/mL) (González et al., 2010).

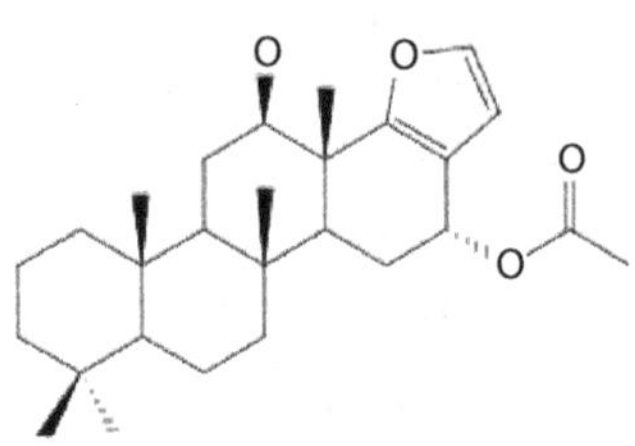

Acetylfuroscalarol

Anti-inflammatory activity: The scalarane sesterterpene, scalaradial, isolated from this species showed anti-inflammatory activity by acting as a phospholipase A2 inhibitor (Sipkema et al., 2005).

Antibacterial activity: The ethanolic extract of this species exerted antibacterial activity against cerebrospinal fluid (CSF) disease (neurological disease)–causing pathogenic bacteria (Sujath et al., 2014). The inhibition zone diameter values for different species of bacteria are given next.

Inhibition Zone Diameter (in mm)

E.coli	*S. aureus*	*H. influenzae*	*N. meningitides*	*K. pneumoniae*	*Streptococcus agalactiae*	*E. faecalis*
No inhibition	6	No inhibition	No inhibition	3	10	No inhibition

Source: Sujatha, S. et al., 2014, *Pharma Innovation*, 3: 81–86.

Smenospongia aurea (Hyatt, 1875)

Order: Dictyoceratida Family: Thorectidae

Common name: Pitted tube sponge, Jamaican sponge

Distribution: Western Central Atlantic, Caribbean, Bahamas

Ecology: This sessile species is found in brackish waters and in mangroves.

Biology: It is a massive sponge, often forming groups of adjacent oscular mounds or volcanos. Color of the sponge varies from brown skin-yellow to light greenish skin-yellow green. Dermis resting on fibers is honeycomb patterned. This sponge darkens and exudes mucus when handled.

Compound(s) and Activities

Cytotoxic activity: A novel phenyl alkene compound, (E)-10-benzyl-5,7-dimethylundeca-1,5,10-trien-4-ol, isolated from this species showed *in vitro* cytotoxic activity (through the inhibition of microtubule activity) against HL-60 human leukemia cancer cells with an IC50 value of 8.1 µM (Hwang et al., 2013).

Antimicrobial activity: The compound 5-bromo-N,N-dimethyltryptamine isolated from this species showed antimicrobial activity (Djura et al., 1980).

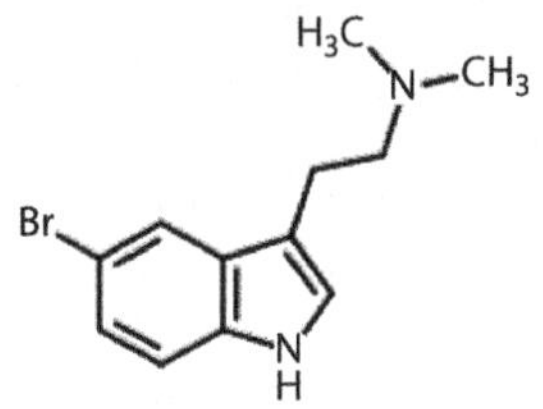

5-bromo-N, N-dimethyltryptamine

Smenospongia cerebriformis (Duchassaing and Michelotti, 1864)

Order: Dictyoceratida Family: Thorectidae

Common name: Not designated

Distribution: Eastern Caribbean, Eastern Sea of Vietnam, St. Thomas

Ecology: It inhabits coral reefs and the Belize barrier reef.

Biology: It is a conical sponge 8 cm high and 7 cm in diameter at the base. It is crowned by a composite, star-shaped vent 12 mm in diameter. It possesss a thick organic dermis. In the absence of dermis, the sponge shows sharp, crude conules with deep valleys between them. It is rigid, brittle, and incompressible. The main fibers are fasciculated, especially at the periphery. Secondary fibers branch from the primaries at right angles. Color of the fibers is yellow red.

Compound(s) and Activities

Cytotoxic activity: A novel phenyl alkene compound, (E)-10-benzyl-5,7-dimethylundeca-1,5,10-trien-4-ol, isolated from this species showed *in vitro* cytotoxic activity (through the inhibition of microtubule activity) against HL-60 human leukemia cancer cells with an IC50 value of 8.1 μM (Hwang et al., 2013). Two new naphtoquinones, namely, smenocerones A and B, and four known sesquiterpene cyclopentenones, namely, dactylospongenones A–D, have been isolated from this species. Among them, the compound smenocerone B significantly exhibited cytotoxic activity toward LU-1, HepG-2, HL-60, MCF-7, and SK-Mel-2 human cancer cells with IC50 values of 5.5, 3.2, 4.0, 4.1, and 5.7 μg/mL, respectively (Huyen et al., 2017).

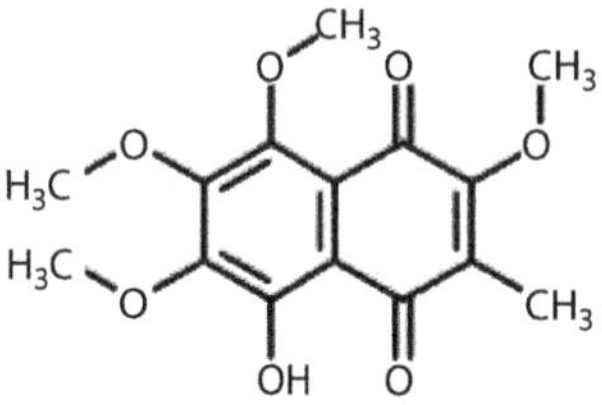

Smenocerone B

Smenospongia echina (de Laubenfels, 1934) (= *Polyfibrospongia echina*)

Order: Dictyoceratida Family: Thorectidae

Common name: Not designated

Distribution: Greater Antilles, Belize barrier reef near Carrie Bow Cay

Ecology: It inhabitats coral reefs.

Biology: This sponge comprises a basal mass (5 × 7 cm) from which primary fibers (processes) of 1.5 cm diameter and 2–4 cm high arise. Surface is black exteriorly and gray interiorly. Consistency is very spongy. Surface has a number of tubercles or cornules that are 2 mm high and 4 mm apart. Pores are not visible. There is only one osculum, which is terminal with 2 mm diameter. Endosome is densly reticulated with small fibers of spongin.

Compound(s) and Activities

Antimicrobial activity: The compound, 5,6-dibromo-N, N-dimethyltryptamine isolated from this species showed antimicrobial acitivty (Djura et al., 1980).

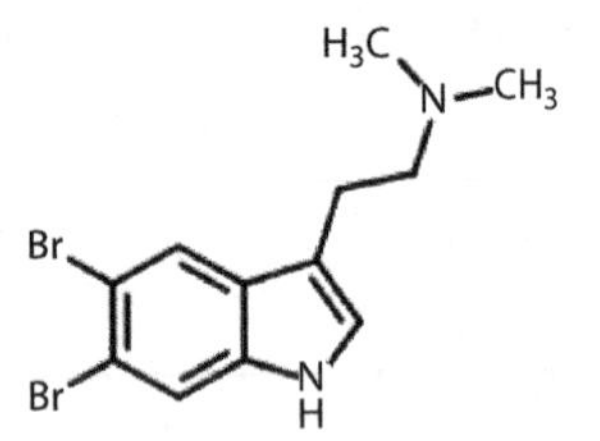

5,6-dibromo-N, N-dimethyltryptamine

Smenospongia sp.

Order: Dictyoceratida Family: Thorectidae

Compound(s) and Activities

Cytotoxicity and antibacterial activity: The compound 12-deacetoxy-22-hydroxy-25-acetoxyscalarin isolated from this sponge showed moderate cytotoxicity against the human chronic myelogenous leukemia K562 cell line (LC50, 22.5 μg/mL). In addition, this compound exhibited antibacterial activity against both Gram-positive and -negative bacteria (González et al., 2010).

The compound 12-dehydroxy-16-deacetoxy-22-hydroxyscalarafuran (1) and the corresponding acetate 12-dehydroxy-16-deacetoxy-22-acetoxy-scalarafuran (2) exhibited antibacterial activity (compound 1: *Bacillus subtilis* MIC = 1.56 μg/mL, *Micrococcus luteus* MIC = 6.25 μg/mL, *Salmonella typhimurium* MIC = 12.5 μg/mL; compound 2: *Bacillus subtilis* MIC= 3.12 μg/mL, *Micrococcus luteus* MIC=25.0 μg/mL), as well as cytotoxicity against human chronic myelogenous leukemia K562 cell line (LC50, 0.97 and 2.3 μg/mL, respectively) (González et al., 2010).

The compound 25α-methoxy-12-deacetoxy-22-hydroxydeoxoscalarin displayed antibacterial activity against *Bacillus subtilis* (MIC = 50.0 μg/mL) and *Salmonella typhimurium* (MIC = 6.25 μg/mL), as well as potent cytotoxicity against the human chronic myelogenous leukemia K562 cell line (LC50 = 0.11 μg/mL) (González et al., 2010).

Cytotoxicity: The compound, 12-deacetoxy-22-hydroxyheteronemin displayed significant cytotoxicity toward the human leukemia cell line K562 (LC50, 0.02 μg/mL, (González et al., 2010).

The compounds, 12-deacetoxy-22-acetoxyheteronemin, 12-deacetoxy-16-Odeacetylheteronemin and 12-deacetoxy-22-formylheteronemin showed moderate cytotoxicity toward the human leukemia cell line K562 (LC50 0.13, 6.8, 5.8 and 17.5 μg/mL, respectively) (González et al., 2010).

Antitumor activity: The compounds smenothiazoles A and B of this species were active and selective against A2780 (ovarian) cancer cells. Further, its smenamides exhibited selectivity and nanomolar cytotoxic activity toward Calu-1 (lung) cancer cells (Gribble, 2015).

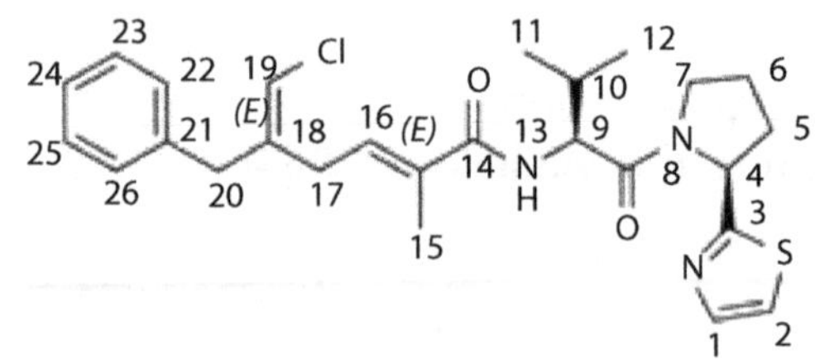

Smenothiazole A

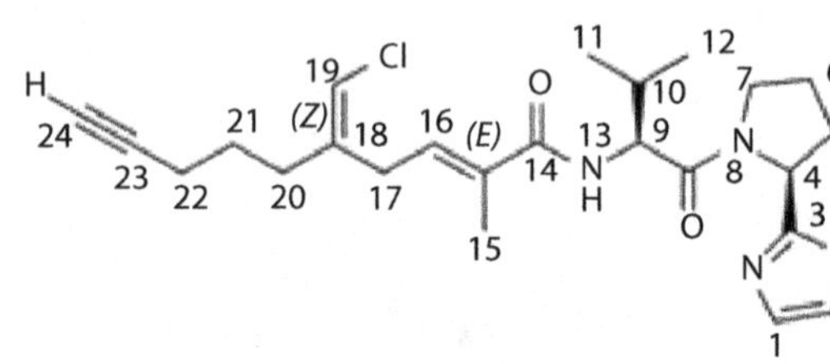

Smenothiazole B

Antibacterial activity: While the compound 12-deacetoxy-22-hydroxyheteronemin displayed antibacterial activity against *Bacillus subtilis* (MIC = 0.78μg/mL) and *Salmonella typhimurium* (MIC = 6.25 μg/mL), the compound 12-deacetoxy-16-odeacetylheteronemin was against *Bacillus subtilis* (MIC= 3.12 μg/mL) and *Salmonella typhimurium* (MIC = 12.5 μg/mL) (González et al., 2010).

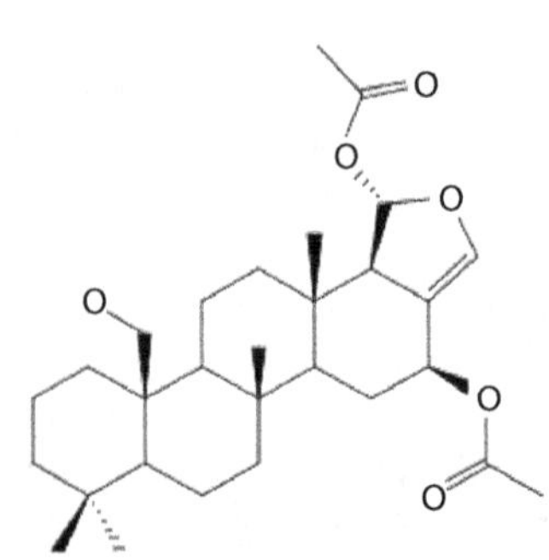

12-deacetoxy-22-hydroxyheteronemin

Others: The compound smenospongine, a sesquiterpene aminoquinone, isolated from this unidenitifed spong showed cytotoxic, antimicrobial, antiproliferetive, and antiangiogenic activities (Perdicaris et al., 2013).

Strepsichordaia aliena (Wilson, 1925)

Order: Dictyoceratida Family: Thorectidae
Common name: Indonesian sponge
Distribution: Indonesia, Singapore
Ecology: Not reported
Biology: Not available

Compound(s) and Activities

Cytotoxicity: The compounds honulactones C and D exhibited cytotoxicity against P-388, A-549, HT-29, and MEL-28 cancer cells (IC50 = 1 μg/mL) (González et al., 2010).

Strepsichordaia lendenfeldi (Bergquist, Ayling, and Wilkinson, 1988)

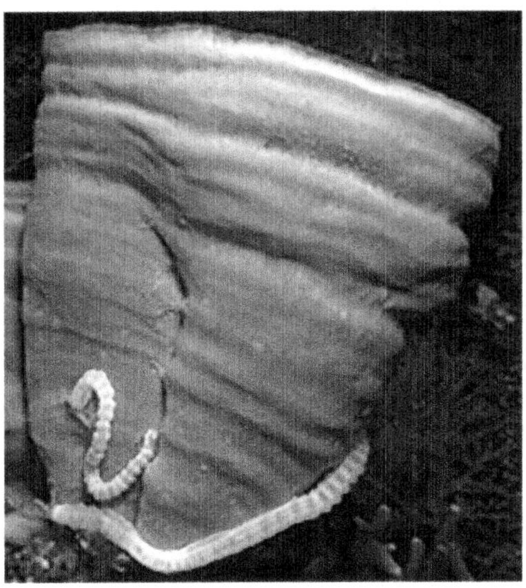

Order: Dictyoceratida Family: Thorectidae
Common name: Vase sponge, Australian sponge
Distribution: Indo-West Pacific Ocean, Central and Southern Great Barrier Reef, Australia
Ecology: It inhabits intertidal and subtidal reef flats.
Biology: Not available

Compound(s) and Activities

Cytotoxicity: The compound 12α-acetoxy-16β-hydroxy-20,24-dimethyl-24-oxoscalaran-25-al of this species exhibited potent cytotoxicity against 13 human tumor cell lines (IC50, 1.13–1.99 μg/mL) and 2 additional tumor cell lines (mouse lymphoma P388 and human lung carcinoma A549, IC50, 0.1 μg/mL), and exhibited 30% to 95% inhibition of the growth of KB cells at 10 μg/mL (González et al., 2010).

Among the compounds 20,24-dimethylscalaranes (1–4) compounds isolated from this sepcies, compounds 1–3 exhibited potent cytotoxicity against two tumor cell lines (mouse lymphoma P388, IC50 = 0.23, 0.5, and 0.67 μg/mL, respectively; human lung carcinoma A549, IC50 = 0.66, 0.5, and 0.67 μg/mL, respectively) (González et al., 2010).

Antibacterial activity: The compound 20,24-dimethylscalarane was active against *Bacillus subtilis* and *Staphylococcus aureus* (González et al., 2010).

Three 20,24-bishomoscalarane members isolated from this sepcies exhibited potent cytotoxicity against two tumor cell lines (mouse lymphoma P388, IC50 = 0.12, 0.12, and 0.2 μg/mL; human lung carcinoma A549, IC50 = 0.25, 0.21, and 0.2 μg/mL, respectively). Two of these compounds were also active against *Bacillus subtilis* (González et al., 2010).

Thorectandra excavatus (Ridley, 1884)

Order: Dictyoceratida Family: Thorectidae
Common name: Not designated
Distribution: Queensland, Australia; Arafura Sea
Ecology: Not reported
Biology: This small sponge has a head that arises from a short pedicel. It is broad and semitruncate above and cuplike. The upper surface is excavated by four pits, the deepest occupying a great part of the thickness of the sponge. Each pit contains a vent. The color of thesponge in spirit is grayish white (putty color). The dermis conceals all of the skeleton. Both the secondary and primary fibers enclose more or less foreign matter. Greatest height and breadth of the single specimen is 31 and 25 mm, respectively.

Compound(s) and Activities

Anti-inflammatory activity: The sesterterpene manoalide monoacetate isolated from this species has shown anti-inflammatory activity against bee venom PLA2 (IC50, 0.3 μM) (Keyzers, 2005).

FAMILY: APLYSINELLIDAE

Aplysinella strongylata (Bergquist, 1980)

Subclass: Verongimorpha Order: Verongiida
Family Aplysinellidae
Common name: Indonesian marine sponge, Balinese marine sponge
Distribution: Indonesia
Ecology: Coral reefs
Biology: It is a pedunculate sponge and is knobbed or lamellate with 6.5 cm height. Texture is rubbery and compressible. Surface is cornulose and the osucles are apical in position. Color of the animal is deep maroon red to flesh pink externally.

Compound(s) and Activities

Antimalarial activity: The psammaplysin derivative 19-hydroxypsammaplysin E isolated from this species has shown *in vitro* antimalarial activity against chloroquine-sensitive (3D7) *P. falciparum* malaria parasites (Perdicaris et al., 2013).

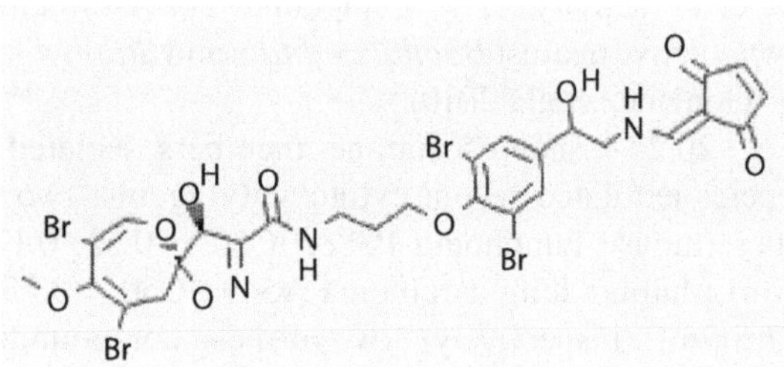

19-hydroxypsammaplysin E

Antiparasitic activity: The compound 19-hydroxypsammaplysin E displayed the highest antimalarial activity against the 3D7 *Plasmodium falciparum* parasite with an IC50 value of 6.4 μM (Gribble, 2015).

Aplysinella sp.

Order: Verongiida Family: Aplysinellidae

Compound(s) and Activities

Enzyme inhibiting activity: The compound aplysamine-1 displayed ApoE-modulating activity by increasing by two-fold the secretion of ApoE from human astrocytoma cells at a concentration of 30 μM (Gribble, 2015).

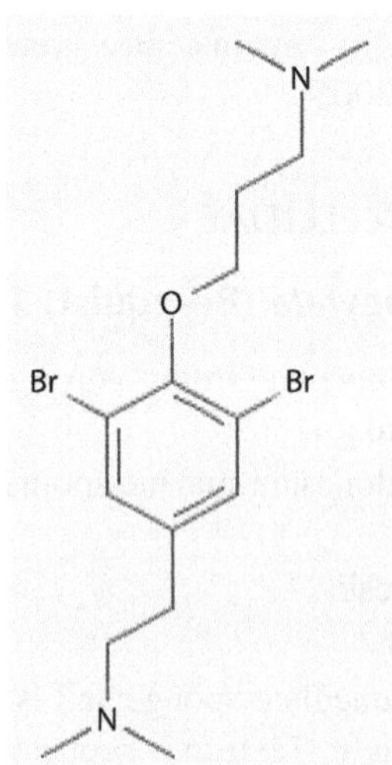

Aplysamine-1

Suberea clavata (Pulitzer-Finali, 1982) (= *Pseudoceratina clavata*)

Order: Verongiida Family: Aplysinellidae

Common name: Great Barrier Reef marine sponge

Distribution: Australia, Queensland, Capricorn Group, Heron Island, northwest end of northern slope

Ecology: It is a reef (Central and Southern Great Barrier Reef)-associated species.

Biology: In fresh condition, this sponge has a height of 14 cm with yellow verging to green color. In formalin-preserved specimen, this color changes to black and the contracted size is 8 cm high and 3 cm wide. Shape of the sponge is claviform and expands from a long peduncle that is 9 mm wide at the abse of attachment. Surface is blunt with irregular tubercles that are irregularly spaced. Consistency is rough and incompressible. Vents are not visible. It sponging is dark amber colored and markedly laminated. The pith occupies about half of the fiber.

Compound(s) and Activities

Antifungal and antibiotic activities: The bacteria *Salinospora* sp. derived from this sponge species yielded the compounds rifamycin B and rifamycin SV, which showed significant cancer cell cytotoxicities as well as antifungal and antibiotic activities (Thomas et al., 2010).

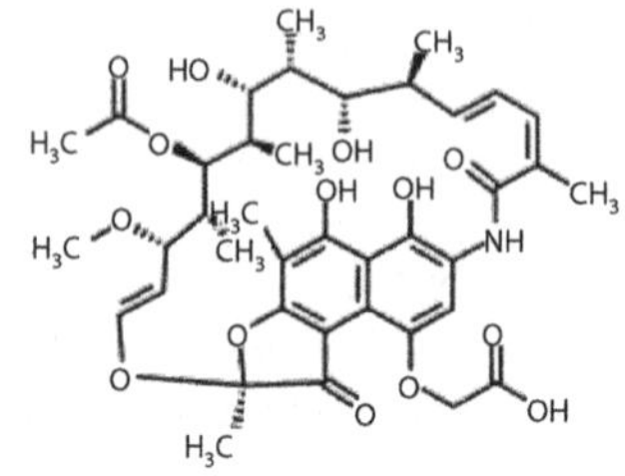

Rifamycin B

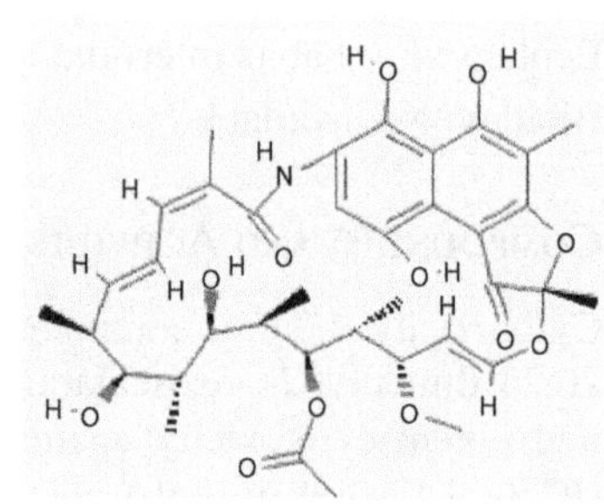

Rifamycin SV

Suberea creba (Bergquist, 1995)

Order: Verongiida Family: Aplysinellidae

Common name: Not known

Distribution: New Caledonia

Ecology: It spreads over stable coral substrates on reef fronts and on cliff walls at depths of 30 to 55 m.

Biology: It is a massive, spreading sponge with an irregular shape and a size of 10 × 15 × 4 cm (thickness). Oscula (3 mm dia.) are scattered and are surrounded by a contractile collagenous rim. Color is bright yellow in life and purple black in alcohol. Texture is hard and corklike. Surface is smooth and the body has many lumps or folds giving a tuberculate impression. Skeleton is dendritic, sparse, and made up of fibers. Ectosome is strongly collagenous and choanocyte chambers are spherical.

Compound(s) and Activities

Antibacterial activity: The compounds dibromo verongoaquin, subereamolline A–B, aerothionin, homoaerothionin, 11,19-dideoxyfistularin-3, and (+)-aeroplysinin-1 of this species showed antibacterial activity (Amr et al., 2017).

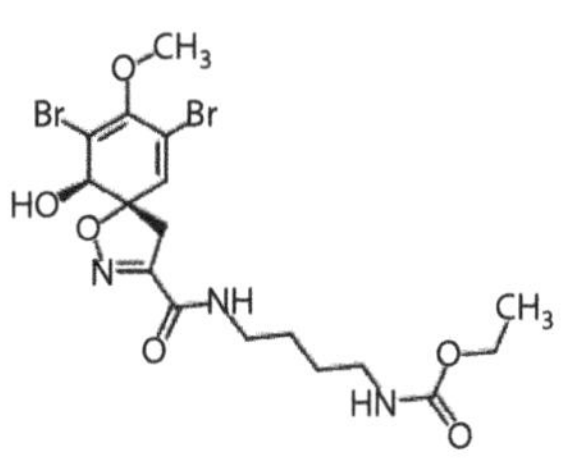

Subereamolline A

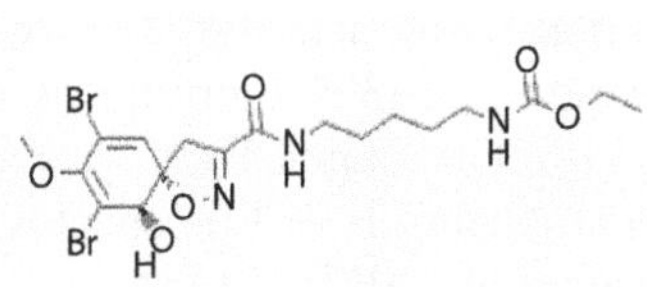

Subereamolline B

Suberea ianthelliformis (Lendenfeld, 1888) (= *Verongia ianthelliformis*)

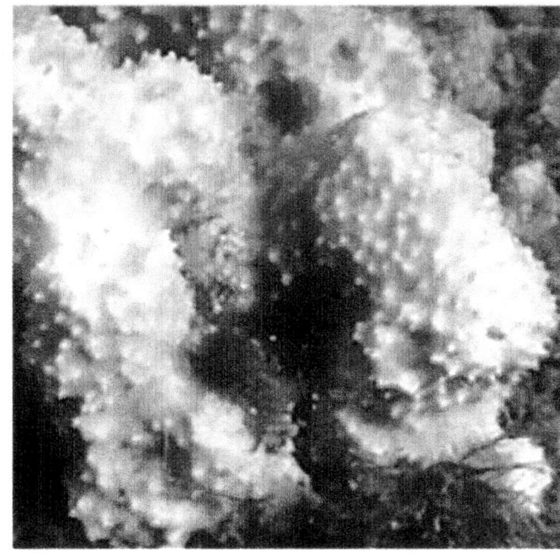

Order: Verongiida Family: Aplysinellidae

Common name: French Polynesian marine sponge

Distribution: Australia

Ecology: It inhabits intertidal areas.

Biology: It is an erect, lamellate sponge with cactiform appearance and a size of 7 cm high and 7 mm thick. Color in life is dark violet to black and very dark brown in spirit. Texture is firm but elastic. Surface is smooth except in areas of the dermal membrane with sharp cornules. These cornules are irregularly distributed over the entire surface except toward the base where they are reduced into small lumps. Skeleton is a very open reticulation of spongin fibers.

Compound(s) and Activities

Antibacterial activity: The compound lanthelliformisamine A of this species showed antibacterial activity against the Gram-negative bacterium *Pseudomonas aeruginosa* with an IC50 value of 6.8 mM (Amr et al., 2017).

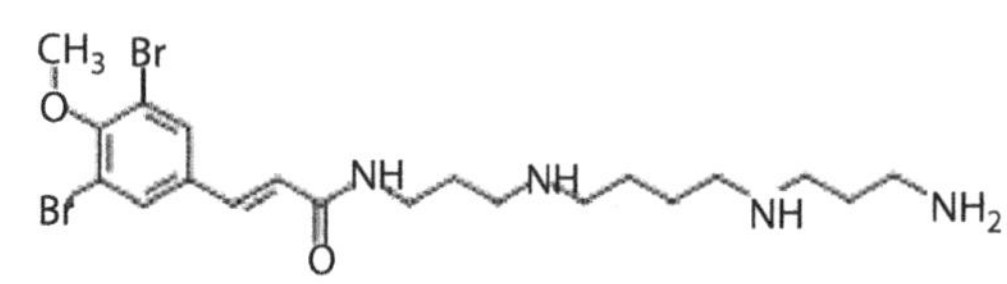

Lanthelliformisamine A

Antiplasmodial activity: The bromotyrosines araplysillin N20 formamide and araplysillin N20 formamide N-oxide showed moderate activity against both chloroquine-resistent (FcB-1) strain of *Plasmodium falciparum* (IC50 3.6 and 5.0 µM, respectively) and the chloroquine-sensitive (3D7) strain (IC50 7.0 and 4.1 µM, respectively) (Amr et al., 2017; Gribble, 2015).

Suberea mollis (Row, 1911) (= *Aplysina mollis; Verongia mollis*)

Order: Verongiida Family: Aplysinellidae

Common name: Red Sea sponge

Distribution: Western Central Pacific, Indonesia

Ecology: This sessile sponge inhabits coral reefs, 15 to 35 m in depth.

Biology: It is an upright mass from which three cylindrical processes arise at the summit. Protuberances and depressions are arranged irregularly in the entire surface. Cornules (2 mm high and 8 mm apart) are arranged regularly on the surface. Oscules are few and are irregularly scattered. Pores are in small depressions of the surface. Color of the sponge is very dark green and deep violet in alcohol. Skeleton is composed of few large pithed fibers only.

Association: Commensalism exists in this species as it harbors the lamellibranch *Avicula* inside. Further it is dark green due to its symbiotic algae.

Compound(s) and Activities

Cytotoxic activity: Among the the compounds subereaphenol A (1), 2-(3′,5′-dibromo-2′-hydroxy-4′-methoxyphenyl) acetamide (2), dibromoverongiaquinol (3), bromochloroverongiaquinol (4), and 2-(3′,5′-dibromo-4′-ethoxy-1′-hydroxy-4′-methoxy-2′,5′-cyclohexadien-1-yl)acetamide (5) isolated from this species, compounds 1 and 4 showed cytotoxicity against the cancer lines P388 and HeLa, respectively, and the IC50 values recorded are given next.

Cytotoxicity of Compounds 1–5 against P388 and Hela Cells (IC50, μg/mL)

	1	2	3	4	5
P388	2.2	>10	NT	NT	NT
HeLa	NT	>10	>10	5	>10

Source: Shaala, L.A. et al., 2008, *Nat. Prod. Commun.*, 3: 219–222.
Note: NT, not tested.

Antimicrobial and antioxidant activities: The compound aeroplysinin-1 of this species displayed significant antimicrobial activity against *S. aureus*, *P. aerugenosa*, and *K. pneumoniae*. Further, the compunds subereaphenols B and C displayed significant DPPH antioxidant activity (Abou-Shoer et al., 2008).

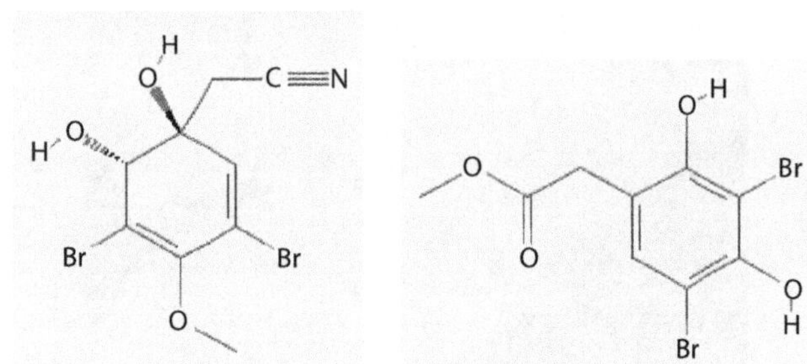

Aeroplysinin Subereaphenol

The compounds bromochloroverongiaquinol, 2-(3′,5′-dibromo-4′-ethoxy-1′-hydroxy-4′methoxy-2′,5′-cyclohexadien-1-yl) acetamide, aeroplysinin-2, and suberaphenol B of this species showed antibacterial activity (Amr et al., 2017).

Amr et al. (2017) reported that the brominated arginine and tyrosine-derived alkaloids subereamines A–B and subereaphenol D,of this species showed antimicrobial activities.

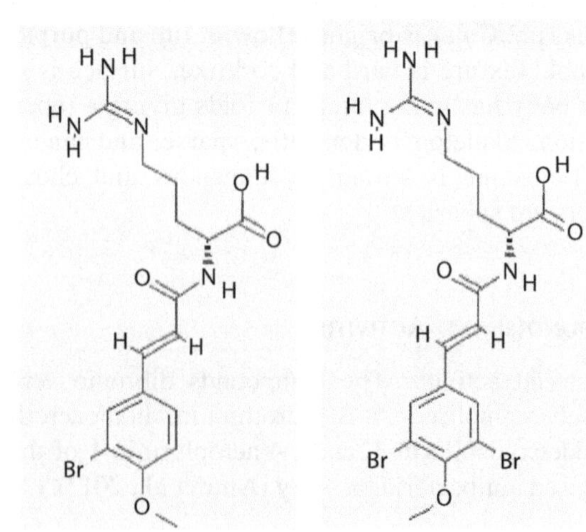

Subereamines A-B

Antibacterial activity: The compounds bromochloroverongiaquinol, 2-(3′,5′-dibromo- 4′-ethoxy-1′-hydroxy-4′methoxy-2′,5′-cyclohexadien-1-yl) acetamide, aeroplysinin-2, and suberaphenol B of this species showed antibacterial activity (Amr et al., 2017).

Lira et al. (2011) reported that the ether extract of this species showed antibacterial activity against *Staphylococcus aureus* at a concentration of 0.2 μL/disc.

Hyperglycemic activity: The 95% ethanol extract of this species showed mouse-intragastric hyperglycemic activity at a dose of 200 mg/kg (Lira et al., 2011).

Suberea sp.

Order: Verongiida Family: Aplysinellidae

Compound(s) and Activities

Antiproliferative activity: The MeOH extract of this sponge produced compounds such as subereamollines C (1) and D (2) aerothionin (3), homoaerothionin (4), aeroplysinin-1 (5), aeroplysinin-2 (6), and subereaphenol C (7). Among these compounds, aerothionin (3) and subereaphenol C (7) showed strong antiproliferative activity against the HeLa cell line with IC50 values of 29 and 13 μM, respectively. Further, the compound aeroplysinin-2 displayed potent antimigratory activity against the human breast cancer cell line MDA-MB-231 with an IC50 value of 18 μM (Shaala et al., 2015).

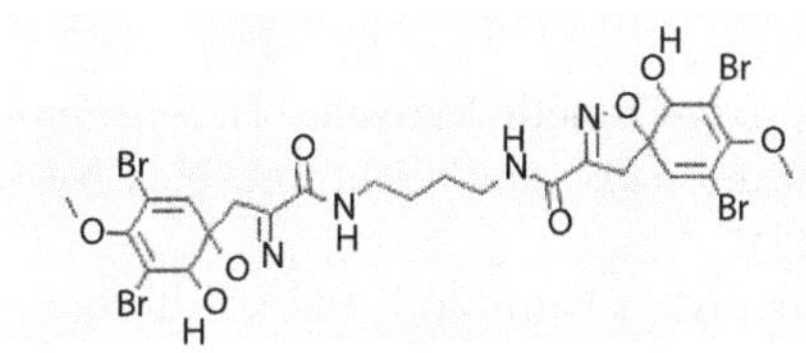

Aerothionin

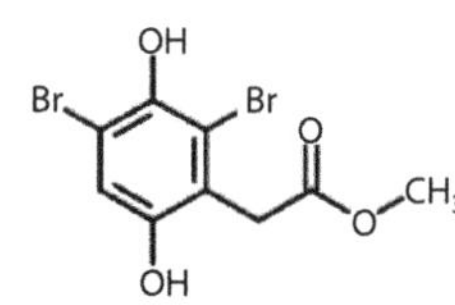

Subereaphenol C

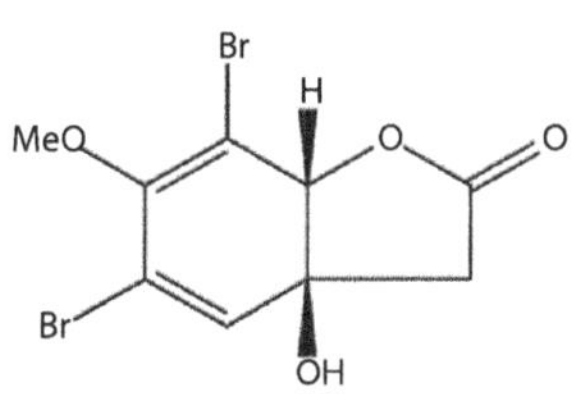

Aeroplysinin-2

Cytotoxicity: A novel spiro polycyclic aromatic alkaloid, lihouidin, isolated from this sponge displayed moderate cytotoxicity against P388D1 mouse lymphoma cells with an IC50 value of 3 mg/mL (Amr et al., 2017).

Antitumor activity: The alkaloids of this species, ma'edamines A and B, displayed significant cytotoxicity against COLO 205 (human colon cancer), MCF-7 (human breast cancer), and A549 (human lung) with IC50 values of 7.9/10.3, 6.9/10.5, and 12.2/15.4 for ma'edamines A/B, respectively (Gribble, 2015).

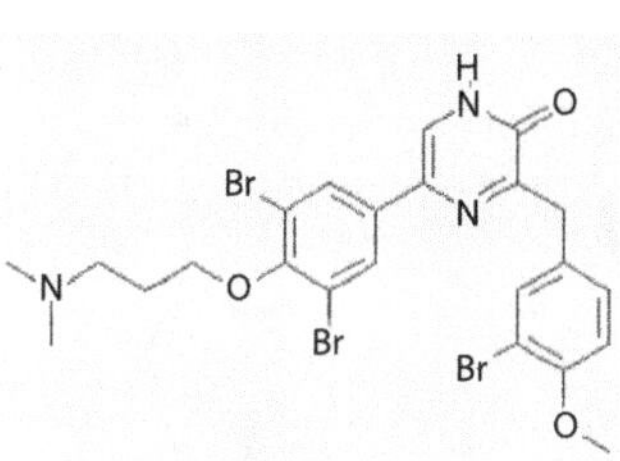

Ma'edamine A

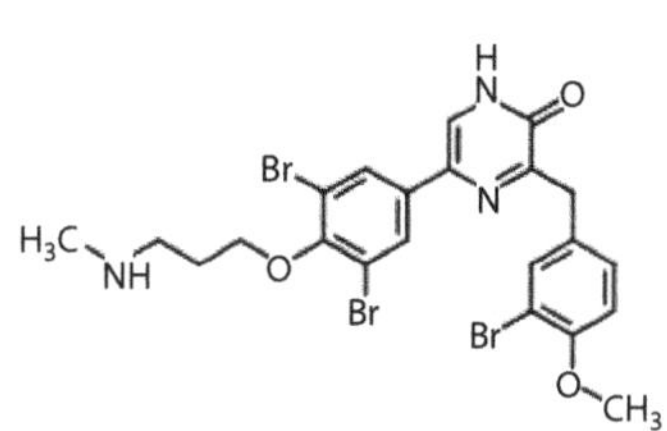

Ma'edamine B

The compounds psammaplysins and bromotyrosine analogs of this species displayed cytotoxicity against a panel of human cancer cell lines. While all the compounds, namely, moloka'iamine, 7-hydroxymoloka'iamine, ceratinamine, and hydroxyceratinamine, showed GI50 values of >70 μM,other compounds showed varying values as shown in the following table (Gribble, 2015).

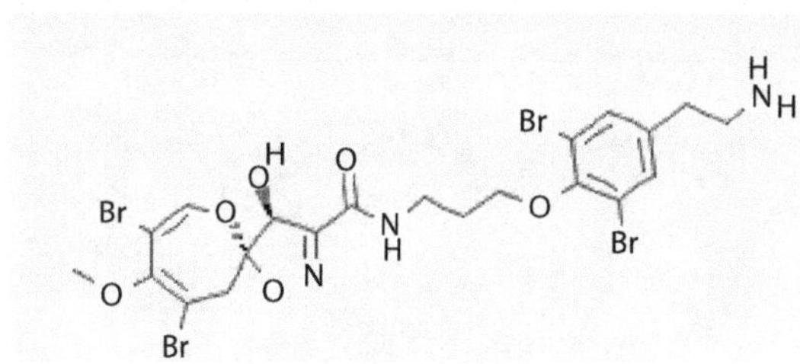

Psammaplysin

Human Cancer Cell Lines (GI50, μM)

Compound	HCT-15	PC-3	ACHN	MDA-MB-231	NUGC-3	NIC-H23
Psammaplysin X (327)	3	2	3	1	4	6
10-hydroxy-psammaplysin X	4	2	3	1	4	4
Psammaplysin A	4	7	5	4	4	12
Psammaplysin B	4	3	2	1	3	4
Psammaplysin D	24	25	27	21	26	27
Psammaplysin E	7	4	10	4	4	7
19-hydroxy-psammaplysin E	4	1	2	1	2	4

Source: Gribble, G.W., 2015, *Mar. Drugs*, 13: 4044–4136.

Anti-inflammatory activity: The compound subersic acid of this species showed anti-inflammatory activity with lipoxygenase inhibitory activity (Sipkema et al., 2005).

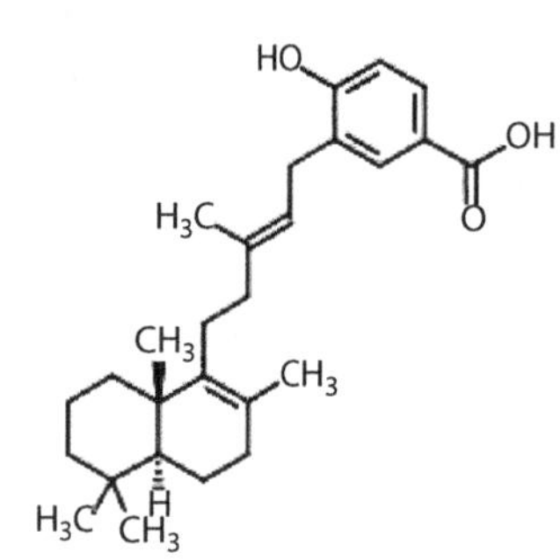

Subersic acid

FAMILY: APLYSINIDAE

Aiolochroia crassa (Hyatt, 1875)

Order: Verongiida Family: Aplysinidae

Common name: Branching tube sponge

Distribution: Western Atlantic, Caribbean, Bahamas, and South Florida

Ecology: This species is known almost exclusively from reef localities of a depth range 3–35 m.

Biology: This species is mainly massive and is frequently lobate to rarely ramose. Ramose outcrops are up to 15 cm long and the lobes are up to 6 cm high. Total size is up to 17 cm in diameter in massive specimens. Surface is irregularly conulose, with large smooth areas showing fields of pores. Interconnecting ridges between the conules may enclose polygonal depressions. Oscules are of 2 to 4 mm in diameter on lobal outcrops. The consistency of the body is rubbery, cheeselike, and somewhat compressible. Color of the animal is normally yellow with blue or green spots, but total bluish specimens may also be found.

Compound(s) and Activities

Antitrypanosomal activity: The dichloromethane and methanol crude extracts of this species showed antitrypanosomal activity and the IC50 values recorded for these extracts in 48 h were 3.95 and 16.7 g/mL, respectively (Tabares, 2011).

Antiparasitic activity: Six bromotyrosine-derived compounds—3-bromo-5-hydroxy-O-methyltyrosine (1), 3-bromo-N,N,N-trimethyltyrosinium (2), 3-bromo-N,N,N,O-tetramethyltyrosinium (3), 3,5-dibromo-N,N,N-trimethyltyrosinium (4), 3,5-dibromo-N,N,N,O-tetramethyltyrosinium (5), and aeroplysinin-1 (6)—have been isolated from this species. Compound 4 showed high activity against *Leishmania (Vianna) panamensis* and a moderate activity against *Plasmodium falciparum*, with 42% and 24% of growth inhibition at 20 and 5 μM, respectively. Compound 6 was active against *P. falciparum* and *T. cruzi* showing an inhibition in growth of 32% and 27% at 5 and 10 μM, respectively. Compound 1 was weakly active against *L. panamensis* and *P. falciparum* with 16.6% and 11.7% at 20 and 5 μM, respectively. Compounds 2 and 3 were only weakly active against *P. falciparum* with an inhibition of 11.6% and 16.5% at 5 μM (Galeano et al., 2012).

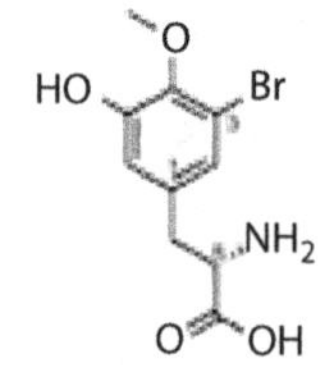

1

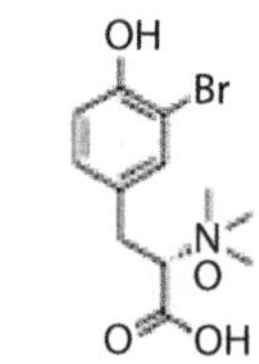

2

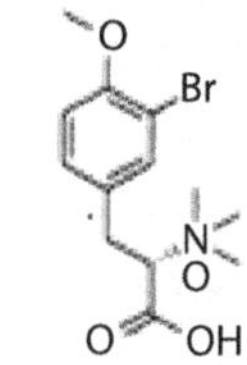

3

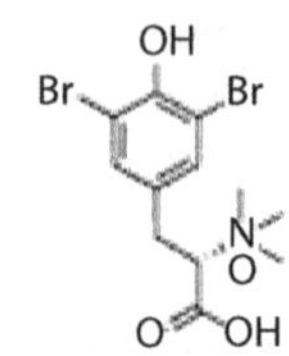

4

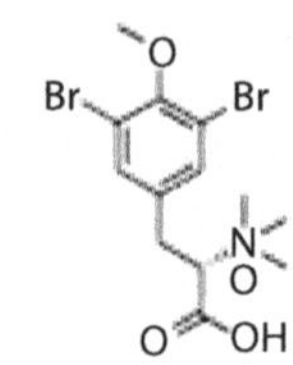

5

From left: 3-bromo-5-hydroxy-O-methyltyrosine (1), 3-bromo-N,N,N-trimethyltyrosinium (2), and 3-bromo-N,N,N,O-tetramethyltyrosinium (3)

3,5-dibromo-N,N,N-trimethyltyrosinium (4) and 3,5-dibromo-N,N,N,O-tetramethyltyrosinium (5)

Antiviral activity: The compounds of this species, aeroplysinin-1, purealidin B, and 3-bromo-5-hydroxy-O-methyltyrosine, inhibited the HIV-1 replication in a dose-dependent fashion, with a median maximum inhibition percentage of 74% at 20 μM and 47% respectively at 80 μM (Gribble, 2015).

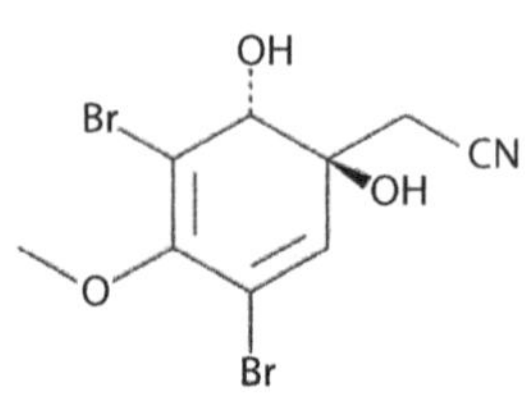

Aeroplysinin-1

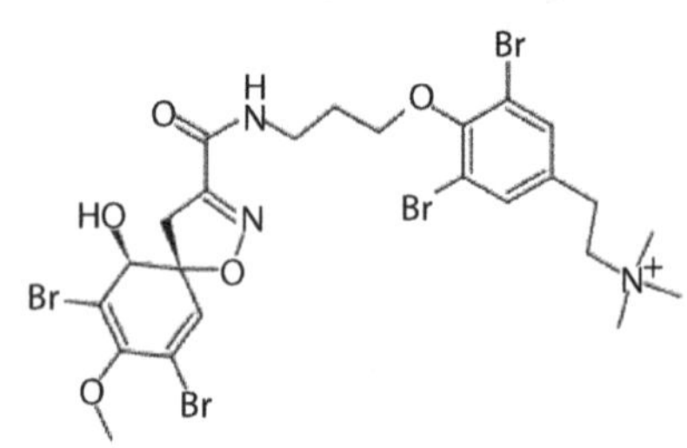

Purealidin B

Aplysina aerophoba (Nardo, 1833)

Order: Verongiida Family: Aplysinidae

Common name: Yellow tube sponge

Distribution: Mediterranean, Portuguese, and northwest Spanish waters

Ecology: It inhabits caves and rocky/detritic/muddy bottoms at a bathymetric range from 10 cm to 100 m.

Biology: It has a massive base from which irregular tubular digitations arise. These digitations are 3–4 cm high, 1–2.5 cm in diameter, and end in a flattened top in the center of which is an oscule with an iris-type of diaphragm. The surface is slippery to the touch and is irregularly covered with small conules. Consistency is firm and fleshy. Color of the animal is bright yellow in life, turning characteristically dark greenish blue-black when taken out of the water.

Compound(s) and Activities

Antiprotozoal activity: The associated actinobacterium, *Streptomyces* sp., and fungal strain, *Micromonospora* sp. RV115, of this species produced bioactive compounds valinomycin and diazepinomicin, respectively. While the former compound showed antiprotozoal activity against *L. major* with an IC50 value of <0.12 μg/mL, diazepinomicin showed activity against *Trypanosoma brucei* with an IC50 value of 6.3 μg/mL (Indraningrat et al., 2016).

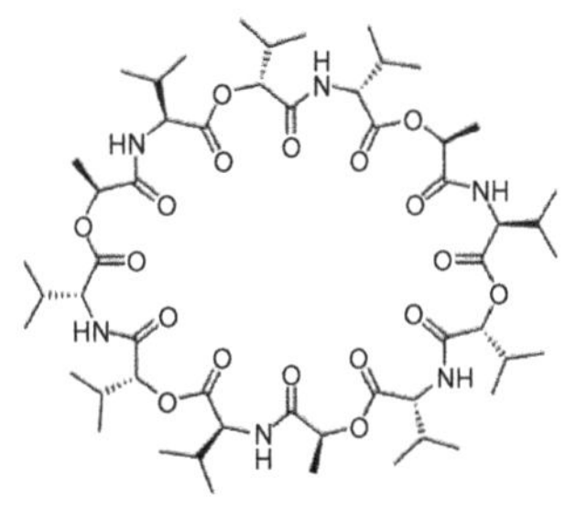

Valinomycin

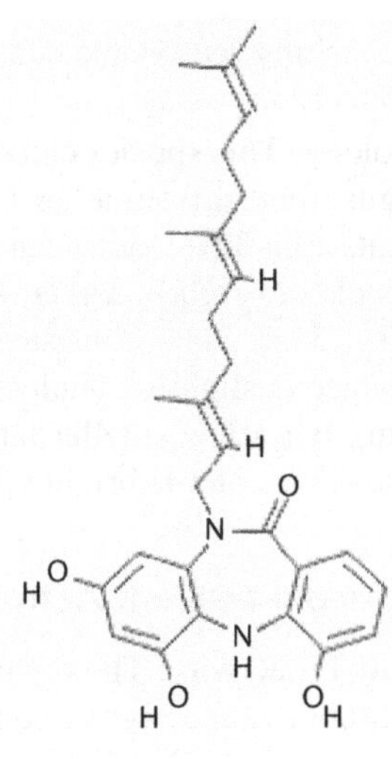

Diazepinomicin

Antimicrobial activity: The associated *Bacillus subtilis* strains A184, A190, and A202 of this species exhibited strong activity against the fungus *Candida albicans* (Thomas et al., 2010). Hentschel et al. (2001) reported that the bacterial isolates of this sponge species showed antibacterial activities against Gram-positive and Gram-negative reference strains. A general pattern was observed in that Gram-positive bacteria inhibited Gram-positive strains while Gram-negative bacteria inhibited Gram-negative isolates.

Antibacterial activity: The compound aeroplysinin-1 isolated from this species has shown growth inhibition potency against *Bacillus cereus*, *Bacillus subtilis*, *Staphylococcus aureus*, *Staphylococcus albus*, *Vibrio anguillarum*, *Flexibacter* sp., and *Moraxella* sp. (García-Vilas eet al., 2016).

Antiviral activity: (+)-Aeroplysinin-1 of this species has been reported to inhibit the key stages of the HIV-1 infection (García-Vilas et al., 2016).

Anti-inflammatory activity: *In vitro* assays with human umbilical vein endothelial cells (HUVEC) and the human monocytic leukemia cell line (THP-1) revealed that (+)-aeroplysinin-1 treatment was able to modulate key proteins of the inflammatory process. These findings strongly suggest that (+)-aeroplysinin-1 could be a novel anti-inflammatory compound with potential pharmacological action on inflammation-dependent diseases (García-Vilas et al., 2016).

Anti-angiogenic activity: (+)-Aeroplysinin-1 has been reported to inhibit the growth of endothelial cells. This effect can be due, at least in part, to the apoptogenic effects of this compound. All these findings suggest that (+)-aeroplysinin-1 is a potent inhibitor of both *in vitro* and *in vivo* angiogenesis (García-Vilas eet al., 2016).

Anticancer activity: The compound 10-hydroxy-18-methoxylbetaenone isolated from the associated, unidentified fungus *Microsphaeropsis* sp. of this species has been reported to inhibit the enzyme, protein kinase C, which plays an important role in neoplastic transformation, carcinogenesis, and tumor cell invasion (Thomas et al., 2010).

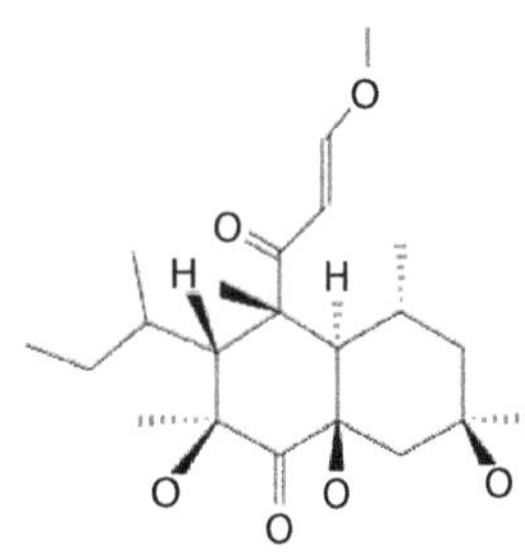

10-hydroxy-18-methoxybetaenone

Cytotoxic activity: The brominated alkaloids of the associated bacterial communities of this species showed cytotoxicity (Thomas et al., 2010).

Anti-tumor activity: (+)-Aeroplysinin-1 has also shown cytostatic and cytotoxic effects on different types of tumor cells lines (García-Vilas eet al., 2016). The IC50 values recorded for the different cell lines are given next.

Tumor Cell Line	IC50 (μM)
HT-1080	2.3
HTC-116	4.7
HeLa	3.0
THP-1	10.0
NOMO-1	17.0
HL-60	5.0

Source: García-Vilas, J.A. et al., 2016, *Mar. Drugs*, 14: 12 pp.

Note: HeLa, human cervix carcinoma; THP-1, human acute monocyte cell line; NOMO-1, human acute myeloid leukemia; HL-60, human promyelocytic leukemia cells; HTC-116, colorectal carcinoma cell line; HT-1080, human fibrosarcoma cell line.

Aplysina archeri (Higgin, 1875)

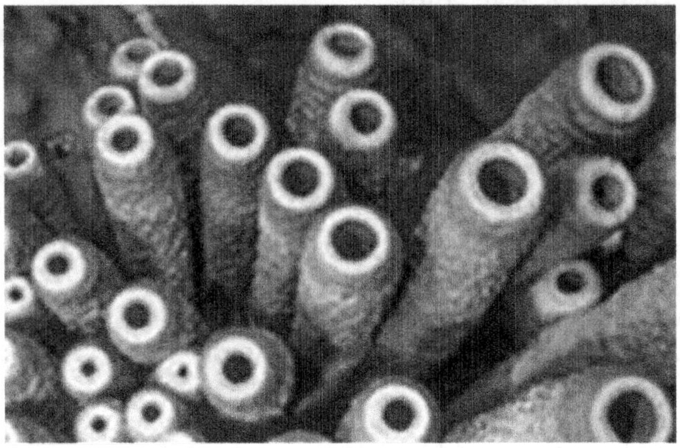

Order: Verongiida Family: Aplysinidae

Common name: Stove-pipe sponge

Distribution: It is rare in the Bahamas, and is more common in the south to Venezuela.

Ecology: It inhabits from the drop-off to deeper water.

Biology: This species forms large tubes, up to 150 cm long with a diameter of 10 cm near the end. Sometimes they grow single, but more often they are in large groups of up to 22 tubes, which are in contact with each other only at their base. The shallow depressions of the surface between the elevations contain rows of pores. The surface is generally finely conulose and does not show the deep and meandering grooves. Color of the body is pink or purplish gray. The tubes' interiors are often cream colored, but always lighter than the exterior color.

Compound(s) and Activities

Antibacterial activity: The organic extracts of this species showed considerable activity toward Gram-negative bacterial strains. The activity of the unheated aqueous extract (MIC = 48 μg/mL) was about 100-fold more potent than the heated extracts, indicating the active compound of this extract may be a peptide or a protein (Sepčić et al., 2010).

Antiviral activity: The methanol extract containing the compound toluene has been reported to possess antiviral activity against feline leukemia (Lira et al., 2011).

Aplysina caissara (Pinheiro and Hajdu, 2001)

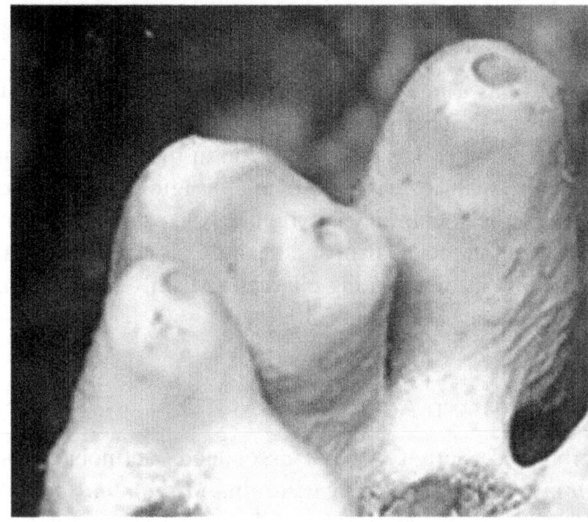

Order: Verongiida Family: Aplysinidae

Common name: Not designated

Distribution: It occurs between the coasts of Sao Paulo and Santa Catarina states of Brazil.

Ecology: The species has a typically patchy distribution and is reaching reasonable densities at a few spots at depths from 0.5 to 12 m.

Biology: This species can combine a collection of small erect digits, that is, from a few (3–4, in young specimens) to many digits (25–30). Digits can be single or anastomosed, cylindrical, or slightly vulcaniform, and are 2–6 cm high and 0.6–1.5 cm wide. Consistency is rather compressible (cheesy). Surface is slightly conulose. Oscula are mostly apical (1.5–4 mm), but a few smaller lateral ones (ca. 1 mm in dia) occur also. Live color is bright yellow externally and internally.

Compound(s) and Activities

Anti-TB activity: The organic extracts of this species showed antitubercular activity and the values of MIC varied from 200 to 400 μg/mL (Azevedo et al., 2008).

Antibacterial activity: The crude extract of this species led to the isolation of five new dibromotyrosine derivatives named agelocaissarines A1, A2, B1, B2, and caissarine C, along with the already known fistularin-3 and 11-hydroxyaerothionin. Among these compounds, fistularin-3 and 11-hydroxyaerothionin displayed moderate antibiotic activity against several pathogenic bacteria (Lira et al., 2006).

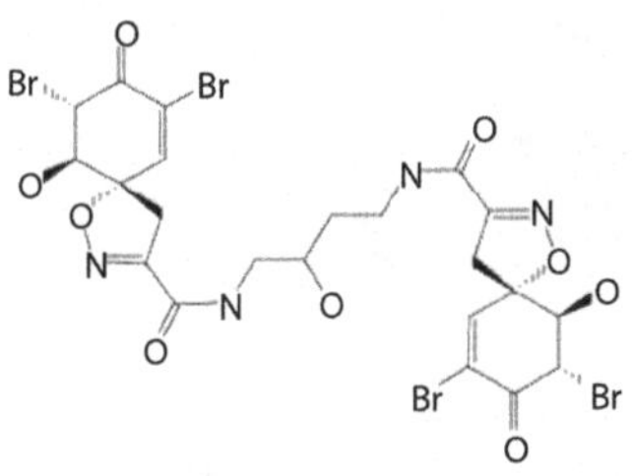

Agelocaissarine A1

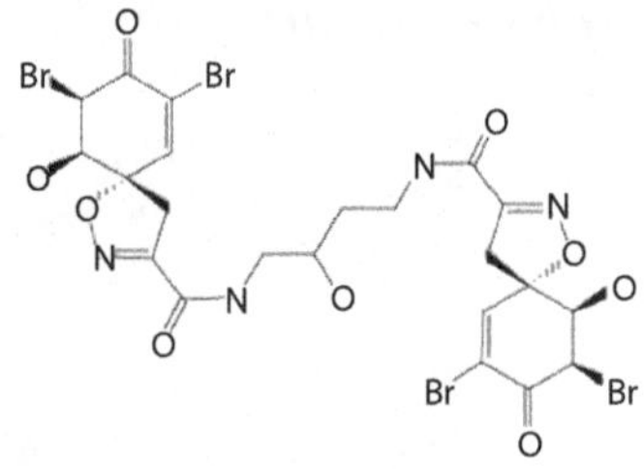

Agelocaissarine A3

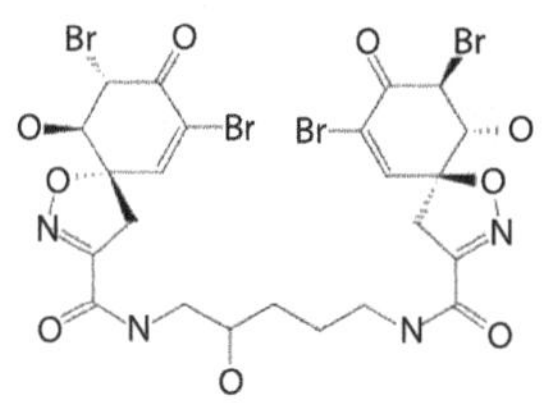

Agelocaissarine B1

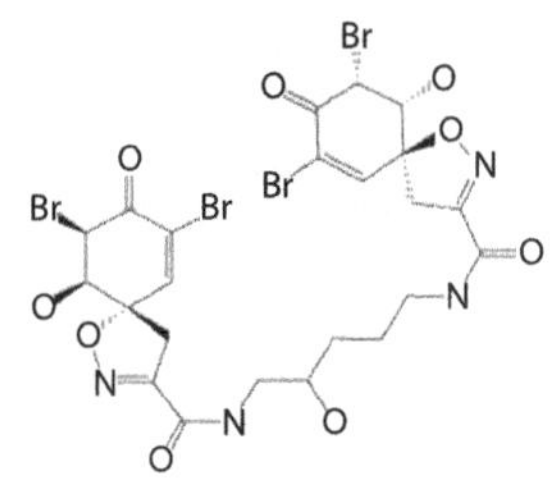

Agelocaissarine B2

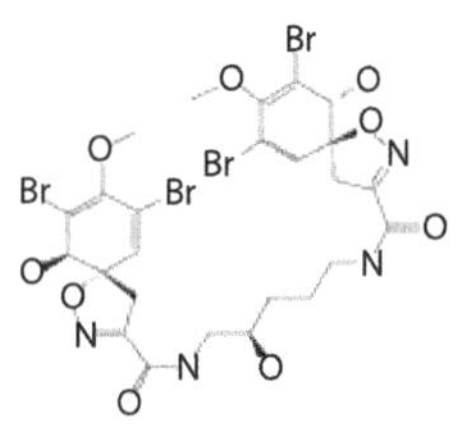

Caissarine C

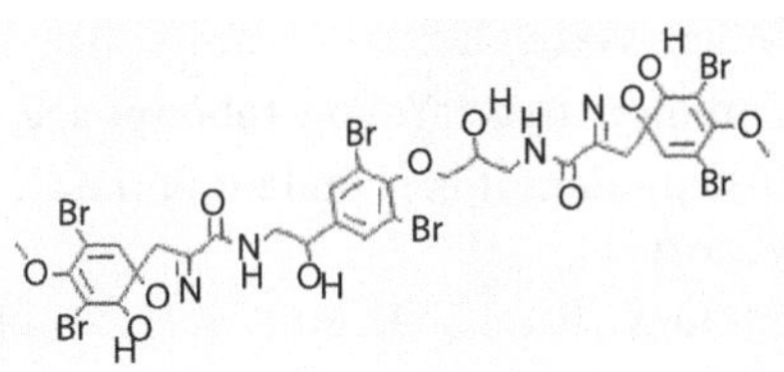

Fistularin-3

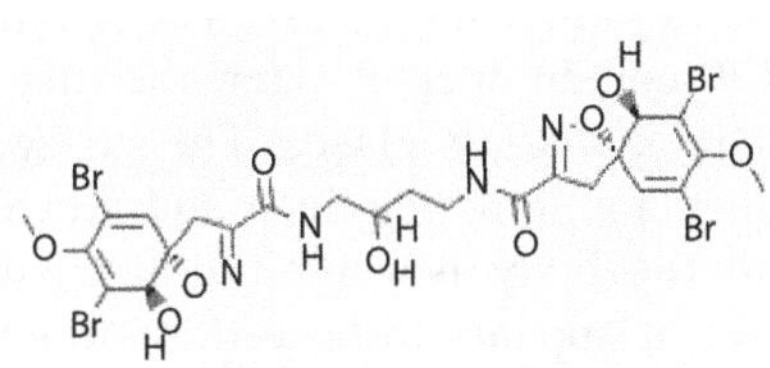

11-hydroxyaerothionin

Cytotoxicity: Two extracts of this species showed cytotoxicity against the human colon tumor (HCT 116) cell line. While the first extract (Ac05) was cytotoxic only at concentrations of 1000 µg/mL and 2000 µg/mL at 72 h, the second extract (Ac06) showed toxicity after 48 h at an extract concentration of 400 µg/mL (Azevedo et al., 2008).

Aplysina cauliformis (Carter, 1882)

Order: Verongiida Family: Aplysinidae

Common name: Row pore rope sponge

Distribution: Florida to Curacao

Ecology: It grows in lagoons and deep sloping reefs.

Biology: It is long, branching, and ropelike and grows up to 60 cm. It is generally hanging downward with the tip of the branches curling upward. Excurrent openings form long rows and are 1 to 4 mm in diameter. The oscules have thin, protruding lips, often of a lighter color. The surface is very finely conulose. The consistency is stiff and hardly compressible. The color of the body is highly variable and not distinctive, it includes red, purple and lavender.

Compound(s) and Activities

Cytotoxic activity: The CHCl3-MeOH extract (1:1) of this species showed cytotoxicity against Cho-K1 cells (Lira et al., 2011).

Anti-TB activity: The compound 2-(3-amino-2,4-dibromo-6-hydroxyphenyl) acetic acid isolated from this species showed antitubercular activity against *Mycobacterium tuberculosis* H37Rv with an MIC value of 49 µM (Kim, 2013).

Aplysina cavernicula (Vacelet, 1959) (= *Verongia cavernicola*)

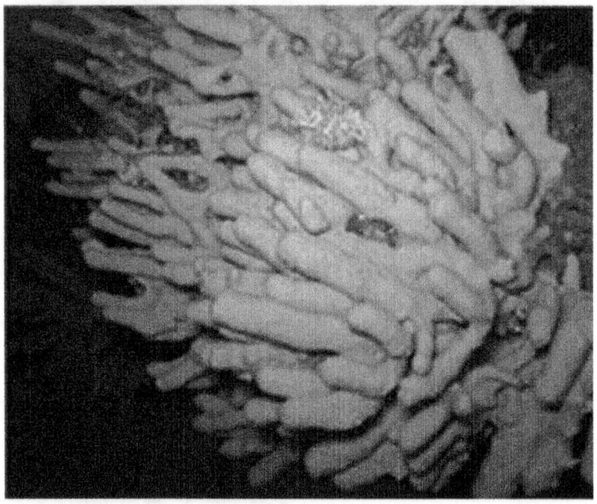

Order: Verongiida Family: Aplysinidae

Common name: Yellow cave sponge

Distribution: Mediterranean Sea and Atlantic coasts of southern Europe

Ecology: It inhabits rocky basins, and shadows, cracks, and caves at depths of 25 to 40 m under low light conditions.

Biology: The body shape is constantly digitate (1–2 cm in dia. and 5–10 cm in height). Each digitation bears one oscule (1–3 mm) at the center of an apical depression. Digitations are regularly arranged on a basal encrusting plate. No exogenous detritus appears in its skeleton. Color of the animal is yellow or pale.

Compound(s) and Activities

Cytotoxic activity: The brominated alkaloids of the associated bacterial communities of this species showed cytotoxic activity (Thomas et al., 2010).

Antimicrobial activity: The bacterial isolates associated with this species showed antimicrobial activities against Gram-positive and Gram-negative reference strains. A general pattern was observed in that Gram-positive bacteria inhibited Gram-positive strains, while Gram-negative bacteria inhibited Gram-negative isolates (Hentschel et al., 2001). Further, the associated *Bacillus subtilis* strains A184, A190, and A202 of this species exhibited strong antifungal activity against the fungus *Candida albicans* (Thomas et al., 2010).

Others: The compound (+)-aeroplysinin-1 extracted from this species exhibited antimicrobial, antiviral, antitumor, anti-angiogenic, and anti-inflammatory properties (García-Vilas et al., 2016).

Aplysina fistularis (Pallas, 1766)

Order: Verongiida Family: Aplysinidae

Common name: Yellow tube sponge

Distribution: It is abundant in the Caribbean, from Florida to Venezuela.

Ecology: It inhabits coral reefs in open water areas and are often growing on vertical walls.

Biology: It is a spicule-less sponge with soft-walled tubes, which are usually in clusters joined at the base. The length of tubes may be up to 120 cm. In deeper water, the tubes tend to become longer than at shallow places. The surface is finely conulose and does not show the deep and meandering grooves. Color of the body is yellow, orange, to brownish. In deeper water, it appears to be yellow green, which is due to its fluorescent pigments. The physiology of this species has been studied in detail.

Behavior: Yellow tube sponges are immobile, but they interact with other organisms when they come close to them. They do not have a specific time of day during which they are active. They do not move around, as they are fixed to one location, the sea floor or on coral reefs. Juveniles grow rapidly after the larva implants itself on a sturdy surface. These sponges have an antipredator response system. They produce chemicals that give their tissues an unappealing flavor. These chemicals prevent other creatures such as barnacles and algae from attaching themselves to their tubes. The yellow tube sponge has a highly specialized behavior. These creatures will alter their cells to create an upright tube when they are disturbed.

Food and feeding: These sponges feed by filtering water though their body wall and by trapping fragments of food such as algae, plankton, dead organic material, and bacteria. The water enters the sponge via microscopic pores along the body of the tube. Tiny hairlike structures can be found on the inside of the tubes of the sponge, which keep the water circulating by beating back and forth rhythmically. This movement allows the sponge to filter the maximum amount of food out of the water. The filtered wastewater leaves the sponge via the large opening at the top of the sponge called the osculum. The sponge absorbs oxygen from the water during the feeding process.

Reproduction: It reproduces sexually by giving off a large volume of sperm into a body of water. The sperm comes into contact with the female reproductive organ that houses eggs in another individual sponge. It is located in the tissue wall of the yellow tube sponge. Once fertilization occurs, tiny larvae are formed. The larvae are released into the water and a new yellow tube sponge begins to grow on the surface that the larva anchors itself to. On an average, there are 100 larvae released. Asexual reproduction also occurs, via budding. A piece of the yellow tube sponge simply attaches itself to a hard stable surface and it begins to grow again. The cells of the yellow tube sponges are very specialized. In laboratory, the cells of the shredded yellow tube sponge may grow within 8 hours into a tube.

Predators: Many species of reef fish eat the yellow tube sponges, but the main predator is the hawksbill turtle.

Environmental uses: The yellow tube sponge is home to microbial symbionts. They are the primary organisms responsible for carbonate bio-erosion on coral reefs. Further, these yellow tube sponges are important in maintaining the framework of the coral reef.

Compound(s) and Activities

Antifungal activity: The antimycotic compounds saadamycin and 5,7-dimethoxy-4-methoxylphenylcoumarin isolated from its endophytic, *Streptomyces* sp., inhibited various fungi that cause skin infection (Kim, 2013).

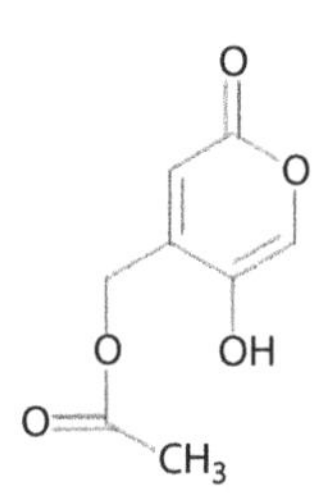

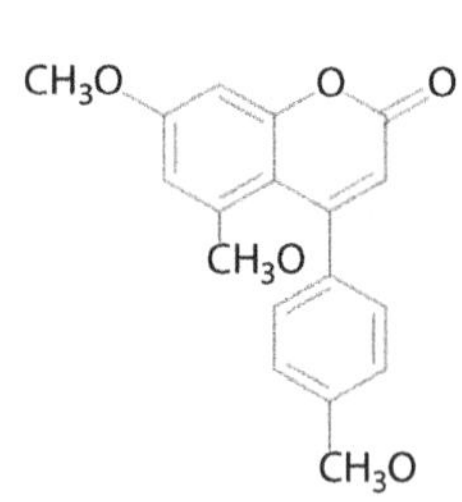

Saadamycin 5, 7-dimethoxy-4-methoxylphenylcoumarin

Antibacterial activity: The MeOH extract of this species possessed antibacterial activity against *Bacillus subtilis*, *Escherichia coli*, *Staphylococcus aureus*, *Sarcina lutea*, *Klebsiella pneumonia*, *Proteus vulgaris*, *Bacteroides fragilis*, *Clostridium perfringens*, and *Mycobacterium aviun* (Lira et al., 2011).

Cytotoxic activity: Two MeOH extracts of this species possessed cytotoxicity against Leuk L-1210 with an ED50 value of 50 mcg/mL and IC50 of 0.14 mcg/mL, respectively. Further this extract containing the compound toluene was active against CA-9KB cells (Lira et al., 2011).

Aplysina fulva (Pallas, 1766)

Order: Verongiida Family: Aplysinidae

Common name: Scattered pore rope sponge

Distribution: From Florida to Venezuela

Ecology: It inhabits the bottom of sandy lagoons and reefs to a depth of 45 m.

Biology: It is long, branching and ropelike, up to 200 cm. It hangs downward with the tip of the branches curling upward. The branches rise from a solid, massively encrusting base. Excurrent openings are found irregularly scattered over the branches and are 2 to 4 mm in dia. The oscules have thin, protruding lips, and are often of a lighter color. The surface is finely conulose. There are no marked grooves, folds, or depressions on the surface. The consistency is toughly compressible. The color of the animal is highly variable, including tan, yellow green, brown, and purple.

Compound(s) and Activities

Cytotoxic activity: The isopropanol extract of this species showed cytotoxicity against CA-9KB and Leuk L-1210 cell lines with an ED50 value of 20 µg/mL (Lira et al., 2011).

Antibacterial activity: The organic extract of this species showed antibacterial activity against the *M. pneumoniae* strain FH at an extract concentration of 1000 µg/mL (Bianco et al., 2015).

Antioxidant activity: The organic extract of this species showed DPPH antioxidant activity with IC50 values of >1000 µg/mL (Bianco et al., 2015).

Anticholinesterase activity: The organic extract of this species showed anticholinesterase activity with an IC50 value of 66 µg/mL (Bianco et al., 2015).

Aplysina gerardogreeni (Gómez and Bakus, 1992)

Order: Verongiida Family: Aplysinidae
Common name: Not known
Distribution: Mexico
Ecology: It inhabits shallow marine habitats.
Biology: Not studied

Compound(s) and Activities

Cytotoxic activity: Four new dibromotyrosine-derived metabolites, aplysinones A–D, isolated from this species have shown cytotoxic activity against human tumor cell lines (Hernández-Guerrer et al., 2007).

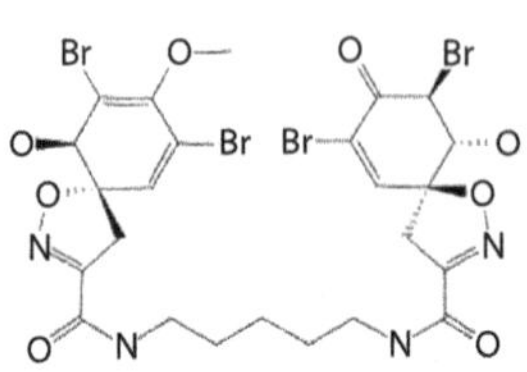

Aplysinone A

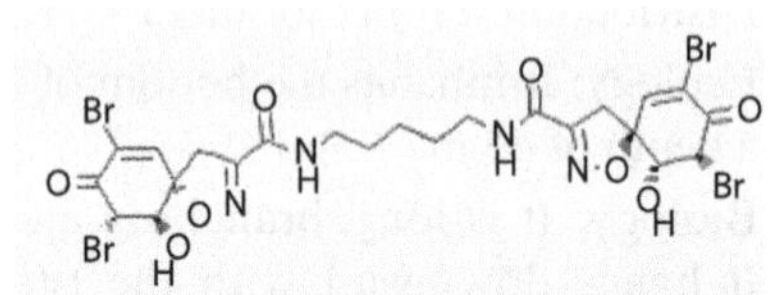

Aplysinone B

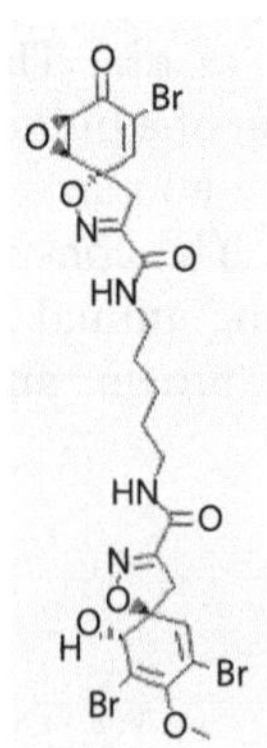

Aplysinone C

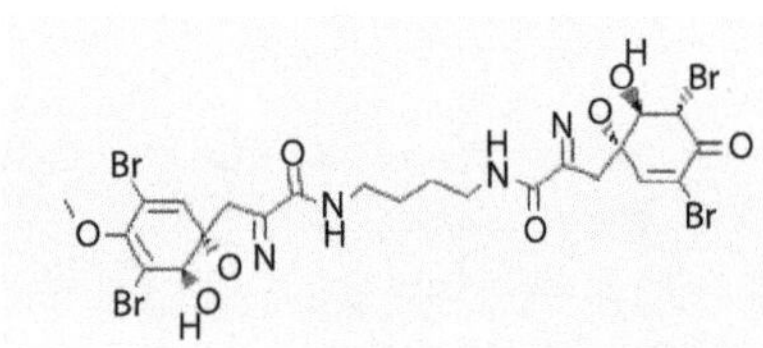

Aplysinone D

Antimycobacterial activity: Among the two bromotyrosine derivatives, namely, aerothionin and calafianin, isolated from this species, only aerothionin was found to be active against all the monoresistant variants of *Mycobacterium tuberculosis* H37 Rv at a concentration of 12.5 µg/mL. Further, this compound was active against all the drug-resistant *Mycobacterium tuberculosis* clinical isolates, regardless of their resistance patterns, with minimum inhibitory concentrations from 6.5 to 25 µg/mL (Encarnación-Dimayuga et al., 2003).

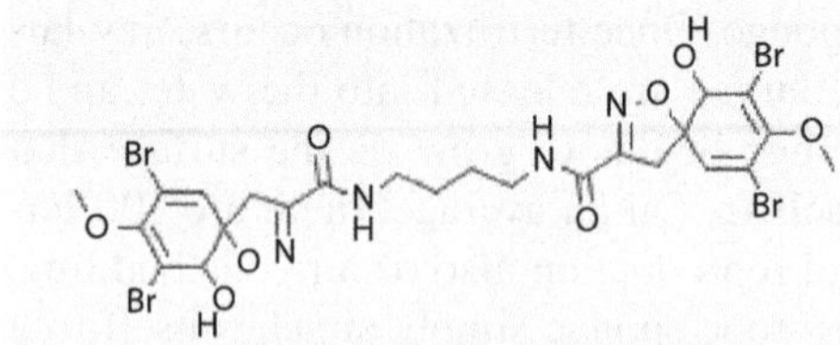

Aerothionin

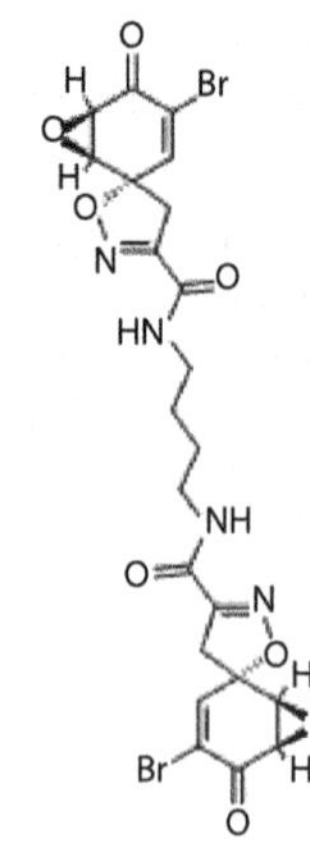

Calatianin

Aplysina lacunosa (Lamarck, 1814)

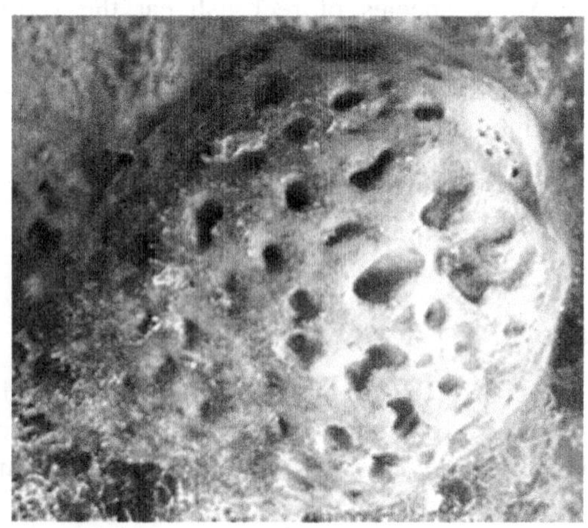

Order: Verongiida Family: Aplysinidae
Common name: Caribbean marine sponge
Distribution: Bahamas, Jamaica, and Caribbean
Ecology: This species inhabits reef and lagoon localities and sandy flats between the reefs.
Biology: This species is single or clustered and has thick-walled tapering tubes of up to 35 cm high and 10 cm in diameter. The tubes taper more strongly toward the base. Tube walls are deeply intended by irregular meandering grooves or disc-shaped depressions of up to 10 mm deep and wide. Consistency is tough and hardly compressible. Hard and soft form animals are also seen. Sediment, algae and various organisms are often seen on the surface. While the ridges of the animal are dark green to yellowish green,

valleys and pits are yellow green to yellow. Interior walls have yellow excurrent pores.

Compound(s) and Activities

Antibacterial and hemagglutinating activity: The aqueous extract of this species showed antibacterial activity and inhibited the growth of *Enterococcus faecalis*, *Bacillus cereus*, *Escherichia coli*, and *Salmonella enteritidis*, with inhibition halos of 24, 20, 24, and 22 mm, respectively. The aqueous extract also revealed the presence of several secondary metabolites with hemagglutinating activity, which is mediated by agglutinative proteins (Kazanjian and Fariñas, 2006).

Antibacterial activity: The MeOH extract of this species showed antibacterial activity of *Bacillus subtilis* and *Escherichia coli* at a concentration of 1.0 mg/Disc (Lira et al., 2011).

Antiyeast activity: The MeOH extract of this species showed weak antiyeast activity against *Saccharomyces cerevisiae* at a concentration of 1.0 mg/Disc (Lira et al., 2011).

Cytotoxicity: The bromotyrosine-derived alkaloid 11-oxoaerothionin of this species showed pronounced and selective antitumor activity against the human colon (HCT 116) cell line within the limited concentration range of 0.01 to 0.1 μg/mL (Acosta and Rodríguez, 1992).

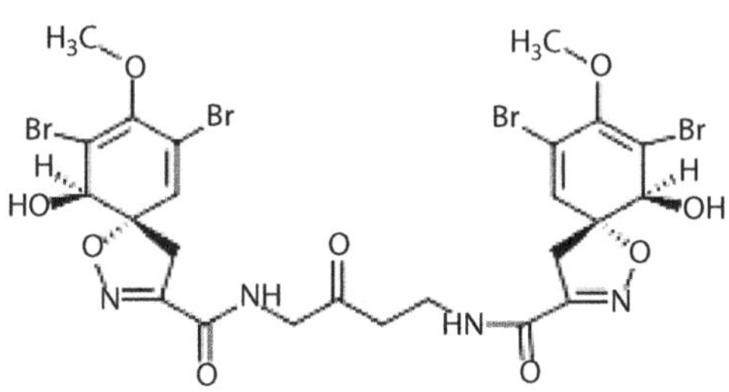

11-Oxoaerothionin

Aplysina muricyana (Pinheiro, Hajdu, and Custodio, 2007)

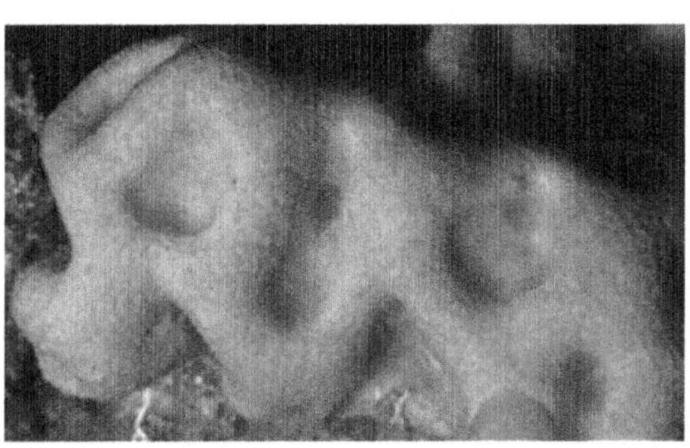

Order: Verongiida Family: Aplysinidae

Common name: Not known, as it is a new species.

Distribution: Atol das Rocas, Brazil

Ecology: It inhabits shallow waters and large crevices, in depths of 5 to15 m. This species is considered to be sciaphilous (shade tolerant).

Biology: This species is of irregular polygonal tubes, which are laterally anastomosed, and predominantly beige *in vivo*. Color of the animal ranges from bright pale yellow to dark reddish brown. It has a maximum of 36 cm in length and 3 cm in width. Some short tubes are projected from the base in varied directions. The surface is finely conulose. The sponge possesses large pseudoscula, frequently apical or pseudo-apical (eccentric), varying from 0.5 cm to 1 cm in diameter. Small oscula are also also seen on the tubes' outer sides. Consistency is soft. Specimens in darker areas are completely beige, and those more exposed to light are more intensely pigmented with darker colors, which may due to its association with cyanobacteria.

Compound(s) and Activities

Cytotoxic activity: The hydroethanolic extracts of this species exhibited cytotoxicity against tumor cell lines at a concentration of 20 mg/mL (Ferreira et al., 2007). The mean inhibitory concentration (IC50) recorded for the different cell lines are given next.

IC50 (μg/mL) Values of Tumor Cell Lines

HL-60	HCT-8	MDA-MB435	SF-295
30	>100	>100	>100

Source: Ferreira, E.G. et al., 2007, *Porifera Res.: Biodiversity, Innovation Sustainability*, 313–318.

Aplysina sp.

Order: Verongiida Family: Aplysinidae

Compound(s) and Activities

The indole alkaloid isoaplysin A isolated from this unidentified species showed weak cytotoxic activity against murine lymphoma L-1210 (IC50 = 11.5 μg/mL) and KB (31% inhibition at 20 μg/mL) cells (Gul et al., 2005).

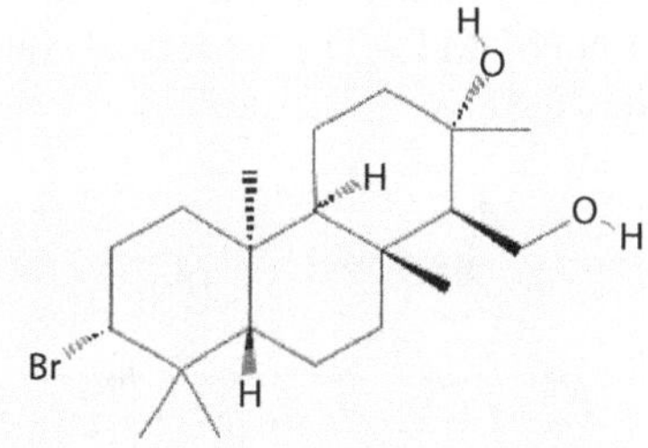

Isoaplysin A

Verongula gigantea (Hyatt, 1875)

Order: Verongiida Family: Aplysinidae

Common name: Netted barrel sponge

Distribution: Caribbean, Bahamas

Ecology: It is a reef-associated species. It is found on overhangs or along hanging walls, between 10 and 40 m.

Biology: This vase-shaped sponge resembles a large barrel, but often it is very elongated or oval. The walls are thin relative to the size of the sponge. It is yellow to green in color with a height of 15 cm. The surface has a honeycombed aspect. The outer honeycombed surface has a distinctive patterning of a raised net or lattice, especially clear in younger specimens. There are no jagged projections on the surface or lip.

Compound(s) and Activities

Antimicrobial activity: The acetone extract of this species showed antimicrobial activity with an MIC value of 50 ng/mL (Sepčić et al., 2010) from *V. gigantea* (#44, MIC = 50 ng/mL) and the butanol extract from *Vibrio rigida* (#38, MIC = 80 ng/mL).

Sedative action: The compound 5-bromo-N,N-dimethyltryptamine isolated from this species caused significant reduction of locomotor activity indicative of a potential sedative action ("5,6-dibromo-N,N-DMT – The 'undersea' equation …," n.d.; Ciminiell et al., 2000).

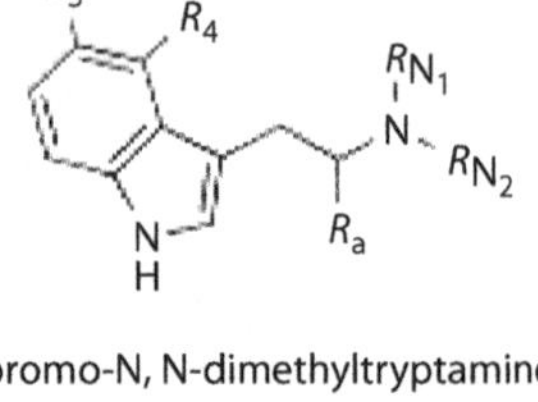

5-bromo-N, N-dimethyltryptamine

Verongula rigida (Esper, 1794)

Order: Verongiida Family: Aplysinidae

Common name: Pitted sponge, Caribbean marine sponge

Distribution: Florida, Bahamas, Gulf of Mexico, Caribbean

Ecology: This species inhabits reefs and lagoons.

Biology: It is a massive lobate sponge, with numerous pits and ridges over the surface. The lobes or tubes bear a conspicuous apical oscule in a depression with perforated diaphragms. The size of the specimens may be up to 20 cm in diameter and 15 cm high. Individual tubes are up to 8 cm in height and 4 cm in dia. Oscules are up to 2 cm in dia. The surface bears meandering grooves of up to 15 mm in width and up to 20 mm in depth. Lamellar ridges are smooth or conulose. The consistency is cheeselike and slightly compressible. The color of the sponge is yellow, yellowish green, green, brown, or tan.

Compound(s) and Activities

Antiparasitic and cytotoxic activities: Nine bromotyrosine-derived compounds—aeroplysinin-1 (1) and dihydroxyaerothionin (2), 3,5-dibromo-N,N,N-trimethyltyraminium (3), 3,5-dibromo-N,N,N,O-tetramethyltyraminium (4), purealidin R (5), 19-deoxyfistularin 3 (6), purealidin B (7), 11-hydroxyaerothionin (8), and fistularin-3 (9)—have been isolated from this species. Among these compounds, compounds 7 and 8 showed selective antiparasitic activity at 10 and 5 μM against *L. panamensis* and *P. falciparum* parasites, respectively. Compound 8 showed 12.6% inhibition of intracellular amastigotes of *Leishmania* at 10 μM. Compound 7 showed 23.2% inhibition *in vitro* over *P. falciparum* at 5 μM. Compound 1 showed 29.1% of parasite growth inhibition *in vitro* over *T. cruzi* at 10 μM. Apart from this, compound 1 exhibited a high cytotoxicity of human promonocytic cell line (94.8%) at 20 μM. On the other hand, compounds 2 and 4 showed moderate cytotoxicity at 20 μM. Compounds 5 and 9 were found to be weak antiparasitic compounds (inhibiting less than 11% at 10 μM) and they produced the 45.3% and 58.2% growth inhibition over U-937 cells at 20 μM. Compound 1 is considered to be the most cytotoxic agent evaluated. Compounds 3 and 6 however, showed no cytotoxic or antiparasitic activity (Galeano et al., 2011; Gribble, 2015; Perdicaris et al., 2013).

Cytotoxic activity: A novel phenyl alkene compound, (E)-10-benzyl-5,7-dimethylundeca-1,5,10-trien-4-ol, isolated from this species showed *in vitro* cytotoxic activity (through the inhibition of microtubule activity) against HL-60 human leukemia cancer cells with an IC50 value of 8.1 μM (Hwang et al., 2013).

Eight 4,9-friedodrimane-type sesquiterpenoids (1–8) were acquired using the oxidative potential of *Verongula rigida* on bioactive metabolites from two *Smenospongia* sponges. Among them, compounds 2 and 8 and the mixture of 3 and 4 were found to suppress β-catenin response transcription (CRT) via degrading β-catenin and exhibited cytotoxic activity on colon cancer cells. Colorectal cancer has emerged as a major cause of death in Western countries (Hwang et al., 2015).

Antimicrobial activity: The butanol extract of this species showed antimicrobial activity with an MIC value of 80 ng/mL (Sepčić et al., 2010).

Antiviral activity: The compounds of this species, namely, aeroplysinin-1, purealidin B, and 3-bromo-5-hydroxy-O-methyltyrosine, inhibited the HIV-1 replication in a dose-dependent fashion, with a median maximum inhibition percentage of 74% at 20 μM and 47%, respectively, at 80 μM (Gribble, 2015).

FAMILY: IANTHELLIDAE

Ianthella basta (Pallas, 1766)

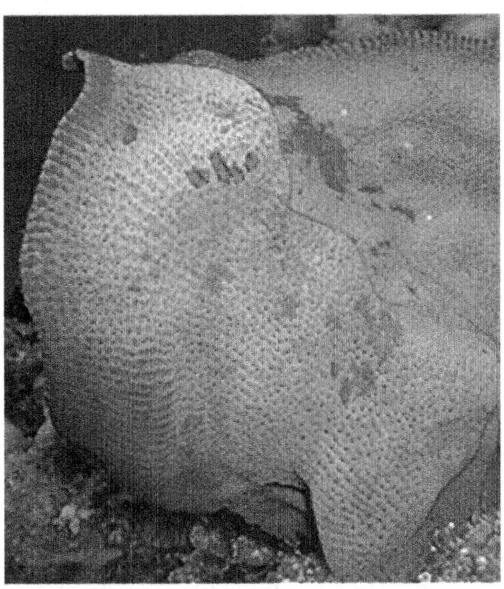

Order: Verongiida Family: Ianthellidae

Common name: Elephant ear sponge, paper sponge, or scroll sponge

Distribution: Indo-Pacific, Kenya, Mozambique Channel, Indonesia, Great Barrier Reef

Ecology: It is found singly or in groups on inshore silty patch and fringing reef slopes with tidal and current flow. Depth range is 0 to 100 m.

Biology: It is an elongate fan or vase-shaped sponge with occasional fanlets. The sponge grows up to 1500 cm in height and width and are 1 to 3 mm thick. It is attached to the substrate at a constricted base. Surface is extremely regular with radiating rows of sharply pointed cornules. Inhalent pores are separated on opposite sides of the spong lamella. The osculates are 0.5 to 1 mm diameter. Sponge is flexible and surface is fleshy. Skeleton is complex reticulation of fibers oriented within one plane, radiating from the base to the margin of the sponge. Both the exterior and interior are bluish violet in color. Sponges that are mimosa yellow, copper green, mandarin orange, opal green, and vivid blue are also common. The two dominant color morphotypes are, however, yellow and purple. Although both color morphs are dominated by only two bacterial symbionts (residing within the Gamma and Alphaproteobacteria), the purple morph also contains minor representatives of the Cyanobacteria, Chloroflexi, and Verrucomicrobia.

Association: The sea cucumber *Synaptula lamperti* is closely associated with the sponge and makes use of certain nutrients exuded by it.

Compound(s) and Activities

Neurotrophic/neuroprotective activity: New compounds bastadin 20 (1), 15,34-O-disulfatobastadin 7 (2), and 10-O-sulfatobastadin 3 (3) have been isolated from this species. Among them, compounds 2 and 3 exhibited moderate differential activity as SR Ca2+ channel agonists (EC50 13.6 and 100 μM, respectively) of the Ry1R FKBP12 complex, which is related to the neurotrophic activity of the nervous system. The potency of compound 1 was only half (EC50 20.6 μM) that of compound 2 (Franklin et al., 1996).

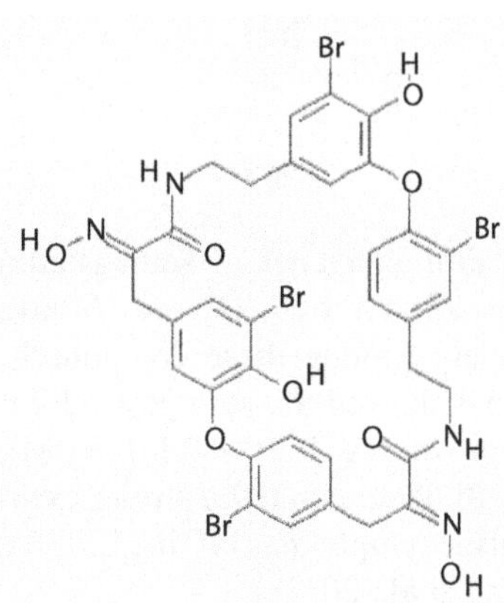

Bastadin 20

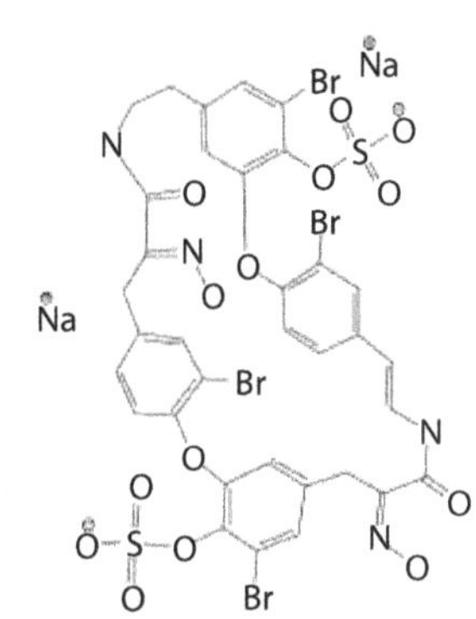

15, 34-O-disulfatobastadin 7

Antitumor activity: The compounds bastadins 6, 7, 11, and 16 of this species inhibited cell proliferation, with IC50 values of 1.5, 5.3, 3.7, and 1.9 μM, respectively (Gribble, 2015).

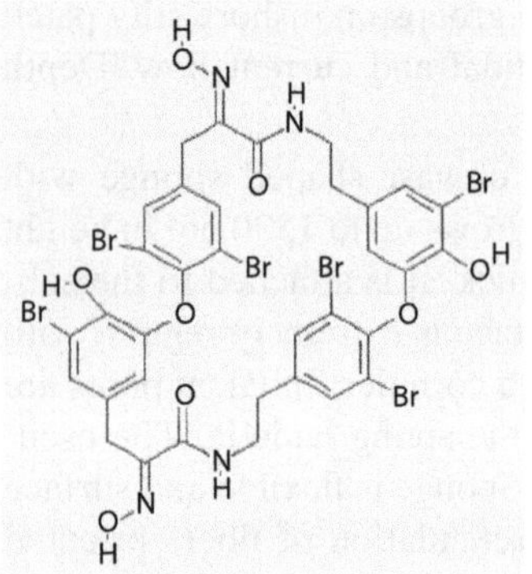

Bastadin 6

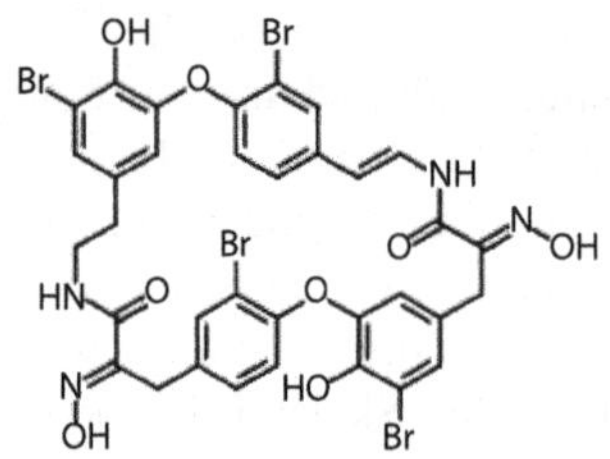

Bastadin 7

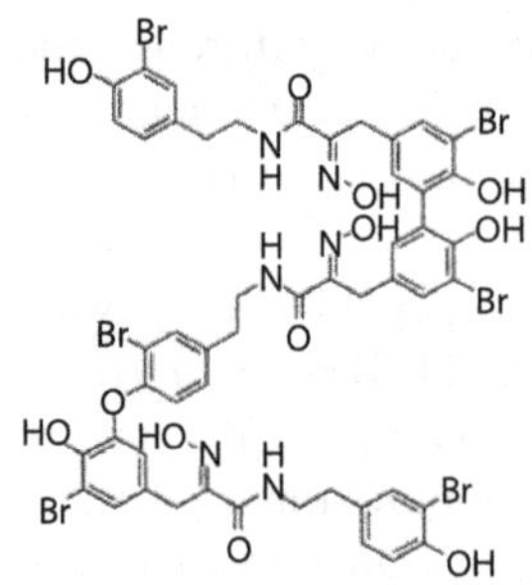

Sesquibastadin 1

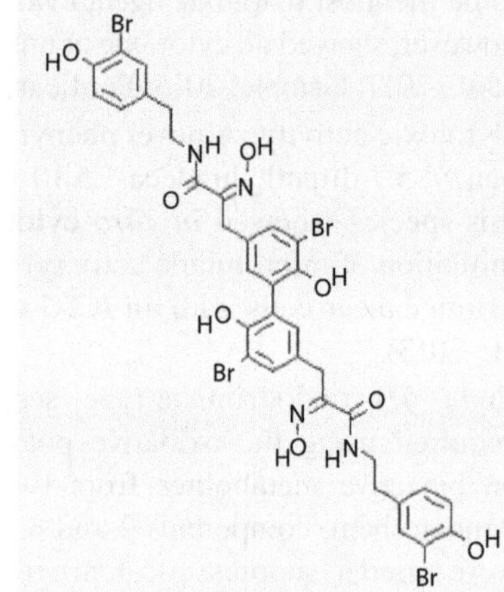

Bastadin 3

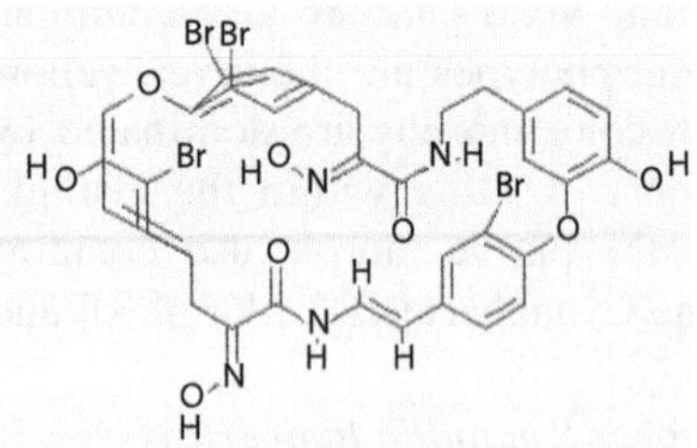

Bastadin 11

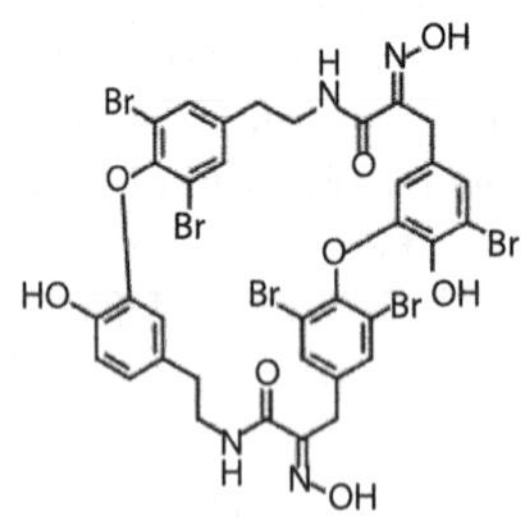

Bastadin 16

Two new cytotoxic metabolites, bastadin 8 and bastadin 9, isolated from this species showed cytotoxicity (Miao et al., 1990).

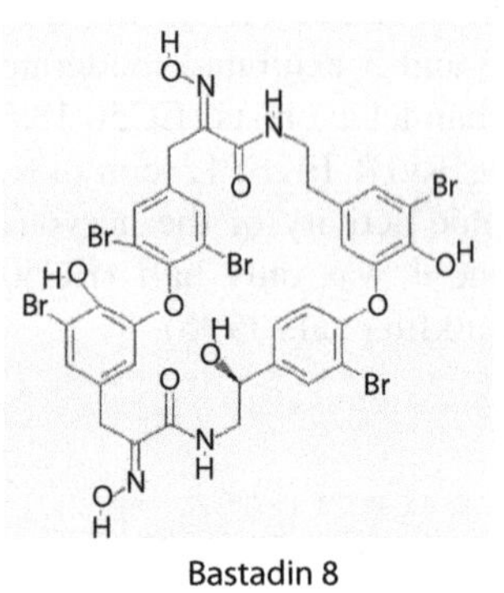

Bastadin 8

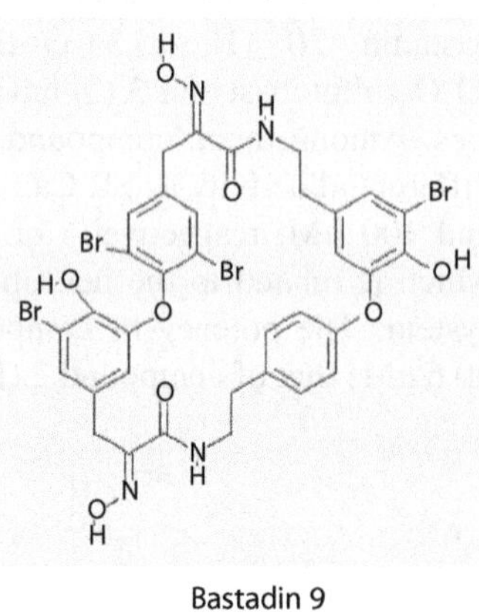

Bastadin 9

The new trimeric hemibastadin congener, sesquibastadin 1, and the already known bastadins 3, 6, 7, 11, and 16 have been isolated from this species. Among these compounds, sesquibastadin 1 and bastadin 3 showed the strongest inhibition rates for at least 22 protein kinases (IC50 = 0.1–6.5 μM), while bastadins 6, 7, 11, and 16 demonstrated a strong cytotoxic potential against the murine lymphoma cell line L5178Y (IC50 = 1.5–5.3 μM) (Niemann et al., 2013).

Anti-inflammatory/cytotoxic activity: The alkaloids bastadins 4, 8, and 9 that have been isolated from this species showed anti-inflammatory activity by inhibiting inflammation in the mouse ear assay. The percentage values of inhibition were found to be 89%, 93%, and 94%, respectively, at 50 μg per ear. These compounds have also shown cytotoxic activity against the murine P388 cell line (Keyzers, 2005).

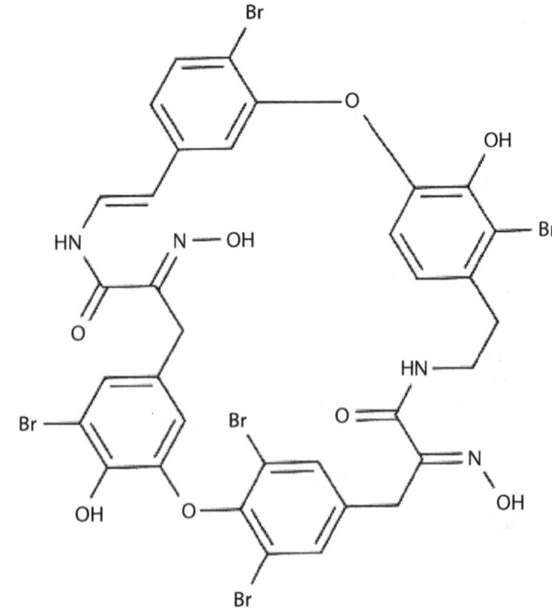

Bastadin 4

Others: The chitin isolated from this species is a promising, highly biocompatible biomaterial for stem cell–based tissue-engineering applications (Mutsenko et al., 2017).

Ianthella quadrangulata (Bergquist and Kelly-Borges, 1995)

Order: Verongiida Family: Ianthellidae

Common name: Australian marine sponge

Distribution: Lord Howe and Norfolk Islands, Tweed-Moreton

Ecology: It is common within harbors and in turbid coastal waters down to 50 m.

Biology: It is an elongate fan 10 to 15 cm high and weide and 3 to 5 mm thick. It has basal fanlets. It has a slightly constricted base of attachment. Surface is smooth and lamella is flexible. External and internal coloration is mimosa yellow. Skeleton is simple reticulation of robust vertical fibers. They are anastomosing to produce a netlike skeleton.

Compound(s) and Activities

Cytotoxic activity: A new cytotoxic bastadin, bastadin 24 (1), and the already known bastadins 4, 5, 6, 7, 12, 13, and 21 have been isolated from this species. All these bastadins were found to be moderately cytotoxic against 36 human tumor cell lines. Bastadin 24 exhibited selective cytotoxic activity toward 5 of the 36 tumor cell lines. Bastadins 7 and 12 significantly inhibited the serum + hEGF-induced (human epithelial growth factor) tubular formation of human umbilical vein endothelial cells (HUVEC) at a concentration of 1 mug (Greve et al., 2008).

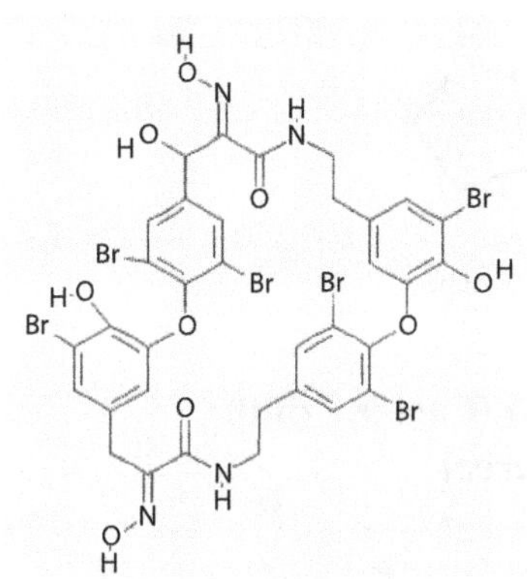

Bastadin 24

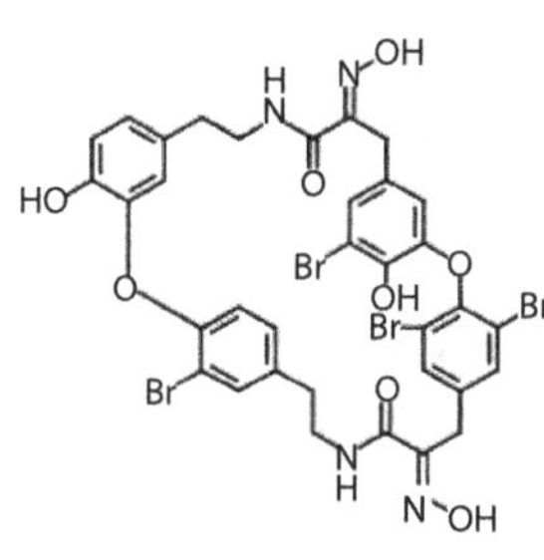

Bastadin 12

Neuroprotective activity: The octopamine derivative ianthellamide A isolated from this species inhibited the activity of kynurenine 3-hydroxylase with an IC50 value of 1.5 μM. It also significantly increased the level of endogenous kynurenic acid in rat brain. This compound may, therefore, have the potential as a neuroprotective agent in the treatment of neurodegenerative disorders (Feng et al., 2012).

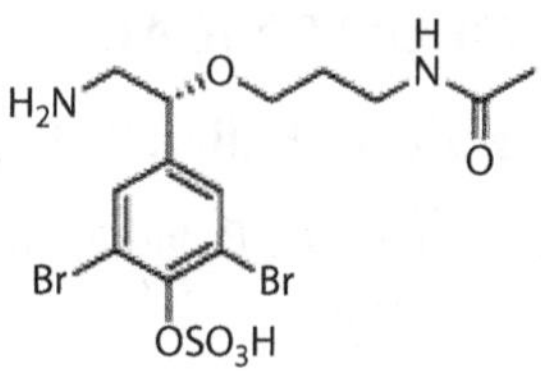

Ianthellamide A

Immunosuppressive activity: The polyketide compound iso-iantheran A isolated from this species has shown immunosuppressive activity (in human dendritic cells) by activating the ionotropic P2Y11 receptor (Anjum et al., 2016).

Three new iantherans—iso-iantheran A, 8-carboxy-iso-iantheran A, iso-iantheran B—and two new brominated tyrosine-derived metabolites 5 and 6 have been isolated from the extract of this species. Among these compounds, iso-iantheran A and 8-carboxy-iso-iantheran A exhibited immunosuppressive activity with agonist activity of P2Y11 receptors with EC50 values of 1.3 μM and 0.5 μM, respectively (Greve et al., 2007).

Ianthella sp.

Order: Verongiida Family: Ianthellidae

Compound(s) and Activities

Antibacterial activity: The compounds dictyodendrins F, H, and I of this species showed antibacterial activity against the Gram-positive *B. subtilis* (ATCC 6051 and 6633): with IC50 values of 2.7 and 2.3 μM (compound F); 1.2 and 3.1 μM (compound H); and 2.5 and 2.8 μM (compound I) (Gribble, 2015).

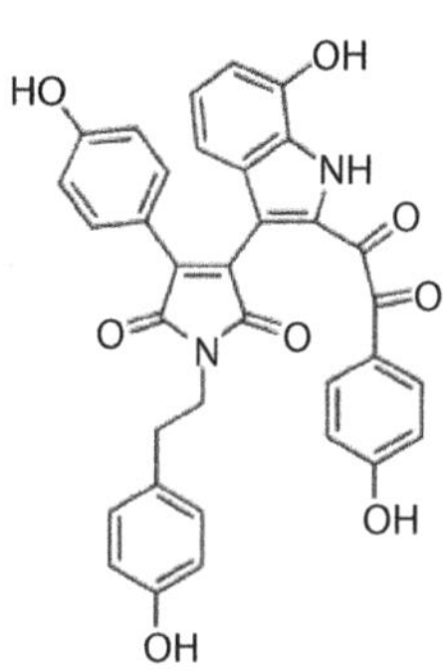

Dictyodendrin J

Enzyme inhibiting activity: The compounds of this species, namely, dictyodendrin F, H, I, and J, were the potent (protease) inhibitors of BACE 1 (β-secretase 1) with IC50 values of 1.0–2.0 μM (Gribble, 2015).

FAMILY: PSEUDOCERATINIDAE

Pseudoceratina arabica (Keller, 1889)

Order: Verongiida Family: Pseudoceratinidae

Common name: Not known

Distribution: Gulf of Aden, Southern Red Sea

Ecology: It inhabits shallow sandy lagoons and seagrass beds.

Biology: It is a sprawling, branching sponge with fingers, and it measures 10–15 cm long and 1–3 cm wide, although highly variable in size. It arises from an encrusting base and is covered in closely spaced, low, rounded conules. Oscules are 1 mm wide and membranous, and are raised and scattered. Body is dense, very rubbery, flexible, shiny, and waxy. Color of the sponge is deep bright yellow with greenish tinges. It may become deep violet blue on exposure to air.

Compound(s) and Activities

Antiproliferative activity: Among the brominated alkaloids, ceratinines A–E (2–6), the sponge also afforded the known alkaloids moloka'iamine (1), hydroxymoloka'iamine (7), and moloka'iakitamide (8). The active fraction of the sponge *Suberea mollis* afforded the three known alkaloids subereamolline A (9), aerothionin (10), and homoaerothionin (11) isolated from this species. Among them the compound subereamolline A (9) potently inhibited the migration and invasion of the highly metastatic human breast cancer cells MDA-MB-231. Further, only this compound showed cytotoxic activity against MDA-MB-231 cells at 30 μM (Shaala et al., 2012).

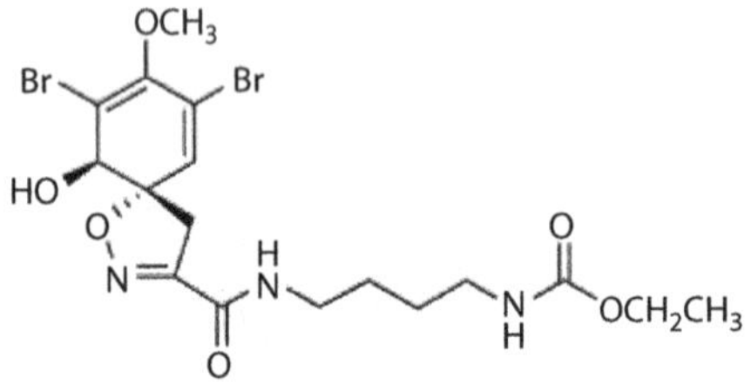

Subereamolline A

Antitumor activity: The compound subereamolline A of this species was highly active, showing IC50 1.7 μM against the highly metastatic MDA-MB-251 human breast cancer cell line (Gribble, 2015).

Pseudoceratina durissima (Carter, 1885) (= *Aplysina laevis*)

Order: Verongiida Family: Pseudoceratinidae

Common name: Not designated

Distribution: Red Sea, Indo-Pacific as far east as Hawaii, Australia, tropical and southeastern temperate waters, Bowl Reef, Great Barrier Reef

Ecology: It is a Great Barrier Reef sponge. Most frequently the sponges are solitary mounds, but multilobate specimens have also been recorded.

Biology: It is a massive rounded sponge, and is hard and barely compressible in life. It is rock hard when dry. It has a smooth surface with scattered low mounds marking the points where fiber brushes intersect the surface. The overall dimensions of this sponge are 5 cm high, and 10 cm by 5 cm basally. However, larger specimens up to 20 cm high have also been collected. Color in life is dull yellow with patchy green-purple areas. In this sponge, the pigment change takes place very quickly (a characteristic feature of its family) and the sponge is becoming a uniform dark purple. Oscules are small, 2–3 mm in diameter, and are scattered over the upper half of the body. The pores are tiny and scattered. Fibers are sparse in relation to the soft tissue, irregular in profile, and are composed only of pith elements. Choanosome is densely collagenous and it may incorporate debris.

Compound(s) and Activities

Antibacterial activity: The acetone extract of this specis has shown antibacterial activity against *Bacillus subtilis* (Lira et al., 2011).

Anticancer activity: The isoxazoline derivatives of this species have sown anticancer activity (Kaur et al., 2014).

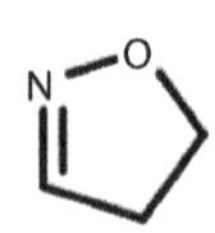

Isoxazoline

Pseudoceratina purpurea (Carter, 1880) (= *Psammaplysilla purpurea*)

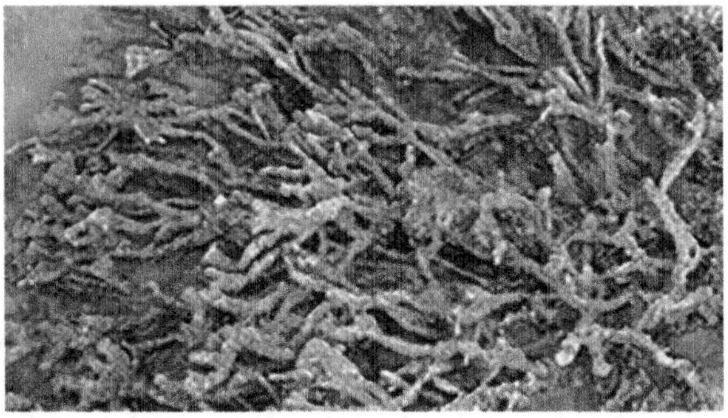

Order: Verongiida Family: Pseudoceratinidae

Common name: Yellow prickly branching sponge

Distribution: Indian Ocean

Ecology: It is commonly seen on coral rubble or small stones on the seashores.

Biology: This sponge is extremely variable in shape from knobs and lobes to slender branching stems (8–12 cm) emerging from an encrusting base (20–30 cm). The sponge can cover large areas. Tips of the sponge are rounded or flattened. The surface is rubbery, prickly, and slippery with small bumps that appear spiky and prickly when the sponge is out of water. There are many tiny holes sprinkled all over the sponge. Colors range from bright yellow, green yellow to olive. It is often dark or olive green with yellowish tips. A dead sponge, however, may be dark brown or deep purple.

Compound(s) and Activities

Anticancer activity: The compound psammaplin A isolated from this species showed activity against several cancer cells by preventing them from growing and dividing ("Anti-cancer drugs," 2011).

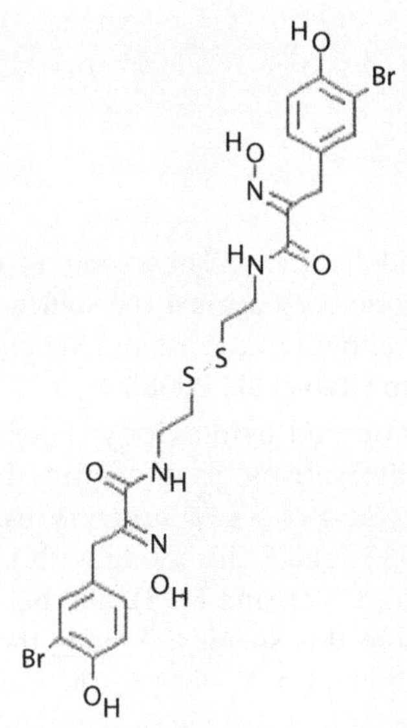

Psammaplin A

The extract of this species showed anticancer activity against A549, HeLa, and HT-29 cancer cell lines with the following GI50 values.

GI50 (μg/mL) Values of Cancer Cell Lines

A549	HeLa	HT-29
10.1	11.9	15.8

Source: Sponge crude extraction and primary screening, n.d, http://shodhganga.inflibnet.ac.in/bitstream/10603/64131/10/10_chapter%205.pdf.

The compound bastadin isolated from this species was foud to be mildly cytotoxic against several cancer cell lines and inhibited the enzymes topoisomerase II (inhibitor in cancer therapy) and dehydrofolate reductase (inhibit DNA synthesis and cell proliferation) (Carney et al., 1993).

Antitumor/cytotoxic activity: The N-methylated cyclic depsipeptides destruxins A, B, and B2, and desmethyl B, E, and E2 chlorohydrin isolated from the extracts of the associated *Metarrhizium* sp. 001103 of this sponge species displayed selective inhibition of human tumor cell lines. E2 chlorohydrin showed cytotoxicity toward the murine c38 cell line (Thomas et al., 2010).

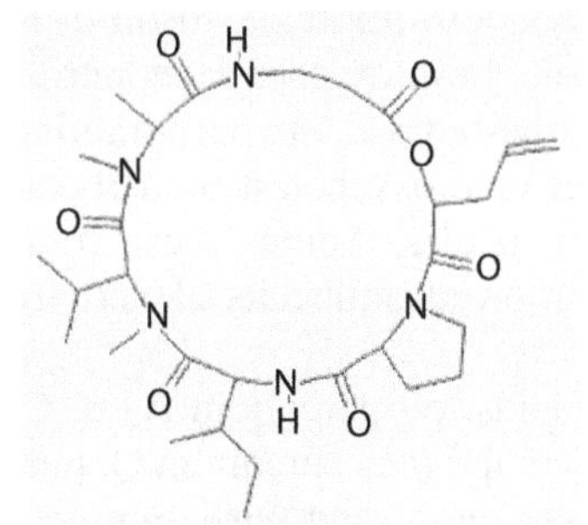

Destruxin A

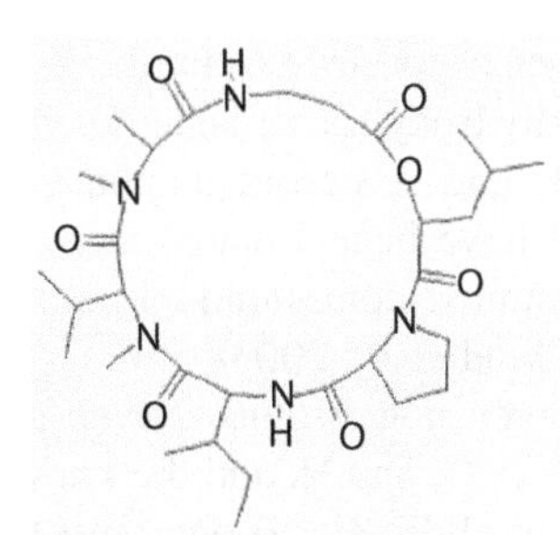

Destruxin B

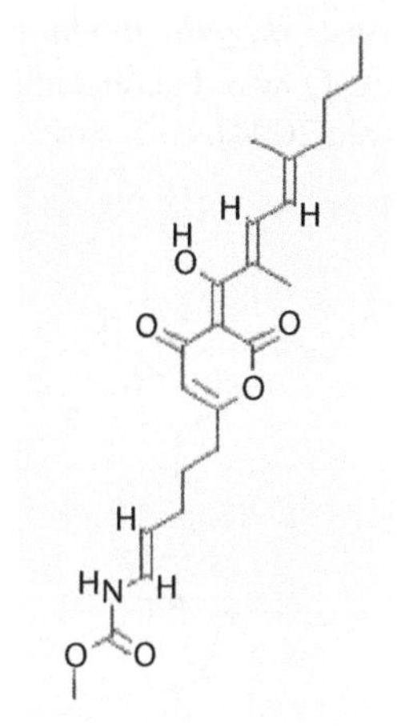

Desmethyl B

Antiviral activity: The mololipids of this species have shown antiviral activity against HIV (Braga, 2014).

Antimicrobial activity: The bromotyrosine-cysteine derivative psammaplin B of this species showed antimicrobial activity against Gram-positive *Staphylococcus aureus* and mild tyrosine kinase inhibition potency (IC50 = 2800 μM) (Emsermann et al., 2016).

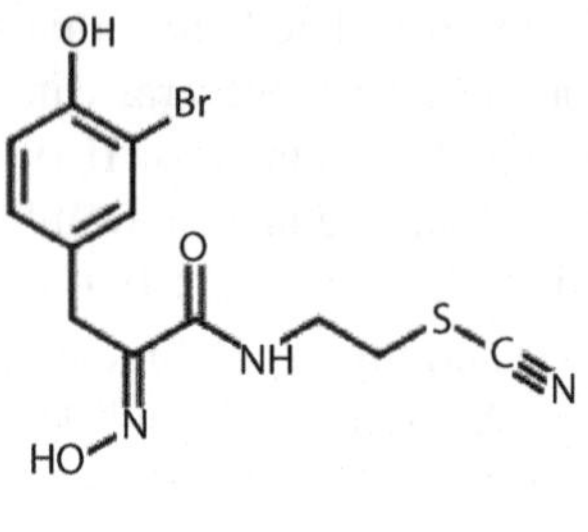

Psammaplin B

Antibacterial activity: Two new bromotyrosine-derived metabolites (1, 2) along with the known compounds 3,5-dibromo-4-methoxyphenylacetonitrile, 3-bromo-4-methoxyphenylacetonitrile, 3-bromo-4-hydroxyphenylacetonitrile, 1-hydroxyuracil, 1-methoxyhemibastadin 2, and purpuramine H, and a steroid 5alpha,8alpha-epidioxycholest-6-en-3beta-ol have been isolated from this species. Among these compounds, compounds 1 and 2 showed antibacterial activity (Goud et al., 2003).

Seven new bromotyrosine alkaloids, purpurealidin A, B, C, D, F, G, and H, and the known compounds purealidin Q, purpurealidin E, 16-debromoaplysamine-4, and purpuramine I have been isolated from this species. Among them, purpurealidin B, 16-debromoaplysamine-4, and purpuramine I exhibited *in vitro* antimicrobial activities against *E. coli*, *S. aureus*, and *V. cholerae*. In addition, purpurealidin B and 16-debromoaplysamine-4 were also active against *Shigella flexineri* and *Salmonella typhi*, while Purealidin Q was bactericidal only against *Salmonella typhi* (Tilvi et al., 2004).

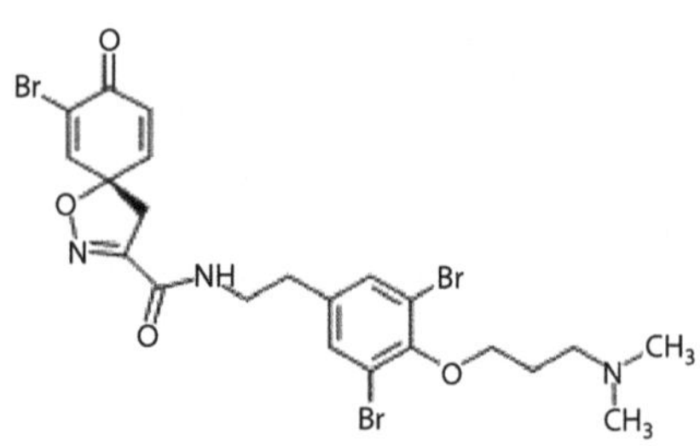

Purpurealidin B

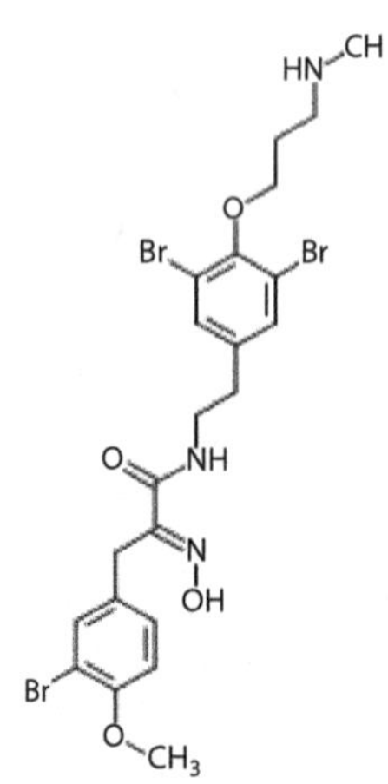

Purpuramine I

Gotsbacher and Karuso (2015) reported that the compound aplysamine 8 displayed notable antibacterial activity against *E. coli* (MIC, 125 μg/mL) and *S. aureus* (MIC, 31 μg/mL). Further, other compounds of this species, namely, hexadellin, aplysamine 2, and 16-debromoaplysamine 4 had mild activity against *S. aureus* (MIC, 125–250 μg/mL).

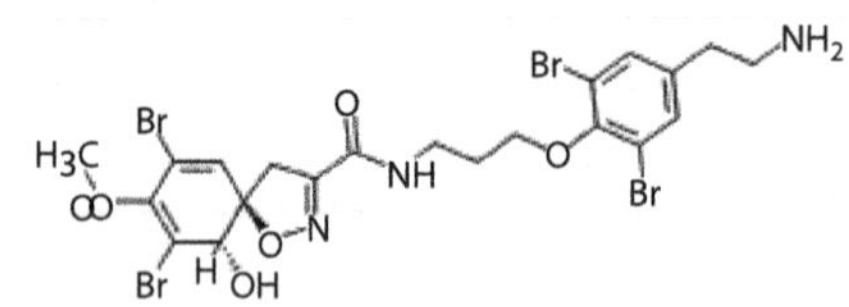

Hexadellin

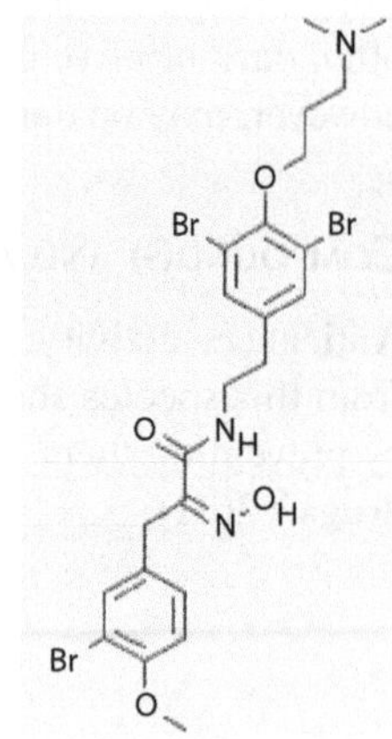

Aplysamine 2

Mosquito larvicidal activity: The extract of this species displayed larvicidal activity against the fourth instar larvae of *Aedes aegypti* (yellow fever mosquito vector) with an LC50 value of 25.9 ppm (Rao et al., 2008).

Enzyme inhibition and cytotoxicity: Four novel bisulfide bromotyrosine derivatives, psammaplins E (9), F (10), G (11), and H (12), and two new bromotyrosine derivatives, psammaplins I (13) and J (14) along with known psammaplins A (4), B (6), C (7), and D (8), and bisaprasin (5) have been isolated from this species. Among these compounds, compounds psammaplin A, bisaprasin, and psammaplin F were found to be potent histone deacetylase inhibitors and also showed mild cytotoxicity. Furthermore, compounds 4, 5, and 11 were potent DNA methyltransferase inhibitors (Piña et al., 2003).

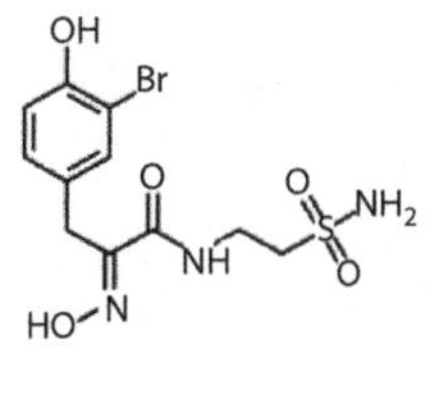

Psammaplin C

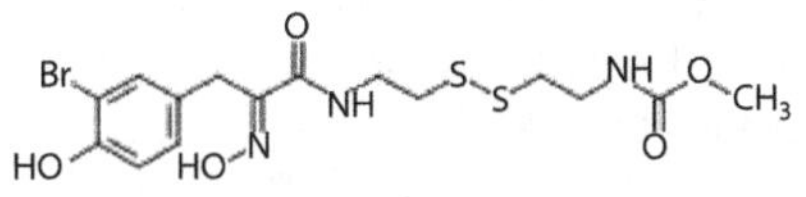

Psammaplin D

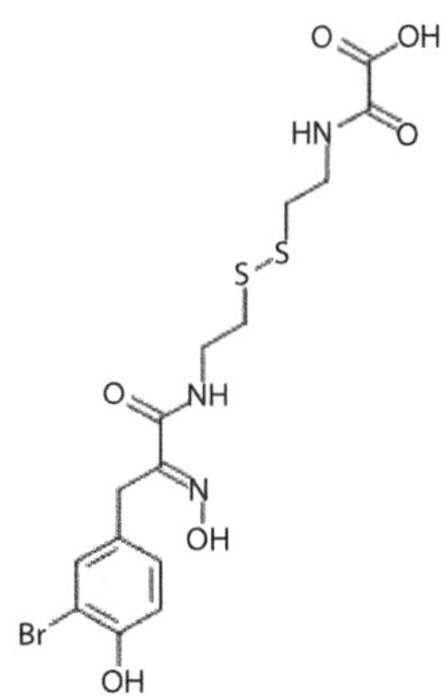

Psammaplin F

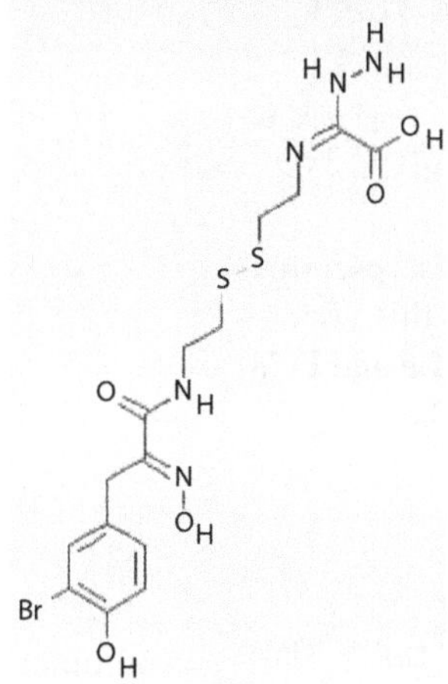

Psammaplin G

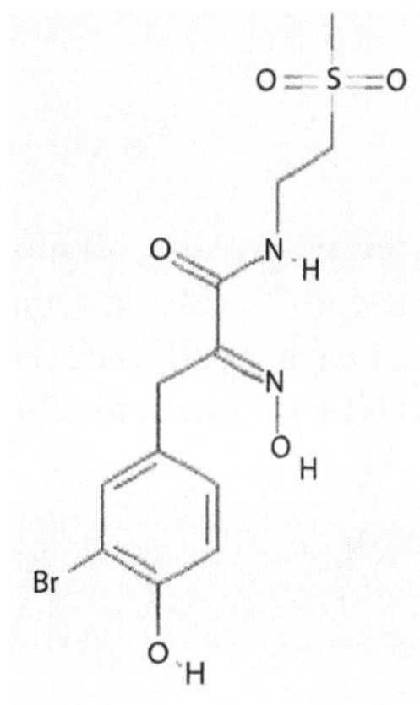

Psammaplin I

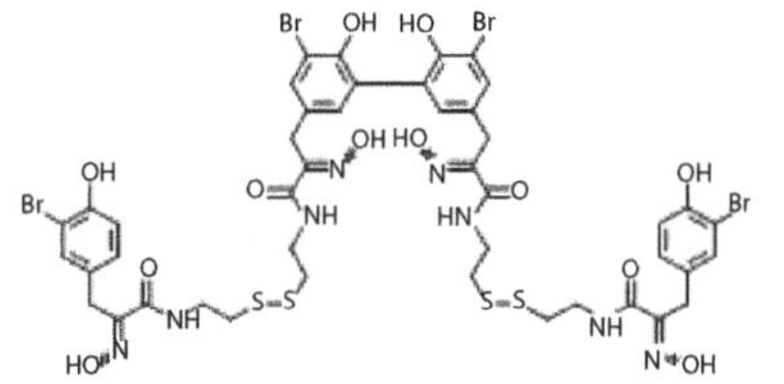

Bisaprasin

Pseudoceratina verrucosa (Bergquist, 1995)

Order: Verongiida Family: Pseudoceratinidae
Common name: Butter pad sponge

Distribution: Eastern Philippines, New Caledonia, Micronesia

Ecology: It occurs in the near-shore areas as a benthic, sessile species at depths of 15 to 18 m in a steep reef front with abundant corals.

Biology: This sponge is massive and repent (fragment: 60 mm long, 30 mm wide, and 30 mm thick) with thick branches. Surface is uneven and verrucose containing small, rounded projections (1–2 mm high). Texture is hard and rubbery and is incompressible. Its color is yellowish brown with greenish patches when out of water. In preservative, it is deep purple almost black. There is a dendritic arrangement of irregular fibers that are composed of pith elements. A large portion of the deeper region of the choanosome is devoid of skeleton. The ectosome contains a layer of collagen. The high growth potential and high survival rate of this sponge species suggest that it is a promising candidate for mariculture development (de Voogd, 2007).

Compound(s) and Activities

Antifungal activity: The bromotyrosine alkaloids, ceratinadins A and B, isolated from this species showed antifungal activity against *C. neoformans* and *C. albicans* (Cheung et al., 2014).

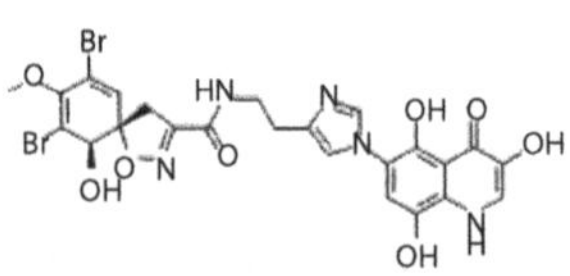
Ceratinadin A

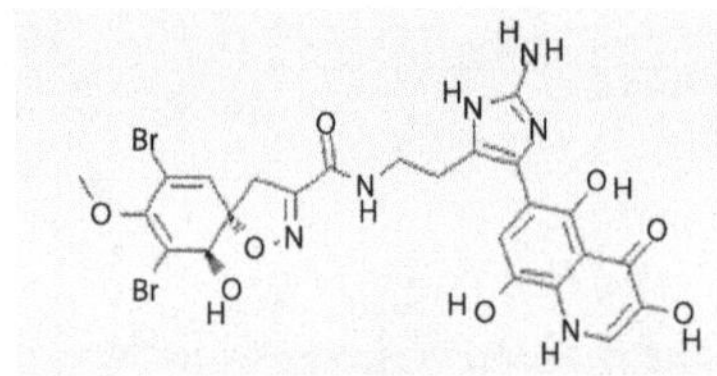
Ceratinadin B

Cytotoxicity: Two new bromotyrosine alkaloids, pseudoceralidinone A and aplysamine 7, isolated from this species showed cytotoxicity against cancer cell lines (HeLa and PC3) and a noncancer cell line (NFF) (Tran et al., 2013).

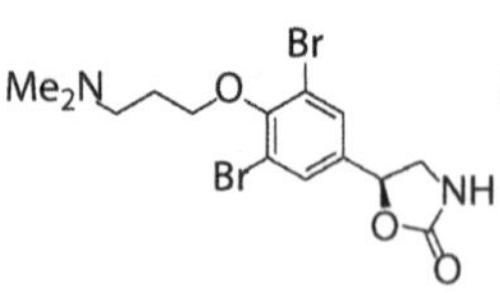
Pseudoceralidinone A

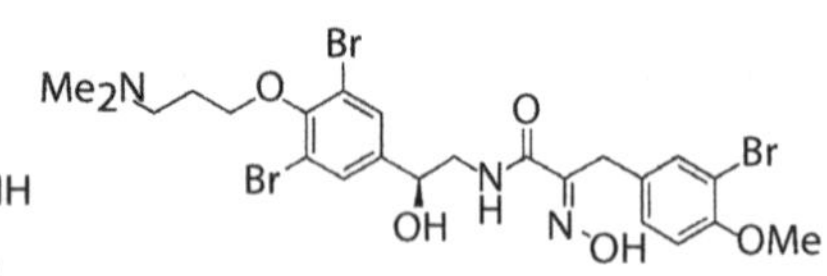
Aplysamine 7

Antitumor activity: The bromotyrosine compound aplysamine 7 of this species showed cytotoxicity toward PC3 (prostate) cancer cells with an IC50 value of 4.9 μM (Gribble, 2015).

Pseudoceratina sp.

Order: Verongiida Family: Pseudoceratinidae

Compound(s) and Activities

Antibacterial activities: The compound, pseudoceramine B of this species inhibited the bacterial growth with IC50 40 μM. Further, the bromotyrosine compound purealin C showed a broad spectrum of activity against two strains each of the Gram-positive *S. aureus* (IC50 2.6 and 6.2 μM) and *B. subtilis* (IC50 2.6 and 2.8 μM), while (–)-purealin B was only active against *B. subtilis* (IC50 3.4 and 3.8 μM) (Gribble, 2015).

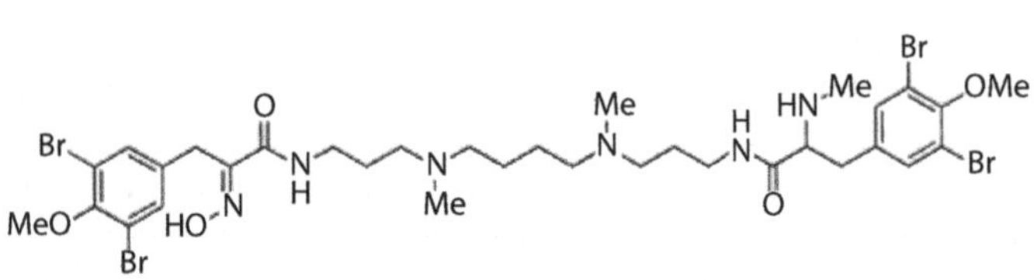
Pseudoceramine B

Antiparasitic activity: The compounds psammaplysin H (1), G (2), and F (3) of this species displayed potent *in vitro* antimalarial activity against the chloroquine-sensitive (3D7) line of *Plasmodium falciparum* (IC50 μM, 0.4, 5.2, and 1.9, respectively) and minimal toxicity toward the mammalian cell lines HEK293 and HepG2 (Gribble, 2015).

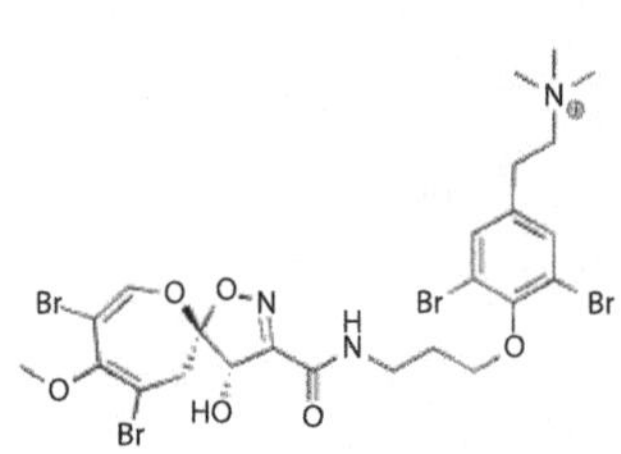
Psammaplysin H

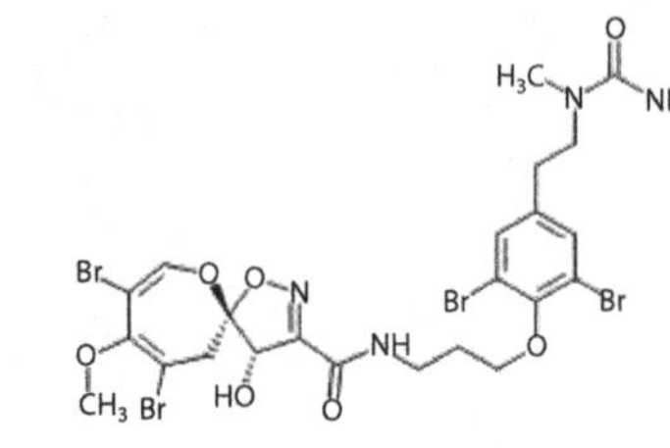
Psammaplysin G

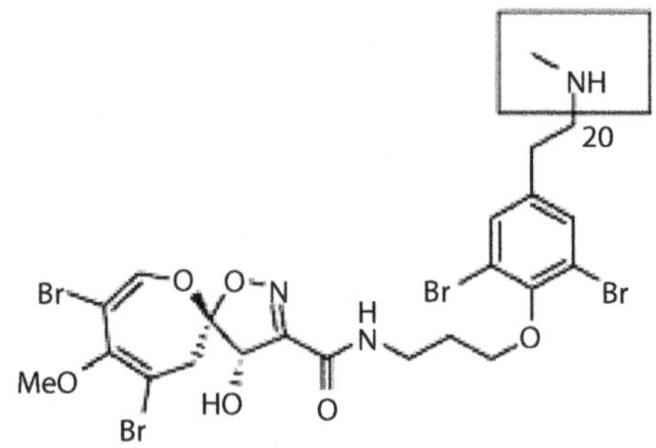
Psammaplysin F

Antitumor activity: The compound dibromo-dihydroxyoxocyclohexenyl acetonitrile of this species had excellent activity against the K562 leukemia cell line (IC50, 1.4 μg/mL) (Gribble, 2015).

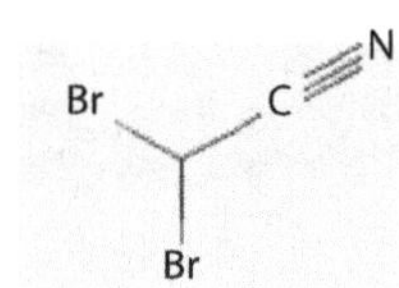
Dibromoacetonitrile

FAMILY: CHONDRILLIDAE

Chondrilla nucula (Schmidt, 1862)

Order: Chondrillida Family: Chondrillidae

Common name: Chicken liver sponge, potato sponge

Distribution: Throughout the Caribbean and Mediterranean

Ecology: It occurs on a wide variety of substrates including reefs and mangrove roots. It is extremely common in shallow inshore waters and can survive in dark habitats such as dark caves and illuminated habitats such as reefs at any depth between 0.5 and 5 m. It is also known to be an aggressive species and overgrows many other organisms, commonly sessile corals.

Biology: It is an encrusting species appearing as a thin sheet over the substrate, but can also be found in globular morphology as well. It can grow up to 15 cm. This sponge occurs in two distinct types, which are also separated by color and habitat. Type 1 is amorphous, with globular lobes, or thickly incrusting, up to about 1 cm thick, spreading horizontally, with pronounced, deeply incised and lacunose, meandering lobes. This type never covers large areas. The oscules are inconspicuous, few, and 0.5 to 1.5 mm in diameter. The surface is smooth, glabrous, and shiny. The consistency is soft, elastic, and tough. Type 2 is thinly encrusting on rocky substrates, 1 to 2 mm thick, with incised margins and lacunae, but cover larger, more continuous areas than type 1. The oscules in this type are conspicuous, more numerous than in type 1, up to 3 mm wide, and are usually slightly elevated. The surface is smooth, glabrous, and shiny. The consistency of this type is somewhat tougher and more cartilaginous than in type 1. The upper surface of type 1 is dark brown to walnut brown or beige; often mottled. The base and inner tissue are light beige to buff. Specimens growing in dark areas may be pale. The color of type 2 is commonly yellowish brown, fulvous, with dark-brown rims. These sponges have toxins that deter fish predation and allow them to compete for space on reefs.

Food and feeding: *Chondrilla nucula* is capable of retaining 80% of suspended particles in the surrounding water column within 24 hours. It is considered to be an extremely efficient feeder because of its high retention rate of food particles.

Reproduction: Fertilization takes place internally. Regeneration rates are also very fast in this species so that they can recover easily.

Associated life: It plays host to a variety of endosymbiotic bacteria, which are distinct to the sponge and very different from the bacteria found in the surrounding water column.

Commercial importance: This species is considered important in bioremediation of seawater in many aquaculture situations. By utilizing this species, aquaculture can maintain seawater that is generally devoid of particulate matter, which this species absorbs as food.

Compound(s) and Activities

Antibacterial activity: The strains of bacteria associated with this sponge species exhibited antibacterial activity. The percent values of inhibition are given next

Staphylococcus aureus	58
Bacillus subtilis	7
Enterococcus faecalis	12
Escherichia coli	12
Pseudoalteromonas atlantica	38
Pseudomonas elongate	33

Source: Chelossi, E. et al., 2007, *Aquat. Microb. Ecol.*, 49: 157–163.

Antiviral activity: The lectin isolated from this species displayed a moderate mitogenic effect on spleen lymphocytes from mice and on CD4-positive human H9 cells. Further, this lectin has the ability to stimulate the (2′-5′)oligoriboadenylate [(2′-5′)A] metabolic pathway in noninfected and human immunodeficiency virus (HIV-1)–infected H9 cells. Moreover, HIV-1–infected H9 cells showed an increase in the time period of HIV-1 release in response to the lectin treatment. That is, the lectin of this species caused a retardation of HIV-1 release due to further increase of the elevated intracellular level of (2′-5′)A (Schroder et al., 1990).

FAMILY: CHONDROSIIDAE

Chondrosia reniformis (Nardo, 1847)

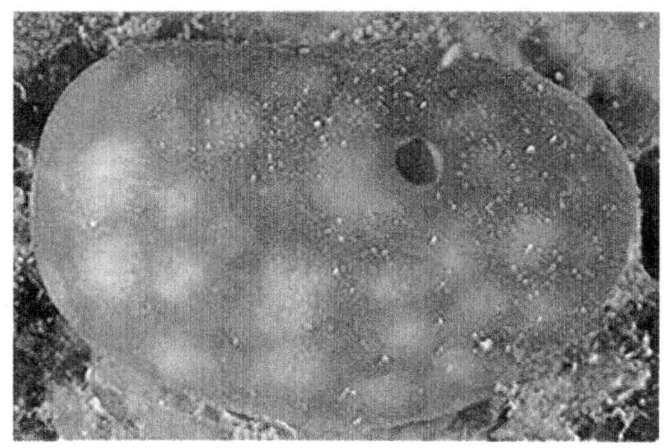

Order: Chondrosiida Family: Chondrosiidae

Common name: Kidney-shaped sponge

Distribution: Australia, Indian Ocean, North Atlantic Ocean, Portugal, Mediterranean

Ecology: It grows in shallow water, exposed habitats.

Biology: It is a massive-lobate species, forming rounded knolls or elongately rounded shapes. Size is up to 5 cm high, 10 or more cm in lateral expansion. Surface is smooth, glistening, and slippery. Oscules are simple, with slightly elevated rims. Consistency of the sponge is cartilaginous. Spicules are absent. Color of ther animal is mottled brown violet to gray, darker on the upper parts and lighter on the sides.

Compound(s) and Activities

Anti-infective agents: The compounds derived from the associated fungus, *Penicillium rugulosum*, namely, polyketides and prugosenes A1–A3, B1, B2, C1, and C2 can be used as templates for new anti-infectives (Thomas et al., 2010).

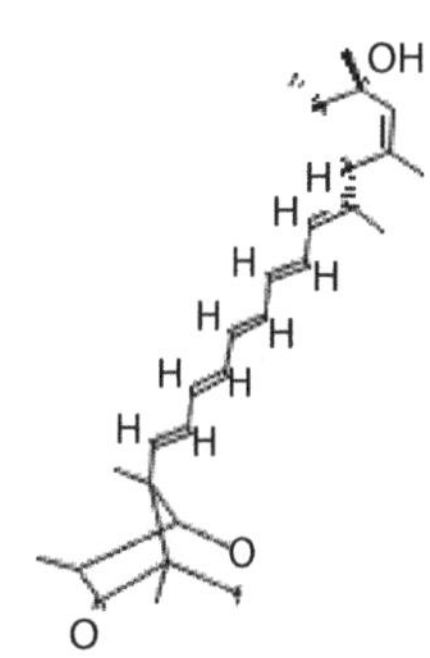

Prugosen A1

Others: The extracts of this species were found to be composed of highly pure mixtures of collagen and gelatin with similar properties to collagen isolated from other marine sources. These compounds may be of great use in biomedical applications (Silva et al., 2016).

FAMILY: PLAKINIDAE

Corticium niger (Pulitzer-Finali, 1996)

Class: Homoscleromorpha Subclass: Homoscleromorpha

Order: Homosclerophorida Family: Plakinidae

Common name: Not designated

Distribution: Bismarck Sea

Ecology: Not reported

Biology: This sponge is cushion shaped and roundish with about 5 cm in diameter and 1 cm thickness. Color of the animal is black. Consistency is tough and scarcely it is resilient.

Compound(s) and Activities

Cytotoxicity: Two new steroidal alkaloids, plakinamines N (1) and O (2), along with two known compounds of the plakinamine class (3, 4) have been isolated from this species. Of these compounds, plakinamines N (1), O (2), and J (4) showed enhanced inhibitory effects against all of the colon cell lines mean GI50 values of 11.5, 2.4, and 1.4 μM, respectively (Sunassee et al., 2014).

Four new steroidal alkaloids, plakinamine I–K (1–3) and dihydroplakinamine K (4), isolated from this species exhibited significant *in vitro* cytotoxicity (Ridley and Faulkner, 2003).

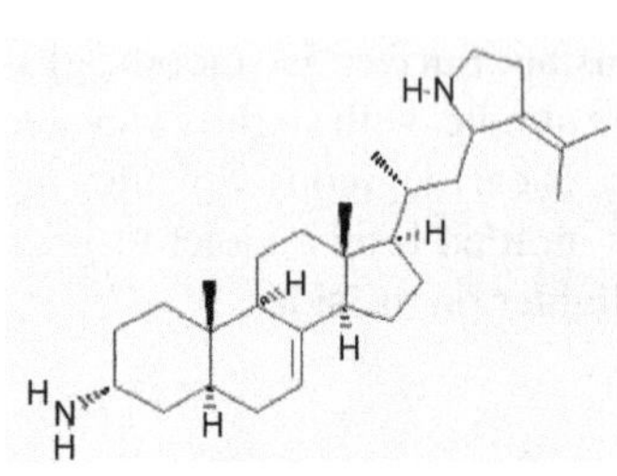

Plakinamine N

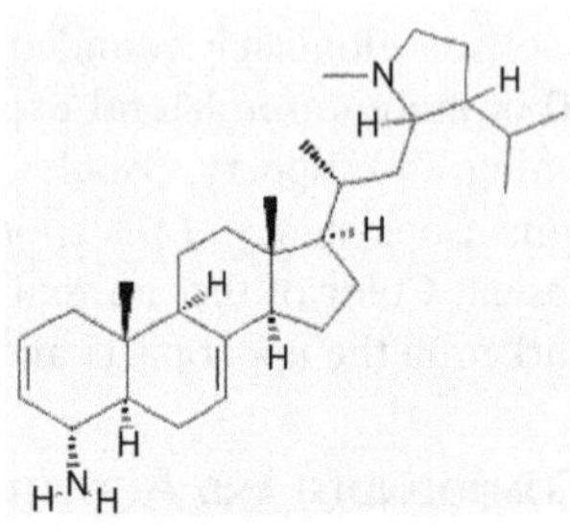

Plakinamine J

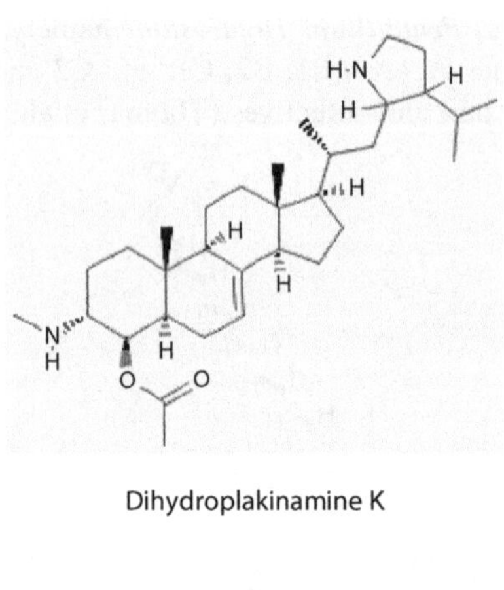

Dihydroplakinamine K

Corticium simplex (Lendenfeld, 1907)

Order: Homosclerophorida Family: Plakinidae

Common name: Indonesian marine sponge

Distribution: Western Australia

Ecology: It occurs in coral reef habitats.

Biology: Not studied

Compound(s) and Activities

Cytostatic antiproliferative activity: Of the steroidal alkaloids, cortistatin J, cortistatin K, and cortistatin L, isolated from this species, the compound, cortistatin J showed cytostatic antiproliferative activity against human umbilical vein endothelial cells (HUVECs) with an IC50 value of 8 nM (Putra and Jaswir, 2014).

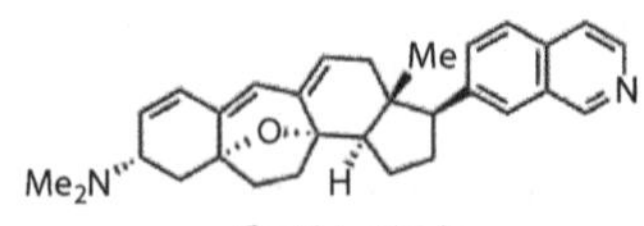

Cortistatin J

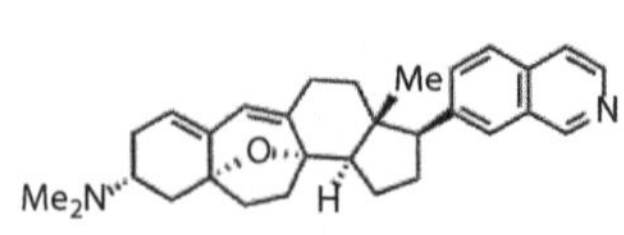

Cortistatin K

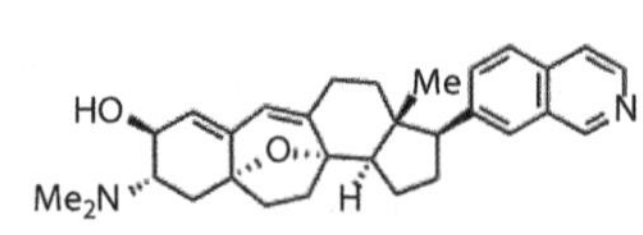

Cortistatin L

Anti-angiogenic and antitumor activities: The steroidal alkaloids, named cortistatins A, B, C, and D, isolated from this species served as specific inhibitors of angiogenesis and promising antitumor agents. Furher, these compounds inhibited the proliferation of HUVECs with high selectivity. Interestingly, cortistatin A showed the strongest antiproliferative activity (IC50 = 0.002 muM) against HUVECs, in which the selective index was more than 3000-fold (Aoki et al., 2006).

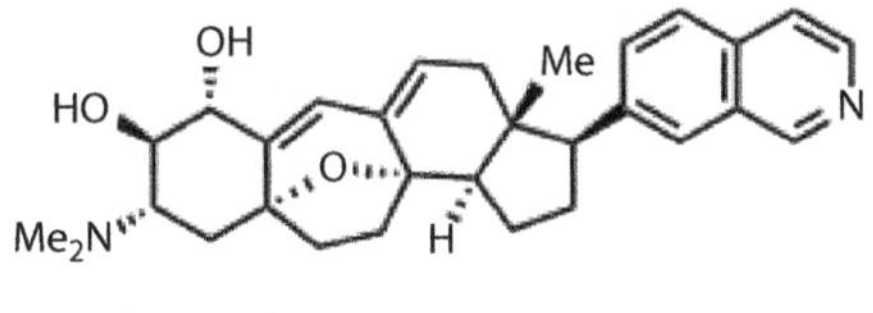

Cortistatin A

Corticium sp.

Order: Homosclerophorida Family: Plakinidae

Compound(s) and Activities

Neuroprotective activity: The alkaloid, 4-acetoxy-plakinamine B isolated from this species significantly inhibited acetylcholinesterase with an IC50 value of 3.8 μM. The acetylcholinesterase, as an inflammatory component, is the causative factor to the pathology of neurodegeneration and associated Alzheimer's disease, Parkinson's disease, and motor neuron disease (Grosso et al., 2014; Senthilkumar and Kim, 2013).

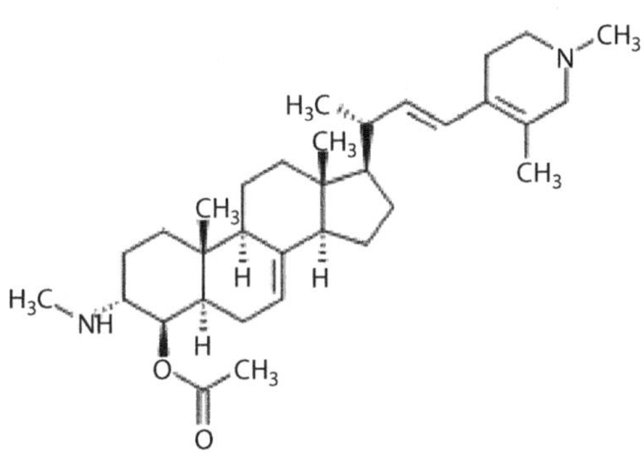

4-acetoxy-plakinamine B

Plakinastrella clathrata (Kirkpatrick, 1900)

Order: Homosclerophorida Family: Plakinidae

Common name: Australian marine sponge

Distribution: South Tropical Pacific

Ecology: It is a common fore-reef sponge; also in exposed habitats.

Biology: *Plakinastrella clathrata* differs from other related species in the overall larger size of all spicule categories

Compound(s) and Activities

Anti-inflammatory activity: The glycolipid 1,6-anhydropyranose isolated from this sponge species exhibited moderate anti-inflammatory activity (Katavic et al., 2013).

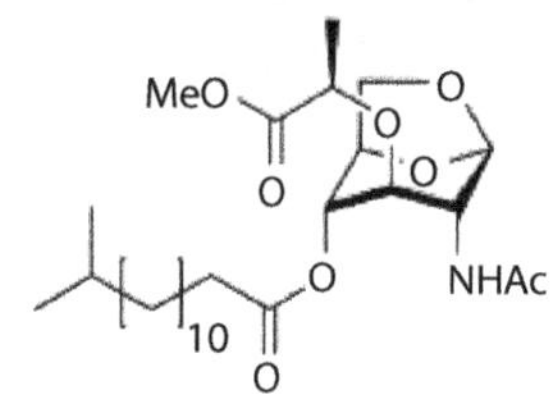

1, 6-anhydropyranose

Plakinastrella mamillaris (Lendenfeld, 1907)

Order: Homosclerophorida Family: Plakinidae

Common name: Not designated

Distribution: Western Australia

Ecology: Not reported

Biology: Not available

Compound(s) and Activities

Anti-inflammatroy activity: The compounds gracilioether B and plakilactone C isolated from this species showed anti-inflammatroy activity by regulating the expression of PPARγ-dependent genes in the liver and inhibited the generation of inflammatory mediators by macrophages (Festa et al., 2012).

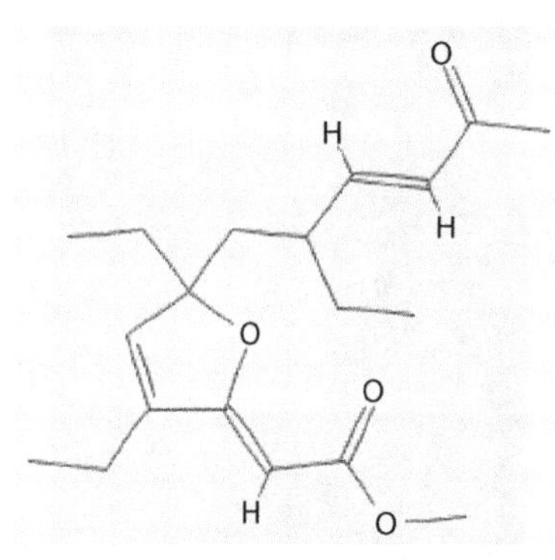

Gracilioether B

Antiplasmodial activity: Among the four new polyketide endoperoxides, plakortides R–U, isolated from this species, plakortide U showed *in vitro* antiplasmodial activity against a chloroquine-resistant strain (Festa et al., 2013).

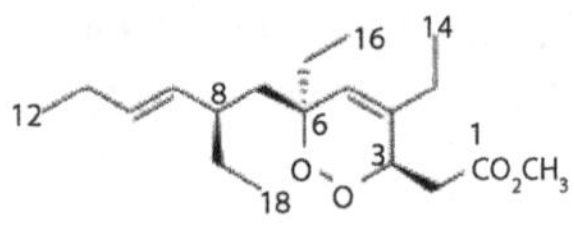

Plakortide R

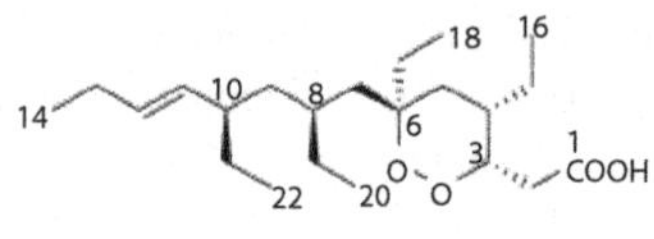

Plakortide S

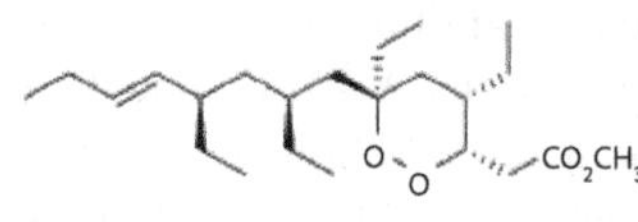

Plakortide T

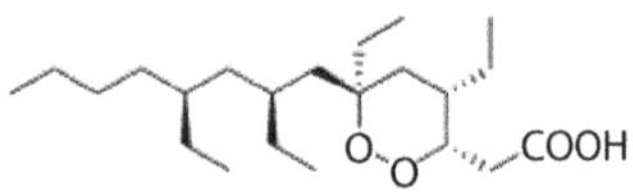

Plakortide U

Among the six new oxygenated polyketides, gracilioethers E–J, isolated from the extracts of this species, the compound gracilioether H was found to show *in vitro* antiplasmodial activity against a chloroquine resistant strain (Festa et al., 2012).

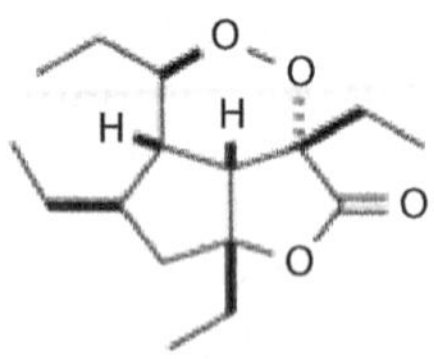

Gracilioether H

Plakinastrella onkodes (Uliczka, 1929)

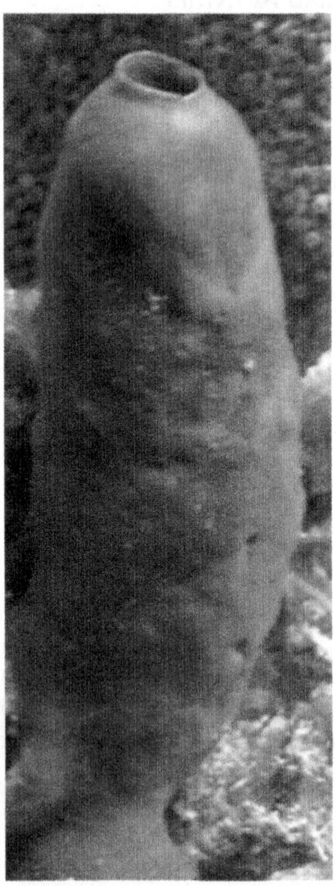

Order: Homosclerophorida Family: Plakinidae
Common name: Jamaican sponge

Distribution: Subtropical, Eastern Caribbean, Western Atlantic, Brazil, Belize, Colombia, and Panama

Ecology: This sessile sponge occurs at a depth of 9 to 11 m. It may also occur in mangrove ponds.

Biology: It is a thick encrusting to massive-lobate sponge. It is brown to gray black externally and tan internally. Some specimens show darker spots. Oscules flush with the smooth surface or are slightly raised. Surface is smooth, but it is slightly rough surface to the touch. It is only slightly compressible and is easy to tear.

Compound(s) and Activities

Antiprotozoal activity: The compound artemisinin analog isolated from the sponge showed antiprotozoal activity against *Toxoplasma gondii* with an IC50 value of 64 nM (Muraleedharan and Avery, n.d.).

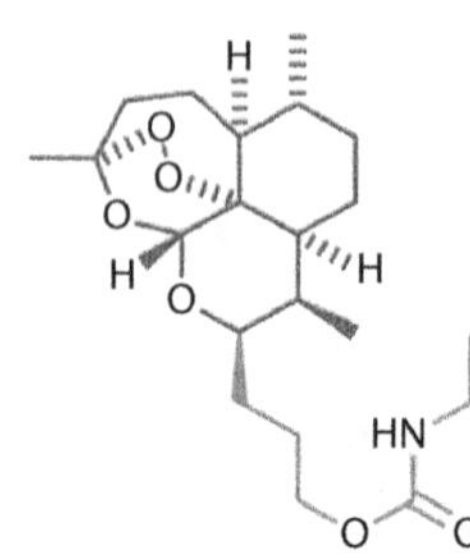

Artemisinin analog

Plakinastrella sp.

Order: Homosclerophorida Family: Plakinidae

Compound(s) and Activities

Antitumor activity: The alkylphenol compound elenic acid isolated from this species showed antitumor activity by inhibiting the enzyme topoisomerase II (Anjum et al., 2016; Sipkema et al., 2005).

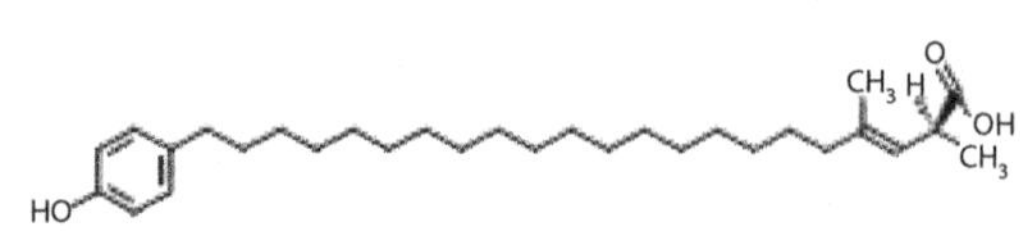

Elenic acid

Plakortis angulospiculatus (Carter, 1879)

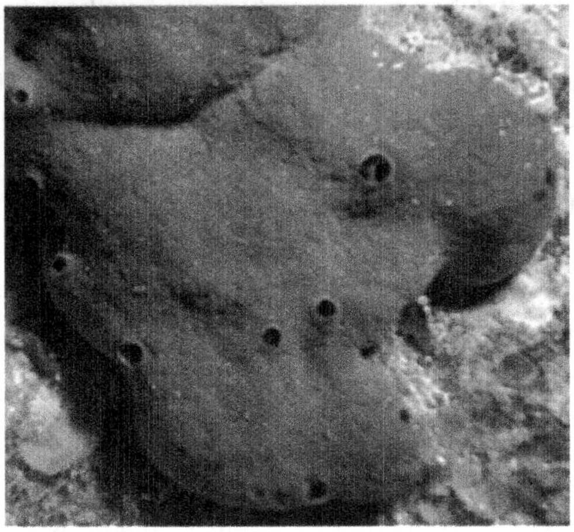

Order: Homosclerophorida Family: Plakinidae

Common name: Viscous sponge, Caribbean marine sponge, Brazilian sponge

Distribution: Florida, Bahamas, and Caribbean

Ecology: It inhabits reefs and walls. It often grows on faces of undercuts, walls of canyons, and in caves at depths from 10 to 30 m.

Biology: These sponges are 8 cm long by 3 cm thick, but they can reach up to 20 cm long in situ. Color *in vivo* is light brown, either homogeneous or with dark brown or greenish patches; sometimes with white dots. It is light brown to cream internally and in spirit. Surface is smooth and even. Oscules (2–10 mm dia.) are conspicuous, circular, and flush with the surface. Consistency is compressible and is easy to tear. Choanosome is compact and is with few cavities.

Compound(s) and Activities

Antitumor activity: The polyketide compound plakortide P isolated from this species showed antitumor activity by inhibiting thromboxane B2 (TxB2) (Anjum et al., 2016).

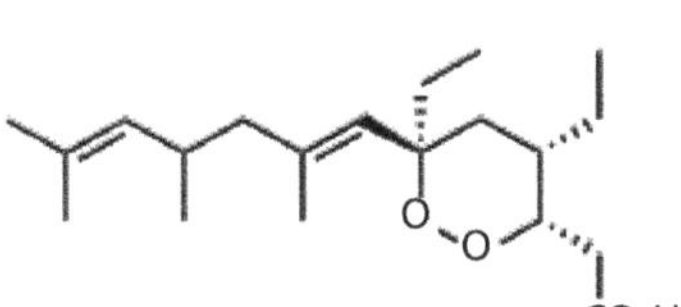

Plakortide P

Cytotoxicity: Of 10 compounds isolated from this species, 6 compounds (a plakortide and other natural products) displayed cytotoxic activities with IC50 values ranging from 0.2 to 10 μM (Santos et al., 2015).

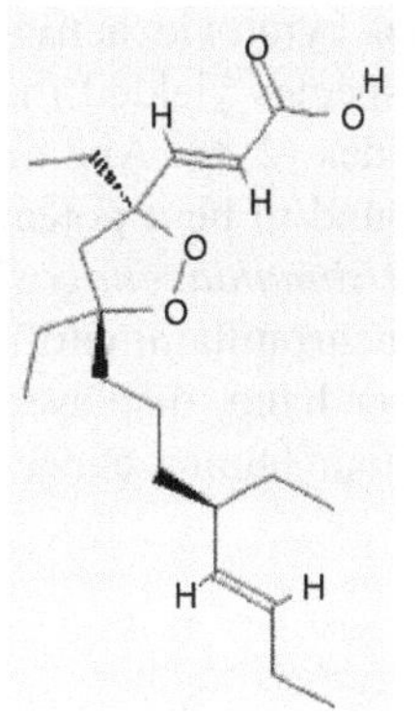

7, 8-dihydroplakortide E

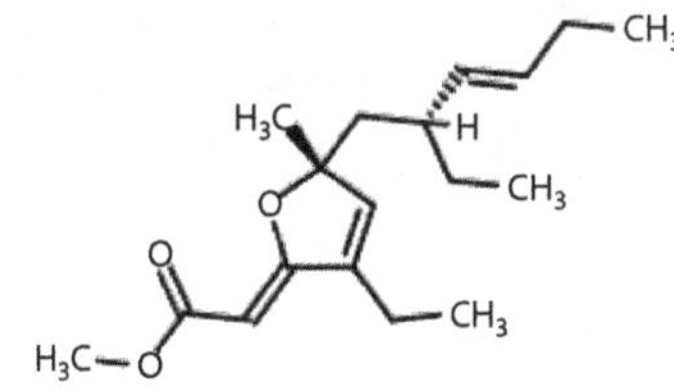

Spongosoritin A

Among the two polyketides, spiculoic acids A and B, isolated from this speices, spiculoic acid A showed *in vitro* cytotoxicity against human breast cancer MCF-7 cells (Huan et al., 2004).

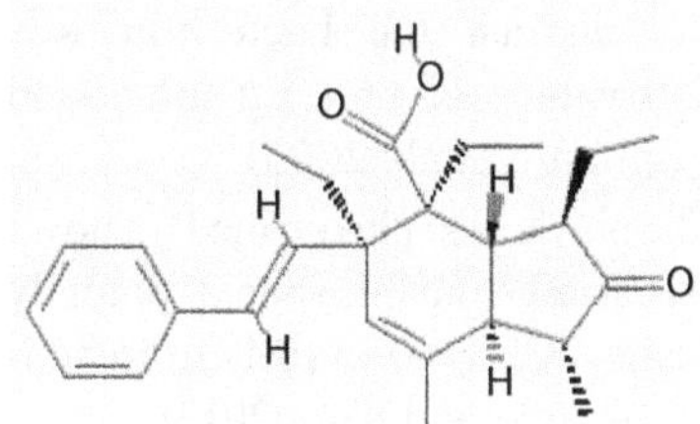

Spiculoic acid A

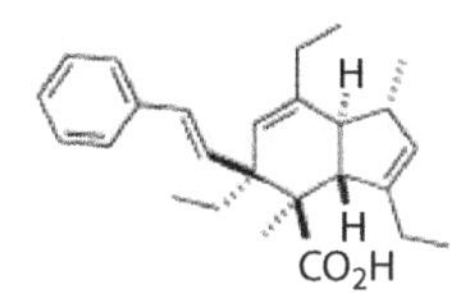

Spiculoic acid B

Antiprotozoal activity: The two cyclic peroxides isolated from this species showed strong *in vitro* antiproliferative effects on promastigotes of *Leishmania mexicana*, a flagellate protozoan that causes leishmaniasis (Compagnone et al., 1998).

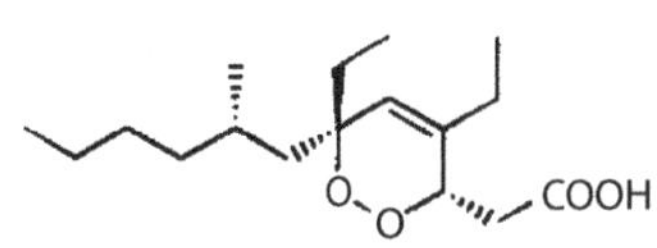

Peroxide 1

Antineuroinflammatory activity: The compound plakortide N isolated from this species showed antineuroinflammatory activity by potently inhibiting thromboxane B2 (TxB2) (IC50 = 0.93 μM) but with low lactic dehydrogenase release (LDH50 >10 μM). The thromboxane inhibitors have also been reported to prevent the progressive dermal ischemia associated with thermal injury (Mayer et al., 2008).

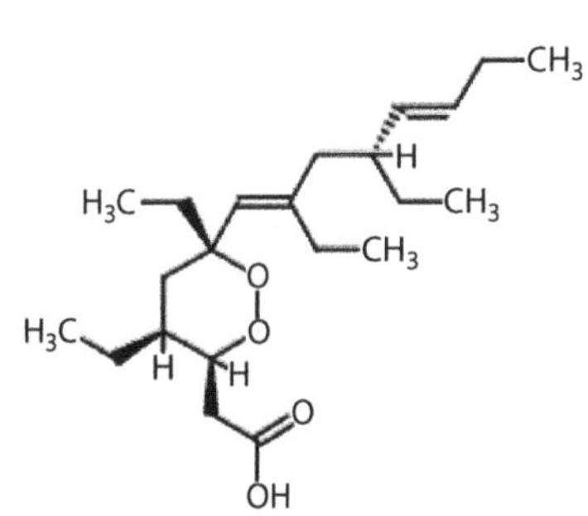

Plakortide N

Antiparasitic, antineuroinflammatory, and cytotoxic activities: The bioactive crude extract of this species yielded the compounds plakortenone (1) and polyketides (2–6). Among these compounds, plakortide P (3) was found to be a potent antiparasitic compound, against both *Leishmania chagasi* and *Trypanosona cruzi*, and exhibited antineuroinflammatory activity. The polyketides 2–6, on the other hand, displayed only moderate cytotoxic activity against four human cancer cell lines (Kossuga et al., 2008).

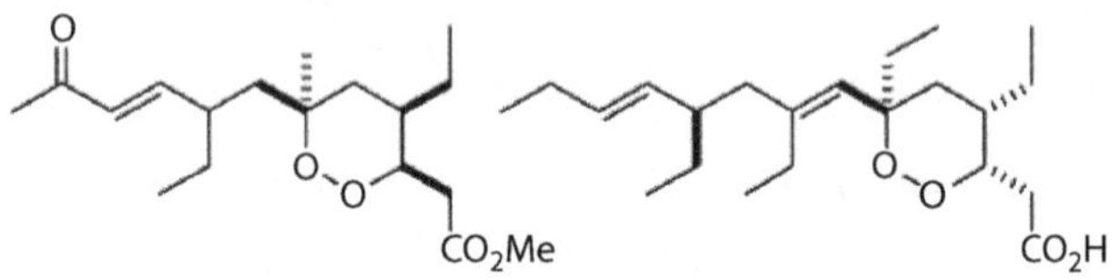

Plakortenone Plakortide

Antibacterial activity: Two crude extracts of this species showed antibacterial activity against the bacterium was *Leucothrix mucor* with LC50 values of 66.1 μg mL^{-1} and 174.2 μg mL^{-1}, respectively (Epifanio et al., 2005).

Anti-inflammatory activity: The polyketide plakortide P (a novel antineuroinflammatory agent) isolated from this species potently inhibited thromboxane B2 release (IC50 = 0.93 μM) from activated rat brain microglia (Senthilkumar and Kim, 2013).

Plakortis halichondrioides (Wilson, 1902) (= *Petrosia halichondrioides*)

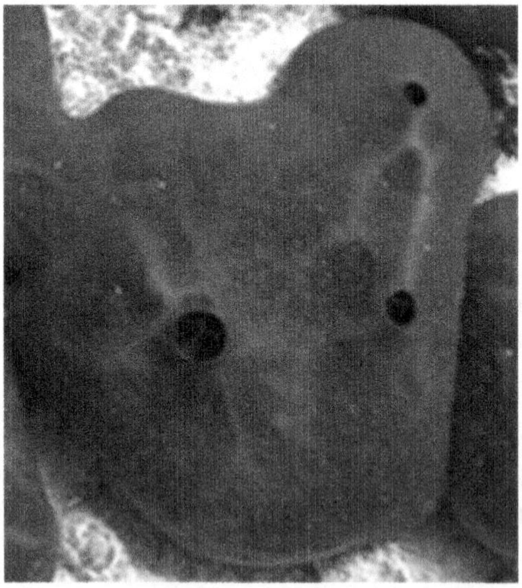

Order: Homosclerophorida Family: Plakinidae

Common name: Caribbean sponge

Distribution: Bocas del Toro region, Panama

Ecology: It occurs in shallow marine habitats.

Biology: It is a massively encrusting sponge that is up to 22 cm long and 1.5 cm thick. Color of the animal is black or dark brown externally and internally. It releases a dark exudate in spirit. Surface is even and slightly pilose. Consistency is very dense and firm, though not hard. The sponge becomes hard and brittle upon drying. Dermal membrane is indistinctly differentiated from subjacent tissue. It presents significantly longer spicules. But they are so loosely packed, without perceptible spongin. Oxea of the spicules are smooth and slightly curved. They are very commonly somewhat irregularly bent in the middle, or sometimes with a slight prominence at that point.

Compound(s) and Activities

Antiplasmodial activity: The compounds artemisinin analogs R = H of the methyl ester extract of this species showed antiplasmodial activity against *P. falciparum* with an IC50 value of 8 μg/mL (Muraleedharan and Avery, n.d.).

Plakortis nigra (Lévi, 1953)

Order: Homosclerophorida Family: Plakinidae

Common name: Indonesian marine sponge

Distribution: Occurs worldwide, including Indo-West Pacific, East African Coral Coast, Southern Red Sea, Mauritius

Ecology: This sessile, deepwater sponge may occur at a depth from 28 m.

Biology: This thickly encrusting sponge has firm consistency and black color in life. Surface is smooth or porous.

Compound(s) and Activities

Antibacterial and antioxidant activities: The different fractions of this species showed antibactierial activity with MIC values ranging from 0.10 to 0.21 mg/mL.

MIC of *P. nigra* Extracts and Positive Control against Test Organisms

	MIC (mg/mL)		
Sample	***E. coli***	***S. aureus***	***P. mirabilis***
Crude extract	2.46	2.46	ND
Hexane fraction	1.28	1.28	1.28
EtOAc fraction	0.10	0.21	0.21
BuOH fraction	0.11	0.21	0.21

Source: Tangnam et al., 2015. *JCLM*, 3: 44–51.

Further, the hexane fractions showed the highest DPPH scavenging activity with the value of 27.5% followed by the crude extract (16.6%), while the EtOAc and BuOH fractions exhibited lower free radical scavenging activity (12.9% and 4.17%), respectively (Tangman et al., 2015).

Anticancer activity: Four beta-carbolines, plakortamines A–D, two cyclic peroxides, epiplakinic acids G and H, and two related gamma-lactones, (2S,4R)- and (2R,4R)-2,4-dimethyl-4-hydroxy-16-phenylhexadecanoic acid 1,4-lactones, have been isolated from this species. The structures of the eight new metabolites were elucidated by interpretation of spectroscopic data. Most of these metabolites inhibited the HCT-116 human colon tumor cell line (Sandler et al., 2002).

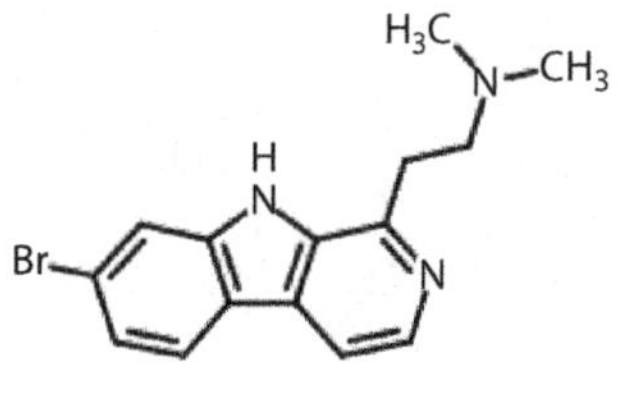

Plakortamine A

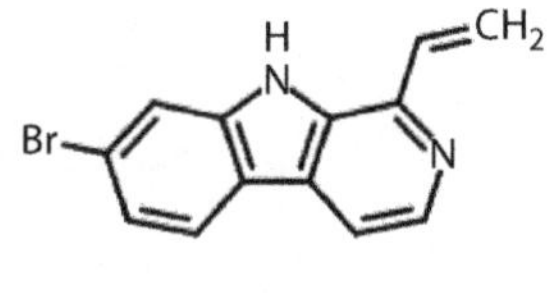

Plakortamine B

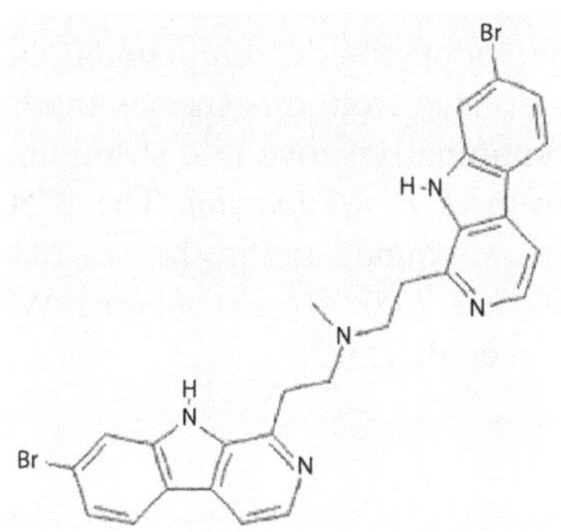

Plakortamine C

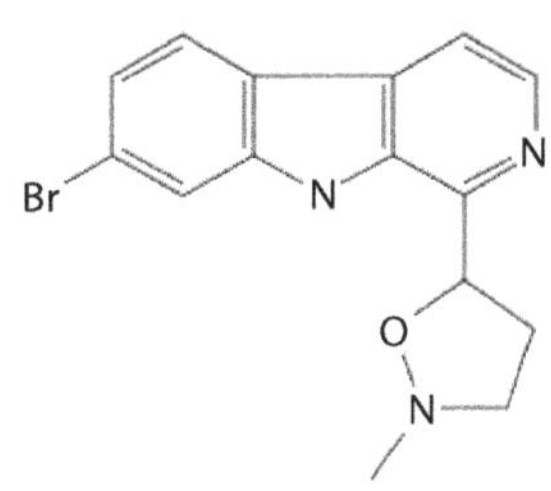

Plakortamine D

Two new cancer cell growth-inhibiting cyclic peroxides, plakorstatins 1 and 2, isolated from this species exhibited moderate cancer cell growth inhibition against the murine P388 lymphocytic leukemia cell line with ED50 values of 1.1 and 0.91 μg/mL, respectively (Pettit et al., 2004).

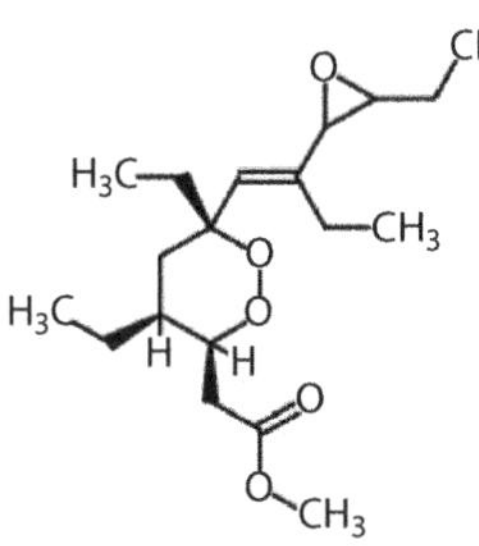

Plakorstatin 1

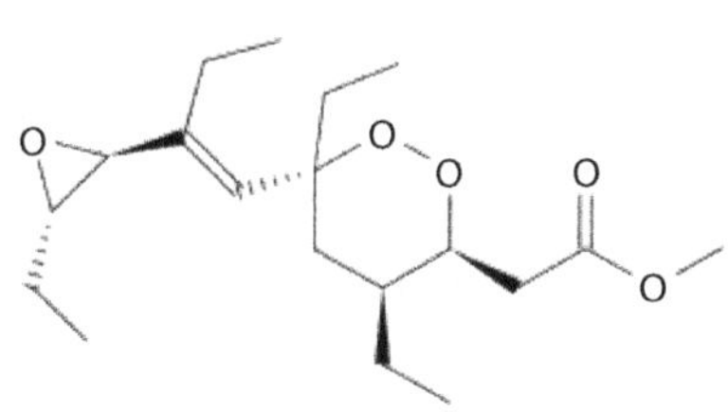

Plakorstatin 2

Plakortis simplex (Schulze, 1880)

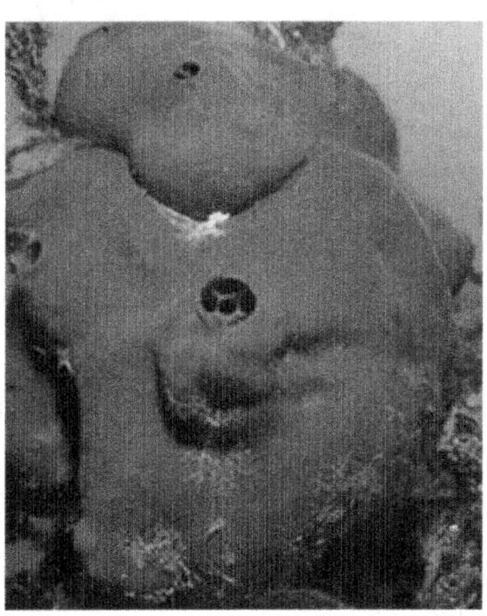

Order: Homosclerophorida Family: Plakinidae

Common name: South China Sea sponge

Distribution: North Atlantic Ocean, Indian Ocean, Kenya, Mozambique, Cape Verde

Ecology: It occurs on and under rocks and boulders from the littoral to at least 15 m depth.

Biology: It is a thinly encrusting sponge with 2 to 5 mm in thickness. Surface is smooth and consistency is compressible. The surface is regularly pierced by small punctiform inhalant openings. Oscules are about 1 mm in diameter and low in frequency. Consistency is compressible. Color of the animal is tan outside and whitish inside.

Compound(s) and Activities

Immunosuppressive activity: The glycolopids simplexides isolated from this species showed immunosuppressive activity by inhibiting T-cell proliferation (Anjum et al., 2016; Sipkema et al., 2005).

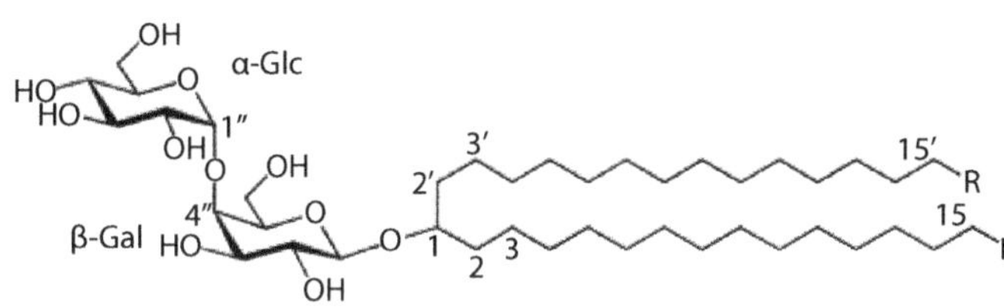

Simplexide

Antifungal, cytotoxic, and PTP1B inhibitory activities: Among the three new polyketides, woodylides A–C (1–3), isolated from the ethanol extract of this species, compounds 1 and 3 showed antifungal activity against fungi *Cryptococcus neoformans* with IC50 values of 3.67 and 10.85 μg/mL, respectively. In the cytotoxicity test, compound 1 exhibited moderate activity against the HeLa cell line with an IC50 value of 11.2 μg/mL, and compound 3 showed cytotoxic activity against the HCT-116 human colon tumor cell line and PTP1B inhibitory activity with IC50 values of 9.4 and 4.7 μg/mL, respectively (Yu et al., 2012).

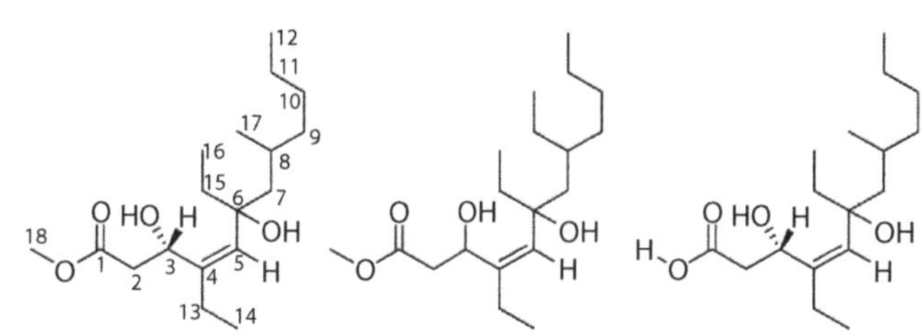

Woodylides A–C

Cytotoxicity: The compound simplexidine of this species exhibited very weak cytotoxicity toward murine macrophages (RAW 264-7) with 30% of growth inhibition at an extract concentration of 80 μg/mL (Fattorusso et al., 2008).

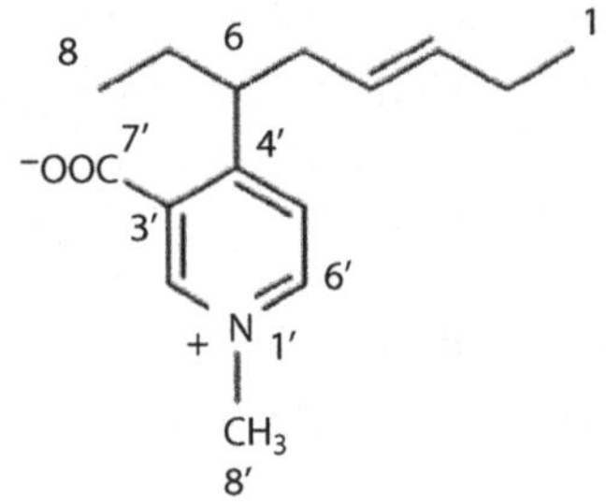

Simplexidine

The MeOH–EtOAc (1:1) extract of this sponge species yielded two new 1,2-dioxane peroxylactones, namely, plakortolides H and I, and a new 1,2-dioxolane, andavadoic acid (3). Among these compounds, andavadoic acid (3) showed significant activity against 13 tumor cells with GI50 values in the submicromolar range (Rudi et al., 2003).

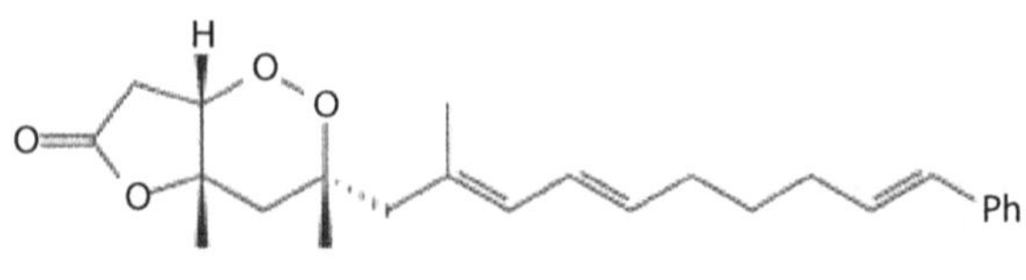

Plakortolide H

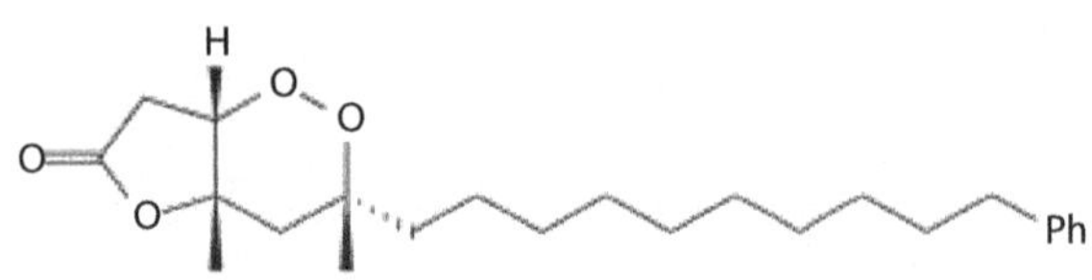

Plakortolide I

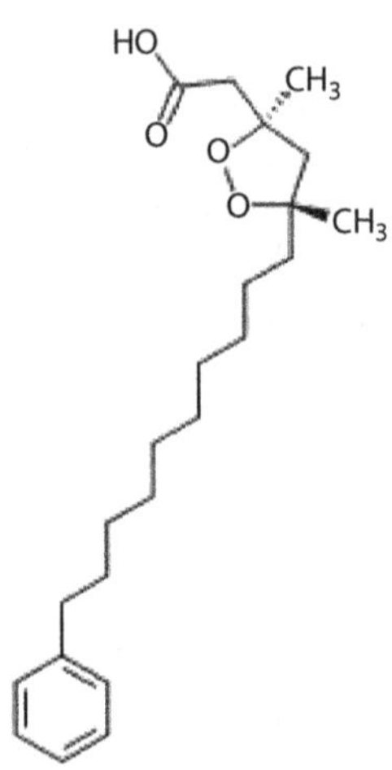

Andavadoic acid

Chianese et al. (2015) reported that the compound spiroplakortone of this species showed a moderate cytotoxicity against the L5178Y (mouse lymphoma) cell line with an IC50 value of 37.5 μM.

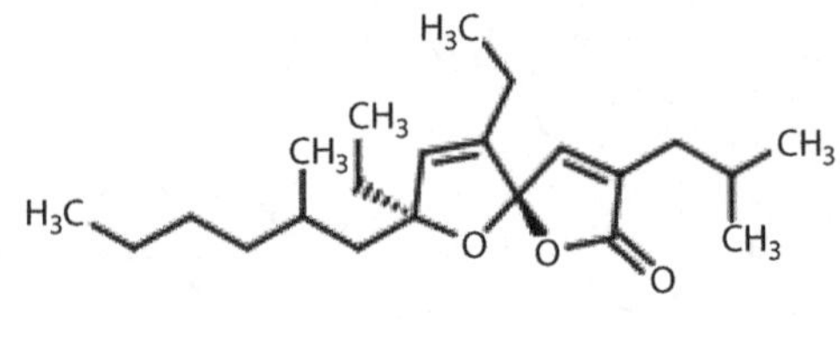

Spiroplakortone

Antiplasmodial activity: The endoperoxide compounds plakortin and dihydroplakortin isolated from this species inhibited the growth of both chloroquine-sensitive D10 strain and chloroquine-resistant W2 strain of *P. falciparum*. The IC50 values were similar for both compounds and in the range of 1263–1117 nM against D10, and 735–760 nM against W2 (Anjum et al., 2016; Fattorusso et al., 2002).

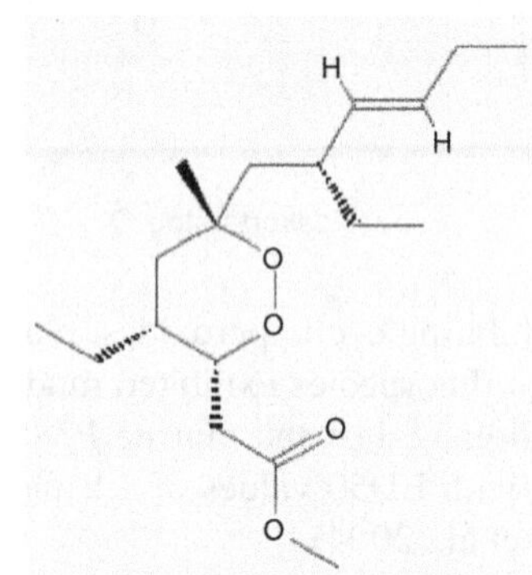

Plakortin

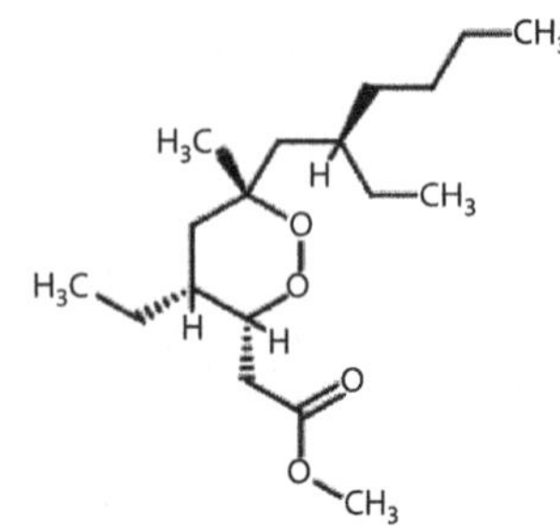

Dihydroplakortin

Plakortis lita (de Laubenfels, 1954)

Order: Homosclerophorida Family: Plakinidae

Common name: Not designated

Distribution: West-Central Pacific; South Korea; Vanuatu; Indonesia; Fiji; Cebu and Mactan Island, Philippines

Ecology: It is encrusting on empty bivalve shells and coral rubbles at depths of 5 to 15 m.

Biology: It is thickly encrusting, toxic sponge (51 mm long, 25 mm wide, and 20 mm thick). Surface is smooth with contractile oscules, which are hard to detect when the sponge is taken out of the water. Texture is soft and compressible, fleshy and easy to tear. In life, the sponge is reddish brown. It is a little brighter and darker at the ectosome than the endosome. In spirit, its color is brown with the ectosome darker than the endosome. Mesoscleres are diods, with straight and sinuous rays. Triods are occasionally present. Spicules are densely packed all throughout, without differential location of spicules.

Compound(s) and Activities

Antiparasitic activity: While the manadoperoxides I and K of this species displayed the greatest activity against *Trypanosoma brucei rhodesiense* (IC50 0.06 and 0.09 µg/mL, respectively) and *Leishmania donovani* (IC50 0.63 and 1.9 µg/mL, respectively), its manadoperoxide B has shown activity with IC50 values of 0.003 and 0.59 µg/mL, respectively. Further, its bromophycoic acids displayed activity against the malaria parasite (3D7 strain) *Plasmodium falciparum*. Among them, the peroxy bromophycoic acid C was found to be most active, with an IC50 value of 8.7 µM. A known macrolide of this species, bromophycolide A, showed an IC50 value of 0.5 µM (Gribble, 2015).

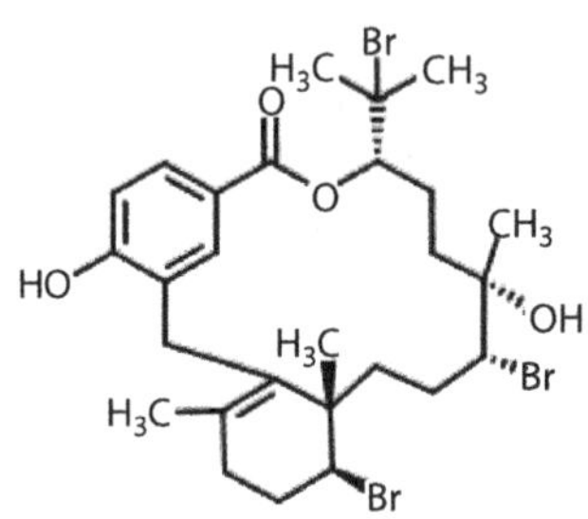

Bromophycolide A

Antimalarial activity: Four novel thiazine-derived alkaloids, thiaplakortones A–D, have been isolated from this species. These compounds displayed significant growth inhibition against chloroquine-sensitive (3D7) and chloroquine-resistant (Dd2) *Plasmodium falciparum* (IC50 values <651 nM) and only moderate cytotoxicity against HEK293 cells (IC50 values >3.9 µM). Thiaplakortone A was the most active natural product, with IC50 values of 51 and 6.6 nM against 3D7 and Dd2 lines, respectively (Davis et al., 2013).

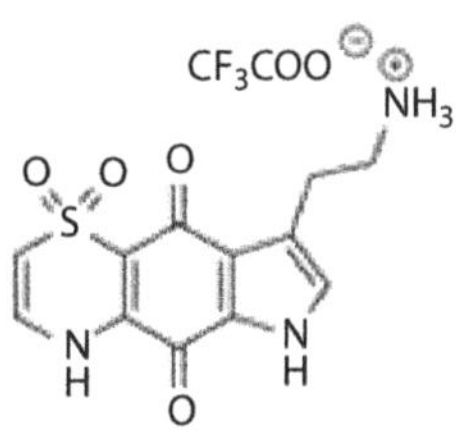

Thiaplakortone A

Plakortis sp.

Order: Homosclerophorida Family: Plakinidae

Compound(s) and Activities

Anti-inflammatory activity: At an extract concentration of 1 µg/mL, the carboxylic acid plakotenin isolated from this unidentified sponge strongly inhibited the proliferation of rheumatoid synovial fibroblasts with an inhibition percentage value of 77 (Keyzers, 2005).

Antiplasmodial activity: The compounds peroxyplakoric acids A3 and B3 (isolated as esters) isolated from this sponge showed antiplasmodial activity against *P. falciparum* with IC50 values of 1.5×10^{-7} M and 1.2×10^{-7} M, respectively (Muraleedharan and Avery, n.d.).

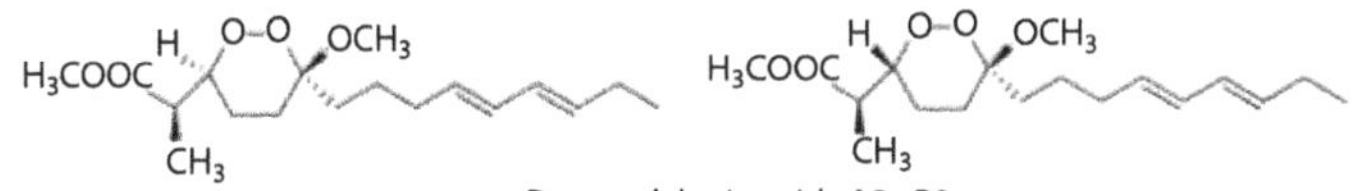

Peroxyplakoric acids A3 , B3

FAMILY: ROSSELLIDAE

Bathydorus spinosus (Schulze, 1886)

Class: Hexactinellida Subclass: Hexasterophora
Order: Lyssacinosida Family: Rossellidae

Common name: Not available

Distribution: It is a polar species. Antarctic Indian Ocean: Kerguelen and Crozet Islands.

Ecology: It is a sessile species occurring at a depth of more than 2000 m.

Biology: The specimen of this species is a thin-walled funnel of 90 mm tall and 150 mm wide when flattened, with a distinct marginal fringe of large diactins projecting 8 mm. The thickness of this sponge is only 6 mm.

Compound(s) and Activities

Cytotoxic activity: The organic dried extract of this species exhibited cytotoxicity against lung fibroblast (V-79) and human colon epithelial (Caco-2) cancer cell lines at a concentration of 100 µg/mL. The percentage inhibition values recorded for these cell lines were found to be 95 and 92, respectively (Turk et al., 2013).

Rossella podagrosa (Kirkpatrick, 1907) (= *Rossella racovitzae*)

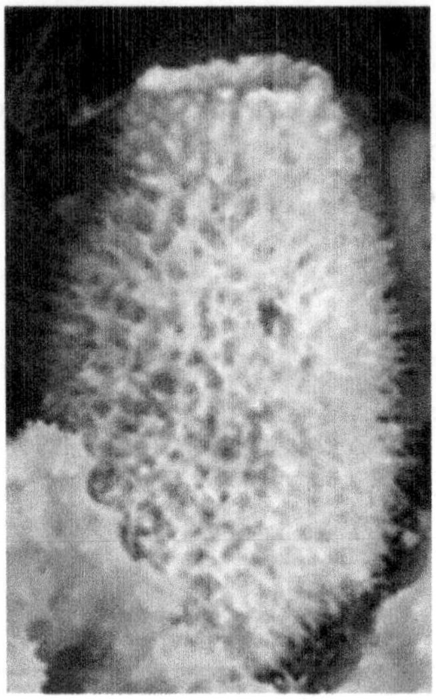

Class: Hexactinellida Order: Lyssacinosida

Family: Rossellidae

Common name: Glass sponge

Distribution: Antarctica; Amundsen/Bellingshausen Sea; Ekström Shelf in the southeastern Weddell Sea; from Sub-Antarctica and South Georgia to the Antarctic Peninsula and Continent.

Ecology: This sessile species occurs on hard or soft substrates at depths of 18 to 2000 m. These sponges deposit spicules, which eventually form hard mats that other sponges, unable to colonize on soft surfaces, can settle on.

Biology: The body of this sponge is of a slender, upright shape with one round osculum at the top. It may take the shape of egg, vase, or barrel. Color of the animal is white, yellow, or orange. It has numerous buds that originate from the basal region of the mother. Colonies are often densely covered by sediment. A single specimen may reach a height of approximately 20 cm. The surface has no conules, but it can be covered by a narrow veil of protruding pentactines. The sponge has a basal root tuft of long pentactines. The inner cavity has a dense surface; its basal part may bear large round cavernous openings. The texture of the sponge is soft. Size (height) of the specimens varies from 15 cm to 50 cm.

Food: It is a suspension feeder. Photosynthetic diatoms are present within its cells, but their role is still unclear.

Reproduction: It is both sexual and asexual. These sponges show an unusual fast growth rate by increasing their length by 16 cm in a period of 10 years and have been shown to increase their volume by almost 300% in 3 years. Further, newly separated buds may grow new buds within 1 year. This mode of reproduction leads to a typical appearance of dense colonies with several specimens growing close to each other.

Associated fauna: The sponge-assocated infauna of this species include foraminiferans, nematodes, polychaetes, arthropods, and mollusks.

Predators: The main predators of this species are seastars and the dorid nudibranch *Austrodoris kerguelenensis*.

Compound(s) and Activities

Cytotoxicity: The organic extracts of this species exhibited moderate cytotoxic potential (Turk et al., 2013).

Rossella sp.

Class: Hexactinellida Order: Lyssacinosida Family: Rossellidae

Compound(s) and Activities

Cytotoxic activity: The organic dried extract of this species exhibited cytotoxicity against lung fibroblast (V-79) and human colon epithelial (Caco-2) cancer cell lines at a concentration of 100 μg/mL. The percentage inhibition values recorded for these cell lines were found to be 101 and 100, respectively (Turk et al., 2013).

Antibacterial activity: The extracts of this species showed antibacterial activity against the laboratory, commensal, and clinically relevant bacterial strains with the MIC values.

Bacterial Strain	MIC (μg/mL)
Staphylococcus aureus (MRSA) S-943 A	350
S. pseudintermedius (MRSP) S-043 A	30
Staphylococcus epidermidis EXB-V55	300
Staphylococcus aureus 10F	300
Bacillus subtilis EXB-V68	30
Macrococcus 1F	200
Micrococcus 2F	150

Source: Turk, T. et al., 2013, *Mar. Drugs*, 11: 1126–1139.

The extracts of this species showed antibacterial activity against the environmental bacterial strains *Exignuobacterium* sp., *Pseudoalteromonas sp.*, *Alteromonas* sp., *Vibrio ruber*, and *Janthinobacterium svalbardensis* with the MIC values of 95, 100, 80, 90, and 100 μg/mL respectively (Turk et al., 2013).

FAMILY: JENKINIDAE

Leucascandra caveolata (Borojevic and Klautau, 2000)

Class: Calcarea Subclass: Calcaronea

Order: Leucosolenida Family: Jenkinidae

Common name: Calcareous sponge

Distribution: New Caledonia

Ecology: It inhabits deepwater habitats.

Biology: This sponge forms large cormi (up to 10 cm high) of a rather loose meshwork that is composed of delicate tubes, the larger ones attaining nearly 1 cm in diameter. The smaller ones (not more than 2 mm in dia.) rise perpendicularly from the larger ones. They are ramified at the distal part, and occasionally anastomosed, bearing a naked osculum at the end. Color of the sponge is brown red in life, but the alcohol-preserved specimen is white.

Compound(s) and Activities

Antitumor activity: The compound leucascandrolide A of this species showed promising activity against certain tumor cells, as well as a pathogenic yeast that can infect AIDS patients (Crow, 2008). Further, this compound has also shown powerful antifungal activity (D'Ambrosio et al., 1996; Sipkema et al., 2005).

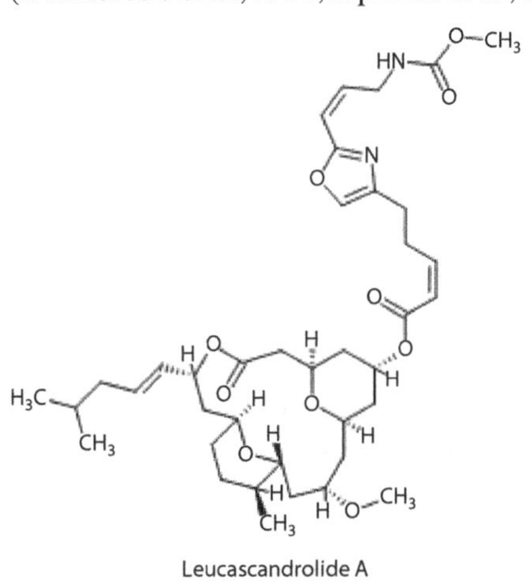

Leucascandrolide A

FAMILY: LEUCOSOLENIIDAE

Leucosolenia sp.

Order: Leucosolenida Family: Leucosoleniidae

Compound(s) and Activities

Cytotoxicity: The compound leucosolenamine A of this sponge was found to be cytotoxic against the murine colon adenocarcinoma C-38 cell line (Putra and Jaswir, 2014).

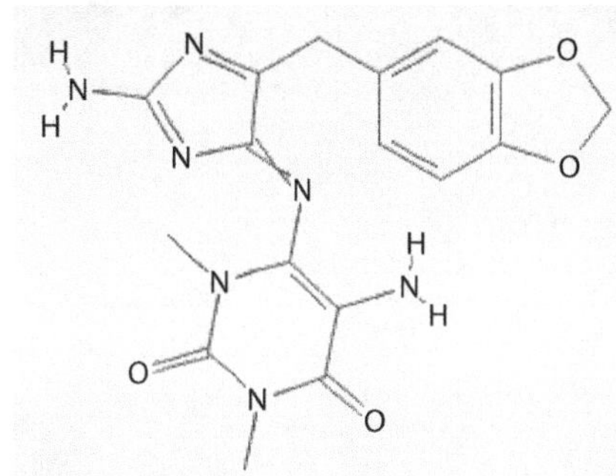

Leucosolenamine A

FAMILY: LEUCETTIDAE

Leucetta avocado (de Laubenfels, 1954)

Subclass: Calcinea Order: Clathrinida

Family: Leucettidae

Common name: Not designated

Distribution: Palau, Panama, East Malaysia, Philippines, Papua New Guinea, Solomon Islands

Ecology: It is conspicuous on the tops of reefs, typically 5–30 m depth.

Biology: It is a volcano- to sack-shaped sponge, hollow, and billowy. Several sponges may join basally. Each sponge has a size of 4–5 cm dia, 15 cm height, and wall thickness of 3–5 mm. Surface is smooth, deeply furrowed, and indented. Interior atrium is deep, voluminous, and with a thin, sharp rim. Small oscules line the atrium. Texture is slightly compressible, tearable, and finely granular. Color in life is golden olive to dark green with a characteristic dark mesh surface pattern. Internal color is vivid neon green when exposed on reef top and yellow in deeper, shaded areas.

Compound(s) and Activities

Antimicrobial activity: The compound N,N-dimethyl naamine D isolated from this species was mildly active against an antimicrobial panel consisting of *E. Coli, S. aureus., B. subtilis,* and *C. albicans* (Putra and Jaswir, 2014).

Leucetta chagosensis (Dendy, 1913)

Order: Clathrinida Family: Leucettidae

Common name: Lemon sponge

Distribution: Indo-Pacific

Ecology: It inhabits coral reef lagoons.

Biology: It is a massive, subspherical or pyriform sponge with a single or a few naked oscules of 1–2 mm in diameter. They generally have a single or only a few naked oscula. The surface is either very regular and smooth, or with small protuberances. The sponge wall is about 1 cm thick, surrounding a wide central atrium, with a smooth surface bearing the large openings of the exhalant cavities. The inhalant pores are small, opening to larger inhalant cavities that run under the thin cortex. The inhalant canals are tubular, running perpendicular to the surface. The color in life is bright yellow, but the preserved specimens are white or brown.

Compound(s) and Activities

Antitumor and immunosuppressive activities: The imidazole alkaloid naamine D isolated from this species has shown antitumor activity by inhibiting the enzyme nitric oxide synthetase. Further, it has also shown immunosuppressive activity (Anjum et al., 2016; Sipkema et al., 2005).

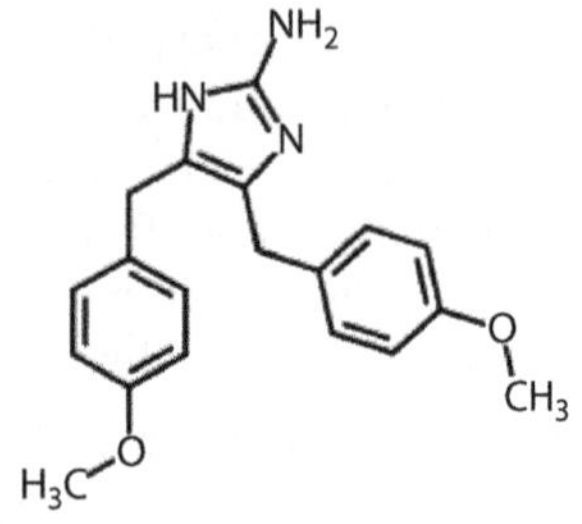

Naamine D

Cytotoxicity: Its compound isonaamine C was found to be cytotoxic to a variety of cell lines, with GI50s of HMO2 (5.3 μg/mL), HepG2 (2.2 μg/mL) and Huh7 (2.1 μg/mL). On the other hand, its naamidine H showed moderate cytotoxicity against HeLa cells (IC50 = 5.6 μg/mL), and naamidine I showed slightly weaker cytotoxicity (IC50 = 15 μg/mL). Naamine G was also found to be cytotoxic to both mouse lymphoma cell lines (46% growth inhibition at 10 μg/mL) and human cervix carcinoma (HeLa) lines (29% growth inhibition at 10 μg/mL) (Putra and Jaswir, 2014).

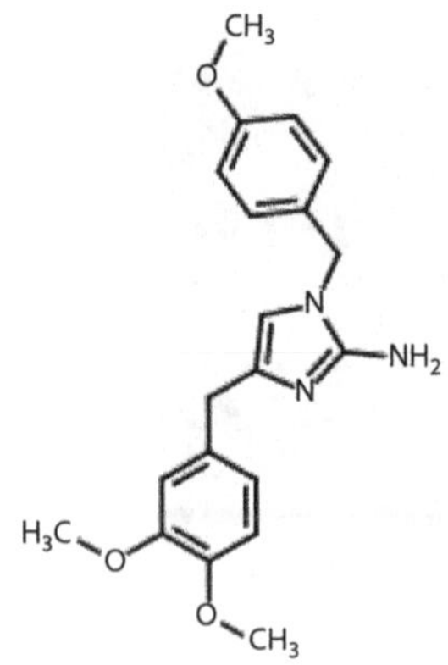

Isonaamine C

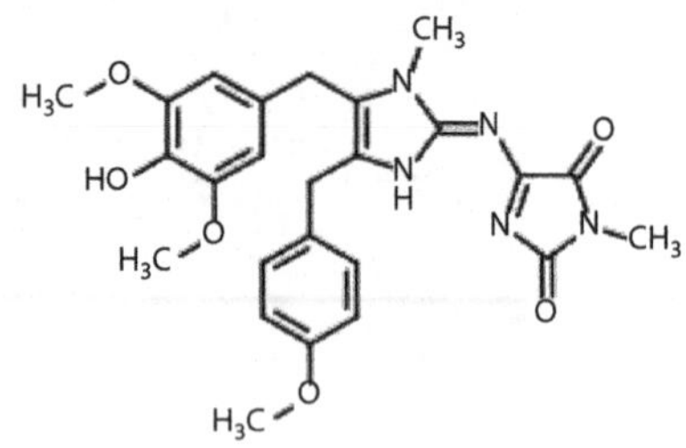

Naamidine H

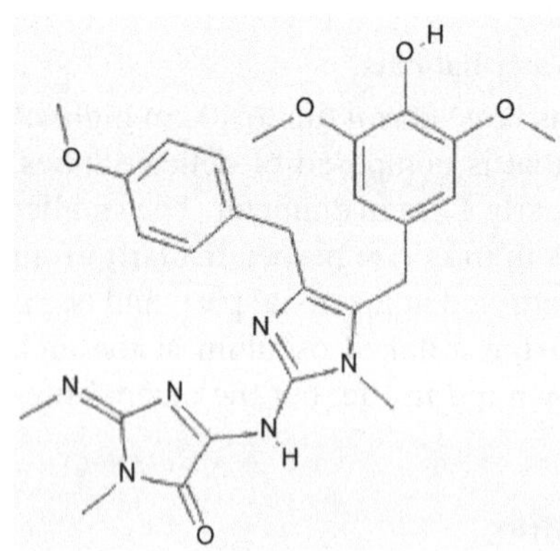

Naamidine I

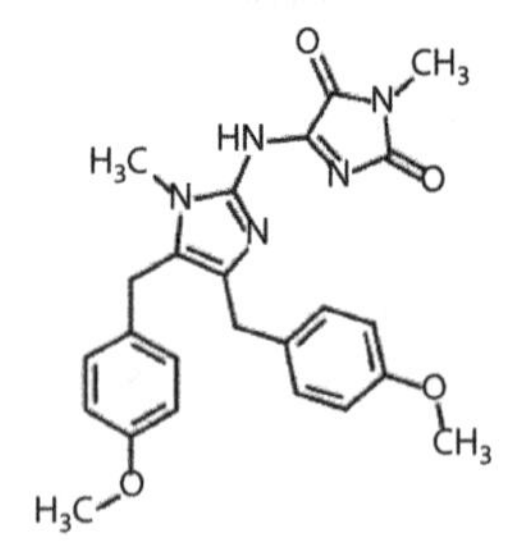

Naamidine G

Antifungal activity: The compound clathridine B of this species showed significant antimycotic activity against *C. albicans* and *S. cerevisiae* at 40 μg/disk. Its compound naamine G was found to be strongly antifungal against *Cladosporium herbarum* (20 mm zone of inhibition at 20 μg/disk). The compund naamine D was active against the opportunistic fungal pathogen *Cryptococcus neoformas* (MIC = 6.25 μg/mL) in AIDS patients (Putra and Jaswir, 2014).

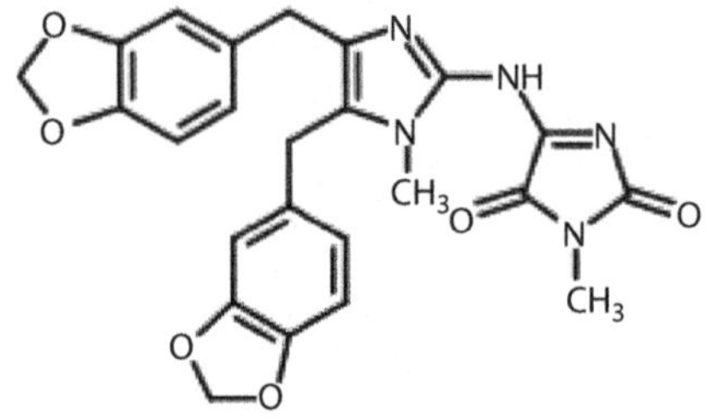

Clathridine B

Leucetta microraphis (Haeckel, 1872)

Order: Clathrinida Family: Leucettidae

Common name: Okinawan marine sponge, Palauan sponge

Distribution: Indo-Pacific, Atlantic, Mediterranean, Red Sea, and Antarctica

Ecology: It is found in exposed and light-protected areas, such as rocky slopes, crevices, and steep walls.

Biology: The body of this sponge is massive and amorphous with single or multiple apical naked oscula. Surface is hispid due to the presence of many colossal triactines and texture is friable. The atrial cavity is reduced and the aquiferous system is leuconoid, with rounded choanocytary chambers. There is no differentiation between the cortical and the choanosomal skeletons. The color of the animal is white, green and brownish yellow or greenish yellow in life.

Compound(s) and Activities

Antitumor activity: The heptapeptide leucamide A isolated from this species has shown antitumor activity (Joseph et al., 2012). Further, the associated cyanobacteria of this species have shown activity by inhibiting the growth of three tumor cell lines, namely, stomach carcinoma, liver carcinoma, and liver carcinoma (Thomas et al., 2010).

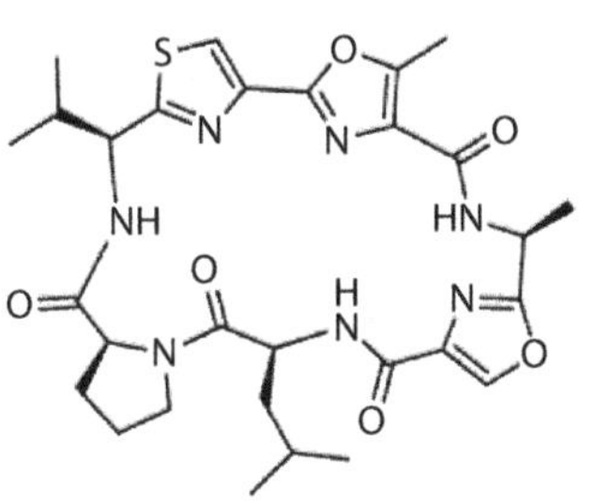

Leucamide A

Cytotoxicity: The compounds leucettamine A and leucettamidine were found to be capable of antagonizing LTB4 binding to its receptor (IC50 = 4.0 and 15.6 μM, respectively). The compound was shown to block the LTB4-induced $Ca2+$ transient (IC50 = 4.6 μM) (Putra and Jaswir, 2014).

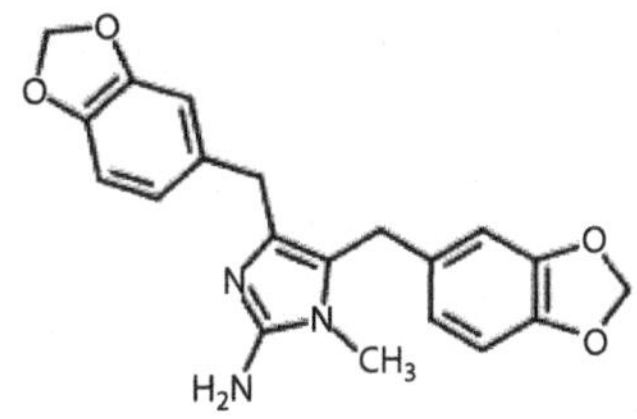

Leucettamine A

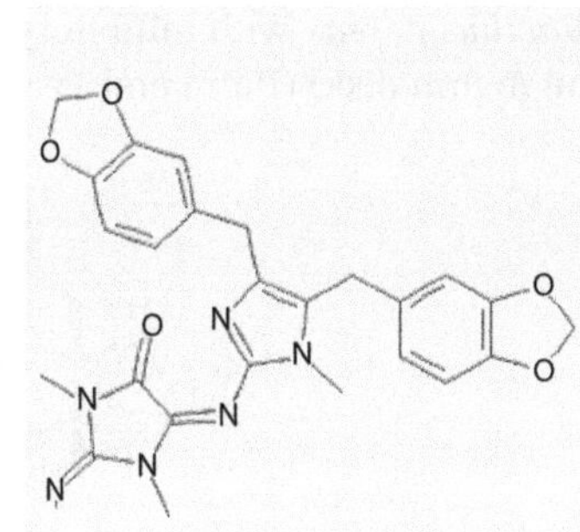

Leucettamidine

Anticancer activity: A new triene aldehyde (2E,6Z,9Z)-2-methyl-2,6,9-icosatrienal isolated from the MeOH extract of this species showed moderate growth inhibitory activity toward HeLa S3 cells (Watanabe et al., 2000).

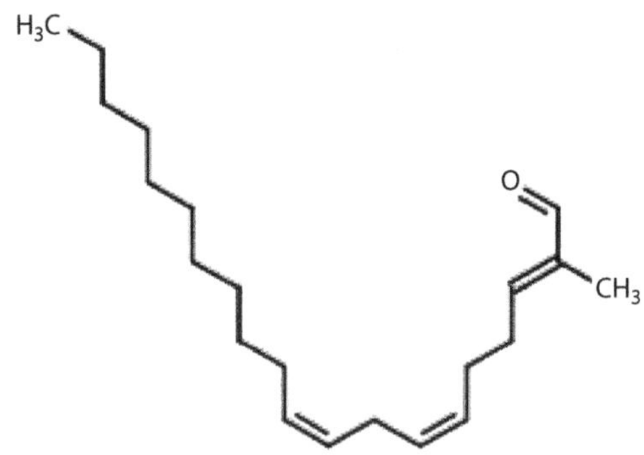

(2E, 6Z, 9Z)-2-methyl-2, 6, 9-icosatrienal

The cyclic hexapeptide compound leucamide A isolated from this species displayed anticancer activity against tumor cell lines HM02, HepG2, and Huh7 tumor cell lines with IC50 values of 5.2 μg/mL, 5.9 μg/mL and 5.1 μg/mL, respectively (Agrawal et al., 2016).

Antimicrobial and anti-inflammatory activities: The compounds leucettamines A and B isolated from this species have shown antimicrobial activity (Kong and Faulkner, 1993). Chan et al. (1993) reported on the isolation of leucettamines A and B (1,2), and leucettamidine (3) from this species. Among them, compound 1 had anti-inflammatory activity, that is, it showed showed potent leukotriene B4 receptor binding activity (pure LTB4 receptor antagonist) (K(i) = 1.3 microM) which can find use in inflammation therapy.

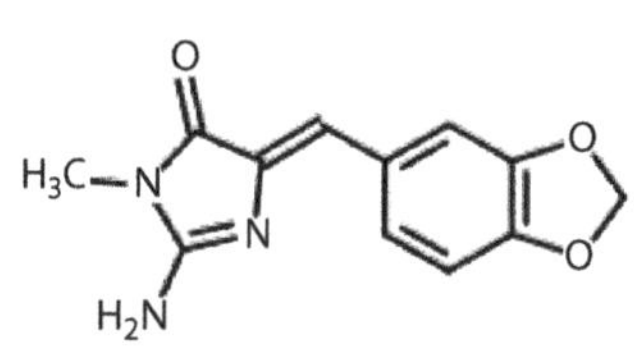

Leucettamine B

The unique spiroquinone-containing alkaloid, spironaamidine, of this species showed antimicrobial activity against

Bacillus cereus with inhibitory zones of 12 by disk assay (10 mg/6 mm disk) (Putra and Jaswir, 2014).

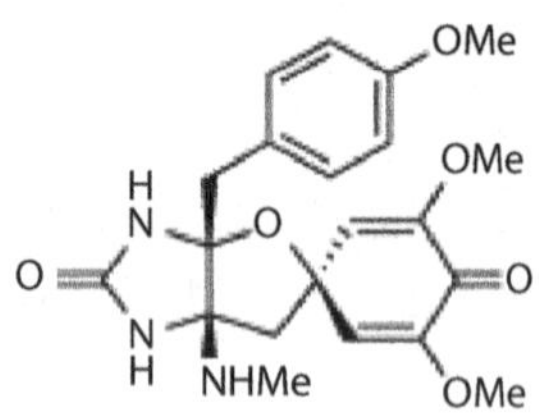

Spironaamidine

Antibacterial activity: Of the three new imidazole alkaloids, namely, leucettamine A, leucettamine B, and leucettamine C, isolated from this species, leucettamine C exhibited only a mild antibiotic activity against *Eschericia coli*, *Staphylococcus aureus*, *Bacillus subtilis*, and *Candida albicans* (Sugumaran and Robinson, 2010).

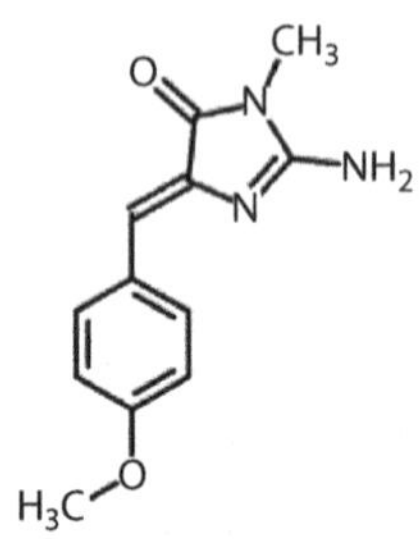

Leucettamine C

The unique spiroquinone-containing alkaloid, spironaamidine, isolated from this species showed antibacterial activity against *Bacillus cereus* with inhibitory zones of 12 by disk assay (10 mg/6 mm diskm) (Putra and Jaswir, 2014).

Leucetta sp.

Order: Clathrinida Family: Leucettidae

Compound(s) and Activities

Cytotoxic activity: The compound pyronaamidine of this species was cytotoxic to KB cells (MIC = 5 µg/mL). Isonaamidine E showed activity to a variety of cell lines with GI50 values of HMO2 (7.0 µg/mL), HepG2 (7.0 µg/mL), and Huh7 (1.3 µg/mL) (Putra and Jaswir, 2014).

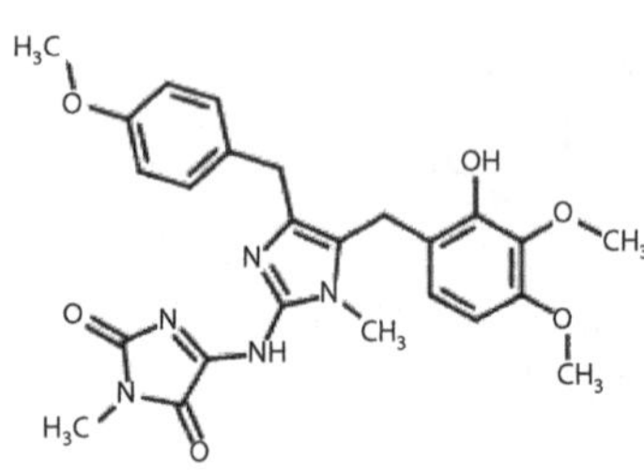

Pyronaamidine

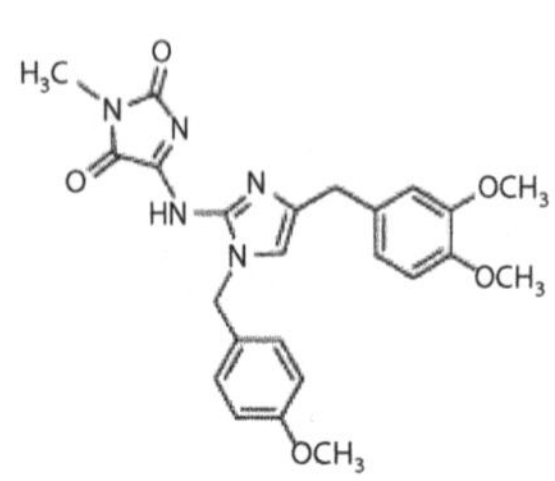

Isonaamidine E

Antifungal activity: The compounds isonaamidine C, isonaamidine D, and isonaamine B showed antifungal activity against *Aspergillus niger* (MIC = 100 µg/mL) (Putra and Jaswir, 2014).

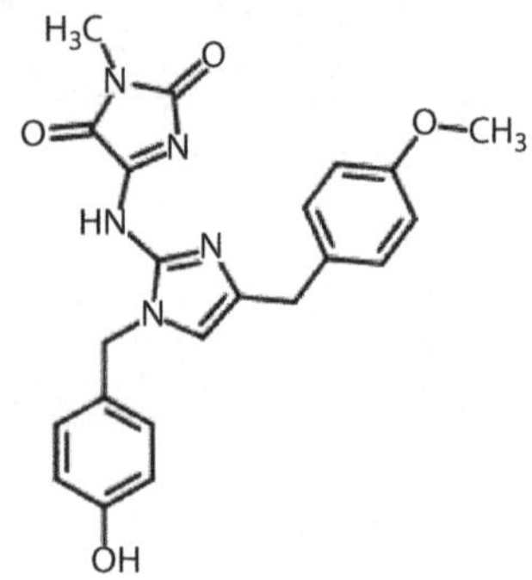

Isonaamine B

Pericharax orientalis (Van Soest and De Voogd, 2015) (= *Pericharax heteroraphis*)

Order: Clathrinida Family: Leucettidae

Common name: South China Sea sponge

Distribution: Indonesia, Papua New Guinea, northeast Australia, New Caledonia, Philippines, Seychelles, Southern Japan (not: Tristan da Cunha)

Ecology: It inhabits roral reefs at depths of 4–36 m.

Biology: It is an upright thick-walled tubular sponge with central vent. Occasionally several individuals may be grouped together. In mature condition, the sides of the tubes are sharply ridged and colors are a mottled reddish brown, green brown, gray green, and grayish yellow. In young specimens, which are usually lighter colored, the walls often lack ridges and at most show rounded folds. Size in mature condition may be 15–30 cm high and 8–20 cm wide. Surface of the sponge is smooth, but it may also be prickly because of large triactines.

Consistency is firm to hard, usually somewhat compressible. At the surface, there is a cortex of small, predominantly slightly sagittal triactines.

Compound(s) and Activities

Cytotoxic and antiviral activities: The imidazole alkaloids naamidine J and naamidine H isolated from this species showed modest cytotoxicities against the K562 cell line, with IC50 values of 11.3 and 9.4 μM, respectively. Further, the compound naamidine H exhibited weak inhibitory activities against the HeLa and A549 cell lines with IC50 values of 21.4 and 22.4 μM, respectively. On the other hand, the compound leucettamine C exhibited weak anti-H1N1 virus activity with an inhibition rate of 33% (Gong et al., 2016).

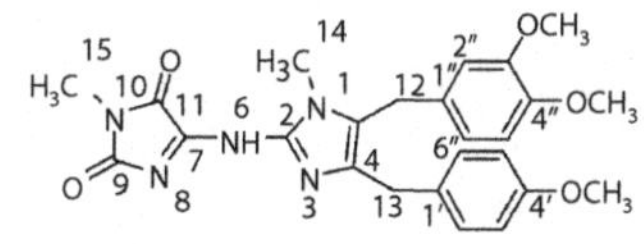

Naamidine J

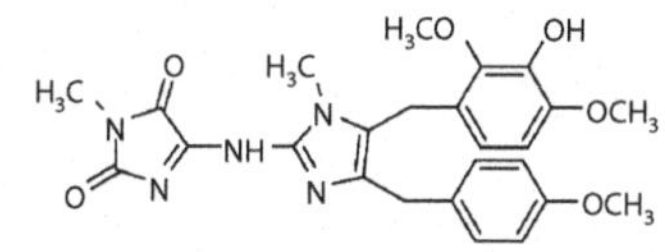

Naamidine H

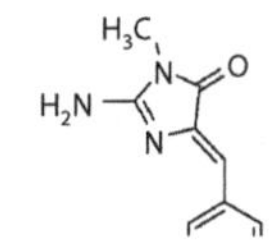

Leucettamine C

Neuroprotective activity/cholinesterases (ChEs) activity: The ethyl acetate extracts of this species rich in terpenoid compounds showed strong AChE inhibition (90%) at a concnentration of 0.1 mg/mL (Grosso et al., 2014).

Consistency is firm to hard, usually somewhat compress-
ible. At the surface, there is a cortex of ..., predominantly
slightly [illegible]

Compound(s) and Activities

[illegible]

... the compound ... [illegible] weak inhibitory
activities against ... cell lines with IC50
values of ... respectively). On the other
[illegible]

References

5,6-dibromo-N,N-DMT – The "undersea" equation …. n.d. https://www.dmt-nexus.me/forum/default.aspx?g=posts&t=24054

Abdelmohsen, U.R., C. Cheng, A. Reimer, V. Kozjak-Pavlovic, A.K. Ibrahim, T. Rudel, U. Hentschel, R. Edrada-Ebel, and S.A. Ahmed. 2015. Antichlamydial sterol from the Red Sea sponge *Callyspongia* aff. *implexa*. *Planta Med.*, 81(05): 382–387.

Abdillah, S., A. Puji, D. Nurhayati, Sri Nurhatika, E. Setiawan, W.L. Heffen. 2013. Cytotoxic and antioxidant activities of marine sponge diversity at Pecaron Bay Pasir Putih Situbondo East Java, Indonesia. *J.Pharm.Res.*, 6: 685–689.

Abdillah, S., F. Zuraiha, and Syarmalina. 2014. Evaluation of antibacterial activities of secondary metabolites produced by *Aaptos suberiptoides* associated bacteria. *IOSR J. Pharm. Biol. Sci.*, 9: 34–38.

Abdjul, D.B. 2016. Studies on bioactive substances from marine sponges collected in Okinawa and Indonesia (Abstract). Tohoku Pharmaceutical University.

Abdjul, D.B., H. Yamazaki, S. Kanno, O. Takahashi, R. Kirikoshi, K. Ukai, and M. Namikoshi. 2015. Structures and biological evaluations of agelasines isolated from the Okinawan marine sponge *Agelas nakamurai*. *J. Nat. Prod.*, 78: 1428–1433.

Abdjul, D.B., H. Yamazaki, O. Takahashi, R.Kirikoshi, K. Ukai, and M. Namikoshi. 2016. Isopetrosynol, a new protein tyrosine phosphatase 1B inhibitor, from the marine sponge *Halichondria* cf. *panacea* collected at Iriomote Island. *Chem. Pharm. Bull.*, 64: 733–736.

Abfa, I.K., O.K. Radjasa, A.B. Susanto, H. Nuryadi, and F.F. Karwur. 2017. Exploration, isolation, and identification of carotenoid from bacterial symbiont of sponge *Callyspongia vaginalis*. *Indonesian J. Mar. Sci.*, 22: 49–58.

Abou-Shoer, M.I., L.A. Shaala, D.T.A. Youssef, J.M. Badr, and A.M. Habib. 2008. Bioactive brominated metabolites from the Red Sea sponge *Suberea mollis*. *J. Nat. Prod.*, 71: 1464–1467.

Acosta, A.L., and A.D. Rodríguez. 1992. 11-Oxoaerothionin: A cytotoxic antitumor bromotyrosine-derived alkaloid from the Caribbean marine sponge *Aplysina lacunosa*. *J. Nat. Prod.*, 55: 1007–1012.

Afifi, A.H., A.H. El-Desoky, H. Kato, R.E.P. Mangindaan, N.J. de Voogd, N.M. Ammar, M.S. Hifnawy, and S. Tsukamoto. 2016. Carteritins A and B, cyclic heptapeptides from the marine sponge *Stylissa carteri*. *Tetrahedron Lett.*, 57: 1285–1288.

Afifi, A.H.H. 2017. The development of new anticancer agents, which inhibit the ubiquitin-proteasome system from marine organisms. Ph.D. thesis, Cairo University.

AFSC News. 2017. Small, deep-water Alaska sponge has molecules that selectively target and kill pancreatic tumor cells. NOAA Fisheries. July 26. https://www.afsc.noaa.gov/News/Green_Sponge.htm

Agrawal, S., A. Adholeya, and S.K. Deshmukh. 2016. The pharmacological potential of nonribosomal peptides from marine sponge and tunicates. *Front. Pharmacol.*, 7: 333.

Aiello, A., M. D'Esposito, E. Fattorusso, M. Menna, W.E. Müller, S. Perović-Ottstadt, and H.C. Schröder. 2006. Novel bioactive bromopyrrole alkaloids from the Mediterranean sponge *Axinella verrucosa*. *Bioorg. Med. Chem.*, 14: 17–24.

Aiello, A., M. D'Esposito, E. Fattorusso, M. Menna, W.E.G. Müller, S. Perović-Ottstadt, H. Tsuruta, T.A.M. Gulder, and G. Bringmann. 2005. Daminin, a bioactive pyrrole alkaloid from the Mediterranean sponge *Axinella damicornis*. *Tetrahedron*, 61: 7266–7270.

Alam, N., W. Wang, J.H. Chonk-Ok, K. Siklm, J.H. Jung. 2013. Cytotoxic sphingosine-4-sulphate from the sponge *Spirastrella abata*. *Bioorg. Med. Chem. Lett.*, 22: 1078–1081.

Alonso, E., R. Alvariño, M. Leirós, J.N. Tabudravu, K. Feussner, M.A. Dam, M.E. Rateb, M. Jaspars, and L.M. Botana. 2016. Evaluation of the antioxidant activity of the marine pyrroloiminoquinone makaluvamines. *Mar. Drugs*, 14: 197.

AlTarabeen, M., G. Daletos, W. Ebrahim, W.E.G. Müller, R. Hartmann, W. Lin, and P. Proksch. 2015. Ircinal E, a new manzamine derivative from the Indonesian marine sponge *Acanthostrongylophora ingens*. *Nat. Prod. Commun.*, 10: 1951–1953.

Alvariño, R., E. Alonso, M. Tribalat, S. Gegunde, O.P. Thomas, and L.M. Botana. 2017. Evaluation of the protective effects of sarains on H2O2-induced mitochondrial dysfunction and oxidative stress in SH-SY5Y neuroblastoma cells. *Neurotox. Res.*, 32: 368–380.

Alvi, K.A., L. Tenenbaum, and P. Crews.1991. Anthelmintic polyfunctional nitrogen-containing terpenoids from marine sponges. *J. Nat. Prod.*, 54: 71–78.

Amagata, T., M. Doi, M. Tohgo, K. Minoura, and A.I. Numata. 1999. Dankasterone, a new class of cytotoxic steroid produced by a *Gymnascella* species from a marine sponge. *Chem. Commun.*, 14: 1321–1322.

Amagata, T., T.A. Johnson, R.H. Cichewicz, K. Tenney, S.L. Mooberry, J. Media, M. Edelstein, F.A. Valeriote, and P. Crews. 2008. Interrogating the bioactive pharmacophore of the latrunculin chemotype by investigating the metabolites of two taxonomically unrelated sponges. *J. Med. Chem.*, 51: 7234–7242.

Amagata, T., K. Minoura, and A. Numata. 2006. Gymnastatins F–H, cytostatic metabolites from the sponge-derived fungus *Gymnascella dankaliensis*. *J. Nat. Prod.*, 69: 1384–1388.

Amr, E.D., A.M. Mamdouh, and A.M. Ali. 2017. Suberea: A Promising marine sponge genus; Chemical and biological remarks. *Org. Med. Chem. J.*, 2: 555–599.

Anjum K., S.Q. Abbas, S.A.A. Shah, N. Akhte, S. Batool, and S.S. Hassan. 2016. Marine sponges as a drug treasure. *Biomol. Ther.*, 24: 347–362.

Ankisetty, S., C.D. Amsler, J.B. McClintock, and B.J. Baker. 2004. Further membranolide diterpenes from the Antarctic sponge *Dendrilla membranosa*. *J. Nat. Prod.*, 67: 1172–1174.

Annamalai, P., M. Thayman, S. Rajan, L.S. Raman, S. Ramasubbu, and P. Perumal. 2015. Ethyl acetate extract from marine sponge *Hyattella cribriformis* exhibit potent anticancer activity by promoting tubulin polymerization as evidenced mitotic arrest and induction of apoptosis. *Pharmacogn. Mag.*, 11: 345–355.

Anti-cancer drugs from a sponge commonly seen on our shores. 2011, February 25. Wild Shores of Singapore. http://wildshores.blogspot.in/2011/02/anti-cancer-drugs-from-sponge-commonly.html#.WfnEFY-CzIU.

Antibacterial activity. n.d. http://shodhganga.inflibnet.ac.in/bitstream/10603/130412/9/09_chapter%205.pdf.

Antiviral activity and characterization of bioactive compounds from Sponge *Sigmadocia pumila* and *Holothuria atra* extracts. n.d. http://shodhganga.inflibnet.ac.in/bitstream/10603/27919/12/12_chapter%205.pdf.

Aoki, S., Y. Watanabe, M. Sanagawa, A. Setiawan, N. Kotoku, and M. Kobayashi. 2006. Cortistatins A,B,C and D and anti-angiogenic steroidal alkaloids from the marine sponge *Corticium simplex*. *A. Am. Chem. Soc.*, 128: 3148–3149.

Asif, M. 2017. Recent developments of some natural products against *Mycobacterium tuberculosis* infection. *J. Pharm. Appl. Chem.*, 3: 225–251.

Atta-ur-Rahman. 2017. *Studies in Natural Products Chemistry: Bioactive Natural Products*, Part 1. Elsevier.

Azevedo, L.G., A.L. Muccillo-Baisch, D.M.V.B. Filgueira, R.T. Boyle, D.F. Ramos, A.D. Soares, C. Lerner, P.A. Silva, and G.S. Trindadea. 2008. Comparative cytotoxic and anti-tuberculosis activity of *Aplysina caissara* marine sponge crude extracts. *Comp. Biochem. Physiol.*, (Part C) 147: 36–42.

Azizah, N., A. Suwarti, and Rosidah. 2017. Antioxidant activity of N-hexane and ethanol extract of sponge (*Pseudosuberites andrewsi*). Lecture paper, Institutional Repository, University of Sumatera Utara, Indonesia.

Bailly, C. 2015. Anticancer properties of lamellarins. *Mar. Drugs*, 13: 1105–1123.

Balakrishnan, D., A.S. Bibiana, A. Vijayakumar, R.S. Santhosh, K. Dhevendaran, and P. Nithyanand. 2014. Antioxidant activity of bacteria associated with the marine sponge *Tedania anhelans. Indian J.Microbiol.*, 55: 13–18.

Barde, S.R., R.S. Sakhare, S.B. Kanthale, P.G. Chandak, and P.G. Jamkhande. 2015. Marine bioactive agents: A short review on new marine antidiabetic compounds. *Asian Pac. J. Trop. Dis.*, 5: S209–S213.

Bary, K., B. Elamraoui, and T. Bamhaoud. 2016. Chemical characterization of *Cliona viridis*: Sponge of Atlantic Moroccan Coast. *IJISR*, 26: 14–22.

Berne, S., M. Kalauz, M. Lapat, L. Savin, D. Janussen, D. Kersken, J.A. Avguštin et al. 2016. Screening of the Antarctic marine sponges (Porifera) as a source of bioactive compounds. *Polar Biol.*, 39: 947–959.

Berrué, F., M.W. McCulloch, P. Boland, S. Hart, M.K. Harper, J. Johnston, and R. Kerr. 2012. Isolation of steroidal glycosides from the Caribbean sponge *Pandaros acanthifolium. J. Nat. Prod.*, 75: 2094–2100.

Bertin, M.J., S.L. Schwartz, J. Lee, A. Korobeynikov, P.C. Dorrestein, L. Gerwick, W.H. Gerwick. 2015. Spongosine production by a Vibrio harveyi strain associated with the sponge *Tectitethya crypta. J Nat. Prod.*, 78: 493–499.

Bhimba, B.V., V. Vinod, and C. Beulah. 2013. Marine sponge *Sigmadocia pumila* a potential supply for drug findings. *J. Pharm. Res.*, 6: 401–403.

Bhimba, B.V., V. Vinod, and M.C. Beulah. 2011. Biopotential of secondary metabolites isolated from marine sponge *Dendrilla nigra. Asian Pac. J. Trop. Dis.*, 1: 299–303.

Biabani, M.F., S.P. Gunasekera, R.E. Longley, A.E. Wright, and S.A. Pomponi. 2002. Tubercidin, a cytotoxic agent from the marine sponge *Caulospongia biflabellata. Pharm. Biol.*, 40: 302–303.

Bianco, E.M., J.L. Krug, P.L. Zimath, A. Kroger, C.J. Paganelli, A.M. Boeder, L. dos Santos et al. 2015. Antimicrobial (including antimollicutes), antioxidant and anticholinesterase activities of Brazilian and Spanish marine organisms—Evaluation of extracts and pure compounds. *Revista Brasileira de Farmacognosia*, 25: 668–676.

Biegelmeyer, R., R. Schröder, D.F. Rambo, R.R. Dresch, J.L.F. Carraro, B. Mothes, J.C.F. Moreira, M.L.C. da Frota Jr., and A.T. Henriques. 2015. Sphingosines derived from marine sponge as potential multi-target drug related to disorders in cancer development. *Mar. Drugs*, 13: 5552–5563.

Blasiak, L.C., and J. Clardy. 2010. Discovery of 3-formyl-tyrosine metabolites from *Pseudoalteromonas tunicata* through heterologous expression. *J. Am. Chem. Soc.*, 132: 926–927.

Blunt, J.W., B.R. Copp, R.A. Keyzers, M.H.G. Munroa, and M.R. Prinsep. 2015. Marine natural products. *Nat. Prod. Rep.*, 32: 116–211.

Bondu, S., G. Genta-Jouve, M. Lrós, C. Vale, J.M. Guigonis, L.M. Botana, and O.P. Thomas. 2012. Additional bioactive guanidine alkaloids from the Mediterranean sponge *Crambe crambe. RSC Adv.*, 2: 2828–2835.

Bonnington, L.S., J. Tanaka, and T. Higa. 1997. Cupolamide A: A cytotoxic cyclic heptapeptide from two samples of the sponge *Theonella cupola. J. Org. Chem.*, 62: 7765–7767.

Botić, T., A. Defant, P. Zanini, M.C. Zuzek, R. Frangez, D. Janussen, D. Kersken, Z. Knez, I. Mancini, and K. Sepcic. 2017. Discorhabdin alkaloids from Antarctic *Latrunculia* spp. sponges as a new class of cholinesterase inhibitors. *Eur. J. Med. Chem.*, 136: 294–304.

Botić, T., D. Cör, A. Anesi, G. Guella, K. Sepčić, D. Janussen, D. Kersken, and Z. Knez. 2015. Fatty acid composition and antioxidant activity of Antarctic marine sponges of the genus *Latrunculia. Polar Biol.*, 38: 1605–1612.

Bourguet-Kondracki, M., C. Debitus, and M. Guyot. 1996. Biologically active sesterterpenes from a New Caledonian marine sponge *Hyrtios* sp. *J. Chem. Res.*, (S), 1996: 192–193.

Boury-Esnault, N., M. Klautau, C. Bézac, J. Wulff, and A.M. Solé-Cava. 1999. Comparative study of putative conspecific sponge populations from both sides of the Isthmus of Panama. *J. Mar. Biol. Ass. U.K.*, 79: 39–50.

Braga, T.F.R. 2014. Evaluation of the antioxidant activity and antitumor activity of marine invertebrates extracts. M.Sc. thesis, Universidade do Algarve.

Brahmachari, G. 2017. *Neuroprotective Natural Products: Clinical Aspects and Mode of Action*. Wiley-VCH.

Brammavidhya, S., and G. Usharani. 2013. Bioactive potential of sponge associated *Bacillus cereus* SBS02 isolated from *Hyattela cribriformis. IJEST*, 3: 61–64.

Breitinger, H.-G., U. Breitinger, D.T.A. Youssef, K. Raafat, and L.A.E. Youssef. n.d. Extract from the sponge amorphinopsis. https://www.google.com/patents/EP2836222A1?cl=en.

British Antarctic Society. n.d. *Kirkpatrickia varialosa* (Kirkpatrick, 1907). http://afg.biodiversity.aq/species/102-kirkpatrickia-variolosa.

Cafieri, F., E. Fattorusso, and O. Taglialatela-Scafati. 1999. Ectyoplasides A–B: Unique triterpene oligoglycosides from the Caribbean sponge *Ectyoplasia ferox. Eur. J. Org. Chem.*, 1999(1): 231–238.

Campagnuolo, C., E. Fattorusso, and O. Taglialatela-Scafati. 2001. Feroxosides A–B, two norlanostane tetraglycosides from the Caribbean sponge *Ectyoplasia ferox. Tetrahedron*, 57: 4049–4055.

Cao, F., Z. Wu, C. Shao, S. Pang, X. Liang, N.J. de Voogd, and C. Wang. 2015. Cytotoxic scalarane sesterterpenoids from the South China Sea sponge *Carteriospongia foliascens. Org. Biomol. Chem.*, 13: 4016–4024.

Capon, R.J., C. Peng, and C. Doomsa. 2008. Trachycladindoles A–G: Cytotoxic heterocycles from an Australian marine sponge, *Trachycladus laevispirulifer. Org. Biomol. Chem.*, 6: 2765–2771.

Carletti, I., B. Banaigs, and P. Amade. 2000. Matemone, a new bioactive bromine-containing oxindole alkaloid from the Indian Ocean Sponge *Iotrochota purpurea. J. Nat. Prod.*, 63: 981–983.

Casapullo, A., A. Cutignano, I. Bruno, G. Bifulco, C. Debitus, L. Gomez-Paloma, and R. Riccio. 2001. Makaluvamine P, a new cytotoxic pyrroloiminoquinone from *Zyzzya* cf. *fuliginosa. J. Nat. Prod.*, 64: 1354–1356.

Casapullo, A., L. Minale, F. Zollo, C. Roussakis, and J.F. Verbist. 1995. New cytotoxic polyoxygenated steroids from the sponge *Dysidea incrustans. Tetrahedron Lett.*, 36: 2669–2672.

Cavalcanti, B.C., C.M. Sombra, J.H. de Oliveira, R.G. Berlinck, M.O. de Moraes, and C. Pessoa. 2008. Cytotoxicity and genotoxicity of ingenamine G isolated from the Brazilian marine sponge *Pachychalina alcaloidifera*. *Comp. Biochem. Physiol. C. Toxicol. Pharmacol.*, 147: 409≠415.

Cayman Chemical. 2013. Product information: Luffariellolide. https://www.caymanchem.com/pdfs/14902.pdf.

Chan, G.W., S. Mong, M.E. Hemling, A.J. Freyer, P.H. Offen, C.W. DeBrosse, H.M. Sarau, and J.W. Westley. 1993. New leukotriene B4 receptor antagonist: Leucettamine A and related imidazole alkaloids from the marine sponge *Leucetta microraphis*. *J. Nat. Prod.*, 56: 116–121.

Chandraraj, S., B. Prakash, and K. Navanath. 2010. Immunomodulatory activities of ethyl acetate extracts of two marine sponges *Gelliodes fibrosa* and *Tedania anhelans* and brown algae *Sargassum ilicifolium* with reference to phagocytosis. *Res. J. Pharm. Biol. Chem. Sci.*, 1: 302–307.

Chang, L.C., S. Otero-Quintero, J.N.A. Hooper, and C.A. Bewl. 2002. Batzelline D and isobatzelline E from the Indopacific sponge *Zyzzya fuliginosa*. *J. Nat. Prod.*, 65: 776–778.

Chang, N., A.C. Lin, C. Hsu, J. Liu, L. Tsui, C. Chen, T. Jayakumar, and T. Fong. 2013. Okadaic acid, a bioactive fatty acid from *Halichondria okadai*, stimulates lipolysis in rat adipocytes: The pivotal role of perilipin translocation. *J. Evid. Based Complementary Altern. Med.*, 2013: Article ID 545739, 10 p.

Chang, Y., S. Tseng, L. Liu, Y. Chou, Y. Ho, M. Lu, and J. Su. 2012. Cytotoxic sesterterpenoids from a sponge *Hippospongia* sp. *Mar. Drugs*, 10: 987–997.

Chapter I: Bioactive metabolites of marine organisms. n.d. http://shodhganga.inflibnet.ac.in/bitstream/10603/28761/3/03_chapter%201.pdf.

Chelossi, E., M. Milanese, A. Milano, R. Pronzato, and G. Riccardi. 2004. Characterisation and antimicrobial activity of epibiotic bacteria from *Petrosia ficiformis* (Porifera, Demospongiae). *J. Exp. Mar. Biol. Ecol.*, 309: 21–33.

Chelossi, E., R. Pantile, R. Pronzato, M. Milanese, and U. Hentschel. 2007. Bacteria with antimicrobial properties isolated from the Mediterranean sponges *Chondrilla nucula* and *Petrosia ficiformis*. *Aquat. Microb. Ecol.*, 49: 157–163.

Cheng, C. 2012. Secondary metabolites investigation of sponge *Callyspongia* aff. *implexa*. https://prezi.com/vr82c-t55ygs/secondary-metabolites-investigation-of-sponge-callyspongia-aff-implexa/.

Cheruku, P., A. Plaza, G. Lauro, J. Keffer, J.R. Lloyd, G. Bifulco, and C.A. Bewley. 2012. Discovery and synthesis of namalide reveals a new anabaenopeptin scaffold and peptidase inhibitor. *J. Med. Chem.*, 55: 735–742.

Cheung, R.C.F., J.H. Wong, W.L. Pan, Y.S. Chan, C.M. Yin, X.L. Dan, H.X. Wang et al. 2014. Antifungal and antiviral products of marine organisms. *Appl. Microbiol. Biotechnol.*, 98: 3475–3494.

Chianese, G., J. Silber, P. Luciano, C. Merten, D. Erpenbeck, B. Topaloglu, M. Kaiser, and D. Tasdemir. 2017. Antiprotozoal linear furanosesterterpenoids from the marine sponge *Ircinia oros*. *J. Nat. Prod.*, 80: 2566–2571.

Chu, M., X. Tang, G. Qin, Y. Sun, L. Li, N.J. de Voogd, P. Li, and G. Li. 2017. Pyrrole derivatives and diterpene alkaloids from the South China Sea sponge *Agelas nakamurai*. *Chem. Biodivers.*, 14: e1600446.

Ciminiello, P., C. Dell'Aversano, E. Fattorusso, S. Magno, and M. Pansini. 2000. Chemistry of verongida sponges. 10: Secondary metabolite composition of the Caribbean sponge *Verongula gigantea*. *J. Nat. Prod.*, 63: 263–266.

Cita, Y.P., F.K. Muzaki, O.K. Radjasa, and P. Sudarmono. 2017. Screening of antimicrobial activity of sponges extract from Pasir Putih, East Java (Indonesia). *J. Marine Sci. Res. Dev.*, 7: 237.

Coello, L., M.J. Martín, and F. Reyes. 2009. 1,5-Diazacyclohenicosane, a new cytotoxic metabolite from the marine sponge *Mycale* sp. *Mar. Drugs*, 7: 445–450.

Coello, L., F. Reyes, M.J. Martín, C. Cuevas, and R. Fernández. 2014. Isolation and structures of pipecolidepsins a and b, cytotoxic cyclic depsipeptides from the Madagascan sponge *Homophymia lamellose*. *J. Nat. Prod.*, 77: 298–303.

Coleman, J.E., E.D. de Silva, F. Kong, R.J. Andersen, and T.M. Allen. 1995. Cytotoxic peptides from the marine sponge *Cymbastela* sp. *Tetrahedron*, 51: 10653–10662.

Collection, extraction of compounds and evaluation of bioactivity profile of chosen marine sponge and holothuria. n.d. http://shodhganga.inflibnet.ac.in/bitstream/10603/27919/8/08er%201.pdf.

Colorado, J., D. Muñoz, D. Marquez, M.E. Marquez, J. Lopez, O.P. Thomas, and A. Martinez. 2013. Ulososides and urabosides—Triterpenoid saponins from the Caribbean marine sponge *Ectyoplasia ferox*. *Molecules*, 18: 2598–2610.

Compagnone, R.S., I.C. Pina, H.R. Rangel, F. Dagger, A.I. Suarez, M.V. Reddy, and D.J. Faulkner. 1998. Antileishmanial cyclic peroxides from the Palauan sponge *Plakortis* aff. *angulospiculatus*. *Tetrahedron*, 54: 3057–3068.

Concepcion, G.P., A.R.J. Anas, and M.A. Azcuna. 2014. Anticancer compounds from Philippine marine organisms act on major pathways in cancer. *Philipp. Sci. Lett.*, 7: 207–227.

Costantini, S., G. Romano, F. Rusolo, F. Capone, E. Guerriero, G. Colonna, A. Ianora, G. Ciliberto, and M. Costantini. 2015. Anti-inflammatory effects of a methanol extract from the marine sponge *Geodia cydonium* on the human breast cancer MCF-7 cell line. *Mediators Inflamm.*, Article ID 204975, 9 p.

Crews, P., L.V. Manes, and M. Boehler. 1986. Jasplakinolide, a cyclodepsipeptide from the marine sponge *Jaspis* sp. *Tedrahedron Lett.*, 27: 2797–2800.

Crow, J.M. 2008. A new hope for anticancer agent. *Chemistry World*, June 11. https://www.chemistryworld.com/news/new-hope-for-anticancer-agent/3002835.article.

Cryptotethya crypta and other sea sponge wonderes: Medicinal uses. 2007. http://bioweb.uwlax.edu/bio203/2011/szotkows_pete/facts.htm.

Dahiya, R., and H. Gautam. 2010. Total synthesis and antimicrobial activity of a natural cycloheptapeptide of marine origin. *Mar. Drugs*, 8: 2384–2394.

Dai, J., Y. Liu, Y. Zhou, and D.G. Nagle. 2007. Hypoxia-selective antitumor agents: Norsesterterpene peroxides from the marine sponge *Diacarnus levii* preferentially suppress the growth of tumor cells under hypoxic conditions. *J. Nat. Prod.*, 70: 130–133.

Daletos, G., N.J. de Voogd, W.E. Müller, V. Wray, W. Lin, D. Feger, M. Kubbutat, A.H. Aly, and P. Proksch. 2014. Cytotoxic and protein kinase inhibiting nakijiquinones and nakijiquinols from the sponge *Dactylospongia metachromia*. *J. Nat. Prod.*, 77: 218–226.

Daletos, G., R. Kalscheuer, H. Koliwer-Brandl, R. Hartmann, N.J. de Voogd, V. Wray, W. Lin, and P. Proksch. 2015. Callyaerins from the marine sponge *Callyspongia aerizusa*: Cyclic peptides with antitubercular activity. *J. Nat. Prod.*, 78: 1910–1925.

D'Ambrosio, M., A. Guerriero, E. Deharo, C. Debitus, V. Munoz, and F. Pietra. 1998. New types of potentially antimalarial agents: Epidioxy-substituted norditerpene and norsesterpenes from the marine sponge *Diacarnus levii*. *Helvetica*, 81: 1285–1292.

D'Ambrosio, M., A. Guerriero, F. Pietra, and C. Debitus. 1996. Leucascandrolide A, a new type of macrolide: The first powerfully bioactive metabolite of calcareous sponges (*Leucascandra caveolata*, a new genus from the coral sea). *Helvetica*, 79: 51–60.

D'Auria, M.V., A. Zampella, L.G. Paloma, L. Minale, C. Debitus, C. Roussakis, and V.L. Bert. 1996. Callipeltins B and C: Bioactive peptides from a marine Lithistida sponge *Callipelta* sp. *Tetrahedron* 52: 9589–9596.

Datta, D., S.N. Talapatra, and S. Swarnakar. 2015. Bioactive compounds from marine invertebrates for potential medicines—An overview. *Intern. Lett. Nat. Sci.*, 34: 42–61.

Davis, R.A., M.S. Buchanan, S. Duffy, V.M. Avery, S.A. Charman, W.N. Charman, K.L. White et al. 2012. Antimalarial activity of pyrroloiminoquinones from the Australian marine sponge *Zyzzya* sp. *J. Med. Chem.*, 55: 5851–5858.

Davis, R.A., S. Duffy, S. Fletcher, V.M. Avery, and R.J. Quinn. 2013. Thiaplakortones A–D: Antimalarial thiazine alkaloids from the Australian marine sponge *Plakortis lita. J. Org. Chem.*, 78: 9608–9613.

Davis, R.A., G.C. Mangalindan, Z.P. Bojo, R.R. Antemano, N.O. Rodriguez, G.P. Concepcion, S.C. Samson et al. 2004. Microcionamides A and B, bioactive peptides from the Philippine sponge *Clathria (Thalysias) abietina. J. Org. Chem.*, 69: 4170–4176.

de Oliveira, J.H., A.M. Nascimento, M.H. Kossuga, B.C. Cavalcanti, M.O.P.C. Moraes, M.L. Macedo, A.G. Ferreira, E. Hajdu, U.S. Pinheiro, and R.G. Berlinck. 2007. Cytotoxic alkylpiperidine alkaloids from the Brazilian marine sponge *Pachychalina alcaloidifera. J. Nat. Prod.*, 70: 538–543.

de Voogd, N.J. 2004. *Callyspongia (Euplacella) biru* (Porifera: Demospongiae: Haplosclerida) from Indonesia. *Zoölogische Mededelingen Leiden*, 78: 477–483.

de Voogd, N.J. 2007. An assessment of sponge mariculture potentials in the Spermonde Archipelago, Indonesia. *J. Mar. Biol. Ass. U.K.*, 87: 1–8.

De Rosa, S., S. De Stefano, and N. Zavodnik. 1988. Cacospongionolide. A new antitumoral sesterterpene, from the marine sponge *Cacospongia mollior. J. Org. Chem.*, 53: 5020–5023.

de Voogd, N.J., J.J.H. Haftka, and B.W. Hoeksema. 2005. Evaluation of the ecological function of amphitoxin in the reef-dwelling sponge *Callyspongia (Euplacella) biru* (Haplosclerida: Callyspongiidae) at southwest Sulawesi, Indonesia. *Contrib. Zool.*, 74: 51–59.

de Voogd, N.J., F.J. Parra-Velandia, and R.W.M. Van Soest. 2008. A new Agelas (Demospongiae: Agelasida: Agelasidae) from the Thousand Islands, West-Java, Indonesia. *Zoölogische Mededelingen Leiden*, 82: 235–243.

Debitus, C., M. Cesario, J. Guilhem, C. Pascard, and M. Païs. 1989. Corallistine, a new polynitrogen compound from the sponge, *Corallistes fulvodesmus. Tetrahedron Lett.*, 30: 1534–1538.

Dellai, A., H. Mansour, A. Clary-Laroche, M. Deghrigue, and A. Bouraoui. 2012. Anticonvulsant and analgesic activities of crude extract and its fractions of the defensive secretion from the Mediterranean sponge, *Spongia officinalis. Cancer Cell Intern.*, 12: 15, 5 p.

Dembitsky, V.M., D.O. Levitsky, T.A. Gloriozovac, and V.V. Poroikov. 2006. Acetylenic aquatic anticancer agents and related compounds. *Nat. Prod. Commun.*, 1: 773–812.

Deng, S., C. Tian, D. Xiao, and H. Wu. 2001. Iotroridoside-A, a novel cytotoxic glycosphingolipid from the marine sponge *Iotrochota ridley. Chin. J. Chem.*, 19: 362–364.

Dhayanithi, N.B., T.T. Ajith Kumar, M. Kalaiselvam, T. Balasubramanian, and N. Sivakumar. 2012. Anti-dermatophytic activity of marine sponge, *Sigmadocia carnosa* (Dendy) on clinically isolated fungi. *Asian Pac. J. Trop. Biomed.*, 2: 635–639.

Dhinakaran, D.I., and A.P. Lipton. 2016. Anti-inflammatory, analgesic and antipyretic potentials of marine sponge *Sigmadocia pumila. J. Coast. Life Med.*, 4: 273–276.

Dhinakaran, D.I., V. Manohari, B. Atchya, K. Tamilselvi, and A.P. Lipton. 2012a. Antifungal and cytotoxic activities of some marine sponges collected from the South East Coast of India. *J. Appl. Pharm. Sci.*, 2: 52–55.

Dhinakaran, D.I., D.R.D. Prasad, R. Gohila, and A.P. Lipton. 2012b. Screening of marine sponge-associated bacteria from *Echinodictyum gorgonoides* and its bioactivity. *Afr. J. Biotechnol.*, 11: 15469–15476.

Ding, L., B. Gu, W. Jiao, W.I. Yuan, Y. Li, W. Tang, H. Yu, X. Liao, B. Han, Z. Li, S. Xu, and H. Lin. 2015. New furan and cyclopentenone derivatives from the sponge-associated fungus *Hypocrea koningii* PF04. *Mar. Drugs*, 13: 5579–5592.

Diyabalanage, T., R. Ratnayake, H.R. Bokesch, T.T. Ransom, C.J. Henrich, J.A. Beutler, J.B. McMahon, and K.R. Gustafson. 2012. Flabelliferins A and B, sesterterpenoids from the South Pacific sponge *Carteriospongia flabellifera. J. Nat. Prod.*, 75:1490–1494.

Djura, P., D.B. Stierle, B. Sullivan, D.J. Faulkner, E.V. Arnold, and J. Clardy. 1980. Some metabolites of the marine sponges *Smenospongia aurea* and *Smenospongia* (.ident. *Polyfibrospongia) echina. J. Org. Chem.*, 45: 1435–1441.

Dresch, R.R., G.D. Zanetti, F.J. Irazoqui, V.G. Sendra, N. Zlocowski, A. Bernardi, R.M. Rosa, A.M.O. Battastini, A.T. Henriques, and M.M. Vozári-Hampe. 2013. Staining tumor cells with biotinylated ACL-I, a lectin isolated from the marine sponge, *Axinella corrugata. Biotech. Histochem.*, 88: 1–9.

Dupont, S., A. Carre-Mlouka, F. Descarrega, A. Ereskovsky, A. Longeon, E. Mouray, I. Florent, and M.L. Bourguet-Kondracki. 2013. Diversity and biological activities of the bacterial community associated with the marine sponge *Phorbas tenacior* (Porifera, Demospongiae). *Lett. Appl. Microbiol.*, 58: 42–52.

Duraikannu, K., D. Edupalli, G. Rameshkumar, and S. Ravichandran. 2009. Antimicrobial peptide from marine sponge *Clathria indica*, (Dendy,1889). *AEJSR (Am.-Eurasian J. Sci. Res.)*, 4: 47–53.

Dyshlovoy, S.A., K.M. Tabakmakher, J. Hauschild, R.K. Shchekaleva, K. Otte, A.G. Guzii, T.N. Makarieva et al. 2016. Guanidine alkaloids from the marine sponge *Monanchora pulchra* show cytotoxic properties and prevent EGF-induced neoplastic transformation in vitro. *Mar. Drugs*, 14: 133.

Ebada, S.S., M.H. Linh, A. Longeon, N.J. de Voogd, E. Durieu, L. Meijer, M.L. Bourguet-Kondracki, A.N. Singab, W.E. Müller, and P. Proksch. 2015. Dispacamide E and other bioactive bromopyrrole alkaloids from two Indonesian marine sponges of the genus *Stylissa. Nat. Prod. Res.*, 29: 231–238.

Ebada, S.S., V. Wray, N.J. DeVoogd, Z. Deng,W. Lin, and P. Proksch. 2009. Two new jaspamide derivatives from the marine sponge *Jaspis splendens. Mar. Drugs*, 7: 435–444.

Eder, C., P. Proksch, V. Wray, R.W. van Soest, E. Ferdinandus, L.A. Pattisina, and S.Sudarsono. 1999. New bromopyrrole alkaloids from the Indopacific sponge *Agelas nakamurai. J. Nat. Prod.*, 62: 1295–1297.

Einarsdottir, E., H.B. Liu, J. Freysdottir, C.H. Gotfredsen, and S. Omarsdottir. 2016. Immunomodulatory N-acyl dopamine glycosides from the Icelandic marine sponge *Myxilla incrustans* collected at a hydrothermal vent site. *Planta Med.*, 82: 903–909.

El Sayed, K.A., M.T. Hamann, N.E. Hashish, W.T. Shier, M. Kelly, and A.A. Khan. 2001. Antimalarial, antiviral, and antitoxoplasmosis norsesterterpene peroxide acids from the Red Sea sponge *Diacarnus erythraeanus*. *J. Nat.Prod.*, 64: 522–524.

El-Amraoui, B., J. Biard, and A. Fassouane. 2013. Haliscosamine: A new antifungal sphingosine derivative from the Moroccan marine sponge *Haliclona viscosa*. *SpringerPlus*, 2: 252.

El-Damhougy, K.A., H.A. El-Naggar, H.A.H. Ibrahim, M.A.E. Bashar, and F. M.A. Senna. 2017. Biological activities of some marine sponge extracts from Aqaba Gulf, Red Sea, Egypt. *Intern. J. Fish. Aquat. Stud.*, 5: 652–659.

El-Gamal, A.A., S.M. Al-Massarani, L.A. Shaala, A.M. Alahdald, M.S. Al-Said, A.E. Ashour, A. Kumar, M.S. Abdel-Kader, W.M. Abdel-Mageed, and D.T.A. Youssef. 2016. Cytotoxic compounds from the Saudi Red Sea sponge *Xestospongia testudinaria*. *Mar. Drugs*, 14: 82.

El-Shitany, N.A., L.A. Shaala, A.T. Abbas, U.A. Abdel-dayem, E.I. Azhar, S.S. Ali, R.W.M. van Soest, and D.T.A. Youssef. 2015. Evaluation of the anti-inflammatory, antioxidant and immunomodulatory effects of the organic extract of the Red Sea marine sponge *Xestospongia testudinaria* against carrageenan induced rat paw inflammation. *PLoS One*, 10: e0138917.

Emsermann, J., U. Kauhl, and T. Opatz. 2016. Marine isonitriles and their related compounds. *Mar. Drugs*, 14: 16.

Encarnación-Dimayuga, R., M.R. Ramírez, and J. Luna-Herrera. 2003. Aerothionin, a bromotyrosine derivative with antimycobacterial activity from the marine sponge *Aplysina gerardogreeni* (Demospongia). *Pharm. Biol.*, 41: 384–387.

Epifanio, R.A., L.S. Pinheiro, and N.C. Alves. 2005. Polyketides from the marine sponge *Plakortis angulospiculatus*. *J. Braz. Chem. Soc.*, 16(6b).

Esteves, A.I.S., M. Nicolai, M. Humanes, and J. Goncalves. 2011. Sulfated polysaccharides in marine sponges: Extraction methods and anti-HIV activity. *Mar. Drugs*, 9: 139–153.

Fathi-Afshar, R., and T.M. Allen. 1988. Biologically active metabolites from *Agelas mauritiana*. *Can. J. Chem.*, 66: 45–50.

Fattorusso, E., A. Romano, F. Scala, and O. Taglialatela-Scafati. 2008. Simplexidine, a 4-alkylpyridinium alkaloid from the Caribbean sponge *Plakortis simplex*. *Molecules*, 13: 1465–1471.

Fattorusso, E., and O. Taglialatela. 2008. *Modern Alkaloids: Structure, Isolation, Synthesis, and Biology*. Wiley-VCH.

Fattorusso, E., S. Parapini, C. Campagnuolo, N. Basilico, O. Taglialatela-Scafati, and D. Taramelli. 2002. Activity against *Plasmodium falciparum* of cycloperoxide compounds obtained from the sponge *Plakortis simplex*. *J. Antimicrob. Chemother.*, 50: 883–888.

Feng, Y., B.F. Bowden, and V. Kapoor. 2012. Ianthellamide A, a selective kynurenine-3-hydroxylase inhibitor from the Australian marine sponge *Ianthella quadrangulata*. *Bioorg. Med. Chem. Lett.*, 22: 3398–3401.

Fernández, D.M., E.M. Fernández, O.P. Thomas, and A.M. Martínez. 2013. In vitro antiproliferative effect of fractions from the caribbean marine sponge *Myrmekioderma gyroderma*. *Revista Cubana de Farmacia*, 47: 368–378.

Ferreira, E.G., D.V. Wilke, P.C. Jimenez, T.A. Portela, E.R. Silveira, E. Hajdu, C. Pessoa, M.O. de Moraes, and L.V. Costa-Lotufo. 2007. Cytotoxic activity of hydroethanolic extracts of sponges (Porifera) collected at Pedra da Risca do Meio Marine State Park, Ceará State, Brazil. *Porifera Res.: Biodiversity, Innovation Sustainability*, 313–318.

Ferretti, C., B. Marengo, C. De Ciucis, M. Nitti, M.A. Pronzato, U.M. Marinari, R. Pronzato, R. Manconi, and C. Domenicotti. 2007. Effects of *Agelas oroides* and *Petrosia ficiformis* crude extracts on human neuroblastoma cell survival. *Int. J. Oncol.*, 30: 161–169.

Field, J.J., A.J. Singh, A. Kanakkanthara, T. Halafihi, P.T. Northcote, and J.H. Miller. 2009. Microtubule-stabilizing activity of zampanolide, a potent macrolide isolated from the Tongan marine sponge *Cacospongia mycofijiensis*. *J. Med. Chem.*, 52: 7328–7332.

Fiorini, L., M. Tribalat, L. Sauvard, J. Cazareth, E. Lalli, I. Broutin, O.P. Thomas, and I. Mus-Veteau. 2015. Natural paniceins from mediterranean sponge inhibit the multidrug resistance activity of patched and increase chemotherapy efficiency on melanoma cells. *Oncotarget*, 6: 22282–22297.

Fouad, M., K. Al-Trabeen, M. Badran, V. Wray, R.A. Edrada, P. Proksch, and R. Ebela. 2004. New steroidal saponins from the sponge *Erylus lendenfeldi*. *ARKIVOC*, 13: 17–27.

Fouad, M., R.A. Edrada, R. Ebel, V. Wray, W.E. Müller, W. H. Lin, and P. Proksch. 2006. Cytotoxic isomalabaricane triterpenes from the marine sponge *Rhabdastrella globostellata*. *J. Nat. Prod.*, 69: 211–218.

Franco, L.A., J.L. Mmacareno, Y.C. Ocampo, I.B. Pájaro, and R. Gaitán. 2012. Marine sponges of the genus *Neopetrosia* with anti-inflammatory activity. *Lat. Am. J. Pharm.*, 31: 976–983.

Franklin, M.A., S.G. Penn, C.B. Lebrilla, T.H. Lam, I.N. Pessah, and T.F. Molinski. 1966. Bastadin 20 and bastadin O-sulfate esters from *Ianthella basta*: Novel modulators of the Ry1R FKBP12 receptor complex. *J. Nat. Prod.*, 59: 1121–1127.

Fujita, M., Y. Nakao, S. Matsunaga, M. Seiki, Y. Itoh, J. Yamashita, R.W. Van Soest, N. Fusetani. 2003a. Ageladine A: An anti-angiogenic matrixmetalloproteinase inhibitor from the marine sponge *Agelas nakamurai*. *J. Am. Chem. Soc.*, 125: 15700–15701.

Fujita, M.,Y. Nakao, S. Matsunaga, R.W. van Soest, Y. Itoh, M. Seiki, and N. Fusetani. 2003b. Callysponginol sulfate A, an MT1-MMP inhibitor isolated from the marine sponge *Callyspongia truncata*. *J. Nat. Prod.*, 66: 569–571.

Fusetani, N., S. Matsunaga, and S. Konosu. 1981. Bioactive marine metabolites II. Halistanol sulfate, an antimicrobial novel steroid sulfate from the marine sponge *Halichondria* cf. *moorei* Bergquist. *Tetrahedron Lett.*, 22: 1985–1988.

Fusetani, N., N. Sata, N. Asai, and S. Matsunaga. 1993. Isolation and structure elucidation of erylusamine B, a new class of marine natural products, which blocked an IL-6 receptor, from the marine sponge *Erylus placenta* Thiele. *Tetrahedron Lett.*, 34: 4067–4070.

Fusetani, N., M. Sugano, S. Matsunaga, and K. Hashimoto. 1987. (+)-Curcuphenol and dehydrocurcuphenol, novel sesquiterpenes which inhibit H,K-ATPase, from a marine sponge *Epipolasis* sp. *Experientia*, 43: 1234–1235.

Fusetani, N., T. Sugawara, S. Matsunaga, and H. Hirota. 1991. Bioactive marine metabolites. Part 35. Cytotoxic metabolites of the marine sponge *Mycale adhaerens* Lambe. *J. Org. Chem.*, 56: 4971–4974.

Fusetani, N., K. Yasumuro, H. Kawaia, T. Natoria, L. Brinenb, and J. Clardy. 1990. Kalihinene and isokalihinol B, cytotoxic diterpene isonitriles from the marine sponge *Acanthella klethra*. *Tetrahedron Lett.*, 31: 3599–3602.

Galeano, E., A. Martínez, O.P. Thomas, S. Robledo, and D. Munoz. 2012. Antiparasitic bromotyrosine derivatives from the Caribbean marine sponge *Aiolochroia crassa*. *Quim. Nova*, 35: 1189–1193.

Galeano, E., O.P. Thomas, S. Robledo, D. Munoz, and A. Martinez. 2011. Antiparasitic bromotyrosine derivatives from the marine sponge *Verongula rigida*. *Mar Drugs*, 9: 1902–1913.

Galzitskaya, O., and A. Caflisch. 1999. Solution conformation of phakellistatin 8 investigated by molecular dynamics simulations. *J. Mol. Graph. Model.*, 17: 19–27.

García-Ruiz, C., and F. Sarabia. 2014. Chemistry and biology of bengamides and bengazoles, bioactive natural products from jaspis sponges. *Mar. Drugs*, 12: 1580–1622.

García-Vilas, J.A., B. Martínez-Poveda, A.R. Quesada, and M.Á. Medina. 2016. Aeroplysinin-1, a sponge-derived multi-targeted bioactive marine drug. *Mar. Drugs*, 14: 12 p.

Gaspar, H., A. Cutignano, L. Grauso, N. Neng, V. Cachatra, A. Fontana, J. Xavier, M. Cerejo, H. Vieira, and S. Santos. 2016. Erylusamides: Novel atypical glycolipids from *Erylus* cf. *deficiens*. *Mar. Drugs*, 14: 179.

Gaspar, H., S.S. Feio, A.I. Rodrigues, and R. van Soest. 2004. Antifungal activity of (+)-curcuphenol, a metabolite from the marine sponge *Didiscus oxeata*. *Mar. Drugs*, 2: 8–13.

Gauvin, A., J. Smadja, M. Aknin, R. Faure, and E. Gaydou. 2000. Isolation of bioactive 5α,8α-epidioxy sterols from the marine sponge *Luffariella* cf. *variabilis*. *Can. J. Chem.*, 78: 986–992.

Goemans, B. 2012. *Stylissa flabelliformis*. http://www.saltcorner.com/AquariumLibrary/browsespecies.php?CritterID=3257.

Gong, K., X. Tang,Y. Liu, P. Li, and G. Li. 2016. Imidazole alkaloids from the South China Sea sponge *Pericharax heteroraphis* and their cytotoxic and antiviral activities. *Molecules*, 21: 150.

González, M.A. 2010. Scalarane sesterterpenoids. *Curr. Bioact. Compd.*, 6: 178–206.

Göthel, Q., and M. Köck. 2014. New sesquiterpene hydroquinones from the Caribbean sponge *Aka coralliphagum*. *Beilstein J. Org. Chem.*, 10: 613–621.

Gotsbacher, M.P., and P. Karuso. 2015. New antimicrobial bromotyrosine analogues from the sponge *Pseudoceratina purpurea* and its predator *Tylodina corticalis*. *Mar. Drugs*, 13: 1389–1409.

Goud, T.V., M. Srinivasulu, V.L. Reddy, A.V. Reddy, T.P. Rao, D.S. Kumar, U.S. Murty, and Y. Venkateswarlu. 2003. Two new bromotyrosine-derived metabolites from the sponge *Psammaplysilla purpurea*. *Chem. Pharm. Bull.* (Tokyo), 51: 990–993.

Goud, T.V., N.S. Reddy, N.R. Swamy, S.T. Ram, and Y. Venkateswarlu. 2003. Anti-HIV active petrosins from the marine sponge *Petrosia similis*. *Biol. Pharma. Bull.*, 26: 1498–1501.

Govinden-Soulange, J., D. Marie, S. Kauroo, R. Beesoo, and A. Ramanjooloo. 2014. Antibacterial properties of marine sponges from Mauritius waters. *Trop. J.Pharm. Res.*, 13: 249–254.

Graça, A.P., J. Bondoso, H. Gaspar, J.R. Xavier, M.C. Monteiro, M. de la Cruz, D. Oves-Costales, F. Vicente, and O.M. Lage. 2013. Antimicrobial activity of heterotrophic bacterial communities from the marine sponge *Erylus discophorus* (Astrophorida, Geodiidae). *PLoS One*, 8: e78992.

Graça, A.P., F. Viana, J. Bondoso, M.I. Correia, L. Gomes, M. Humanes, A. Reis, J.R. Xavier, H. Gaspar, and O.M. Lage. 2015. The antimicrobial activity of heterotrophic bacteria isolated from the marine sponge *Erylus deficiens* (Astrophorida, Geodiidae). *Front Microbiol.*, 6: 389.

Gravalos, D.G., Kashman, Y., Rudi, A., and J.A. De La Fuente. 1995. Oxazole derivatives as antitumoral agents. EP 0687673A1. http://www.google.mk/patents/EP0687673A1?cl=en.

Green, D., I. Goldberg, Z. Stein, M. Ilan, and Y. Kashman. 1992. Cyanthiwigin A-D, novel cytotoxic diterpenes from the sponge *Epipolasis reiswigi*. *Nat. Prod. Lett.*, 1: 193–199.

Gregson, R.P., B.A. Baldo, P.G. Thomas, R.J. Quinn, P.R. Bergquist, J.F. Stephens, and A.R. Horne. 1979. Fluorine is a major constituent of the marine sponge *Halichondria moorei*. *Science*, 206: 1108–1109.

Greve, H., S. Kehraus, A. Krick, G. Kelter, A. Maier, H.H. Fiebig, A.D. Wright, and G.M. König. 2008. Cytotoxic bastadin 24 from the Australian sponge *Ianthella quadrangulata*. *J. Nat. Prod.*, 71: 309–312.

Greve, H., S. Meis, M.U. Kassack, S. Kehraus, A. Krick, A.D. Wright, and G.M. König. 2007. New iantherans from the marine sponge *Ianthella quadrangulata*: Novel agonists of the P2Y(11) receptor. *J. Med. Chem.*, 50: 5600–5607.

Gribble, G.W. 2015. Biological activity of recently discovered halogenated marine natural products. *Mar. Drugs*, 13: 4044–4136.

Grkovic, T., J.S. Blees, M.M. Bayer, N.H. Colburn, C.L. Thomas, C.J. Henrich, M.L. Peach, J.B. McMahon, T. Schmid, and K.R. Gustafson. 2014. Tricyclic guanidine alkaloids from the marine sponge *Acanthella cavernosa* that stabilize the tumor suppressor PDCD4. *Mar Drugs*, 12: 4593–4601.

Grosso, C., P. Valentão, F. Ferreres, and P.B. Andrade. 2014. Bioactive marine drugs and marine biomaterials for brain diseases. *Mar. Drugs*, 12: 2539–2589.

Guimarães, T.R., C.G. Quiroz, C. Rigotto, S.Q. de Oliveira, M.T.R. de Almeida, E.M. Bianco, M.I.G. Moritz et al. 2013. Anti HSV-1 activity of halistanol sulfate and halistanol sulfate C isolated from Brazilian marine sponge *Petromica citrina* (Demospongiae). *Mar. Drugs*, 11: 4176–4192.

Gul, W., and M.T. Hamann. 2005. Indole alkaloid marine natural products: An established source of cancer drug leads with considerable promise for the control of parasitic, neurological and other diseases. *Life Sci.*, 78: 442–453.

Gunasekera, S.P., M. Gunasekera, R.E. Longley, and G.K. Schulte. 1990. Discodermolide: A new bioactive polyhydroxylated lactone from the marine sponge *Discodermia dissoluta*. *J. Org. Chem.*, 55: 4912–4915.

Gunasekera, S.P., S.H. Sennett, M. Kelly-Borges, and R.W. Bryant. 1994. Ophirapstanol trisulfate, a new biologically active steroid sulfate from the deep water marine sponge *Topsentia ophiraphidites*. *J. Nat. Prod.*, 57: 1751–1754.

Gupta, P., U. Sharma, T.C. Schulz, A.B. McLean, A.J. Robins, and L.M. West. 2012. Bicyclic C21 terpenoids from the marine sponge *Clathria compressa*. *J. Nat. Prod.*, 75: 1223–1227.

Hadasa, E., M. Shpigelb, and M. Ilan. 2005. Sea ranching of the marine sponge *Negombata magnifica* (Demospongiae, Latrunculiidae) as a first step for latrunculin B mass production. *Aquaculture*, 244: 159–169.

Hadisaputri, Y.E. 2017. Potential cytotoxic effects of sponge *Callyspongia aerizusa* extract in several cancer cells. *2nd ISEJ*, 1(1): 40.

Hallock, Y.F., J.H. Cardellina, M.S. Balaschak, M.R. Alexander, T.R. Prather, R.H. Shoemaker, and M.R.J. Boyd. 1995. Antitumor activity and stereochemistry of acetylenic alcohols from the sponge *Cribrochalina vasculum*. *Nat. Prod.*, 58: 1801–1807.

Han, Y., B. Yang, F. Zhang, X. Miao, and Z. Li. 2009. Characterization of antifungal chitinase from marine *Streptomyces* sp. DA11 associated with South China Sea sponge *Craniella australiensis*. *Mar. Biotechnol.* (NY), 11: 132–140.

Handayani, D., R.F. Ahdinur, and R. Rustini. 2015. Antimicrobial activity of endophytic fungi from marine Sponge *Haliclona fascigera*. *J. App. Pharm. Sci.*, 5: 154–156.

Handayani, D., and I. Aminah. 2017. Antibacterial and cytotoxic activities of ethyl acetate extract of symbiotic fungi from West Sumatra marine sponge *Acanthrongylophora ingens*. *J. Appl. Pharm. Sci.*, 7: 237–240.

Handayani, D., R.A. Edrada, P. Proksch, V. Wray, L. Witte, R.W.M. Van Soest, and A. Kunzmann. 1997. Four new bioactive polybrominated diphenyl ethers of the sponge *Dysidea herbacea* from West Sumatra, Indonesia. *J. Nat. Prod.*, 60: 1313–1316.

Handayani, D., M. Murniati, and R. Rustini. 2016a. In vitro inhibitory activity of ethyl acetate extract of symbiotic bacteria isolated from the marine sponge *Haliclona fascigera* against multidrug resistant organism (MDRO). *J. App. Pharm. Sci.*, 6: 218–222.

Handayani, D., R. Ornando, and Rustini. 2016b. Antimicrobial activity screening of symbiotic fungi from marine sponge *Petrosia nigrans* collected from South Coast of West Sumatera, Indonesia. *IJPPR* 8: 623–626.

Hanssen, K., J.H. Andersen, T. Stiberg, R.A. Engh, J. Svenson, A. Geneviève, and E. Hansen. 2012. Antitumoral and mechanistic studies of ianthelline isolated from the Arctic sponge *Stryphnus fortis*. *Anticancer Res.*, 32: 4287–4297.

Hardoim, C.C.P., and R. Costa. 2014. Microbial communities and bioactive compounds in marine sponges of the family Irciniidae—A review. *Mar. Drugs*, 12: 5089–5122.

Hassan, M.H.A., M.E. Rateb, M. Hetta, T.A. Abdelaziz, M.A. Sleim, M. Jaspars, and R. Mohammed. 2015. Scalarane sesterterpenes from the Egyptian Red Sea sponge *Phyllospongia lamellose*. *Tetrahedron*, 71: 577–583.

Hayes, M. 2012. *Marine Bioactive Compounds: Sources, Characterization and Applications*. Springer.

He, F., L.H. Mai, A. Longeon, B.R. Copp, N. Loaëc, A. Bescond, L. Meijer, and M. Bourguet-Kondracki. 2015. Novel adociaquinone derivatives from the Indonesian sponge *Xestospongia* sp. *Mar. Drugs*, 13: 2617–2628.

He, W., D. Xue, L. Yao, J. Li, H. Liu, and Y. Guo. 2016. A new bioactive steroidal ketone from the South China Sea sponge *Xestospongia testudinaria*. *J. Asian Nat. Prod. Res.*, 18: 195–199.

Hentschel, H., M. Schmid, M. Wagner, L. Fieseler, C. Gernert, and J. Hacker. 2001. Isolation and phylogenetic analysis of bacteria with antimicrobial activities from the Mediterranean sponges *Aplysina aerophoba* and *Aplysina cavernicola*. *Microb. Ecol.*, 35: 305–312.

Hernández-Guerrero, C.J., E. Zubía, M.J. Ortega, and J.L. Carballo. 2007. Cytotoxic dibromotyrosine-derived metabolites from the sponge *Aplysina gerardogreeni*. *Bioorg. Med. Chem.*, 15: 5275–5282.

Hirade, H., T. Haruyama, N. Kobayashi, N.J. de Voogd, and J. Tanaka. 2017. A new imidazole from the sponge *Dercitus (Halinastra) japonensis*. *Nat. Prod. Commun.*, 12: 19–20.

Hirano, S., H.T. Quach, T. Watanabe, N. Kanoh, Y. Iwabuchi, T. Usui, and T. Kataoka. 2015. Irciniastatin A, a pederin-type translation inhibitor, promotes ectodomain shedding of cell-surface tumor necrosis factor receptor 1. *J. Antibiot.* (Tokyo), 68: 417–420.

Hong, L.L., J. Sun, F. Yang, M. Liu, J. Tang, F. Sun, W. Jiao, S. Wang, W. Zhang, and H. Lin. 2017a. New diterpene alkaloids from the marine sponge *Agelas mauritiana*. *RSC Adv.*, 7: 23970–23976.

Hong, L.L., H.B. Yu, J. Wang, W.H. Jiao, B.H. Cheng, F. Yang, Y.J. Zhou, B.B. Gu, S.J. Song, and H.W. Lin. 2017b. Unusual anti-allergic diterpenoids from the marine sponge *Hippospongia lachne*. *Sci. Rep.*, 7: 43138, 7 p.

Hoshino, A., H. Mitome, H. Miyaoka, A. Shintani, Y. Yamada, R.W. van Soest. 2003. New strongylophorines from the Okinawan marine sponge *Petrosia (Strongylophora) corticata*. *J. Nat. Prod.*, 66: 1600–1605.

Huang, R., W. Ma, J. Dong, X. Zhou, T. Xu, K. Lee, X. Yang, S. Xu, and Y. Liu. 2010. A new 1,4-diazepine from South China Sea marine sponge *Callyspongia Species*. *Molecules*, 15: 871–877.

Huyen, L.T., D.T.T. Hang, N.X. Nhiem, P.H. Yen, H.L.T. Anh, T.H. Quang, B.H. Tai, N.V. Dau, and P.V. Kiem. 2017. Naphtoquinones and sesquiterpene cyclopentenones from the sponge *Smenospongia cerebriformis* with their cytotoxic activity. *Chem. Pharm. Bull.* (Tokyo), 65: 589–592.

Hwang, B.S., K. Lee, C. Yang, E.J. Jeong, and J. Rho. 2013. Characterization and anti-inflammatory effects of iodinated acetylenic acids isolated from the marine sponges *Suberites mammilaris* and *Suberites japonicus*. *J. Nat. Prod.*, 76: 2355–2359.

Hwang, I.H., O.H. Joonseok, W. Zhou, S. Park, J. Kim, A.G. Chittiboyina, D. Ferreira et al. 2015. Cytotoxic activity of rearranged drimane meroterpenoids against colon cancer cells via down-regulation of β-catenin expression. *J. Nat. Prod.*, 78: 453–461.

Ibrahim, S.R., and G.A. Mohamed. 2016. Ingenines C and D, new cytotoxic pyrimidine-β-carboline alkaloids from the Indonesian sponge *Acanthostrongylophora ingens*. *Phytochem. Lett.*, 18: 168–171.

Ibrahim, S.R.M., and G.A. Mohamed. 2017. Ingenine E, a new cytotoxic β-carboline alkaloid from the Indonesian sponge *Acanthostrongylophora ingens*. *J. Asian Nat. Prod. Res.*, 19: 504–509.

Ibrahim, S.R., G.A. Mohamed, M.F. Zayed, and H.M. Sayed. 2015. Ingenines A and B, two new alkaloids from the Indonesian sponge *Acanthostrongylophora ingens*. *Drug Res.* (Stuttg). 65: 361–365.

Ibrahim, S.R.M., R.A. Edrada-Ebel, G.A. Mohamed, D.T.A. Youssef, V. Wray, and P. Proksch. 2008. Callyaerin G, a new cytotoxic cyclic peptide from the marine sponge *Callyspongia aerizusa*. *Arkivoc (Archiv. Org. Chem.)*, 12: 164–171.

Ibrahim, S.R.M., C.C. Min, F. Teuscher, R. Ebel, C.Kakoschke, W. Lin, V. Wray, R. Edrada-Ebel, and P.Proksch. 2010. Callyaerins A–F and H, new cytotoxic cyclic peptides from the Indonesian marine sponge *Callyspongia aerizusa*. *Bioorg. Med. Chem.*, 18: 4947–4956.

Ibrahim, S.R.M., G.A. Mohamed, E.S.E. Mostaf, and A.F.P. Proksch. 2013. Sagitol C, a new cytotoxic pyridoacridine alkaloid from the sponge *Oceanapia* sp. *Bull. Faculty Pharm., Cairo Univ.*, 51: 229–232.

Inbaneson, S.J., and S. Ravikumar. 2012. In vitro antiplasmodial activity of marine sponge *Clathria indica* associated bacteria against *Palsmodium falciparum*. *Asian Pac. J. Trop. Biomed.*, 2: S1090–S1095.

Indraningrat, A.A.G., H. Smidt, and D. Sipkema. 2016. Bioprospecting sponge-associated microbes for antimicrobial compounds. *Mar. Drugs*, 14: 87.

Inman, W.D., W.M. Bray, N.C. Gassner, R.S. Lokey, K. Tenney, Y.Y. Shen, K. Tendyke, T. Suh, and P. Crews. 2010. A beta-carboline alkaloid from the Papua New Guinea marine sponge *Hyrtios reticulatus*. *J. Nat. Prod.*, 73: 255–257.

Ishibashi, M., Y. Ohizumi, J.F. Cheng, H. Nakamura, Y. Hirata, T. Sasaki, and J. Kobayashi. 1988. Metachromins A and B, novel antineoplastic sesquiterpenoids from the Okinawan sponge *Hippospongia* cf. *metachromia*. *J. Org. Chem.*, 53: 2855–2858.

Jadulco, R., R.A. Edrada, R. Ebel, A. Berg, K. Schaumann, V. Wray, K. Steube, and P. Proksch. 2004. New communesin derivatives from the fungus *Penicillium* sp. derived from the Mediterranean aponge *Axinella verrucosa. J. Nat. Prod.*, 67: 78–81.

Jang, K.H., Y. Lee, C.J. Sim, K.B. Oh, and J. Shin. 2012. Bioactive lipids from the sponge *Spirastrella abata. J. Bioorg.Med. Chem. Lett.*, 22: 1078–1081.

Jares-Erijman, E.A., R. Sakai, and K.L. Rinehart. 1991. Crambescidins: New antiviral and cytotoxic compounds from the sponge *Crambe crambe. J. Org. Chem.*, 56: 5712–5715.

Jasplakinolide, *Jaspis johnstonii*. n.d. http://search.cosmobio.co.jp/cosmo_search_p/search_gate2/docs/CBC_/420107.20040716.pdf.

Jeong, S.J., M. Inagaki, R. Higuchi, T. Miyamoto, M. Ono, M. Kuwano, R.W. Van Soest. 2003. 1,3-dimethylisoguaninium, an antiangiogenic purine analog from the sponge *Amphimedon paraviridis. Chem. Pharm. Bull. (Tokyo)*, 51: 731–733.

Jha, R.K., and X. Zi-rong. 2004. Biomedical compounds from marine organisms. *Mar. Drugs*, 2: 123–146.

Jiao, W., B. Cheng, G. Shi, G. Chen, B. Gu, Y. Zhou, L. Hong et al. 2017a. Dysivillosins A–D, unusual anti-allergic meroterpenoids from the marine sponge *Dysidea villosa. Sci. Rep.*, 7: 8947.

Jiao, W., L.L. Hong, J.B. Sun, S.J. Piao, G.D. Chen, H. Deng, S.P. Wang, F. Yang, and H. W. Lin. 2017b. (±)-Hippolide J—A pair of unusual antifungal enantiomeric sesterterpenoids from the marine sponge *Hippospongia lachne. Eur. J. Org. Chem.*, 2017: 3421–3426.

Jiao, W., J. Li, Q. Liu, T. Xu, Q. Shi, H. Yu, F. Yang, B. Han, M. Li, and H. Lin. 2014. Dysidinoid A, an unusual meroterpenoid with anti-MRSA activity from the South China Sea sponge *Dysidea* sp. *Molecules*, 19: 18025–18032.

Jiao, W., T. Xu, B. Gu, G. Shi, Y. Zhu, F. Yang, B. Han et al. 2015. Bioactive sesquiterpene quinols and quinones from the marine sponge *Dysidea avara. RSC Advances*, 106. http://pubs.rsc.org/-/content/articlelanding/2015/ra/c5ra18876h#!div(Abstract).

Jin, Y., S. Fotso, Z. Yongtang, M. Sevvana, H. Laatsch, and W. Zhang. 2006. Halichondria sulfonic acid, a new HIV-1 inhibitory guanidino-sulfonic acid, and halistanol sulfate isolated from the marine sponge *Halichondria rugosa* Ridley & Dendy. *Nat. Prod. Res.*, 20: 1129–1135.

Johnson, T.A., T. Amagata, K.V. Sashidhara, A.G.O. Tenney, T. Matainaho, K.K. Ang, J.H. McKerrow, and P. Crews. 2009. The aignopsanes, a new class of sesquiterpenes from selected chemotypes of the sponge *Cacospongia mycofijiensis. Org. Lett.*, 11: 1975–1978.

Joseph, B., V.M. Nair, and S. Sujatha. 2012. Pharmacological perception of peptides from marine sponge: A review. *IJPSR*, 3: 4689–4696.

Jyothi, M., Y.H.I. Mohammed, Zabiulla, S.V. Mamatha, F.H. Al-Ostoot, and S.A. Khanum. 2016. Study of marine-derived anti-mycobacterium compounds: A review. *J. Bio. Innov.*, 5: 647–662.

Kakou, Y., P. Crews, and G.J. Bakus. 1987. Dendrolasin and latrunculin A from the Fijian sponge *Spongia mycofijiensis* and an associated nudibranch *Chromodoris lochi. J. Nat. Prod.*, 50: 482–484.

Karyawati, A.T., W. Manalu, I.H. Suparto, and R.S. Mauboy. 2015. Antioxidant and anticancer activity of the marine sponge *Clathria Basilana. J. Nat. Sci. Res.*, 5: 10–16.

Kashman, Y., G. Koren-Goldshlager, M.D.G. Gravalos, and M. Schleyer. 1999. Halitulin, a new cytotoxic alkaloid from the marine sponge *Haliclona tulearensis. Tetrahedron Lett.*, 40: 997–1000.

Kato, Y., N. Fusetani, S. Matsunaga, and K. Hashimoto. 1988. Calyculins, potent antitumour metabolites from the marine sponge *Discodermia calyx*: Biological activities. *Drugs Exp. Clin. Res.*, 14: 723–728.

Kaur, K., V. Kumar, A.K. Sharma, and G.K. Gupta. 2014. Isoxazoline containing natural products as anticancer agents: A review. *Eur. J. Med. Chem.*, 77: 121–133.

Kazanjian, A., and M. Fariñas. 2006. Actividades biológicas del extracto acuoso de la esponja *Aplysina lacunosa* (Porifera: Aplysinidae). *Rev. Biol. Trop. (Int. J. Trop. Biol.)*, 54: S189–S200.

Kelly, M., C. Sim-Smith, R. Stone, T. Samaai, H. Reiswig, and W. Austin. 2016. New taxa and arrangements within the family Latrunculiidae (Demospongiae, Poecilosclerida). *Zootaxa*, 4121: 1–48.

Keyzers, R.A., and M.T. Davies-Coleman. 2005. Anti-inflammatory metabolites from marine sponges. *Chem. Soc. Rev.*, 34: 355–365.

Khakshoor, M.S., and J. Pazooki. 2014. Bactericidal and fungicidal activities of different crude extracts of *Gelliodes carnosa* (sponge, Persian Gulf). *Iran. J. Fish. Sci.*, 13: 776–784.

Kijjoa, A., R. Wattanadilok, N. Campos, M.S.J. Nascimento, M. Pinto, and W. Herz. 2007. Anticancer activity evaluation of kuanoniamines A and C isolated from the marine sponge *Oceanapia sagittaria*, collected from the Gulf of Thailand. *Mar. Drugs*, 5: 6–22.

Kilimnik, A., D.V. Kuklev, and V.M. Dembitsky. 2016. Antitumor acetylenic lipids. *Matthews J. Phar. Sci.*, 1: 005.

Kim, S. 2013. *Marine Microbiology: Bioactive Compounds and Biotechnological Applications*. Wiley-VCH.

Kitamura, H., K. Iwakabe, T. Yahata, S. Nishimura, A. Ohta, Y. Ohmi, M. Sato et al. 1999. The natural killer T (NKT) cell ligand alpha-galactosylceramide demonstrates its immunopotentiating effect by inducing interleukin (IL)-12 production by dendritic cells and IL-12 receptor expression on NKT cells. *J. Exp. Med.*, 189: 1121–1128.

Kobayashi, J., K. Yuasa, T. Kobayashi, T. Sasaki, and M. Tsuda. 1996. Jaspiferals A ~ G, new cytotoxic isomalabaricane-type nortriterpenoids from Okinawan marine sponge *Jaspis stellifera. Tetrahedron*, 52: 5745–5750.

Kobayashi, M. 1991. Mechanism of inotropic action of xestoquinone, a novel cardiotonic agent isolated from a sea sponge. *J. Pharmacol. Exp. Ther.*, 257: 82–89.

Kobayashi, M., K. Higuchi, N. Murakami, H. Tajima, and S. Aoki. 1997. Callystatin A, a potent cytotoxic polyketide from the marine sponge, *Callyspongia truncata. Tetrahedron Lett.*, 38: 2859–2862.

Kon, K., and M. Rai. 2016. *Antibiotic Resistance: Mechanisms and New Antimicrobial Approaches*. Academic Press.

Kong, F., and D.J. Faulkner. 1993. Leucettamines A and B, two antimicrobial lipids from the calcareous sponge *Leucetta microraphis. J. Org. Chem.*, 58: 970–971.

König, G.M., and A.D. Wright. 1996. Novel potent antimalarial diterpene isocyanates, isothiocyanates, and isonitriles from the tropical marine sponge *Cymbastela hooperi. J. Org. Chem.*, 61: 3259–3267.

Kossuga, M.H., A.M. Nascimento, J.Q. Reimão, A.G. Tempone, N.N. Taniwaki, K. Veloso, A.G. Ferreira et al. 2008. Antiparasitic, antineuroinflammatory, and cytotoxic polyketides from the marine sponge *Plakortis angulospiculatus* collected in Brazil. *J. Nat. Prod.*, 71: 334–339.

Kumar, M.S., D. Desai, and A.K. Pal. 2012. Immunomodulatory potential of a marine sponge *Spongosorites halichondrioides* (Dendy, 1905). *Turk. J. Fish. Aquat. Sci.*, 12: 67–71.

Kumar, M.S., and S. Gopalkrishnan. 2014. Evaluation, partial characterization and purification of acetylcholine esterase enzyme and antiangiogenic activity from marine sponges. *J. Coast. Life Med.*, 2: 849–854.

Kumar, M.S., and A.K. Pal. 2016. A review of bioactive compounds from marine organisms with special mention on the potential of marine sponges in pharmacological applications. *J. Mar. Biol. Ass. India*, 58: 87–96.

Kumar, M.S., K.K. Vijaylaxmi, and A.K. Pal. 2014. Anti-inflammatory and antioxidant properties of *Spongosorites halichondriodes*, a marine sponge. *Turk. J. Pharm. Sci.*, 11: 285–294.

Kumar, S.P., and S. Ravikumar. 2014. In vitro antiplasmodial activity of marine sponge *Clathria vulpina* extract against chloroquine sensitive *Plasmodium falciparum*. *Asian. Pac. J. Trop. Dis.*, 4: S162–S166.

Kutluay, V.M., B. Konuklugil, and I. Saracoglu. 2016. Cytotoxic activity screening of some Turkish marine sponge species. *Planta Med.*, 82: S1–S381.

Lai, K., Y. Liu, J. Su, M. El-Shazly, C. Wu, Y. Du, Y. Hsu et al. 2016. Antileukemic scalarane sesterterpenoids and meroditerpenoid from *Carteriospongia (Phyllospongia)* sp., induce apoptosis via dual inhibitory effects on topoisomerase II and Hsp90. *Sci. Rep.*, 6: 36170.

Lai, Y., M. Lu, L. Wang, J. Chen, L. Fang, Y. Wu, and P. Sung. 2015. New scalarane sesterterpenoids from the formosan sponge *Ircinia felix*. *Mar. Drugs*, 13: 4296–4309.

Lakshmi, V., P. Khare, P. Misra, M.N. Srivastava, and A. Dube. 2015. Antileishmanial potential of a marine sponge *Haliclona oculata* against experimental visceral leishmaniasis. *J. Coast. Life Med.*, 3: 187–192.

Lakshmi, V., S.K. Mishra, S. Srivastava, A. Chaturvedi, M.N. Srivastava, and P.K. Shukl. 2010. Antifungal activity of marine sponge *Haliclona exigua* (Kirkpatrick). *J. Mycol. Med.*, 20: 31–35.

Lakshmi, V., A. Saxena, S.K. Mishra, M. Mishra, S. Srivastava, and S. Ghoshal. 2009. Antiamebic activity of marine sponge *Haliclona exigua* (Krikpatrick). *Bangladesh J. Pharmacol.*, 4: 55–59.

Lakshmi, V., and P.K. Shukla. 2013. Antifungal property of marine sponge *Haliclona oculata* (Krikpatrick). *Nat. Prod.*, 9: 339–344.

Lauritano, C., and A. Lanora. 2016. Marine organisms with anti-diabetes properties. *Mar. Drugs*, 14: 220.

Lee, H., K. Yoon, Y. Han, K.J. Lee, S. Chung, T. Kim, S. Lee, J. Shin, and K. Oh. 2009. 5-Hydroxyindole-type alkaloids, as *Candida albicans* isocitrate lyase inhibitors, from the tropical sponge *Hyrtios* sp. *Bioorg. Med. Chem. Lett.*, 19: 1051–1053.

Lefranc, F., G. Nuzzo, N.A. Hamdy, I. Fakhr, L.M.Y. Banuls, G. Van Goietsenoven, G. Villani et al. 2013. In vitro pharmacological and toxicological effects of norterpene peroxides isolated from the Red Sea sponge *Diacarnus erythraeanus* on normal and cancer cells. *J. Nat. Prod.*, 76: 1541–1547.

Li, J., L. Du, M. Kelly, Y. Zhou, and D.G. Nagle. 2013. Structures and potential antitumor activity of sesterterpenes from the marine sponge *Hyrtios communis*. *J. Nat. Prod.*, 76: 1492–1497.

Li, J., B. Xu, J. Cui, Z. Deng, N.J. de Voogd, P. Proksch, and W. Lin. 2010. Globostelletins A–I, cytotoxic isomalabaricane derivatives from the marine sponge *Rhabdastrella globostellata*. *Bioorg. Med. Chem.*, 18: 4639–4647.

Li, W., Y. Yi, H. Wu, Q. Xu, H. Tang, D. Zhou, H. Lin, and Z. Wang. 2003. Isolation and structure of the cytotoxic cycloheptapeptide phakellistatin 13. *J. Nat. Prod.*, 66: 146–148.

Li, Y., S.W.A. Himaya, and S. Kim. 2013. Triterpenoids of marine origin as anti-cancer agents. *Molecules*, 18: 7886–7909.

Li, Y., H. Tang, X. Tian, H. Lin, M. Wang, and M. Yao. 2015. Three new cytotoxic isomalabaricane triterpenes from the marine sponge *Stelletta tenuis*. *Fitoterapia*. 106: 226–230.

Li, Y., Y. Zhang, X. Shen, and Y. Guo. 2009. A novel sesquiterpene quinone from Hainan sponge *Dysidea villosa*. *Bioorganic Med. Chem. Lett.*, 19: 390–392.

Li, Z.Y., and Y. Liu. 2006. Marine sponge *Craniella austrialiensis*-associated bacterial diversity revelation based on 16S rDNA library and biologically active Actinomycetes screening, phylogenetic analysis. *Lett. Appl. Microbiol.*, 43: 410–416.

Lima, B.A., S.P. de Lirac, M.H. Kossugad, R.B. Gonçalvese, R.G.S. Berlinckd, and R.U. Kamiya. 2014. Halistanol sulfate A and rodriguesines A and B are antimicrobial and antibiofilm agents against the cariogenic bacterium *Streptococcus mutans*. *Rev. Bras. Farmacogn.*, 24: 651–659.

Lin, H., Z. Wang, J. Wu, N. Shi, H. Zhang, W. Chen, S.L. Morris-Natschke, and A. Lin 2007. Stellettins L and M, Cytotoxic isomalabaricane-type triterpenes, and sterols from the marine sponge *Stelletta tenuis*.. *J. Nat. Prod.*, 70: 1114–1117.

Lin, K., P. Yang, H. Yang, A.H. Liu, L.G. Yao, Y.W. Guo, and S.C. Mao. 2015. Lysophospholipids from the Guangxi sponge *Spirastrella purpurea*. *Lipids*, 50: 697–703.

Linington, R.G., M. Robertson, A. Gauthier, B.B. Finlay, J.B. MacMillan, T.F. Molinski, R. van Soest, and R.J. Andersen. 2006. Caminosides B–D, antimicrobial glycolipids isolated from the marine sponge *Caminus sphaeroconia*. *J. Nat. Prod.*, 69: 173–177.

Lira, N.S., R.C. Montes, J.F. Tavares, M.S. da Silva, E.V.L. da Cunha, P.F. de Athayde-Filho, L.C. Rodrigues, C.S. Dias, and J.M. Barbosa-Filho. 2011. Brominated compounds from marine sponges of the genus *Aplysina* and a compilation of their 13C NMR spectral data. *Mar. Drugs*, 9: 2316–2368.

Lira, T.O., G.F.F. Nascimento, E. Hajdu, and R.G.S. Berlinck. 2006. Further dibromotyrosine-derived metabolites from the marine sponge *Aplysina caissara*. *J. Braz. Chem. Soc.*, 17: 1233–1240.

Liu, C., X. Tang, P. Li, and G. Li. 2012. Suberitine A–D, four new cytotoxic dimeric aaptamine alkaloids from the marine sponge *Aaptos suberitoides*. *Org. Lett.*, 14: 1994–1997.

Liu, H., M. Namikoshi, S. Meguro, H. Nagai, H. Kobayashi, and X. Yao. 2004. Isolation and characterization of polybrominated diphenyl ethers as inhibitors of microtubule assembly from the marine sponge *Phyllospongia dendy*i collected at Palau. *J. Nat. Prod.*, 67: 472–474.

Liu, R., Y. Liu, Y. Zhou, and D.G. Nagle. 2007. Molecular-targeted antitumor agents 15: Neolamellarins from the marine sponge *Dendrilla nigra* inhibit hypoxia-inducible factor-1 (HIF-1) activation and secreted vascular endothelial growth factor (VEGF) production in breast tumor cells. *J. Nat. Prod.*, 70: 1741–1745.

Liu, W.K., F.W.K. Cheung, Y.H. Ling, and C.T. Che. 2013. Anti-melanoma activity of stellettin A. *Planta Med.*, 79–PE4.

Liu, W.K., J.C.K. Ho, and C.T. Che. 2005. Apoptotic activity of isomalabaricane triterpenes on human promyelocytic leukemia HL60 cells. *Cancer Let.*, 230: 102–110.

Liu, Y., H. Ji, J. Dong, S. Zhang, K.J. Lee, and S. Matthew. 2008. Antioxidant alkaloid from the South China Sea marine sponge *Iotrochota* sp. *Z. Naturforsch C.*, 63: 636–638.

Longeon, A., B.R. Copp, E. Quévrain, M. Roué, B. Kientz, T. Cresteil, S. Petek, C. Debitus, and M. Bourguet-Kondracki. 2011. Bioactive indole derivatives from the South Pacific marine sponges *Rhopaloeides odorabile* and *Hyrtios* sp. *Mar Drugs*, 9: 879–888.

Lu, Z., R.M. Van Wagoner, M.K. Harper, H.L. Baker, J.N. Hooper, C.A. Bewley, and C.M. Ireland. 2011. Mirabamides E-H, HIV-inhibitory depsipeptides from the sponge *Stelletta clavosa. J. Nat. Prod.*, 74: 185–193.

Lv, F., M. Xu, Z. Deng, N.J. de Voogd, R.W.M. van Soest, P. Proksch, and W. Lin. 2008. Rhabdastrellins A–F, isomalabaricane triterpenes from the marine sponge *Rhabdastrella* aff. *distincta. J. Nat. Prod.*, 71: 1738–1741.

Ly, B.T.K., N.T.P. Dung, N.Q. Huy, P.T.K. Tram, H.T. Chi, and N.H. Dang. 2017. Antibacterial activity of crude methanolic and fractionated extracts of *Aaptos suberitoides. J. Sci., Ho Chi Minh City Open Univ.*, 7: 66–71.

Lysek, N., R. Kinscherf, R. Claus, and T. Lindel. 2003. L-5-Hydroxytryptophan: Antioxidant and anti-apoptotic principle of the intertidal sponge *Hymeniacidon heliophila. Z. Naturforsch C.*, 58: 568–572.

Machida, K., T. Abe, D. Arai, M. Okamoto, I. Shimizu, and N.J. de Voogd, N. Fusetani, and Y. Nakao. 2014. Cinanthrenol A, an estrogenic steroid containing phenanthrene nucleus, from a marine sponge *Cinachyrella* sp. *Org. Lett.*, 16: 1539–1541.

Makarieva, T.N., K.M. Tabakmaher, A.G. Guzii, V.A. Denisenko, P.S. Dmitrenok, A.S.Kuzmich, H. Lee, and V.A. Stonik. 2012. Monanchomycalins A and B, unusual guanidine alkaloids from the sponge *Monanchora pulchra. Tetrahedron Lett.*, 53: 4228–4231.

Makarieva, T.N., K.M. Tabakmaher, A.G. Guzii, V.A. Denisenko, P.S. Dmitrenok, L.K. Shubina, A.S. Kuzmich, H. Lee, and V.A. Stonik. 2011. Monanchocidins B–E: Polycyclic guanidine alkaloids with potent antileukemic activities from the sponge *Monanchora pulchra. J. Nat. Prod.*, 74: 1952–1958.

Marinho, P.R., G.R.S. Muricy, M.F.L. Silva, M.G. de Marvall, and M.S. Laport. 2010. Antibiotic-resistant bacteria inhibited by extracts and fractions from Brazilian marine sponges. *Braz. J. Pharmacog.*, 20: 267–275.

Marques, J., E. Vilanova, P.A.S. Mourão, and X. Fernàndez-Busquet. 2016. Marine organism sulfated polysaccharides exhibiting significant antimalarial activity and inhibition of red blood cell invasion by *Plasmodium. Sci. Rep.* 6: 24368.

Márquez, D.M., S.M. Robledo, and A. Martínez. 2007. Antileishmanial epidioxysterols from extracted sterols of the Colombian marine sponge *Ircinia campana. Porifera Res.: Biodiversity, Innovation Sustainability*, 433–437.

Martin, J., F. Berrue, P. Amade, R. Fernandez, A. Fracesch, F. Reyes, and C. Cunevas. 2005. Halogenated helinane derivatives from the maria sponge *Spirastrellahartmani, J. Nat. Prod.*, 68: 1554–1555.

Martínez, J.M., C.E.R. Veloz, A.F. Valdés, D.A. Rodríguez, C.L. Rivero, and O.V. Iglesias. 2012. In vivo antiplasmodial activity of *Mycale laxissima* and *Clathria echinata* sponges. *Rev. Cubana Med. Trop.*, 64: 244–255.

Mauti, G.O., E.M. Mauti, and J. Onguso. 2014. *Oceanic Findings Applied In Destruction of Diseases: Bioactive Secondary Metabolites from Sea Sponge (Clathria indica) vs urinary Tract Pathogens.* LAP LAMBERT Academic Publishing.

Mayer, A.M., K.B. Glaser, C. Cuevas, R.S. Jacobs, W. Kem, R.D. Little, J.M. McIntosh, D.J. Newman, B.C. Potts, and D.E. Shuster. 2010. The odyssey of marine pharmaceuticals: A current pipeline perspective. *Trends Pharmacol. Sci.*, 31: 255–265.

Mayer, A.M., M.H. Kossuga, K. Veloso, A.G. Ferreira, E. Hajdu, and R.G. Berlinck. 2008. Antineuroinflammatory polyketide plakortide N from the Brazilian marine sponge *Plakortis angulospiculatus. FASEB J.*, 22(Suppl. 1).

McCulloch, M.W.B., T.S. Bugni, G.P. Concepcion, G.S. Coombs, M.K. Harper, S. Kaur, G.C. Mangalindan, M.M. Mutizwa, C.A. Veltri, D.M. Virshup, and C.M. Ireland. 2009. Carteriosulfonic acids A–C, GSK-3β inhibitors from a *Carteriospongia* sp. *J. Nat. Prod.*, 72: 1651–1656.

McNamara, C.E., L. Larsen, N.B. Perry, J.L. Harper, M.V. Berridge, E.W. Chia, M. Kelly, and V.L. Webb. 2005. Anti-inflammatory sesquiterpene-quinones from the New Zealand sponge *Dysidea* cf. *cristagalli. J. Nat. Prod.*, 68: 1431–1433.

Medeirosa, M.A., A. Lourenc, M.R. Tavares, M.J.M. Curto, S.S. Feio, and J.C. Roseiro. 2006. (–)-Agelasidine A from *Agelas clathrodes. Z. Naturforsch C.*, 61: 472–476.

Meesala, S., V.R. Lakwal, D.S. Kharate, S.S. Mokashe, and M.G. Watve. 2016. Isolation and characterization of bioactive benzofuran sesquiterpene from marine sponge *Ircinia fusca* (Carter, 1880). *J. Chem. Bio. Phy. Sci.* (Sec. A), 6: 1400–1418.

Mehbub, M.F., J. Lei, C. Franco, and W. Zhan. 2014. Marine sponge derived natural products between 2001 and 2010: Trends and opportunities for discovery of bioactives. *Mar. Drugs*, 12: 4539–4577.

Mehbub, M.F., J.E. Tanner, S.J. Barnett, C.M. Franco, and W. Zhang. 2016. The role of sponge-bacteria interactions: The sponge *Aplysilla rosea* challenged by its associated bacterium Streptomyces ACT-52A in a controlled aquarium system. *Appl. Microbiol. Biotechnol.*, 100: 10609–10626.

Miao, S., R.J. Andersen, and T.M. Allen. 1990. Cytotoxic metabolites from the sponge *Ianthella basta* collected in Papua New Guinea. *J.Nat. Prod.*, 53: 1441–1446.

Millipore Sigma. 2018. 208851: Calyculin A, *Discodermia calyx* - CAS 101932-71-2 – Calbiochem. http://www.merckmillipore.com/IN/en/product/Calyculin-A-Discodermia-calyx-CAS-101932-71-2-Calbiochem,EMD_BIO-208851?ReferrerURL=https%3A%2F%2Fwww.google.co.in%2F&bd=1.

Minale, L., R. Riccio, and G. Sodano. 1974. Acanthellin-1, an unique isonitrile sesquiterpene from the sponge *Acanthella acuta. Tetrahedron*, 30: 1341–1343.

Mioso, R., F.J.T. Marante, R.S. Bezerra, F.V.P. Borges, B.V.O. Santos and I.M.B. de Laguna. 2017. Cytotoxic compounds derived from marine sponges: A review (2010–2012). *Molecules*, 22: 208.

Mishra, P.M., A. Sree, and S. Baliarsingh. 2009. Antibacterial study and fatty acid analysis of lipids of the sponge *Myrmekioderma granulata. Chem. Nat. Compd.*, 45: 621.

Mohammed, R., J. Peng, M. Kelly, and M.T. Hamann. 2006. Cyclic heptapeptides from the Jamaican sponge *Stylissa caribica. J. Nat. Prod.*, 69: 1739–1744.

Mokhlesi, A., R. Hartmann, T. Kurtán, H. Weber, W. Lin, C. Chaidir, W.E.G. Müller, G. Daletos, and P. Proksch. 2017. New 2-methoxy acetylenic acids and pyrazole alkaloids from the marine sponge *Cinachyrella* sp. *Mar. Drugs*, 15: 356.

Mol, R.S.R., A.P. Lipton, and V. Thankamani. 2016. Antibacterial and antifouling activity of the marine sponge *Callyspongia diffusa* collected from south-west coast of India. *IJBB*, 12: 33–42.

Moles, J., A. Torrent, M.J. Alcaraz, R. Ruh, and C. Avila 2014. Anti-inflammatory activity in selected Antarctic benthic organisms. *Front. Mar. Sci.*, 1: 1–5.

Molinski, T.F., and D.J. Faulkner. 1988. An antibacterial pigment from the *sponge Dendrilla membranosa. Tetrahedron Lett.*, 29: 2137–2138.

Moodie, L.W.K., M.C. Žužek, R. Frangež, J.H. Andersen, E. Hansen, E.K. Olsen, M. Cergolj, K. Sepčić, K. Hansen, and J. Svenson. 2016. Synthetic analogs of stryphnusin isolated from the marine sponge *Stryphnus fortis* inhibit acetylcholinesterase with no effect on muscle function or neuromuscular transmission. *Org. Biomol. Chem.*, 14: 11220–11229.

Morais, S.R., and K. Chitra. 2016. Investigation of bio-active compound and in-vitro antioxidant activity of *Spirastrella pachyspira. Int. J. Pharm. Sci. Rev. Res.*, 38: 162–166.

Morinaka, M.I., and T.F. Molinski. 2011. Mollenyne A, a long-chain chlorodibromohydrin amide from the sponge *Spirastrella mollis. Org. Lett.*, 13: 6338–6341.

Motuhi, S., M. Mehiri, C.E. Payri, S.L. Barre, and S. Bach. 2016. Marine natural products from New Caledonia—A review. *Mar. Drugs*, 14: 58.

Moura, R.M., A.F. Queiroz, J.M. Fook, A.S. Dias, N.K. Monteiro, J.K. Ribeiro, G.E. Moura, L.L. Macedo, E.A. Santos, and M.P. Sales. 2006. CvL, a lectin from the marine sponge *Cliona varians*: Isolation, characterization and its effects on pathogenic bacteria and *Leishmania* promastigotes. *Comp. Biochem. Physiol. A Mol. Integr. Physiol.*, 145: 517–523.

Müller, W.E.G. 2003. *Sponges (Porifera)*. Springer.

Muraleedharan, K.M., and M.A. Avery. n.d. Progress in the development of peroxide-based antiparasitic agents. ars.els-cdn.com /content/image/1-s2.0-S1359644609001731-mmc1.doc.

Murayama, S., Y. Imae, K. Takada, J. Kikuchi, Y. Nakao, R.W.M. van Soest, S. Okada, and S. Matsunaga. 2011. Shishicrellastatins, inhibitors of cathepsin B, from the marine sponge *Crella (Yvesia) spinulata. Bioorg. Med. Chem.*, 19: 6594–6598.

Muricy, G., E. Hajdu, F.V. Araujo, and A.N. Hagler. 1993. Antimicrobial activity of South-western Atlantic shallow-water marine sponges (Porifera). *Sci. Mar.*, 57: 427–432.

Musman, M., I.I. Ohtani, D. Nagaoka, J. Tanaka, and T. Higa. 2001. Hipposulfates A and B, new sesterterpene sulfates from an Okinawan sponge, *Hippospongia cf. metachromia. J. Nat. Prod.*, 64: 350–352.

Mutsenko, V.V., O. Gryshkov, L. Lauterboeck, O. Rogulska, D.N. Tarusin, V.V. Bazhenov, K. Schütz et al. 2017. Novel chitin scaffolds derived from marine sponge *Ianthella basta* for tissue engineering approaches based on human mesenchymal stromal cells: Biocompatibility and cryopreservation. *Int. J. Biol. Macromol.*, 104: 1955–1965.

Nadeem, F., M. Oves, H.A. Qari, and I.M.I. Ismail. 2015. Red Sea microbial diversity for antimicrobial and anticancer agents. *J. Mol. Biomark. Diagn.*, 7: 267.

Nagasawa, Y., R. Ueoka, R. Yamanokuchi, N. Horiuchi, T. Ikeda, H. Rotinsulu, R.E. Mangindaan et al. 2011. Isolation of salsolinol, a tetrahydroisoquinoline alkaloid, from the marine sponge *Xestospongia* cf. *vansoesti* as a proteasome inhibitor. *Chem. Pharm. Bul.* (Tokyo), 59: 287–290.

Nakao, Y., M. Fujita, K. Warabi, S. Matsunaga, and N. Fusetani. 2000. Miraziridine A, a novel cysteine protease inhibitor from the marine sponge *Theonella aff. mirabilis. J. Am. Chem. Soc.*, 122: 10462–10463.

Nandhagopal, S., A. Muthamil Iniyan, R.R. Kannan, and S.G.P. Vincent. 2017. In vivo evaluation of anti-MRSA compound from *Streptomyces collinus* ICN1 in zebrafish embryos. *Indian J. Geo-Mar. Sci.*, 46: 1155–1161.

Nazemi, M., S.T. Jahromi, Z. Salari, and M. Gozaei. 2017a. Comparison of antibacterial activity in methanol extract of sea cucumber (*Holothuria leucospilota*) and sponge *Niphates furcata* from Hengam Island, Persian Gulf. *J. Mar. Bio.*, 8: 65–72.

Nazemi, M., Y. Moradi, F.R. Gilkolai, M.A. Ahmaditaba, M. Gozari, and Z. Salari. 2017b. Antimicrobial activities of semi polar-nonpolar and polar secondary metabolites of sponge *Dysidea pallescens* from Hengam Island, Persian Gulf. *Iran. J. Fish. Sci.*, 16: 200–209.

Nèeman, I., L. Fishelson, and Y. Kashman. 1975. Isolation of a new toxin from the sponge *Latrunculia magnifica* in the Gulf of Aquaba (Red Sea). *Mar. Biol.*, 30: 293–296.

Newman, D.J., and G.M. Cragg. 2016. Natural products as sources of new drugs from 1981 to 2014. *J. Nat. Prod.*, 79: 629–661.

Nguyen, H.Y.X. 2015. Phylogeny, bioactivity, and physiology of unique bacterial isolates from Hawaiian marine sponges from Hawaii Island: Insights into the bacteria-sponge relationship. M.Sc. thesis, University of Hawaii.

Nhiem, N.X., N.V. Quang, C.V. Minh, D.T. Hang, T.A. Hle, B.H. Tai, P.H. Yen, N.T. Hoai, C. Thung do, and P.V. Kiem. 2013. Biscembranoids from the marine sponge *Petrosia nigricans. Nat. Prod. Commun.*, 8: 1209–1212.

Nicholas, G.M., T.W. Hong, T.F. Molinski, M.L. Lerch, M.T. Cancilla, and C.B. Lebrilla. 1999. Oceanapiside, an antifungal bis-α,ω-amino alcohol glycoside from the marine sponge *Oceanapia phillipensis. J. Nat. Prod.*, 62: 1678–1681.

Niemann, H., W. Lin, W.E.G. Müller, M. Kubbutat, D. Lai, and P. Proksch. 2013. Trimeric hemibastadin congener from the marine sponge *Ianthella basta. J. Nat. Prod.*, 76: 121–125.

Northcote, P.T., J.W. Blunt, and M.H.G. Munro. 1991. Pateamine: A potent cytotoxin from the New Zealand marine sponge, *Mycale* sp. *Tetrahedron Lett.*, 32: 6411–6414.

Nurhayati, A.P.D., R. Pratiwi, S. Wahyuono, Istriyati, and S. Abdillah. 2014. Cytotoxic activity of ethanolic extract and fractions of the marine sponge *Cinachyrella* sp. *J. Pharm. Sci. Res.*, 6: 404–410.

Nuzzoa, G., M.L. Ciavatta, G. Villania, E. Manzoa, A. Zanfardinob, M. Varcamonti, and M. Gavagnina. 2012. Fulvynes, antimicrobial polyoxygenated acetylenes from the Mediterranean sponge *Haliclona fulva. Tetrahedron*, 68: 754–760.

Oclarit, J.M., S. Ohta, K. Kamimura, Y. Yamaoka, and S. Ikegami. 1994. Production of an antibacterial agent, o-aminophenol, by a bacterium isolated from the marine sponge, *Adocia* sp. *Fish Sci.*, 60: 559–562.

Oh, K.B., W. Mar, S. Kim, J.Y. Kim, T.H. Lee, J.G. Kim, D. Shin, C.J. Sim, and J. Shin. 2006. Antimicrobial activity and cytotoxicity of bis(indole) alkaloids from the sponge *Spongosorites* sp. *Biol. Pharm. Bull.*, 29: 570–573.

Ohno, O., T. Chiba, S. Todoroki, H. Yoshimura, N. Maru, K. Maekawa, H. Imagawa et al. 2011. Halichonines A, B, and C, novel sesquiterpene alkaloids from the marine sponge *Halichondria okadai* Kadota. *Chem. Commun.*, 47: 12453–12455.

Okada, Y., S. Matsunaga, R.W. van Soest, and N. Fusetani. 2006. Sokodosides, steroid glycosides with an isopropyl side chain, from the marine sponge *Erylus placenta. J. Org. Chem.*, 71: 4884–4888.

Oliveira, M., L. Barreira, K.N. Gangadhar, M.J. Rodrigues, T. Santos, J. Varela, and L.Custódio. 2016. Natural products from marine invertebrates against *Leishmania* parasites: A comprehensive review. *Phytochem. Rev.*, 15: 663–697.

Olsen, E.K., C.K. de Cerf, G.A. Dziwornu, E. Puccinelli, I.J. Ansorge, T. Samaai, L.M.K. Dingle, A.L. Edkins, and S.N. Sunasseel. 2016a. Cytotoxic activity of marine sponge extracts from the sub-Antarctic Islands and the Southern Ocean. *S. Afr. J. Sci.*, 112: 1–5.

Olsen, E.K., E. Hansen, J. Isaksson, L. Moodie, J. Svenson, and J.H. Andersen. n.d. Natural products isolated from *Geodia barretti*; a biological and synthetic investigation. Arctic University of Norway. http://bioprosp.com/sites/bioprosp.com/files/Poster_ElisabethOlsen.pdf.

Olsen, E.K., E. Hansen, L. Moodie, J. Isaksson, K. Sepčić, M. Cergolj, J. Svenson, and J.H. Andersen. 2016b. Marine AChE inhibitors isolated from *Geodia barretti*: Natural compounds and their synthetic analogs. *Org. Biomol. Chem.*, 14: 1629–1640.

Orhan, I.E., B. Ozcelik, B. Konuklugil, A. Putz, U.G. Kaban, and P. Proksch. 2012. Bioactivity screening of the selected Turkish marine sponges and three compounds from *Agelas oroides. Rec. Nat. Prod.*, 6: 356–367.

Ortega, M.J., E. Zubía, J.L. Carballo, and J. Salvá. 1996. Fulvinol, a new long-chain diacetylenic metabolite from the sponge *Reniera fulva. J. Nat. Prod.*, 59: 1069–1071.

Ortega, M.J., E. Zubía, M.C. Sánchez, J. Salvá, and J.L. Carballo. 2004. Structure and cytotoxicity of new metabolites from the sponge *Mycale cecilia. Tetrahedron*, 60: 2517–2524.

Ottinger, S., V. Rausch, L. Liu, G. Kallifatidis, A. Klöppel, F. Brümmer, A.V. Salnikov, M.W. Büchler, and I. Herr. n.d. Marine sponge *Crambe crambe* for efficient elimination of pancreatic cancer stem cells. https://www.klinikum.uni-heidelberg.de/fileadmin/MolOnkoChir/Poster/APA_Chicago.pdf.

Pailee, P., C. Mahidol, S. Ruchirawat, and V. Prachyawarakorn. 2017. Sterols from Thai marine sponge *Petrosia (Strongylophora)* sp. and their cytotoxicity. *Mar. Drugs*, 15: 54.

Pallela, R., and H. Ehrlich. 2016. *Chemicobiological and Biomedical Applications*. Springer.

Palma, M., and F.C. Sumera. 1995. Isolation and purification of some bioactive constituents from a marine sponge *Dysidea herbacea. KIMIKA*, 11: 33–38.

Patil, A.D., A.J. Freyer, L. Killmer, P. Offen, B. Carte, A. J. Jurewicz, and R.K. Johnson. 1997. Frondosins, five new sesquiterpene hydroquinone derivatives with novel skeletons from the sponge *Dysidea frondosa*: Inhibitors of interleukin-8 receptors. *Tetrahedron*, 53: 5047–5060.

Patil, A.D., N.V. Kumar, W.C. Kokke, M.F. Bean, A.J. Freyer, C. De Brosse, S. Mai, A. Truneh, and B. Carte. 1995. Novel alkaloids from the sponge *Batzella* sp.: Inhibitors of HIV gp120-human CD4 binding. *J. Org. Chem.*, 60: 1182–1188.

Peng, C.S., G. Gunaherath, A.M. Piggott, Z. Khalil, M. Conte, and R.J. Capon. 2010. 9-(5′-Deoxy-5′-thio-β-D-xylofuranosyl)adenine disulfide isolated from this species possessed from the Southern Australian Marine Sponge *Trachycladus laevispirulifer*: The first natural occurrence of a nucleoside disulfide. *Aust. J. Chem.*, 636: 873–876.

Perdicaris, S., T. Vlachogianni, and A. Valavanidis. 2013. Bioactive natural substances from Marine sponges: New developments and prospects for future pharmaceuticals. *Nat. Prod. Chem. Res.*, 1: 3.

Perry, N.B., L. Ettouati, M. Litaudon, J.W. Blunt, M.H.G. Munro, S. Parkin, and Hako. 1994. Alkaloids from the antarctic sponge *Kirkpatrickia varialosa*: Part 1: Variolin b, a new antitumour and antiviral compound. *Tetrahedron,* 50: 3987–3992.

Petrichtcheva, N.V., C. Duque, A. Dueñas, S. Zea, N. Hara, and Y. Fujimoto. 2002. New nitrogenous eudesmane-type compounds isolated from the Caribbean sponge *Axinyssa ambrosia. J. Nat. Prod.*, 65: 851–855.

Pettit, G.R., J.C.Z. Barkoczy, A.C. Dorsaz, D.L. Herald, M.D. Williams, D.L. Doubek, J.M. Schmidt et al. 1993a. Isolation and structure of the marine sponge cell growth inhibitory cyclic peptide phakellistatin. *J. Nat. Prod.*, 56: 260–267.

Pettit, G.R., F. Gao, and R. Cerny. 1993b. Isolation and structure of axinastatin 4 from the western Indian Ocean marine sponge *Axinella cf. carteri. Heterocycles*, 35: 711–718.

Pettit, G.R., F. Gao, J.M. Schmidt, J.C. Chapuis, and R.L. Cerny. 1994a. Isolation and structure of axinastatin 5 from a republic of comoros marine sponge. *Bioorg. Med. Chem. Lett.*, 4: 2935–2940.

Pettit, G.R., C.L. Herald, Z.A. Cichacz, F. Gao, J.M. Schmidt, M.R. Boyd, N.D. Christie, and F.E. Boettner. 1993c. Isolation and structure of the powerful human cancer cell growth inhibitors spongistatins 4 and 5 from an African *Spirastrella spinispirulifera* (Porifera*). J. Chem. Soc. Chem. Commun.*, 24: 1805–1807.

Pettit, G.R., R.T. Michael, D.W.L.T. Jean, M.S. Ronald, L.C. John, and N.A. Hooper. 1993d. Isolation and structure of phakellistatin 2 from the eastern indian ocean marine sponge *Phakellia carteri. Bioorg. Med. Chem. Lett.*, 3: 2869–2874.

Pettit, G.R., T. Nogawa, J.C. Knight, D.L. Doubek, and J.N.A. Hooper. 2004. Antineoplastic agents. 535. Isolation and structure of plakorstatins 1 and 2 from the Indo-Pacific sponge *Plakortis nigra. J. Nat. Prod.*, 67: 1611–1613.

Pettit, G.R., and R. Tan. 2005. Isolation and structure of phakellistatin 14 from the Western Pacific Marine Sponge *Phakellia* sp. *J. Nat. Prod.*, 68: 60–63.

Pettit, G.R., J.P. Xu, Z. Cichacz, J.M. Schmidt, A.C. Dorsaz, M.R. Boyd, and R.L. Cerny. 1995. Isolation and structure of the human cancer cell growth inhibitory phakellistatin 4 from the western Pacific sponge *Phakellia costata. Heterocycles*, 40: 501–506.

Pettit, G.R., J. Xu, Z.A. Cichacz, M.D. Williams, J. Chapuis, and R.L. Cerny. 1994b. Antineoplastic agents 323. Isolation and structure of phakellistatin 6 from a Chuuk archipelago marine sponge. *Bioorg. Med. Chem. Lett.*, 4: 2677–2682.

Pham, C., R. Hartmann, W.E.G. Müller, N. de Voogd, D. Lai, and P. Proksch. 2013. Aaptamine derivatives from the Indonesian sponge *Aaptos suberitoides. J. Nat. Prod.*, 76: 103–106.

Phuwapraisirisan, P., S. Matsunaga, R.W.M. van Soest, and N. Fusetani. 2002. Isolation of a new mycalolide from the marine sponge *Mycale izuensis. J. Nat. Prod.*, 65: 942–943.

Piao, S., Y. Song, W. Jiao, F. Yang, X. Liu, W. Chen, B. Han, and H. Lin. 2013. Hippolachnin A, a new antifungal polyketide from the South China Sea sponge *Hippospongia lachne. Org. Lett.*, 15: 3526–3529.

Piao, S., H. Zhang, H. Lu, F. Yang, W. Jiao, Y. Yi, W. Chen, and H. Lin. 2011. Hippolides A–H, acyclic manoalide derivatives from the marine sponge *Hippospongia lachne. J. Nat. Prod.*, 74: 1248–1254.

Pietra, F. 2002. *Biodiversity and Natural Product Diversity*. Tetrahedron Organic Chemistry Series, Vol. 21. Elsevier.

Piña, C., J.T. Gautschi, G.Y. Wang, M.L. Sanders, F.J. Schmitz, D. France, S. Cornell-Kennon et al. 2003. Psammaplins from the sponge *Pseudoceratina purpurea*: Inhibition of both histone deacetylase and DNA methyltransferase. *J. Org. Chem.*, 68: 3866–3873.

Plaza, A., H.L. Baker, and C.A. Bewley. 2008. Mirabalin, an antitumor macrolide lactam from the marine sponge *Siliquariaspongia mirabilis. J. Nat. Prod.*, 71: 473–477.

Plaza, A., E. Gustchina, H.L. Baker, M. Kelly, and C.A. Bewley. 2007. Mirabamides A–D. depsipeptides from the sponge *Siliquariaspongia mirabilis* that inhibit HIV-1 fusion. *J. Nat. Prod.*, 70: 1753–1760.

Plaza, A., G. Bifulco, M. Masullo, J.R. Lloyd, J.L. Keffer, P.L. Colin, J.N.A. Hooper, L.J. Bell, and C.A. Bewl. 2010. Mutremdamide A and koshikamides C–H, peptide inhibitors of HIV-1 entry from different *Theonella* species. *J. Org. Chem.*, 75: 4344–4355.

Pulitzer-Finali, G. 1969. *Reniera sarai*, new name for *Haliclona viscosa Sarà. Bollettino dei Musei e degli Istituti Biologici dell'Universitá di Genova*, 37: 97–101.

Pulitzer-Finali, G. 1993. A collection of marine sponges from East Africa. Annales Museo Civico Storia Naturale. Giacomo Doria, 89: 247–350.

Purushottama, G.B., K. Venkateshvaran, K. Pani Prasad, and P. Nalini. 2009. Bioactivities of extracts from the marine sponge *Halichondria panacea. J. Venom. Anim. Toxins incl. Trop. Dis.*, 15: 444–459.

Putra, M.Y., and T.A. Hadi. 2017. Chemical composition, antimicrobial, cytotoxic and antiplasmodial activities of three sponges from Buton Islands, Indonesia. *Ilmu Kelautan*, 22: 147–154.

Putra, M.Y., and I. Jaswir. 2014. The alkaloids from Indonesian marine sponges. *Oceanography* 2: 125.

Qaralleh, H. 2016. Chemical and bioactive diversities of marine sponge *Neopetrosia. Bangladesh J. Pharmacol.*, 11: 433–452.

Qiu, Y., and X.M. Wang. 2008. A new sesquiterpenoid hydroquinone from the marine sponge *Dysidea arenaria. Molecules*, 13: 1275–1281.

Queiroz, A.F.S., R.A. Silva, R.M. Moura, J.L. Dreyfuss, E.J. Paredes-Gamero, A.C.S. Souza, I.L.S. Tersariol et al. 2009. Growth inhibitory activity of a novel lectin from *Cliona varians* against K562 human erythroleukemia cells. *Cancer Chemother. Pharmacol.*, 63: 1023–1033.

Quévrain, E., M. Roué, I. Domart-Coulon, and M. Bourguet-Kondracki. 2014. Assessing the potential bacterial origin of the chemical diversity in calcareous sponges. *J. Mar. Sci. Technol.*, 22: 36–49.

Quiao, M.A.D., and M.M. Uy. 2013. Pyrimidines from the Philippine marine sponge *Aaptos suberitoides. IJSER*, 4: 1.

Rabelo, L., N. Monteiro, R. Serquiz, P. Santos, R. Oliveira, A. Oliveira, H. Rocha, A.H. Morais, A. Uchoa, and E. Santo. 2012. A lactose-binding lectin from the marine sponge *Cinachyrella apion* (Cal) induces cell death in human cervical adenocarcinoma cells. *Mar. Drugs*, 10: 727–743.

Radwan, M.M., A.S. Wanas, F.R. Fronczek, M.R. Jacob, and S.A. Ross. 2015. Polybrominated diphenyl ethers from the marine organisms *Lendenfeldia dendyi* and *Sinularia dura* with anti-MRSa activity. *Med. Chem. Res.*, 24: 3398–3404.

Rady, H. 2014. Sponge mesohyl induces anti-proliferation activity and cell cycle arrest in colon cancer in-vitro. *Res. J. Pharm. Biol. Chem. Sci.*, 5: 1070–1075.

Raj, S., and D. Inbakandan. 2013. Anti cancer activity prediction of secondary metabolites from marine sponge *Discodermia calyx*: An in silico approach. *Indian J. Geo-Mar. Sci.*, 42: 653–658.

Rajendran, I., K.S. Shobana, G.A.S. Sonia, K. Chakraborty, K.K. Vijayan, and P. Vijayagopal. 2011. Antibacterial and antifungal properties of southeast Indian coastal sponges. *J. Mar. Biol. Ass. India*, 53: 272–274.

Rakotondraibe, H., J. Li, L.C. Blasiak, R.T. Hill, and M.B. Cassera. 2014. Antimalarial dipeptide from a *Streptomyces* sp. associate of the sponge *Xestospongia muta. L. Planta Med.*, 80–PB15.

Randazzo, A., F.D. Piaz, S. Orrù, C. Debitus, C. Roussakis, P. Pucci, and L. Gomez-Paloma. 1998. Axinellins A and B: New proline-containing antiproliferative cyclopeptides from the Vanuatu sponge *Axinella carteri. Eur. J. Org. Chem.*, 1998: 2659–2665.

Rao, J.V., P.K. Usman, and J.B. Kumar. 2008. Larvicidal and insecticidal properties of some marine sponges collected in Palk Bay and Gulf of Mannar waters. *Afr. J. Biotechnol.*, 7: 109–113.

Rao, K.V., M.S. Donia, J. Peng, E. Garcia-Palomero, D. Alonso, A. Martinez, M. Medina, S. et al. 2006. Manzamine B and E and ircinal A related alkaloids from an Indonesian *Acanthostrongylophora* sponge and their activity against infectious, tropical parasitic, and Alzheimer's Diseases. *J. Nat. Prod.*, 69: 1034–1040.

Rao, M., and D.J. Faulkner. 2002. Isotactic polymethoxydienes from the Philippines sponge *Myriastra clavosa. J. Nat. Prod.*, 65: 1201–1213.

Rao, T.S.P., N.S. Sarma, Y.L.N. Murthy, V.S.S.N. Kantamreddi, C.W. Wright, and P.S. Parameswaran. 2010. New polyhydroxy sterols from the marine sponge *Callyspongia fibrosa* (Ridley & Dendly). *Tetrahedron Lett.*, 51: 3583–3586.

Rashid, M.A., K.R. Gustafson, J.L. Boswell, and M.R. Boyd. 2000. Haligramides A and B, two new cytotoxic hexapeptides from the marine sponge *Haliclona nigra. J. Nat. Prod.*, 63: 956–959.

Rashid, M.A., K.R. Gustafson, L.K. Cartner, N. Shigematsu, L.K. Pannell, and M.R. Boyd. 2001. Microspinosamide, a new HIV-inhibitory cyclic depsipeptide from the marine sponge *Sidonops microspinosa. J. Nat. Prod.*, 64: 117–121.

Ravichandran, S., S. Wahidullah, L. D'Souza, and R.M. Anbuchezhian. 2011. Antimicrobial activity of marine sponge *Clathria indica* (Dendy, 1889). *Bioorg. Khim.*, 37: 483–489.

Reddy, A.V., K. Ravinder, M. Narasimhulu, A. Sridevi, N. Satyanarayana, A.K. Kondapi, and Y. Venkateswarlu. 2006. New anticancer bastadin alkaloids from the sponge *Dendrilla cactos. Bioorg. Med. Chem.* 14: 4452–4457.

Reegan, A.D., A.V. Kinsalin, M.G. Paulraj, and S. Ignacimuthu. 2015. Larvicidal, ovicidal, and repellent activities of marine sponge *Cliona celata* (Grant) extracts against *Culex quinquefasciatus* Say and *Aedes aegypti* L. (Diptera: Culicidae). *Asian Pac. J. Trop. Med.*, 8: 29–34.

Regalado, E.L., D. Tasdemir, M. Kaiser, N. Cachet, P. Amade, and O.P. Thomas. 2010. Antiprotozoal steroidal saponins from the marine *Sponge Pandaros acanthifolium. J. Nat. Prod.*, 73: 1404–1410.

Regalado, E.L., T. Turk, D. Tasdemir, M. Gorjan, and M.K.P. Olivier. 2011. Cytotoxic and haemolytic steroidal glycosides from the Caribbean sponge *Pandaros acanthifolium. Steroids*, 76: 1389–1396.

Reimer, A., A. Blohm, T. Quack, C.G. Grevelding, V. Kozjak-Pavlovic, T. Rudel, U. Hentschel, and U.R. Abdelmohsen. 2015. Inhibitory activities of the marine streptomycete-derived compound SF2446A2 against *Chlamydia trachomatis* and *Schistosoma mansoni. J. Antibiot. (Tokyo)*, 68: 674–679.

Reyes, F., R. Martín, A. Rueda, R. Fernández, D. Montalvo, C. Gómez, and J.M. Sánchez-Puelles. 2004. Discorhabdins I and L, cytotoxic alkaloids from the sponge *Latrunculia brevis. J. Nat. Prod.*, 67: 463–465.

Ridley, C.P., and D.J. Faulkner. 2003. New cytotoxic steroidal alkaloids from the Philippine sponge *Corticium niger. J. Nat. Prod.*, 66: 1536–1539.

Ridley, S.O. 1884. *Spirastrella vagabunda* Ridley, 1884, Spongiida. Report on the Zoological Collections made in the Indo-Pacific Ocean during the Voyage of H.M.S. 'Alert', 1881-2. *British Museum (Natural History)*: London, 366–482.

Ridley, S.O., and A. Dendy. 1886. Preliminary report on the Monaxonida collected by H.M.S. Challenger. *Part I. Annals and Magazine of Natural History*, 18: 325–351.

Rogers, E.W., and T.F. Molinski. 2005. A cytotoxic carotenoid from the marine sponge *Prianos osiros. J. Nat. Prod.*, 2005, 68: 450–452.

Rogoza, L.N., N.F. Salakhutdinov, and G.A. Tolstikov. 2011. 3. Antitubercular activity of natural products: Recent developments. *Opportunity, Challenge, Scope Nat. Prod. Med. Chem.*, 2011: 103–120.

Roll, D.M., C.M. Ireland, H.S.M. Lu, and J. Clardy. 1988. Fascaplysin, an unusual antimicrobial pigment from the marine sponge *Fascaplysinopsis* sp. *J. Org. Chem.*, 53: 3276–3278.

Rudi, A., R. Afanii, L.G. Gravalos, M. Aknin, E. Gaydou, J. Vacelet, and Y. Kashman. 2003. Three new cyclic peroxides from the marine sponge *Plakortis* aff *simplex. J. Nat. Prod.*, 66: 682–685.

Rudi, A., T. Yosief, S. Loya, A. Hizi, M. Schleyer, and Y. Kashman. 2001. Clathsterol, a novel anti-HIV-1 RT sulfated sterol from the sponge *Clathria* species. *J. Nat. Prod.*, 64: 1451–1453.

Rueda, A., E. Zubía, M.J. Ortega, J.L. Carballo, and J. Salvá. 1998. New metabolites from the sponge *Spongia agaricina. J. Nat. Prod.*, 61: 258–261.

Rungprom, W., W. Chavasiri, U. Kokpol, A. Kotze, and M.J. Garson. 2004. Bioactive chromodorolide diterpenes from an Aplysillid sponge. *Mar. Drugs*, 2: 101–107.

Ryu, G., S. Matsunaga, and N. Fusetani. 1994a. Discodermin E, a cytotoxic and antimicrobial tetradecapeptide, from the marine sponge *Discodermia kiiensis. Tetrahedron Lett.*, 35: 8251–8254.

Ryu, G., S. Matsunaga, and N. Fusetani. 1994b. Discodermins F–H, cytotoxic and antimicrobial tetradecapeptides from the marine sponge *Discodermia kiiensis*: Structure revision of discodermins A-D. *Tetrahedron*, 50: 13409–13416.

Sa'adah, N.N., A.P.D. Nurhayati, and P.T.E.K.M. Shovitri. 2016. The anticancer activity of the marine sponge *Aaptos suberitoides* to protein profile of fibrosarcoma mice (*Mus musculus*). *J. Technol. Sci.*, 27: 53–58.

Sagar, S., M. Kaur and K.P. Minneman. 2010. Antiviral lead compounds from marine sponges. *Mar. Drugs*, 8: 2619-2638.

Samoylenko, V., S.I. Khan, M.R. Jacob, B.L. Tekwani, L.A. Walker, C.D. Hufford, and I. Muhammad. 2009. Bioactive (+)-manzamine A and (+)-8-hydroxymanzamine A tertiary bases and salts from *Acanthostrongylophora ingens* and their preparations. *Nat. Prod. Commun.*, 4: 185–192.

Sandler, J.S., P.L. Colin, J.N. Hooper, and D.J. Faulkner. 2002. Cytotoxic beta-carbolines and cyclic peroxides from the Palauan sponge *Plakortis nigra. J. Nat. Prod.*, 65: 1258–1261.

Santafé, G., V. Paz, J. Rodríguez, and C. Jiménez. 2002. Novel cytotoxic oxygenated C29 sterols from the Colombian marine sponge *Polymastia tenax. J.Nat. Prod.*, 65: 1161–1164.

Santos, E.A., A.L. Quintela, E.G. Ferreira, T.S. Sousa, P. Fd, E. Hajdu, M.S. Carvalho, S. Salani, D.D. Rocha, D.V. Wilke et al. 2015. Cytotoxic plakortides from the Brazilian marine sponge *Plakortis angulospiculatus. J. Nat. Prod.*, 78: 996–1004.

Sata, N.U., S. Wada, S. Matsunaga, S. Watabe, R.W.M. van Soest, and N. Fusetani. 1999. Rubrosides A–H, new bioactive tetramic acid glycosides from the marine sponge *Siliquariaspongia japonica. J. Org. Chem.*, 64: 2331–2339.

Sauleau, P., P. Retailleau, S. Nogues, I. Carletti, L. Marcourt, R. Raux, A.A. Mourabit, and C. Debitus. 2011. Dihydrohymenialdisines, new pyrrole-2-aminoimidazole alkaloids from the marine sponge *Cymbastela cantharella. Tetrahedron Lett.*, 52: 2676–2678.

Saurav, K., R. Bar-Shalom, M. Haber, I. Burgsdorf, G. Oliviero, V. Costantino, D. Morgenstern, and L. Steindler. 2016. In search of alternative antibiotic drugs: Quorum-quenching activity in sponges and their bacterial isolates. *Front Microbiol.*, 7: article 416, 18 p.

Schneemann, I., I. Kajahn, B. Ohlendorf, H. Zinecker, A. Erhard, K. Nagel, J. Wiese and J.F. Imhoff. 2010a. Mayamycin, a cytotoxic polyketide from a *Streptomyces* strain isolated from the marine sponge *Halichondria panacea. J. Nat. Prod.*, 73: 1309–1312.

Schneemann, I., K. Nagel, I. Kajahn, A. Labes, J. Wiese, and J.F. Imhof. 2010b. Comprehensive investigation of marine actinobacteria associated with the sponge *Halichondria panicea. Appl. Envtl. Microb.*, 76: 3702–3714.

Schroder, H.C., Z. Kljajic, B.E. Weiler, M. Gasic, G. Uhlenbruck, B. Kurelec, and W.E.G. Muller. 1990. The galatose-specific lectin from the sponge *Chondria nucula* displays anti-human immunodeficiency virus in vitro via stimulation of the (2′-5′) oligoadenylate metabolism. *Antivir. Chem. Chemother.*, 1: 99–105.

Schwartsmann, G., A.B. da Rocha, R.G. Berlinck, and J. Jimeno. 2001. Marine organisms as a source of new anticancer agents. *Lancet Oncol.*, 2: 221–225.

Schyschka, L., A. Rudy, I. Jeremias, N. Barth, G.R. Pettit, and A.M. Vollmar. 2008. Spongistatin 1: A new chemosensitizing marine compound that degrades XIAP. *Leukemia*, 22: 1737–1745.

Segraves, N.L., and P. Crews. 2005. A Madagascar sponge *Batzella* sp. as a source of alkylated iminosugars. *J. Nat. Prod.*, 68: 118–121.

Selvin, J., S. Joseph, K.R. Asha, W.A. Manjusha, V.S. Sangeetha, D.M. Jayaseema, M.C. Antony, and A.J.D. Vinitha. 2004. Antibacterial potential of antagonistic *Streptomyces* sp. isolated from marine sponge *Dendrilla nigra. FEMS Microbiol. Ecol.*, 50: 117–122.

Senthilkumar, K., and S. Kim. 2013. Marine invertebrate natural products for anti-inflammatory and chronic diseases. *Evid. Compl. Alter. Med.*, 2013: article ID 572859, 10 p.

Sepčić, K., S. Kauferstein, D. Mebs, and T. Turk. 2010. Biological activities of aqueous and organic extracts from tropical marine sponges. *Mar. Drugs*, 8: 1550–1566.

Seradj, H., M. Moein, M. Eskandari, and F. Maaref. 2012. Antioxidant activity of six marine sponges collected from the Persian Gulf. *Iranian J. Pharm. Sci.*, 8: 249–255.

Setyowati, E.P., and S.S. Wahyuono. 2004. Cytotoxicity and antimicrobial test of the bioactive compound isolated from *Stylissa flabelliformis* sponge. *Majalah Farmasi Indonesia*, 15: 50–56.

Seydi, E., A. Motallebi, M. Dastbaz, S. Dehghan, A. Salimi, M. Nazemi, and J. Pourahmad. 2015. Selective toxicity of Persian Gulf sea cucumber (*Holothuria parva*) and sponge (*Haliclona oculata*) Methanolic extracts on liver mitochondria isolated from an animal model of hepatocellular carcinoma. *Hepat. Mon.*, 15: e33073.

Shaala, L.A., S.I. Khalifa, M.K. Mesbah, R.W.M. van Soest, and D.T.A. Youssef. 2008. Subereaphenol A, a new cytotoxic and antimicrobial dibrominated phenol from the Red Sea sponge *Suberea mollis. Nat. Prod. Commun.*, 3: 219–222.

Shaala, L.A., D.T.A. Youssef, J.M. Badr, M. Sulaiman, and A. Khedr. 2015. Bioactive secondary metabolites from the Red Sea marine verongid sponge *Suberea Species. Mar. Drugs*, 13: 1621–1631.

Shaala, L.A., D.T.A. Youssef, M. Sulaiman, F.A. Behery, A.I. Foudah, and K.A. El Sayed. 2012. Subereamolline A as a potent breast cancer migration, invasion and proliferation inhibitor and bioactive dibrominated alkaloids from the Red Sea sponge *Pseudoceratina arabica. Mar. Drugs*, 10: 2492–2508.

Shaari, K., K.C. Ling, Z.M. Rashid, T.P. Jean, F. Abas, S.M. Raof, Z. Zainal, N.H. Lajis, H. Mohamad, and A.M. Ali. 2009. Cytotoxic aaptamines from Malaysian *Aaptos aaptos. Mar. Drugs*, 7: 1–8.

Shen, S., D. Liu, C. Wei, P. Proksch, and W. Lin. 2012. Purpuroines A–J, halogenated alkaloids from the sponge *Iotrochota purpurea* with antibiotic activity and regulation of tyrosine kinases. *Bioorg. Med. Chem.*, 20: 6924–6928.

Shen, Y., C. Lu, R. Chakraborty, and Y. Kuo. 2003. Isolation of sesquiterpenoids from sponge *Dysidea avara* and chemical modification of avarol as potential antitumor agents. *Nat. Prod. Res.*, 17: 83–89.

Shen, Y.C., C.Y. Chen, and Y.H. Kuo. 2001. New sesquiterpene hydroquinones from a Taiwanese marine sponge, *Hippospongia metachromia. J. Nat. Prod.*, 64: 801–803.

Shin, J., H. Lee, L. Woo, J. Rho, Y. Seo, K.W. Cho, and C.J. Sim. 2001. New triterpenoid saponins from the sponge *Erylus nobilis. J. Nat. Prod.*, 64: 767–771.

Silva, J.C., A.A. Barros, I.M. Aroso, D. Fassini, T.H. Silva, R.L. Reis, and A.R.C. Duarte. 2016. Extraction of collagen/gelatin from the marine demosponge *Chondrosia reniformis* (Nardo, 1847) using water acidified with carbon dioxide—Process optimization. *Ind. Eng. Chem. Res.*, 55: 6922–6930.

Sionov, E., D. Roth, H. Sandovsky-Losica, Y. Kashman, A. Rudi, L. Chill, I. Berdicevsky, and E. Segal. 2005. Antifungal effect and possible mode of activity of a compound from the marine sponge *Dysidea herbacea*. http://dx.doi.org/10.1016/j.jinf.2004.07.014.

Sipkema, D., M.C.R. Franssen, R.Osinga, J. Tramper, and R.H. Wijffels. 2005. Marine sponges as pharmacy. *Mar. Biotechnol. (NY)*, 7: 142–162.

Sobahi, T.R.A., S.N. Ayyad, A. Abdel-Lateff, M.M. Algandaby, H.S. Alorfi, and A.B. Abdel-Naim. 2017. Cytotoxic metabolites from *Callyspongia siphonella* display antiproliferative activity by inducing apoptosis in HCT-116 Cells. *Pharmacogn. Mag.*, 13(Suppl. 1): S37–S40.

Soest, R.W.M.V., R. Desqueyroux-Faündez, A.D. Wright, and G.M. König. 1996. *Cymbastela hooperi* sp. nov. (Halichondrida: Axinellidae) from the Great Barrier Reef, Australia. *Bull. Inst. R. Sci. Nat. Belg. Biol.*, 66 Suppl: 103–108.

Somerville, M.J., J.N. Hooper, and M.J. Garson. 2006. Mooloolabenes A–E, norsesterterpenes from the Australian sponge *Hyattella intestinalis. J. Nat. Prod.*, 69: 1587–1590.

Song, J., J. Jeon, T.H. Won, C.J. Sim, D. Oh, K. Oh, and J. Shin. 2014. New cyclic cystine bridged peptides from the sponge *Suberites waedoensis.. Mar. Drugs*, 12: 2760–2770.

Sonia, G.A.S., and A.P. Lipton. 2012. Mosquito larvicidal activity of marine sponge metabolites. *GJP*, 6: 1–3.

Sorres, J., M. Martin, S. Petek, H. Levaique, T. Cresteil, S. Ramos, O. Thoison, C. Debitus, and A. Al-Mourabit. 2012. Pipestelides A–C: Cyclodepsipeptides from the Pacific Marine Sponge *Pipestela candelabra. J. Nat. Prod.*, 75: 759–763.

Souza, T.M., J.L. Abrantes, R. de A Epifanio, C.F.L. Fontes, and I.C. Frugulhetti. 2007. The alkaloid 4-methylaaptamine isolated from the sponge *Aaptos aaptos* impairs *Herpes simplex* virus type 1 penetration and immediate-early protein synthesis. *Planta Med.*, 73: 200–205.

Sponge crude extraction and primary screening. n.d. http://shodhganga.inflibnet.ac.in/bitstream/10603/64131/10/10_chapter%205.pdf.

Stankevicins, L., C.A.F. Aiub, J.L. Mazzei, G. Lobo-Hajdu, and I. Felzenszwalb.2008. Cytotoxic, mutagenic and antimutagenic screening of *Arenosclera brasiliensis* acetone and ethanol extracts. *Genet. Mol. Res.*, 7: 542–548.

Stanojković, T., S. Milović, I. Matić, N. Grozdanić, Z. Kljajić, and L. Sirovine. 2015. In vitro antitumoral activity of the extract of sponge *Acanthella acuta. LEK. SIROV*, 35: 89–101.

Su, J., S. Tseng, M. Lu, L. Liu, Y. Chou, and P. Sung. 2011. Cytotoxic C21 and C22 terpenoid-derived metabolites from the sponge *Ircinia* sp. *J. Nat. Prod.*, 74: 2005–2009.

Suarez-Jimenez, G., A. Burgos-Hernandez, and J. Ezquerra-Brauer. 2012. Bioactive peptides and depsipeptides with anticancer potential: Sources from marine animals. *Mar. Drugs*, 10: 963–986.

Subugade, S., S. Mokashe, S.G. Gupta, V. Lakwal, and D. Kharate. 2015. Antimicrobial potential of the marine intertidal sponge, *Ircinia fusca* from west coast of India. *Indian J. Pharm. Biol. Res.*, 3: 12–14.

Sugumaran, M., and W.E. Robinson. 2010. Bioactive dehydrotyrosyl and dehydrodopyl compounds of marine origin. *Mar. Drugs*, 8: 2906–2935.

Sujatha, S., G.P. Vincent, and M.J. Dhas. 2014. Marine sponge and its potential bioactive spectrum against CSF affected bacterial organisms. *Pharma Innov.*, 3: 81–86.

Sullivan, J.D., R.L. Giles, and R.E. Looper. 2009. 2-Aminoimidazoles from *Leucetta* sponges: Synthesis and biology of an important pharmacophore. *Curr. Bioact. Compd.*, 5: 39–78.

Sun, J., W. Cheng, N.J. de Voogd, P. Proksch, and W. Lin. 2016. Stylissatins B-D, cycloheptapeptides from the marine sponge *Stylissa massa. Tetrahedron Lett.*, 57: 4288–4292.

Sun, S., C. Canning, K. Wang, W. Zhu, F.Yang, Y. Zhang, and K. Zhou. 2017. Antibacterial activity of 2-(3′,5′-Dibromo-2′-methoxyphenoxy)-3,5-dibromophenol isolated from *Phyllospongia papyracea. Nat. Product Comm.*, 12: 567–569.

Sunassee, S.N., T. Ransom, C.J. Henrich, J.A. Beutler, D.G. Covell, J.B. McMahon, and K.R. Gustafson. 2014. Steroidal alkaloids from the marine sponge *Corticium niger* that inhibit growth of human colon carcinoma cells. *J. Nat. Prod.*, 77: 2475–2480.

Supriyono, A., B. Schwarz, V. Wray, L. Witte, W.E.G. Müller, R. van Soest, W. Sumaryono, and P. Proksch. 1995. Bioactive alkaloids from the tropical marine sponge *Axinella carteri. Zeitschrift für Naturforschung C*, 50: 669–674.

Suzuki, M., R. Ueoka, K. Takada, S. Okada, S. Ohtsuka, Y. Ise, and S. Matsunaga. 2012. Isolation of spirastrellolides A and B from a marine sponge *Epipolasis* sp. and their cytotoxic activities. *J. Nat. Prod.*, 75: 1192–1195.

Tabares, P. 2011. Antimicrobial, anti-protease and immunomodulatory activities of secondary metabolites from Caribbean sponges and their associated bacteria. Ph.D. thesis, University of Würzburg.

Tada, H., and F. Yasuda. 1985. Metabolites from the marine sponge *Epipolasis kushimotoensis. Chem.Pharm. Bull.*, 33: 1941–1945.

Takada, K., Y. Nakao, S. Matsunaga, R.W.M. van Soest, and N. Fusetani. 2002. Nobiloside, a new neuraminidase inhibitory triterpenoidal saponin from the marine sponge *Erylus nobilis*. *J. Nat. Prod.*, 65: 411–413.

Tan, K.C., T. Wakimoto, and I. Abe. 2014. Lipodiscamides A–C, new cytotoxic lipopeptides from *Discodermia kiiensis*. *Org. Lett.*, 16: 3256–3259.

Tan, K.C., and T. Wakimoto, and I. Abe. 2016. Sulfoureido lipopeptides from the marine sponge *Discodermia kiiensis*. *J. Nat. Prod.*, 79: 2418–2422.

Tang, S.A., Q. Zhou, W.Z. Guo, Y. Qiu, R. Wang, M. Jin, W. Zhang, K. Li, T. Yamori, S. Dan, and D. Kong. 2014. In vitro antitumor activity of stellettin B, a triterpene from marine sponge *Jaspis stellifera*, on human glioblastoma cancer SF295 cells. *Mar. Drugs*, 12: 4200–4213.

Tangman, S., J.G. Soulange, and D. Marie. 2015. Bioactive profile of *Plakortis nigra*, a sea sponge from Mauritius Islands. *JCLM*, 3: 44–51.

Tasdemir, D., R. Mallon, M. Greenstein, L.R. Feldberg, and S.C. Kim. 2002. Aldisine alkaloids from the Philippine sponge *Stylissa massa* are potent inhibitors of mitogen-activated protein kinase kinase-1 (MEK-1). *J. Med. Chem.*, 45: 529–532.

Tasdemir, D., B. Topaloglu, R. Perozzo, R. Brun, R. O'Neill, N.M. Carballeira, X. Zhang, P.J. Tonge, A. Linden, and P. Rüedi. 2007. Marine natural products from the Turkish sponge *Agelas oroides* that inhibit the enoyl reductases from *Plasmodium falciparum, Mycobacterium tuberculosis* and *Escherichia coli. Bioorg. Med.Chem.*, 15: 6834–6845.

Thomas, T.R., D.P. Kavlekar, and P.A. LokaBharathi. 2010. Marine drugs from sponge-microbe association—A review. *Mar. Drugs*, 8: 1417–1468.

Thompson, M.N., and W. Gallimore. 2013. Antileishmanial, antimalarial and antimicrobial activity of the Jamaican "Touch-me-not" sponge *Neofibularia nolitangere. J. Appl. Pharm.* Sci., 3: 80–83.

Tilvi, S., M. Majik, and C.G. Naik. 2005. A tendem mass spectrometric approach for determining the structure of molecular species of ceramide in the marine sponge, *Haliclona cribricutis. Eur. J. Mass Spectrom.*, 11: 345–351.

Tilvi, S., C. Rodrigues, C.G. Naik, P.S. Parameswaran, and S. Wahidullah. 2004. New bromotyrosine alkaloids from the marine sponge *Psammaplysilla purpurea. Tetrahedron*, 60: 10207–10215.

Torres, Y.R., R.G.S. Berlinck, G.G.F. Nascimento, S.C. Fortier, C. Pessoa, and M.O. Moraes. 2002. Antibacterial activity against resistant bacteria and cytotoxicity of four alkaloid toxins isolated from the marine sponge *Arenosclera brasiliensis. Toxicon*, 40: 885–891.

Tovar-Barrios, G., A.L.S. Ramírez, R. Valle-Molinares, and J.W.A. Valencia. 2016. Antioxidant activity of poriferans from the Colombian Caribbean rocky coast by DPPH and ABTS method. *J. Archives*, 1: 64–71.

Tran, T.D., N.B. Pham, G. Fechner, D. Zencak, H.T. Vu, J.N. Hooper, and R.J. Quinn. 2012. Cytotoxic cyclic depsipeptides from the Australian marine sponge *Neamphius huxleyi. J. Nat. Prod.*, 75: 2200–2208.

Tran, T.D., N.B. Pham, G. Fechner, J.N. Hooper, and R.J. Quinn. 2013. Bromotyrosine alkaloids from the Australian marine sponge *Pseudoceratina verrucosa. J. Nat. Prod.*, 76: 516–5123.

Tran, T.D., N.B. Pham, G.A. Fechner, J.N. Hooper, and R.J. Quinn. 2014. Potent cytotoxic peptides from the Australian marine sponge *Pipestela candelabra. Mar. Drugs*, 12: 3399–33415.

Trevisi, L., S. Bova, G. Cargnelli, D. Danieli-Betto, M. Floreani, E. Germinario, M.V. D'Auria, and S. Luciani. 2000. Callipeltin A, a cyclic depsipeptide inhibitor of the cardiac sodium-calcium exchanger and positive inotropic agent. *Biochem. Biophys. Res. Commun.*, 279: 219–222.

Trimurtulu, B.G., D.J. Faulkner, N.B. Perry, L. Ettouati, M. Litaudon, and J.W. Blunt. 1994. Alkaloids from the antarctic sponge *Kirkpatrickia varialosa*. Part 2: Variolin A and N(3′)-methyl tetrahydrovariolin. *Tetrahedron*, 50: 3993–4000.

Tsuda, M., M. Ishibashi, K. Agemi, T. Sasaki, and J. Kobayashi. 1991. Stelliferins A–F, new antineoplastic isomalabaricane triterpenes from the Okinawan marine sponge *Jaspis stellifera. Tetrahedron*, 47: 2181–2194.

Tsukamoto, S., R. Yamanokuchi, M. Yoshitomi, K. Sato, T. Ikeda, H. Rotinsulu, R.E. Mangindaan, N.J. de Voogd, R.W. van Soest, and H. Yokosawa. 2010. Aaptamine, an alkaloid from the sponge *Aaptos suberitoides*, functions as a proteasome inhibitor. *Bioorg. Med. Chem. Lett.*, 20: 3341–3343.

Turk, T., J.A. Avguštin, U. Batista, G. Strugar, R. Kosmina, S. Čivović, D. Janussen, S. Kauferstein, D. Mebs, and K. Sepčić. 2013. Biological activities of ethanolic extracts from deep-sea Antarctic marine sponges. *Mar. Drugs*, 11: 1126–1139.

Tziveleka, L., E. Ioannou, D. Tsiourvas, P. Berillis, E. Foufa, and V. Roussis. 2017. Collagen from the marine sponges *Axinella cannabina* and *Suberites carnosus*: Isolation and morphological, biochemical, and biophysical characterization. *Mar. Drugs.*, 15: 152.

Uchida, K., H. Watanabe, and K. Mori. 1997. Synthesis of (+)-hiburipyranone, a cytotoxic metabolite of marine sponge *Mycale adhaerens. Biosci. Biotech. Biochem.*, 61: 1564–1567.

Ueda, T., Y. Nakamura, C.M. Smith, B.A. Copits, A. Inoue, T. Ojima, S. Matsunaga, G.T. Swanson, and R. Sakai. 2013. Isolation of novel prototype galectins from the marine ball sponge *Cinachyrella* sp. guided by their modulatory activity on mammalian glutamate-gated ion channels. *Glycobiology*, 23: 412–425.

Urban, S., M.S. Butler, and R.J. Capon. 1994. Lamellarins O and P: New aromatic metabolites from the Australian marine sponge *Dendrilla cactos. Aust. J. Chem.*, 47: 1919–1924.

Utkina, N.K. 2013. Antioxidant activity of zyzzyanones and makaluvamines from the marine sponge *Zyzzya fuliginosa. Nat. Prod. Commun.*, 8: 1551–1152.

Utkina, N.K., A.E. Makarchenko, and V.A. Denisenko. 2005. Zyzzyanones B-D, dipyrroloquinones from the marine sponge *Zyzzya fuliginosa, J. Nat. Prod.*, 68: 1424.

Vasanthabharathi, V., and S. Jayalakshmi. 2012. Bioactive potential of symbiotic bacteria and fungi from marine sponges. *Afr. J. Biotechnol.*, 11: 7500–7511.

Velho-Pereira, S., and I. Furtado. 2012. Antibacterial activity of halophilic bacterial bionts from marine invertebrates of Mandapam-India. *Indian J. Pharm. Sci.*, 74: 331–338.

Venables, D.A., G.P. Concepción, S.S. Matsumoto, L.R. Barrows, and C.M. Ireland. 1997. Makaluvamine N: A new pyrroloiminoquinone from *Zyzzya fuliginosa. J. Nat. Prod.*, 60: 408–410.

Vera, B., J. Vicente, and A.D. Rodríguez. 2009. Isolation and structural elucidation of euryjanicins B–D, proline-containing cycloheptapeptides from the Caribbean marine sponge *Prosuberites laughlini. J. Nat. Prod.*, 72: 1555–1562.

Viegelmann, C., J. Parker, T. Ooi, C. Clements, G. Abbott, L. Young, J. Kennedy, A.D.W. Dobson, and R. Edrasa-Ebel. 2014. Isolation and identification of antitrypanosomal and antimycobacterial active steroids from the sponge *Haliclona simulans. Mar. Drugs*, 12: 2937–2952.

von Salm, J.L., C.G. Witowski, R.M. Fleeman, J.B. McClintock, C.D. Amsler, L.N. Shaw, and B.J. Baker. 2016. Darwinolide, a new diterpene scaffold that inhibits methicillin-resistant *Staphylococcus aureus* biofilm from the Antarctic sponge *Dendrilla membranosa*. *Org. Lett.*, 18: 2596–2599.

Vuong, D., R.J. Capon, E. Lacey, J.H. Gill, K. Heiland, and T. Friedel. 2001. Onnamide F: A new nematocide from a southern Australian marine sponge, *Trachycladus laevispirulifer*. *J. Nat. Prod.*, 64: 640–642.

Wakimoto, T., Y. Egami, Y. Nakashima, Y. Wakimoto, T. Mori, T. Awakawa, T. Ito, H. Kenmoku, Y. Asakawa, J. Piel, and I. Abe. 2014. Calyculin biogenesis from a pyrophosphate protoxin produced by a sponge symbiont. *Nat. Chem. Biol.*, 10: 648–655.

Warabi, K., S. Matsunaga, R.W. van Soest, and N. Fusetani. 2003. Dictyodendrins A–E, the first telomerase-inhibitory marine natural products from the sponge *Dictyodendrilla verongiformis*. *J. Org. Chem.*, 68: 2765–2770.

Warabi, K., L.M. McHardy, L. Matainaho, R. van Soest, C.D. Roskelley, M. Roberge, and R.J. Andersen. 2004. Strongylophorine-26, a new meroditerpenoid isolated from the marine sponge *Petrosia (Strongylophora) corticata* that exhibits anti-invasion activity. *J. Nat. Prod.*, 67: 1387–1389.

Warad, V.B., P. Habbu, and R. Shastri. 2017. Antimicrobial activity of *Callyspongia diffusa* (marine sponge) associated endophytic bacteria l strains. *Int. J. Pharm. Pharm. Sci.*, 9: 90–96.

Watanabe, K., Y. Tsuda, M. Iwashima, and K. Iguchi. 2000. A new bioactive triene aldehyde from the marine sponge *Leucetta microraphis*. *J. Nat. Prod.*, 63: 258–260.

Weinheimer, A.J., C.W. Chang, J.A. Matson, and P.N. Kaul. 1978. Marine cardioactive agents. Adenosine and 2′-deoxyadenosine from *Dasychalina cyathina*. *Lloydia*, 41: 488–490.

Wellington, K.D., R.C. Cambie, P.S. Rutledge, and P.R. Bergquist. 2000. Chemistry of sponges. 19. Novel bioactive metabolites from *Hamigera tarangaensis*. *J. Nat. Prod.*, 63: 79–85.

Wikipedia. Petrosia (Petrosia) ficiformis. 2018, March 22. https://en.wikipedia.org/wiki/Petrosia_ficiformis.

Wikipedia. Tectitethya crypta. 2018, March 22. https://en.wikipedia.org/wiki/Tectitethya_crypta.

Wikipedia. Xestospongia testudinaria. 2018, March 22. https://en.wikipedia.org/wiki/Xestospongia_testudinaria.

Wilkins, S.P., A.J. Bluma, D.E. Burkepile, T.J. Rutland, A. Wierzbicki, M. Kelly, and M.T. Hamanna. 2002. Isolation of an antifreeze peptide from the Antarctic sponge *Homaxinella balfourensis*. *Cell. Mol. Life Sci.*, 59: 2210–2215.

Williams, D.E., M. Roberge, R. van Soest, and R.J. Andersen. 2003. Spirastrellolide A, an Antimitotic macrolide isolated from the Caribbean marine sponge *Spirastrella coccinea*. *J. Am. Chem. Soc.*, 125: 5296–5297.

Williams, D.E., A. Steinø, N.J. de Voogd, A.G. Mauk, and R.J. Andersen. 2012. Halicloic acids A and B isolated from the marine sponge *Haliclona* sp. collected in the Philippines inhibit indoleamine 2,3-dioxygenase. *J. Nat. Prod.*, 75: 1451–1458.

Williams, D.E., K. Yu, H.W. Behrisch, R. van Soest, and R.J. Andersen. 2009. Rolloamides A and B, cytotoxic cyclic heptapeptides Isolated from the Caribbean marine sponge *Eurypon laughlini*. *J. Nat. Prod.*, 72: 1253–1257.

Winder, P.L. 2009. Therapeutic potential, mechanism of action, and ecology of novel marine natural products. Ph.D. thesis, Florida Atlantic University.

Witowski, C.G. 2015. Investigation of bioactive metabolites from the Antarctic sponge *Dendrilla membranosa* and marine microorganisms. Ph.D. thesis, University of South Florida.

Woo, J.K., C.K. Kim, S.H. Kim, H. Kim, C.H. Ahn, D.C. Oh, K.B. Oh, and J. Shin. 2015. Sesterterpenes and a nortriterpene saponin from the sponge *Clathria gombawuiensis*. *Planta Med.*, 81–PQ15.

Wright, A.D., and N. Lang-Unnasch. 2009. Diterpene formamides from the tropical marine sponge *Cymbastela hooperi*, and their antimalarial activity *in vitro*. *J. Nat. Prod.*, 72: 492–495.

Wright, A.D., A. McCluskey, M.J. Robertson, K.A. MacGregor, C.P. Gordon, and J. Guenther. 2011. Anti-malarial, anti-algal, anti-tubercular, anti-bacterial, anti-photosynthetic, and anti-fouling activity of diterpene and diterpene isonitriles from the tropical marine sponge *Cymbastela hooperi*. *Org. Biomol. Chem.*, 9: 400–407.

Xiong, F., S.J. Francis, Kelly-Borges, M. McCready, L. Tara, and F.B.H. Charles. 1998. Clavosines A-C from the marine sponge *Myriastra clavosa*: Potent cytotoxins and inhibitors of protein phosphatases 1 and 2A. *JOC*, 63: 7957–7963.

Xu, N.J., X. Sun, and X.J. Yan. 2007. A new cyclostellettamine from sponge *Amphimedon compressa*. *Chin. Chem. Lett.*, 18: 947–950.

Xu, Y., J. Lang, W. Jiao, R. Wang, Y. Peng, S. Song, B. Zhang, and H. Lin. 2012. Formamido-diterpenes from the South China Sea sponge *Acanthella cavernosa*. *Mar. Drugs*, 10: 1445–1458.

Xuea, D., H. Liua, S. Chena, E. Mollob, M. Gavagninb, J. Lia, X. Lia, and Y. Guo. 2017. 5-Alkylpyrrole-2-carboxaldehyde derivatives from the Chinese sponge *Mycale lissochela* and their PTP1B inhibitory activities. *Chin. Chem. Lett.*, 28: 1190–1193.

Yamaguchi, M., M. Miyazaki, M.P. Kodrasov, H. Rotinsulu, F. Losung, R.E. Mangindaan, N.J. de Voogd, H. Yokosawa, B. Nicholson, and S. Tsukamoto. 2013. Spongiacidin C, a pyrrole alkaloid from the marine sponge *Stylissa massa*, functions as a USP7 inhibitor. *Bioorg. Med. Chem. Lett.*, 23: 3884–3886.

Yamanokuchi, R., K. Imada, M. Miyazaki, H. Kato, T. Watanabe, M. Fujimuro, Y. Saeki, S. Yoshinaga, H. Terasawa, N. Iwasaki, and H. Rotinsulu. 2012. Hyrtioreticulins A–E, indole alkaloids inhibiting the ubiquitin-activating enzyme, from the marine sponge *Hyrtios reticulata*. *Bioorg. Med. Chem.*, 20: 4437–4442.

Yang, F., M.T. Hamann, Y. Zou, M. Zhang, X. Gong, J. Xiao, W. Chen, and H. Lin. 2012. Antimicrobial metabolites from the Paracel Islands sponge *Agelas mauritiana*. *J. Nat. Prod.*, 75: 774–778.

Yoo, H.D., D. Leung, J. Sanghara, D. Daley, R. Van Soest, and R.J. Andersen. 2003. Isoarenarol, a new protein kinase inhibitor from the marine sponge *Dysidea arenaria*. *Pharm. Biol.*, 41: 223–225.

Youssef, D.T., A.K. Ibrahim, S.I. Khalifa, M.K. Mesbah, A.M. Mayer, and R.W. van Soest. 2010. New anti-inflammatory sterols from the Red Sea sponges *Scalarispongia aqabaensis* and *Callyspongia siphonella*. *Nat. Prod. Commun.*, 5: 27–31.

Youssef, D.T., L.A. Shaala, G.A. Mohamed, J.M. Badr, F.H. Bamanie, and S.R. Ibrahim. 2014. Theonellamide G, a potent antifungal and cytotoxic bicyclic glycopeptide from the Red Sea marine sponge *Theonella swinhoei*. *Mar. Drugs*, 12: 1911–1923.

Youssef, D.T.A. 2005. Hyrtioerectines A–C, cytotoxic alkaloids from the Red Sea sponge *Hyrtios erectus*. *J. Nat. Prod.*, 68: 1416–1419.

Youssef, D.T.A., J.M. Badr, L.A. Shaala, G.A. Mohamed, and F.H. Bamanie. 2015a. Ehrenasterol and biemnic acid; new bioactive compounds from the Red Sea sponge *Biemna ehrenbergi*. *Phytochem. Lett.*, 12: 296–301.

Youssef, D.T.A., L.A. Shaala, and K.Z. Alshali. 2015b. Bioactive hydantoin alkaloids from the Red Sea marine sponge *Hemimycale arabica*. *Mar. Drugs*, 13: 6609–6619.

Youssef, D.T.A., R. K. Yamaki, M. Kelly, and P.J. Scheuer. 2002. Salmahyrtisol A, a novel cytotoxic sesterterpene from the Red Sea sponge *Hyrtios erecta*. *J. Nat. Prod.*, 65: 2–6.

Yu, H.-B., X. Liu, Y. Xu, J.W. Jiao, Y. Shen, and H. Lin. 2012. Woodylides A–C, new cytotoxic linear polyketides from the South China Sea sponge *Plakortis simplex*. *Mar. Drugs*, 10: 1027–1036.

Yu, H.-B., F. Yang, F. Sun, J. Li, W. Jiao, J. Gan, W. Hu, and H. Lin. 2014a. Aaptamine derivatives with antifungal and anti-HIV-1 activities from the South China Sea sponge *Aaptos aaptos*. *Mar. Drugs*, 12: 6003–6013.

Yu, H.-B., F. Yang, F. Sun, G.Y. Ma, J.H. Gan, W.Z. Hu, B.N. Han, W.H. Jiao, and H.W. Lin. 2014b. Cytotoxic aaptamine derivatives from the South China Sea sponge *Aaptos aaptos*. *J. Nat. Prod.*, 77: 2124–2129.

Yu, Z., G. Lang, I. Kajahn, R. Schmaljohann, and J.F. Imhoff. 2008. Scopularides A and B, cyclodepsipeptides from a marine sponge-derived fungus, *Scopulariopsis brevicaulis*. *J. Nat. Prod.*, 71: 1052–1054.

Zabriskie, T.M., and C.M. Ireland. 1989. The isolation and structure of modified bioactive nucleosides from *Jaspis johnstoni*. *J. Nat. Prod.*, 52: 1353–1356.

Zampella, A., M.V. D'Auria, C. Debitus, and J. Menou. 2000. New isomalabaricane derivatives from a new species of *Jaspis* sponge collected at the Vanuatu Islands. *J. Nat. Prod.*, 63: 943–946.

Zampella, A., M.V. D'Auria, L.G. Paloma, A. Casapullo, L. Minale, C. Debitus, and Y. Henin. 1996. Callipeltin A, an anti-HIV cyclic depsipeptide from the New Caledonian lithistida sponge *Callipelta* sp. *J. Am. Chem. Soc.*, 118: 6202–6209.

Zampella, A., C. Giannini, C. Debitus, C. Roussakis, and M.V. D'Auria. 1999. Isolation and structural elucidation of crellastatins B–H: Cytotoxic bis(steroid) derivatives from the Vanuatu marine sponge *Crella* sp. *Eur. J. Org. Chem.*, 1999: 949–953.

Zampella, A., V. Sepe, P. Luciano, F. Bellotta, M.C. Monti, M.V. D'Auria, T. Jepsen et al. 2008. Homophymine A, an anti-HIV cyclodepsipeptide from the sponge *Homophymia* sp. *J. Org. Chem.*, 73: 5319–5327.

Zeng, Z., J. Zhao, C. Ke, and D. Wang. 2013. Antimicrobial activities of novel cultivable bacteria isolated from marine sponge *Tedania anhelans*. *Chin. J. Oceanol. Limnol.*, 31: 581–590.

Zhang, B., R. Higuchi, T. Miyamoto, R.W.M. van Soest. 2008. Neuritogenic activity-guided isolation of a free base form manzamine A from a marine sponge, *Acanthostrongylophora* aff. *ingens* (Thiele, 1899). *Chem. Pharm. Bull.*, 56: 866–869.

Zhang, H., P. Crews, K. Tenney, and F.A. Valeriotec. 2017a. Cytotoxic phyllactone analogs from the marine sponge *Phyllospongia papyrecea*. *Med. Chem.*, 13: 295–300.

Zhang, H., M. Dong, H. Wang, and P. Crews, 2017b. Secondary metabolites from the marine sponge genus *Phyllospongia*. *Mar. Drugs*, 15: 12.

Zhang, H., J. Sun, H. Lin, Z. Wang, H. Tang, P. Cheng, W. Chen, and Y. Yi. 2007. A new cytotoxic cholesterol sulfate from marine sponge *Halichondria rugosa*. *Nat. Prod. Res.*, 21: 953–958.

Zhang, H., Z. Zhao, and H. Wang. 2017c. Cytotoxic natural products from marine sponge-derived microorganisms. *Mar. Drugs*, 15: 68.

Zhang, L., and A.L. Demain. 2005. *Natural Products: Drug Discovery and Therapeutic Medicine*. Humana Press, New York.

Zhang, W., F. Zhang, Z. Li, X. Miao, Q. Meng, and X. Zhang. 2009. Investigation of bacteria with polyketide synthase genes and antimicrobial activity isolated from South China Sea sponges. *J. Appl. Microbiol.*, 107: 567–575.

Zhang, W.H., and C.T. Che. 2001. Four isomalabaricane triterpenes were isolated from marine sponge Geodia japonica. Isomalabaricane-type nortriterpenoids and other constituents of the marine sponge *Geodia japonica, J. Nat. Prod.*, 64: 1489–1492.

Zhang, Z., C. Zong, G. Song, G. Lv, Y. Chun, P. Wang, N. Ding, and Y. Li. 2010. Total synthesis of caminoside B, a novel antimicrobial glycolipid isolated from the marine sponge *Caminus sphaeroconia*. *Carbohydr. Res.*, 345: 750–760.

Zovko, A., M. Novak, P. Hååg, D. Kovalerchick, T. Holmlund, K. Färnegårdh, M. Ilan, S. Carmeli, R. Lewensohn, and K. Viktorsson. 2016. Compounds from the marine sponge *Cribrochalina vasculum* offer a way to target IGF-1R mediated signaling in tumor cells. *Oncotarget*, 19 p.

Zovko, A., K. Viktorsson, P. Hååg, D. Kovalerchick, K. Färnegårdh, A. Alimonti, M. Ilan, S. Carmeli, and R. Lewensohn. 2014. Marine sponge *Cribrochalina vasculum* compounds activate intrinsic apoptotic signaling and inhibit growth factor signaling cascades in non-small cell lung carcinoma. *Mol. Cancer Ther.*, 13: 2941–2954.

Zubía, E., M.J. Ortega, C.J. Hernández-Guerrero, and J.L. Carballo. 2008. Isothiocyanate sesquiterpenes from a sponge of the genus *Axinyssa*. *J. Nat. Prod.*, 71: 608–614.

Compounds

A

B

bis-1,3-dialkylpyridinium
bisaprasin
bis-oxaquinolizidine
blanchaquinone
blastamycin
bolinaquinone
bovinocidin
brevianamide A
brocaenols A–C
bromoalterochromide A
bromochloroverongiaquinol
bromoconicamin
bromotopsentin
bromotyrosine
butenolide

C

C17 acetylenic acid
C27-polyacetylene
C27-polyacetylene
cacospongin D
cacospongionolide B
cadlinolide C
caissarine C
calafianin
callimplexen A
callipeltins A–C; F–I; J, K
callipeltosides A–C
callyaerins A–G; I–M
callyspongamide A
callyspongendiol
callyspongenols A-D
callysponginol sulfate A
callyspongiolide
callyspongynic acid
callysponine
callystatin A
callysterol (ergosta-5,11-dien-3beta-ol)
calyculin A
cambescidins 1–4
caminosides A-D
carteramine A
carteriofenones A–K
carteriosulfonic acids A–C
carteritin A
celebesides A–C
celenamides
ceramides
ceratinadins A, B
ceratinamine
ceratinines A–E
C-glucoside iso-D8646-2-6; 8646-2-6
chartarutinea B, G, H
chelonin A
chloriolines A–C
chlorodibromohydrin
chondropsins A, B
chromodorolides A–C
chujamides A, B
ciliatamides A–C
cinachylenic acid A
cinachyrella galectins (CchGs)
cinanthrenol A
cis, cis-ceratospongamide
cis-4-hydroxysalsolinol
citronamide A
clathric acid
clathridine
clathriols A, B
clathsterol
clavamycin D
clavosines A, B
clionamide
clionasterol
communesins B–D
contignasterol
coriolin B
corticatic acids A–E
cortistatins A–D; J–L
crambescidins 816, 826, 830, 844, 860
crambescin C1
crellastatins A–H
criamides A, B
cryptosporiopsinol
cupolamide A
curcudiol
curcuphenol
cyanthiwigins AA-AA; A–D
cyclo-(glycyl-L-prolyl-Lglutamyl),
cyclo-(glycyl-Lseryl-L-prolyl-L-glutamyl)
cyclo-(L-proline-L-methionine)
cyclo-(L-Pro-L-Phe)
cyclostellettamine
cytarabine
cytoskyrin A

D

dactylospongenones A–D
daminin
damirones A, B
dankastatins A, B
dankasterones A, B
darwinolide
daryamide C
debromohamacanthin B
debromosceptrin
dechlorogriseofulvin
decumbenones A, B
dehydrocurcuphenol
dehydrofoliaspongin
dehydrofurodendin
dehydroisophonochalynol

E

F

G

H

I

J

O

P

theonellamide G
theopapuamides B–D
thiaplakortones A–D
thiocyanatoneopupukeanane
thorectidaeolide A
thorectolide monoacetate
thorectolide monoacetate
thymidine
title acid
tobramycin
tokaramide A
toluene
topsentiasterol sulfates A–E
topsentins B1, B2
toyocamycin
trachycladindoles A–G
trans, trans-ceratospongamide
trans-4-hydroxysalsolinol
triangulynes A, E
trichodenones A–C
trichodermol
trichoverrins A, B
trisindoline
tsitsikammamine C
tubercidin

U

ulocladol
ulososide A, B, F
urauchimycins A, B
urupocidin A

V

validamycin D
valinomycin
valistatin
variabines A,B
variolins A, B
vasculyne
venturicidin A
verrucarins A, M
versiol
vidarabine

W

wainunuamide
weinbersterols A, B
woodylides A–C

X

X 14931A
xanthobaccin A
xestosterol
xestoadociaminals A–D
xestoadociaquinones A, B
xestobergsterols A, B
xestodecalactones A–C
xestokerols A, B
xestolactone A
xestoquinone
xestosapro C
xestosaprol O
xestospongins A–D
xestosterol ester of 18′-bromooctadeca-7′E,9′E-diene-7′,15′-diynoic acid
xestosterol palmitate
xestovanin A
xylosyl esters A–C

Z

zampanolide
zooanemonin
zyzzyanones A–D
(1H-indol-3-yl) oxoacetamide
(1H-indol-3-yl) oxoacetic acid methyl ester
(1R*, 4S*, 6R*, 7S*)-4-isothiocyanato-9-amorphene
(1R, 6S, 7S, 10S)-10-isothiocyanato-4-amorphene
(1S*, 3S*, 4R*, 7S*, 8S*, 12S*, 13S*)-7-formamido-cyclo amphilect-11(20)-ene
(1S*,3S*,4R*,7S*,8S*,12S*,13S*)-7-isocyano-amphilecta-10,14-diene
(1S, 3S, 4R, 7S, 8S, 11S, 12S, 13S, 15R, 20R)-7, 20-diformamido-isocycloamphilectane
(1S, 3S, 4R, 7S, 8S, 11S, 12S, 13S, 15R, 20R)-7-formamido-20-isocyano-isocycloamphilectane
(22E)-ergosta-22-ene-8,14-epoxy-3,7-dione
(22E)-ergosta-5,8,22-trien-7-one-3β-ol
(24R) 5α,8α-epidioxy-24(ξ)-methylcholesta-6,9(11),22-trien-3β-ol
(24R), 5α,8α-epidioxy-24(S)-ethylcholest-6-en-3β-ol
(24R)-ergosta-6,22-diene-5,8-epidioxy-3-ol
(24S)-ergostane 3beta, 5alpha, 6beta, 25 tetraol-25-monoacetate
(24S)-ergostane-1beta, 3beta, 5alpha, 6beta, 25-pentaol-25-monoacetate
(2E,6Z,9Z)-2-methyl-2,6,9-icosatrienal
(2R,3R,6R)-2-aminooctadec-1,3,6-triol (halisphingosine B)
(2R,3R,6R,7Z)-2-aminooctadec-7-ene-1,3, 6-triol (halisphingosine A)
(2S,4R)- and (2R,4R)-2,4-dimethyl-4-hydroxy-16-phenyl hexadecanoic acid 1,4-lactones
(3,7-dimethyl-2-(methylamino)- 3H-purin-6(7H)-one
(3R)-6-methoxy-7-chloromellein
(3R)-6-methoxymellein
(3R,3′R,5S)-3,3′,5,19′-tetrahydroxy-7′,8′-didehydro-γ,ε-carotene-8-one
(3R,4R)-hydroxymellein
(3R,4S)-hydroxymellein

4-acetoxythorectidaeolide A
4a-methyl-5a-cholest-8-en-3~-ol
4-bromo-N-(butoxymethyl)-1H-pyrrole-2-carboxamide
4E, 6E-dehydromanoalide
4-formamidoeudesm-7-ene
4-hydroxythreonine
4-isothiocyanatoeudesm-11-ene
4-methylaaptamine
4-thiouracil
5α-choletan-3β-ol, dihydrocholesterol, cholesterol
5-(6′-Methyl-7′-oxo-octyl)-SH-furan-2-one
5-(methoxycarbonyl)tubercidin
5,6-dibromo-l-hypaphorine
5,6-dibromo-N,N-dimethyltryptamine
5,7-dimethoxy-4-methoxylphenylcoumarin
5alpha,6alpha-epoxy-24R-ethylcholest-8(14)-en-3beta, 7alpha-diol
5alpha,6alpha-epoxy-24R-ethylcholest-8-en-3beta, 7alpha-diol
5alpha,8alpha-epidioxycholest-6-en-3beta-ol
5-benzoyldemethyaaptamine
5-bromo-N,N-dimethyltryptamine
5-debromomidpacamide
5-hydroxy-16-methyleicos- (3Z)-en-1-yne
5-hydroxy-1H-indole-3-carboxylic acid methyl ester
5-hydroxy-3-(1-hydroxy-2-methylpropyl)-4-methyl-2(5H)-furanone
5-hydroxy-4-oxonorvaline
5-hydroxyindole-3-aldehyde
5-hydroxyindole-3-carbaldehyde
5α,8α-epidioxy sterols
5α,8α-epidioxy-24(S)-ethylcholesta-6,22-dien-3β-ol
5α,8α-epidioxy-24(S)-methylcholest-6-en-3β-ol and its epimer (24R)
5α,8α-epidioxy-24(ξ)-methylcholesta-6,22-dien-3β-ol
5α,8α-epidioxycholesta-6,22-dien-3β-ol, 5α,8α-epidioxy cholest-6-en-3β-ol
6-hydroxy-3,4-dihydro-1-oxo-b-carboline
6‴-O-α-d-Mannopyranosylmannosido-streptomycin
6-bromoindole-3-carbaldehyde
6-Hydroximino-4-en-3-one
6-hydroxymanzamine E
7,8-dihydroplakortide E
7-formamidoamphilect- 11(20),15-diene
7-hydroxymoloka'iamine
7-hydroxyneolamellarin A
7-isocyanoneoamphilecta-1(14),15-diene
7-methyl-7-((E)-4-methylpenta-1,3-dien-1-yl)-5,6,7,7a-tetra hydrobenzofuran-3a(4H)-ol (Iricinol A),
7-oxopetrosterol
7α,12β,20β-trihydroxycholesta-22E-en-3-one
7α-hydroxyaragusterol I
8,8′-dienecyclostellettamine
8,9-dihydrobarettin
8′-oxo-agelasine D
80 -oxo-agelasines B -E
8-carboxy-iso-iantheran A
8-deoxytrichothecin
8-hydroxymanzamine
8-hydroxymanzamine A
8-hydroxymanzamine A hydrochloride
8-isocyanato-15-formamidoamphilect-11(20)-ene
8-isocyanatocycloamphilect-10-ene
8-isocyano-15- formamidoamphilect-11(20)-ene
8-isothiocyanato-15-formamidoamphilect-11(20)-ene,
8-isothiocyanato-amphilect-10-ene
9-(5′-deoxy-5′-thio-β-D-xylofuranosyl)adenine disulfide
10,11-dihydrosodwanone B
10-acetoxy-15,21-dideacetyl-4-oxo-28-hydroraspacionin
10-acetoxy-15-deacetyl-4-21-dioxo-28-hydroraspacionin 1
10-acetoxy-15-deacetyl-4-oxo-28-hydroraspacionin
10-acetoxy-21-deacetyl-28- hydroraspacionin
10-acetoxy-21-deacetyl-4-acetyl-28-hydroraspacionin
10-acetoxy-21-deacetyl-4-oxo-28-hydroraspacionin
10-acetoxy-4,21-dioxo-28-hydroraspacionin 1
10-acetoxy-4-acetyl-15-deacetyl-28-hydroraspacionin
10-acetoxy-4-acetyl-28-hydroraspacionin
10-acetylirciformonin B
10E-hymenialdisine
10-formamido-kalihinene
10-hydro-9-hydroxyagelasine F
10-hydroxy-18- methoxylbetaenone
10-hydroxypsammaplysin X
10-isothiocyanato-4,6-amorphadiene, and (4S*, 10S*)-10-isothiocyanato-5-amorphen-4-ol
10-isothiocyanato-4-amorphene
10-O-sulfatobastadin 3
10Z-debromohymenialdisine
10Z-hymenialdisine
11 β-acetoxyspongi-12-en-16-one
11,19-dideoxyfistularin-3
11-hydroxyaerothionin
11-isothiocyano-7β-H-eudesm-5-ene
11-methoxy-3H-[1,6] naphthyridino [6,5,4-def] quinoxalin-3-one
11-Oxoaerothionin
11β-hydroxyspongi-12-en-16-one
12-deacetoxyscalarin 19-acetate
12-deacetoxyscalarin acetate
12-deacetylhyrtial
12,16-di-epi-12-deacetyl-16-acetylfuroscalarol
12,16-di-epi-12-O-deacetyl-16-O-acetylfuroscalarol
12,18-di-epi-scalaradial
12-acetoxytetrahydroaplysulfurin-1
12-acetyl-12,16-diepi-16-deacetylscalarolbutenolide
12-acetyl-16-deacetyl-16-episcalarolbutenolide
12-acetyl-16-epi-scalarolbutenolide
12-deacetoxy-16-Odeacetylheteronemin
12-deacetoxy-19-acetoxyscalarin
12-deacetoxy-22-acetoxyheteronemin
12-deacetoxy-22-formylheteronemin

12-deacetoxy-22-hydroxy-25-acetoxyscalarin
12-deacetoxy-22-hydroxyheteronemin
12-deacetyl-17-hyrtial
12-deacetyl-12,18-diepi-scalaradial
12-deacetyl-12-epi-19-deoxyscalarin
12-deacetyl-12-epi-scalaradial
12-deacetylhyrtial and methyl 18-hydroxy-25-norscalar-16-en-24-carboxylate
12-deacetylnorscalaral B
12-dehydroxy-16-deacetoxy-22-acetoxy-scalarafuran
12-epi-19-deoxyscalarin
12-epideacetyl-24α-methoxy-25α-acetoxy-deoxoscalarin
12-epi-deoxoscalarin
12-epi-heteronemin acetate
12-epi-scalaradial
12-epi-scalarin
12-N-methyl stevensine
12-Oacetyl-16-O-methylhyrtiolide
12-O-deacetyl-12-epi-scalarin
12-O-deacetyl-scalarin
12α, 24-dihydroxy-20, 24-dimethyl-15, 17-scalaradien-25, 24-olides
12α-acetoxy-13β,18β-cyclobutan-20,24-dimethyl-24-oxoscalar-16-en-25β-ol
12α-acetoxy-16β-hydroxy-20,24-dimethyl-24-oxoscalaran-25-al
12α-Acetoxy-16β-hydroxy-20,24-dimethyl-25-norscalar-17-en-24-one
12α-Acetoxy-16β-hydroxy-24- methyl-24-oxoscalaran-25-al
12α-acetoxy-20,24-dimethyl-25-norscalar-16-en-24-one
12α-acetoxy-24,25-epoxy-24-hydroxy-20,24-dimethylscalarane
12α-Acetoxy-24-methyl-24-oxoscalar-16-en-22,25-dial
12β-(3′β-hydroxybutanoyloxy)-20,24-dimethyl-24-oxo-scalara-16-en-25-al
13(E)-29-hydroxystelliferin E
13(E)-3-epi-29-hydroxystelliferin E
13(E)-stelliferin G
13′-acetyltrichoverrin B
13-deoxytedanolide
14,15-secocurvularin
14-carboxy-xestoquinol sulfate
14-hydroxyaragusterol
14-methoxyhalenaquinone
15,34-O-disulfatobastadin 7
15-acetylirciformonin B
15-formamido-kalihinene
16-bromo (7E,11E,15E)hexadeca-7,11,15-triene-5,13-diynoic acid
16-deacetoxyscalara-furan
16-deacetyl-16-episcalarolbutenolide
16-debromoaplysamine-4
16-epi-scalarolbutenolide
16β,22-Dihydroxy- 24-methyl-24-oxoscalaran-25,12β-olactone
16β-hydroxy-24-methyl-12, and 24-dioxoscalaran-25-al
18,18-dibromo-(9E)-octadeca-9,17-diene-5,7-diynoic acid
18-bromooctadeca-(9E,13E,17E)-triene-7,15-diynoic acid
18-bromooctadeca-(9E,17E)-diene-7,15-diynoic acid
18-epi-latrunculol A
18-hydroxy-5-epi-hyrtiophenol
19-acetoxydeoxoscalarin
19-acetyl sesterstatin 3
19-butanoyloxy-mytiloxanthin
19-deoxyfistularin 3
19-hexanoyl-oxymytiloxanthin
19-hydroxy-polyfibrospongol B
19-hydroxypsammaplysin E
20,24-bishomo-25-norscalarane
20,24-bishomoscalarane
20,24-dimethylscalarane
20-methyl-pregn-6-en-3β-ol,5a,8a-epidioxy
21-deacetylraspacionin
21-dehydroxybolinaquinone
21-hydroxypetrosaspongiolide K
21-hydroxypetrosaspongiolide P
21-oxoraspacionin, 15-deacetyl-21-dioxoraspacionin
22(S)-hydroxyingamine A
22-acetoxy-16β-hydroxy-24-methyl-24-oxoscalaran-25,12β-olactone
22-acetylisocyclocitrinol A
22-hydroxy-25-methoxy-24-methyldeoxoscalarin
24 acetoxytetrahydroaplysulfurin-1
24 α-methylcholestanol
24-ethyl-cholest-5alpha-7-en-3-beta-ol
24-hydroxy-scalarolide
24-methyl-12,24,25-trioxoscalar-16-en-22-oate
24-methyl-12,24,25-trioxoscalar-16-en-22-oic acid
24-methyl-24,25-dioxoscalar-16-en-12β-yl 3-hydroxybutanoate
24-methylenecholesterol
24-methylscalarane
24-methylscalarane, 12α-Acetoxy-24-methyl-24-oxoscalar-16-en-22,25-dial
24-vinyl-cholest-9-ene-3β,24-diol
24ξ,25-dimethyl-3α-hydroxyl-cholest-5-ene-2β-ol sodium sulfate
24ξ,25-dimethyl-cholest-5-ene-2β,3α-diol disodium sulfate
24ξ,25-dlmethylcholestane-2β,3α,6α-triyl trisodium sulfate
25-dehydroxy-12-epi-scalarin
25-dehydroxy-12-epi-deacetylscalarin
25α-methoxy-12-deacetoxy-22-hydroxydeoxoscalarin
26,27-bisnorcholest-5,16-dien-23-yn-3β,7α-diol
26,27-bisnorcholest-5,16-dien-23-yn-3β,7α-diol, gelliusterol E
26,27-cyclo-24,27-dimethylcholest-5-ene-3β,7α,12β,20β-tetraol
26,27-cyclo-24,27-dimethylcholestan-3,3-dimethoxy-12β,21,20α,22α-tetraol
29-hydroxystelliferins A,B,E
30,32-dihydroxymycalolide A

Index

D

E

F

G

Q

R

S

T

U

V

W

X

Y

Z